Soilless Culture

SOILLESS CULTURE: THEORY AND PRACTICE

MICHAEL RAVIV
J. HEINRICH LIETH

ELSEVIER

Amsterdam • Boston • Heidelberg • London • New York • Oxford
Paris • San Diego • San Francisco • Singapore • Sydney • Tokyo

Elsevier
84 Theobald's Road, London WC1X 8RR, UK
Radarweg 29, PO Box 211, 1000 AE Amsterdam, The Netherlands
30 Corporate Drive, Suite 400, Burlington, MA 01803, USA
525 B Street, Suite 1900, San Diego, CA 92101-4495, USA

First edition 2008

Copyright © 2008 Elsevier BV. All rights reserved

No part of this publication may be reproduced, stored in a retrieval system or transmitted in any form or by any means electronic, mechanical, photocopying, recording or otherwise without the prior written permission of the publisher

Permissions may be sought directly from Elsevier's Science & Technology Rights Department in Oxford, UK: phone (+44) (0) 1865 843830; fax (+44) (0) 1865 853333; email: permissions@elsevier.com. Alternatively you can submit your request online by visiting the Elsevier web site at http://elsevier.com/locate/permissions, and selecting *Obtaining permission to use Elsevier material*

Notice
No responsibility is assumed by the publisher for any injury and/or damage to persons or property as a matter of products liability, negligence or otherwise, or from any use or operation of any methods, products, instructions or ideas contained in the material herein. Because of rapid advances in the medical sciences, in particular, independent verification of diagnoses and drug dosages should be made

British Library Cataloguing in Publication Data
A catalogue record for this book is available from the British Library

Library of Congress Cataloging-in-Publication Data
A catalog record for this book is available from the Library of Congress

ISBN: 978-0-444-52975-6

For information on all Elsevier publications
visit our web site at books.elsevier.com

Printed and bound in the United States of America
08 09 10 11   10 9 8 7 6 5 4 3 2 1

Working together to grow
libraries in developing countries

www.elsevier.com | www.bookaid.org | www.sabre.org

ELSEVIER   BOOK AID International   Sabre Foundation

UAA CONSORTIUM LIBRARY
Please return items on the
date below. For renewal before
the due date you may call
786-1871 during business hours
RESERVE MATERIALS CANNOT BE
RENEWED OVER THE PHONE

Title: UAA ILL/SOILLESS CULTURE: THEORY AN
D PRACTICE
Item ID: 171707
Date due: 4/3/2013,23:59

Total checkouts for session:1
Total checkouts:4

CONTENTS

LIST OF CONTRIBUTORS XVII
PREFACE XIX

1

SIGNIFICANCE OF SOILLESS CULTURE IN AGRICULTURE
MICHAEL RAVIV AND J. HEINRICH LIETH

1.1 Historical Facets of Soilless Production 1
1.2 Hydroponics 6
1.3 Soilless Production Agriculture 6
References 10

2

FUNCTIONS OF THE ROOT SYSTEM
UZI KAFKAFI

2.1 The Functions of the Root System 13
2.2 Depth of Root Penetration 17
2.3 Water Uptake 18
2.4 Response of Root Growth to Local Nutrient Concentrations 22
    2.4.1 Nutrient Uptake 22
    2.4.2 Root Elongation and P Uptake 22
    2.4.3 Influence of N Form and Concentration 25

2.5 Interactions Between Environmental Conditions and Form of N Nutrition   26
  2.5.1 Temperature and Root Growth   26
  2.5.2 Role of Ca in Root Elongation   30
  2.5.3 Light Intensity   31
  2.5.4 pH   32
  2.5.5 Urea   32
  2.5.6 Mycorrhiza–Root Association   33
2.6 Roots as Source and Sink for Organic Compounds and Plant Hormones   33
  2.6.1 Hormone Activity   33
References   34
Further Readings   40

3

Physical Characteristics of Soilless Media

Rony Wallach

3.1 Physical Properties of Soilless Media   41
  3.1.1 Bulk Density   42
  3.1.2 Particle Size Distribution   42
  3.1.3 Porosity   44
  3.1.4 Pore Distribution   45
3.2 Water Content and Water Potential in Soilless Media   46
  3.2.1 Water Content   46
  3.2.2 Capillarity, Water Potential and its Components   50
  3.2.3 Water Retention Curve and Hysteresis   58
3.3 Water Movement in Soilless Media   65
  3.3.1 Flow in Saturated Media   65
  3.3.2 Flow in an Unsaturated Media   67
  3.3.3 Richards Equation, Boundary and Initial Conditions   71
  3.3.4 Wetting and Redistribution of Water in Soilless Media – Container Capacity   73
3.4 Uptake of Water by Plants in Soilless Media and Water Availability   76
  3.4.1 Root Water Uptake   76
  3.4.2 Modelling Root Water Uptake   79
  3.4.3 Determining Momentary and Daily Water Uptake Rate   84
  3.4.4 Roots Uptake Distribution Within Growing Containers   88
  3.4.5 Water Availability vs. Atmospheric Demand   90

3.5 Solute Transport in Soilless Media 95
  3.5.1 Transport Mechanisms – Diffusion, Dispersion, Convection 95
  3.5.2 Convection–Dispersion Equation 99
  3.5.3 Adsorption – Linear and Non-linear 99
  3.5.4 Non-equilibrium Transport – Physical and Chemical Non-equilibria 101
  3.5.5 Modelling Root Nutrient Uptake – Single-root and Root-system 102
3.6 Gas Transport in Soilless Media 104
  3.6.1 General Concepts 104
  3.6.2 Mechanisms of Gas Transport 105
  3.6.3 Modelling Gas Transport in Soilless Media 107
References 108

4

Irrigation in Soilless Production

J. Heinrich Lieth and Lorence R. Oki

4.1 Introduction 117
  4.1.1 Water Movement in Plants 119
  4.1.2 Water Potential 119
  4.1.3 The Root Zone 122
  4.1.4 Water Quality 124
4.2 Root Zone Moisture Dynamics 126
  4.2.1 During an Irrigation Event 126
  4.2.2 Between Irrigation Events 126
  4.2.3 Prior to an Irrigation Event 127
4.3 Irrigation Objectives and Design Characteristics 128
  4.3.1 Capacity 128
  4.3.2 Uniformity 128
4.4 Irrigation Delivery Systems 130
  4.4.1 Overhead Systems 132
  4.4.2 Surface Systems 134
  4.4.3 Subsurface 137
4.5 Irrigation System Control Methods 141
  4.5.1 Occasional Irrigation 141
  4.5.2 Pulse Irrigation 141
  4.5.3 High Frequency Irrigation 142
  4.5.4 Continuous Irrigation 142
4.6 Irrigation Decisions 143
  4.6.1 Irrigation Frequency 143
  4.6.2 Duration of Irrigation Event 144

4.7   Approaches to Making Irrigation Decisions   145
      4.7.1   'Look and Feel' Method   145
      4.7.2   Gravimetric Method   146
      4.7.3   Time-based Method   146
      4.7.4   Sensor-based Methods   147
      4.7.5   Model-based Irrigation   151
4.8   Future Research Directions   153
References   155

5

TECHNICAL EQUIPMENT IN SOILLESS PRODUCTION SYSTEMS

ERIK VAN OS, THEO H. GIELING AND J. HEINRICH LIETH

5.1   Introduction   157
5.2   Water and Irrigation   158
      5.2.1   Water Supply   158
      5.2.2   Irrigation Approaches   161
      5.2.3   Fertigation Hardware   167
5.3   Production Systems   178
      5.3.1   Systems on the Ground   178
      5.3.2   Above-ground Production Systems   186
5.4   Examples of Specific Soilless Crop Production Systems   192
      5.4.1   Fruiting Vegetables   192
      5.4.2   Single-harvest Leaf Vegetables   194
      5.4.3   Single-harvest Sown Vegetables   195
      5.4.4   Other Speciality Crops   195
      5.4.5   Cut Flowers   197
      5.4.6   Potted Plants   199
5.5   Discussion and Conclusion   201
References   204

6

CHEMICAL CHARACTERISTICS OF SOILLESS MEDIA

AVNER SILBER

6.1   Charge Characteristics   210
      6.1.1   Adsorption of Nutritional Elements to Exchange Sites   216

6.2 Specific Adsorption and Interactions Between Cations/Anions and Substrate Solids 217
    6.2.1 Phosphorus 218
    6.2.2 Zinc 223
    6.2.3 Effects of P and Zn Addition on Solution Si Concentration 224
6.3 Plant-induced Changes in the Rhizosphere 225
    6.3.1 Effects on Chemical Properties of Surfaces of Substrate Solids 225
    6.3.2 Effects on Nutrients Availability 230
    6.3.3 Assessing the Impact of Plants: The Effect of Citric Acid Addition on P Availability 233
6.4 Nutrient Release from Inorganic and Organic Substrates 236
References 239

7

ANALYTICAL METHODS USED IN SOILLESS CULTIVATION

CHRIS BLOK, CEES DE KREIJ, ROB BAAS AND GERRIT WEVER

7.1 Introduction 245
    7.1.1 Why to Analyse Growing Media? 245
    7.1.2 Variation 248
    7.1.3 Interrelationships 248
7.2 Physical Analysis 249
    7.2.1 Sample Preparation (Bulk Sampling and Sub-sampling) 249
    7.2.2 Bulk Sampling Preformed Materials 249
    7.2.3 Bulk Sampling Loose Material 249
    7.2.4 Sub-sampling Pre-formed materials 250
    7.2.5 Sub-sampling Loose Materials 250
7.3 Methods 250
    7.3.1 Bulk Density 250
    7.3.2 Porosity 253
    7.3.3 Particle Size 254
    7.3.4 Water Retention and Air Content 255
    7.3.5 Rewetting 257
    7.3.6 Rehydration Rate 258
    7.3.7 Hydrophobicity (or Water Repellency) 259
    7.3.8 Shrinkage 260
    7.3.9 Saturated Hydraulic Conductivity 261
    7.3.10 Unsaturated Hydraulic Conductivity 262
    7.3.11 Oxygen Diffusion 264
    7.3.12 Penetrability 267
    7.3.13 Hardness, Stickiness 269

7.4 Chemical Analysis 270
    7.4.1 Water-soluble Elements 272
    7.4.2 Exchangeable, Semi- and Non-water Soluble Elements 275
    7.4.3 The pH in Loose Media 276
    7.4.4 Nitrogen Immobilization 277
    7.4.5 Calcium Carbonate Content 277
7.5 Biological Analysis 277
    7.5.1 Stability (and Rate of Biodegradation) 278
    7.5.2 Potential Biodegradability 279
    7.5.3 Heat Evolution (Dewar Test) 279
    7.5.4 Solvita Test™ 279
    7.5.5 Respiration Rate by CO_2 Production 280
    7.5.6 Respiration Rate by O_2 Consumption (The Potential Standard Method) 280
    7.5.7 Weed Test 282
    7.5.8 Growth Test 283
References 286

8

NUTRITION OF SUBSTRATE-GROWN PLANTS

AVNER SILBER AND ASHER BAR-TAL

8.1 General 291
8.2 Nutrient Requirements of Substrate-grown Plants 292
    8.2.1 General 292
    8.2.2 Consumption Curves of Crops 295
8.3 Impact of N Source 300
    8.3.1 Modification of the Rhizosphere pH and Improvement of Nutrient Availability 303
    8.3.2 Cation-anion Balance in Plant and Growth Disorders Induced by NH_4^+ Toxicity 307
8.4 Integrated Effect of Irrigation Frequency and Nutrients Level 310
    8.4.1 Nutrient Availability and Uptake by Plants 311
    8.4.2 Direct and Indirect Outcomes of Irrigation Frequency on Plant Growth 315
8.5 Salinity Effect on Crop Production 318
    8.5.1 General 318
    8.5.2 Salinity-nutrients Relationships 319
    8.5.3 Yield Quality Induced by Salinity 324

8.6 Composition of Nutrient Solution 325
    8.6.1 pH Manipulation 326
    8.6.2 Salinity Control 327
References 328

9

FERTIGATION MANAGEMENT AND CROPS RESPONSE TO SOLUTION RECYCLING IN SEMI-CLOSED GREENHOUSES

BNAYAHU BAR-YOSEF

9.1 System Description 343
    9.1.1 Essential Components 343
    9.1.2 Processes and System Variables and Parameters 344
    9.1.3 Substrate Considerations 346
    9.1.4 Monitoring 354
    9.1.5 Control 355
9.2 Management 359
    9.2.1 Inorganic Ion Accumulation 359
    9.2.2 Organic Carbon Accumulation 365
    9.2.3 Microflora Accumulation 367
    9.2.4 Discharge Strategies 367
    9.2.5 Substrate and Solution Volume Per Plant 369
    9.2.6 Effect of Substrate Type 373
    9.2.7 Water and Nutrients Replenishment 374
    9.2.8 Water Quality Aspects 380
    9.2.9 Fertigation Frequency 381
    9.2.10 pH Control: Nitrification and Protons and Carboxylates Excretion by Roots 383
    9.2.11 Root Zone Temperature 391
    9.2.12 Interrelationship Between Climate and Solution Recycling 393
    9.2.13 Effect of N Sources and Concentration on Root Disease Incidence 395
9.3 Specific Crops Response to Recirculation 397
    9.3.1 Vegetable Crops 397
    9.3.2 Ornamental Crops 405
9.4 Modelling the Crop-Recirculation System 409
    9.4.1 Review of Existing Models 409
    9.4.2 Examples of Closed-loop Irrigation System Simulations 410
9.5 Outlook: Model-based Decision-support Tools for Semi-Closed Systems 416

Acknowledgement 417
Appendix 418
References 419

10

Pathogen Detection and Management Strategies in Soilless Plant Growing Systems

Joeke Postma, Erik van Os and Peter J. M. Bonants

10.1 Introduction 425
    10.1.1 Interaction Between Growing Systems and Plant Pathogens 425
    10.1.2 Disease-Management Strategies 426
    10.1.3 Overview of the Chapter 426
10.2 Detection of Pathogens 427
    10.2.1 Disease Potential in Closed Systems 427
    10.2.2 Biological and Detection Thresholds 428
    10.2.3 Method Requirements for Detection and Monitoring 430
    10.2.4 Detection Techniques 430
    10.2.5 Possibilities and Drawbacks of Molecular Detection Methods for Practical Application 432
    10.2.6 Future Developments 433
10.3 Microbial Balance 434
    10.3.1 Microbiological Vacuum 434
    10.3.2 Microbial Populations in Closed Soilless Systems 435
    10.3.3 Plant as Driving Factor of the Microflora 437
    10.3.4 Biological Control Agents 438
    10.3.5 Disease-suppressive Substrate 440
    10.3.6 Conclusions 441
10.4 Disinfestation of the Nutrient Solution 442
    10.4.1 Recirculation of Drainage Water 442
    10.4.2 Volume to be Disinfected 442
    10.4.3 Filtration 444
    10.4.4 Heat Treatment 446
    10.4.5 Oxidation 447
    10.4.6 Electromagnetic Radiation 449
    10.4.7 Active Carbon Adsorption 450
    10.4.8 Copper Ionisation 451
    10.4.9 Conclusions 451
10.5 Synthesis: Combined Strategies 452
    10.5.1 Combining Strategies 452
    10.5.2 Combining Biological Control Agents and Disinfestation 452
    10.5.3 Non-pathogenic Microflora After Disinfestation 452

10.5.4 Addition of Beneficial Microbes to Sand Filters   453
10.5.5 Detection of Pathogenic and Beneficial Micro-organisms   453
10.5.6 Future   453
Acknowledgements   454
References   454

11

ORGANIC SOILLESS MEDIA COMPONENTS

MICHAEL MAHER, MUNOO PRASAD AND MICHAEL RAVIV

11.1 Introduction   459
11.2 Peat   460
    11.2.1 Chemical Properties   463
    11.2.2 Physical Properties   464
    11.2.3 Nutrition in Peat   466
11.3 Coir   468
    11.3.1 Production of Coir   468
    11.3.2 Chemical Properties   469
    11.3.3 Physical Properties   472
    11.3.4 Plant Growth in Coir   473
11.4 Wood Fibre   473
    11.4.1 Production of Wood Fibre   473
    11.4.2 Chemical Properties   474
    11.4.3 Physical Properties   476
    11.4.4 Nitrogen Immobilization   476
    11.4.5 Crop Production in Wood Fibre   477
    11.4.6 The Composting Process   477
11.5 Bark   479
    11.5.1 Chemical Properties   479
    11.5.2 Nitrogen Immobilization   481
    11.5.3 Physical Properties   481
    11.5.4 Plant Growth   481
11.6 Sawdust   482
11.7 Composted Plant Waste   482
11.8 Other Materials   486
11.9 Stability of Growing Media   487
    11.9.1 Physical and Biological Stability   487
    11.9.2 Pathogen Survival in Compost   489
11.10 Disease Suppression by Organic Growing Media   490
    11.10.1 The Phenomenon and its Description   490

11.10.2 Suggested Mechanisms for Suppressiveness of Compost Against Root Diseases   490
11.10.3 Horticultural Considerations of Use of Compost as Soilless Substrate   494
References   496

12

Inorganic and Synthetic Organic Components of Soilless Culture and Potting Mixes

Athanasios P. Papadopoulos, Asher Bar-Tal, Avner Silber, Uttam K. Saha and Michael Raviv

12.1 Introduction   505
12.2 Most Commonly Used Inorganic Substrates in Soilless Culture   506
    12.2.1 Natural Unmodified Materials   507
    12.2.2 Processed Materials   511
    12.2.3 Mineral Wool   516
12.3 Most Commonly Used Synthetic Organic Media in Soilless Culture   518
    12.3.1 Polyurethane   518
    12.3.2 Polystyrene   520
    12.3.3 Polyester Fleece   521
12.4 Substrates Mixtures — Theory and Practice   523
    12.4.1 Substrate Mixtures — Physical Properties   523
    12.4.2 Substrate Mixtures — Chemical Properties   531
    12.4.3 Substrate Mixtures — Practice   532
12.5 Concluding Remarks   536
Acknowledgements   537
References   537

13

Growing Plants in Soilless Culture: Operational Conclusions

Michael Raviv, J. Heinrich Lieth, Asher Bar-Tal and Avner Silber

13.1 Evolution of Soilless Production Systems   545
    13.1.1 Major Limitation of Soilless- vs. Soil-growing Plants   546

- 13.1.2 The Effects of Restricted Root Volume on Crop Performance and Management 547
- 13.1.3 The Effects of Restricted Root Volume on Plant Nutrition 548
- 13.1.4 Root Confinement by Rigid Barriers and Other Contributing Factors 550
- 13.1.5 Root Exposure to Ambient Conditions 552
- 13.1.6 Root Zone Uniformity 552

13.2 Development and Change of Soilless Production Systems 553
- 13.2.1 How New Substrates and Growing Systems Emerge (and Disappear) 553
- 13.2.2 Environmental Restrictions and the Use of Closed Systems 554
- 13.2.3 Soilless 'Organic' Production Systems 555
- 13.2.4 Tailoring Plants for Soilless Culture: A Challenge for Plant Breeders 557
- 13.2.5 Choosing the Appropriate Medium, Root Volume and Growing System 557

13.3 Management of Soilless Production Systems 561
- 13.3.1 Interrelationships Among Various Operational Parameters 561
- 13.3.2 Dynamic Nature of the Soilless Root Zone 562
- 13.3.3 Sensing and Controlling Root-zone Major Parameters: Present and Future 566

References 567

INDEX OF ORGANISM NAMES 573
SUBJECT INDEX 579

LIST OF CONTRIBUTORS

Rob Baas FytoFocus, The Netherlands.

Asher Bar-Tal Agricultural Research Organization, Institute of Soil, Water and Environmental Sciences, Volcani Center, Bet Dagan, P.O.B. 6, 50250, Israel.

Bnayahu Bar-Yosef Agricultural Research Organization, Institute of Soil, Water and Environmental Sciences, Volcani Center, Bet Dagan, P.O.B. 6, 50250, Israel.

Chris Blok Wageningen UR Greenhouse Horticulture Postbus 20, 2665 ZG Bleiswijk, The Netherlands.

Peter J.M. Bonants Wageningen UR Greenhouse Horticulture P.O. Box 16, 6700 AA Wageningen, The Netherlands.

Theo H. Gieling Plant Research International B.V., P.O. Box 16, 6700 AA Wageningen, The Netherlands.

Uzi Kafkafi Faculty of Agriculture, Hebrew University of Jerusalem, P.O.B. 12 Rehovot, 71600, Israel.

Cees de Kreij Research for Floriculture and Glasshouse Crops, Pater Damiaanstraat 48, 2131 EL Hoofddorp, The Netherlands.

J. Heinrich Lieth Department of Plant Sciences, University of California, Davis, Mailstop 2, Davis, CA, 95616 USA.

Michael Maher Teagasc, Kinsealy Research Centre, Dublin 17, Ireland.

Lorence R. Oki Department of Plant Sciences, University of California, Davis, Mailstop 6, Davis, CA, 95616 USA.

Athanasios P. Papadopoulos Greenhouse and Processing Crops Research Centre, Agriculture and Agri-Food Canada, Harrow, Ontario N0R 1G0, Canada.

Joeke Postma Plant Research International B.V., P.O. Box 16, 6700 AA Wageningen, The Netherlands.

Munoo Prasad Research Centre, Bord na Mona Horticulture, Main Street Newbridge, Co. Kildare, Ireland.

Michael Raviv Agricultural Research Organization, Institute of Plant Sciences, Newe Ya'ar Research Center, P.O.B 1021, Ramat Yishay, 30095, Israel.

Uttam K. Saha Soil and Water Science Department, University of Florida, 2169 McCarty Hall, Gainesville, Florida 32 611, USA.

Avner Silber Agricultural Research Organization, Institute of Soil, Water and Environmental Sciences, Volcani Center, Bet Dagan, P.O.B. 6, 50250, Israel.

Erik van Os Greenhouse Technology, Plant Research International B.V., P.O. Box 16, 6700 AA Wageningen, The Netherlands.

Rony Wallach Faculty of Agriculture, Hebrew University of Jerusalem, P.O.B. 12, Rehovot, 71600, Israel.

Gerrit Wever Stichting RHP, Galgeweg 38, 2691 MG's-Gravenzande, The Netherlands.

PREFACE

Since the onset of the commercial application of soilless culture, this production approach has evolved at a fast pace, gaining popularity among growers throughout the world. As a result, a lot of information has been developed by growers, advisors, researchers, and suppliers of equipment and substrate. With the rapid advancement of the field, an authoritative reference book is needed to describe the theoretical and practical aspects of this subject. Our goal for this book is to describe the state-of-the-art in the area of soilless culture and to suggest directions in which the field could be moving. This book provides the reader with background information of the properties of the various soilless media, how these media are used in soilless production, and how this drives plant performance in relation to basic horticultural operations such as irrigation and fertilization.

As we assemble this book, we are aware that many facets of the field are rapidly changing so that the state-of-the-art is continuing to advance. Several areas in particular are in flux. Two of these are (1) the advent of governmental pressures to force commercial soilless production systems to include recirculation of irrigation effluent and (2) a desire for society to use fewer agricultural chemicals in food production. The authors that have contributed to this book are all aware of these factors, and their contributions to this book attempt to address the state-of-the-art.

This book should serve as reference book or textbook for a wide readership including researchers, students, greenhouse and nursery managers, extension specialists; in short, all those who are involved in the production of plants and crops in systems where the root-zone contains predominantly of soilless media or no media at all. It provides information concerning the fundamental principles involved in plant production in soilless culture and, in addition, may serve as a manual that describes many of the useful techniques that are constantly emerging in this field.

In preparing this book, we were helped by many authorities in the various specialized fields that are covered. Each chapter was reviewed confidentially by prominent scholars in the respective fields. We take this opportunity to thank these colleagues who contributed their time and expertise to improve the quality of the book. The responsibility, however, for the content of the book rests with the authors and editors.

For both of us, the assembly of this book has been an arduous task in which we have had numerous discussions about the myriad of facets that make up this field. This has served to stimulate in us a more in-depth respect for the field and a deeper appreciation for our many colleagues throughout the world. We are very appreciative of all the work that our authors invested to make this book the highest quality that we could achieve, and hope that after all the repeated requests from us for various things, that they are still our friends.

We also note that while no specific agency or company sponsored any of the effort to assemble this book, we are in debt to some extent to various funding sources that supported our research during the time of this book project. This includes BARD (especially Project US-3240-01) and the International Cut Flower Growers Association. Our own employers (The Agricultural Research Organization of Israel and the University of California), of course, supported our efforts to create this work and for that we are deeply grateful.

We also thank our wives, Ayala Raviv and Sharyn Lieth for their understanding and support.

1

SIGNIFICANCE OF SOILLESS CULTURE IN AGRICULTURE

Michael Raviv and J. Heinrich Lieth

1.1 Historical Facets of Soilless Production
1.2 Hydroponics
1.3 Soilless Production Agriculture
    References

1.1 HISTORICAL FACETS OF SOILLESS PRODUCTION

Although we normally think about soilless culture as a modern practice, growing plants in containers aboveground has been tried at various times throughout the ages. The Egyptians did it almost 4000 years ago. Wall paintings found in the temple of Deir el Bahari (Naville 1913) showed what appears to be the first documented case of container-grown plants (Fig. 1.1). They were used to transfer mature trees from their native countries of origin to the king's palace and then to be grown this way when local soils were not suitable for the particular plant. It is not known what type of growing medium was used to fill the containers, but since they were shown as being carried by porters over large distances, it is possible that materials used were lighter than pure soil.

Starting in the seventeenth century, plants were moved around, especially from the Far and Middle East to Europe to be grown in orangeries, in order to supply aesthetic value, and rare fruits and vegetables to wealthy people. An orangery is 'a sheltered place, especially a greenhouse, used for the cultivation of orange trees

FIGURE 1.1 Early recorded instance of plant production and transportation, recorded in the temple of Hatshepsut, Deir el-Bahari, near Thebes, Egypt (Naville, 1913; Matkin et al., 1957).

in cool climates' (American Heritage Dictionary), so it can be regarded as the first documented case of a container-growing system, although soil was mostly used to fill these containers. Orangeries can still be found today throughout Europe. An exquisite example of an organery from Dresden, Germany, is shown in Fig 1.2.

1.1 HISTORICAL FACETS OF SOILLESS PRODUCTION

FIGURE 1.2 The organery at Pillnitz Palace near Dresden, Germany (see also Plate 1).

The orangery at Pillnitz Palace near Dresden Germany was used to protect container-grown citrus trees during the winter. Large doors at the east side allowed trees to be moved in and out so that they could be grown outdoors during the summer and brought inside during the winter. Large floor-to-ceiling windows on the south side allowed for sunlight to enter.

As suggested by the name, the first plants to be grown in orangeries were different species of citrus. An artistic example can be seen in Fig. 1.3.

Two major steps were key to the advancement of the production of plants in containers. One was the understanding of plant nutritional requirements, pioneered by French and German scientists in the nineteenth century, and later perfected by mainly American and English scientists during the first half of the twentieth century. As late as 1946, British scientists still claimed that while it is possible to grow plants in silica sand using nutrient solutions, similarly treated soil-grown plants produced more yield and biomass (Woodman and Johnson, 1946 a,b). It was not until the 1970s that researchers developed complete nutrient solutions, coupled their use to appropriate rooting media and studied how to optimize the levels of nutrients, water and oxygen to demonstrate the superiority of soilless media in terms of yield (Cooper, 1975; Verwer, 1976).

The second major step was the realization that elimination of disease organisms that needed to be controlled through disinfestation was feasible in container-grown production while being virtually impossible in soil-grown plants. In the United States,

FIGURE 1.3 Orangery (from *The Nederlanze Hesperides* by Jan Commelin, 1676).

a key document was the description of a production system that provided a manual for the use of substrates in conjunction with disease control for production of container-grown plants in outdoor nursery production. Entitled *The U.C. System for Producing Healthy Container-grown Plants through the use of clean Soil, Clean Stock, and Sanitation* (Baker, 1957), it was a breakthrough in container nursery production in the 1950s and 1960s and helped growers to such an extent that it became universally adopted since growers using the system had a dramatic economic advantage over competitors that did not use it. This manual described several growing media mixes consisting of sand and organic matter such as peat, bark or sawdust in various specific percentages (Matkin and Chandler, 1957). These became known as 'UC mixes'. It should be noted that in this manual these mixes are called 'soil' or 'soil mixes', largely because prior to that time most container media consisted of a mix of soil and various other materials. That convention is not used in this book; here we treat the term 'soil' as meaning only a particular combination of sand, silt, clay and organic matter found in the ground. Thus, when we talk about soilless substrates in this book, they may include mineral components (such as sand or clay) that are also found in soil, but not soil directly. The term 'compost' was also used as a synonym to 'soil mix' for many years, especially in Europe and the United Kingdom (Robinson and Lamb, 1975), but also in the United States (Boodley and Sheldrake, 1973). This term included what is now usually termed 'substrate' or 'growing medium' and, in most cases, suggests the use of mix of various components, with at least one of them being of organic origin. In this book, we use the term 'growing medium' and 'substrate' interchangeably.

These scientific developments dispelled the notion that growing media can be assembled by haphazardly combining some soil and other materials to create 'potting soil'. This notion was supported in the past by the fact that much of the development of ideal growing media was done by trial and error. Today we have a fairly complete picture of the important physical and chemical characteristics (described in Chaps. 3 and 6, respectively), which are achieved through the combination of specific components (e.g. UC mix) or through industrial manufacture (e.g. stone wool slabs).

Throughout the world there are many local and regional implementations of these concepts. These are generally driven by both horticultural and financial considerations. While the horticultural considerations are covered in this book, the financial considerations are not. Yet this factor is ultimately the major driving force for the formulation of a particular substrate mix that ends up in use in a soilless production setting. The financial factor manifests itself through availability of materials, processing costs, transportation costs and costs associated with production of plants/crops as well as their transportation and marketing. Disposal of used substrates is, in some cases, another important consideration of both environmental and economical implications. For example, one of the major problems in the horticultural use of mineral wool (stone- and glass-wool) is its safe disposal, as it is not a natural resource that can be returned back to nature. Various methods of stone wool recycling have been developed but they all put a certain amount of financial burden on the end-user.

In countries where peat is readily available, perhaps even harvested locally, growers find this material to be less expensive than in countries where it has to be imported from distant locations. As prices of raw materials fluctuate, growers must evaluate whether to use a 'tried and true' component (e.g. peat) or a replacement (e.g. coconut coir) in a recipe that may have proven to work well over the years. In some years the financial situation may force consideration of a change. Since the properties of all substrates and mixes differ from each other, replacement of one particular component (such as peat) with another component might result in other costs or lower quality crops (which may be valued less in the market place), especially if the substitution is with a material with which the grower has less experience. Thus growers throughout the world face the challenge of assembling mixes that will perform as desired at the lowest possible overall cost.

The result of this is that the substrates used throughout the world differ significantly as to their make-up, while attempting to adhere to a specific set of principles. These principles are quite complex, relating to physical and chemical factors of solids, liquids and gasses in the root zone of the plant.

Today the largest industries in which soilless production dominates are greenhouse production of ornamentals and vegetables and outdoor container nursery production. In urban horticulture, virtually all containerized plants are grown without any field soil.

1.2 HYDROPONICS

Growing plants without soil has also been achieved through water culture without the use of any solid substrates. This type of soilless production is frequently termed 'hydroponics'. While this term was coined by Gericke (1937) to mean water culture without employing any substrate, currently the term is used to mean various things to various persons. Many use the term to refer to systems that do include some sort of substrate to anchor or stabilize the plant and to provide an inert matrix to hold water. Strictly speaking, however, hydroponics is the practice of growing plants in nutrient solutions. In addition to systems that use exclusively nutrient solution and air (e.g. nutrient film technique (NFT), deep-flow technique (DFT), aerohydroponics), we also include in this concept those substrate-based systems, where the substrate contributes no nutrients nor ionic adsorption or exchange. Thus we consider production systems with inert substrates such as stone wool or gravel to be hydroponic. But despite this delineation, we have in this book generally avoided the use of the term 'hydroponics' due to the fact that not every one agrees on this delineation.

Initially scientists used hydroponics mainly as a research tool to study particular aspects of plant nutrition and root function. Progress in plastics manufacturing, automation, production of completely soluble fertilizers and especially the development of many types of substrates complemented the scientific achievements and brought soilless cutivation to a viable commercial stage. Today various types of soilless systems exist for growing vegetables and ornamentals in greenhouses. This has resulted with a wide variety of growing systems; the most important of these are described in Chap. 5.

1.3 SOILLESS PRODUCTION AGRICULTURE

World agriculture has changed dramatically over the last few decades, and this change continues, since the driving forces for these changes are still in place. These forces consist of the rapid scientific, economic and technological development of societies throughout the world. The increase in worlds' population and the improvement in the standard of living in many countries have created a strong demand for high-value foods and ornamentals and particularly for out-of-season, high-quality produce. The demand for floricultural crops, including cut flowers, pot plants and bedding plants, has also grown dramatically. The result of these trends was the expanded use of a wide variety of protected cultivation systems, ranging from primitive screen or plastic film covers to completely controlled greenhouses. Initially this production was entirely in the ground where the soil had been modified so as to allow for good drainage. Since the production costs of protected cultivation are higher than that of open-field production, growers had to increase their production intensity to stay competitive. This was achieved by several techniques; prominent among these is the rapid increase in soilless production relative to total agricultural crop production.

The major cause for shift away from the use of soil was the proliferation of soil-borne pathogens in intensively cultivated greenhouses. Soil was replaced by various substrates, such as stone wool, polyurethane, perlite, scoria (tuff) and so on, since they are virtually free of pests and diseases due to their manufacturing processes. Also in reuse from crop to crop, these materials can be disinfested between uses so as to kill any microorganisms. The continuing shift to soilless cultivation is also driven by the fact that in soilless systems it is possible to have better control over several crucial factors, leading to greatly improved plant performance.

Physical and hydraulic characteristics of most substrates are superior to those of soils. A soil-grown plant experiences relatively high water availability immediately after irrigation. At this time the macropores are filled with water followed by relatively slow drainage which is accompanied by entry of air into the soil macropores. Oxygen, which is consumed by plant roots and soil microflora, is replenished at a rate which may be slower than plant demand. When enough water is drained and evapotranspired, the porosity of the soil is such that atmospheric oxygen diffuses into the root zone. At the same time, some water is held by gradually increasing soil matric forces so that the plant has to invest a considerable amount of energy to take up enough water to compensate for transpiration losses due to atmospheric demand. Most substrates, on the other hand, allow a simultaneous optimization of both water and oxygen availabilities. The matric forces holding the water in substrates are much weaker than in soil. Consequently, plants grown in porous media at or near container capacity require less energy to extract water. At the same time, a significant fraction of the macropores is filled with air, and oxygen diffusion rate is high enough so that plants do not experience a risk of oxygen deficiency, such as experienced by plants grown in a soil near field capacity. This subject is quantitatively discussed in Chap. 3 and its practical translation into irrigation control is described in Chap. 4.

Another factor is that nutrient availability to plant roots can be better manipulated and controlled in soilless cultivation than in most arable soils. The surface charge and chemical characteristics of substrates are the subjects of Chap. 6, while plant nutrition requirement and the methods of satisfying these needs are treated in Chap. 8. Chapter 7 is devoted to the description of the analytical methods, used to select adequate substrate for a specific aim, and other methods, used to control the nutritional status during the cropping period, so as to provide the growers with recommendations, aimed at optimizing plant performance.

Lack of suitable soils, disease contamination after repeated use and the desire to apply optimal conditions for plant growth are leading to the worldwide trend of growing plants in soilless media. Most such plants are grown in greenhouses, generally under near-optimal production conditions. An inherent drawback of soilless vs. soil-based cultivation is the fact that in the latter the root volume is unrestricted while in containerized culture the root volume is restricted. This restricted root volume has several important effects, especially a limited supply of nutrients (Dubik et al., 1990; Bar-Tal, 1999). The limited root volume also increases root-to-root competition since there are more roots per unit volume of medium. Chapter 2 discusses the main functions of the root system while Chap. 13 quantitatively analyses the limitations imposed by a restricted root volume. Various substrates of organic origin are described in Chap. 11, while Chap. 12 describes substrates of inorganic origin and the issue of potting mixes. In both the chapters, subjects such as production, origin, physical and chemical characteristics, sterilization, re-use and waste disposal are discussed.

Container production systems have advantages over in-ground production systems in terms of pollution prevention since it is possible, using these growing systems, to minimize or eliminate the discharge of nutritional ions and pesticide residues thus conserving freshwater reservoirs. Simultaneously, water- and nutrient-use efficiencies are typically significantly greater in container production, resulting in clear economic benefits. Throughout the developed countries more and more attention is being directed to reducing environmental pollution, and in the countries where this type of production represent a large portion of agricultural productivity, regulations are being created to force recirculation so as to minimize or eliminate run-off from the nurseries and greenhouses. The advantages and constraints of closed and semi-closed systems in an area that is currently seeing a lot of research and the state-of-the-art is described in Chap. 9. The risk of disease proliferation in recirculated production and the methods to avert this risk are described in Chap. 10.

The book concludes with a chapter (Chap. 13) dealing with operational conclusions. In many cases practitioners are treating irrigation as separate from fertilization, and in turn as separate from the design and creation of the substrate in which the plants are grown. This chapter addresses the root-zone as a dynamic system and shows how such a system is put together and how it is managed so as to optimize crop production, while at the same time respecting the factors imposed by society (run-off elimination, labour savings, etc.). Another subject which is mentioned in this chapter is the emerging trend of 'Organic hydroponics' which seems to gain an increasing popularity in some parts of the world.

One of the main future challenges for global horticulture is to produce adequate quantities of affordable food in less-developed countries. Simple, low-cost soilless production systems may be part of the solution to the problems created by the lack of fertile soils and know-how. The fact that a relatively small cultivated area can provide food for a large population can stimulate this development. This, in turn, should stimulate professionals to find alternatives to current expensive and high-tech pieces of equipment and practices, to be suitable and durable for the needs of remote areas. One of the most important advantages of soilless cultivation deserves mentioning in this context: in most of the developing countries, water is scarce and is of low quality. By superimposing the FAO's hunger map (Fig. 1.4) on the aridity index map (Fig. 1.5), it is clear that in many regions of the world such as sub-Sahalian Africa, Namibia,

1.3 SOILLESS PRODUCTION AGRICULTURE

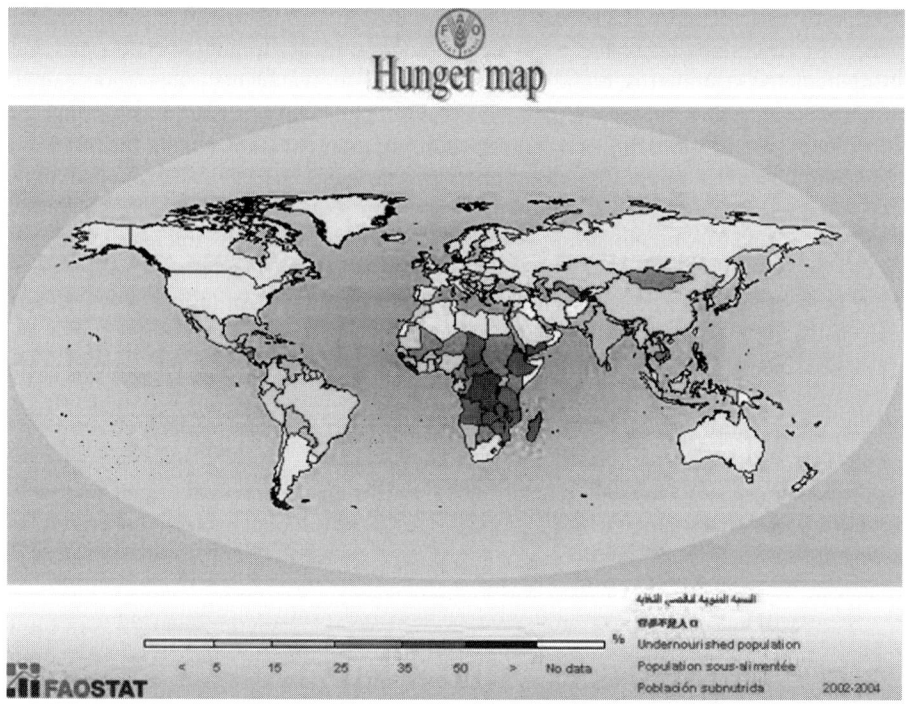

FIGURE 1.4 Percentage of undernourished population around the globe (see also Plate 2; with kind permission of the FAO).

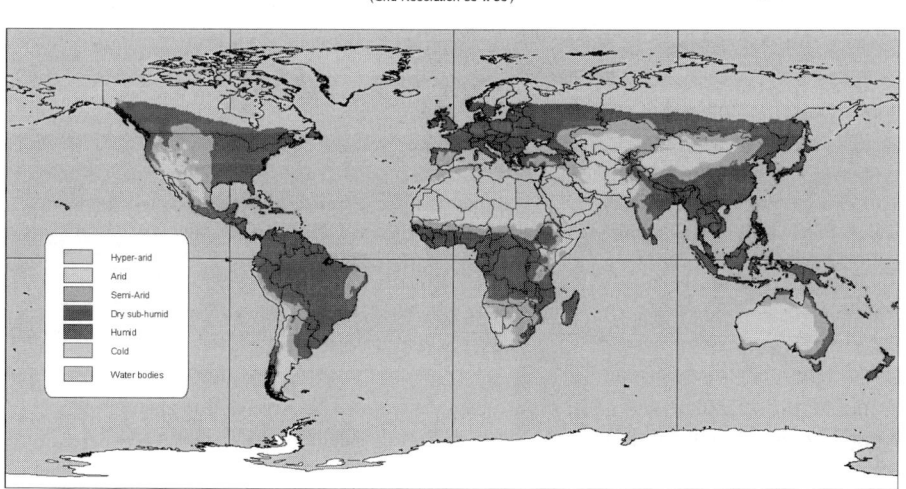

FIGURE 1.5 Aridity index around the globe (see also Plate 3; with kind permission of the FAO).

Mongolia and so on, a large part of the population suffers hunger mainly due to water scarcity. Since water-use efficiency of soilless plant production (and especially in recirculated systems) is higher than that of soil-grown plants, more food can be produced with such systems with less water. Also, plants growing in such systems can cope better with higher salinity levels than soil-grown plants. The reason for this is the connection between ample oxygen supply to the roots and their ability to exclude toxic ions such as Na^+ and to withstand high osmotic pressure (Kriedemann and Sands, 1984; Drew and Dikumwin, 1985; Drew and Lauchli, 1985). It is interesting to note, in this respect, that soilless cultivation is practised in large scale in very arid regions such as most parts of Australia, parts of South Africa, Saudi Arabia and the southern part of Israel. In none of these countries, hunger is a problem.

The science of plant production in soilless systems is still young, and although much work has been done, many questions still remain unanswered. One of the purposes of this book is to focus on the main issues of the physical and chemical environment of the rhizosphere and to identify areas where future research is needed so as to take further advantage of the available substrates and to propose desirable characteristics for future substrates and growing practices to be developed by next generation of researchers.

REFERENCES

Baker, K.F. (ed.) (1957). *The U.C. System for Producing Healthy Container-Grown Plants through the use of Clean Soil, Clean Stock, and Sanitation*. University of California, Division of Agricultural Sciences, p. 332.

Bar-Tal, A. (1999). The significance of root size for plant nutrition in intensive horticulture. In *Mineral Nutrition of Crops: Fundamental Mechanisms and Implications* (Z. Rengel, ed.). New York: Haworth Press, Inc., pp. 115–139.

Boodley, J.W. and Sheldrake, R. Jr. (1973). Boron deficiency and petal necrosis of 'Indianapolis White' chrysanthemum. *Hort. Science*, **8**(1), 24–26.

Cooper, A.J. (1975). Crop production in recirculating nutrient solution. *Sci. Hort.*, **3**, 251–258.

Drew, M.C. and Dikumwin, E. (1985). Sodium exclusion from the shoots by roots of *Zea mays* (cv. LG 11) and its breakdown with oxygen deficiency. *J. Exp. Bot.*, **36**(162), 55–62.

Drew, M.C. and Lauchli, A. (1985). Oxygen-dependent exclusion of sodium ions from shoots by roots of *Zea mays* (cv. Pioneer 3906) in relation to salinity damage. *Plant Physiol.*, **79**(1), 171–176.

Dubik, S.P., Krizek, D.T. and Stimart, D.P. (1990). Influence of root zone restriction on mineral element concentration, water potential, chlorophyll concentration, and partitioning of assimilate in spreading euonymus (*E. kiautschovica Loes.* 'Sieboldiana'). *J. Plant Nutr.*, **13**, 677–699.

Gericke, W.F. (1937). Hydroponics – crop production in liquid culture media. *Science*, **85**, 177–178.

Kriedemann, P.E. and Sands, R. (1984). Salt resistance and adaptation to root-zone hypoxia in sunflower. *Aust. J. Pl. Physiol.*, **11**(4): 287–301.

Matkin, O.A. and Chandler, P.A. (1957). The U.C.-type soil mixes. In *The U.C. System for Producing Healthy Container-Grown Plants Through the Use of Clean Soil, Clean Stock, and Sanitation* (K.F. Baker, ed.). University of California, Division of Agricultural Sciences, pp. 68–85.

Matkin, O.A., Chandler, P.A. and Baker, K.F. (1957). Components and development of mixes. In *The U.C. System for Producing Healthy Container-grown Plants through the Use of Clean Soil, Clean Stock, and Sanitation* (K.F. Baker, ed.). University of California, Division of Agricultural Sciences, pp. 86–107.

Naville, E.H. (1913). *The Temple of Deir el-Bahari (Parts I–III)*, Vol. 16. London: Memoirs of the Egypt Exploration Fund. pp. 12–17.

Robinson, D.W. and Lamb, J.G.D. (1975). *Peat in Horticulture*. Academic Press, London, xii, 170pp.

Verwer, F.L.J.A.W. (1976). Growing horticultural crops in rockwool and nutrient film. In Proc. 4th Inter. Congr. On Soilless Culture. ISOSC, Las Palmas, pp. 107–119.

Woodman, R.M. and Johnson, D.A. (1946a). Plant growth with nutrient solutions. II. A comparison of pure sand and fresh soil as the aggregate for plant growth. *J. Agric. Sci.*, **36**, 80–86.

Woodman, R.M. and Johnson, D.A. (1946b). Plant growth with nutrient solutions. III. A comparison of sand and soil as the aggregate for plant growth, using an optimum nutrient solution with the sand, and incomplete supplies of nutrients with 'once-used' soil. *J. Agric. Sci.*, **36**, 87–94.

2

FUNCTIONS OF THE ROOT SYSTEM

UZI KAFKAFI

2.1 The Functions of the Root System
2.2 Depth of Root Penetration
2.3 Water Uptake
2.4 Response of Root Growth to Local Nutrient Concentrations
2.5 Interactions Between Environmental Conditions and Form of N Nutrition
2.6 Roots as Source and Sink for Organic Compounds and Plant Hormones
References
Further Readings

2.1 THE FUNCTIONS OF THE ROOT SYSTEM

The root is the first organ to emerge from the germinating seed. In fact, it is packed in the seed in an emerging position (Fig. 2.1)

Root elongation is a continuous process that is essential for healthy plant growth. It allows the plant to explore new soil volumes for water and nutrients and as a support for the growing plant. Any reduction in the rate of root elongation negatively affects the growth and function of aerial organs which, eventually, is translated into restricted plant development. Continuous root elongation is needed for mechanical anchoring, water uptake, nutrient uptake and the avoidance of drought conditions. Both touch

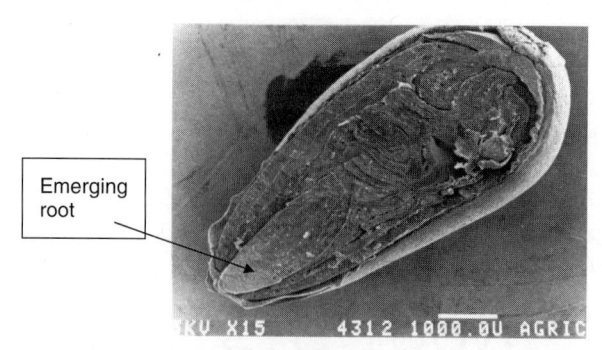

FIGURE 2.1 Root starting to emerge from a cotton seed. Picture taken 3 h after imbibition at 33°C (picture taken by A. Swartz and U. Kafkafi, unpublished).

and gravity are essential stimuli for normal root growth, engaging thigmotropic and gravitropic response mechanisms, respectively. Thigmotropism is the response of a plant organ to mechanical stimulation. Intuitively, one can imagine that the gravitropic and thigmotropic responses in roots are intimately related. In fact, a recent study by Massa and Gilroy (2003) has suggested that proper root-tip growth requires the integration of both the responses. Environmental conditions known to impair root growth involve physical factors such as soil compaction, shortage of water, insufficient soil aeration and extreme soil temperatures, and chemical factors such as saline and sodic soils, soils with low pH (which causes toxicity and an excess of exchangeable aluminium), shortage or excess of plant macronutrients and shortage or excess of heavy metals. Oxygen plays a critical role in determining root orientation, as well as root metabolic status. Oxytropism enables roots to avoid oxygen-deprived soil strata and may also be a physiological mechanism designed to reduce the competition between roots for water and nutrients, as well as oxygen (Porterfield and Musgrave, 1998).

In container-grown plants, the role of the roots in maintaining water and nutrient uptake and production of growth-regulating hormones is essentially the same as in field-grown plants. The main difference is that in containers, the entire root system is exposed to every environmental change, whereas in the field, deep roots sense changes in daily root temperature and moisture more slowly than the surface roots. Therefore, in container-grown plants, there is no room to escape human-imposed mistakes, especially those involving critical temperature and moisture values, or nutrient deficiencies and salt accumulation: the smaller the root container, the higher the risk of root damage due to human mismanagement in the greenhouse. The most extreme example is the Bonsai (literally 'plant in a tray'), a plant-growth system that is based on severe limitation of plant root growth by confining the roots to a small container with rigid walls. Common consequences of mistakes in container-grown plants are as follows: root death due to oxygen deficiency as a result of over-irrigation, especially during hot growing periods, salt accumulation when the root zone is not sufficiently leached by irrigation water, ammonium toxicity due to high concentrations of fertilizer during periods of high temperature, or exposure of the plant container to direct radiation from the sun which can cause over-heating, and consequently root death (Kafkafi, 1990).

Seedlings growing in containers, especially tree seedlings confined to containers for long periods, frequently develop roots in the space between the medium and the container wall and at the bottom of the container. This is due to compaction of the growing medium, which causes oxygen deficiency and root death at the centre of the container (Asady et al., 1985). This phenomenon can be even more pronounced when the medium contains organic matter which is subject to decomposition by oxygen-consuming microorganisms. Downward root growth is a natural response to gravitropism and hydrotropism, typical to all active roots. In containers, however, this frequently results in a root mat developing at the bottom, where it may be exposed to oxygen deficiency due to competition among the roots for oxygen associated with the frequent accumulation of a water layer at the bottom of the container.

The container material and its colour affect the absorbed radiation and have an important effect on the temperature to which the roots are exposed. Clay pots keep roots cool due to evaporative cooling from the container walls. Plastic or metal containers cause root temperature to rise above ambient air temperature, with devastating implication on hot days, especially when high ammonium-N is present (Kafkafi, 1990). Sand used as a growing medium ingredient may cause aeration and compaction problems in container-grown plants. Each physical impact on the container, during frequent handling, causes the sand to compress and reduces air spaces, increasing the mechanical resistance to root penetration. The successful use of light-weight growing media, for example peat, pumice, artificial stone- or glass-wool, is due to their high water-retention capacity while maintaining sufficient aeration for the root zone. The limited commercial distribution of plant-growth systems such as the nutrient film growing technique (NFT; Cooper, 1973) is probably due to the demand for continuous care and maintenance. It has been shown that even a 10-min. shortage of oxygen supply can stop root growth, and a 30-min. shortage results in death of the elongation zone above the root tip (Huck et al., 1999).

This chapter presents early observations on the importance of root growth and elongation as well as recent work that has unveiled the reasons underlying the field observations.

Root architecture in the soil profile is determined by two main factors: (1) genetic architecture of the root system, and (2) local soil and water constraints facing roots during their propagation. Schroeder-Murphy et al. (1990) reported how RHIZO-GEN, a two-dimensional root graphic model, visually demonstrates root demographic responses to soil bulk density, water content and relative soil fertility. A review of the different shapes and topologies of plant root systems is described by Fitter (2002). Quick root penetration and distribution in the soil provide the plant with anchorage and prevent its lodging at later stages of plant growth. Continuous observation of root growth is difficult and may interfere with the natural environment for normal root growth (Voorhees, 1976; Böhm, 1979; Smucker, 1988). However, the use of modern rhizotrons enables close, non-destructive observation, allowing very intricate root studies (Majdi et al., 1992). Examples of such studies include root architecture and spatial and temporal development (Walter et al., 2002; Gautam et al., 2003), the effects of soil moisture and bulk density (Asady et al., 1985; Asady and Smacker, 1989;

Smucker et al., 1991; Smucker and Aiken, 1992; Kuchenbuch and Ingram, 2002), nutrition (Wang et al., 2004, 2005), real-time formation of root-mycorrhizal associations (Schubert et al., 2003), ambient conditions (Norby et al., 2004), root pathogens including parasitic weeds (Wang et al., 2003; Eizenberg et al., 2005) and even root and mycorrhizal effects on mineral weathering and soil formation (Arocena et al., 2004). Although most of the work conducted in rhizotrons is meant to be relevant to broader soil conditions, much of it can be considered relevant to soilless conditions as well. The mathematical tool Fractal has been used to predict the expected direction and architecture of plant root systems. Early efforts presented two-dimensional (Tatsumi et al., 1989) and, later, three-dimensional descriptions of root proliferation in field crops (Ozier-Lafontaine et al., 1999). This mathematical tool is of great importance and is very helpful compared to the destructive methods of direct observation (Weaver, 1926). However, the Fractal method assumes that the resistance to root penetration is uniform in all directions, which is far from the reality of field conditions (Asady and Smucker, 1989). Soil control of root penetration is relevant to growing medium conditions, and special care should be taken to prevent compaction while filling the growing pot and before the root reaches the container wall, which mechanically changes the direction of its growth.

The most popular explanation for how plants perceive gravitational changes in their environment is the starch/statolith hypothesis, whereby starch-filled amyloplasts are displaced when gravitational stimulation changes (Hasenstein et al., 1988; Kiss et al., 1989). These amyloplasts are found in the columellar cells of the root cap (statoliths) and in the endodermal cells of the shoot (statocytes). When laser ablation was used to remove the central root columellar cells in *Arabidopsis*, a large inhibitory effect was seen with respect to root curvature in response to gravitational stimulation (Blancaflor et al., 1998). When a tap root encounters resistance in the soil, it elongates along the compacted layer until it finds a crack through which the root cap can continue to penetrate downward, as dictated by the statoliths within it (Sievers et al., 2002). An example of this is shown in Fig. 2.2 for a cotton (*Gossypium hirsutum*) root in a sand dune soil profile.

FIGURE 2.2 Field-grown cotton root deflected by subsoil compaction (Figure 3c in Bennie, 1991).

Once a sandy soil is compacted, it is very hard to reverse the compaction stage (Bennie, 1991). In containers, sand tends to become compacted due to careless handling of the pots, physical impact on growing surfaces when moving the pots and over-irrigation. Once sand is compacted in a pot, the process is essentially irreversible and root growth is restricted.

2.2 DEPTH OF ROOT PENETRATION

When soil compaction is not a limiting factor, root systems of crop plants vary with their botanical origin. Corn (*Zea mays* L.), carrot (*Daucus carota* L.) and white cabbage (*Brassica oleracea* L. convar. *Capitata* L. Alef. var *alba* DC) (Kristensen and Thorup-Kristensen, 2004) demonstrate the general principles: the monocots, with their multiple parallel roots, penetrate to relatively shallow depths, while dicotyledonous plants have a tap root that may reach 2.5 m into the soil, much deeper than any feasible mechanical agricultural practice. Cabbage stores carbohydrate products in the root for the following year's flower growth, while corn transfers all of its reserves to the above-ground grains. André et al. (1978) showed that the rate of root growth in corn reaches its maximum when the plant reaches its flowering growth stage (at day 60–65 after emergence in their cultivar), and that root growth rate then declines in parallel to the growth-rate decline of the aerial part of the plant. The depth of the rooting system has important biological and agronomic consequences: the deeper the roots, the better the plant's ability to withstand environmental extremes such as long periods of drought and short frost events, and to access, for example, leached nitrogen compounds (Shani et al., 1995).

Technically, the root consists of three main sections, starting from its tip: the root cap, the elongation zone (with root hairs) and the mature part of the root from which the lateral roots emerge.

The root cap enables easier root penetration through the soil by means of its secretion of slimy mucilage and sloughing of outer cells (Iijima et al., 2003). To estimate the contribution of the root cap to facilitating root elongation and overcoming soil resistance to root progress, these authors measured root-penetration resistance in decapped primary roots of maize (*Zea mays* L.) and compared it with that of intact roots in a loose soil (bulk density $1.0\,g\,cm^{-3}$) and in a compact soil ($1.4\,g\,cm^{-3}$), with penetration resistances of 0.06 and 1.0 MPa, respectively. While root elongation rate and diameter were the same for decapped and intact roots when the plants were grown in the loose soil, in the compacted soil, the elongation rate of the decapped roots was only about half of that of intact ones. However, when a root's elongation is restricted, Iijima et al. (2003) reported that its radial growth is 30 per cent higher than that of non-restricted roots due to the flow-down of photosynthetic products and their accumulation in the root tissue. The presence of a root cap alleviates much of the mechanical impedance to root penetration, and enables roots to grow faster in compacted soils. The lubricating effect of the root cap was about 30 per cent and was unaffected by the degree of soil compaction at a constant penetrometer resistance of 0.52 MPa (Iijima et al., 2004).

The effects of soil compaction on the growth and seed yield of pigeon pea (*Cajanus cajan* L.) in a coastal oxisol were studied in a field experiment in Australia by Kirkegaard et al. (1992), and for dry edible bean (*Phaseolus vulgaris* L.) by Asady et al. (1985). Plant response was related to the ability of the root system to overcome the soil-strength limitations of the compacted soil. Under dry soil conditions, root penetration was restricted by high soil strength. Root restriction resulted in reduced water uptake and shoot growth.

Roots exert a profound influence on the soils with which they are in contact (Bowen and Rovira, 1991; McCully, 1999). This influence includes exertion of physical pressure, localized drying and rewetting (Aiken and Smucker, 1996), changes in pH and redox potential (Marschner et al., 1986), mineralogical changes, nutrient depletion and the addition of a wide variety of organic compounds (including root-cap mucilage and surfactants). Roots also affect the soil indirectly through the activities of the specific microbial communities that become established in the rhizosphere (McCully, 1999).

The acquisition of water and nutrients is one of the major functions of roots. In agricultural soils, the dimensions of the root's absorbing surface as well as the ability to explore non-depleted soil horizons are important factors for mineral nutrient uptake (Silberbush and Barber, 1983). These factors are used to explain genotypic differences in growth and yield under conditions of low soil fertility (Sattelmacher and Thoms, 1991). Mineral nutrient supply influences the size and morphology of the whole plant as well as the root system. These effects are due to the type of nutrient, its concentration range near the root, the type of field application used, the soil type and the soil environmental conditions.

2.3 WATER UPTAKE

After germination, the main purpose of root elongation is to penetrate the soil as quickly as possible to secure a supply of water for the emerging cotyledons. The direction of early root growth is not coincidental and has been defined as hydrotropism, or growth (or movement) of an organ towards water or towards soil with a higher water potential (Takahashi et al., 1997). In watermelon (*Citrulus lanatus* (Thunb.) Matsum. & Nakai), 2 weeks after seeding, root length had already reached 20 cm, while the cotyledons where at the opening stage (Fig. 2.3). All of the energy needed for root development during the initial establishment stage is derived from storage compounds found in the seeds. In dicotyledonous plants, a tap root develops and lateral roots start to appear a few centimetres from the root cap, while in monocotyledons such as wheat (*Triticum aestivum* L.), several roots develop simultaneously to support the developing plant.

The rate of growth of the tap root varies among plants. In watermelon, rates of 2–3 cm per day have been observed at a soil temperature of 23°C (Fig. 2.3).

The downward direction of plant roots is controlled by gravity sensors in the root cap (Sievers et al., 2002). The depth of root proliferation varies with soil environmental factors such as temperature, compaction, moisture and zone of nutrient abundance.

FIGURE 2.3 Watermelon seedlings 6 days after germination. At the cotyledon-opening stage, roots are already 20 cm long (Kafkafi, unpublished).

Ozanne et al. (1965) studied the root distribution of 12 annual pasture species. Most of the roots were found in the soil's top 10 cm with an exponential decline of root density with depth. However, variations were found between species. Increasing soil compaction at 10–12 cm reduced the yield of cotton (Taylor and Burnett, 1964). However, irrigation and maintaining moisture in the upper part of the soil resulted in no loss of yield. This suggests that deep root penetration is a necessary trait for survival under natural soil and climatic environmental conditions.

Improvement of root-penetration ability in durum wheat (*Triticum turgidum* L.) has become an important breeding target to overcome yield losses due to soil compaction and drought (Kubo et al., 2004). Eight weeks after planting, no genotype penetrated through a Paraffin–Vaseline disc of 0.73 MPa hardness. However, the number of roots penetrating through a disc of 0.50 MPa hardness showed significant differences among genotypes, with the highest in an Ethiopian landrace genotype and the lowest in a North American genotype, indicating large genotypic variation for root-penetration ability in durum wheat. Increasing soil bulk density decreased the total length of primary and of the lateral roots of 17-day-old eucalyptus seedlings by 71 and 31 per cent, respectively, with an increase in penetrometer resistance from 0.4 to 4.2 MPa, respectively (Misra and Gibbons, 1996). The authors concluded that primary roots are more sensitive to high soil strength than lateral roots, most probably due to the differences in root diameter between them. Deep rooting is essential for securing water in relatively dry soils, but when water and nutrient supply is secure, plants can be satisfied with shallow rooting. Such conditions are frequently found in acid soils, usually in wet climatic zones, where deep soil layers are usually high in exchangeable aluminium which restricts root growth (Pearson et al., 1973).

In a classic study, Weaver (1926) showed that root growth responds to local nutrient conditions and especially to phosphorus (P) concentration. The radioactive ^{32}P has been used to non-destructively estimate root proliferation in the soil (Böhm, 1979). However, any detection method, including the most accurate field method (Jacobs et al., 1969), has its limitations. Any interference with soil structure by penetrating to various soil depths produces a tunnel through which the instrument used to introduce the chemical isotope to a particular depth was inserted. Roots that encounter such a tunnel penetrate faster downwards due to lesser soil compaction.

The effect of local water availability on root distribution was described by Shani et al. (1995). When the irrigation was supplied from a trickle source on the soil surface, the main root system developed in the upper 20 cm. At the same site, with the same plant cultivar, when the water was supplied at a depth of 50 cm, the roots developed deep in the soil with a very fine root system.

During the early stages of seed development, before the first leaf opens and transpiration begins, water uptake by the root must be controlled by osmotic factors. Once transpiration starts through the stomatal openings, the volume of water flow through the plant exceeds any amount of flow that could be explained by osmotic forces. Moreover, it is well established that water transport through root cell membranes towards the xylem tubes is through aquaporins (Daniels et al., 1994) which enable a large flux of water to penetrate the lipid layers of the cells.

Deep rooting is necessary for the plant to make use of deep moisture reserves in the soil (Pearson et al., 1973). The ability of the plant to take up water from deep locations is a function of root distribution in the soil profile. Any factor that prevents or retards root propagation will affect the plant's ability to draw water from deep soil layers, a crucial characteristic for plants grown in arid and semiarid climates. A common factor stopping root growth is shortage of oxygen. Cotton root stopped growing immediately upon replacement of soil oxygen with nitrogen gas (Huck et al., 1970). In a field experiment, continuous sprinkler irrigation, supplying 70 mm of water over an 8-h period on a swelling clay soil, during active root growth of corn resulted in huge losses of nitrate by denitrification (Bar Yosef and Kafkafi, 1972), most probably due to the few hours of oxygen shortage in the soil. It is most probable that in field and greenhouse practices, excess irrigation involve short-term deficiencies in oxygen supply that causes extended root damage which is seldom noticed by farmers.

As soil dries out, the roots growing in it may shrink and retain only partial contact with the soil. The effect of root shrinkage on the water inflow across soils and roots with various hydraulic conductivities was modelled by Nye (1994), and the effect of changes in stem and root radius was modelled by Genard et al. (2001). The water inflow was very sensitive to root radius, root shrinkage, root hydraulic conductivity and water potential in the bulk soil and at the root endodermis relative to standard conditions of a secondary root in loam at field capacity. The inflow was insensitive to the density of rooting and to soil hydraulic conductivity, except in dry sandy soil, where the inflow was lower than the root hydraulic conductivity. Genard et al. (2001) calculated that loss of full root contact with the soil might decrease the inflow by a factor of up to about three. In very dry soils, water-vapour transfer across the air gaps between the root and soil surfaces can contribute usefully to the total water inflow. Their model could explain the field observation of stem flow in field-grown chickpea (*Cicer arietinum* L.).

When soil water moisture is below 'field capacity', water loss from the plants might exceed supply from the soil. Our measurements on chickpeas grown in a heavy

2.3 WATER UPTAKE

clay soil (montmorillonite-type clay) (Fig. 2.4) show that in replenishing irrigation by only 75 per cent of evaporation for field-grown chickpea, the root must have shrunk, as sap flow shows a decline during midday while at the same time, at 100 per cent irrigation, a peak in water transport is observed. These measurements can be explained by midday root shrinkage (Huck et al., 1970) that causes water films around the roots to show discontinuities, which brings about a decrease in water movement towards the root, and results in closure of the stomata at midday, as shown in Figure 1. The night observations support this explanation, since the roots that shrank during the day slowly swell again and start to deliver water that moves through the xylem, even during the night. In the early morning hours, close to about 10 a.m., the stem

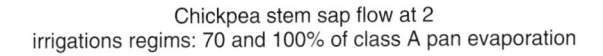

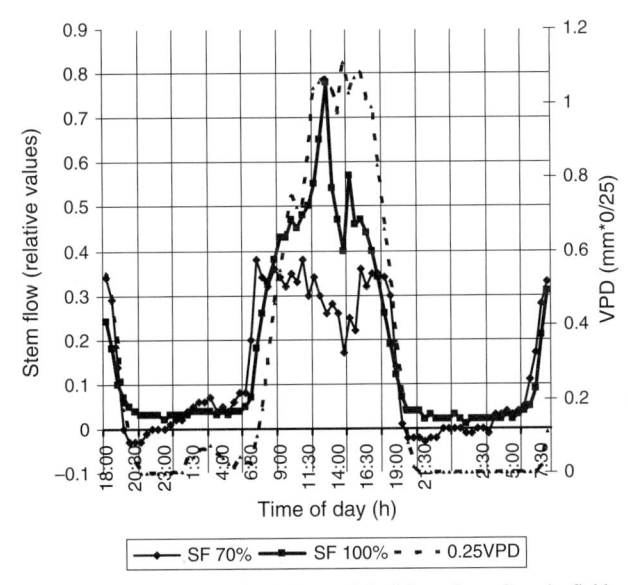

FIGURE 2.4 Continuous monitoring for 1 day and 2 nights of sap flow in field-grown chickpea at two levels of irrigation: 100 and 70 per cent of evaporation from a class A pan. The vapor-pressure deficit (VPD) is zero during the night and starts to increase at 0700h, while sap flow starts about half an hour earlier. Under the wet conditions – SF100% – flow does not change during the night and following VPD, reaches its first peak at about 1300h. VPD shows a drop at 1330h and a new peak at 1400h due to a passing cloud. The stem flux follows the same pattern and starts to decline faster than the VPD from 1500h. The drier irrigation treatment deviates from the behaviour of the wet irrigation treatment. During the first night, sap flow reaches a minimum at 2030h, but a slow positive increase of sap flow is monitored in the stem during the first and second nights. At about 0600h, the stem sap flow in the dry treatment increases faster during the first hour of daylight and then starts to fluctuate, reaching a minimum at 1400h, exactly at maximum VPD. This inverse relationship to sap flow behaviour in the wet treatment can be explained by assuming that water transport to the root is slower than water losses from the plant. As a result, the root must shrink and the water-film continuity around it is temporarily disrupted. When VPD becomes zero during the night, the roots regain water and restore turgor pressure (Raw data were kindly supplied by Phytech Inc., Rehovot Israel).

flow in both treatments is the same, due to similar evaporative demand; however, with an increase in evaporative demand, the rate of water flow through the stem starts to decline towards midday and then to increase with the decrease in evaporative demand in the afternoon. In container-grown plants, where the soil volume is restricted, special care must be taken to maintain root-volume moisture without drying out or creating over-saturation which may cause root death. Figure 2.4 implies that the morning hours are the most important in terms of water supply to the roots and tops. This subject is further discussed in Chaps. 3 and 4.

2.4 RESPONSE OF ROOT GROWTH TO LOCAL NUTRIENT CONCENTRATIONS

2.4.1 NUTRIENT UPTAKE

Mineral nutrients usually enter the plant through the root cells. The root membranes exhibit distinctive selectivity in terms of which chemical elements are allowed entry. Among the cations, potassium (K) is positively selected while sodium (Na) entry is selectively prevented or, in many plants, secreted (Marschner, 1995), despite the fact that Na concentration in the soil solution is many times higher than that of K. Recently, the gene that controls secretion of Na from the root cells was identified in rice (Ren et al., 2005). Glycophytes have developed mechanisms to exclude Na from the xylem flow (Marschner, 1995); plants adapted to growing in salt marshes have developed ways to transport the Na to the leaves, where special glands are used to excrete the salt (Liphschitz and Waisel, 1974). Salt (NaCl) is usually abundant in nature, in saline soils and in water. Inside the plant, chloride (Cl) moves quite freely in the xylem flow (Xu et al., 2000), while Na flow in the plant is restricted. As a result of the relative charge enrichment of Na in the soil, it is balanced by negative charge on the clay particles and by the HCO_3^- anion in the soil solution.

The basic difference between Na and K ions lies in their hydrated ionic radius, which is reflected in the higher hydration energy of Na relative to K (Eisenman, 1960). These variations between Na and K were explored in detail by Hille (1992). Garofoli et al. (2002) calculated the selectivity for K over Na by using the hydration energy of the ions, and membrane-channel size and composition to explain the K selectivity. The K channels must have a specific opening that fits the size of K^+ but not the hydrated Na^+. The reason for the variability in Na transport through plant roots is still not clear (Flowers and Flowers, 2005). There is a remarkable difference among plants with regard to Na uptake and transport to different organs (Marschner, 1995). This explains why some plants can grow under saline conditions, such as in sea water (mangroves), while other land plants vary in their degree of salt sensitivity.

2.4.2 ROOT ELONGATION AND P UPTAKE

Phosphate ions have a very short diffusion pathway in the soil (Black, 1968). Early hypotheses suggested that diffusion of P towards the root is the main way in which roots acquire P (Barber, 1962). By definition, a nutrient is regarded as deficient if its addition results in an increase in plant yield and growth. When P is deficient in the soil, roots elongate to further depths in an effort to find it (Weaver, 1926). If a source of available P is encountered on the way, the roots respond by local proliferation around the P source (Weaver, 1926; Blanchar and Coldwell, 1966). However, the mechanism by which a root senses and reacts to P is not fully understood. Kafkafi and Putter (1965) suggested that when the response curves of a plant to added fertilizer P are concave upward (sigmoid shape), assuming uniform P-source distribution, one can assume that the average distance the root has to move from one point source of fertilizer P to the next is not enough to supply the plant's need for P. Another explanation that might account for this situation is when root exploration does not provide the required P. Once the distances between P sources start to overlap, the plant response to additional P sources follows the regularly diminishing increment response curve.

Beck et al. (1989), using ^{33}P labelling, demonstrated that in corn roots, the distance from which P is withdrawn from the soil around the root does not exceed the radius of the root plus 1-mm long root hairs. They showed that the primary root of corn is active throughout the life span of the plant; however, the root hairs last only 2–3 days. They observed that most of the plant's P uptake is due to the development of lateral roots. The main P-uptake region was found around the root tip, where proton extrusion was observed (Schaller and Fischer, 1985). They concluded that P is predominantly acquired during the early developmental stages of the corn plant.

However, when corn plants were grown in a nutrient solution that was replaced daily, André et al. (1978) showed that P is taken up continuously by the roots during almost all of the growing period and in a relatively large proportion during the grain-filling stage. The question remains: Is it the plant's physiology or the soil conditions that control P uptake? From the above studies, it is safe to assume that if the top soil in the field, which usually contains the highest amount of P, is maintained under moist and aerated conditions, roots will continue to take up P until the end of grain-filling in corn. Wheat, on the other hand, when the level of P in the soil is high at germination, will take all the P needed by the plant by about day 100 of growth (Williams, 1948), and in the last 40 days of growth, P in the grain is derived from the vegetative parts. However, when a plant is deficient in P, the roots continue to take up P from the soil until the seeds finish developing. It is therefore safe to conclude that it is the internal metabolism of the plant that controls the root's uptake of P from the soil. Glass (2002) came to a similar conclusion based on many further studies.

In the field, Stibbe and Kafkafi (1973) showed that when P is placed at a depth of more than 20 cm in a P-deficient soil, the plant reaches the P source about 20 days later than plants seeded in soil in which P fertilizer was applied to the soil surface. A 20-d delay is crucial to rain-fed wheat crops, where an early shortage of P cannot be compensated for by late root excavation of P. The response of the plant to P should therefore include the timing of supply in relation to physiological stages of plant development. In wheat, the size of the panicle is determined in the first 3 weeks after germination (Sibony and Pinthus, 1988). Wheat grown on soil rich in available P has strong vigour and, late in the season, produces non-fertile tillers that compete with the limited water in the soil. As a result of late water shortage, the grain filling shrinks, resulting in low grain weight and a high yield of stubble (Kafkafi and Halevy, 1974).

In legumes, Walk et al. (2006) explained the differences between bean genotypes in P uptake from the upper soil by means of genotypic differences in adventitious root development.

Barber (1962) tried to identify the mechanisms by which nutrients move in the soil towards roots. He based his calculations on total nutrient uptake by the plant at harvest on the one hand, and equilibrium concentrations of nutrients in the soil solution on the other. His calculations suggested that nitrate, calcium (Ca) and magnesium (Mg) move towards the root with the mass flow of the soil solution while K and P move to the root by diffusion. Jungk (2002) examined the 40 years of work on nutrient transport towards roots and concluded that even with soil nitrate (which is present only in the soil solution, being a monovalent anion), movement towards the roots does not follow Barber's assumption and cannot a priori be assumed to be supplied to the root by mass flow with the soil solution. The problem with Barber's simple approach, in contrast to his nutrients flow model, was that nutrients are not homogeneously distributed in the soil (Glass, 2002) and no specific role, except as a diffusion sink, was given to the root in mobilizing soil minerals towards it. It is well known that roots excrete protons (Römheld, 1986) and organic anions (Imas et al., 1997a,b). The fact that plant roots excrete citric acid was known to Dyer in 1894 (as cited by Russell, 1950). Dyer suggested using citric acid extraction of soils to estimate their level of available P. The ability of sweet clover (*Melilotus officinalis* L.) to dissolve non-water-soluble rock phosphate and make it ready for use by the next crop was described by Bray and Kurtz (1945); the same effect was found for lupin (*Lupinus albus* L.) by Russell (1950). This fact has been known to farmers and used in practice for many years, although the exact organic acid excretions from the roots were not known at the time. P may become phytotoxic when accumulated by plants to high concentrations. Certain plant species, such as *Verticordia plumosa* L., suffer from P toxicity at solution concentrations far lower than those tolerated by most other plant species (Silber et al., 2002). Exposure of *V. plumosa* plants to a solution containing as low as 3 mg P l^{-1} resulted in growth inhibition and symptoms of P toxicity. Observations on transgenic tomato (*Lycopersicon esculantum* Mill.) plants that showed high hexokinase activity due to over-expression of Arabidopsis hexokinase (AtHXK1) revealed senescence symptoms similar to the symptoms of P toxicity in *V. plumosa*. The resemblance of these two plants' symptoms suggested a role for increased sugar metabolism in P toxicity in plants. To test this hypothesis, Silber et al. (2002) determined the amount of hexose phosphate, the product of hexokinase, in *V. plumosa* leaves grown at various P levels in the nutrient solution. Positive correlations were reported between P concentration in the medium, P concentration in the plant, hexose phosphate concentration in leaves and P-toxicity symptoms. At the same time, foliar zinc (Zn) application suppressed P-toxicity symptoms, and reduced the level of hexose phosphate and inhibited hexokinase activity in leaves. Based on these results, they suggested that P toxicity involves sugar metabolism via increased activity of hexokinase, which accelerates senescence (Silber et al., 2002). Such plant sensitivity to high soil P content is of global interest, as recently suggested by Wassen et al. (2005). They observed the disappearance of endangered plant species on a worldwide scale due to P application in agricultural soils. Future gene transfer might use this trait in grain crops that are able to excavate soil P which has accumulated in the soil in unavailable forms to present-day crops. Lewis and Quirk (1965, 1967), working in potted plants with very high concentrations of P in the soil, concluded that only a few meters of roots were enough to supply all the P needed by a wheat plant, while in the field the root length per plant is much greater. Silber et al.'s (2002) study demonstrated that P acquisition by the root cannot be regarded only as a mechanical sink of P flow towards the root.

P uptake by the root is not only a function of soil-solution P concentration as measured by soil extraction. The type of N source near the root has a profound effect on P uptake. A fertilizer band that contained ammonium (NH_4) with P fertilizer contributed more P to the root than a nitrate fertilizer (Tisdale and Nelson, 1966). In practice, band applications of fertilizers create high local concentrations that influence plant uptake of P. After 23 days of growth, corn took up six times more P when given in the band with NH_4 sulphate, as compared with mixing in the band with Ca-nitrate (Tisdale and Nelson, 1966).

When NH_4-N is taken up, proton excretion results in P solubilization, even from non-soluble P compounds in the soil (Marschner, 1995).

Under natural aerobic conditions, most of the N is converted to nitrate which is usually the main source of N for plant roots. Under oxygen deficiency in the soil, the oxidized NO_3-N is reduced to N_2, which escapes as a gas to the atmosphere. During mineralization, the organic N is released as an NH_4^+ cation that adsorbs to the clay particles in the soil. Plants can use both NH_4 and nitrate forms of N but general plant performance, dry matter production and the ionic balance of nutrient uptake are affected by the form of N available in the root zone (Kafkafi, 1990; Sattelmacher et al., 1993; Marschner, 1995). The uptake of P, iron (Fe), molybdenum (Mo) and Zn is dependent on the pH around the root (Marschner et al., 1986). This pH is a function of the proportion of ammonium-to-nitrate uptake by the plant.

2.4.3 INFLUENCE OF N FORM AND CONCENTRATION

Control of the N form that is taken up by the roots has a major effect on root growth and their uptake of water and nutrients. The nitrate concentration range in soil solution that is optimal for plant uptake is 5–15 mM l^{-1}. In the case of NH_4, only a much lower concentration is considered safe for plants (Moritsugu et al., 1983). NH_4 concentration value is valid only for plants grown in nutrient solutions or on chemically inert media such as stone wool, silica sand or perlite, since NH_4 is strongly adsorbed to the soil and its concentration in the soil solution is low in clay-containing soils. In soils with low cation-exchange capacity, N supplied in NH_4 form might cause considerable damage when applied close to the roots. The pH of the soil solution controls the equilibrium of NH_4 in solution according to the general equation: $K_w(NH_3)\,(aq)(H^+) = K_b\,(NH_4^+)$, where K_w is the water constant and K_b is the base constant. Using this equation,

Bennet and Adams (1970) demonstrated the effect of solution pH on the ratio of NH_3 to NH_4^+ and the toxic effect of ammonia on root growth as affected by solution pH.

The forms of mineral N that might be present near a growing root are mainly NH_4 or nitrate and, in transit, urea and nitrite during urea hydrolysis. All commercial fertilizers contain one or more of these forms of nitrogen in the final formula. In the field, all N fertilizers end up as nitrate, provided time is allowed for the oxidizing soil bacteria to convert NH_4 and urea to nitrate.

Positive effects of NH_4 at very low concentrations (Cox and Resienauer, 1973) have been reported for wheat. Moritsugu et al. (1983) carefully maintained a constant pH in the root zone and showed that at 5 mM, NH_4 has a detrimental effect on spinach (*Spinacia oleracea* L.) and Chinese cabbage (*Brassica rapa*), a very big negative effect in tomato, cucumber (*Cucumis sativum* L.) and carrot, but a relatively lesser negative effect in rice (*Oryza sativa*), corn and sorghum (*Sorghum bicolor* L.). However, when very low NH_4 concentrations (0.05 mM) were automatically maintained near the roots, even very sensitive plants such as Chinese cabbage grew very well (Moritsugu and Kawasaki, 1983). Their results suggested that the concentration of NH_4 near the root has a vital effect on root growth. Kafkafi (1990) suggested that the variation in carbohydrate allocation between leaves and roots is the main factor in plants' sensitivity to NH_4 nutrition. Plants that allocate a large proportion of their carbohydrates to the roots [e.g. corn (Smucker, 1984) and rice] can stand higher external NH_4 concentrations near the root. A negative effect of high NH_4 concentrations on root growth has been observed in many investigations (Maynard et al., 1968; Wilcox et al., 1985). Growth retardation in the presence of high NH_4 concentrations in the nutrient solution has been related to several mechanisms: induced nutrient deficiency of other ions (Barker et al., 1967; Wilcox et al., 1985) and interference in ionic balance. Furthermore, depletion of soluble sugars due to detoxification of NH_4 (Breteler, 1973) and the uncoupling of electron transport and ATP synthesis have also been suggested (Puritch and Barker, 1967).

2.5 INTERACTIONS BETWEEN ENVIRONMENTAL CONDITIONS AND FORM OF N NUTRITION

2.5.1 TEMPERATURE AND ROOT GROWTH

The relationship between root-zone temperature and N nutrition has been considered in many studies. When a nutrient solution contains both NH_4 and nitrate ions, NH_4 uptake is higher the lower the temperature (Clarkson et al., 1986; MacDuff and Jackson, 1991). These results suggest that, in the temperature range of 3–11°C, NH_4 is preferred over nitrate. Cox and Resienauer (1973) argued that the lower energy demand for NH_4 assimilation than for nitrate could explain the stimulatory effect of NH_4 on plant growth in the low temperature range. At high root-zone temperatures (>25°C), however, plant tolerance to high NH_4 concentrations is often reduced (Kafkafi, 1990). This has been attributed to the low carbohydrate concentration in the cytoplasm available for detoxification of cytoplasmic ammonia (NH_3). The pH in the cytoplasm is about 7.3 to 7.5 (Marschner, 1995). This means that the instantaneous concentration of non-dissociated NH_3 could accumulate to toxic levels if sugar is not present nearby. At high root temperatures, sugar in the root is rapidly consumed by cell respiration (André et al., 1978). NH_4 metabolism is restricted to the root, where the sugar supply detoxifies the free NH_3 produced in the cytoplasm (Marschner, 1995). The combination of low sugar concentration at high root temperature and increasing concentrations of NH_3 inside the cell is dangerous to cell survival, since a temperature point is reached at which all the sugar in the root is consumed and nothing is left to prevent NH_3 toxicity (Ganmore-Newmann and Kafkafi, 1985; Kafkafi, 1990). Ali et al. (1994, 1996) showed that at low root temperatures, nitrate transport to the leaves is restricted while NH_4-fed plants are able to transfer NH_4 metabolites to the leaves.

The optimum temperature for root growth is species, and even cultivar, specific (Cooper, 1973). This author summarized early studies on plant growth and nutrient-uptake responses to root temperatures. For most commercially grown crops, the response curve to root temperature shows an optimum with a slow decreasing slope towards the cold zone and a sharp negative response when root temperatures are above the plant's optimum as shown in Fig. 2.5.

The knowledge of such a response curve for any specific crop is essential to nursery and glasshouse container-grown crops, especially in hot climates. As mentioned earlier, in a container, all of the roots are exposed to the temperature extremes; the plant has no room to escape and plant death is observed with the combination of high temperature and NH_4 (Ganmore-Newmann and Kafkafi, 1980, 1983).

Under low temperature conditions, sensitive plants such as cucumber and melon (*Cucumis melo* L.) start losing water from their leaves at dawn, while the roots are still cold. This results in total loss of plants due to morning desiccation (Shani et al., 1995).

Figure 2.6 shows that when light starts at 0600 h, water flow starts immediately in the stem to maximum values at about 1200 h. A general observation is that water flux starts to decrease after midday and stops almost completely during the dark hours.

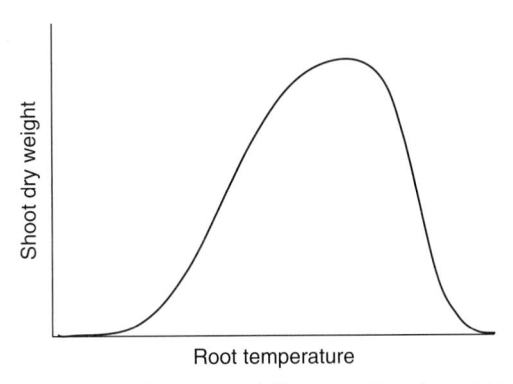

FIGURE 2.5 Generalized shape of temperature influence on shoot dry weight (following Cooper, A.J. (1973), by permission).

Water flow rate in tomato at cold and normal root temperatures

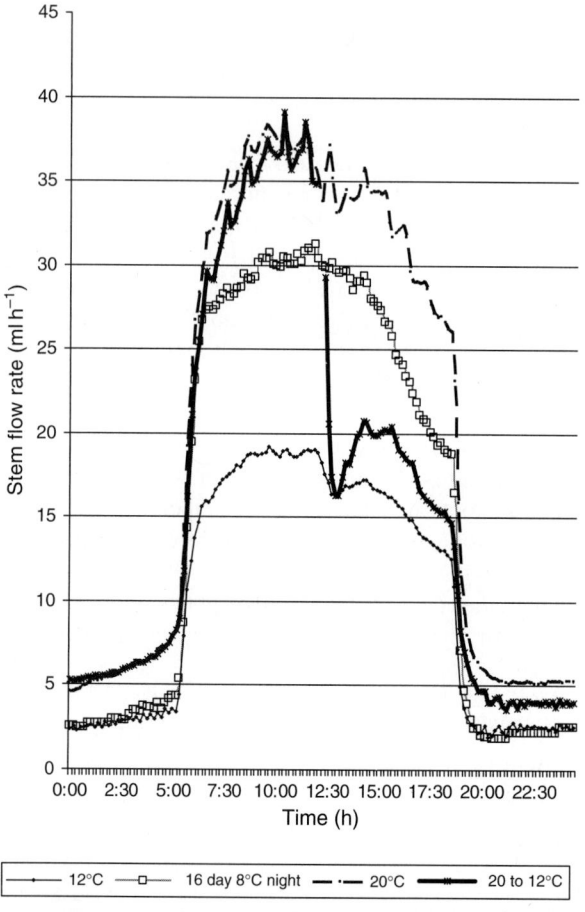

FIGURE 2.6 Root temperature influence on water transport through tomato plant stems. The roots were kept at four different regimes: a constant 12°C; 8°C at night and 16°C during the daylight hours; constant 20°C; 20°C at night until 12:00 a.m. and then transfer to 12°C (based on Ali et al. (1996), with kind permission from Marcel Dekker Inc.).

The fourth treatment (heavy black line in Fig. 2.6) shows how quickly the root responds to a decrease in root temperature. When the roots were cooled from 20 to 12°C, water transport decreased within 30 min to the same value as that of a parallel plant that was kept at a constant 12°C (Ali et al., 1996).

A rapid plant response to lowering root temperature was also reported for cucumbers by Lee et al. (2002). They examined root tips from cucumber seedlings grown at 8°C for varying periods ranging from 15 min to 96 h. Marked changes in the ultrastructure of cortical cells were observed within only 15 min of exposure. The effect of chilling injury included alterations in cell walls, nuclei, mitochondria, plastids and ribosomes. The extent of the alterations varied greatly among cells, moderate to severe alterations in cellular components being observed among adjoining cells. Measurements of root pressure using the root-pressure probe showed a sudden, steep drop in response to lowering the temperature of the bathing solution from 25 to 8°C. The effect of low temperature on fatty-acid unsaturation and lipoxygenase activity was examined in cucumber and figleaf gourd (*Cucurbita ficifolia* Bouché) cells by Lee et al. (2005). Their results suggested that the degree of non-saturation of root plasma membrane lipids correlates positively with chilling tolerance. Water transport across root systems of young cucumber seedlings was measured following exposure to low temperature (8–13°C) for varying periods of time by Lee, Singh, Chung et al. (2004). These authors also evaluated the amount of water transported through the stems using a heat-balance sap-flow gauge. Following the low temperature treatment, hydrogen peroxide (H_2O_2) was localized cytochemically in root tissues, as measured by the oxidation of $CeCl_3$. The effects of H_2O_2 on the hydraulic conductivity of single cells in root tissues, and on the H^+-ATPase activity of isolated root cells' plasma membranes, were also measured. Cytochemical evidence suggested that exposure of roots to low temperature stress causes the release of H_2O_2 in the millimolar range around the plasma membranes.

In response to a low root temperature (13°C), the hydraulic conductivity of the root decreased by a factor of four. Decreasing root temperatures from 25 to 13°C increased the half-times of water exchange in a cell by a factor of six to nine. Lee, Singh, Chung et al. (2004) showed that only a small amount of water is transported when cucumber roots are exposed to 8°C. In the field, such conditions brought about melon plant desiccation in the early morning after a chilling night (Shani et al., 1995). When root temperature dropped below 25°C, there was a sharp drop in the root pressure and hydraulic conductivity of excised roots of young cucumber seedlings, as measured with a root-pressure probe (Lee, Singh, Chung et al., 2004). A detailed analysis of root hydraulics provided evidence for a larger reduction in the osmotic component of the root water potential (77 per cent) in comparison with the hydrostatic component (34 per cent) in response to exposure of the root system to 13°C. They concluded that the rapid drop in water permeability in response to low temperature is largely caused by a reduction in the activity of the plasma membrane H^+-ATPase, rather than loosening of the endodermis which would cause substantial solute losses. They related water permeability of the root cell membrane at low temperatures to changes in the activity (open/closed state) of the water channels.

The effects of high root temperature on plant growth, leaf elemental composition, root respiration and sugar content in cucumber plants were studied by Du and Tachibana (1994). Root dry weight, leaf area and leaf concentrations of most nutrients were all reduced as the root temperature was raised to 35°C, and in particular to 38°C. High root temperature markedly enhanced intact root respiration. This increase was due to stimulation of an alternative respiration pathway. On the other hand, cytochrome respiration deteriorated at high root temperatures. They suggested that disturbance of carbohydrate metabolism in the root is a primary factor responsible for the growth inhibition of cucumber roots grown at supra-optimal root temperatures.

The main interactions of root temperature with NH_4 concentration in the nutrient solution can be summarized as follows: the higher the root temperature, the higher the consumption of sugar in the root by respiration (Marschner, 1995). The concentration of NH_4 in a nutrient solution affects the rate of NH_4 ion uptake. Inside the cytoplasm, the pH is about 7.3 and therefore NH_4^+ is transformed into NH_3 within a very short time, causing a reduction in root growth (Bennet and Adams, 1970). The metabolic detoxification of NH_3 consumes sugar in the root cell. At low root temperatures, the consumption of sugar by respiration in the root cells is low, and strawberry (*Fragaria* × *ananassa* Duchense) plants were found to be able to withstand 7 mM NH_4^+ in the nutrient solution. However, at 32°C, the plants died because the high rate of root respiration consumed all the sugar, leaving none to detoxify the NH_4^+ (7 mM in the nutrient solution) (Ganmore-Newmann and Kafkafi, 1985).

Since in potted plants, all of the roots sense the variation in external temperature, controlling the ammonium concentration in nutrient-solution composition according to variations in root temperature is vital to greenhouse crop production.

2.5.2 ROLE OF Ca IN ROOT ELONGATION

The role of Ca in ameliorating crop performance in acid soils rich in exchangeable aluminum was clearly demonstrated by Pearson et al. (1973). The use of $CaSO_4$ to ameliorate saline and sodic soils has been practised for many years (Richards et al., 1954). Kafkafi (1991) suggested that the basic differences in plant tolerance to salinity were due to the specific surface-charge density of root cells and that the plant's selectivity for Ca increases with increases in surface-charge density of the root membranes. To test this suggestion, Yermiyahu et al. (1994) studied root elongation in four melon cultivars that differ in their salt sensitivity to find out whether the plasma membrane was the site of salinity tolerance in the root cell. They concluded that salt tolerance is not due to variations in the plasma membrane, but rather to variations in the cell wall of the elongation zone (Yermiyahu et al., 1999).

At an external concentration of 40 mM NaCl, the increase in root elongation with increasing Ca in solution is obvious. Figure 2.7 shows that increasing Ca concentration from 1 to 10 mmol l^{-1} increases root growth. As the sum of concentrations increased, no osmotic effects could be related to these results. The authors concluded that in this concentration range, the cell-wall exchange complex was responding to the competition of Ca with Na and prevented a negative effect on root growth.

Na toxicity in the concentration range of 0–40 mM cannot therefore be described as an osmotic effect (Fig. 2.7); above 40 mM NaCl, osmotic effects start to influence root growth as seen from the negative slope of root elongation with increasing salt concentration. However, even at a very high concentration of saline solution, Ca helps the root to overcome the increasing Na concentrations. These characteristics of the cell wall are most likely responsible for Ca's amelioration of plant tolerance to salinity.

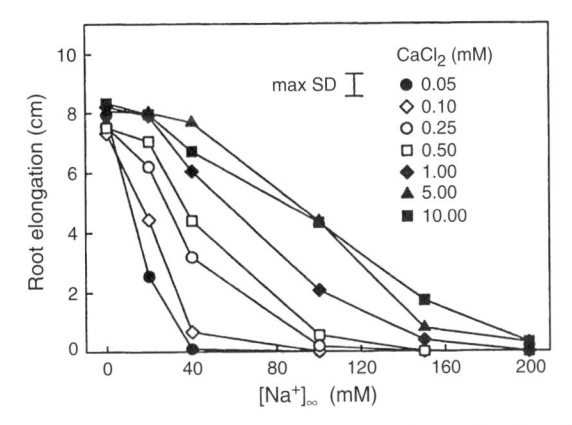

2.5 ENVIRONMENTAL CONDITIONS AND FORM OF N NUTRITION

FIGURE 2.7 Elongation of melon roots 5 days after germination as a function of NaCl concentration in the presence of various $CaCl_2$ concentrations in the growing solution (Fig. 4, Yermiyahu, U. et al. (1997), by permission, Heidelberg@springer.com).

Demidchik et al. (2004) measured the effects of free oxygen radicals on plasma membrane Ca^{2+}- and K^+-permeable channels in plant root cells. They proposed two functions for the cation-channel activation by free oxygen radicals: (1) initialization/amplification of stress signals and (2) control of cell elongation in root growth. In an independent work, Liszkay et al. (2004) hypothesized that the cell-wall-loosening reaction controlling root elongation is affected by the production of reactive oxygen intermediates, initiated by the NAD(P)H-oxidase-catalysed formation of superoxide radicals ($O_2^{\cdot B}$) at the plasma membrane and culminating in the generation of polysaccharide-cleaving hydroxyl radicals ($^{\cdot}OH$) by cell-wall peroxidase. Their results showed that juvenile root cells transiently express the ability to generate $^{\cdot}OH$ and to respond to these radicals by wall loosening, in passing through the growing zone. Moreover, studies with inhibitors indicated that $^{\cdot}OH$ formation is essential for normal root growth (Liszkay et al., 2004).

2.5.3 LIGHT INTENSITY

High light intensity on growing plants generally stimulates root growth and increases root-to-shoot ratios. Using various N sources (nitrate vs. NH_4), Precheur and Maynard (1983) proposed a beneficial effect of high light intensities for NH_4-grown plants, as the carbohydrate pool necessary for NH_4 detoxification is enhanced. However, Zoronoza et al. (1987) reported lower root dry weight in pepper (*Capsicum annum* L.) plants grown with 20 per cent NH_4 when compared to all nitrate-treated plants at high light intensity, whereas roots were stimulated by NH_4 at low light intensity. Magalhaes and Wilcox (1983) found that tomato shoots are also severely affected and the detrimental effect of high light intensity on the plant's NH_4 tolerance was attributed to the accelerated translocation of NH_4 from the roots to the shoots, and the consequent accumulation of toxic NH_4 levels there. However, these conclusions should be regarded with caution as it was shown by Kafkafi and Ganmore-Neumann (1997) that the levels of free NH_4 reported in various papers for vegetative parts of plant tissues are most probably artefacts due to the disintegration of short-chain amino compounds in the leaves during the analytical procedure for total N determination in leaf tissues. Sattelmacher et al. (1993) clearly demonstrated the importance of light intensity on root growth. The differences in root dry weight between similar N-level treatments increased with increasing light intensity. The form of the N seemed to be of minor importance in their experiment. However, they stressed that it should be kept in mind that root dry weight is not the most suitable parameter to describe root growth. Based on their results, root morphology is usually more responsive to nutritional and environmental treatments.

2.5.4 pH

Nitrate or NH_4 uptake by plant roots changes the pH of the medium (Marschner, 1995). However, different plants respond differently to the same external pH. To stress the importance of plant metabolism on the pH of the root rhizosphere, the cover of Marschner's book (1995, 2nd edn) shows that in the same nitrate medium, chickpea acidifies the root surroundings while corn increases the pH. The susceptibility of NH_4-treated roots to low pH may result from the lower cytoplasmic and vacuolar pH of their tips, as demonstrated by Gerendas et al. (1991).

2.5.5 UREA

Urea is the most common N fertilizer everywhere in the world except the United States (FAO statistics). It undergoes substantial changes and transformations in the soil that, given enough time, produce nitrate in its final form. The classical work by Court et al. (1962) demonstrates the sequence of events in urea transformation. Urea's transformation to NH_4 results in an immediate rise in soil pH to toxic levels if applied at relatively high concentrations, as in the case of fertilizer bands placed near plant roots. At pH values above 7.7, nitrobacter activity is inhibited (Russell, 1950), resulting in the accumulation of toxic levels of nitrite in the soil. When applied near emerging seedlings, total plant death was reported (Court et al., 1962). At a high local concentration, after application of a band of urea fertilizer, it may take 4–5 weeks for the nitrite to be completely oxidized to nitrate and become a safe N source for plants. The toxicity of ammonia $(NH_3)_{aq}$ to plant roots was studied in detail by Bennet and Adams (1970), who clearly specified the toxicity hazards to cotton roots and the precautions to take when NH_3 gas or liquid NH_3 are given as fertilizers.

While the uptake rate of different forms of N depends on the pH of the medium, it can also influence the pH of the medium (Marschner and Röhmheld, 1983). Poor growth of NH_4-treated roots has been attributed to acidification of the rhizosphere (Kirkby and Hughes, 1970). The effect of N form on rhizospheric pH affects the availability of P and several micronutrients (Marschner et al., 1986) which can significantly affect root and total plant growth.

2.5.6 MYCORRHIZA–ROOT ASSOCIATION

In nature, plants may obtain N and P via their association with an array of microorganisms from the arctic region to the tropics; one such association is with fungal species, known as mycorrhizal association (Harley, 1969). The mycorrhizal association is usually specific to soil site and specific plants, enabling the plant survival in their natural habitats (Harley, 1969) and supplying the plant roots with P when soluble P concentration in the soil is very low. Usually, this plant-fungus association disappears when the external soil P concentration increases due to heavy application of P fertilizers. Therefore, in conventional container-grown plants, the role of mycorrhizal associations is negligible. However, in the production of organically certified transplants, mycorrhizal inoculation is beneficial (Raviv et al., 1998). Mulligan et al. (1985) reported that root-mycorrhizal associations are reduced by soil compaction associated with soil tillage.

2.6 ROOTS AS SOURCE AND SINK FOR ORGANIC COMPOUNDS AND PLANT HORMONES

Leaves are the 'source' of carbohydrate compounds in the plant, while roots and propagation organs consume these compounds, thereby functioning as 'sinks'. The amount of total fixed carbon (C) moving from the leaves to the roots varies with plant species (Smucker, 1984). In legume plants, C translocation to the root might reach 60–70 per cent of total photosynthetic C fixation (Pate et al., 1974). The variation among plants is significant. When roots play a function in carbohydrate storage, to enable the following year's propagation, roots become the main C storage organ for the plant. Austin (2002) reviewed roots as a source for human food. Fifteen species of plants accumulate C in their roots which are used as food for man, in all parts of the world. The huge biosynthetic potential of roots as a source of food was recently described (Vivanco et al., 2002) and the use of root-derived metabolites in medicine has been reviewed by Yaniv and Bachrach (2002). The wide range of chemicals produced in roots is not only beneficial for use by man, they are mainly there to protect the plants from the numerous soil microorganisms and other soil dwellers.

2.6.1 HORMONE ACTIVITY

Plant roots respond to water or salt stress by increasing the synthesis of abscisic acid (ABA). However, mechanical impedance increases the concentration of indole acetic acid (IAA) (Lachno, 1983), the hormone that initiates root formation. Root elongation and growth is an irreversible process resulting from an irreversible increase in cell volume. Pilet (2002) specified five critical parameters that are essential for root cell growth: (1) the extensibility of their cell walls, (2) their permeability to water, (3) osmotic potential, (4) their proton extrusion, and (5) Ca availability. The first four are properties of the root cells themselves, and their responses to external changes occur without any human intervention. The Ca concentration around the root plays a critical role in cell-wall extension and serves as an antidote to increasing Na concentrations in that area, as previously described (Yermiyahu et al., 1997; Fig. 2.6). The balance of hormones present in the growing and elongating roots- IAA, ABA, cytokinins, gibberellins and ethylene controls root growth. Any change in environmental conditions near the growing root tip is translated into an increase or decrease in these internal compounds and, as a result, root elongation rate is affected.

Tanimoto (2002) suggested that gibberellic acid (GA) may inhibit root elongation by increasing endogenous levels of auxin and/or increasing the sensitivity to auxin as suggested for stems (Ockerse and Galston, 1967). The mechanical extensibility of cell walls is thought to involve GA-mediated root elongation. GA was found to increase cell-wall extensibility of pea (*Pisum sativum* L.) root cells (Tanimoto, 1994; Tanimoto and Yamamoto, 1997). It also modified the sugar composition and molecular mass of cell-wall polysaccharides in pea root cells (Tanimoto, 1995).

Root growth precedes leaf growth in the emerging seed. The information transported from the root to the growing centres of the upper plant influences the development and growth of the entire plant. The root is exposed to many deficiencies and conditions that affect root as well as whole-plant growth. Shortage of water affects all plant activities while excess water induces oxygen deficiency and growth arrest. Transport of minerals to the root and their uptake by the root are influenced by a wide array of physical, chemical and biological activities of soil microorganisms. Knowledge of all the hurdles to root growth is an important tool for increasing world food production.

REFERENCES

Aiken, R.M. and Smucker, A.J.M. (1996). Root system regulation of whole plant growth. *Annu. Rev. Phytopathol.*, **34**, 325–346.

Ali, I.E., Kafkafi, U., Sugimoto. Y. and Inanaga, S. (1994). Response of sand-grown tomato supplied with varying ratios of NO$_3$/NH$_4$ at constant and variable root temperatures. *J. Plant Nutr.*, **17**, 2001–2024.

Ali, I.E., Kafkafi, U., Yamaguchi, I., et al. (1996). Effect of low root temperature on sap flow rate, soluble carbohydrates, nitrate contents and on cytokinins and gibberellin levels in root xylem exudates of sand grown tomato. *J. Plant Nutr.*, **19**, 619–634.

André, M., Massimino, D. and Daguenet, A. (1978). Daily patterns under the life cycle of a maize crop II. Mineral nutrition, root respiration and root excretion. *Physiol. Plant.*, **44**, 197–204.

Arocena, J.M., Göttlein, A. and Raidl, S. (2004). Spatial changes of soil solution and mineral composition in the rhizosphere of Norway-spruce seedlings colonized by *Piloderma croceum*. *J. Plant Nutrition and Soil Science*, **167**(4), 479–486.

Asady, G.H. and Smucker, A.J.M. (1989). Compaction and root modifications of soil aeration. *Soil Sci. Soc. Am. J.*, **53**, 251–254.

Asady, G.H., Smucker, A.J.M. and Adams M.W. (1985). Seedling test for the quantitative measurement of root tolerances to compacted soil. *Crop Sci.*, **25**, 802–806.

Austin, D.F. (2002). Roots as a source of food. In *Plant Roots the Hidden Half* (Y. Waisel, A. Eshel and U. Kafkafi, eds). New York: Marcel Dekker, pp. 1025–1044.

Barber, S.A. (1962). A diffusion and mass flow concept of soil nutrients availability. *Soil Sci.*, **93**, 39–49.

Barker, A.V., Maynard, D.N. and Lachman, W.H. (1967). Induction of tomato stem and leaf lesions and potassium deficiency by excessive ammonium nutrition. *Soil Sci.*, **103**, 319–327.

Bar-Yosef, B. and Kafkafi, U. (1972). Rates of growth and nutrient uptake of irrigated corn as affected by N and P fertilization. *Soil Sci. Soc. Am. J.*, **36**, 931–936.

References

Beck, E., Fussedder, A. and Kraus, M. (1989). The maize root system in situ: Evaluation of structure and capability of utilization of phytate and inorganic soil phosphates. *Z. Pflanzenernähr. Bodenk.*, **15**, 2159–2167.

Bennet, A.C. and Adams, F. (1970). Concentration of NH_3 (aq) required for incipient NH_3 toxicity to seedlings. *Soil Sci. Soc. Amer. J.*, **34**, 259–263.

Bennie, A.T.P. (1991). Growth and mechanical impedance. In *Plant Roots the Hidden Half* (Y. Waisel, A. Eshel and U. Kafkafi, eds). New York: Marcel Dekker, pp. 393–414.

Bennie, A.T.P. (1996). Growth and mechanical impedance. In *Plant Roots: the Hidden Half* (Y. Waisel et al. ed.). 2nd ed. New York: Marcel Dekker, pp. 453–470.

Black, C.A. (1968). *Soil-Plant Relationships* (2nd edn). New York: John Wiley & Sons, Inc.

Blancaflor, E.B., Fasano, J.M. and Gilroy, S. (1998). Mapping the functional roles of cap cells in the response of *Arabidopsis* primary roots to gravity. *Plant Physiol.* **116**, 213–221.

Blanchar, R.W. and Coldwell, A.C. (1966). Phosphate ammonium moisture relationships in soils: II. Ion concentrations in leached fertilizer zones and effects on plants. *Soil Sci. Soc. Am. J.*, **30**, 43–48.

Böhm, W. (1979). *Methods of Studying Root Systems.* Berlin: Springer-Verlag, 188pp.

Bowen, G.D. and Rovira, A.D. (1991). The rhizosphere. In *Plant Roots the Hidden Half* (Y. Waisel, A. Eshel and U. Kafkafi, eds). New York: Marcel Dekker, p. 641.

Bray, R.H. and Kurtz, L.T. (1945). Determination of total, organic and available forms of phosphorus in soils. *Soil Sci.*, **59**, 39–45.

Breteler, H. (1973). A comparison between ammonium and nitrate nutrition of young sugar beet plants grown in nutrient solutions at constant acidity. 2. Effect of light and carbohydrate supply. *Neth. J. Agric. Sci.*, **21**, 297–307.

Clarkson, D.T., Hopper, M.J. and Jones, L.H.P. (1986). The effect of root temperature on the uptake of nitrogen and the relative size of their root system in Lolium perenne. 1. Solutions containing both NH_4 and NO_3. *Plant Cell Environ.*, **9**, 535–545.

Cooper, A.J. (1973). Root temperature and plant growth. Research Review No. 4 C.A.B. Slough, England: Farnham Royal.

Court, M.N., Stephen, R.C. and Waid, J. S. (1962). Nitrite toxicity arising from the use of urea of a fertilizer. *Nature*, **194**, 1263–1265.

Cox, W.J. and Resienauer, H.M. (1973). Growth and ion uptake by wheat supplied nitrogen as nitrate or ammonium or both. *Plant Soil*, **38**, 363–380.

Daniels, M.J., Mirkov, T.E. and Chrispeels, M.J. (1994). The plasma-membrane of *Arabidopsis thaliana* contains a mercury-insensitive aquaporin that is a homolog of the tonoplast water channel protein tip. *Plant Physiol.*, **106**(4), 1325–1333.

Demidchik, V., Shabala, S.N., Coutts, K.B., et al. (2004). Free oxygen radicals regulate plasma membrane Ca^{2+}- and K^+-permeable channels in plant root cells. *J. Cell Sci.* **116**, 81–88.

Du, Y.C. and Tachibana, S. (1994). Effect of supraoptimal root temperature on the growth, root respiration and sugar content of cucumber plants. *Scientia Horticulturae*, **58**(4), 289–301.

Eisenman, G. (1960). On the elementary atomic origin of equilibrium atomic specificity. In *Membrane Transport and Metabolism* (A. Kleinzeller and A. Kotic, eds). London: Academic Press, pp. 163–179.

Eizenberg, H., Shtienberg, D., Silberbush, M. and Ephrath, J.E. (2005). A new method for *in-situ* monitoring of the underground development of *Orobanche cumana* in Sunflower (*Helianthus annuus*) with a mini-rhizotron. *Annals of Botany*, **96**(6), 1137–1140.

Fisher, P.R., Huang, J. and Argo, W.R. (2006). Modeling lime reaction in peat-based substrates. *Acta Hort.* (ISHS), *718*: III International Symposium on Models for Plant Growth, Environmental Control and Farm Management in Protected Cultivation (HortiModel 2006).

Fitter, A. (2002). Characteristic and functions of root systems. In *Plant Roots the Hidden Half* (Y. Waisel, A. Eshel and U. Kafkafi, eds). New York: Marcel Dekker, pp. 33–48.

Flowers, T.J. and Flowers, S.A. (2005). Why does salinity pose such a difficult problem for plant breeders? *Agric. Water Management*, **78**, 15–24.

Ganmore-Newmann, R. and Kafkafi, U. (1980). Root temperature and ammonium/nitrate ratio effect on tomato. II Nutrient uptake. *Agron. J.*, **72**, 762–766.

Ganmore-Newmann, Ruth and Kafkafi, U. (1983). The effect of root temperature and NO_3^-/NH_4^+ ratio on strawberry plants. I. Growth, flowering and root development. *Agron. J.*, **75**, 941–947.

Ganmore-Newmann, R. and Kafkafi, U. (1985). The effect of root temperature and NO_3^-/NH_4^+ ratio on strawberry plants. II. Nutrient uptake and plant composition. *Agron. J.* **77**, 835–840.

Garofoli, S., Miloshevsky, G., Dorman, V.L. and Jordan, P.C. (2002). Permeation energetics in a model potassium channel. *Novartis Found Symp.*, **245**, 109–122; discussion 122-6, 165–8.

Gautam, M.K., Mead, D.J., Clinton, P.W. and Chang, S.X. (2003). Biomass and morphology of Pinus radiata coarse root components in a sub-humid temperate silvopastoral system. *Forest Ecology and Management*, **177**(1), 387–397.

Genard, M., Fishman, S., Vercambre, G., et al. (2001). A biophysical analysis of stem and root diameter variations in woody plants. *Plant Physiol.*, **126**, 188–202.

Gerendas, J., Ratcliffe, R.G. and Sattelmacher, B. (1991). ^{31}P nuclear magnetic resonance evidence for differences in intracellular pH in the roots of maize seedlings grown with nitrate and ammonium. *J. Pl. Physiol.*, **137**, 125–128.

Glass, A.D.M. (2002). Nutrient absorption by plant roots: Regulation of uptake to match plant demand. In *Plant roots the hidden half* (Y. Waisel, A. Eshel and U. Kafkafi, eds). New York: Marcel Dekker, pp. 571–586.

Harley, J.L. (1969). The biology of mycorrhiza. *Plant Science Monographs*. London: Leonard Hill.

Hasenstein, K.H., Evans, M.L., Stinemetz, C.L., et al. (1988). Comparative effectiveness of metal ions in inducing curvature of primary roots of Zea mays. *Plant Physiol.*, **86**, 885–889.

Hille, B. (1992). *Ionic Channels of Excitable Membranes* (2nd edn). Sunderland, MA: Sinauer Associates Inc.

Huck, M.G., Klepper, B. and Taylor, H.M. (1970). Diurnal variations in root diameter. *Plant Physiol.*, **45**, 529–530.

Huck, M.G., Klepper, B. and Taylor, H.M. (1999). Cotton Root Growth & Time Lapse Photography of Root Growth (VHS), VHS, ASA, CSSA, and SSSA. Item number: B30463. Wisconsin USA.

Iijima, M., Higuchi, T. and Barlow, P.W. (2004). Contribution of root cap mucilage and presence of an intact root cap in maize (Zea mays L.) to the reduction of soil mechanical impedance. *Ann. Bot.-(London)*. **94**(3), 473–477.

Iijima, M., Higuchi, T., Barlow, P.W. and Bengough, A.G. (2003). Root cap removal increases root penetration resistance in maize (Zea mays L.). *J. Exp. Bot.*, **54**(390), 2105–2109.

Imas, P., Bar-Yosef, B., Kafkafi, U. and Ganmore-Neumann, R. (1997a). Release of carboxylic anions and protons by tomato roots in response to ammonium nitrate ratio and pH in solution culture. *Plant and Soil*, **191**, 27–34.

Imas, P., Bar-Yosef, B., Kafkafi, U. and Ganmore-Neumann, R. (1997b). Phosphate induced carboxylate and proton release by tomato roots. *Plant and Soil*, **191**, 35–39.

Jacobs, E., Atzmon, D. and Kafkafi, U. (1969). A convenient method of placing radioactive substances in the soil for studies of root development. *Agron. J.*, **62**, 303–304.

Jungk, A.O. (2002). Dynamic of nutrient movement at the soil root interface. In *PlantRroots the Hidden Half* (Y. Waisel, A. Eshel and U. Kafkafi, eds). New York: Marcel Dekker, pp. 587–616.

Kafkafi, U. (1990). Root temperature, concentration and ratio of NO_3^-/NH_4^+ effect on plant development. *J. Plant Nutr.*, **13**(10), 1291–1306.

Kafkafi, U. (1991). Root growth under stress: Salinity. In *Plant Roots the Hidden Half* (Y. Waisel, A. Eshel and U. Kafkafi, eds). New York: Marcel Dekker, pp. 375–393.

Kafkafi, U. and Ganmore-Neumann, R. (1997). Ammonium in plant material, real or artefact? *J. Plant Nutr.*, **20**(1), 107–118.

Kafkafi, U. and Halevy, J. (1974). Rates of growth and nutrients consumption of semidwarf wheat. *Trans 10th Int. Congr. Soil Sci.*, Moscow. **4**, pp. 137–143.

Kafkafi, U. and Putter, J. (1965). Some aspects of sigmoidal yield response curves related to the geometry of granule fertilizer availability. *Israel J. Agric. Res.*, **15**, 169–178.

Kirkby, E.A. and Hughes, A.D. (1970). Some aspects of ammonium and nitrate nutrition in plant metabolism. In *Nitrogen Nutrition of the Plant* (E.A. Kirkby, ed.). Leeds: University of Leeds Press, pp. 69–77.

References

Kirkegaard, J.A., Troedson. R.J., So, H.B. and Kushwaha, B.L. (1992). The effect of compaction on the growth of pigeonpea on clay soils ii. Mechanisms of crop response and seasonal effects on an oxisol in a humid coastal environment. *Soil Tillage Res.*, **24**(2), 129–147.

Kiss, J.Z., Hertel, R. and Sack, F.D. (1989). Amiloplasts are necessary for full gravitropic sensitivity in *Arabidopsis* roots. *Plant*, **177**, 198–206.

Kristensen, H.L. and Thorup-Kristensen, K. (2004). Uptake of ^{15}N labeled nitrate by root systems of sweet corn, carrot and white cabbage from 0.2–2.5 meters depth. *Plant and soil*, **265**, 93–100.

Kubo, K., Jitsuyama, Y., Iwama, K., et al. (2004). Genotypic difference in root penetration ability by durum wheat (*Triticum turgidum* L. var. durum) evaluated by a pot with paraffin-Vaseline discs. *Plant Soil*, **262**, 169–177.

Kuchenbuch R.O. and Ingram, K.T. (2002). Image analysis for non-destructive and non-invasive quantification of root growth and soil water content in rhizotrons. *J. Plant Nutrition and Soil Science*, **165**(5), 573–581.

Lachno, D.R. (1983). Abscisic acid and indole-3-acetic acid in maize roots subject to water, salt and mechanical stress. In *Growth Regulators in Root Development Monograph 10* (M.B. Jackson and A.D. Stead, eds). Wantage: British plant growth regulator group, pp. 37–71.

Lee, S.H., Ahn, S.J., Im, Y.J., et al. (2005). Differential impact of low temperature on fatty acid unsaturation and lipoxygenase activity in figleaf gourd and cucumber roots. *Biochem. Biophys. Res. Commun.*, **330**(4), 1194–1198.

Lee, S.H., Singh, A.P. and Chung, G.C. (2004). Rapid accumulation of hydrogen peroxide in cucumber roots due to exposure to low temperature appears to mediate decreases in water transport. *J. Exp. Bot.*, **55**(403), 1733–1741.

Lee, S.H., Singh, A.P., Chung, G., et al. (2004). Exposure of roots of cucumber (*Cucumis sativus*) to low temperature severely reduces root pressure, hydraulic conductivity and active transport of nutrients. *Physiologia Plantarum*, **120**, 413–420.

Lee, S.H., Singh, A.P., Chung, G.C., et al. (2002). Chilling root temperature causes rapid ultrastructural changes in cortical cells of cucumber (*Cucumis sativus* L.) root tips. *J. Exp. Bot.*, **53**(378), 2225–2237.

Lewis, D.G. and Quirk, J.P. (1965). Diffusion of phosphate to plant roots. *Nature*, **205**, 765–766.

Lewis, D.G. and Quirk, J.P. (1967). Phosphate diffusion in soil and uptake by plants: III. ^{32}P movement and uptake by plants as indicated by ^{32}P autoradiography. *Plant and Soil*, **26**, 445–453.

Liphschitz, N. and Waisel, Y. (1974). Existence of salt glands in various genera of the *Gramineae*. *New Phytol.*, **73**, 507–513.

Liszkay, A., van der Zalm, E. and Schopfer, P. (2004). Production of reactive oxygen intermediates (O_2^{AB}, H_2O_2, and ^AOH) by maize roots and their role in wall loosening and elongation growth. *Plant Physiol.*, **136**, 3114–3123.

MacDuff, J.H. and Jackson, S.B. (1991). Growth and preference for ammonium or nitrate uptake by barley in relation to root temperature. *J. Exp. Bot.*, **42**, 521–530.

Magalhaes, J.S. and Wilcox, G.E. (1983). Tomato growth and nutrient uptake patterns as influenced by nitrogen form and light intensity. *J. Plant. Nutr.*, **6**, 941–956.

Majdi, H., Smucker, A.J.M. and Persson, H. (1992). A comparison between minirhizotron and monolith sampling methods for measuring root growth of maize (Zea mays L.). *Plant and Soil*, **147**, 127–134.

Marschner, H. (1995). *Mineral Nutrition of Higher Plants* (2nd edn). London: Academic Press.

Marschner, H. and Röhmheld, V. (1983). In vivo measurement of root induced changes at the soil root interface: effect of plant species and nitrogen source. *Z. Pflanzenphysiol.*, **111**, 241–251.

Marschner, H., Röhmheld, V., Horst, W.J. and Martin, P. (1986). Root induced changes in the rhizosphere: importance for the mineral nutrition of plants. *Z. Pflanzenernähr. Bodenk.*, **149**, 441–456.

Massa, G.D. and Gilroy, S. (2003). Touch modulates gravity sensing to regulate the growth of primary roots of *Arabidopsis thaliana*. *Plant J.*, **33**, 435–445.

Maynard, D.N., Barker, A.V. and Lachman, W.H. (1968). Influence of potassium on the utilization of ammonium by tomato plants. *Proc. Am. Soc. Hortic. Sci.*, **92**, 537–542.

McCully, M.E. (1999). Roots in soil: unearthing the complexities of roots and their rhizospheres. *Annu. Rev. Plant Physiol. Plant Mol. Biol.*, **50**, 695–718.

Misra, R.K. and Gibbons, A.K. (1996). Growth and morphology of eucalypt seedling-roots, in relation to soil strength arising from compaction. *Plant and Soil*, **182**(1), 1–11.

Moritsugu, M. and Kawasaki, T. (1983). Effect of nitrogen source on growth and mineral uptake in plants under nitrogen restricted culture condition. Berichte des Ohara Instituts fuer Landwirtschaftliche Biologie, *Okayama Universitaet*, **18**(3), 145–158.

Moritsugu, M., Suzuki, T. and Kawasaki, T. (1983). Effect of nitrogen source on growth and mineral uptake in plants under constant pH and conventional culture condition. Berichte des Ohara Instituts fuer Landwirtschaftliche Biologie, *Okayama Universitaet*, **18**(3), 125–145.

Mulligan, M.F., Smucker, A.J.M. and Safir, G.F. (1985). Tillage modifications of dry edible bean root colonization by VAM fungi. *Agron. J.*, **77**, 140–144.

Norby, R.J., Ledford Joanne, Reilly Carolyn, D., Miller Nicole, E. and O'Neill Elizabeth G. (2004). Fine-root production dominates response of a deciduous forest to atmospheric CO_2 enrichment. *PNAS* **101**(26), 9689–9693.

Nye, P.H. (1994).The effect of root shrinkage on soil-water inflow. *Phil. Trans. R. Soc. Lond., B, Biol. Sci.*, **345**(1314), 395–402.

Ockerse, R, and Galston, A.W. (1967). Gibberellin-auxin interaction in pea stem elongation. *Plant Physiol.*, **42**, 47–54.

Ozanne, P.G., Asher, C.J. and Kirton, D.J. (1965). Root distribution in a deep sand and its relationship to the uptake of added potassium by pasture plants. *Aust. J. Agric. Res.*, **16**, 785–800.

Ozier-Lafontaine, H., Lecompte, F. and Sillon, J.F. (1999). Fractal analysis of the root architecture of *Gliricidia sepium* for the spatial prediction of root branching, size and mass: model development and evaluation in agroforestry. *Plant and Soil*, **209**, 167–180.

Pate, J.S., Sharkey, P.J. and Lewis, O.A.M. (1974). Xylem to phloem transfer of solutes in fruiting shoots of legumes, studied by a phloem bleeding technique. *Planta*, **122**, 11–26.

Pearson, R.W., Childs, J. and Lund, Z.F. (1973). Uniformity of limestone mixing in acid subsoil as a factor in cotton root penetration. *Soil Sci. Soc. Am. J.*, **37**, 727–732.

Pilet, P.E. (2002). Root growth and gravireaction: a critical study of Hormone and regulator implication. In *Plant Roots the Hidden Half* (Y. Waisel, A. Eshel and U. Kafkafi, eds). New York: Marcel Dekker, pp. 489–504.

Porterfield, D.M. and Musgrave, M.E. (1998). The tropic response of plant root to oxygen: oxitropism in *Pisum Sativum L. Planta*, **206**, 1–6.

Precheur, R.J. and Maynard, D.N. (1983). Growth of asparagus transplants as influenced by nitrogen form and lime. *J. Am. Soc. Hortic. Sci.*, **108**, 169–172.

Puritch, G.S. and Barker, A.V. (1967). Structure and function of tomato leaf chloroplasts during ammonium toxicity. *Pl. Physiol.*, **42**, 1229–1238.

Raviv, M., Zaidman, B. and kapulnik, Y. (1998). The use of compost as a peat substitute for organic vegetable transplant production. *Compost Science and Utilization*, **6**, 46–52.

Ren, Z.H., Gao, J.P., Li, L.G., et al. (2005). A rice quantitative trait locus for salt tolerance encodes a sodium transporter. *Nature Gene.*, **37**, 1141–1146.

Richards, L.A. (1954). *Diagnosis and Improvement of Saline and Alkali Soils.* Agriculture handbook 60. USDA.

Römheld, V. (1986). pH changes in the rhizosphere of various crops in relation to the supply of plant nutrients. In Potash Review 11 Berne Switzerland international Potash Institute.

Russell, E.J. (1950). *Soil Conditions and Plant Growth* (8th edn). London: Longmans Green & Co.

Sattelmacher, B. and Thoms, K. (1991). Morphology of maize root systems as influenced by a local supply of nitrate or ammonia. In *Plant Roots and their Environment* (B.L. Mc Michael and H. Persson, eds). Amsterdam: Elsevier, pp. 149–156.

Sattelmacher, B., Gerendas, J., Thoms, K., et al. (1993). Interaction between root growth and mineral nutrition. *Environ.l and Experim. Bot.*, **33**, 63–73.

Schaller, G. and Fischer, W.R. (1985). pH-Änderungen in der rhizosphere von Mais- und Erdnushwurzeln. *Z. Pflanzenernähr. Bodenk.*, **148**, 306–320.

Schubert, R., Raidl, S., Funk, R., et al. (2003). Quantitative detection of agar-cultivated and rhizotron-grown Piloderma croceum Erikss. and Hjortst. by ITS1-based fluorescent PCR. *Mycorrhiza*, **13**, 159–165.

References

Schroeder-Murphy, S.L., Huang, B., King, R.L. and Smucker, A.J.M. (1990). Measurement of whole plant responses to compacted and flooded soil environments in the teaching laboratory. *J. Agron. Ed.*, **19**(2), 171–175.

Shani, U., Waisel, Y. and Eshel, A. (1995). The development of melon roots under trickle irrigation: Effects of the location of the emitters. In *Structure and Function of Roots* (F. Baluska, M. Ciamporova, O. Gasparikova and P.W. Barlow, eds). Netherlands: Kluwer Academic Publishers, pp. 223–225.

Sibony, M. and Pinthus, M.J. (1988). Floret initiation and development in spring wheat (*Triticum aestivum* L.). *Ann. Bot.*, **61**, 473–479.

Sievers, A., Broun, M. and Monshausen, G.B. (2002). The root cap: structure and function. In *Plant Roots the Hidden Half* (Y. Waisel, A. Eshel and U. Kafkafi, eds). New York: Marcel Dekker, pp. 33–48.

Silber, A., Ben-Jaacov, J., Ackerman, A., et al. (2002). Interrelationship between phosphorus toxicity and sugar metabolism in Verticordia plumosa L. *Plant Soil*, **245**, 249–260.

Silberbush, M. and Barber, S.A. (1983). Sensitivity analysis of parameters used in stimulating K-uptake with a mechanistic mathematical model. *Agron. J.*, **75**, 851–854.

Smucker, A.J.M. (1984). Carbon utilization and losses by plant root systems. In *Roots, Nutrient and Water Influx, and Plant Growth* (S.A. Barber and D.R. Boulden, eds). Madison, WI: American Society Agronomy, pp. 27–46.

Smucker, A.J.M. (1988). Video recording and image analyses of the rhizosphere. *Yearbook of Science and Technology*. McGraw-Hill, 3pp.

Smucker, A.J.M. and Aiken, R.M. (1992). Dynamic root responses to water deficits. *Soil Science*, **154**(4), 281–289.

Smucker, A.J.M., S.L. McBurney and A.K. Srivastava (1982). Quantitative separation of roots from compacted soil profiles by the hydropneumatic elutriation system. *Agron. J.*, **74**, 500–503.

Smucker, A.J.M., Srivastava, A.K., Adams, M.W. and Knezek, B.D. (1991). Secondary tillage and traffic compaction modifications of the growth and production of dry edible beans and soybeans. *Amer. Soc. of Ag. Engineers*, **7**(2), 149–152.

Stibbe, E. and Kafkafi, U. (1973). Influence of tillage depths and P-fertilizer application rates on the yields of annual cropped winter-grown wheat. *Agron. J.*, **65**, 617–620.

Takahashi, H. (1997). Hydrotropism: The current state of our knowledge. *J. Plant Res.*, **110**(2), 163–169.

Tanimoto, E. (1994). Interaction of gibberellin A_3 and ancymidol in the growth and cell-wall extensibility of dwarf pea roots. *Plant Cell Physiol.*, **29**, 269–280.

Tanimoto, E. (1995). Effect of gibberellin and ancymidol on the growth and cell wall components of pea (*Pisum sativum* L.) roots. In *Structure and Function of Roots* (F. Baluska, M. Ciamporova, O. Gasparikova and P.W. Barlow, eds). Netherlands: Kluwer Academic Publishers, pp. 91–98.

Tanimoto, E. (2002). Gibberellins. In *Plant Roots the Hidden Half* (Y. Waisel, A. Eshel and U. Kafkafi, eds). New York: Marcel Dekker, pp. 405–416.

Tanimoto, E. and Yamamoto, R. (1997). Change in cell wall extensibility during gibberellin-regulated growth of pea roots. *Zemedelska Technika*, **43**, 15–19.

Tatsumi, J., Yamauchi, A. and Kono, Y. (1989). Fractal analysis of plant root systems. *Ann. Bot.*, **64**, 499–503.

Taylor, H.M. and Burnett, E. (1964). Influence of soil strength on root growth habits of plants. *Soil Science*, **98**, 174–180.

Tisdale, S.L. and Nelson, W.L. (1966). *Soil Fertility and Fertilizers*. New York: MacMillan.

Vivanco, J.M., Guimaraes, R.L. and Flores, H.E. (2002). Underground plant metabolism: The biosynthetic potential of roots. In *Plant Roots the Hidden Half* (Y. Waisel, A. Eshel and U. Kafkafi, eds). New York: Marcel Dekker, pp. 1045–1070.

Walk, T.C., Jaramill, O.R. and Lynch, J.P. (2006). Architectural tradeoffs between adventitious and basal roots for phosphorus acquisition. *Plant and Soil*, **279**(1–2), 347–366.

Wang, R., Okamoto M., Xing X., Crawford, N.M. (2003). Microarray analysis of the nitrate response in Arabidopsis roots and shoots reveals over one thousand rapidly responding genes and new linkages to glucose, trehalose-6-P, iron and sulfate metabolism. *Plant Physiol.*, **132**, 556–567.

Wang, R., Tischner, R., Gutiérrez, R.A., Hoffman Maren, Xing, Xiujuan Chen, M., Coruzzi Gloria and Crawford N.M. (2004). Genomic analysis of the nitrate response using a nitrate reductase-null mutant of Arabidopsis. *Plant Physiol.*, **136**, 2512–2522.

Walter, S. and H. Bürgi. (1996). Report on the project ROOT DIRECTOR: Computer aided evaluation of scanned images of roots. ETH Zurich, *Inst. of Plant Sci.*, Zurich: Switzerland.

Wassen, M.J., Venterink, H.O., Lapshina, E.D. and Tanneberger, F. (2005). Endangered plants persist under phosphorus limitation. *Nature*, **437**(Sept.), pp. 347–350.

Weaver, J.E. (1926). *Root Development of Field Crops*, New York: McGraw-Hill.

Wilcox, G.E., Magalhaes, J.R. and Silva, F.L.I.M. (1985). Ammonium and nitrate concentration as factors in tomato growth and nutrient uptake. *J. Plant Nutr.*, **8**, 989–998.

Williams, R.F. (1948). The effects of phosphorus supply on the rates of intake of phosphorus and nitrogen and upon certain aspects of phosphorus metabolism in gramineous plants. *Austr. J. Sci. Res. Series B – Biol. Sci.*, **1**, 333.

Xu, G.H., Magen, H., Tarchitzky, J. and Kafkafi, U. (2000). Advances in Chloride Nutrition of Plants. *Adv. Agron.*, **68**, 97–150.

Yaniv, Z. and Bachrach, U. (2002). Roots as a source of metabolites with medicinal activity. In *Plant Roots the Hidden Half* (Y. Waisel, A. Eshel and U. Kafkafi, eds). New York: Marcel Dekker, pp. 1071–1093.

Yermiyahu, U., Nir, S., Ben-Hayyim, G.L. and Kafkafi, U. (1994). Quantitative competition of calcium with sodium or magnesium for sorption sites on plasma membrane vesicles of melon (*Cucumis melo* L.) root cells. *J. Memb. Biol.*, **138**, 55–63.

Yermiyahu, U., Nir, S., Ben-Hayyim G., et al. (1997). Root elongation in saline solution related to calcium binding to root cell plasma membranes. *Plant and Soil*, **191**, 67–76.

Yermiyahu, U., Nir, S., Ben-Hayyim, G., et al. (1999). Surface properties of plasma membrane vesicles isolated from melon (*Cucumis melo* L.) root cells differing in salinity tolerance. *Colloids and Surf.*, **14**, 237–249.

Zoronoza, P., Casselles, J. and Carpena, O. (1987). Response of pepper plants to NO_3:NH_4 ratio and light intensity. *J. Plant Nutr.*, **10**, 773–782.

FURTHER READINGS

The following two movies show roots responding to different unfavourable soil conditions. Cotton Root Growth (15:52 minutes) describes the elongation of the taproot of cotton seedlings with cold or compacted soils and under different soil chemical conditions including pH and osmotic strength. Time Lapse Photography of Root Growth (14:49 minutes) shows details of roots growing in compacted subsoils and under anaerobic conditions. Both are suitable for the general biology student and provide an excellent introduction to how roots grow and explore the soil. VHS, 1999; ASA, CSSA, and SSSA. Item number: B30463.

3

PHYSICAL CHARACTERISTICS OF SOILLESS MEDIA

RONY WALLACH

3.1 **Physical Properties of Soilless Media**
3.2 **Water Content and Water Potential in Soilless Media**
3.3 **Water Movement in Soilless Media**
3.4 **Uptake of Water by Plants in Soilless Media and Water Availability**
3.5 **Solute Transport in Soilless Media**
3.6 **Gas Transport in Soilless Media**
    References

3.1 PHYSICAL PROPERTIES OF SOILLESS MEDIA

Growing substrates and soil are both porous media and the physical principles of both are similar. Research in soil physics is ahead of that in substrates since research in this field started many years before the onset of soilless cultivation and up to now more efforts are devoted to soil physics, compared to physics of substrates. An appropriate adaptation is needed when soil-related knowledge is transferred to substrates due to the differences in structure and limited root zone volume. In this chapter, many examples will be presented based on current soil-related knowledge but only when they are relevant to substrate issues.

All media are composed of three phases: solid, aqueous and gaseous. In the following sections, the physical characteristics of these three phases will be discussed separately and in combination.

3.1.1 BULK DENSITY

Bulk density (BD) of a medium is defined as its dry mass per unit of volume (in a moist state) and is measured in $g\,cm^{-3}$. Numerous methods for the measurement of BD (as well as other physical parameters) can be found in the literature (e.g. De Boodt and Verdonck, 1972; Wever, 1995; Raviv and Medina, 1997; Gruda and Schnitzler, 1999; Morel et al., 2000). Some methods are used primarily for research purposes (e.g. the standard ISHS method, as described by Verdonck and Gabriels, 1992). Others are used as industrial standards in certain countries or regions of the world (e.g. BS EN 12580:2001 in the UK, both the LUFA and DIN methods in Germany and the CEN method in the EU). All of them, however, are based on one principle: Wet material is allowed to settle within or compressed using known pressure into a cylinder of a known volume. It is then dried down completely and weighed. For specific details, see Chap. 7.

As many media are composed of more than one ingredient, the characteristics of each ingredient contribute to the total BD of the medium. These are individual and combined particles' arrangements, BD of the ingredients and compaction qualities. In particular, media components that differ significantly in particle size have higher BDs as a mix (Pokorny et al., 1986). Similarly, they have lower total porosity (TP), water holding capacity and air-filled porosity (AFP) than media composed of similar particle sizes.

The BD affects the choice of media in various ways. For example, outdoor production of tree saplings requires high BD media to prevent container instability under windy conditions. This can be achieved by the inclusion of heavy mineral constituents such as sand, soil, clay or tuff in the mix. On the other hand, high-intensity greenhouse crops, which are frequently irrigated and may be exposed to oxygen deficiency if hydraulic conductivity and AFP are not high, require media of low BD. Another consideration is that the mixing and transportation of low BD-media are easier than those of high BD-media.

3.1.2 PARTICLE SIZE DISTRIBUTION

Particle size distribution (PSD) is the most fundamental physical property of a porous medium and defines its texture. The particle sizes present and their relative abundance have a significant influence on most of its physical properties. Particle size analysis consists of isolating various particle sizes or size increments and then measuring the abundance of each size. The material of which the medium solid phase is composed includes discrete particles of various shapes and sizes, as well as amorphous compounds such as colloidal organic matter. The particle size and mineral composition largely determine the nature and behaviour of the medium: its internal geometry and porosity, its interactions with fluids and solutes, as well as its compressibility and strength.

3.1 PHYSICAL PROPERTIES OF SOILLESS MEDIA

The term 'medium texture' refers to the size range of particles in the medium; that is, whether the particles composing a particular medium are of a wide or relatively narrow range of sizes, and whether they are mainly large, small or of some intermediate size. As such, the term carries both qualitative and quantitative connotations. Qualitatively, it indicates whether the material is coarse and gritty, or fine and smooth. Quantitatively, it denotes the precisely measured distribution of particle sizes and the proportions of the various size ranges of particles within a given medium. As such, medium texture is an intrinsic attribute of the medium and the one most often used to characterize its physical make-up.

The PSD attempts to divide what in nature is generally a continuous array of particle sizes into discrete fractions. The PSD curves for two types of tuff (scoria) (data from Wallach et al., 1992a) and perlite (data from Orozco and Marfa, 1995) are shown in Fig. 3.1. The shape of particle size distribution is related to the pore size (radii)

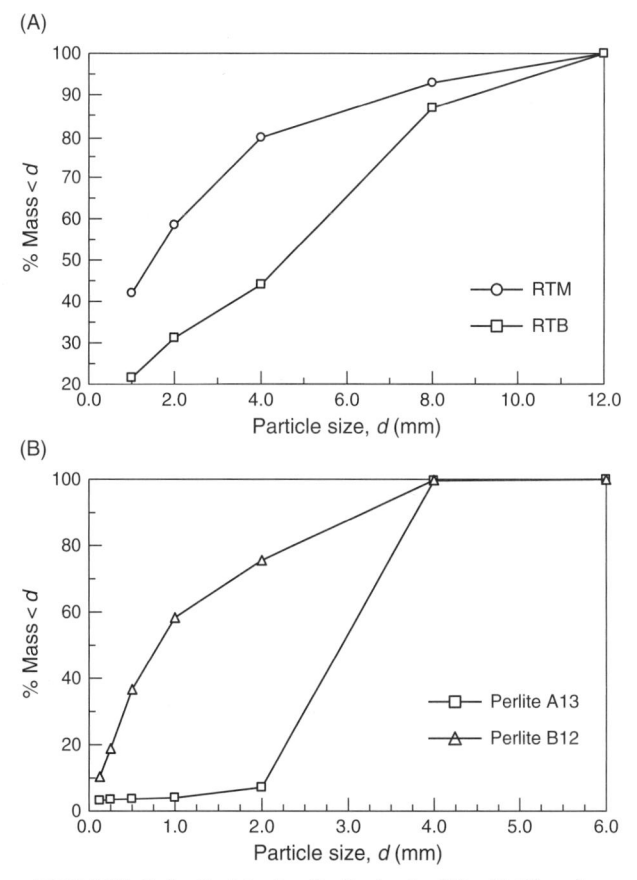

FIGURE 3.1 Particle size distribution for (A) tuff, (B) perlite.

distribution which was related, as will be discussed in the following, to the water retention characteristics and the hydraulic conductivity.

The information obtainable from this representation of PSD includes the diameter of the largest grains in the assemblage and the grading pattern; that is, whether the substrate is composed of distinct groups of particles each of uniform size, or of a more or less continuous array of particle sizes. Poorly graded media have a dominance of particles of several distinct sizes and are represented by a step-like distribution curve. A medium with a flattened and smooth distribution curve is termed 'well graded'. Based on the shape of tuff and perlite particle size distribution (Fig. 3.1), it can be concluded that RTB tuff and A13 perlite, whose distribution curves are more step-like, are poorly graded relative to RTM tuff and B12 perlite, respectively.

The PSD is used in soil science for estimating soil hydraulic properties, such as the water retention curve $\theta(\psi)$ (Sect. 3.2.3) and saturated as well as unsaturated hydraulic conductivities (Gupta and Larson, 1979; Arya and Paris, 1981; Campbell, 1985; Schuh and Bauder, 1986; Vereecken et al., 1989). Mathematical relationships between the particle-size and hydraulic properties tend to be fairly good for sandy soils, but not as accurate for soils with larger fractions of clay (Cornelis et al., 2001). Significant contributions were made by Arya and Paris (1981) to predict water retention curves $\theta(\psi)$, using the PSD. Their physico-empirical approach is based mainly on the similarity between shapes of the cumulative PSD and $\theta(\psi)$ curves. Various authors have developed similar models (Haverkamp and Parlange, 1986; Wu et al., 1990; Smettem and Gregory, 1996; Zhuang et al., 2001). To date, no attempts have been made to find such relationships for soilless media.

Arya et al. (1999) also derived a model to compute the hydraulic conductivity function, $K(\psi)$ (Sect. 3.3.2), directly from the PSD. Unlike other models, the need for both measured $\theta(\psi)$ and the saturated hydraulic conductivity, K_s, is eliminated.

3.1.3 POROSITY

Total porosity (TP) and its components are expressed as a percentage of the total volume of the medium. The combined volume of the aqueous and the gaseous phases of the medium are defined as its total pore space or total porosity. TP is related to the shape, size and arrangement of media particles. The various methods of measuring BD are also relevant to the determination of porosity and its components. Some of these methods are described in Chap. 7.

In many cases, BD is inversely related to TP (Bugbee and Frink, 1983; Bunt, 1988). However, the medium BD cannot accurately determine TP if components that have closed pores, such as perlite or pumice, are used (Bunt, 1988; Bures et al., 1997). Not only mineral substrates contain inaccessible pores. Evans et al. (1996) sampled different coir batches. Contrary to their expectations, the samples exhibited a lower BD with lower TP. This, of course, can only happen if a significant fraction of the pores was not saturated with water during the saturation stage of the measurement, suggesting that those pores were probably of a strong hydrophobic nature. Unlike the case of closed pores, unsaturation of organic substrates may be regarded as an inherent drawback of the testing method, as can be expected during the practical use of such substrates, their hydrophobicity will be lost and all non-accessible pores will be frequently filled up with water.

The volumetric amount of water, θ, which saturates a given volume of a substrate, is defined as its effective pore space. The difference between total pore space and effective pore space constitutes the volume of closed pores that are not accessible to water. In horticulture, the air-filled porosity (AFP) is defined as the volumetric percentage of the medium filled with air at the end of free (gravitational) drainage. Availability of oxygen to the roots depends on the rate of gaseous exchange between the atmosphere and the growing medium. Growing medium aeration is positively related to AFP and negatively to water content. For mineral soils, an AFP of 10 per cent is usually presented as the minimum limit for gaseous diffusion and 10–15 per cent for root respiration and growth.

Most media and mixes have an AFP of 10–30 per cent. Given that perched water table rests at the bottom of the growing containers after the end of free drainage, optimal AFPs may vary greatly according to the size of the container and the irrigation frequency. A wider discussion of this topic is given in Sect. 3.2.3. For the rooting of cuttings under intermittent mist, AFPs of > 20 per cent are essential. A somewhat lower AFP is required for bedding plants grown in shallow trays or plugs. On the other hand, an AFP as low as 10 per cent may suffice for deep containers with slow growing, infrequently irrigated tree saplings. For all types of containers and media, it is important to consider the tendency of most root systems to grow gravitropically and to form a dense layer of roots at the bottom of the container. This may be the reason for less than optimal performance of plants in media that are otherwise considered adequate in terms of AFP (Raviv et al., 2001).

3.1.4 PORE DISTRIBUTION

Porous medium structure consists of a 3-D network of pores. Large pores play an important role in allowing roots, gas and water to penetrate into the medium. The higher the large pore (macropore) density, the more the soil can be exploited by plant roots (Scott et al., 1988). Similarly, the more continuous the macropores are, the more freely gases can interchange with the atmosphere. Continuous macropores also have a direct effect on water infiltration and solute transport in soil. Although small pores restrict the penetration of roots, the macropores provide favourable conditions for root growth. Several studies have shown that the presence of continuous macropores significantly benefits root growth (Bennie, 1991). One of the most important factors influencing soil fertility, besides water and nutrient content, is soil aeration (Glinski and Stepniewski, 1985; Hillel, 1998). Large pores are the paths available for gas exchange between the growing medium and the atmosphere. Intuitively, large and continuous pores facilitate water transport. It is now well known that the size and connectivity of soil pores play a major role in the flow characteristics of water and the transport of solutes through soil (Ma and Selim, 1997). Jury and Flühler (1992) stated that fluid transport through well-defined structural voids is not predictable unless the distributions of the voids, aperture sizes and shapes, depths of penetration and interconnectivity are known.

The capacity of a medium to store water and air as well as its ability to provide them to the plant (via its hydraulic conductivity and rate of gas exchange) are determined by its total porosity and pore size distribution, tortuosity and connectivity. Tortuosity is one of the most meaningful 3-D parameters of pore structure. It is defined as the ratio of the effective average path in the porous medium to the shortest distance measured along the direction of the pore (Jury et al., 1991). Tortuosity is a dimensionless factor always greater than one, which expresses the degree of complexity of the sinuous pore path. Tortuosity can easily be related to the conductivity of a porous medium since it provides an indication of increased resistance to flow due to the pore system's greater path length (Dullien, 1992). Connectivity is a measure of the number of independent paths between two points within the pore space (Dullien, 1992). In other words, connectivity is the number of non-redundant loops enclosed by a specific geometrical shape. Each macropore network has a connectivity which is a positive integer equal to the number of different closed circuits between two points in the network. If there is only one open circuit, the connectivity is equal to 0; the connectivity is 1, if the circuit is closed.

An important problem associated with the characterization of soilless media pores is the lack of standard terminology related to their classification into distinct size ranges. Several researchers have identified the need for a standard classification scheme, and suggestions for such a classification have been made. Typical examples are the proposed index for soil pore size distribution by Cary and Hayden (1973) and the suggested classification of micro-, meso-, and macroporosity by Luxmoore (1981). Although the pore size distribution of many porous materials could be measured by different techniques, for example water desorption and mercury intrusion methods (Danielson and Sutherland, 1986), none of them have been applied to soilless container media.

It is recognized by soil physicists that the water retention curve (Sect. 3.2.3) is essentially the pore-size distribution curve. The expression relating pore size to the equivalent suction of water in porous media is the capillarity equation (Eq. [8]). Pore size distributions have been used to develop expressions for soils' water retention curves (Sect. 3.2.3) by Arya and Paris (1981), Haverkamp and Parlange (1986), Kosugi (1994), Assouline et al. (1998) and Or and Tuller (1999), but not to growing substrates so far.

3.2 WATER CONTENT AND WATER POTENTIAL IN SOILLESS MEDIA

3.2.1 WATER CONTENT

Water content or wetness of a porous medium is the volume or mass of water V_w occupying space within the pores (Gardner, 1986). Water content is typically expressed in one of the three ways: mass basis, volume or depth basis and percent or degree saturation.

3.2 WATER CONTENT AND WATER POTENTIAL IN SOILLESS MEDIA

Mass wetness (sometimes referred to as gravimetric water content) is determined by extracting a soilless growing medium sample, oven-drying (generally 105°C for 24 h), and determining the amount of water lost through the drying process. Oven drying is necessary to remove hygroscopic water adhering to particles that cannot be removed by air drying. This is sometimes a significant amount, depending on small particles of the growing medium and specific surface area; however, not all hygroscopic water can be removed. Mass wetness, ω, is described by

$$\omega = \frac{V_w \rho_w}{V_s \rho_s} \tag{1}$$

where V_s and V_w are the volume fractions of the solid and liquid phases respectively, ρ_s is the density of the solid phase, and ρ_w is the density of water at measured temperature (for analytical processes of bulk densities, see Chap. 7).

Volumetric water content is generally more useful for field and laboratory studies, because it is the form in which growing medium water content is usually expressed (as a fractional basis, e.g., 0.34, 0.48), and is reported in the results from gamma attenuation, neutron probe and time domain reflectometry (TDR) water content measuring devices. Like mass wetness, it is generally reported as a percentage, but is compared to total volume rather than the volume of solids present. Because both the V_w and the V_t units of measure are cm^3, they cancel in the equation, leaving θ dimensionless. Volumetric water content is given by

$$\theta = \frac{V_w}{V_t} \tag{2}$$

where V_t is the total growing medium volume. Volumetric water content may also be obtained by its relation to mass wetness and bulk density, as follows:

$$\theta = \omega \frac{\rho_b}{\rho_w} \tag{3}$$

where ρ_w and ρ_b are the wet and dry bulk densities respectively.

Equivalent depth of water is a measure of the ratio of depth of water per unit depth of porous media, described by volumetric water content:

$$d_w = \theta d_t \tag{4}$$

where d_w is the equivalent depth of growing medium water if it were extracted and ponded over the surface (cm), and d_t is the total depth of growing medium under consideration (cm). Depth of water is a very important concept in agricultural and irrigation practices.

Degree of saturation, S, expresses the volume of water in relation to the volume of pores. Therefore, S can be as high as 100 per cent in porous medium that remain constantly moist. In most soils and growing media, S will not reach 100 per cent because of air entrapped in closed pores. The degree of saturation is expressed by

$$S = \frac{V_w}{V_w + V_a} \tag{5}$$

where V_a is the volume fracture of the air phase.

Effective Saturation, S_e, or reduced water content, is defined by

$$S_e = \frac{\theta - \theta_r}{\theta_s - \theta_r} \tag{6}$$

where θ_s is the volumetric water content at saturation and θ_r is the residual water content. The residual water content is somewhat arbitrarily defined as the water content at which the corresponding hydraulic conductivity is appreciably zero, but very often it is used as an empirical constant when fitting hydraulic functions. When $\theta_r = 0$, S_e approaches S (Eq. 5). Note that S_e varies between zero and one.

Water content measurement techniques are often classified as either direct or indirect. Direct methods involve some form of removal or separation of water from the soilless growing medium matrix with a direct measurement of the amount of water removed. Indirect methods measure some physical or chemical property of a soilless growing medium that are related to its water content such as the dielectric constant, electrical conductivity, heat capacity and magnetic susceptibility. In contrast to the direct methods, the indirect methods are less destructive or non-destructive, that is the water content of the sample is not necessarily altered during measurement. Thus, these methods allow continued, real-time measurement of water content, which enables to increase water availability to the plant and ensures more efficient irrigation.

The most known and simple direct and destructive method to measure the weight-based moisture content in soil science is the thermogravimetry method. The sample is weighed at its initial wetness and then dried to remove interparticle absorbed water in an oven at 105°C until the soil mass becomes stable; this usually requires 24–48 h or more, depending on the sample size, wetness and soil characteristics (texture, aggregation, etc.). The difference between the wet and the dry weights is the mass of water held in the original soil sample (Eq. [1]). The gravimetric method is considered the standard against which many indirect techniques are calibrated. The primary advantages of gravimetry are the direct and relatively inexpensive processing of samples. The shortcomings of this method are its labour- and time-intensive nature, the time delay required for drying (although this may be shortened by the use of a microwave rather than a conventional oven, the methodology has not yet been standardized), and the fact that the method is destructive, thereby prohibiting repetitive measurements within the same soil volume. These shortcomings are intensified in soilless growing media where changes in moisture content are fast compared to field soils and its volume within the container is limited so that intensive sampling may use significant parts of the total medium volume. The gravimetric method is rarely practised in soilless culture to monitor moisture content variation.

Many indirect methods to measure moisture content are used in soil science. A detailed description can be found in Gardner (1986). The indirect method that has been widely used for soilless culture so far is the TDR method. This method is based on the measurement of reflected electromagnetic signals travelling in transmission cables and waveguides embedded in the growing medium to measure the apparent dielectric constant of the medium surrounding the waveguide. The propagation velocity of an electromagnetic wave along a transmission line (waveguide) of length embedded in the medium is determined from the time response of the system to a pulse generated by

3.2 WATER CONTENT AND WATER POTENTIAL IN SOILLESS MEDIA

the TDR cable tester. The medium bulk dielectric constant is governed by the dielectric constant of liquid water ($\varepsilon_w \sim 80$), as the dielectric constants of other medium constituents are much smaller, for example, growing medium's minerals ($\varepsilon_s \sim 3$ to 5), and air ($\varepsilon_a = 1$). This large disparity of the dielectric constants makes the method relatively insensitive to medium composition and texture (other than organic matter and some clays), and thus, a good method for liquid water measurement.

The TDR calibration establishes the relationship between the growing medium bulk dielectric constant, ε_b and θ. The empirical relationship for mineral soils as proposed by Topp et al. (1980)

$$\theta = -5.3 \times 10^{-2} + 2.92 \times 10^{-2} \varepsilon_b - 5.5 \times 10^{-4} \varepsilon_b^2 + 4.3 \times 10^{-6} \varepsilon_b^3 \qquad (7)$$

provides adequate description for many soils and for the water content range $\theta < 0.5$ (which covers the entire range of interest in most mineral soils), with an estimation error of about 0.013 for θ. However, Eq. (7) fails to adequately describe the $\varepsilon_b - \theta$ relationship for water contents exceeding 0.5, and for organic rich soils, mainly because Topp's calibration was based on experimental results for mineral soils and concentrated in the range of $\theta < 0.5$.

Numerous studies have been conducted during the past 20 years, focusing on TDR and capacitive methods for water content determination in growing media (Pépin et al., 1991; Paquet et al., 1993; Anisko et al., 1994; Kipp and Kaarsemaker, 1995; Lambiny et al., 1996; da Silva et al., 1998; Morel and Michel, 2004; Naasz et al., 2005). Figure 3.2 presents a calibration of TDR for different growing media (da Silva et al., 1998).

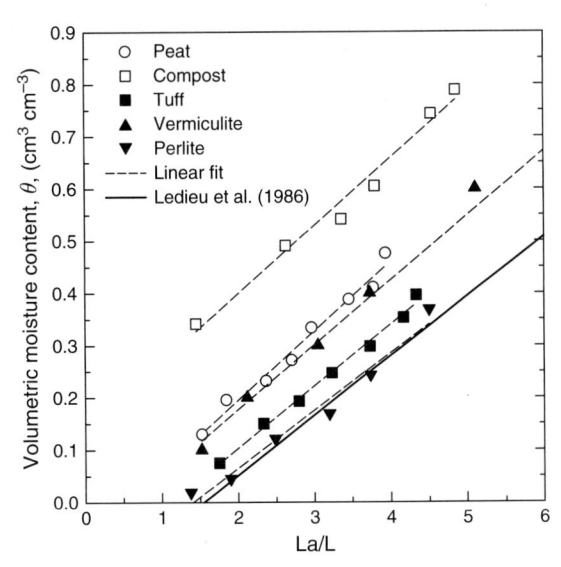

FIGURE 3.2 TDR calibration curves: measured data and corresponding linear regression for the different soil substitutes. The solid line represents Ledieu's equation.

In situ description of water storage and flow characteristics using TDR was made *inter alia* by Paquet et al. (1993), Caron et al. (2002), Nemati et al. (2002) and Naasz et al. (2005).

3.2.2 CAPILLARITY, WATER POTENTIAL AND ITS COMPONENTS

Water that has entered a porous medium but has not drained deep out of the sample bottom will be retained within pores or on the surface of individual medium particles. The particles are typically surrounded by thin water films that are bound to solid surfaces within the medium by the molecular forces of adhesion and cohesion. Because of this, a simple measurement of water content is not sufficient to enumerate the complete status of water in medium. An example of this may be observed in two different media that have been treated in the same manner: though the same amount of water has been applied to each, they will have different water content and varying abilities to contain water due to individual physical and chemical properties. As a result, defining water content alone will not give any indication of the osmotic or other water potentials in the medium and their availability to the plant roots.

While the quantity of water present in the growing medium is very important (affecting such processes as diffusion, gas exchange with the atmosphere, soil temperature), the potential or affinity with which water is retained within the medium matrix is perhaps more important. This potential may be defined as the amount of work done or potential energy stored, per unit volume, in moving that mass, m, from the reference state (typically chosen as pure free water). In this manner, one may think of matric potential as potential energy per unit volume, E ($J\,m^{-3}$). Energy is work with units N (Newton) per distance (N · m). Consequently, $J\,m^{-3} = N \cdot m \cdot m^{-3}$ or $N \cdot m^{-2}$, which is also expressed as Pascal (Pa). A Pascal is a force per unit area or pressure, which explains the use of the term pressure potential, and is the reason why soil physicists refer to matric potential as soil pressure or if it is divided by the bulk density as pressure head.

When a water droplet is formed or placed on a clean surface, the size of the droplet will depend on the attractive forces associated with the air–liquid, solid–liquid, and solid–air interfaces. Forces, for bulk water, consist primarily of London–van der Waals forces and hydrogen bonding; one-third London–van der Waals and two-thirds hydrogen bonding (Stumm, 1992). Molecules at an air–water interface are not subjected to attractive forces from without, but are attracted inwards to the bulk phase. This attraction tends to reduce the number of molecules on the surface region of the droplet due to an increase in intermolecular distance. For the area of the interface to be enlarged, energy must be expended. For water at 20°C, this force, per unit of new area, may be expressed as a force per unit length that has the value $73\,N \cdot m^{-1}$. This stretched surface in tension is called surface tension.

As can be seen in Fig. 3.3, the curvature of the liquid surface is dependent on the contact angle. The angle that the tangent to the fluid surface makes at the point of contact with the solid surface (inside the fluid) is known as the contact angle, and is generally represented by the symbol θ (not to be confused with volumetric

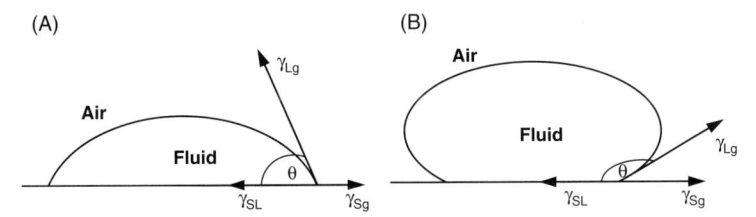

FIGURE 3.3 Interfacial tension components for the equilibrium of a droplet of fluid on a smooth surface in contact with air. The subscripts, S, L and g refer to solid, liquid and gas, respectively. The contact angle is shown by θ.

water content). Normally, this angle is curved, but can range from 0 to 180°. For the majority of fluids on glass, it is less than 90° (Fig. 3.3A). If the contact angle is greater than 90°, $\cos\theta$ is negative and the liquid does not wet the solid (Fig. 3.3B). For water, this condition is termed 'hydrophobic'; drops tend to move about easily, but not enter capillary pores. It is commonly accepted that soil and soilless growing medium water repellency is caused by organic compounds derived from living or decomposing plants or microorganisms. Sources of hydrophobic substances are vegetation, fungi and microorganisms, growing medium organic matter and humus. The identification of the specific compounds causing water repellency has continued to be a focus of soil research in the last decade. However, despite advances in analytical techniques, identifying the exact substances, responsible in a given soil, has yet to be achieved. Furthermore, how these compounds are bonded to soil particles also remains unclear (Doerr et al., 2000).

Soil water repellency (hydrophobicity) reduces the affinity of soils to water such that they resist wetting for periods ranging from a few seconds to hours (e.g. King, 1981; Dekker and Ritsema, 1994; Wallach et al., 2005). Water repellency of soil has substantial hydrological repercussions. These include the reduced infiltration capacity, uneven wetting patterns, development of preferential flow (e.g. Ritsema and Dekker, 1994) and the accelerated leaching of agrochemicals. In a porous medium surface, like soils, water infiltration is inhibited, forming a water pond on the soil surface prior to infiltration. For hydrophobic porous media with sufficiently large pore openings, water might occupy the openings and lead to preferential flow but will not cover the individual grains, whereas hydrophilic particles will be covered by a film of water (Anderson, 1986). A review on soil water repellency, its causes, characteristics and hydro-geomorphological significance can be found in Doerr et al. (2000).

The influence of organic matter on water repellency is even more important in horticulture due to its widespread use as substrate (e.g. peat, coir, bark and other composts). These potentially hydrophobic substrates can cause problems since, after drying out, they require a long time to rewet and the physical properties of the substrate are impaired (Valat et al., 1991; Michel et al., 2001). The evaluation of wettability, not only in an air-dried state but during desiccation, constitutes an important characteristic of organic matter used in horticulture. Non-uniform wetting of organic substrates within containers may take place due to their hydrophobic nature. This topic is currently under intensive research in water repellent soils and less in organic substrates.

Capillarity is also referred to as surface tension (force per unit length). Capillarity deals with both the macroscopic and the statistical behaviour of interfaces, rather than their molecular structure. This phenomenon is extremely important in water retention in porous media. Surface tension, normally expressed by the symbol γ, occurs at the molecular level and involves two types of molecular forces: adhesive forces, which are the attractive forces of molecules of dissimilar substances, and cohesive forces, which are the attractions between molecules in similar substances. Cohesive forces decrease rapidly with distance, and are strongest in the order: solids, liquids, gases. Being work per unit area, the units of γ are dynes cm^{-1}, J m^{-2} or N m^{-1}.

As water rises in a cylindrical tube (Fig. 3.4), the meniscus is spherical in shape and concave upwards. By letting r equal the tube radius, the excess pressure above the meniscus compared to the pressure directly below it can be described under various assumptions by $2\gamma/r$. As the pressure on the water surface outside the capillary tube is atmospheric, the pressure in the liquid below the meniscus will be less than the atmospheric pressure above the meniscus by $2\gamma/r$. This will force the fluid up the tube until the hydrostatic pressure of the fluid column within the tube equals the excess pressure of $2\gamma/r$. Because the circumference of the tube is $2\pi r$, the total force (upward)

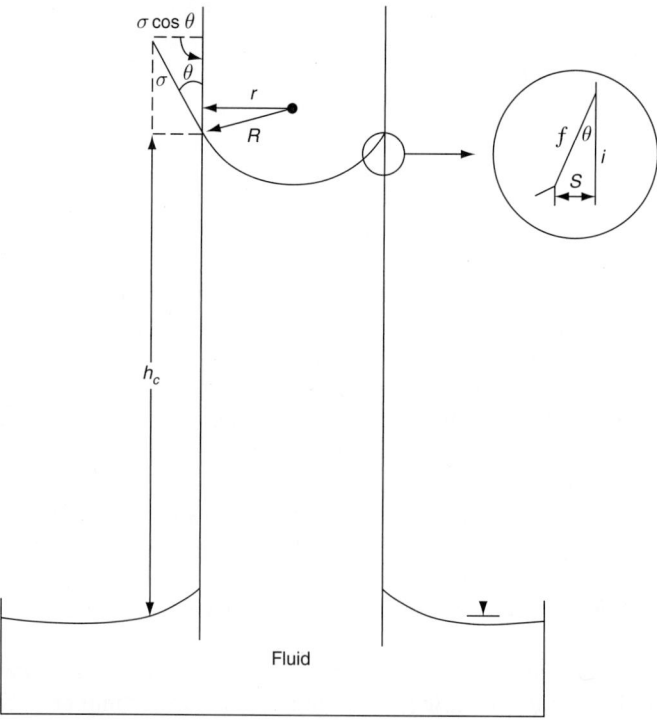

FIGURE 3.4 Capillary rise in a cylindrical tube, where θ is the contact angle, r is the tube radius, R is radius of curve of the gas–liquid interface and h_c is the height of capillary rise. i is the area of the solid–water interface associated with the triangular volume, and f is the air–water interface.

on the fluid is $2\pi r \gamma \cdot \cos\theta$. This force supports the weight of the fluid column to the height h_c. The height of capillary rise can be given as

$$h_c = \frac{2\gamma \cos\theta}{\rho g r} \qquad (8)$$

The radius of curvature, R, of the capillary meniscus in Fig. 3.4 can be determined by $R = r/\cos\theta$. It should also be noted that for a contact angle of zero, $r = R$.

Equation (8) can be used to calculate the height of water rise in a specific porous medium for which the largest effective pore size is known, and also to calculate the diameter of the largest effective pore (assuming h_c is known). These equations are exact for capillary tubes; however, because growing media do not behave as a single capillary such as a glass tube, these equations are at best approximations of water behaviour in them.

The pressure potential above the concave portion of the meniscus in the tube and at the bottom of the capillary is zero (Fig. 3.4). The pressure potential of the water in the capillary tube, ψ_p, decreases with height to offset the increasing gravitational potential, ψ_g (both potential components are discussed in the following). As a result, the pressure of the water within the capillary tube is less than atmospheric pressure, creating a pressure difference on both sides of the meniscus. Because $r = R \cdot \cos\theta$, a general relation for the pressure difference across the meniscus interface with radius of curvature R is $\Delta p = 2\sigma/R$. It should be noted here that the highest pressure is on the concave side of the meniscus. This results in the pressure difference across the interface being inversely proportional to R.

Capillary potential relates directly to the air entry value of porous media. If water is expelled from a capillary by a positive gas pressure, that pressure is termed 'the bubbling pressure'. Because h_c is the pressure head at the air–water interface of the meniscus, this concept can be extended to growing medium pores. For example, if the pressure potential in a porous medium pore is lower than the air entry value, both cohesive and adhesive forces cannot hold the water any longer. This will result in water draining from the pore until the pressure potential at the air–water interface is equal to that of the air-entry value. Since porous media are of many irregularly shaped capillaries and not a clean glass tube, the concepts discussed here can only be applied qualitatively to capillary water in growing media. Being a threshold of desaturation, air-entry value is generally small in coarse-textured and in well-aggregated soils having large pores, but tends to be larger in dense, poorly aggregated, medium-textured or fine-textured soils. The physical properties of growing media are significantly different to field soils in the sense that their texture is coarse with low air-entry value that enables fast free drainage of most of the water following irrigation. The fast and intensive drainage enables air to replace the draining water and prevent water clogging. Note that the coarseness of the growing media and the resulting air-entry value should be adapted to the height of the growing container, namely, it should be lower than the container height.

The total potential of growing medium water has been defined as the amount of work that must be done per unit quantity of pure water in order to transport reversibly and isothermally an infinitesimal quantity of water from a pool of pure water at a specified elevation at atmospheric pressure to the medium water at the point of consideration. This definition was retained in the *Glossary of Terms* of the Soil Science Society of America (1996). The transformation of pure water from the reference state to the medium-water state is actually broken into a series of steps. These steps are generally reversible and isothermal. Because of this, the total potential is actually a sum of each sequential step. Growing medium water is subjected to a number of possible forces, each of which may cause its potential to differ from that of pure, free water at the reference elevation. Such force fields result from the mutual attraction between the solid matrix and water, from the presence of solutes in the medium solution, as well as from the action of external gas pressured and gravity. Accordingly, the total potential of growing medium water can be considered as the sum of the separate contributions of these various factors.

$$\psi_t = \psi_g + \psi_m + \psi_o + \cdots \tag{9}$$

where ψ_t is the total potential, ψ_g is the gravitational potential, ψ_m the matric potential and ψ_o the osmotic potential. Additional terms are theoretically possible in Eq. (9).

All objects on Earth are attracted downwards due to gravitational force. This force is equal to the weight of an object, which is a product of mass and gravitational acceleration. Gravitational potential is the energy of water (on a unit volume basis) that is required to move a specific amount of pure, free water from an arbitrary reference point to the medium-water elevation. If the medium-water elevation is above the reference point, z is positive; if the medium-water elevation is below the reference, z is negative, where z is the vertical distance from the reference point to the point of interest. This means that the gravitational potential is independent of medium properties, and is solely dependent on the vertical distance between the arbitrary reference point and the medium-water elevation or elevation in question. Gravitational potential has the value

$$\psi_g = \rho_l g z \tag{10}$$

Here, ψ_g has the units of $J\,m^{-3}$, and is assumed to be positive upwards.

The primary effect of salts in the growing medium water is that of lowering the vapour pressure. However, this does not directly affect the mass flow of water in the medium, except in the presence of a membrane barrier of some type and in the case of vapour diffusion. Plants make up part of a dynamic medium-water-plant system, and their roots are often affected by the presence of salts in the soil solution. The membranes of plant roots transmit water more readily than salts, and are influenced by osmotic forces. Osmotic potential may be thought of more as a suction than a pressure. In the presence of a selective permeable membrane (as found in plant roots), the pressure or energy potential (a suction) on the solution side is less than that of pure, free water, and the water first will pass through the membrane to the solution side. One can prevent flow across the membrane and maintain static equilibrium by exerting gas pressure, P, on the medium solution side in excess of atmospheric pressure. This will result in the total potential at the surface of the solution, ψ_t, being

$$\psi_t = \frac{1}{\rho_w}P + \psi_g + \psi_o \tag{11}$$

where P is the air pressure (Pa) in excess of atmospheric. The term $1/\rho_w$ is included due to the small difference between the density of the solution and that of pure water, and is expressed in units of $J\,kg^{-1}$. However, in practice, it is often neglected, and the simple term P is used (in which case the units are expressed in $J\,m^{-3}$). At the surface of the pure water, the total potential is

$$\psi_t = \psi_g + \psi_m + \psi_o = \psi_g \tag{12}$$

Because the solution is pure water, we can ignore the matric potential, ψ_m, and osmotic potential, ψ_o. At static equilibrium, $\psi_t = \psi_g$. Consequently, expressed in $J\,m^{-3}$,

$$P + \psi_o = 0 \tag{13}$$

To maintain equilibrium, the excess air pressure needed must be equal to what is commonly referred to as the osmotic pressure, Π, of the solution, resulting in an expression for osmotic potential of

$$\psi_o = -\frac{\Pi}{\rho_w} \tag{14}$$

Equation (14) (in $J\,kg^{-1}$) could be expanded to include the total potential of growing medium water components, and is written as

$$\psi_t = \frac{\Pi - \psi_m}{\rho_w} + gz \tag{15}$$

The matric potential of a porous medium, ψ_m, results primarily from both the adsorptive and the capillary forces due to medium matrix properties and is, thus, a dynamic medium property. The matric potential is often referred to as capillary or pressure potential, and is usually expressed with a negative sign. By convention, pressure is generally expressed in positive terms. Assume a pot, 26-cm high, filled with substrate with free water at its bottom as a result of free drainage after saturation or irrigation (Fig. 3.5). The matric potential above the free water would be negative; the potential at the free water would be zero. If the pot was initially saturated with water table at the medium surface, the pressure potential below the water table varies with height and is equal to ρgh, where h is the distance below the water table. It should be noted that an unsaturated medium has no pressure potential, but only a matric potential with negative units.

The matric potential of growing media is measured by tensiometers. A tensiometer consists of a porous cup, usually made of porous ceramic connected to a vacuum gauge through a water-filled tube. The porous cup is placed in a close contact with the bulk medium at the depth of measurement. When the measured matric potential is lower (more negative) than the equivalent pressure inside the tensiometer cup, water moves from the tensiometer along a potential energy gradient to the medium through the

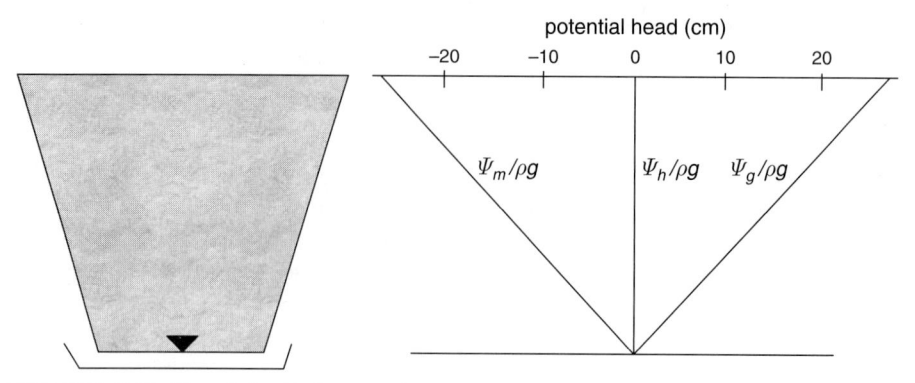

FIGURE 3.5 The relationship between the components potential in a container, 26-cm high, filled with growing medium after free drainage (container capacity).

saturated porous cup, thereby creating suction sensed by the gauge. Water flow into the medium continues until equilibrium is reached and the suction inside the tensiometer equals the soil matric potential. When the medium is wetted, flow may occur in the reverse direction, that is soil water enters the tensiometer until a new equilibrium is attained. The tensiometer equation is

$$\psi_m = \psi_{gauge} + (z_{gauge} - z_{cup}) \tag{16}$$

The vertical distance from the gauge plane (z_{gauge}) to the cup (z_{cup}) must be added to the matric potential measured by the gauge to obtain the matric potential at the depth of the cup, when potential is expressed per unit of weight (named as suction head). This accounts for the positive head at the depth of the ceramic cup exerted by the overlying tensiometer water column.

Electronic sensors called pressure transducers often replace the mechanical vacuum gauges. The transducers convert mechanical pressure into an electric signal which can be more easily and precisely measured. The combination of electronic sensors with data logging equipment provides continuous measurements of soil matric potential. A problem that one should be aware of is the time required for matric potential sensors to reach equilibrium. There is a significant difference between tension (and moisture content) variation in field soils and limited-volume containers that are filled with coarse soilless materials. The rate of change of these variables in a limited-volume container is much higher than that in field soils. As such, tensiometers with ceramic cups of larger pores with lower air-entry values should be used in order to follow in real-time the fast changes in matric potential. Owing to the relatively low matric-potential values that are usually practiced in soilless culture, air does not invade the pores of the ceramic cup and their fast response to changes in the containerized growing medium is attained.

An example of matric potential variation, during a typical summer day in a growing container, is illustrated in Fig. 3.6. These tensiometeric data were taken from an experiment showing the relationship between irrigation frequency and water availability (Wallach and Raviv, 2005). The containers, cylinders of 20 cm diameter and

3.2 WATER CONTENT AND WATER POTENTIAL IN SOILLESS MEDIA

(A) Perlite 'dry' treatment

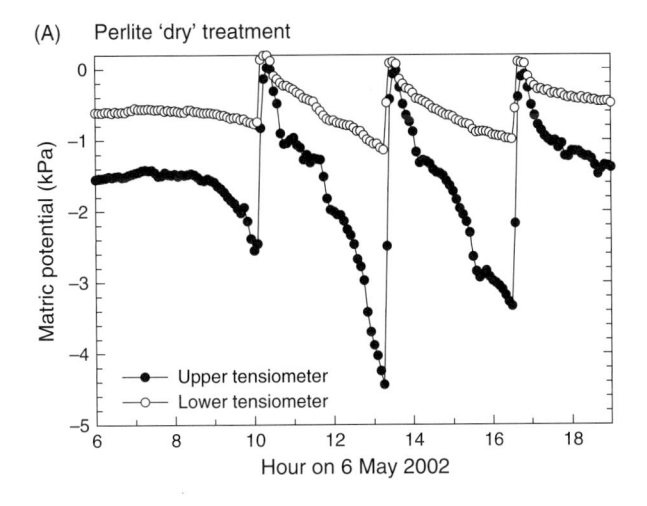

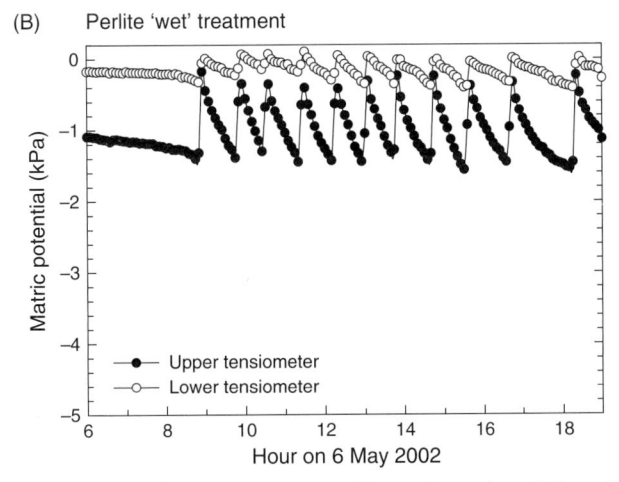

(B) Perlite 'wet' treatment

FIGURE 3.6 Matric potential variation during an arbitrary chosen day within an irrigated pot filled with perlite. The tensiometers were located 7 (lower) and 14 (upper) cm above the base of a 21-cm high pot. The grown rose plants were irrigated at two irrigation frequencies (Wallach and Raviv, unpublished).

21 cm height (volume of 10 liters), were filled with RTM tuff 0–8 mm (volcanic ash, scoria, Merom Golan, Israel). Rose plants were grown in these containers and were drip irrigated with four l/hr drippers per container. The matric potential was measured by two high-flow electro-tensiometers that were installed horizontally in each container at 7 and 14 cm from the bottom. The tensions were measured every 10 s and averaged for 5-min intervals. Two irrigation treatments are shown in Fig. 3.6: dry (A) and wet (B). The irrigation scheduling in this experiment was automatically controlled according to predefined container-weight thresholds. The dry treatment was obtained by lower weight thresholds at which irrigations were turned on. Matric potential decreases upon

the medium wetting by irrigation and increases during periods when water is depleted by root water extraction (Fig. 3.6). A similar threshold high tension was obtained in each of the two treatments owing to the weight threshold that started the irrigation. Although container weight was used here to control irrigation, tensiometers can be an alternative controller for irrigation scheduling (Karlovitch and Fonteno, 1986; Hansen and Pasian, 1999; Lebeau et al., 2003). The rate at which matric potential decreases (tension increases) depends on the water depletion rate or transpiration rate, which themselves depend on the momentary balance between the atmospheric demand and the water availability in the medium. This issue will be elaborated later in this chapter.

3.2.3 WATER RETENTION CURVE AND HYSTERESIS

The water characteristic curve of porous media (abbreviated in the soil physics literature by SWC), also known as the moisture retention curve (RC), describes the functional relationship between the water content, θ, and the matric potential, ψ_m, under equilibrium conditions. This curve is an important property related to the distribution of pore space (sizes, interconnectedness), which is strongly affected by texture and structure, as well as related factors including organic matter content. The RC indicates the amount of water in the porous medium at a given matric potential and is frequently used in soilless culture to estimate the water availability to plants and for irrigation management. It is also a primary hydraulic property required for modelling water flow in growing media. RCs are highly non-linear functions and are relatively difficult to obtain accurately (Figs. 3.7 and 3.8).

When suction is applied incrementally to a saturated porous medium, the first pores to be emptied are the relatively large ones that cannot retain water against the suction applied. From the capillary equation (Eq. [8]), it can be readily predicted that a gradual increase in suction will result in the emptying of progressively smaller pores, until, at high suction values, only the very narrow pores retain water. Similarly, an increase in medium-water suction is associated with decreasing thickness of the hydration envelopes adsorbed to the particle surfaces. Increasing suction is thus associated with decreasing medium wetness (and a decrease in the osmotic potential of the aqueous phase). The amount of water remaining in the medium at equilibrium is a function of the sizes and volumes of the water-filled pores and the amount of water adsorbed to the particles; hence it is a function of water suction.

The traditional method of determining the water RC involves establishing a series of equilibria between water in the porous medium sample and a body of water at known suctions. The medium-water system is in hydraulic contact with the body of water via a water-wetted porous plate or membrane. At each equilibrium, the volumetric water content, θ, of the medium is determined and paired with a value of the matrix pressure head, ψ, determined from the pressure in the body of water and the gas phase pressure in the substrate. The data pair, (θ, ψ), forms one point on a RC. A summary of methods used in soil science to determine the RC can be found in Klute (1986). For analysis procedure mostly used for soilless growing media research, see Chap. 7. Owing to its unique texture, the RC of a soilless growing medium is usually determined for very

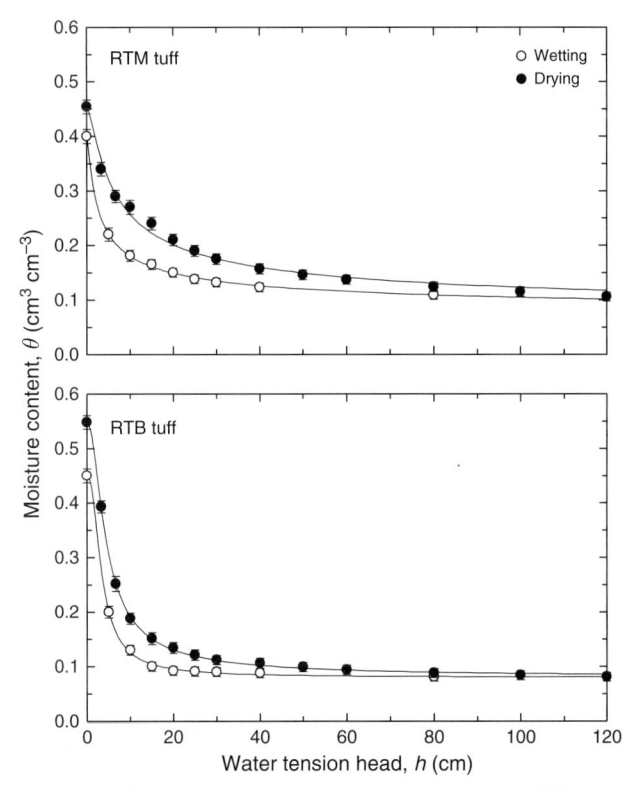

3.2 WATER CONTENT AND WATER POTENTIAL IN SOILLESS MEDIA

FIGURE 3.7 Measured and fitted water retention curves for tuff RTM and RTB.

low suctions (up to 300 cm) using a suction funnel (Goh and Mass, 1980), suction table (Ball and Hunter, 1988) or a pressure plate system (Fonteno et al., 1981). A record of measured RCs for different soilless growing media is given in Table 3.1.

In situ measurement of RCs may provide supplementary information on the dependence of RC on water dynamics in the substrates (including the effect of water depletion by root uptake) – dynamic non-equilibrium (Smiles et al., 1971; Vachaud et al., 1972; Plagge et al., 1999), and variation of the physical properties of the substrates with time (including the effect of root growth on pore-size distribution). The *in situ* RC measurement is obtained by simultaneous measurement of the transient moisture content and tensions by adjacent pairs of TDR probes and tensiometers. The effect of irrigation frequency on the *in situ* measured RC is demonstrated in Wallach and Raviv (2005).

The RCs of soilless growing media have an intrinsically different shape than field soils, a property that is needed to properly manage the air–water relationship within containers at low suctions range. As an illustration, RCs for RTM and RTB tuff (Wallach et al., 1992a), stone wool (da Silva et al., 1995) and peat (da Silva et al. (1993a) that were measured over a 0–120 cm suction range are presented in Figs. 3.7 and 3.8, respectively. The tension head (cm) shown in this figure is the matric potential

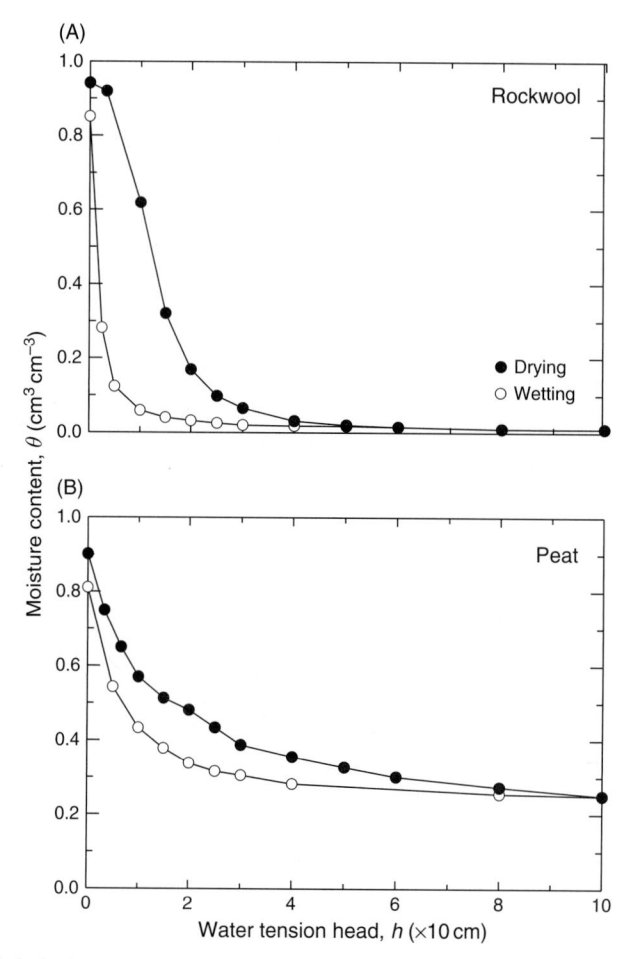

FIGURE 3.8 Measured and fitted water retention curves for stone wool (A), and peat (B).

divided by ρg. A notable characteristic of soilless growing media is their high value of saturated water content, θ_s, and low value of air-entry value, for which the medium remains saturated and air does not replace water. Low air-entry values are needed to keep growing containers, usually with height ranging from 5 to 20 cm, aerated soon after irrigation. Temporary shortage of oxygen can reduce root and shoot growth and anaerobic conditions for only a few days will result in the death of some roots. As can be seen in Figs. 3.7 and 3.8, and in other measured RCs for container media (list appears in Table 3.1), the air-entry value for these media can be practically taken as zero. Beyond the air-entry suction, any suction increase (up to 25–40 cm) is associated with a sharp decrease in water content until almost constant moisture content is reached (for this measured suction range). The rate at which moisture content decreases per unit suction increase ($C = d\theta/d\psi$, noted as the specific water capacity) is significantly

TABLE 3.1 Sources for Retention Curves [θ(ψ)] and Unsaturated Hydraulic Conductivities [K(ψ)] of Representative Media

Material	Reference	Comments
Sand	Riviere (1992)	θ(ψ)
Stone wool	da Silva et al. (1995)	θ(ψ) and K(ψ)
Stone wool	Riviere (1992)	θ(ψ)
Perlite	Orozco and Marfa (1995)	θ(ψ) and K(ψ) for different types available in Spain
Vermiculite	Fonteno and Nelson (1990)	θ(ψ)
Vermiculite	Riviere (1992)	θ(ψ)
Zeolite	Riviere (1992)	θ(ψ)
Pumice	Raviv et al. (1999)	θ(ψ) and K(ψ)
Tuff – Scoria (RTM, RTB)	Wallach et al. (1992a)	θ(ψ) and K(ψ)
Sphagnum peat moss	da Silva et al. (1993a)	θ(ψ) and K(ψ), including its mixtures with tuff
Sphagnum peat moss	Heiskanen (1995a,b; 1999)	θ(ψ), including its mixture with coarse perlite
Sphagnum peat moss	Heiskanen (1999)	θ(ψ) and K(ψ), including its mixtures with perlite and sand
Canadian sphagnum peat	Fonteno and Nelson (1990)	θ(ψ)
Composted agricultural wastes	Wallach et al. (1992b)	θ(ψ) and K(ψ), including its mixtures with tuff
Composted cow manure	Raviv and Medina (1997) and Raviv et al. (1998)	θ(ψ)
Coir	Raviv et al. (2001)	θ(ψ) and K(ψ)
Pine bark	Fonteno and Nelson (1990)	θ(ψ)
Pine bark mixed with hardwood bark	Bilderback (1985)	θ(ψ)
Sawdust	Goh and Haynes (1977)	θ(ψ)
UC mix	Raviv et al. (2001)	θ(ψ) and K(ψ)
Peat and vermiculite mix (1:1)	Fonteno (1989)	θ(ψ)
Pine bark, peat and sand mix (3:1:1)	Fonteno (1989)	θ(ψ)

different for soilless growing media and soils, being steeper in the former. The sharp decrease in Figs. 3.7 and 3.8 is initiated at very low suction, as the air-entry value is close to zero. The constant θ values that follow a sharp decrease presents a unique shape to the characteristic hydraulic conductivity curve for container media, as will be discussed later.

The difference between RC shapes of soilless growing media and those of regular soils are not solely due to differences in their particle size. This is illustrated in Fig. 3.9, which shows the RCs measured for a 1–2 mm fraction of RTM tuff and similarly textured sand (Wallach et al., 1992a). In spite of the similar particle size, the tuff porosity (0.58 cm^3 cm^{-3}) is much higher than the sand porosity (0.30 cm^3 cm^{-3}). The porosity disparity is attributed to the inner microporosity of the tuff particles (Chen et al., 1980) which is not present in sand. Sand has a typical S-shaped RC,

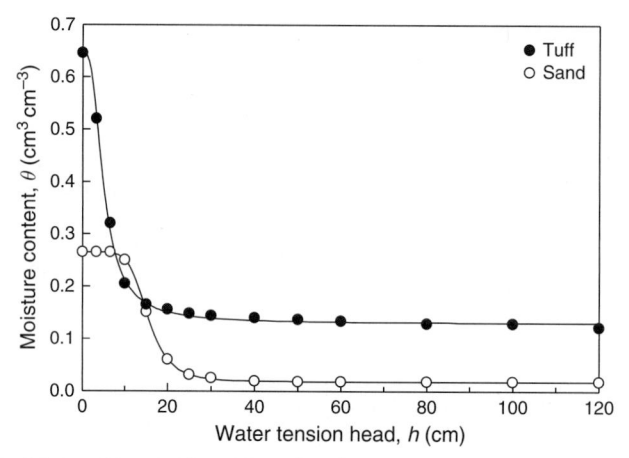

FIGURE 3.9 Measured (symbols) and fitted (lines) drying retention curves for 1–2 mm fractions of red tuff and quartz sand.

which characterizes soils of different types, whilst the tuff RC lacks the first vertical part of the S shape and looks more like a decreasing hyperbolic function, similar to the RCs shown in Fig. 3.9. The value of the air-entry suction of tuff could not be determined from the measured RC. It is probably a few millimetres compared to 10 cm for sand. The difference in air-entry value can be attributed to the particle shapes, which affect their spatial distribution in the container. Sand particles are regular and smooth compared to tuff particles, which are rough and irregular. The water content approaches 0.125 and 0.02 for tuff and sand, respectively, when suction increases beyond 30 cm. The difference in the residual water content values can be attributed to the inner microporosity of the tuff particles.

Measured values of water content and suction ($\theta - \psi$) are often fragmentary, and are usually based on relatively few measurements over the wetness range of interest. For modelling and analysis purposes and for the characterization and comparison of different substrates and scenarios, it is essential to represent the RC in continuous and parametric form. A parametric expression of an RC model should contain as few parameters as possible to simplify its estimation and describe the behaviour of the RC at the limits (wet and dry ends) while closely fitting the non-linear shape of the $\theta - \psi$ data.

An effective and commonly used parametric model for relating water content to the matric potential was proposed by van Genuchten (1980) and is denoted as VG:

$$S_e = \frac{\theta - \theta_r}{\theta_s - \theta_r} \left[\frac{1}{1+(\alpha\psi)^n} \right]^m, \quad S_e = \frac{\theta - \theta_r}{\theta_s - \theta_r} \left[\frac{1}{1+(\alpha h)^n} \right]^m \quad (17)$$

where S_e is the effective saturation (Eq. [6]), $h = \psi_m/\rho g$ is the matric or tension head (−L).

The parameters α, n and m in Eq. (17) determine the shape of the RC and are determined by curve-fitting techniques. The subscripts s and r in Eq. (17) refer to the saturated and residual values of θ, respectively. The latter is defined as the water content at which the gradient $d\theta/d\psi$ becomes zero (excluding the region near θ_s, which may also present a zero gradient). Both θ_s and θ_r can be either measured or estimated along with α, n and m. As opposed to θ_s, that has a clear physical significance, the meaning of θ_r and its estimation has not yet been resolved. Stephens and Rehfeldt (1985) reported improved model accuracy using a measured value of θ_r. Fonteno and Nelson (1990) determined θ_r as the water content at $\psi = 300$ cm. However, Ward et al. (1983) and van Genuchten (1978, 1980) suggested that θ_r should be viewed as a fitting-parameter rather than a soil property. van Genuchten and Nielsen (1985) considered not only θ_r but also θ_s to be empirical parameters that should be fitted to the $\theta - \psi$ data. The VG model can be used in conjunction with predictive models for unsaturated hydraulic conductivity, as will be discussed in the next section.

For small m/n ratios, parameter α in Eq. (17) approximately equals the inverse of the air-entry value. For large values of m/n, this parameter roughly equals the inverse of the suction at the inflection point of the retention curve (van Genuchten and Nielsen, 1985). Parameter n is related to the pore size distribution of the medium and the product $m \cdot n$ determines the slope of the $\theta(\psi)$ curve at large suction values. Therefore, n may be viewed as being mostly affected by the structure of the medium (van Genuchten and Nielsen, 1985).

The VG model (Eq. [17]) was applied to soilless growing media by Milks et al. (1989a,b,c), who found that it fits the measured $\theta - \theta_r$ data better than the cubic polynomial. It has also been used by Fonteno (1989), Wallach et al. (1992a,b), da Silva et al. (1993a, 1995), Orozco and Marfa (1995) and Raviv et al. (2001). Examples of the VG model fit to measured RCs of different growing media are shown in Fig. 3.7 for RTB and RTM tuffs and in Fig. 3.8 for stone wool and peat.

There are two initial stages at which RCs are measured: A drainage curve is obtained by establishing a series of equilibria by drainage from zero suction to higher suctions. A wetting curve is obtained by equilibrating samples wetted from a low water content or high suction to zero suction. The retention curve is hysteretic, that is the water content at a given suction for a wetting medium is less than that for a draining medium (e.g., Fig. 3.7 and 3.8). The drainage curve that starts on complete saturation of the medium is called the initial drainage curve and the main wetting curve is obtained by wetting the medium from low moisture content. The measured initial drainage and main wetting curves were successfully fitted by the VG model, as shown in Figs. 3.7 and 3.8. As the medium is wetted along the main wetting curve and the suction approaches zero, the water content approaches a value that is less than the total porosity, ϕ, due to the presence of entrapped air (Figs. 3.7 and 3.8). This value is about 0.8ϕ to 0.9ϕ and is called the natural saturation of the satiated water content (Klute, 1986). The drainage curve, starting at natural saturation, is called the main drainage curve. This curve merges asymptotically with the initial drainage curve as the suction increases. The values of θ_s (at saturation and at the end of one cycle of drying and wetting), together with the parameters α, n and ($m = 1/n$) for the VG retention model are shown in Table 3.2, for both drying and wetting.

TABLE 3.2 Curve Fitting of Drying (d) and Wetting (w) Retention Data of Tested Media Using the VG Retention Model: Fitted Parameter Values, Standard Error Coefficients (in Parenthesis) and Coefficient of Determination (R^2). θ_s is the Water Content at Saturation, and at the End of One Cycle of Drying and Wetting

Medium	θ_s	α	Error coefficient	n	Error coefficient	θ_f	Error coefficient	R^2
Tuff RTM	d 0.454	d 0.346	(0.038)	1.529	(0.030)	0.063	(0.023)	0.990
	w 0.400	w 1.000	(0.086)	1.455	(0.014)			0.999
Tuff RTB	d 0.548	d 0.324	(0.063)	2.186	(0.034)	0.079	(0.020)	1.000
	w 0.450	w 0.387	(0.026)	2.534	(0.080)			0.999
Coarse sand	d 0.265	d 0.067	(0.001)	6.532	(0.060)	0.018	(0.003)	0.985
	w 0.260	w 0.080	(0.000)	6.615	(0.022)			1.000
Stone wool	d 0.935	d 0.083	(0.001)	3.725	(0.059)	0.007	(0.003)	1.000
	w 0.750	w 0.739	(0.069)	2.157	(0.081)			1.000
Sphagnum peat	d 0.901	d 0.264	(0.014)	1.390	(0.098)	0.0		0.998
	w 0.810	w 0.582	(0.124)	0.345	(0.029)			0.989

Naasz et al. (2005) measured the water retention and hydraulic conductivity of drying–wetting cycle curves for peat and composted pine bark by a transient procedure (IPM). Results showed differences in the physical behaviour of the two substrates studied. Hysteresis phenomena were evident in the water retention and hydraulic conductivity curves for peat, whereas this phenomenon was very limited for pine bark. The use of the VG (van Genuchten, 1980) retention model to describe the water retention characteristics of the two materials revealed a high correlation and seemed to be in agreement with other hydraulic studies of substrates. The differences in hysteretic phenomena among the two substrates could be related to their different structure; fine and fibrous in the case of peat, and coarse for pine bark. With regard to peat, the hysteresis phenomenon can be also related to the change in the solid phase organization (swelling/shrinkage phenomena) and consequently in the pore interconnection, as well as to variations in wettability as was suggested by Michel et al. (2001).

When second and third cycles of drying and wetting are measured, they provide normally different wetting and drying curves. The initial drying curve differs markedly from the second one, while the wetting curves are close to each other. This behaviour is usually attributed to entrapped air. Retention curve measurements in stone wool showed that (data not shown) the secondary and tertiary curves show a considerable reduction of the hysteretic loop in θ(h) curves. The converging drainage and wetting scanning curves at higher cycles may indicate that at frequent irrigation regime, usually applied in horticultural substrates, a single RC can be used. The chosen curve should depend on the irrigation method: bottom flooding or top irrigation.

The main hysteresis mechanism is referred to as the inkbottle effect, due to its similarity with the behaviour of bottles of ink with very narrow spout. Another source of hysteretic behaviour is hysteresis in the contact angle. The contact angle often exhibits a different value in advancing and receding cases. A central question is when and why hysteresis must be included when describing container media processes. The answer to this question is not yet clear, given that few studies have been performed on hysteresis in growing media. The adjustment of the theory used in soil physics, and its application to container media, is not straightforward. This is due to the special conditions that exist in the limited container volume where: (1) wetting and drying frequently occur, (2) special conditions that exist at the container boundaries (the bottom may be open to the atmosphere or in some cases continuously saturated), and (3) the unique textural and structural properties of soilless substrates compared to field soils. Thus, additional research is needed to increase knowledge of the effects of hysteresis on static and dynamic distributions of moisture content vs. suction.

3.3 WATER MOVEMENT IN SOILLESS MEDIA

3.3.1 FLOW IN SATURATED MEDIA

Following the definition of hydraulic potential (Eq. 11), the hydraulic head, H, is defined as the potential per water specific weight, ρg. Water will therefore flow from regions of high to low hydraulic head. The flow equation in saturated media is Darcy's law

$$J_W = -K_s \frac{\Delta H}{\Delta s} \tag{18}$$

With K_s [LT^{-1}], the saturated hydraulic conductivity; H [L], the hydraulic head; s [L], distance along a stream line in the flow field; $\Delta H/\Delta s$, the hydraulic-head gradient along the stream line; and J_w, the flux density or flow per unit area opposite to the direction defined by hydraulic-head gradient. If water flow takes place in the horizontal or vertical directions, s becomes x or z, respectively. Water pressure in saturated porous media is greater than zero (although a medium is also saturated at negative pressure that is higher than the water-entry value). Darcy's law is empirically based in spite of its similarity to classical relationship (e.g. Poiseuille's law). If the streamlines do not superimpose on the Cartesian coordinates (x, y, z), Darcy's law is written in a more general way

$$\vec{J}_W = -K_s \operatorname{grad} H \tag{19}$$

where 'grad' is the vector gradient of H. In Cartesian coordinates, Eq. (19) may be expressed by individual components:

$$\begin{aligned} J_x &= -K_s \frac{\partial H}{\partial x} \\ J_y &= -K_s \frac{\partial H}{\partial y} \\ J_z &= -K_s \frac{\partial H}{\partial z} \end{aligned} \tag{20}$$

We assume that the growing medium is isotropic, namely, the hydraulic conductivity value does not depend on the flow direction (unisotropy). As opposed to many problems in field soils and groundwater, this assumption holds in most of the cases in soilless culture.

Deviations from Darcy's law can occur at high flow velocities, when J_w eventually becomes non-linear with respect to the hydraulic gradient (Fig. 3.10). The departure of the measured curve from the linear line obtained by the Darcy's law, Eq. (18), indicates that the latter does not apply at large flow velocities. As flow velocity increases, especially in systems with large pores, the occurrence of turbulent eddies or non-linear laminar flow results in dissipation of effective energy by the internal mixing of the liquid. As a result, the hydraulic potential gradient becomes less effective in inducing flow. Owing to the coarse texture of soilless container media, high velocities are obtained even at relatively low hydraulic gradients. A criterion for departure from laminar flow is based on the Reynolds number, Re,

$$Re = \frac{J_w \cdot \rho \cdot d}{\eta} \qquad (21)$$

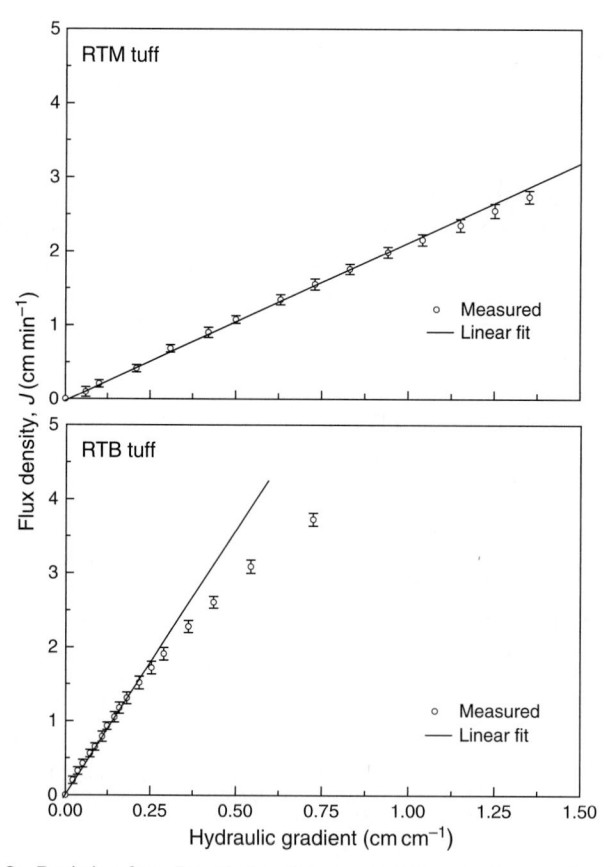

FIGURE 3.10 Deviation from Darcy's law for measured flux density as a function of hydraulic gradient for tuff RTB and RTM.

where $d[L]$ is an effective pore diameter. The critical Reynolds number is much lower for porous media than for straight tubes, such as those assumed for flow in pipes. Wide ranges have been observed, but generally, for $Re < 1$, laminar conditions are expected and the Darcy law is valid (Hillel, 1998). At a higher Reynolds number, the inertial forces become significant relative to viscous forces and Darcy's law cannot be used. The deviations from Darcy's law in Fig. 3.10 can therefore be related to the development of turbulent flow at $Re > 1$.

The value of the hydraulic conductivity of a saturated porous medium, $K_s[LT^{-1}]$, depends on the properties of the medium and the flowing fluid.

$$K_s = \frac{k \cdot \rho g}{\eta} \tag{22}$$

K_s can be separated into two factors: fluidity (defined as $\eta/\rho g$, where ρ is the fluid density and η is its kinematic viscosity) and intrinsic permeability, k. The intrinsic permeability of a medium is a function of pore structure and geometry. Particles of smaller-sized individual grains have a larger specific surface area, increasing the drag on water molecules that flow through the medium which results with a reduced intrinsic permeability and K_s.

The measurement of K_s is based on the direct application of the Darcy's law (Eq. [18]) to a column of uniform cross-sectional area that is filled with the saturated medium. A hydraulic-head difference is imposed on the column and the resulting water flux is measured. A constant or variable head difference is maintained throughout the experiment (Klute and Dirksen, 1986). Note that the applied hydraulic gradient should be limited to the range within which Darcy's law is valid (Fig. 3.10) where measured flux density is a linear function of the hydraulic gradient (Wallach et al., 1992a).

3.3.2 FLOW IN AN UNSATURATED MEDIA

In most cases of irrigated substrates, their pores will be filled by both air and water (unsaturation). Assuming that pressure of the air in the pores is atmospheric, the flux density for the water–air system is

$$\vec{J}_W = -K[\text{grad}(h) + \text{grad}(z)] \tag{23}$$

where $h[L]$ is the tension head (matric potential divided by ρg) and K is the unsaturated hydraulic conductivity that depends on water status expressed as a function of the tension head h or water content θ.

The value of K decreases rapidly as h increases or θ decreases and a tortuous path is necessary to move from position to position. Just as the medium water characteristic curves of θ vs. h can exhibit hysteresis, so can the conductivity function.

Although the saturated hydraulic conductivity, K_s, is usually determined with little difficulty, many technical problems are involved in the measurement of the hydraulic conductivity under unsaturated conditions. It would be desirable to determine $K(\theta)$ or $K(h)$ by direct measurement. However, this is often not possible because the values of the hydraulic conductivity may vary by several orders of magnitude within the water content range of interest. In addition, the measurements are tedious and expensive, and most measurement systems cannot efficiently cover such a wide range of variation.

A method to measure $K - h$ relationship is by the steady-state flux control method (Klute and Dirksen, 1986). According to this method, a constant flux of water, Jw, is established at the upper end of a vertical uniformly packed substrate column, while its lower end is maintained at atmospheric pressure. The flux, Jw, which should be lower than the saturated hydraulic conductivity, K_s, is applied to a previously saturated column, which drains to a condition of steady-state downward flow. Upon reaching this condition, the suction distribution along the column is expected to be relatively constant throughout its upper region. A unit hydraulic gradient ($dh/dz = 0$ in Eq. [23]) should, therefore, be established in that part of the column, and under these conditions K is numerically equal to J_w.

This method was used by Wallach et al. (1992) and da silva et al. (1995) to measure the $K(h)$ of tuff RTM and RTB, and stone wool. The experimental set-up included a 50-cm long column (10 cm i.d.), four tensiometers, a pressure transducer, a manual scanning valve system and a peristaltic pump. Each tensiometer consisted of a high-flow ceramic porous cup which was mounted in a horizontal position, extending about 6 cm across the column. The vertical distance between tensiometers was 10.5 cm. Starting at saturation, a controlled flow was maintained until the tensiometer readings stabilized and the volumetric outflow rate was constant. Due to the unit-gradient, the constant flux yielded K, while the suction was measured to obtain the related suction. This process was then repeated at a series of decreasing flow rates, and each time Jw and h were recorded. For each flux, measurements continued until a steady-state condition was attained. This usually took from several hours to several days. The set-up for $K(h)$ measurement is shown in Figure 2 in Wallach et al., 1992a. The measured $K(h)$ for tuff RTM and stone wool are shown in Figs. 3.11A and 3.11B, respectively.

Given the curve-fitting equation for $\theta(h)$, the unsaturated hydraulic conductivity $K(h)$ can be directly calculated by means of a predictive equation based on the fitted $\theta(h)$ curve, and a single measurement of the saturated hydraulic conductivity, K_s. The combination of a curve-fitting equation for $\theta(h)$ with a predictive model for K produces the structure of a combined model for the determination of the substrate's hydraulic properties. To obtain an accurate predictive equation for the unsaturated hydraulic conductivity, an analytical expression that accurately describes the medium water RC, over the whole relevant suction range, is required. This prerequisite is essential, and any attempt to predict the unsaturated hydraulic conductivity from retention data will fail if the assumed function cannot describe the data over the whole range of interest (van Genuchten and Nielsen, 1985). The accuracy of the predictive equation for unsaturated hydraulic conductivity also depends on the theory and assumptions on which that equation is ultimately based.

Several efforts have been made to relate pore-size distribution to the unsaturated (as well as saturated) hydraulic conductivity. Two of these general relationships are by Mualem (1976) and Burdine (1953). Both of these general relationships are based largely on interconnected capillary bundle. As mentioned previously, the pore-size

3.3 WATER MOVEMENT IN SOILLESS MEDIA

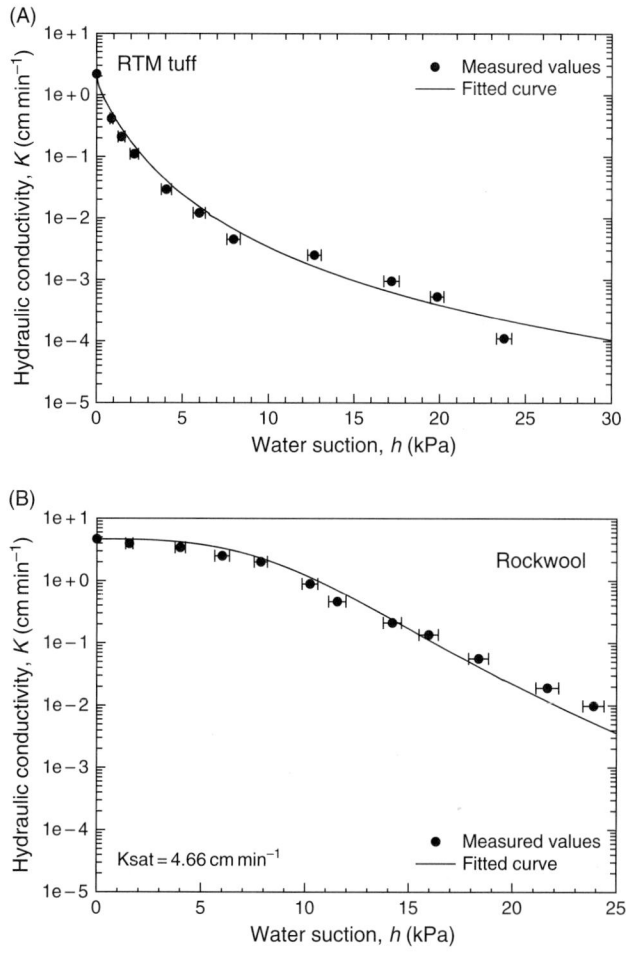

FIGURE 3.11 Measured (symbols) and calculated (lines) hydraulic conductivity, $K(h)$, for RTM tuff (A) and stone wool (B).

distribution curve can be defined in terms of the soil water characteristic curve by the capillary rise equation. Mualem developed his relationship using an interactive capillary bundle theory with the result.

$$K_r = \frac{K}{K_s} = \frac{S_e^p \left[\int_0^{S_e} dx/h(x) \right]^2}{\left[\int_0^1 dx/h(x) \right]^2} \quad (24)$$

where S_e is the effective saturation (Eq. [6]) and x is an integration variable. Introducing the van Genuchten (1980) soil–water characteristic relationship (Eq. [17]) into Eq. (24), under the assumption of $m = 1 - 1/n$ (where α, m, n are positive empirical constants in Eq. (17)) and $0 < m < 1$ results in the relative hydraulic conductivity

$$K_r = S_e^p \left[1 - \left(1 - S_e^{1/m}\right)^m\right]^2 \quad (25)$$

The value $p = 0.5$ in Eq. (2) was proposed by Mualem (1976) and is widely used in Eq. (3). In terms of h, Eq. (3) becomes,

$$K_r(h) = \frac{\left\{1 - (\alpha h)^{n-1}\left[1 + (\alpha h)^n\right]^{-m}\right\}^2}{[1 - (\alpha h)^n]^{m/2}} \quad (26)$$

$K(h)$ that were calculated by Eq. (26) with α, m and n values obtained by fitting Eq. (17) to measured RCs for RTM tuff and stone wool (Table 3.2) were successfully compared to independently measured $K(h)$ data for the drying branch (Fig. 3.4). Predicted hydraulic conductivity curves, $K(h)$, by Eq. (26) for RTB and RTM tuff and

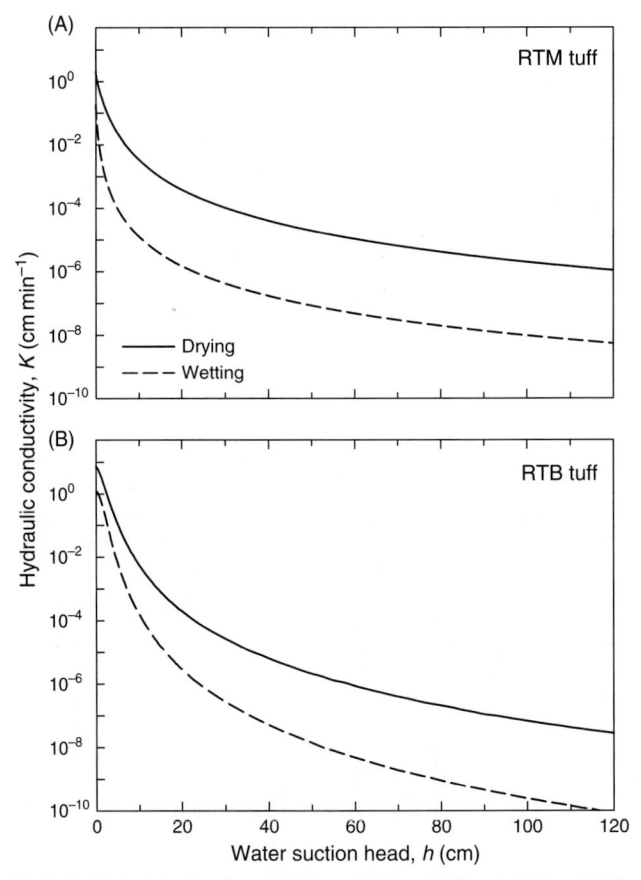

FIGURE 3.12 Predicted $K(h)$ for RTM tuff (A) and RTB tuff (B).

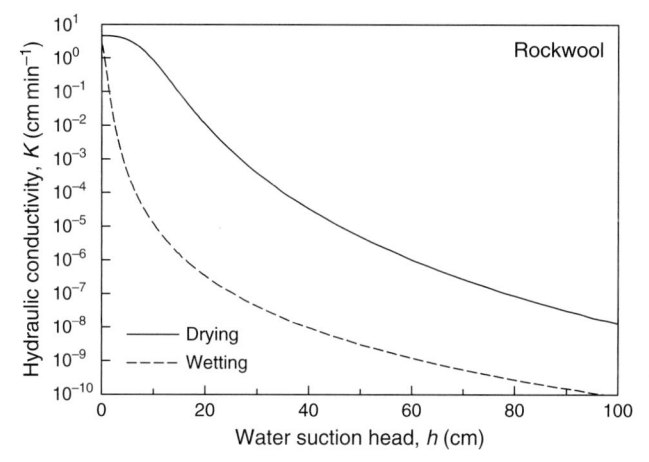

FIGURE 3.13 Predicted drying and wetting $K(h)$ for stone wool.

stone wool are shown in Fig. 3.12A, 3.12B and 3.13 respectively, for both wetting and drying branches. The parameters of Eq. (26) for these $K(h)$ curves were obtained by fitting Eq. (17) to the measured wetting and drying retention curves (Table 3.2) that are shown in Fig. (3.7). Predicted hydraulic conductivity curves for other soilless growing media that were calculated by Eq. (26) can be found in: Wallach et al. (1992b), for composted agricultural wastes and their mixtures with tuff; in da Silva et al. (1993a), for sphagnum peat moss and its mixture with tuff; in Orozco and Marfa (1995), for perlite; in Raviv et al. (1999), for two types of pumice and in Raviv et al. (2001), for coir and UC mix. An overall conclusion that can be made for soilless growing media is that $K(h)$ decreases by several orders of magnitude over a narrow range of suctions. The extreme variation in $K(h)$ at a narrow range of suction that was determined by De Boodt and Verdonck (1972) as easily available water has a tremendous effect on water dynamics in the growing container and its availability to the container-grown plants. This topic will be further discussed in the following.

3.3.3 RICHARDS EQUATION, BOUNDARY AND INITIAL CONDITIONS

Darcy's law can be coupled with the conservation of mass principles to derive a continuity equation. The continuity equation is derived by taking a cube of infinitesimal dimensions, dx, dy, dz and balancing the in- and out-fluxes at each one of the directions. The outcome is (Hillel, 1998)

$$\frac{\partial \theta}{\partial t} = -\nabla \cdot \vec{J}_w \qquad (27)$$

Substituting Darcy's law (Eq. [23]) in Eq. (27) leads to

$$\frac{\partial \theta}{\partial t} = -\nabla \cdot (K\nabla H); \qquad H = h + z \tag{28}$$

or

$$\frac{\partial \theta}{\partial t} = \frac{\partial}{\partial x}\left(K\frac{\partial h}{\partial x}\right) + \frac{\partial}{\partial y}\left(K\frac{\partial h}{\partial y}\right) + \frac{\partial}{\partial z}\left(K\frac{\partial h}{\partial z}\right) + \frac{\partial K}{\partial z} \tag{29}$$

The last two equations, along with their various alternative formulations, are known as the Richards equation. Equation (29) has two dependent variables, θ and h. The number of dependent variables may be reduced from two to one, provided a soil–water characteristic relationship exits, either as $h = h(\theta)$ or as $\theta = \theta(h)$. The θ-based version of Eq. (29) is

$$\frac{\partial \theta}{\partial t} = \frac{\partial}{\partial x}\left(D\frac{\partial \theta}{\partial x}\right) + \frac{\partial}{\partial y}\left(D\frac{\partial \theta}{\partial y}\right) + \frac{\partial}{\partial z}\left(D\frac{\partial \theta}{\partial z}\right) + \frac{\partial K}{\partial z} \tag{30}$$

where $D = K \cdot \partial h/\partial \theta$ is called the hydraulic diffusivity. The h-based version of Eq. (29) is

$$C\frac{\partial h}{\partial t} = \frac{\partial}{\partial x}\left(K\frac{\partial h}{\partial x}\right) + \frac{\partial}{\partial y}\left(K\frac{\partial h}{\partial y}\right) + \frac{\partial}{\partial z}\left(K\frac{\partial h}{\partial z}\right) + \frac{\partial K}{\partial z} \tag{31}$$

where $C(h) = \partial\theta/\partial h$ is the slope of the water RC and is called the specific moisture capacity. The advantage of the θ-based form is that D does not vary with θ nearly as much as K varies with h.

For a well-posed problem, not only the domain boundaries should be specified, but also a description relating the dependent variables along the boundaries. For example, the conditions at the growing medium surface (e.g. flooded or flux-controlled irrigation), the lower part of the root zone or the bottom of the growing container (e.g. free- or controlled-drainage), the walls of the growing container or at a symmetry cross-section or line within the container and so on. For time-dependent problems, the initial conditions within the domain should be specified. A formal specifications of boundary conditions helps not only to assure completeness in a mathematical sense, but also to bring attention to necessary assumption and data limitations.

Some common choices of initial cases include the following.

a. Constant matric potential (suction head) or water content
   $h(x, y, z, 0) = \text{Constant}$
   $\theta(x, y, z, 0) = \text{Constant}$
b. Constant head (static conditions)
   $H(x, y, z, 0) = \text{Constant}$
   $H(z, 0) = -z$ (in one-dimensional problems with z the elevation) that is widely used for container capacity (end of drainage and water redistribution)
c. General matric potential or water content distributions
   $h(x, y, z, 0) = f(x, y, z)$
   $\theta(x, y, z, 0) = f(x, y, z)$.

3.3.3.1 Boundary Conditions

There are two kinds of boundary conditions: 'Dirichlet' (first kind), for which the dependent variable is given; 'Neumann' (second kind), for which the flux or gradient of the dependent variable is given. Common boundary conditions include the following:

a. h or θ are specified on the boundaries (Dirichlet)
b. vertical water flux as from irrigation at the growing media surface (Neumann)
c. zero flux specified at the growing container vertical or inclined surfaces, along symmetry cross sections between plants (Neumann)
d. evaporation through the soil surface (Neumann).

3.3.4 WETTING AND REDISTRIBUTION OF WATER IN SOILLESS MEDIA – CONTAINER CAPACITY

The small volume of growing containers and pots filled with growing substrates, compared to field soils, enables to analyse the average moisture content changes with time during irrigation, moisture redistribution and free drainage afterwards, and water extraction by the plant roots. The container in this type of analysis is an analogue to a weighing lysimeter. The typical container-weight variation prior, during and after an irrigation event is shown in Fig. 3.14. The substrate wetting started at 12:40.

A close look at the container-weight variation in Fig. 3.14 indicates that it may be divided into four stages, dominated by different driving forces for moisture distribution within the container. The first stage (noted as Stage I in Fig. 3.14) is the time period when water is added to the top of the container. The container weight is then markedly increasing during a peak value. The weight peak is determined by the irrigation duration, if it stops before equilibrium between the input (irrigation flux) and the output (the drainage flux through the holes at the container bottom) is reached. If irrigation continues beyond this period (as if frequently the case due to salt discharge

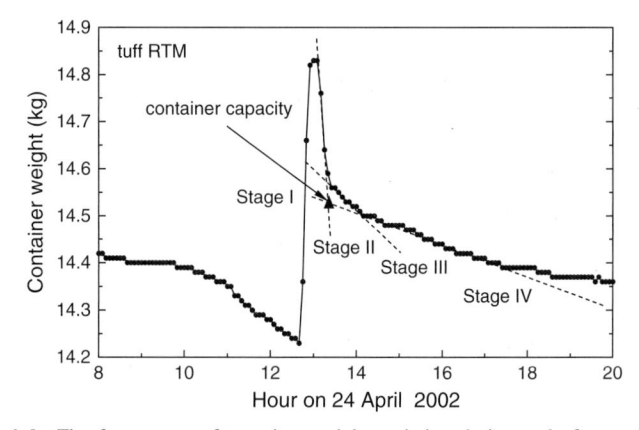

FIGURE 3.14 The four stages of container-weight variation during and after an irrigation event. The data is for a rose plant grown in a container filled with tuff RTM (Wallach and Raviv, unpublished).

considerations), the peak weight sustained until irrigation terminates, as can be seen in Fig. 3.14. The second stage is characterized by a high rate of free drainage flux that leaves the container bottom. This stage starts when irrigation stops and terminates when the container weight loss becomes small, when approaching container capacity. The drainage flux depends on the momentary distribution of the medium hydraulic conductivity, which itself depends in a non-linear manner on the moisture content. Thus, the drainage flux is high when irrigation ends, as the moisture content is high, and then diminishes markedly, in parallel with the continuous decrease in moisture content. The duration of this stage and its intensity depend mainly on the hydraulic properties of the medium and the growing-container volume and geometry. Given that the time interval between any two data points in Fig. 3.14 is 5 min, the duration of this stage for the case shown in this figure (tuff substrate and an 8-l container volume) was between 15 and 20 min. Following the sharp container-weight decrease, the weight continued to decrease, but at a much lower rate, during a few minutes afterwards. This stage, designated as the third stage, is a transition period between the high drainage flux at the second stage, and the low weight decrease rate afterwards owing to evapotranspiration, noted as the fourth stage. If the drainage flux from the container bottom is monitored, for example by tipping buckets (Fig. 3.15), the determination of the initiation and termination of the third stage is obvious. If the container surface is fully covered by the plant canopy or by mulch, the contribution of evaporation to the weight loss is much smaller compared to transpiration (Urban et al., 1994) and can be practically ignored for this type of an analysis. Thus, the rate of container weight decrease during the fourth stage may be considered as almost solely related to the transpiration-driven water uptake rate. The calculation of momentary transpiration rate and its distribution along the day hours can be performed by tracking the change of container weight with time ($\Delta W/\Delta t$).

Figure 3.15 shows the weight variation of two containers filled with perlite and the related drainage that leaves the container through holes at its bottom. This data was taken from an experiment on the relationship between irrigation frequency and water availability, as the tensiometers data shown in Fig. 3.6 (Wallach and Raviv, 2005). The two irrigation regimes, 'wet' and 'dry', were predetermined by threshold lower and upper container weights. These figures indicate that the average moisture content (over time) in the 'wet' treatment is much higher than that in the 'dry treatment', as also indicated by the suction variation in Fig. 3.6, which potentially provide higher water availability. The cumulative drainage in Fig. 3.15 was 650 g (about 220 ml per irrigation) in the 'dry' treatment and 585 g (about 53 ml per irrigation) in the 'wet' treatment. The upper threshold weight that controls the drainage was set up in a way to keep a certain predefined drainage to prevent salt accumulation in the container. The threshold-container-weights method used to control irrigations in Fig. 3.15 provided a higher irrigation frequency during the hours when water demand is high and less frequent during the early morning and afternoon hours, when water demand is lower.

The term 'container capacity' is widely used in the soilless culture literature to designate the upper moisture content threshold of the available water range. This term is analogous to 'field capacity', a term used by soil physicists and irrigation scientists

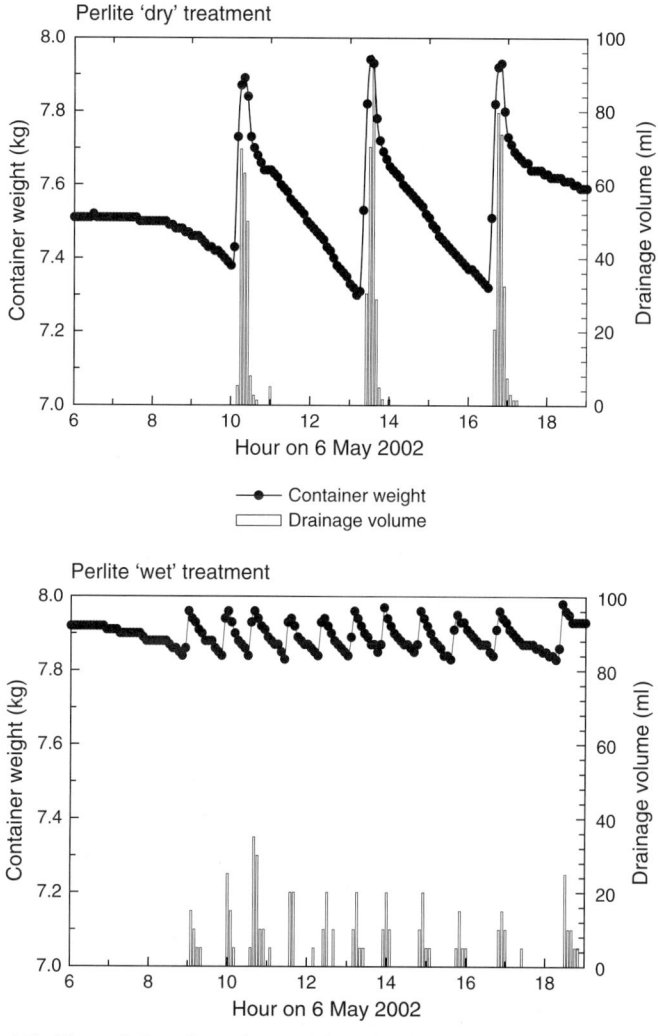

FIGURE 3.15 The variation of container weight and drainage leaving the bottom of the pots during an arbitrary chosen day for an irrigated pot filled with perlite. The grown rose plants were irrigated at two irrigation frequencies. Note that this data matches the tensiometers data in Fig. 3.6 (Wallach and Raviv, unpublished).

for field-irrigated crops. Moisture content at field capacity is usually determined as the moisture content in the root zone at the end of the free drainage following an irrigation event. As the upper threshold of available water moisture content range is dictated by the growing medium type and its hydraulic properties, the lower moisture content threshold depends on the decision when to turn the irrigation on. The issue of water availability will be discussed in detail in the next section while other irrigation-related issues are discussed in Chaps. 4 and 5. Different methods have been suggested for determining the container capacity. White and Mastalerz (1966) defined container capacity as the amount of water retained in a containerized medium after drainage from saturation has ceased, but before evaporation has started. A combination of mathematical functions for the water characteristic curves and container geometry was later shown to provide a more consistent description of container capacity (Bilderback and Fonteno, 1987; Fonteno, 1989). According to this approach, container capacity is the total volume of water in the container, as given by its water RC, divided by the container volume. The assumption made while determining container capacity is that the pressure distribution along the container height when drainage stops is static (Fig. 3.5).

The basic assumption in the RC-related methods to determine container capacity is that the medium drainage starts from saturation. However, owing to the high porosity of soilless media, saturation rarely occurs during irrigation by conventional on-surface methods. In addition, the RC used to determine container capacity by these methods is usually measured independently for smaller medium volumes, without roots, and independent of the container geometry. Referring to Fig. 3.14, container capacity is reached sometime during the third stage when both free drainage and root uptake take place simultaneously. The question is how the free drainage can be separated from the root uptake in order to find the moment when free drainage ceases. It is suggested herein that container capacity is reached at the intersection between the continuation of the lines fitted to the weight variation during the second and fourth stages (Fig. 3.14). The point of intersection between the two lines is signified by a solid triangle in Fig. 3.14. As container weight changes continuously with plant growth, flower cutting, fruit picking, pruning, trimming and so on, its use for container capacity determination is problematic. However, when tensiometers are used together with load cells, the container weight at container capacity can be related to suction that can be further translated to moisture content by using the independently measured RC of the specific growing medium. An implicit assumption in this procedure to determine container capacity is that moisture content is uniformly distributed along the container height during moisture redistribution (throughout the second stage and early third stage).

3.4 UPTAKE OF WATER BY PLANTS IN SOILLESS MEDIA AND WATER AVAILABILITY

3.4.1 ROOT WATER UPTAKE

Water is transported through the growing medium into the roots and plant xylem towards the plant canopy where it eventually transpires into the atmosphere. The continuous uptake of water is essential for the growth and survival of plants because they lose quite large amounts of water throughout the day, and mainly during periods of high evaporative demand. Unless water loss is replaced immediately, dehydration-related stresses may start to affect the performance of the plant. Thus, water availability to the plant roots and its relation to irrigation scheduling deserve a careful attention in general and in soilless growing media in particular owing to the limited root volume and root density.

One key function of plant roots is their ability to link the growing substrate, where water and nutrients reside, to the organs and tissues of the plant, where these resources are used. Hence roots serve to connect the growing substrate environment to the atmosphere by providing a link in the pathway for water fluxes from the medium through the plant to the atmosphere. Fluxes along the growing substrate (or soil)–plant–atmosphere continuum (SPAC) are regulated by above-ground plant properties, for example the leaf stomata, which can regulate plant transpiration through interaction with the atmosphere and with other plant organs, and plant root-system properties like depth and spatial distribution, roots permeability or hydraulic conductance. The root density is often expressed as root length density in centimetres of roots per cubic centimetres of soil. There is a significant difference among plants that are grown in soils and those grown in limited volumes (container, pots, sleeves, etc.) filled with substrates. The limited growing volume affects the spatial root distribution and length density, and other physiological and environmental parameters. Some of these issues are further discussed in Chaps. 2 and 13.

Uptake of water occurs along gradients of decreasing potential from the growing medium to the roots. However, the gradient is produced differently in slowly and rapidly transpiring plants, resulting in two uptake mechanisms. Active or osmotic uptake occurs in slowly transpiring plants where the roots behave as osmometers, whereas passive uptake occurs in rapidly transpiring plants where water is pulled in through the roots, which act merely as absorbing surfaces. When the growing medium is warmer than the ambient air and the air is humid, transpiration is slow, as at night and on cloudy days. Under these conditions, water in the xylem often is under positive pressure (root pressure) as indicated by the occurrence of guttation and exudation of sap from wounds or hydathodes. In moist growing medium, uptake at night and early in the morning is largely by the osmotic mechanism, but as daytime transpiration increases the demand for water in the leaves, uptake increasingly occurs by the passive mechanism and the osmotic mechanism becomes less important (Kramer and Boyer, 1995).

The force bringing about uptake of water by transpiring plants originates in the leaves and is transmitted to the roots or the lower end of cut stems (as in the case of cut flowers) through the sap stream in the xylem. Evaporation of water from leaf cells decreases their water potential, causing water to move from the xylem of the leaf veins. This reduces the potential in the xylem sap, and the reduction is transmitted through the cohesive water columns to the roots where the reduced water potential causes inflow from the growing medium. In this situation, water can be regarded as moving through the plant in a continuous, cohesive column, pulled by the matric or capillary forces developed in the evaporating surfaces of stem and leaf cell walls.

When considering root water uptake, we accept the continuum approach as presented by Van den Honert (1948), assuming that flow through the SPAC is at a steady state for an unspecified time period, and that water potential across the SPAC is continuous and determined by the cohesion theory. For this condition, an Ohm's law analogue between water flow and electrical current is valid and the electric analogue model assumes that the water-flux ($cm^3 \, cm^{-2} \, s^{-1}$) through the root zone and the root–stem–leaf–stomata path, namely transpiration rate, T, is proportional to the total head difference Δh_{total} (cm) and inversely proportional to the total resistance R_{total} (expressed in seconds) of the system. A steady-state condition implies that water does not accumulate/deplete in the different sections of the SPAC, so that water flow within each section of the SPAC pathway can be determined by the ratio of water potential gradient and flow resistance within each section.

$$T = -\frac{\Delta h_{total}}{R_{total}} = -\frac{h_{root} - h_{substrate}}{R_{substrate}} = -\frac{h_{leaf} - h_{root}}{R_{plant}} = \frac{h_{leaf} - h_{substrate}}{R_{substate} + R_{plant}} \quad (32)$$

where T (cm s^{-1}) is the transpiration rate, $h_{substrate}$, h_{root} and h_{leaf} (cm) are pressure heads in the growing medium (substrate), at the root surface and in the leaves, respectively, $R_{substrate}$ and R_{plant} (s) are liquid-flow resistances of the substrate and the plant (Fig. 3.16). Hence R_{plant} does not include stomatal resistance. When the transpiration demand of the atmosphere on the plant system is high or when the substrate is rather dry, $R_{substrate}$ and R_{plant} influence h_{leaf} in such a way that transpiration is reduced by closure of the stomata. Equation (32) can be applied to the root system as a whole by measuring T, $h_{substrate}$, and h_{leaf} during two periods, thus obtaining two equations with two unknowns, from which $R_{substrate}$ and R_{plant} can be computed. The relative magnitude of $R_{substrate}$ and R_{plant} has been an important object of many studies, mainly in soil-grown plants. Under wet conditions, $R_{substrate}$ is small. Generally one can state, except for very dry soil, that $R_{substrate} > R_{plant}$. Most of the plant resistance is concentrated in the roots, to a lesser extent in the leaves, and a minor part in the xylem vessels.

Equation (32) is applied to the macroscopic flow of water across a complete rooting system (Gardner and Ehlig, 1962) but it was also used to quantify water transport across a single root in a microscopic approach, where T denotes the volumetric uptake rate per unit length of root per unit root surface area (Molz, 1981). A review of the simplifications and implications of Eq. (32) was presented by Philip (1966). The constant, time-independent resistances in the electrical analogue theory (Eq. [32]) rarely exist in reality. The plant system is much more complex, resembling more a series-parallel network of flow paths, each characterized by different resistances.

There has been considerable discussion concerning the relative importance of soil and root resistance with respect to water uptake. Early investigators, for example Gardner (1960) and Gardner and Ehlig (1962), concluded that resistance in the soil

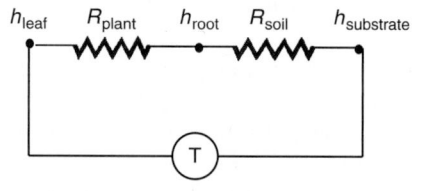

FIGURE 3.16 Potentials and resistances in the substrate-plant system. T is the transpiration rate, h_{leaf} is the leaf-water potential, h_{root} is the root-water potential, $h_{substrate}$ is the substrate-water potential, R_{plan} is the plant resistance and R_{soil} is the soil resistance.

would exceed resistance in the roots at a soil water potential of −100 to −200 kPa, and this view was supported by Cowan (1965) and others. Later studies examined the effect of root density on soil resistance (Newman, 1969) and concluded that root densities used by previous investigators were much lower than those usually found in nature, and if normal root densities were used, soil resistance in the vicinity of roots would not be limiting until the soil water content approached the permanent wilting percentage. Regarding the plant resistance in Eq. (32), it is likely to vary with transpiration rate (Passioura, 1988; Steudle et al., 1988) and water potential gradients, for example, due to reduced plant conductance by cavitation. The steady-state assumption when using Eq. (32) is valid at short time scales, but is less likely to apply at time scales longer than a day in field soils and more than a couple of hours in limited-volume growing substrates.

The validation of the steady-state resistive flow model for the SPAC continuum (Eq. [32]) was investigated by Li et al. (2002a) by studying the effects of soil moisture distribution on water uptake of drip-irrigated corn (*Zea mays* L.) and simultaneously monitoring the diurnal evolution of sap flow rate in stems, of leaf water potential and of soil moisture during intervals between successive irrigations. The results invalidate the steady-state resistive flow model. High hydraulic capacitance of wet soil and low hydraulic conductivity of dry soil surrounding the roots depressed significantly diurnal fluctuations of water flow from bulk soil to root surface. On the contrary, sap flow responded directly to the large diurnal variation of leaf water potential. In wet soil, the relation between the diurnal courses of uptake rates and leaf water potential was linear. Water potential at the root surface remained nearly constant and uniformly distributed. The slope of the lines allowed calculating the resistance of the hydraulic path in the plant. Resistance increased in inverse relation to root length density. Soil desiccation induced a diurnal variation of water potential at the root surface, the minimum occurring in the late afternoon. The increase of root surface water potential with depth was directly linked to the soil desiccation profile. The development of a water potential gradient at the root surface implies the presence of a significant axial resistance in the root hydraulic path that explains why the desiccation of the soil upper layer induces an absolute increase of water uptake rates from the deeper wet layers.

3.4.2 MODELLING ROOT WATER UPTAKE

Models for water transfer have been developed for the SPAC in order to improve the understanding of plant behaviour in the environment. The major difficulty in general and in these models in particular is in estimating the effective water availability to the plant in the growing media, which is a fundamental limiting factor of plant transpiration. In this context, models have been developed to simulate the water budget and water movement in the soil on two different scales with appropriate models for each scale: the *microscopic scale* that considers individual roots and the associated volumes of soil, and the *macroscopic scale* that includes part or the entire root system. For overviews of root water-uptake models, the reader may refer to Feddes (1981), Molz (1981), and Hopmans and Bristow (2002).

The *microscopic* analysis generally considers the convergent radial flow of water towards and into a representative individual root, taken to be a line or narrow-tube sink uniform along its length, that is of constant and definable thickness and absorptive properties. The microscopic root system as a whole can then be described as a set of such individual roots, assumed to be regularly spaced in the growing medium at specified distances that may vary within the root volume in the container profile. The Richards equation (Eq. [29]), written for the microscopic scale, is solved for the distribution of soil water pressure heads, water contents in the growing medium and fluxes from the root outwards. As the flow to a single root has a radial symmetry, the flow equations are written in cylindrical coordinates,

$$\frac{\partial \theta}{\partial t} = \frac{1}{r}\frac{\partial}{\partial r}\left[rK(h)\frac{\partial h}{\partial r}\right], \quad \frac{\partial \theta}{\partial t} = \frac{1}{r}\frac{\partial}{\partial r}\left[rD(\theta)\frac{\partial \theta}{\partial r}\right] \quad (33)$$

where r is the radial coordinate from the centre of the root, $D(\theta) = K(h) \cdot \partial h/\partial \theta$ is the diffusivity (Eq. [30]). The boundaries for Eq. (33) are at $r = r_0$ and $r = r_1$, where r_0 is the radius of the plant root, the internal radius of the equivalent cylindrical shell of growing medium associated with the plant root, and r_1 is the external radius of the equivalent cylindrical shell of growing medium associated with the plant root. The radius r_1 can be half the distance between adjacent roots, or a distance from the root surface where the effect of water extraction at the root surface on the moisture content or tension head variation is approaching zero. Boundary conditions at the two radii and an initial condition for the dependent variable in the entire domain should be specified in order to obtain a specific solution for Eq. (33).

Gardner (1960) was among the first to use Eq. (33) to study the extraction of water from soil. In his analysis, he considered uptake of water by a plant root with radius r_0 surrounded by a cylindrical shell of soil with outer radius r_1. He have considered an approximate solution after linearizing Eq. (33) and assuming that $r_0 << r_1$ and a constant flux into a line sink. Gardner (1960) used the line sink solution to calculate water-depletion patterns around individual roots. But more importantly, he also used it as a point of departure for the formulation of a simpler model, in which the depletion resulting from uptake by a single root is treated as a series of steady flows in the cylindrical shell of soil surrounding the root, with the soil–root interface at the inner edge and the water coming from the outer edge. This simple model has been used ever since for more sophisticated microscale as well as macroscale models of water uptake. In the single-root model of Gardner (1960), the root is viewed as a cylinder of infinite length having uniform water-absorbing properties. For steady-state conditions, $(\partial \theta/\partial t = 0)$, in the soil shell surrounding the root with water flowing from the outer cylindrical surface at $r = r_1$ to the inner cylindrical soil–root interface at $r = r_0$, the solution of Eq. (33) under the assumption of constant hydraulic conductivity K gives an expression for the flux q_r at the soil/growing medium–root interface

$$q_r = \frac{2\pi K}{\ln(r_i/r_0)}(h_1 - h_0) \quad (34)$$

where q_r (cm^3 cm^{-1} d^{-1}) is the rate of water uptake per unit length of root, h_1 (cm) is pressure head of the soil/growing medium at $r = r_1$, h_0 (cm) is the pressure head at the soil/growing medium–root interface. Equation (34) is an analogous steady-state flux towards a well per unit length of well in groundwater hydrology. Cowan (1965) realized that the assumption of constant K used in Eq. (34) can be replaced by

$$\overline{K} = \frac{\int_{h_0}^{h_l} K(h) \mathrm{d}h}{h_l - h_0} \qquad (35)$$

The integral in Eq. (35) can be evaluated from any of the commonly used expressions to represent the $K(h)$ relationship. The sharp decline of $K(h)$ within a small range of h values in containerized substrates (Figs. 3.12 and 3.13) raises doubts about the use of constant or average K in the microscale models.

In spite of the assumptions made in Gardner's studies, they were insightful and simulating and inspired the models that have been developed later. On the other hand, the single-root approach is not practical when a whole rooting system with complex geometries has to be considered. The flow processes in the SPAC can be highly dynamic, thereby requiring transient formulations of root water uptake. Consequently, later studies of water extraction by plants roots have considered the macroscopic approach.

On the microscopic scale, the uptake is represented by a flux across the soil/growing medium–root interface. That flux is a consequence of the interaction of processes in the soil/growing medium and in the plant. In the soil/growing medium, flow of water towards or away from individual plant roots may be described by a non-linear diffusion equation, subject to appropriate initial and boundary conditions. The microscale model involves at least two characteristic lengths describing the root–soil/growing medium geometry, for an individual plant root one characteristic length is the internal radius of the plant root r_0 and the external radius of the soil/growing medium associated with the plant root r_1, and two characteristic times describing, respectively, the capillary flow of water from soil/growing medium to plant roots and the ratio of supply of water in the soil/growing medium and uptake by plant roots. Generally, at a certain critical time, uptake will switch from demand-driven to supply-dependent. The resulting microscopic expressions for the evolution of the average water content can be used as a basis for up-scaling to the macroscopic scale.

On the *macroscopic* scale, the uptake of water by plant roots is represented by a sink term in the volumetric mass balance Eq. (27), representing the rate of water extraction by roots

$$\frac{\partial \theta}{\partial t} = -\frac{\partial J_w}{\partial x_i} - S(x_i, h) \qquad (36)$$

where S (cm^3 cm^{-3} d^{-1}) is the sink term, h is the spatially distributed tension head and x_i is the spatial coordinate ($i = 1, 2, 3$). The volumetric flux J_w is given by Darcy's law (Eq. [23]). The depth-dependent sink term may be represented by different root water uptake functions ranging from linear to non-linear (e.g., Raats, 1974; Hoogland et al., 1981; Prasad, 1988) and from one dimension to three dimension according to the considered dimensionality of the system (Vrugt et al., 2001). Thus, the local root uptake rate, S, depends on location in the root zone, time, the water tension head in the medium, root density or a combination of these variables. Boundary conditions can be included to allow for specified medium water potentials and fluxes at the growing medium surface and the bottom boundary of the growing medium domain, whereas user-specified initial conditions and time-varying source/sink volumetric flow rates should be specified.

The macroscopic models of root water uptake can be further divided into two groups. The first group contains water potential and hydraulic parameters inside plant roots (Nimah and Hanks, 1973; Hillel et al., 1976; Molz, 1981; Kramer and Boyer, 1995), which are difficult to quantify. In the second group, the root water extraction rate is calculated from plant transpiration rate, rooting depth and soil water potential (Molz and Remson, 1970; Feddes et al., 1974, 1978; Raats, 1976; Gardner, 1983; Prasad, 1988). The parameters in the second group are relatively easy to obtain; therefore, the approach is widely used and has been implemented into commonly used numerical models (SWMS, Simunek et al., 1992; HYDRUS, Simunek et al., 1999). The root water extraction models by Molz and Remson (1970) and Raats (1976) ignore the effect of medium water content on the distribution of root water uptake. For example, in Molz and Remson's (1970) model, the root system always extracts 40 per cent of the total transpiration from the top quarter of root zone, even if the top layer is desiccated by evapotranspiration.

The sink term, for a one-dimensional vertical mass balance equation (Eq. [36])

$$C(h)\frac{\partial h}{\partial t} = \frac{\partial}{\partial z}\left[K(h)\left(\frac{\partial h}{\partial z}+1\right)\right] - S(z,h) \tag{37}$$

according to Feddes et al. (1978) is

$$S(z,h) = \gamma(h) \cdot S_{\max}(z) \tag{38}$$

with $S_{\max}(z)$, the maximal root water uptake as a function of depth [T^{-1}], and $\gamma(h)$, a dimensionless reduction function that simulates the effects of medium water deficit on root water extraction. This function is characterized by different tension head values h_0, h_1, h_2 (low and high according to the climatic demand) and h_3, as shown in Fig. 3.17 Above h_0, $\gamma(h)$ is zero, between h_1 and h_{2l} or h_{2h} is 1, and between h_3 and h_3 the values of the reduction function is given by

$$\gamma(h) = \frac{h(z) - h_3}{h_2 - h_3} \tag{39}$$

The actual root water uptake given by Eq. (37) is subsequently integrated from the medium surface downwards until it reaches either the potential transpiration Tp (no water stress) or the rooting depth (water stress if the integrated root water uptake is smaller than Tp). Specifically, the tension head threshold at which water stress initiates reduction in root water uptake is determined by Tp, with water stress occurring earlier at the less negative values h_{2h}, if Tp is high. This type of functional dependence allows for less favourable water-supplying medium moisture conditions with increasing plant transpiration. Simulation of root water uptake using the above-described model with VG models for the RC and hydraulic conductivity function, with a model to calculate

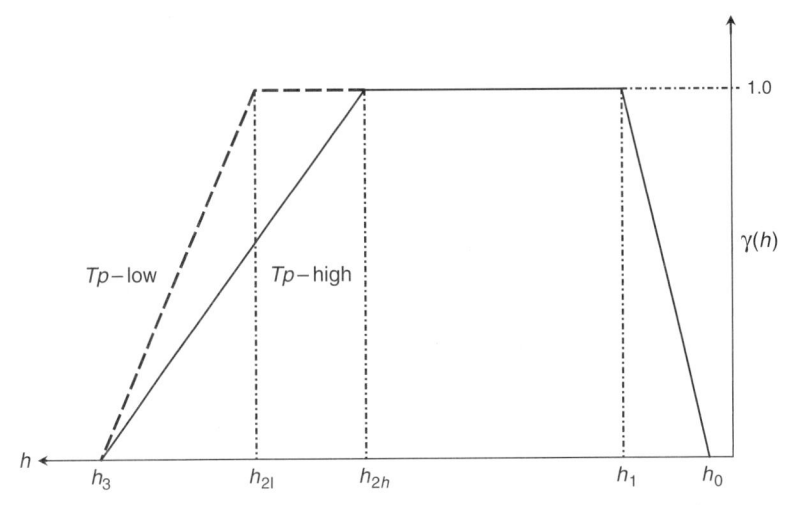

3.4 UPTAKE OF WATER BY PLANTS IN SOILLESS MEDIA AND WATER AVAILABILITY

FIGURE 3.17 Reduction function, $\gamma(h)$ (after Feddes et al., 1978).

Tp during a relatively short period for which the crop is fully developed and crop growth is negligible, requires the specification of about 17 parameters; 6 are medium related and 11 plant related.

The volume and shape of growing containers have a significant effect on the root density distribution (Fig. 3.18). The container's impervious walls and bottom (except

FIGURE 3.18 Root distribution in limited-volume pots, containers and sleeves (two left photos courtesy of Michael Raviv).

holes to allow drainage) increase the root density along these walls. Moreover, the vertical distribution of moisture content in containers, pots and growing sleeves differs substantially from field soils owing to the existence of a lower boundary. This moisture distribution has an effect on the vertical root-density distribution. As such, the use of the one-dimensional model (Eq. [37]) for simulations of water extraction by roots in pots and containers will thus provide erroneous predictions. The complexity in representing the spatial and temporal root-density distribution in growing containers negates the option of using the multi-dimensional (two-dimensional/radial and three-dimensional models) mathematical model as well. If one is interested to predict $\theta(x, y, z, t)$ or $h(x, y, z, t)$ within the growing container, its progress depends on the evaluation of root-density distribution as a function of x, y, z, t. However, if the prediction of water availability under different water demand scenarios is of interest, alternative methods should be used. As an alternative, root water uptake in containerized media can be indirectly inferred from growing medium water balance studies, such as container weight variation or characterizing the moisture content variation at different spatial and temporal scales by tensiometers and/or TDRs. These methods enable to separate the vertical moisture flow and the water depletion curve by root extraction (Figs. 3.6, 3.14 and 3.15) and to evaluate the overall and local momentary root water extraction.

3.4.3 DETERMINING MOMENTARY AND DAILY WATER UPTAKE RATE

The knowledge of the momentary crop transpiration rate over time may serve to improve irrigation control in soilless culture, since an accurate and dynamic control of the water supply is needed to meet the plants' water requirements in the low water-holding capacity and limited volume systems. The crop transpiration rate can be evaluated by various direct and indirect methods. One approach is to extrapolate from porometer measurements on single leaves to values for the plant canopy. This method suffers from the major disadvantage that the measurement changes the microclimate at the leaf surface, which can influence the observed transpiration. Moreover, stomatal patchiness and leaf-to-leaf variability make whole-plant transpiration prediction, based on individual leaf measurements, questionable. The heat-pulse method is used to measure sap flow (Jones et al., 1988; Green et al., 2003; Möller et al., 2004) that can be related to transpiration rate. However, a calibration for each crop is needed and frequent measurements may yield noisy data. Crop transpiration can be modelled using the Penman–Monteith equation (Monteith, 1990) and has been evaluated at all levels – from single leaves to whole canopies – in greenhouses growing both vegetable (Jolliet and Bailey, 1992; Boulard and Jemaa, 1993; Hamer, 1996) and ornamental crops (Bailey et al., 1993; Baille et al., 1994; Montero et al., 2001). This requires the internal, or stomatal, and the external, or boundary layer, resistances to be known along with the environmental variables within the plant canopy. Most of these studies have been conducted in a specific growth phase of the crop, while there is little information available on the transpiration rate over the plant's ontogeny and its modelling over the entire crop cycle. The relation of the evaluated transpiration rates to water availability in the growing containers has not been widely studied.

3.4 Uptake of Water by Plants in Soilless Media and Water Availability

A relatively old and simple method to evaluate the daily and hourly transpiration rate is by using frequent measurements of the weight of a growing container (Teare et al., 1973; Yang et al., 2003; Medrano et al., 2005; Rouphael and Colla, 2005). The weight measurement frequency varies from minutes to hours and even a day (e.g. Ray and Sinclair, 1998). It is often argued that lysimeter-grown plants do not accurately represent plants grown in normal soil (Girona et al., 2002). This argument is not valid for container-grown plants, and the method has found wide use when the transpiration rate was related to other environmental and physiological variables. This method includes a single plant or set of plants grown in a container of specific volume and known initial weight, that is weighed at each of n predetermined time points, $t_1, t_2, \ldots, t_n$. The momentary water uptake rate is then determined by calculation of the numerical derivative of the measured weight

$$T = \frac{dW}{dt} \approx \frac{W_{k+1} - W_k}{t_{k+1} - t_k} \tag{40}$$

where W_k is the measured weight of the container at time t_k.

Differentiation with time (Eq. [40]) significantly amplifies the high-frequency noise in the measured time series. Given that measurement errors can not be avoided, additional attention should be paid when this method is used to calculate the transpiration rate. An additional analysis to the time series of container weight prior to differentiation may improve the accuracy of the calculated momentary water uptake rate.

The calculated momentary transpiration rate depends on the time interval among successive data points, $\Delta t = t_{k+1} - t_k$. Different average rates with different errors are obtained as time intervals among samplings expand. On the one hand, the greater the time interval among samplings is, the smaller the error of calculated average rate. On the other hand, different average rates for different time intervals carry different (often incomparable) information about the studied system. Therefore, the time intervals at which average transpiration is calculated should correspond to the purpose of the sampling.

The time series of the measured perlite-filled pot weight during 7 August 2001 (Fig. 3.19A) is used herein to demonstrate the role of the sampling time on the calculated rate of water uptake. Irrigation events took place at 15:40 on the previous day, August 6, and at 18:20 on August 7 (not shown). Being covered by plant canopy, the evaporation from the substrate surface was neglected and the container weight decrease (Fig. 3.19A) can be related to transpiration (fourth stage). The momentary transpiration rate of a rose plant grown in this pot was calculated first for $\Delta t = 5$ min (shown as dotted lines in Fig. 3.19B), which was the time interval at which data was logged (an average over 5 min of weight sampling every 10 s). The negative derivative values stands for weight loss. The momentary transpiration rate for $\Delta t = 5$ min is very noisy (Fig. 3.19B), which hinders the identification of the actual transpiration pattern. This noisy time derivative was obtained in spite of the relatively smooth measured weight pattern in Fig. 3.19A. This noisy derivative cannot be used for studying the relationship between water uptake rate and other physiological or environmental parameters or as a basis for irrigation management without any further analysis. The

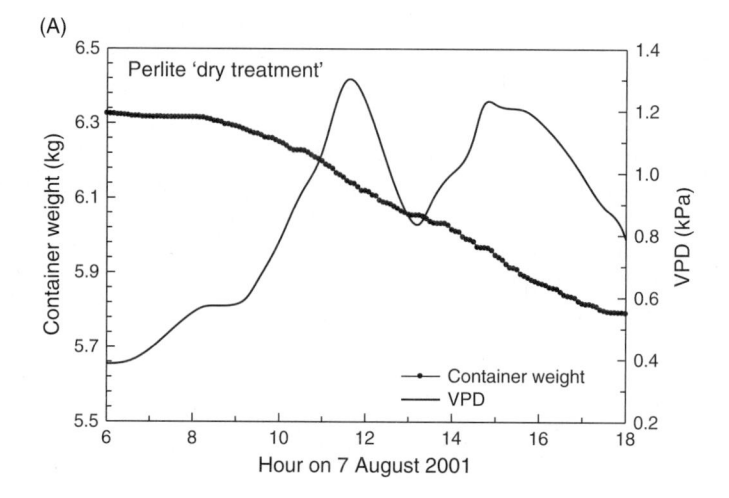

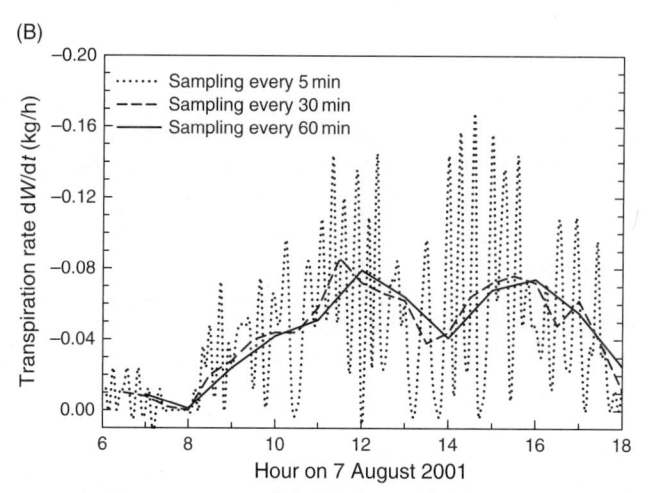

FIGURE 3.19 Perlite container weight and VPD variation during August 7 (A), and transpiration rate during this day, calculated by $\Delta W/\Delta t$, for different sampling interval Δt (B) (Wallach and Raviv, unpublished).

effect of larger sampling intervals on the time derivative is demonstrated by producing new time series for $\Delta t = 30$ and 60 min by omission of measured data points from the original measured time series (with $\Delta t = 5$ min). The transpiration rates calculated by $\Delta W/\Delta t$ for these time series are shown in Fig. 3.19B. The conclusion from this figure is the time series of less-frequent sampled data filter noises associated with numerical derivatives. However, a careful look on the transpiration rate obtained for the increased sampling interval time series reveals that the temporary variation of momentary transpiration rate, its minimum and maximum values, mainly during the midday hours that may be indicative for water shortage and plant stress development, are not unique;

it depends on the sampling time interval (Fig. 3.19B). A comparison between the calculated transpiration rate and the vapour pressure deficit (VPD) variation during the day (Fig. 3.19A) indicates that the deviation between the VPD and transpiration peaks, and the in-between dip timing increases with Δt. Such meaningful dependence on data-sampling intervals raises questions regarding the method of increasing time interval to filter out derivative noise and what is the adequate sampling interval. The shortcomings of using sparsely measured data to 'smooth' noises associated with numerical differentiation while gaining the advantages of the additional information associated with more frequently measured data can be rectified by applying filtering methods to the time series prior to differentiation.

Noises in calculated transpiration rates can be reduced, while preserving the information accompanying frequently measured time series, by smoothing (detrend) the time series prior to differentiation. This smoothing filters out the noise from the measured data while keeping the leading variation patterns. Many methods are available for smoothing noisy data in time-series analysis (Harvey, 1993). These methods can be categorized as non-parametric smoothing (e.g. moving average, Savitzky–Golay, and FFT filtering) and non-parametric regression (fitting polynomials of various orders, exponential functions, symmetrical and asymmetrical transition functions and so on to the measured data). Smoothing should be carried out with care since substantially different results may be obtained after differentiation, in spite of the high R^2 values that are usually obtained for the different smoothing methods. Three procedures are used herein to smooth the container-weight time series prior to differentiation (Fig. 3.19A): fitting a linear and a fourth order polynomial, and noise filtering by Savitzky–Golay method. The transpiration rate patterns obtained by time derivative of the smoothed weight time series are shown in Fig. 3.20. A remarkable difference was obtained

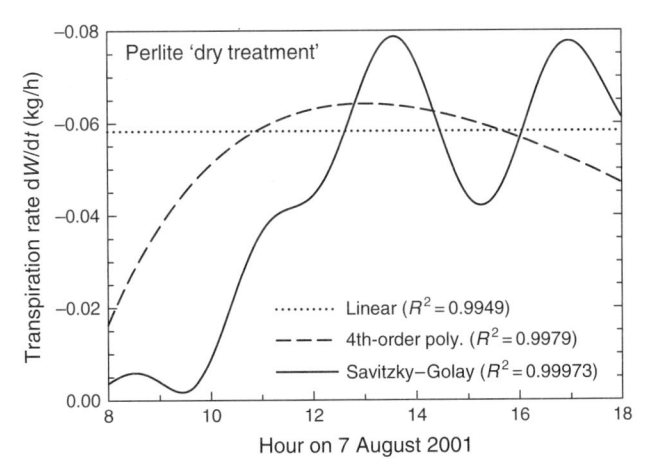

FIGURE 3.20 Comparison of three procedures to smooth the container-weight time series prior to differentiation (Wallach and Raviv, unpublished).

between these patterns, in spite of the extremely high R^2 values. Note that the transpiration rate pattern calculated by the Savitzky–Golay method is the only pattern that follows the VPD variation (Fig. 3.19A). The smoothed time-series provides a smooth transpiration rate that enables to figure out momentary values. Many other smoothing methods that were not demonstrated here can also be used to filter noise from the container-weight time series. However, a pertinent question regarding the smoothing procedure is how intensive should the smoothing be in order to preserve the relevant characteristics of the time series. A further discussion on this issue is beyond the scope of this chapter.

3.4.4 ROOTS UPTAKE DISTRIBUTION WITHIN GROWING CONTAINERS

The calculated or directly measured whole-plant transpiration rate provides information on the temporal variation of the average moisture content within the root zone (container in our context). Yet, it fails to provide information on the spatial variation of moisture content within the root zone/container and to provide an insight on the dynamics of water extraction by roots. Such information may be interpreted to root density variation and temporary water availability at different parts of the container that can be translated to a better irrigation management.

The matric potential at two levels, 7 and 14 cm above the container bottom that was measured by high flow tensiometers (see experimental setup in Wallach and Raviv, 2005), can be transformed into moisture content by the independently measured retention curve. The matric potential and the associated transformed moisture content at these two levels within the container are shown in Fig. 3.21. The moisture content is usually higher at the lower part of the container between frequent subsequent irrigation events. However, this tendency may change at longer periods for less frequent irrigation events, as appears in Fig. 3.21B, for the late afternoon hours. The relative contribution of moisture content at the two levels (7 and 14 cm) above the container bottom to the container weight for the data of August 7 (Fig. 3.21) is shown in Fig. 3.22A. The derivative of the curves in Fig. 3.22A, $d\theta/dW$, was calculated and the derivative vs. the container weight is shown in Fig. 3.22B. This figure highlights the differences in the relative contribution of moisture extraction rate at the two levels to the transpiration rate while the container weight decreases. The curves in Fig. 3.22B indicate that the contribution to the container weight decrease (transpiration rate) was low in the morning when the container weight was high, increased during noon time and intensified during afternoon hours when the container weight was low (note that the subsequent irrigation started when the container reached the lowest weight value). The comparison between Fig. 3.22B and 3.19A indicates that the rate at which water was extracted, as well as the root density, was higher at the lower part of the container, as shown qualitatively in Fig. 3.18. The initial preference of the plant is to extract water from the lower part of the container where moisture content is high and water is easily available. When moisture content decreases and so its availability, the intensive

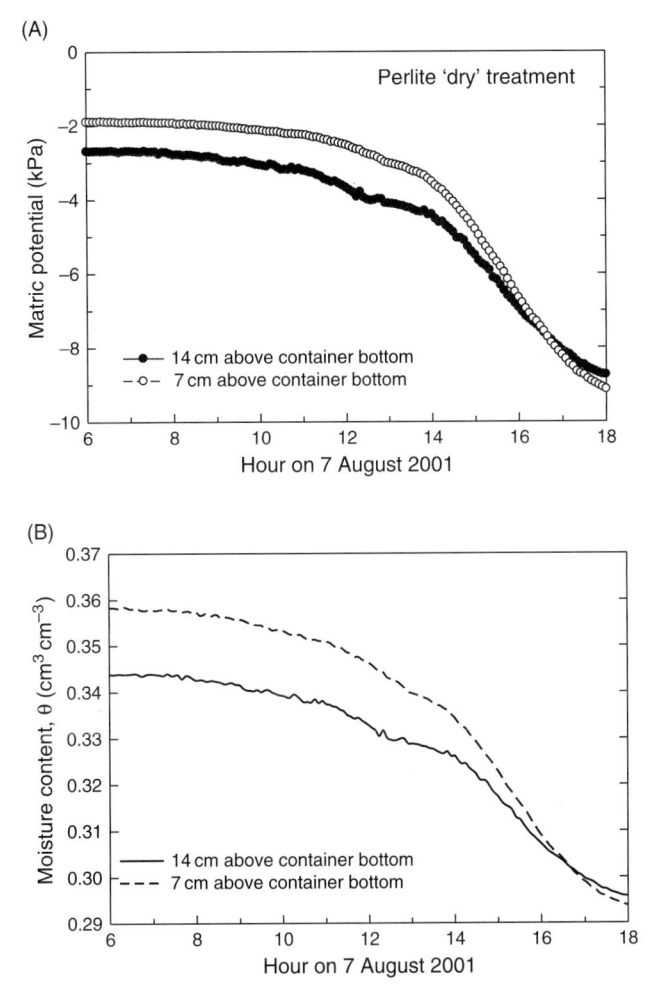

3.4 UPTAKE OF WATER BY PLANTS IN SOILLESS MEDIA AND WATER AVAILABILITY 89

FIGURE 3.21 Matric potential (suction head) (A) and moisture content (B) variation during August 7 in the perlite container for the 'dry' treatment (Wallach and Raviv, unpublished).

water extraction moves upwards (Fig. 3.22B). Note that local water extraction rate depends on the combination of local root density and water availability (moisture content). The migration of the zone where intensive water extraction rate takes place is also known as 'the moving source' and was observed in a field soil by Li et al. (2002b), but at an opposite direction, namely, from the upper layers downwards. The actual transpiration rate deviated from the potential one at given ambient conditions when the moisture content is low at sections where root density is high. The sections with low root density cannot extract the appropriate amount of water, even at relatively high moisture contents.

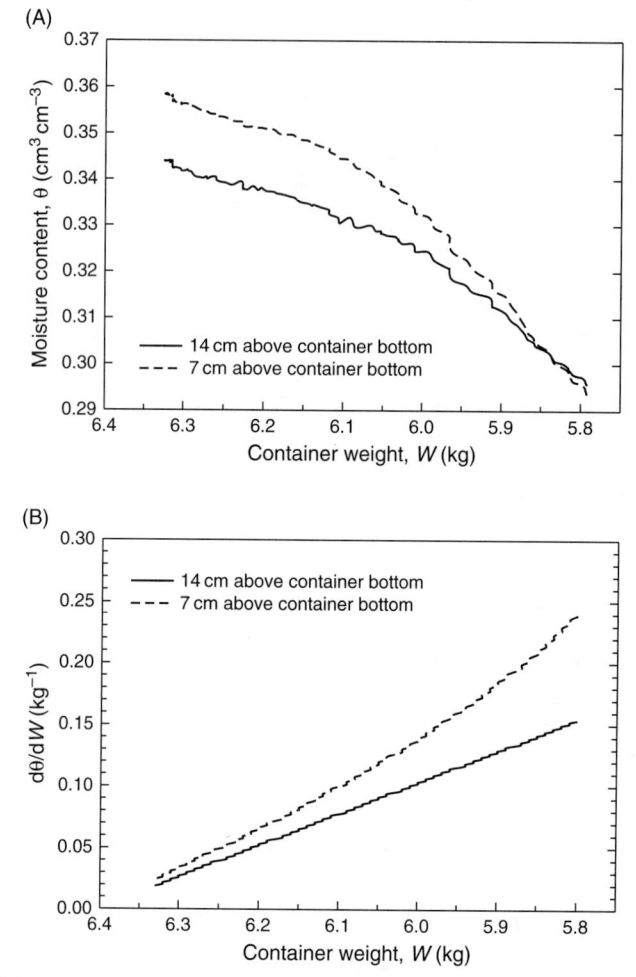

FIGURE 3.22 Moisture content (A) and moisture content derivative by container weight (B) variation with container weight for the perlite container during August 7 (Wallach and Raviv, unpublished).

3.4.5 WATER AVAILABILITY VS. ATMOSPHERIC DEMAND

The term 'available water' is a key parameter in crop irrigation. The amount of water which is 'available' for root water uptake is defined in field soils as the amount of soil water between field capacity and permanent wilting point. A soil matric pressure of -10 or -33 kPa for field capacity and -1500 kPa for wilting point are commonly used values (FAO, 1979). The widely used determination of water availability in soilless culture was introduced by De Boodt and Verdonck (1972), as shown in Fig. 3.23. They introduced the concept of 'easily available water' (EAW), which they defined

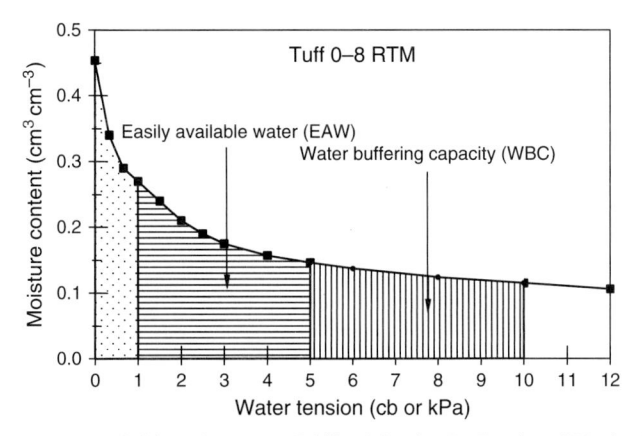

3.4 UPTAKE OF WATER BY PLANTS IN SOILLESS MEDIA AND WATER AVAILABILITY 91

FIGURE 3.23 Definition of water availability following De Boodt and Verdonck (1972).

as the difference between the water content at suctions of 1 and 5 kPa and the 'water buffering capacity' (WBC) as the difference between the water content of the medium at 5 and 10 kPa. These definitions and others, for water availability, have not taken into account the accessibility of the 'available' water to meet the evaporation demand or the potential rate of transpiration. Wallach et al. (1992a,b) and da Silva et al., (1993a,b) hypothesized that $K(\theta)$ of the medium bulk indicates the availability (amounts and rates) of medium-water to plant roots and significantly affects the performance of the plant. The differences in potential water availability for peat moss and tuff substrates at container capacity are demonstrated in Fig. 3.24, where the moisture content and $K(\theta)$ distributions along a 25-cm container height are plotted. As tension head (expressed in the figure as height from the container bottom) increases from 0 to 25 cm and water content decreases accordingly, the hydraulic conductivity decreases by approximately three orders of magnitude for peat and by approximately four orders of magnitude for tuff. As such, the hydraulic conductivity variation in the pot will have a major influence on water availability and should be considered when criteria for water availability are postulated. Naasz et al. (2005) also found that threshold values of water availability obtained for peat and pine bark by direct measurement of their retention curves and hydraulic conductivity functions do not correspond to the common empirical value as proposed by De Boodt and Verdonck (1972) and could provide inaccurate predictions as to potential plant response. Denmead and Shaw (1962) verified that transpiration could be restricted and plants could wilt over a wide range of soil water contents, depending on root density, soil hydraulic properties and on the transpiration demand of the atmosphere. They found that a soil water potential of only −0.1 MPa limited uptake at high rates of transpiration, but −1.0 MPa was not limiting at low rates. The limiting effect of low soil water potential increases as atmospheric conditions favour high potential rates of transpiration.

The momentary balance between the actual water flux from the medium volume surrounding the root and the potential water extraction rate (which depends mainly on

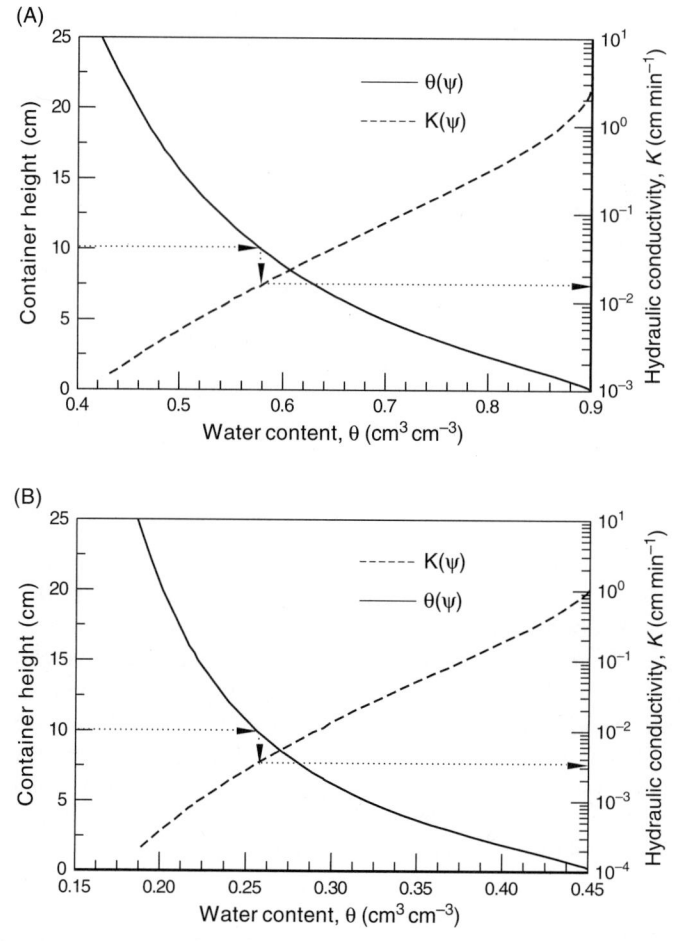

FIGURE 3.24 Water content and hydraulic conductivity distributions throughout a container of 25 cm height filled with (A) peat moss, (B) tuff RTM at container capacity.

the momentary atmospheric conditions) determines whether medium moisture is fully available to the plant. By 'fully available', we mean that the water extracted from the vicinity of the root is fully replenished at a rate that coincides with the potential transpiration. When the moisture within the rhizosphere cannot be immediately replenished, water uptake by the root will be limited by the transport ability of the medium water towards the roots, that is by the medium's conductivity near the roots. Thus, when the transpiration rate is plant-atmosphere demand-driven, the water in the container is fully available. However, when water uptake becomes substrate supply-dependent, water is still available, but not fully. This concept of water availability and its dependence on the momentary hydraulic conductivity of the growing substrate was demonstrated by Raviv et al. (1999, 2001). In the first study (Raviv et al., 1999), rose plants were grown in two substrates, tuff and pumice, and were irrigated frequently. The yield differences between the two substrates were greater in the summer than in the cooler period, in spite of the frequent irrigations, large amounts of irrigation water and average low tension that was maintained in the containers. The transpirational demand in the summer exceeded the maximum water flux that can flow from the pumice media to the roots within the relevant range of hydraulic conductivity, for longer periods than in the winter, accentuating the negative effects of low water availability. In the second study (Raviv et al., 2001), rose plants were grown in UC mix (42 per cent composted fir bark, 33 per cent peat, and 25 per cent sand (by volume)) or in coconut coir. Low specific transpiration rate (transpiration rate per leaf area) values found at tensions between 0 and 1.5 kPa in UC mix suggest that this medium has insufficient free air space for proper root activity within this range. Above 2.3 kPa, unsaturated hydraulic conductivity of UC mix was lower than that of coir, possibly lowering specific transpiration rate values of UC mix-grown plants. As a result of these two factors, specific transpiration rate of plants grown in coir was 20–30 per cent higher than that of plants grown in UC mix. Specific transpiration rate of coir-grown plants started to decline only at tensions around 4.5 kPa. Yield (number of flowers produced) by coir-grown plants was 19 per cent higher than UC mix-grown plants. These studies demonstrated the superiority of hydraulic conductivity over tension as a measure of water availability.

An example for the relationship between atmosphere-demand and substrate-supply driven transpiration is given in the following by comparing the perlite container with 'dry' treatment (Figs. 3.19–3.22) with a perlite container with 'wet' treatments. The only difference between the two treatments is the irrigation frequency, namely the threshold weight at which the irrigation was turned on. Contrary to a single irrigation on August 7, at 18:00, in the 'dry' treatment (Fig. 3.19A), the 'wet' treatment was irrigated three times during that day – 10:20, 14:00 and 21:30. The matric potential and the moisture content (determined as for the 'dry' treatment from the measured matric potential and the independently determined retention curve) variation for the 'wet' treatment are shown in Figs. 3.25A and 3.25B, respectively. Pronounced differences in matric potential and moisture content were obtained between the two treatments (Fig. 3.21 and 3.25). The lower matric potential and higher moisture content in the 'wet' treatment indicate that the momentary hydraulic conductivity of the perlite is high and so is the water availability. The transpiration rate for the two 'wet' and 'dry' treatments are shown in Fig. 3.26. Comparison of these patterns with the VPD variation (Fig. 3.19A) reveals that both transpiration rate patterns followed the VPD pattern, with higher transpiration rate for the 'wet' treatment. The deviation between the two transpiration rate patterns was high during the peak VPD hours and decreased significantly during the period of lower VPD in between the two VPD peaks. The deviation pattern between the transpiration patterns demonstrates the meaning of water availability as a balance between input and output fluxes. When atmospheric demand is high (peak VPD values), the higher moisture content (and water availability) in the

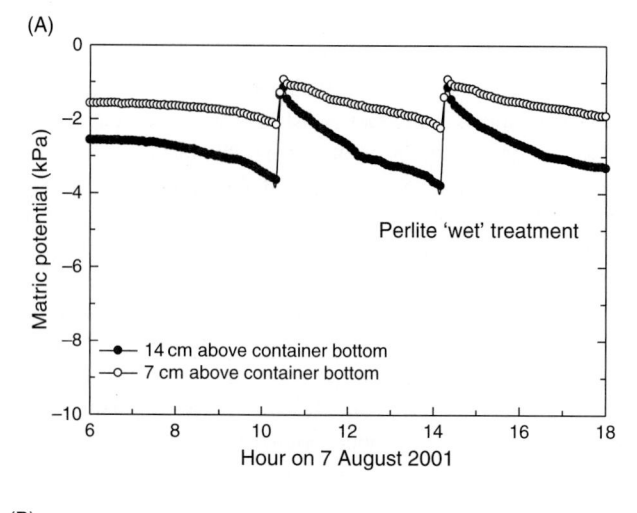

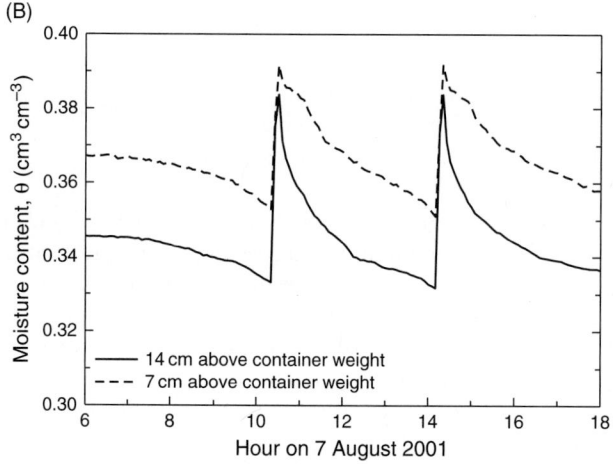

FIGURE 3.25 Matric potential (suction head) (A) and moisture content (B) variation during August 7 in the perlite container for the 'wet' treatment (Wallach and Raviv, unpublished).

wet substrate (Fig. 3.25B) enables a higher transpiration rate than that from the drier substrate. When there is a relief in the atmospheric demand (lower VPD values), water in both treatments were fully available and the two transpiration rates became close, in spite of the significant differences in moisture content between the two treatments. The existing differences in momentary transpiration rate among the two irrigation treatments that still exist during the periods of lower VPD values are probably due to the differences in plant size; the plant in the 'wet' treatment was larger than the one in the 'dry' treatment.

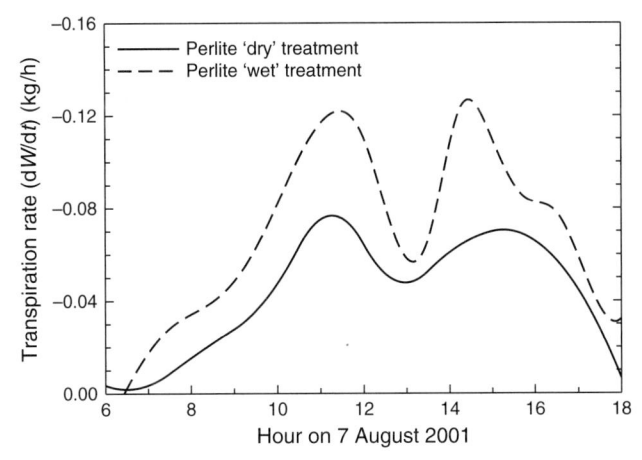

FIGURE 3.26 Transpiration rate for rose grown in perlite containers for the 'dry' and 'wet' treatment (Wallach and Raviv, unpublished).

3.5 SOLUTE TRANSPORT IN SOILLESS MEDIA

3.5.1 TRANSPORT MECHANISMS – DIFFUSION, DISPERSION, CONVECTION

Soil scientists and agricultural engineers have traditionally been interested in the behaviour and effectiveness of agricultural chemicals (fertilizers, pesticides) applied to soils for enhancing crop growth, as well as in the effect of salts and other dissolved substances in the soil profile on plant growth. The movement and fate of solutes in soil is affected by a large number of physical, chemical and microbiological processes requiring a broad array of mathematical and physical sciences to study and describe solute transport. A vast amount of work on solute transport can be found in the soil science literature. Soil scientists have paid much attention to water movement and mainly chemical transport in the absence of roots. However, much less attention has been devoted to solute movement and transport mechanisms that include nutrient uptake by roots in containerized media of limited volume with the unique structure of soilless media, frequent irrigations and chemicals that are continuously added with the irrigation water (fertigation). Consequently, both crop simulation and water flow models tend to treat the solute transport and root uptake mostly by empirical means, thereby limiting their general applicability.

The scope of this chapter permits only an introductory treatment of the subject. First, the standard transport mechanisms relevant to the fundamental convection-dispersion equation (CDE) will be introduced. This equation is most often used to model solute transport in porous media. The movement of a solute that undergoes adsorption by the solid particles requires modifications of the CDE, particularly if several solute species are present that may participate in a number of different reactions. The nutrient uptake mechanisms will be discussed as well, both qualitatively and quantitatively.

The mass balance equation for a solute species at the macroscopic level in a three-phase substrate–air–water system that is subject to arbitrary reactions is given as

$$\frac{\partial \theta C}{\partial t} = -\frac{\partial J_s}{\partial x_i} - \theta R_s(x_i) \tag{41}$$

where θ is the volumetric water content [$L^3\ L^{-3}$], C is the solute concentration expressed as solute mass-per-water volume [$M\ L^{-3}$], t is time [T], x_i is the spatial coordinate ($i=1, 2, 3$) [L], J_s is the solute flux expressed in solute mass-per-cross-sectional area of soil-per-unit time [$M\ L^2\ T^{-1}$], and R_s denotes arbitrary solute sinks (<0) or sources (>0) [$M\ L^{-3}\ T^{-1}$]. The one-dimensional mass balance equation is obtained by replacing x_i by x in Eq. (41). The solute flux is usually distinguished in an advective and a dispersive component according to

$$J_s = J_w C + J_D \tag{42}$$

where J_w is a vector quantifying the water flux [LT^{-1}], namely the Darcian velocity (Eq. [23]) and J_D is a solute flux to quantify transport caused by a gradient in the solute concentration [$M\ L^{-2}\ T^{-1}$], also per unit area of soil/growing medium.

Molecular diffusion is an important mechanism for solute transport in soil/growing medium in directions where there is little or no water flow. A net transfer of molecules of a solute species usually occurs from regions with higher to lower concentrations as the result of diffusion as described by Fick's first law. The one-dimensional mass flux, J_{diff} [$M\ L^2\ T^{-1}$], due to molecular diffusion is given by

$$J_{\text{diff}} = -D_0 \frac{\partial c}{\partial x_i} \tag{43}$$

where D_0 is the coefficient of molecular diffusion for a free or bulk solution [$L^2\ T^{-1}$]. Many publications exist that provide data on D_0 (Kemper, 1986; Lide, 1995). To characterize diffusion in growing media, the diffusion in a free solution is typically adjusted with terms accounting for a reduced solution phase (a smaller cross-sectional area available for diffusion) and an increased path length. A general treatment of diffusion in soils can be found in Olsen and Kemper (1968) and Nye (1979). The macroscopic diffusive flux per unit area of soil is

$$J_{\text{diff}} = -\theta D_{\text{dif}} \frac{\partial c}{\partial x_i} \tag{44}$$

where the molecular diffusion within the soil liquid D_{dif} can be expressed in terms of the diffusion in free water D_0 by

$$D_{\text{dif}} = \xi \theta D_0 \tag{45}$$

where ξ is the tortuosity factor. One form of ξ is (Millington and Quirk, 1961; Jury et al., 1991)

$$\xi = \frac{\theta^{7/3}}{\phi^2} \tag{46}$$

In some publications, the moisture content, θ, in Eq. (45) is combined in the tortuosity factor, providing $\xi = \theta^{10/3}/\phi^2$.

Mechanical dispersion is formed by local variations in water flow in a porous medium. Several mechanisms contribute to mechanical dispersion: (1) the development of a velocity profile within an individual pore such that the highest velocity occurs in the centre of the pore, and presumably little or no flow at the pore walls; (2) different mean flow velocities in pores of different sizes; (3) the mean water flow direction in the porous medium being different from the actual streamlines within individual pores, which differ in shape, size and orientation; and (4) solute particles converging to or diverging from the same pore. All of these processes contribute to the increased spreading, in which initially steep concentration fronts become smoother during movement along the main flow direction.

The effects of dispersion can be measured in a laboratory experiment in which water and a dissolved tracer are applied to an initially tracer-free, or vice versa, uniformly packed growing medium column of length L. The column is subjected to steady-state water flow with uniform water content. As more of the tracer is added, the initially very sharp concentration front near the soil surface becomes spread out because of dispersion. Eventually, a smooth and sigmoidally shaped effluent curve can be monitored at the column exit as shown in Fig. 3.27 for a 30-cm long column filled with 1 mm saturated sand. In the absence of dispersion, the front of a perfectly inert tracer will travel as a square wave through the column (a process often called 'piston flow') to reach the bottom of the column at time $t = L/v$, where v is the average pore water or interstitial velocity (Fig. 3.27). This velocity is the ratio of the Darcian water flux density (J_w), and the volumetric water content (θ). Note that v is given per unit area of fluid, whereas J_w is defined per unit area of (bulk) soil. For piston flow, the

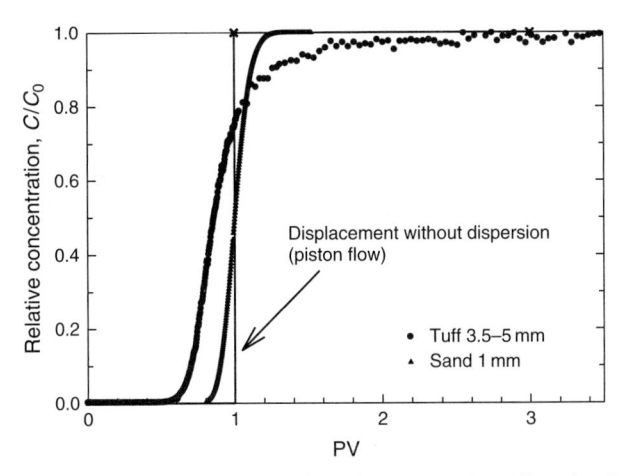

FIGURE 3.27 Measured breakthrough curves (BTCs) for saturated sand (1 mm) and tuff (3.5–5.0 mm) column. Water discharge of 10 ml/min was displacing by 6 g/l NaCl from bottom to top of a 30-cm long column.

tracer reaches the column exit exactly after one pore volume of tracer solution has been injected (or collected at the column exit). Pore volume is defined as the amount of water stored in that column.

The one-dimensional solute flux, due to mechanical dispersion in a uniform isotropic growing medium, may be approximated in a similar way as Fick's law:

$$J_{dis} = -\theta D_{dis} \frac{\partial C}{\partial x} \tag{47}$$

where J_{dis} is the dispersive solute flux [M L^{-2} T^{-1}]. It is commonly assumed that D_{is} is a linear function of the flow velocity

$$D_{dis} = \alpha v \tag{48}$$

where α [L] is the dispersivity. When the above one-dimensional geometry is too simplistic for a given problem, two or three dimensional dispersion is quantified with a dispersion tensor with α_L and α_T as the longitudinal and transverse dispersivity.

Hydrodynamic dispersion is a non-steady, irreversible process (i.e., the initial tracer distribution cannot be obtained by reversing the flow) in which the tracer mass mixes with the non-labelled portion of the solution. This is due to the presence of flow through a complicated system of pathways within the growing medium matrix. Hydrodynamic dispersion consists of mechanical dispersion and molecular diffusion. These two contributions are usually artificially separated, but in actuality are totally inseparable in form because both occur together; however, the dependence of each on different parameters can vary due to changes in physical and chemical conditions. As an example, mechanical dispersions is more prominent at high-water content and greater-flow velocities because it is here that nutrient particles mix more freely with water within the soil pores, as that water meanders around individual particles. Molecular diffusion predominates at low water content and low-flow velocity, since at this stage, chemical phenomena associated with a tracer continue, even though mechanical dispersions due to water movement have ceased. Molecular diffusion alone takes place at the molecular level in the absence of motion while dispersion occurs at the pore level. Hydrodynamic dispersion is also generally associated with early breakthrough of the tracer.

The structure of the substrate affects dispersion and the resultant shape of the breakthrough curve in numerous ways. As particle size increases or the medium is aggregated, the graphic representation of the breakthrough curve has more tailing due to diffusion into stagnant regions, but the effluent appears at the end of the column sooner due to larger pore size. However, to get $C/C_0 = 1$, a larger volume of effluent would be required, due to diffusion into both the stagnant areas and into the tortuous path of large aggregates. Smaller particle sizes yield a more even distribution in pore size, which means there is less stagnant water, so the effluent appears at a later time, and requires less volume for $C/C_0 = 1$ than for a soil with large aggregates.

The macroscopic similarity between diffusion and mechanical dispersion has led to the practice of describing both processes with one coefficient of hydrodynamic dispersion ($D = D_{dis} + D_{dif}$). This practice is consistent with results from experiments which do not permit a distinction between mechanical dispersion and molecular diffusion. The hydrodynamic dispersive flux (J_D) (Eq. [42]) consists of contributions from molecular diffusion (Eq. [43]) and mechanical dispersion (Eq. [47]) as

$$J_D = J_{\text{dif}} + J_{\text{dis}} \tag{49}$$

Hydrodynamic dispersion is often simply referred to as 'dispersion', as will be done in the remainder of this chapter.

Dispersion coefficients may be determined by fitting a mathematical solution to observed concentrations (Toride et al., 1995). Additional procedures to determine dispersion coefficients are given by Fried and Combarnous (1971) and van Genuchten and Wierenga (1976). Values of the longitudinal dispersivity (δ_L) for laboratory experiments typically vary between 0.1 and 10 cm with 6–20 times smaller values for δ_T (Klotz et al., 1980).

3.5.2 CONVECTION–DISPERSION EQUATION

The expressions for the convective and dispersive solute fluxes can be substituted in the mass balance (Eq. [41]). The one-dimensional convection–dispersion equation (CDE) for solute transport in a homogeneous growing medium becomes

$$\frac{\partial(\theta C)}{\partial t} = -\frac{\partial}{\partial x}\left(J_w C - \theta D \frac{\partial C}{\partial x}\right) + \theta R_s \tag{50}$$

In the case where the water content is invariant with time and space, the CDE may be simplified to ($v = J_w/\theta$)

$$\frac{\partial C}{\partial t} = D\frac{\partial^2 C}{\partial x^2} - v\frac{\partial C}{\partial x} + R_s \tag{51}$$

This is a second-order linear partial differential equation, which, like the diffusion equation, is classified as a parabolic differential equation. To complete the mathematical formulation of the transport problem, several concentration types, mathematical conditions as well as analytical and numerical solutions can be found in different textbooks and papers (e.g. van Genuchten and Alves, 1982; Warrick, 2003).

A variety of solute source or sink terms may be substituted for R_s. The most common source/sink term is due to adsorption/desorption and ion exchange stemming from chemical and physical interactions between the solute and the soil solid phase. Many other processes such as radioactive decay, aerobic and anaerobic transformations, volatilization, photolysis, precipitation/dissolution, reduction/oxidation and complexation may also affect the solute concentration. A further refinement of the transport model is necessary in the case of nonuniform interactions between the solute and the soil, or if there is adsorption on moving particles and colloids. In the following, only interactions at the solid–liquid interface will be considered.

3.5.3 ADSORPTION – LINEAR AND NON-LINEAR

Dissolved substances in the liquid phase can interact with solid particles and inorganic or organic colloids. Adsorption of solute by the growing medium is an important phenomenon affecting the fate and movement of solutes. The CDE for one dimensional transport of an adsorbed solute may be written as

$$\frac{\partial C}{\partial t} + \frac{\rho_b}{\theta}\frac{\partial S}{\partial t} = D\frac{\partial^2 C}{\partial x^2} - v\frac{\partial C}{\partial x} \qquad (52)$$

where S is the adsorbed concentration, defined as mass of solute per mass of dry soil [M M^{-1}]. The above equation can be expressed in terms of one dependent variable by assuming a suitable relationship between the adsorbed and the liquid concentrations. This is typically done with a simple adsorption isotherm to quantify the adsorbed concentration as a function of the liquid concentration at a constant temperature. In addition to temperature, the adsorption isotherm is generally also affected by the solution composition, total concentration, the pH of the bulk solution, and sometimes the method used for measuring the isotherm. A mathematically pertinent distinction is often made between linear and non-linear adsorption. Although most adsorption isotherms are non-linear, the adsorption process may often be assumed linear for low solute concentrations or narrow concentration ranges.

There are three common types of adsorption isotherms for which the CDE has been written: linear,

$$S = K_d C \qquad (53)$$

Freundlich,

$$S = k_1 C^n \qquad (54)$$

and Langmuir,

$$S = \frac{k_2 C}{1 + k_3 C} \qquad (55)$$

where K_d is the partition coefficient, often referred to as the distribution coefficient, expressed in volume of solvent per mass of soil [L^3 M^{-1}], k_1, k_2, k_3, and n are empirical constants. Using the derivative chain-rule $\partial S/\partial t = \partial S/\partial C \cdot \partial C/\partial t$, the CDEs for the linear, Freundlich and Langmuir isotherms are

$$\left(1 + \frac{K_d}{\theta}\right)\frac{\partial C}{\partial t} = D\frac{\partial^2 C}{\partial x^2} - v\frac{\partial C}{\partial x} \qquad (56)$$

$$\left(1 + \frac{nk_1}{\theta}C^{n-1}\right)\frac{\partial C}{\partial t} = D\frac{\partial^2 C}{\partial x^2} - v\frac{\partial C}{\partial x} \qquad (57)$$

$$\left[1 + \frac{k_2}{\theta(1+k_3 C)^2}\right]\frac{\partial C}{\partial t} = D\frac{\partial^2 C}{\partial x^2} - v\frac{\partial C}{\partial x} \qquad (58)$$

respectively. The term that multiplies the derivative $\partial C/\partial t$ in Eqs. (55)–(57) is noted as the retardation factor, R. The convective and dispersive fluxes are reduced by a factor R as a result of adsorption. The movement of the solute is said to be retarded with respect to the average solvent movement. If there is no interaction between the solute and the soil, the value for R is equal to unity. The value for R can be readily calculated from the parameters values as obtained from chemical analyses of the solution and adsorbed phases. Alternatively, R can be estimated from solute displacement studies on laboratory soil columns.

3.5.4 NON-EQUILIBRIUM TRANSPORT – PHYSICAL AND CHEMICAL NON-EQUILIBRIA

Solute breakthrough curves for aggregated growth media exhibit asymmetrical distributions or non-sigmoidal concentration fronts, as shown in Fig. 3.27 for the 3–3.5 mm tuff substrate. The concept behind physical non-equilibrium models is that differences between regions of the liquid phase lead to mostly lateral gradients in the solute concentration, resulting in a diffusive type of solute transfer process. Depending upon the exact pore structure of the medium, asymmetry is sometimes enhanced by desaturation when the relative fraction of water residing in the marginally continuous immobile region increases.

Since most of the sorption sites are only accessible after diffusion through the immobile region of the liquid phase, a corresponding delay in adsorption will occur. The delayed adsorption can also be explained with a kinetic description of the adsorption process. Both cases may be described with chemical non-equilibrium models, which distinguish between sites with equilibrium and kinetic sorption. The two-site chemical non-equilibrium model was applied successfully to describe solute breakthrough curves by Selim et al. (1976), van Genuchten (1981) and Nkedi-Kizza et al. (1983), among others. Bi-continuum or dual-porosity non-equilibrium models are the most widely used. Only two concentrations need to be considered and the equilibrium CDE (Eq. [52]) can be readily modified for this purpose. The same dimensionless mathematical formulation can be used for physical and chemical non-equilibrium models. If necessary, the CDE can be modified to incorporate additional non-equilibrium processes and continua.

The physical non-equilibrium approach is based on a partitioning of the liquid phase into a mobile or flowing region and an immobile or stagnant region. Solute movement in the mobile region occurs by both convection and dispersion, whereas solute exchange between the two regions occurs by first-order diffusion (Coats and Smith, 1964). Following van Genuchten and Wierenga (1976), the governing equations for the two region model are

$$(\theta_m + f\rho_b K_d) \frac{\partial C_m}{\partial t} = \theta_m D_m \frac{\partial^2 C_m}{\partial x^2} - \theta_m v_m \frac{\partial C_m}{\partial x} - \alpha (C_m - C_{im}) \quad (59)$$

$$[\theta_m + (1-f)\rho_b K_d] \frac{\partial C_{im}}{\partial t} = \alpha (C_m - C_{im}) \quad (60)$$

where f represents the fraction of sorption sites in equilibrium with the fluid of the mobile region, α is a first-order mass transfer coefficient $[T^{-1}]$ and the subscripts m and im, respectively, refer to the mobile and immobile liquid regions (with $\theta = \theta_m + \theta_{im}$), while ρ_b and K_d are the soil bulk density and distribution coefficient for linear sorption.

Transport equation, Eq. (59), follows directly from the addition of a source/sink term (R_s) to Eq. (52).

Growing media are coarse materials compared to regular soil. Pockets of air, to some extent, are generally present within the pores regardless of water content. The mobile and immobile regions within soils of similar structure have been widely used in direct relation to this phenomenon. However, the use of solute transport models to soilless growing media is exceptional. Fig. 3.27 introduces a breakthrough curve (BTC) that was measured for a saturated tuff growing medium packed in a 30-cm column. Comparing the asymmetrical tuff with the symmetrical sand, BTCs (Fig. 3.27) highlight the role of the immobile water in the tuff substrate on solute transport. The following parameters for Eqs. (59–60) were determined by using the CXTFIT software (Toride et al., 1995) for the measured tuff BTC: $v = 0.012 \text{ cm s}^{-1}$ and $n = 0.61$ (both directly measured), $D_m = 2.9 \times 10^{-3} \text{ cm s}^{-1}$, $\theta_m = 87\%$, $\theta_{im} = 13\%$, and $\alpha = 7.1 \times 10^{-5} \text{ s}^{-1}$. Note that θ_m does not exist in the sand.

3.5.5 MODELLING ROOT NUTRIENT UPTAKE – SINGLE-ROOT AND ROOT-SYSTEM

Uptake of chemicals by the root system depends mainly on two factors. These are the characteristics of the root system (e.g. morphology, growth rate, nutrient absorption rate) and the interaction of the chemical species with the soil and the soil-water supply (e.g. soil adsorption, mass transport, diffusion). In recent years, several mathematical models have been developed to simulate root growth and its interaction with both the soil water and the nutrient supply.

Modelling root nutrient uptake started with simulation of mass flow and diffusion of nutrients to a single root, approximated by a cylinder of constant radius, solving a simple form of the convection–dispersion equation for a circular geometry (Passioura and Frére, 1967) considering water but no nutrient uptake. Nye and Marriott (1969) included nutrient uptake, using a Michaelis–Menten type boundary condition at the soil–root interface, and explored the sensitivity of the solution to changes in some of the equation parameters (i.e. diffusion coefficient, soil buffer capacity, root absorbing power and nutrient flux at the root surface), whereas Barley (1970) simulated uptake in a system of irregular, parallel roots. Later models have included root competition (Baldwin et al., 1973) and root growth (Claassen and Barber, 1976; Barber and Cushman, 1981), but the root system was modelled only in terms of root length per unit depth, and no attention was paid to the interaction between three-dimensional distribution of the roots and water and solute transport patterns. Advances have also been made in the modelling of root growth, including the analytical approach of Hackett and Rose (1972a,b), the numerical simulations of Lungley (1973), the three-dimensional simulation models of Diggle (1988) and Pages et al. (1989) and the stochastic approach of Jourdan et al. (1995).

Coupling of such efforts as presented above has led to models capable of simulating various aspects of root–soil dynamics. Models have been developed focusing on the transport and uptake of highly immobile nutrients transported mainly by diffusion

(Silberbush and Barber, 1984) or mobile nutrients such as nitrate moving through the soil mainly by mass flow. Jones et al. (1991) simulated one-dimensional root growth of a user-selected crop as influenced by soil physical properties, soil water and nutrient content, and temperature. Clausnitzer and Hopmans (1994) simulated three-dimensional root architecture and the interaction between root growth and soil-water movement, emphasizing the effects of soil strength on root growth, whereas Benjamin et al. (1996) modelled the influence of rooting patterns on root-water uptake and leaching in two dimensions. Somma et al. (1998) included the interaction between plant growth and nutrient concentration in their model, thus providing a tool for studying the dynamic relationships between changing soil-water and nutrient status and root activity.

The general transport model for root uptake that follows the development of the mass-balance equation in Sect. 3.5.3 to solve the three dimensional form of the convection–dispersion equation for solute concentration C [M L^{-3}], as fully described in Simunek et al. (1999), is

$$(\theta + \rho K_d) \frac{\partial C}{\partial t} = \frac{\partial}{\partial x_i}\left(\theta D \frac{\partial C}{\partial x_i}\right) - J_{wi} \frac{\partial C}{\partial x_i} - S' \tag{61}$$

where K_d is the linear adsorption coefficient, J_{wi} is the water flux component in the ith direction, and S' [T^{-1}] is the sink term to account for root nutrient uptake. Many more rate constants can be added to the CDE, for example, to allow for reactions of the solute in the dissolved or adsorbed phase such as microbial degradation, volatilization and precipitation. The solution of Eq. (61) yields the spatial and temporal distribution of nutrient concentration and fluxes at the same time resolution as the solution of Eq. (37). Note that the solution of Eqs. (37) and (61) yields macroscopic quantities, that is values of matric potential, concentration, fluxes, in resolution which is usually much larger than root diameter and root spacing.

The Nye–Tinker–Barber model for nutrient uptake by a single cylindrical root in an infinite extent of the soil was first developed in the 1970s (Nye and Tinker, 1977; Barber, 1984; Tinker and Nye, 2000). In that model, the nutrient uptake by the root is considered to occur from the liquid phase of the soil only. The soil is assumed to be homogeneous and isotropic, and the changes in the moisture conditions are assumed to be negligible since the soil is assumed to be fully saturated. The classical Nye–Tinker–Barber model also neglects the effect of root exudates, microbial activity, mycorrhiza and so on the plant nutrient uptake and nutrient movement in the soil. The movement of nutrient to the root surface takes place by convection due to water uptake by the plant, and by diffusion of nutrient ions in the soil pore water.

The relation between the net influx of nutrients into the root and the concentration of nutrients at the root surface is described in the Nye–Tinker–Barber model by an experimentally measured heuristic Michaelis–Menten type uptake law. Classically, the Michaelis–Menten kinetics is derived from the most basic enzyme catalysed reaction. This is compatible with the nutrient 'carrier' theory in which the nutrients are considered to be carried through the root surface cell membranes by the so-called nutrient 'carrier' proteins. Thus, the Michaelis–Menten uptake law appears to include the characteristics of symplastic pathway of nutrients into the root. Alternatively, the Michaelis–Menten type uptake law could also be derived similarly to that of Langmuir adsorption isotherm. This describes the fraction of free and bound nutrient ion-binding sites on a membrane or on a solid surface. This results in what looks like a Michaelis–Menten type equilibrium condition for the number of binding sites that are bound at any given time. However, as mentioned, the Nye–Tinker–Barber model does not consider the exact mechanism of nutrient uptake by the plant root, instead it uses an experimentally measured heuristic Michaelis–Menten nutrient uptake law.

For the specific case of a cylindrical root, the problem can be written in terms of polar radius r, and moreover the Darcy flux (satisfying $\nabla \cdot \vec{u} = 0$) is given as $u = -aV/r$, where V is the water flux to the root and a is the root radius. The CDE for a single root, written in radial coordinates is

$$(\theta + Kd)\frac{\partial C}{\partial t} = \frac{\theta D}{r}\frac{\partial}{\partial r}\left(r\frac{\partial C}{\partial r}\right) \qquad (62)$$

The boundary condition at the root surface $(r = a)$ is an uptake flux of Michaelis–Menten type.

$$\theta D\frac{\partial C}{\partial r} + VC = \frac{F_m C}{K_m + C} \quad \text{on} \quad r = a \qquad (63)$$

where F_m and K_m are properties of the root surface. The boundary condition far from the root surface is the initial concentration, C_0

$$C \to C_0 \quad \text{as} \quad r \to \infty \qquad (64)$$

The initial soil concentration is prescribed, and equal to the far-field concentration away from the root:

$$C = C_0 \text{ at } t = 0 \qquad (65)$$

A combined model for water flow, solute transport and uptake by roots to containerized growing media has been developed by Heinen (2001). In this simulation model, water flow is modelled by the Richards equation and the solute transport by the convection–dispersive equation without adsorption. Water uptake and nutrient uptake were treated analogously. Such models could quantitatively highlight the effect of irrigation frequency and its effect on water uptake, nutrients availability and EC distribution in the root zone at different stages during and between subsequent irrigation events and so on.

3.6 GAS TRANSPORT IN SOILLESS MEDIA

3.6.1 GENERAL CONCEPTS

The understanding of gas transport in growing media is important for the evaluation of soil aeration or movement of O_2 from the atmosphere to the medium. Medium aeration is critical for plant root growth. The aeration conditions of growing media during cultivation are still not fully known in spite of the intensive research that has been conducted. For example, mixtures of peat-based growing media are used worldwide in horticulture and in forest tree nurseries (Bunt, 1988; Landis et al., 1990) and several studies and handbooks on peat-based growing media are available (Puustärvi, 1977; Bunt, 1983, 1988; Verdonck et al., 1983; Landis et al., 1990). However, little is known about the actual aeration conditions in these growth media during cultivation (Heiskanen 1995a,b).

Gas in unsaturated substrates is generally moist air, but has higher CO_2 concentrations than atmospheric air because of plant root respiration and microbial degradation of organic compounds. Oxygen concentrations are generally inversely related to CO_2 concentrations because processes producing CO_2 generally deplete O_2 levels.

The volumetric gas content, θ_g, is defined as

$$\theta_g = \frac{V_g}{V_T} \tag{66}$$

where V_g [L^3] is the volume of the gas and V_T [L^3] is the total volume of the sample. This definition is similar to that used for volumetric water content. In many cases, the volumetric gas content is referred to as 'the gas porosity'. The saturation with respect to the gas phase S_g is

$$S_g = \frac{V_g}{V_v} \tag{67}$$

where V_v [L^3] is the volume of pores.

In unsaturated systems, water wets the solids and is in direct contact with them, whereas gas is generally separated from the solid phase by the water phase. Water fills the smaller pores whereas gas is restricted to the larger pores. Gas and water differ greatly. The gas density is much lower than the liquid density; the density of air ranges generally from 1 to 1.5 kg m^{-3}, depending upon its composition, whereas the density of water is close to 1000 kg m^{-3}. When water flow is considered, it is assumed that its velocity at the solid surfaces is zero (no slip flow), whereas gas velocities are generally nonzero (slip flow). As the dynamic viscosity of air is ~50 times lower than that of water, significant air flow can occur at much smaller pressure gradients. The density and viscosity differences between gas and water induce a lower hydraulic conductivity (an order of magnitude) for gas than for water in the same material. Because gas molecular diffusion coefficients are about four orders of magnitude greater than those of water, gas diffusive fluxes are generally much greater than those of water.

3.6.2 MECHANISMS OF GAS TRANSPORT

As in liquid transport (see details in Sect. 3.5), transport in the gas phase may also be described by convection (advection) and dispersion. Although some studies have found mechanical dispersion or velocity-dependent dispersion to be important for chemical transport in the gas phase (Rolston et al., 1969), in most cases mechanical dispersion is ignored because gas velocities are generally too small and the effects of diffusion are generally much greater than dispersion in the gas phase. The one-dimensional diffusive flux of a given gas can be described by Fick's law, similar to Eq. (43) for solute transport,

$$J_g = -D_{gs}\frac{dC_g}{dx} \tag{68}$$

where J_g is the flux of gases, D_{gs} is the effective gas diffusion coefficient and C_g is the gas concentration [L^3 L^{-3}] in the air phase. Soil gas diffusivity and its dependency on θ_g and growing media type (texture, structure) control gas transport and fate where diffusive gas transport is normally dominant compared with convective gas transport. Accurate predictive models for D_{gs} are needed to evaluate, for example soil aeration (Buckingham, 1904; Taylor, 1949), the diffusion and emission of fumigants at soil fumigation sites (Call, 1957; Jin and Jury, 1995), the diffusion and volatilization of organic chemicals from polluted soil sites (Petersen et al., 1996), and the diffusion and biodegradation of greenhouse gases such as methane (Kruse et al., 1996). Numerous predictive–descriptive models for D_{gs} as a function of ε are available and were divided by Moldrup et al. (2004) into six groups:

1. The first group consists of predictive $D_{gs}(\theta_g)$ models based only on θ_g. The first θ_g-based models were introduced a century ago. Buckingham suggested that the relative oxygen diffusion coefficient in soil was proportional to θ_g^2 (Buckingham, 1904). Other classical $D_{gs}(\theta_g)$ models in the first group are the linear $D_{gs}(\theta_g)$ models by Penman (1940),

$$D_{gs}/D_{g0} = 0.66\theta_g \tag{69}$$

van Bavel (1952), and Call (1957), and the non-linear models by Marshall (1959) and Millington (1959). The latter two can be considered mechanistically based (cutting and randomly rejoining pores) models (Ball et al., 1988; Collin and Rasmuson, 1988).

2. The second group consists of simple, empirically, or mechanistically based, non-linear $D_{gs}(\theta_g)$ models that take into account both θ_g and soil total porosity, ϕ. These predictive models introduce a minor soil type effect through ϕ that is dependent on, for example, soil texture and management. Among the numerous models within this group are the Millington and Quirk (1960) model, as re-introduced by Jin and Jury (1996), and the Millington and Quirk (1961) model

$$D_{gs}/D_{g0} = \theta_g^{10/3}/\phi^2 \tag{70}$$

that is almost universally accepted and applied in unsaturated soil transport and fate models to describe both gas and solute diffusivity (Eq. [46]). The frequent use of the Millington and Quirk (1961) model is noteworthy, since the model has never been validated against gas diffusivity data for undisturbed soils representing a broad interval of soil types and porosities.

3. The models in the third group use the soil water characteristic curve (SWC) as an additional input to take into account soil type effects on gas diffusivity. Moldrup et al. (1996) introduced the Campbell (Campbell, 1974) SWC parameter b as the third

model parameter, together with θ_g and ϕ, in $D_{gs}(\theta_g)$ models. Moldrup et al. (1999) combined the Campbell b dependent $D_{gs}(\theta_g)$ model with the Buckingham (1904) expression for gas diffusivity in dry soil (void of water) to develop the so-called 'Buckingham–Burdine–Campbell model'. Moldrup et al. (2000) further introduced the air-filled porosity at 100 cm H_2O of soil–water matric potential to describe soil structure effects on gas diffusivity.

4. The fourth group of models consists of generalized power law models that introduce additional, empirical model parameters and thereby can provide a good fit to $D_{gs}(\theta_g)$ data within the θ_g interval where measurements are available. The most frequently used within this group is the Troeh et al. (1982) model, where two additional fitting parameters are introduced. The Troeh et al. (1982) model was successfully used in several studies to fit and subsequently represent measured $D_{gs}(\theta_g)$ data in gas transport and fate models (Petersen et al., 1994, 1996). Although one of the Troeh et al. (1982) model parameters can be interpreted as the air-filled porosity where gas diffusion ceases due to interconnected water films (creating blocked pore space), no relationships between the Troeh et al. (1982) model parameters and soil physical properties have been identified. The model at present is descriptive rather than predictive (Moldrup et al., 2003).

5. The fifth group consists of two- or three-region $D_{gs}(\theta_g)$ models that partition the pore space into, for example, easily accessible, difficult accessible, and non-accessible pore space (Arah and Ball, 1994). The models have typically been developed for highly structured or highly aggregated artificial porous media or repacked soil aggregates and include the classical model by Millington and Shearer (1971). The models contain additional pore shape and pore region parameters and mainly descriptive models that can be used to fit detailed $D_{gs}(\theta_g)$ data. Moldrup et al. (2004) introduced a three-porosity model that requires only one point (at -100 cm H_2O) on the SWC and predicts the gas diffusion coefficient as well as those who use the entire SWC and better than typically used soil type independent models.

6. The sixth group of $D_{gs}(\theta_g)$ models consists of macroscopic pore-size distribution models based on equivalent pore radius capillary tube, jointed tubes of different radii or multidimensional capillary tube networks (Ball, 1981; Nielson et al., 1984; Steele and Nieber, 1994). Further, Freijer (1994) established links between this type of model and the multi-parameter Mualem–van Genuchten SWC model (van Genuchten, 1980). The $D_{gs}(\theta_g)$ models in this group at present have several empirical constants that must be fitted to actual $D_{gs}(\theta_g)$ data for the soil. Caron and Nkongolo (2004) have recently introduced a rapid method for the estimation of gas diffusivity *in situ* in peat substrates that is based on water storage and flow measurements.

3.6.3 MODELLING GAS TRANSPORT IN SOILLESS MEDIA

Under the assumption that gas transport in soils and growing media is basically diffusion-dependent, both vertical (long distance) and horizontal (short distance) transport is driven mainly by concentration gradients. In horizontal transport, the geometry of aggregates and the liquid phase are the major components of resistance for diffusion.

Equation (68) describes the flux of gas (oxygen, for example) but it does not account for its consumption. Thus, to account for more realistic cases, the mass balance equation with a sink term should be used in order to account for local concentration changes and gas component depletion. Equation (51) and its radial and cylindrical forms can be used to model different cases of gas transport. Note that the solute concentration and diffusion coefficient should be replaced by the proper terms for the relevant gases.

REFERENCES

Anderson, W.G. (1986). Wettability literature survey: Part 3. The effects of wettability on the electrical properties of porous media. *J. Pet. Tech.*, **38**, 1371–1378.

Anisko, T., NeSmith, D.S. and Lindstrom, O.M. (1994). Time-domain reflectometry for measuring water content of organic growing media in containers. *Hort. Science*, **29**, 1511–1513.

Arah, J.R.M. and Ball, B.C. (1994). A functional model of soil porosity used to interpret measurements of gas diffusion. *Eur. J. Soil Sci.*, **45**, 135–144.

Arya, L.M. and Paris, J.F. (1981). A physicoempirical model to predict the soil moisture characteristic from particle-size distribution and bulk density data. *Soil Sci. Soc. Am. J.*, **45**, 1023–1030.

Arya, L.M., Leij, F.J., Shouse, P.J. and van Genuchten, M.Th. (1999). Relationship between the hydraulic conductivity function and the particle-size distribution. *Soil Sci. Soc. Am. J.*, **63**, 1063–1070.

Assouline, S., Tessier, D. and Bruand, A. (1998). A conceptual model of the soil water retention curve. *Water Resour. Res.*, **34**, 223–231. (Correction, *Water Resour. Res.*, **36**, 3769).

Bailey, B.J., Montero, J.I., Biel, C., et al. (1993). Transpiration of Ficus Benjamina: comparison of measurements with predictions of the Penman–Monteith model and a simplified version. *Agric. For. Meteorol.*, **65**, 229–243.

Baille, M., Baille, A. and Laury, J.C. (1994). A simplified model for predicting evapotranspiration rate of nine ornamental species vs. climate factors and leaf area. *Sci. Hort.*, **59**, 217–232.

Baldwin, J.P., Nye, P.H. and Tinker, P.B. (1973). Uptake of solutes by multiple root systems from soil. III. A model for calculating the solute uptake by a randomly dispersed root system developing in a finite volume of soil. *Plant and Soil*, **38**, 621–635.

Ball, B.C. (1981). Modeling of soil pores as tubes using gas permeabilities, gas diffusivities and water release. *J. Soil Sci.*, **32**, 465–481.

Ball, B.C. and Hunter, R. (1988). The determination of water release characteristics of soil cores at low suctions. *Geoderma*, **43**, 195–212.

Ball, B.C., Sullivan, M.F. and Hunter, R. (1988). Gas diffusion, fluid flow and derived pore continuity indices in relation to vehicle traffic and tillage. *J. Soil Sci.*, **39**, 327–339.

Barber, S.A. (1984). *Soil Nutrient Bioavailability. A Mechanistic Approach*. New York: John Wiley & Sons.

Barber, S.A. and Cushman, J.H. (1981). Nitrogen uptake model for agronomic crops. In *Modeling waste water Renovation-Land treatment*. (I.K. Iskandar, eds). New York: John Wiley & Sons, pp. 382–409.

Barley, K.P. (1970). The configuration of the root system in relation to nutrient uptake. *Adv. Agron.*, **22**, 159–201.

Benjamin, J.G., Ahuja, L.R. and Allmaras, R.R. (1996). Modeling corn rooting patterns and their effects on water uptake and nitrate leaching. *Plant and Soil*, **179**, 223–232.

Bennie, A.T. (1991). Growth and mechanical impedance. In *Plant Roots: The Hidden Half*. Marcel Dekker (Y. Waisel et al., eds). New York: Marcel Dekker, pp. 393–414.

Bilderback, T.E. (1985). Physical properties of pine bark and hardwood bark media and their effects with four fertilizers on growth of Ilex x 'Nelie R. Stevens Holly'. *J. Environ. Hort.*, **3**, 181–185.

Bilderback, T.E. and Fonteno, W.C. (1987). Effects of container geometry and media physical properties on air and water volumes in containers. *J. Environ. Hort.*, **5**, 180–182.

Boulard, T. and Jemaa, R. (1993). Greenhouse tomato crop transpiration model application to irrigation control. In *Irrigation of Horticultural Crops* (López-Gálvez, ed.). Acta Hort. (ISHS), **355**, 381–387.

Buckingham, E. (1904). Contributions to our knowledge of the aeration of soils. USDA Bur. Soil Bul. 25. Washington, DC: U.S. Gov. Print. Office.

Bugbee, G.J. and Frink, C.R. (1983). Quality of potting soils. Bulletin 812. New Haven, CT: The Connecticut Agr. Exp. Station.

Bunt, A.C. (1984). Physical properties of mixtures of peats and minerals of different particle size and bulk density for potting substrates. Acta Hort. (ISHS), **150**, 143–154.

Burdine, N.T. (1953). Relative permeability calculations from pore-size distribution data. *Trans. Am. Inst. Min. Eng., Pet. Dev. Tech., Pet Branch*, **198**, 71–78.

Bures, S., Marfa, O., Perez, T., et al. (1997). Measure of substrates unsaturated hydraulic conductivity. Acta Hort. (ISHS), **450**, 297–304.

Call, F. (1957). Soil fumigation: V. Diffusion of ethylene dibromide through soils. *J. Sci. Food Agric.*, **8**, 143–150.

Campbell, G.S. (1974). A simple method for determining unsaturated conductivity from moisture retention data. *Soil Sci.*, **117**, 311–314.

Campbell, G.S. (1985). *Soil Physics with BASIC: Transport Models for Soil-Plant Systems*. Amsterdam: Elsevier.

Caron, J. and Nkongolo, N.V. (2004). Assessing gas diffusion coefficients in growing media from in situ water flow and storage measurements. *Vadose Zone J.*, **3**, 300–311.

Caron, J., Riviere, L.M., Charpentier, S., et al. (2002). Using TDR to estimate hydraulic conductivity and air entry in growing media and sand. *Soil Sci. Soc. Am. J.*, **66**, 373–383.

Cary, J.W. and Hayden, C.W. (1973). An index for soils pore size distribution. *Geoderma*, **9**, 249–256.

Chen, Y., Banin, A. and Ataman, Y. (1980). Characterization of particles, pores, hydraulic properties and water-air-ratios of artificial growth media and soil. Proc. 5th Intl. Cong. Soilless Culture, Wageningen, pp. 63–82.

Claassen, N. and Barber, S.A. (1976). Simulation model for nutrient uptake from soil by a growing plant root system. *Agron. J.*, **68**, 961–964.

Clausnitzer, V. and Hopmans, J.W. (1994). Simultaneous modeling of transient three-dimensional root growth and soil water flow. *Plant and Soil*, **164**, 299–314.

Coats, K.H. and Smith., B.D. (1964). Dead end pore volume and dispersion in porous media. *Soc. Petrol. Eng. J.*, **4**, 73–84.

Collin, M. and Rasmuson, A. (1988). A comparison of gas diffusivity models for unsaturated porous media. *Soil Sci. Soc. Am. J.*, **52**, 1559–1565.

Cornelis, W.M., Ronsyn, J., Van Meirvenne, M. and Hartmann, R. (2001). Evaluation of pedotransfer functions for predicting the soil moisture retention curve. *Soil Sci. Soc. Am. J.*, **65**, 638–648.

Cowan, I.R. (1965). Transport of water in the soil-plant-atmosphere system.*J. Appl. Ecol.*, **2**, 221–239.

Danielson, R.E. and Sutherland, P.L. (1986). Porosity. In *Methods of Soil Analysis, Part 1. Physical and Mineralogical Methods, Agronomy Monograph no. 9* (2nd edn). Madison, WI: Amer. Soc. Agron. Inc. & Soil Sci. Soc. Am. Inc., pp. 443–461.

da Silva, F.F., Wallach, R. and Chen, Y. (1993a). Hydraulic properties of sphagnum peat moss and Tuff (scoria) and their potential effects on water availability. *Plant and Soil*, **154**, 119–126.

da Silva, F.F., Wallach, R. and Chen, Y. (1993b). A dynamic approach to irrigation scheduling in container media. Proc. 6th Intl. Conf. on Irrigation. Tel Aviv, Israel: Agritech, Ministry of Agriculture, pp. 183–198.

da Silva, F.F., Wallach, R. and Chen, Y. (1995). Hydraulic properties of rockwool slabs used as substrates in horticulture. Acta Hort. (ISHS), **401**, 71–75.

da Silva, F.F., Wallach, R., Polak, A. and Chen, Y. (1998). Measuring water content of soil substitutes with time domain reflectometry (TDR). *J. Amer. Soc. Hort. Science*, **123**, 734–737.

De Boodt, M. and Verdonck, O. (1972). The physical properties of the substrates in horticulture. Acta Hort. (ISHS), **26**, 37–44.

Dekker, L.W. and Ritsema, C.J. (1994). How water moves in a water repellent sandy soil: 1. Potential and actual water repellency. *Water Resour. Res.*, **30**, 2507–2517.

Denmead, O.T. and Shaw, R.H. (1962). Availability of soil water to plants as affected by soil moisture content and meteorological conditions. *Agron. J.,* **54**, 385–390.

Diggle, A.J. (1988). ROOTMAP – A model in three-dimensional coordinates of the growth and structure of fibrous root systems. *Plant and Soil,* **105**, 169–178.

Doerr S.H., Shakesby, R.A. and Walsh, R.P.D. (2000). Soil water repellency: its causes, characteristics and hydro-geomorphological significance. *Earth-Science Rev.,* **51**, 33–65.

Dullien, F.A.L. (1992). *Porous Media – Fluid Transport and Pore structure* (2nd edn). New York: Academic Press.

Evans, M.R., Konduru, S. and Stamps, R.H. (1996). Source variation in physical and chemical properties of coconut coir dust. *Hort. Science,* **31**, 965–967.

FAO (1979). Yield response to water. FAO irrigation and drainage paper 33. Rome: FAO, 193p.

Feddes, R.A. (1981). Water use models for assessing root zone modification. In *Modifying the Root Environment to Reduce Crop Stress* (G.F. Arkin and H.M. Taylor, eds). American Society of Agricultural Engineers, St. Joseph, ASAE Monograph no. 4, pp. 347–390.

Feddes, R.A, Bresler, E. and Neuman, S.P. (1974). Field test of a modified numerical model for water uptake by root systems. *Water Resour. Res.,* **10**, 1199–1206.

Feddes, R.A., Kowalik P.J. and Zaradny, H. (eds) (1978). Water uptake by plant roots. *Simulation of Field Water Use and Crop Yield.* New York: John Wiley & Sons, pp. 16–30.

Fiscus, E.L. (1977). Determination of hydraulic and osmotic properties of soybean root systems. *Plant Physiol.,* **59**, 1013–1020.

Fonteno, W.C. (1989). An approach to modelling air and water status of horticultural substrates. *Acta Hort.* (ISHS), **238**, 67–74.

Fonteno, W.C., Cassel, D.K. and Larson, R.A. (1981). Physical properties of three container media and their effect on poinsettia growth. *J. Am Soc. Hort. Science,* **106**, 736–741.

Fonteno, W.C. and Nelson, P.V. (1990). Physical properties of and plant responses to rockwool-amended media. *J. Amer. Soc. Hort. Science,* **115**, 375–381.

Freijer, J.I. (1994). Calibration of jointed tube models for the gas diffusion coefficient in soil. *Soil Sci. Soc. Am. J.,* **58**, 1067–1076.

Fried, J.J. and Combarnous, M.A. (1971). Dispersion in porous media. *Adv. Hydrorsci.,* **9**, 169–282.

Gardner, W.R. (1960). Dynamic aspects of water availability to plants. *Soil Sci.,* **89**, 63–73.

Gardner, W.R. (1983). Soil properties and efficient water use: A review. In *Limitations to Efficient Water Use in Crop Production* (H.M. Taylor, W.R. Jordan and T.R. Sinclair, eds). Madison: SSSA, pp. 45–64.

Gardner, W.H. (1986). Water content. In *Methods of Soil Analysis. Part I, Physical and Mineralogical Methods,* 2nd edn (A. Klute, ed.). Monograph 9, Madison, Wisconsin: Am. Soc. Agron, pp. 493–541.

Gardner, W.R. and Ehlig, C.F. (1962). Some observations on the movement of water to plant roots. *Agron. J.,* **54**, 453–456.

Girona, J., Mata, M., Fereres, E., et al. (2002). Evapotranspiration and soil water dynamics of peach trees under water deficits. *Agric. Water Management,* **54**, 107–122.

Glinski, J. and Stepniewski, W. (1985). *Soil Aeration and Its Role for Plants.* Boca Raton, FL: CRC Press.

Goh, K.M. and Haynes, R.J. (1977). Evaluation of potting media for commercial nursery production of container-grown plants. I. Physical and chemical characteristics of soil and soilless media and their constituents. *New Zealand Jour. Agr. Res.,* **20**, 363–370.

Goh, K.M. and Mass, E.F. (1980). A procedure for determining air and water capacity of soilless media and a method for presenting the results for easier interpretations. *Acta Hort.* (ISHS), **99**, 81–91.

Green, S., Clothier, B. and Jardine, B. (2003). Theory and practical application of heat pulse to measure sap flow. *Agron. J.,* **95**, 1371–1379.

Gruda, N. and Schnitzler, W.H., (1999a). Determination of volume weight and water content of wood fibre substrates with different methods. *Agribiol. Res.,* **52**, 163–170.

Gupta, S.C. and Larson, W.E. (1979). A model for predicting packing density of soils using particle size distribution. *Soil Sci. Soc. Am. J.,* **43**, 758–764.

Hackett, C. and Rose, D.A. (1972a). A model of the extension and branching of a seminal root of barley, and its use in studying relations between root dimensions. I. The model. *Aust. J. Biol. Sci.,* **25**, 669–679.

Hackett, C. and Rose, D.A. (1972b). A model of the extension and branching of a seminal root of barley, and its use in studying relations between root dimensions. II. Results and inferences from manipulation of the model. *Aust. J. Biol. Sci.*, **25**, 681–690.

Hamer, P.J.C. (1996). Validation of a model used for irrigation control of a greenhouse crop. In *Water Quality and Quantity in Greenhouse Horticulture* (R.M. Carpena, ed.). *Acta Hort.* (ISHS), **458**, 75–81.

Hansen, R.C. and Pasian, C.C. (1999). Using tensiometers for precision microirrigation of container-grown roses. *Applied Engr. in Agric.*, **15**, 483–490.

Harvey, A.C. (1993). *Time Series Models* (2nd edn). Cambridge, MA: The MIT Press.

Haverkamp, R. and Parlange, J.Y. (1986). Predicting the water retention curve from particle size distribution: I. Sandy soils without organic matter. *Soil Sci.*, **142**, 325–339.

Heinen, M. (2001). FUSSIM2: brief description of the simulation model and application to fertigation scenarios. *Agronomie*, **21**, 285–286.

Heiskanen, J. (1995a). Water status of sphagnum peat and a peat-perlite mixture in containers subjected to irrigation regimes. *Hort. Science*, **30**, 281–284.

Heiskanen, J. (1995b). Physical properties of two-component growth media based on *Sphagnum* peat and their implications for plant-available water and aeration. *Plant and Soil*, **172**, 45–54.

Heiskanen, J. (1999). Hydrological properties of container media based on sphagnum peat and their potential implications for availability of water to seedling after outplanting. *Scan. J. Forest Res.*, **14**, 78–85.

Hillel, D. (1998). *Environmental Soil Physics*, Academic Press, San Diego, CA.

Hillel, D., Talpaz H. and Van Keulen, H. (1976). A macroscopic-scale model of water uptake by a nonuniform root system and of water and salt movement in the soil profile. *Soil Sci.*, **121**, 242–255.

Hoogland, J.C., Feddes, R.A. and Belmans, C. (1981). Root water uptake model depending on soil water pressure head and maximum extraction rate. *Acta Hort.* (ISHS), **119**, 123–131.

Hopmans, J.W. and Bristow, K.L. (2002). Current capabilities and future needs of root water and nutrient uptake modeling. *Adv. Agron.*, **77**, 103–183.

Jin, Y. and Jury, W.A. (1995). Methyl bromide diffusion and emission through soil columns under various management techniques. *J. Environ. Qual.*, **24**, 1002–1009.

Jin, Y. and Jury, W.A. (1996). Characterizing the dependency of gas diffusion coefficient on soil properties. *Soil Sci. Soc. Am. J.*, **60**, 66–71.

Jolliet, O. and Bailey, B.J. (1992). The effect of climate on tomato transpiration in greenhouse: measurements and models comparison. *Agric. For. Meteorol.*, **58**, 43–63.

Jones, C.A., Bland, W.L., Ritchie, J.T., and Williams, J.R. (1991). Simulation of root growth. *Modeling Plant and Soil Systems*. ASA-CSSA-SSSA Agronomy Monograph no. 31.

Jones, H.G., Hamer, P.J.C. and Higges, K.H. (1988). Evaluation of various heat-pulse methods for estimation of sap flow in orchard trees: comparison with micrometeorological estimates of evaporation. *Trees Struct. Funct.*, **2**, 250–260.

Jourdan, C., Rey, H. and Guédon, Y. (1995). Architectural analysis and modeling of the branching process of the young oil-palm root system. *Plant and Soil*, **177**, 63–72.

Jury, W.A. and Flühler, H. (1992). Transport of chemicals through soil: Mechanisms, models, and fields applications. *Adv. Agron.*, **47**, 141–201.

Jury, W.A., Gardner, W.R. and Gardner, W.H. (1991). *Soil Physics* (5th edn). New York: John Wiley & Sons.

Karlovitch, P. and Fonteno, W. (1986). Effect of soil moisture tension and soil water content on the growth of *Chrysanthemum* in three container media. *Amer. Soc. Hort. Science*, **111**, 191–195.

Kemper, W.D. (1986). Solute diffusivity. In *Methods in Soil Analysis. I. Physical and Mineralogical Methods* (A. Klute, ed.). Madison, WI: Soil Sci. Soc. Am., pp. 1007–1024.

King, P.M. (1981). Comparison of methods for measuring severity of water repellence of sandy soils and assessment of some factors that affect its measurement. *Aust. J. Soil Res.*, **19**, 275–285.

Kipp, J.A. and Kaarsemaker, S.C. (1995). Calibration of time domain reflectometry water content measurements in growing media. *Acta Hort.* (ISHS), **401**, 49–55.

Klotz, D., Seiler, K.P., Moser, H. and Neumaier, F. (1980). Dispersivity and velocity relationship from laboratory and field experiments. *J. Hydrol.*, **45**, 169–184.

Klute, A. (ed.) (1986). Water retention: Laboratory methods. In *Methods of Soil Analysis. Part I, Physical and Mineralogical Methods* (2nd edn). Monograph 9, Madison, Wisconsin: Am. Soc. Agron., pp. 635–660.

Klute, A. and Dirksen, C. (1986). Hydraulic conductivity and diffusivity: Laboratory methods. In *Methods of Soil Analysis. Part I, Physical and Mineralogical Methods* (2nd edn). (Klute ed.) Monograph 9, Madison, Wisconsin: Am. Soc. Argon., pp. 635–660.

Kosugi, K. (1994). Three-parameter lognormal distribution model for soil water retention. *Water Resour. Res.*, **30**, 891–901.

Kramer, P.J. and Boyer, J.S. (1995). *Water Relations of Plants and Soils*. New York: Academic Press.

Kruse, C.W., Moldrup, P. and Iversen, N. (1996). Modeling diffusion and reaction in soils: II. Atmospheric methane diffusion and consumption in a forest soil. *Soil Sci.*, **161**, 355–365.

Lambiny, G., Robidas, L. and Ballester, P. (1996). Measurement of soil water content in a peat-vermiculite medium using time domain reflectometry (TDR): A laboratory and field evaluation. *Tree Plant.* **47**, 88–93.

Landis, T.D., Tinus, R.W., McDonald, S.E. and Barnett, J.P. (1990). Containers and growing media. The container free nursery manual. Vol 2. USDA, Forest Service, Washington, DC. Agric. Handbook **674**, 87p.

Lebeau, B., Barrington, S. and Bonnell, R. (2003). Micro-tensiometers to monitor water retention in peat potted. *App. Eng. Agric.*, **19**, 559–564.

Ledieu, J., de Ridder, P., de clerck, P. and Dautrebande, S. (1986). A method of measuring soil moisture by time-domain reflectometry. *J. Hydrol.*, **88**, 319–328.

Li, Y., Fuchs, M., Cohen, S., et al. (2002a). Changes of the hydraulic path in roots of drip-irrigated corn during a drying period. *Cell, Plant and Environment*, **25**(4), 491–512.

Li, Y., Wallach, R. and Cohen, Y. (2002b). The role of soil hydraulic conductivity on the spatial and temporal variation of root water uptake in drip-irrigated corn. *Plant and Soil*, **243**, 131–142.

Lide, D.R. (1995). *Handbook of Chemistry and Physics*. Boca Raton, FL: CRC Press.

Lungley, D.R. (1973). The growth of root systems – A numerical computer simulation model. *Plant and Soil*, **38**, 145–159.

Luxmoore, R.J. (1981). Micro-, meso-, and macroporosity of soil. Letter to the editor. *Soil Sci. Soc. Am. J.*, **45**, 671–672.

Ma, L. and Selim, H.M. (1997). Physical non-equilibrium modeling approaches to solute transport in soils. *Adv. Agron.*, **58**, 95–153.

Marshall, T.J. (1959). The diffusion of gases through porous media. *J. Soil Sci.*, **10**, 79–82.

Medrano, E., Lorenzo, P., Sanchez-Guerrero, M.C. and Montero, J.I. (2005). Evaluation and modelling of greenhouse cucumber-crop transpiration under high and low radiation conditions. *Scientia Horticulturae*, **105**, 163–175.

Michel, J.C., Riviere, L.M. and Bellon-Fontaine, M.N. (2001). Measurement of the wettability of organic materials in relation to water content by the capillary rise method. *European J. Soil Sci.*, **52**, 459–467.

Milks, R.R., Fonteno, W.C. and Larson, R. (1989a). Hydrology of horticultural substrates: I. Mathematical models for moisture characteristics of horticultural container media. *J. Am. Soc. Hort. Science*, **144**, 48–52.

Milks, R.R., Fonteno, W.C. and Larson, R. (1989b). Hydrology of horticultural substrates: II. Predicting physical properties of media in containers. *J. Am. Soc. Hort. Science*, **144**, 53–56.

Milks, R.R., Fonteno, W.C. and Larson, R. (1989c). Hydrology of horticultural substrates: III. Predicting air and water content of limited-volume plug cells. *J. Am. Soc. Hort. Science*, **144**, 57–61.

Millington, R.J. (1959). Gas diffusion in porous media. *Science*, **130**, 100–102.

Millington, R.J. and Quirk, J.M. (1960). Transport in porous media. In F.A. Van Beren et al. (eds) Trans. 7th Int. Congr. Soil Sci., Vol. 1, Madison, WI. 14–24 Aug., Amsterdam: Elsevier, pp. 97–106.

Millington, R.J. and Quirk, J.M. (1961). Permeability of porous solids. *Trans. Faraday Soc.*, **57**, 1200–1207.

Millington, R.J. and Shearer, R.C. (1971). Diffusion in aggregated porous media. *Soil Sci.*, **111**, 372–378.

Moldrup, P., Kruse, C.W., Rolston, D.E. and Yamaguchi, T. (1996). Modeling diffusion and reaction in soils: III. Predicting gas diffusivity from the Campbell soil water retention model. *Soil Sci.*, **161**, 366–375.

Moldrup, P., Olesen, T., Schjønning, P., et al. (2000). Predicting the gas diffusion coefficient in undisturbed soil from soil water characteristics. *Soil Sci. Soc. Am. J.*, **64**, 94–100.

Moldrup, P., Olesen, T., Yamaguchi, T., et al. (1999). Modeling diffusion and reaction in soils: IV. The Buckingham-Burdine-Campbell equation for gas diffusivity in undisturbed soil. *Soil Sci.*, **164**, 542–551.

Moldrup, P., Olesen, T., Yoshikawa, S., et al. (2004). Three-porosity model for predicting the gas diffusion coefficient in undisturbed soil. *Soil Sci. Soc. Am. J.*, **68**, 750–759.

Moldrup, P., Yoshikawa, S., Olesen, T., et al. (2003). Gas diffusivity in undisturbed volcanic ash soils: Test of soil-water-characteristic-based prediction models. *Soil Sci. Soc. Am. J.*, **67**, 41–51.

Möller, M., Tanny, J., Li, Y. and Cohen, S. (2004). Measuring and predicting evapotranspiration in an insect-proof screenhouse. *Agric. and Forest Meteorology*, **127**, 35–51.

Molz, F.J. (1981). Models for water transport in the soil-plant system: A review. *Water Resour. Res.*, **17**, 1245–1260.

Molz, F.J. and Remson, I. (1970). Extraction term models of soil moisture use by transpiring plants. *Water Resour. Res.*, **6**, 1346–1356.

Monteith, J.L. (1990). *Principles of Environmental Physics.* London: Edward Arnold, xii + 291pp.

Montero, J.I., Anton, A., Muñoz, P. and Lorenzo, P. (2001). Transpiration from geranium grown under high temperatures and low humidities in greenhouses. *Agric. For. Meteorol.*, **107**, 323–332.

Morel, P. and Michel, J.-C. (2004). Control of the moisture content of growing media by time domain reflectometry (TDR). *Agronomie*, **24**, 275–279.

Morel, P., Riviere, L.M. and Julien, C. (2000). Measurement of representative volumes of substrates: which method, in which objective? *Acta Hort.* (ISHS), **517**, 261–269.

Mualem, Y. (1976). A new model for predicting the hydraulic conductivity of unsaturated porous media. *Water Resources Res.*, **12**, 513–522.

Naasz, R., Michel, J.-C. and Charpentier, S. (2005). Measuring hysteretic hydraulic properties of peat and pine bark using a transient method. *Soil Sci. Soc. Am. J.*, **69**, 13–23.

Nemati, R., Caron, J., Banton, O. and Tardif, P. (2002). Determining air entry in peat substrates. *Soil Sci. Soc. Am. J.*, **66**, 367–373.

Newman, E.I. (1969). Resistance to water flow in soil and plant. I. Soil resistance in relation to amounts of roots: theoretical estimates. *J. Appl. Ecol.*, **6**, 1–12.

Nielson, K.K., Rogers, V.C. and Gee, G.W. (1984). Diffusion of Radon through soils. A pore distribution model. *Soil Sci. Soc. Am. J.*, **48**, 482–487.

Nimah, M.N. and Hanks, R.J. (1973). Model for estimating soil water, plant and atmospheric interrelations: I. Description and Sensitivity. *Soil Sci. Soc. Am. Proc.*, **37**, 522–532.

Nkedi-Kizza, P., Biggar, J.W., van Genuchten, M.Th., et al. (1983). Modeling tritium and chloride 36 transport through an aggregated oxisol. *Water Resour. Res.*, **19**, 691–700.

Nye, P.H. and Marriott, F.H.C. (1969). A theoretical study of the distribution of substances around roots resulting from simultaneous diffusion and mass flow. *Plant and Soil*, **3**, 459–472.

Nye, P.H. and Tinker, P.B. (1977). *Solute Movement in the Soil-Root System.* Oxford: Blackwell Scientific Publications.

Olsen, S.R. and Kemper, W.D. (1968). Movement of nutrients to plant roots. *Adv. Argon.*, **20**, 91–151.

Or, D. and Tuller, M. (1999). Liquid retention and interfacial area in variably saturated porous media: scaling from single-pore to sample-scale model. *Water Resour. Res.*, **35**, 3591–3605.

Orozco, R. and Marfa, O. (1995). Granulometric alteration, air entry potential and hydraulic conductivity in perlites used in soilless cultures. *Acta Hort.* (ISHS), **408**, 147–161.

Pages, L., Jordan, M.O. and Picard, D. (1989). A simulation model of the three-dimensional architecture of the maize root system. *Plant and Soil*, **119**, 147–154.

Paquet, J.M., Caron, J. and Banton, O. (1993). In situ determination of the water desorption characteristics of peat substrates. *Can. J. Soil Sci.*, **73**, 329–339.

Passioura, J.B. (1988). Water transport in and to roots. *Annu. Rev. Plant Physiol. Plant Mol. Biol.*, **39**, 245–265.

Passioura, J.B. and Frére, M.H. (1967). Numerical analysis of convection and diffusion of salutes to roots. *Aust. J. Soil Res.*, **5**, 149–159.

Penman, H.L. (1940). Gas and vapour movements in soil: The diffusion of vapours through porous solids. *J. Agric. Sci. (Cambridge)*, **30**, 437–462.

Pépin, S., Plamondon, A.P. and Stein, J. (1991). Peat water content measurement using time domain reflectometry. *Can. J. For. Res.*, **22**, 534–540.

Petersen, L.W., El-Farhan, Y.H., Moldrup, P., et al. (1996). Transient diffusion, adsorption, and emission of volatile organic vapours in soils with fluctuating low water contents. *J. Environ. Qual.*, **25**, 1054–1063.

Petersen, L.W., Rolston, D.E., Moldrup, P. and Yamaguchi, T. (1994). Volatile organic vapour diffusion and adsorption in soils. *J. Environ. Qual.*, **23**, 799–805.

Philip, J.R. (1966). Plant water relations: Some physical aspects. *Ann Rev. Plant Physiol.*, **17**, 245–268.

Plagge, R., Häupl, P. and Renger, M. (1999). Transient effects on the hydraulic properties of porous media. In *Proc. Int'l. Workshop, Characterization and Measurement of the Hydraulic Properties of Unsaturated Porous Media* (M.Th. van Genuchten et al., eds). Riverside, CA: University of California, pp. 905–912.

Pokorny, F.A., Gibson, P.G. and Dunavent, M.G. (1986). Prediction of bulk density of pine bark and/or sand potting media from laboratory analyses of individual components. *J. Amer. Soc. Hort. Science*, **111**, 8–11.

Prasad, R (1988). A linear root water uptake model. *J. Hydrol.*, **99**, 297–306.

Puustärvi, V. (1977). Peat and its use in horticulture. Turveteollisuuslitto ry. Publication 3. Liikekirjapaino, Helsinki, 160p.

Raats, P.A.C. (1974). Steady flows of water and salt in uniform soil profiles with plant roots. *Soil Sci. Soc. Am. Proc.*, **38**, 717–722.

Raats, P.A.C. (1976). Analytical solutions of a simplified flow equation. *Trans. ASAE*, **19**, 683–689.

Raviv, M., Lieth, J.H., Burger, D.W. and Wallach, R. (2001). Optimization of transpiration and potential growth rates of 'Kardinal' Rose with respect to root-zone physical properties. *J. Amer. Soc. Hort. Science*, **126**, 638–645.

Raviv, M. and Medina, Sh. (1997). Physical characteristics of separated cattle manure compost. *Compost Science & Utilization*, **5**, 44–47.

Raviv, M., Reuveni, R. and Zaidman, B. (1998). Improved medium for organic transplants. *Biol. Agric. and Hort.* **16**, 53–64.

Raviv, M., Wallach, R., Silber, A., et al. (1999). The effect of hydraulic characteristics of volcanic materials on yield of roses in soilless culture. *J. Am. Soc. Hort. Science*, **124**(2), 205–209.

Ray, J.D. and Sinclair, T.R. (1998). The effect of pot size on growth and transpiration of maize and soybean during water deficit stress. *J. Exp. Bot.*, **49**, 1381–1386.

Ritsema, C.J. and Dekker, L.W. (1994). How water moves in a water repellent sandy soil: 2. Dynamics of fingered flow. *Water Resour. Res.*, **30**, 2519–2531.

Riviere, L.M. (1992). Le fonctionnement hydrique du système substrat-plante en culture hors sol. Report submitted to the University of Angers.

Rolston, D.E., Kirkham, D. and Nielsen, D.R. (1969). Miscible displacement of gases through soils columns. *Soil Sci. Soc. Am. Proc.*, **33**, 488–492.

Rouphael, Y. and Colla, G. (2005). Radiation and water use efficiencies of greenhouse zucchini squash in relation to different climate parameters. *Europ. J. Agron.*, **23**, 183–194.

Scott, G.J.T., Webster, R. and Nortcliff, S. (1988). The topology of pore structure in cracking clay soil: I. The estimation of numerical density. *J. Soil Sci.*, **39**, 303–314.

Schuh, W.M. and Bauder, J.W. (1986). Effect of soil properties on hydraulic conductivity-moisture relationships. *Soil. Sci. Soc. Am. J.*, **54**, 1509–1519.

Selim, H.M., Davidson, J.M. and Mansell, R.S. (1976). Evaluation of a two-site adsorption-desorption model for describing solute transport in soils. In proceedings of the summer computer simulation conference, Washington, D.C. Simulation councils, La Joua, CA.

Silberbush, M. and Barber, S.A. (1984). Phosphorus and potassium uptake of field-grown soybean cultivars predicted by a simulation model. *Soil Sci. Soc. Am. J.*, **48**, 592–596.

Simunek, J., Sejna, M. and van Genuchten, M.Th. (1999). *The HYDRUS – 2D Software Package for Simulating Two-Dimensional Movement of Water, Heat, and Multiple Solutes in Variable Saturated Media*. Version 2.0, IGWMC-TPS-53. Golden, CO: International Ground Water Modeling Center, Colorado School of Mines.

Simunek, J., Vogel, T. and van Genuchten, M.Th. (1992). The SWMS-2D code for simulating water flow and solute transport in two-dimensional variably saturated media, V.1.1, Research Report No.126, U.S. Salinity Lab, ARS USDA, Riverside.

Smettem, K.R.J. and Gregory, P.J. (1996). The relation between soil water retention and particle-size distribution parameters for some predominantly sandy Western Australian soils. *Aust. J. Soil Res.*, **34**, 695–708.

Smiles, D.E., Vachaud, G. and Vauclin, M. (1971). A test of the uniqueness of the soil moisture characteristic during transient, non-hysteretic flow of water in a rigid soil. *Soil Sci. Soc. Am. Proc.*, **35**, 535–539.

Soil Science Society of America (SSSA) (1996). *Glossary of Soil Science Terms*. Madison, WI: Soil Science Soc. Am.

Somma, F., Hopmans, J.W. and Clausnitzer, V. (1998). Transient three-dimensional modeling of soil water and solute transport with simultaneous root growth, root water and nutrient uptake. *Plant and Soil*, **202**, 281–293.

Steele, D.D. and Nieber, L.L. (1994). Network modeling of diffusion coefficients for porous media: I. Theory and model development. *Soil Sci. Soc. Am. J.*, **58**, 1337–1345.

Stephens, D.B. and Rehfeldt, K.R. (1985). Evaluation of closed-form analytical modules to calculate conductivity in a fine sand. *Soil Sci. Soc. Am. J.*, **49**, 12–19.

Steudle, E., Oren, R. and Schulze, E.D. (1988). Water transport in maize roots – measurement of hydraulic conductivity, solute permeability, and of reflection coefficients of excised roots using the root pressure probe. *Plant Physiol.*, **84**, 1220–1232.

Stumm, W. (1992). *Chemistry of the Solid-Water Interface: Processes at the Mineral-Water and Particle-Water Interface in Natural Systems*. New York: Wiley, 428p.

Taylor, S.A. (1949). Oxygen diffusion in porous media as a measure of soil aeration. *Soil Sci. Soc. Am. Proc.*, **14**, 55–61.

Teare, I.D., Schimmelpfennig, H. and Waldren, R.P. (1973). Rainout shelter and drainage lysimeters to quantitatively measure drought stress. *Agron. J.*, **65**, 544–547.

Tinker, P.B. and Nye, P.H. (2000). *Solute Movement in the Rhizosphere*. Oxford: University Press.

Topp, G.C., Davis, J.L. and Annan, A.P. (1980). Electromagnetic determination of soil water content: measurements in coaxial transmission lines, *Water Resour. Res.*, **16**, 574–582.

Toride, N., Leij, F.J. and van Genuchten, M.Th. (1995). The CXTFIT code for estimating transport parameters from laboratory of field tracer experiments. Version 2.0. Riverside, CA: U.S. Salinity Lab. Res. Rep. 137, 121pp.

Troeh, F.R., Jabro, J.D. and Kirkham, D. (1982). Gaseous diffusion equations for porous materials. *Geoderma*, **27**, 239–253.

Urban, L., Fabret, C. and Barthelemy, L. (1994). Interpreting changes in stem diameter in rose plants. *Physiol. Plant.*, **92**, 668–672.

Vachaud, G., Vauclin, M. and Wakil, M. (1972). A study of the uniqueness of the soil moisture characteristic during desorption by vertical drainage. *Soil Sci. Soc. Am. Proc.*, **36**, 531–532.

Valat, B., Jouany, C. and Riviere, L.M. (1991). Characterization of the wetting properties of air dried peats and composts. *Soil Sci.*, **152**, 100–107.

van Bavel, C.H.M. (1952). Gaseous diffusion and porosity in porous media. *Soil Sci*, **73**, 91–104.

Van den Honert, T.H. (1948). Water transport in plants as a catenary process. *Disc. Faraday Sot.*, **3**, 146–153.

van Genuchten, M.Th. (1978). Calculating the unsaturated hydraulic conductivity with a new closed-form analytical model. Research Report 78-WR-08. New Jersey: Dept. of Civil Eng. Princeton Univ., 63p.

van Genuchten, M.Th. (1980). A closed-form equation for predicting the hydraulic conductivity of unsaturated soils. *Soil Sci. Soc. Am. J.*, **49**, 12–19.

van Genuchten, M.Th. (1981). *Non-equilibrium Transport Parameters from Miscible Displacement Experiments*. Riverside, CA: S. Salinity Lb. Re. Rep. 119.

van Genuchten, M.Th. and Alves, W.J. (1982). Analytical solutions of the one-dimensional convective-dispersive solute transport equation, Tech. Bull. 1661. Washington, DC: U.S. Dep. of Agric.

van Genuchten, M.Th. and Nielsen, D.R. (1985). On describing and predicting the hydraulic properties of unsaturated soils. *Annales Geophysicae*, **3**, 615–628.

van Genuchten, M.Th. and Wierenga, P.J. (1976). Mass transfer studies in sorbing porous media. I. Analytical solutions. *Soil Sci. Soc. Am. J.*, **40**, 473–480.

Verdonck, O. and Gabriels, R. (1992). Reference method for the determination of physical and chemical properties of plant substrates. *Acta Hort.* (ISHS), **302**, 169–179.

Vereecken, H., Maes, J., Feyen, J. and Darius, P. (1989). Estimating the soil moisture retention characteristic from texture, bulk density, and carbon content. *Soil Sci.*, **148**, 389–403.

Vrugt, J.A., van Wijk, M.T., Hopmans, J.W. and Simunek, J. (2001). One-, two-, and three-dimensional root water uptake functions for transient modeling. *Water Resour. Res.*, **37**, 2457–2470.

Wallach, R., Ben Arie, O. and Graber, E.R. (2005). Soil water repellency induced by long-term irrigation with treated sewage effluent. *J. Environ. Quality*, **34**, 1910–1920.

Wallach, R., da Silva, F.F. and Chen, Y. (1992a). Unsaturated hydraulic characteristics of tuff (scoria) from Israel. *J. Amer. Soc. Hort. Science*, **117**, 415–421.

Wallach, R., da Silva, F.F. and Chen, Y. (1992b). Unsaturated hydraulic characteristics of composted agricultural wastes, tuff and mixtures. *Soil Sci.,* **153**, 434–441.

Wallach, R. and Raviv, M. (2005). The dependence of moisture-tension relationship and water availability on irrigation frequency in containerized growing medium. *Acta Hort.* (ISHS), **697**, 293–300.

Ward, A., Wells L.G. and Philips, R.E. (1983). Characterizing the unsaturated Hydraulic conductivity on Western Kentucky mine spoils and soils. *Soil Sci. Soc. Am J.*, **47**, 847–854.

Warrick, A.W. (2003). *Soil Water Dynamics.* New York: Oxford University Press.

Wever, G. (1995). Physical analysis of peat and peat-based growing media. *Acta Hort.* (ISHS), **401**, 561–567.

White, J.W. and Mastalerz, J.W. (1966). Soil moisture as related to container capacity. *Proc. Am. Soc. Hort. Science*, **89**, 757–765.

Wu, L., Vomocil, J.A. and Childs, S.W. (1990). Pore size, particle size, aggregate size, and water retention. *Soil Sci. Soc. Am. J.*, **54**, 952–956.

Yang, S.L., Aydin, M., Yano, T. and Li, X. (2003). Evapotranspiration of orange trees in greenhouse lysimeters. *Irrigation Sci.*, **21**, 145–149.

Zhuang, J., Jin, Y. and Miyazaki, T. (2001). Estimating water retention characteristic from soil particle-size distribution using a non-similar media concept. *Soil Sci.*, **166**, 308–321.

4

IRRIGATION IN SOILLESS PRODUCTION

J. HEINRICH LIETH AND LORENCE R. OKI

4.1 Introduction
4.2 Root Zone Moisture Dynamics
4.3 Irrigation Objectives and Design Characteristics
4.4 Irrigation Delivery Systems
4.5 Irrigation System Control Methods
4.6 Irrigation Decisions
4.7 Approaches to Making Irrigation Decisions
4.8 Future Research Directions
References

4.1 INTRODUCTION

Irrigation is the process of delivering water to plants so as to meet their needs for several important resources. It is known that providing too little or too much water can reduce crop productivity or, when extreme, can lead to plant death. Improper water management can also lead to excessive run-off. While providing water is certainly a major facet of irrigation, various other materials that are dissolved in the water are also provided to the plant at the same time. These include an array of nutrients as well as oxygen.

Although fertilization is not a specific aspect of this chapter, it should be noted that in soilless plant production, soluble fertilizers are frequently dissolved in the irrigation water using injection equipment. When the irrigation scheme includes soluble fertilizers dissolved in the irrigation water at various concentrations, this is called 'fertigation'. The substances in the water may also include pesticides and oxygen. In many soilless systems, fertigation is the preferred approach to supplying nutrients, especially where the substrate is not capable of holding (bonding) nutrient ions (i.e. due to low CEC). Specific technologies involved in combining fertilizers with irrigation water are covered elsewhere in the book (Chap. 5). In this chapter, we assume that this type of irrigation is being practised. So here, the terms 'fertigation' and 'irrigation' are used interchangeably.

In general, the plant's fresh weight typically consists of more than 90 per cent water and less than 10 per cent everything else. There are, of course, many exceptions reflecting organ function and plant adaptation to the native environment. When comparing the amount of water held in fresh matter of a particular plant with the total amount of water that the plant has taken into the roots, only a small percentage (about 1 per cent) of the total water actually ends up as fresh matter (Raviv and Blom, 2001). The rest of the water is transpired and most of this water use is necessary as part of the plant's need to cool itself.

The total amount of water that flows through a particular plant system is generally called 'the transpiration stream'. Despite what the name implies, this includes all the water that enters the roots and thus also accounts for the fraction of water that ends up tied up in biomass. While transpiration cools the leaves, this flux of the transpiration stream has several other consequences that are very important. One of these is the fact that when this flux is positive, it has to match an equivalent flux in the root zone of liquid moving towards the roots. This liquid, called the root zone solution, carries with it nutrients and gasses, and the motion of this liquid is driven by the transpiration stream of the plant.

It should be noted that dissolved elements can move by either diffusion or bulk (mass) flow. Diffusion is the process whereby an element moves due to a gradient in concentration of that element. Bulk flow (or mass flow) is when the liquid in which the material is dissolved carries the element to another location. Diffusion is a relatively slow process so that movement of an element over 1 cm distance can take hours or days (depending on the magnitude of the gradient); bulk flow can carry elements over this distance in a matter of seconds or less. The biggest implication for this is that when the transpiration stream is actively moving water, or when water is applied through irrigation, bulk flow of nutrients and oxygen is carrying these elements towards the root surface.

Of all resources that the plant needs, water is needed in greatest quantities and needs to be plentiful and readily available so as to sustain a transpiration stream that is actively carrying the nutrients needed for plant growth or production to the surface of the roots.

4.1 INTRODUCTION

4.1.1 WATER MOVEMENT IN PLANTS

Plants are able to capture light energy and use it to transform water and CO_2 into a useful chemical form (carbohydrates) through the photosynthetic reactions that occur in the leaves. In addition to light, the plant also requires water and carbon dioxide for this process to take place. While water is mainly absorbed by the plants through its roots, CO_2 enters the plant through pores in the leaves called 'stomata'. These pores must be open for the CO_2 to enter and, as a consequence, water vapour inside the leaves is able to escape into the atmosphere. This movement of water into the atmosphere is called 'transpiration' and this process drives the movement of water from the soil, through the plant, into the atmosphere.

At the leaf surface there is a relatively still, thin layer of air surrounding the leaf, called the 'boundary layer'. Even with wind or air current flowing past the leaf, this boundary layer persists, but becomes thinner as the rate of air movement increases. The movement of water from the stomata through this boundary layer is driven by diffusion (which is relatively slow) and only works if the water molecules are then carried away by air currents (a type of bulk flow which can be relatively fast). If the air surrounding the leaf is saturated with moisture (i.e. RH>95 per cent), then water movement is slow since the diffusion gradients are small.

Most of the evaporation process at the leaves occurs through the stomata. The plant has the ability to control the degree to which the stomata are open and this mechanism allows the plant to exert control over the amount of water evaporated from the leaves. It should be noted that the surfaces of the plant are generally covered with a waxy layer (cuticle) which limits water loss. This water loss is generally negligible compared to the rate of water loss through the stomata. However, young leaves may not have a substantial cuticle and water loss control through these leaves is not as effective. The total rate of water loss through evaporation is slightly less than (but very close to) the rate of water uptake from the root zone. The total water loss from plants is also affected by the number of stomata that are present on the foliage. Thus, the more leaf area the plant has, the greater the potential rate of transpiration.

The potential rate of removal of water by plants from the root zone is driven by the following factors: temperature, light, relative humidity, wind speed, leaf area, and the degree to which the plant's stomata are open. The potential amount of water removal can be calculated by integrating this rate over time. Various irrigation control strategies (described below) use this information as part of the control strategy.

4.1.2 WATER POTENTIAL

The hydraulic forces of water within both plants and substrate play an important role in how water is managed. These forces can generally be expressed in terms of energy potentials such as matric, osmotic, and gravimetric potentials. When dealing with irrigation issues, such potentials are expressed as negative numbers with units of kilopascal (kPa). 'Matric Potential' or 'moisture potential' are synonymous; they have positive values in conditions of positive pressure and negative values for suction. In contrast, we use the term moisture 'tension' as being the absolute force of suction.

As a consequence, 'high tension' in substrates means 'dry conditions'. In relation to the types of values that are typically recorded in soilless production, a relatively dry condition in most substrates corresponds to a moisture tension between 10 and 20 kPa (or a matric potential between −10 and −20 kPa). In general, all irrigation practices that explicitly attempt to avoid water stress in soilless production are confined to the range of 0–8 kPa of tension. In some substrates, such as stone wool, the range is much narrower, with the onset of water stress problems occurring if the production system is allowed to exceed 5 kPa. In contrast, the types of tensions encountered in field soils may well range as high as 75 kPa; in such systems, we rarely see tensions below 10 kPa, except during or just after an irrigation event.

Matric potential results from the interaction of water and solid materials. It can be thought of as the pressure that would need to be exerted to remove water from a substrate. For example, we understand that a sponge can hold water and we need to squeeze it (impose a pressure) to make the sponge release the water. If we could maintain a constant pressure on the sponge, at some point no more water would be released while some would be retained in the sponge. Matric potential is holding the remainder of the water in the sponge against the squeezing pressure.

Gravimetric potential is related to the height of water above some point of reference. This force of gravity translates to approximately 1 kPa per 10 cm of water column (note that the horizontal size of the layer has no influence on this). For example, take a sponge saturated with water, stand it on one end, and allow it to drain. At some point, water will stop draining from it. If the height of the sponge is 10 cm, then the gravimetric potential at the top of the sponge would be 1 kPa.

Something is holding water in that sponge and that force can be measured as matric potential. Since energy needs to be balanced and we know that the gravimetric potential of the 10 cm sponge is 1 kPa, then the matric potential must be −1 kPa so that there is a net of 0 kPa.

Solute or osmotic potential, Ψ_s, is related to salinity (and thus electrical conductivity) and can be described mathematically with the formula:

$$\Psi_s = -RT(n_s/v_w)$$

where R = gas constant, T = temperature, n_s = number of moles of solutes, and v_w = volume of water. Note that the n_s/v_w portion is a concentration, so solute potential is directly related to the concentration of salt in the substrate solution. In the root zone, this can increase through the addition of salts or through the removal of water. This second process can occur in plant systems whenever water is removed from the growing substrate at a greater rate than the solutes within it. The empirical relationship between electrical conductivity (EC, in dS m^{-1}) and osmotic potential (kPa) of

$$\Psi_s = -36EC$$

has been suggested to estimate how salinity, at levels that plant growth will occur, accounts for the contribution to the overall potential (Richards, 1954). So, as the concentration of solutes in a solution and the EC increases, the solute potential decreases.

Understanding the concept of potential energy can provide insight to how management practices affect plant water status. Water will move passively from a state of high

4.1 INTRODUCTION

energy to a relatively lower energy state, and the difference of these two energy states is related to the rate of water movement. So, increasing the gradient will increase the movement of water down that gradient. When an irrigation event occurs, the matric potential of the substrate increases and the water therein is of higher energy. This facilitates water movement into the roots that might be present since the water in them contains solutes and is lower in energy. If the water that is applied happens to contain a large amount of solutes, which reduces its energy potential, the gradient is reduced and the 'ease' that water can move into the roots decreases.

The moisture content of any substrate is also affected by the various physical characteristics of the substrate (particle size distribution, particle surface properties, pore space, etc.), each affecting the various potentials. For any substrate, there is a relationship between the moisture tension and the moisture content. This relationship can be thought of as a 'signature' for a particular substrate.

This curve has various names such as 'moisture release curve', 'moisture retention curve', and so on; the diagram (Fig. 4.1) shows the relationship between moisture tension and moisture content and indicates a container capacity of 75 per cent for the substrate UC mix (see Chap. 3 for other examples of this curve). In addition to the various potentials, there are also forces due to gravity that affect the moisture status of the root zone, and Fig. 4.1 can be used to interpret how a root zone might behave with regard to moisture dynamics by considering the typical root zone as consisting of multiple layers of substrate 'stacked' on top of each other and each affecting the layers above and below, as the water is held in place by internal forces (capillary forces), with the weight of the water in all the layers below a particular layer pulling down on the water in the layers above. Thus the top layers have more weight pulling on them than the bottom layer so that the moisture tension varies with depth. With field

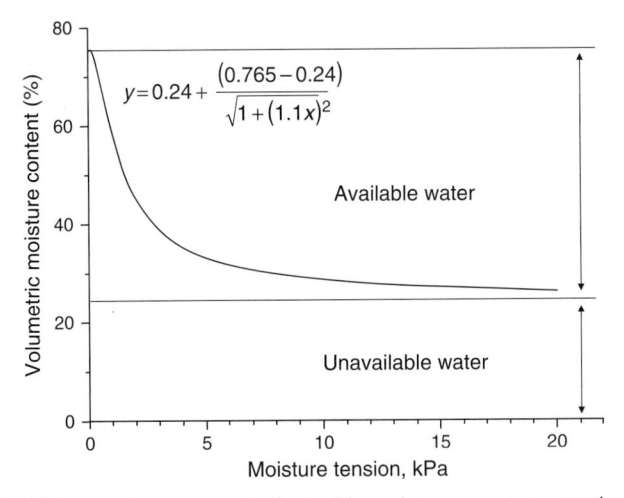

$$y = 0.24 + \frac{(0.765 - 0.24)}{\sqrt{1 + (1.1x)^2}}$$

FIGURE 4.1 Moisture release curve of UC mix. The moisture content at saturation is 76.5 per cent and the unavailable portion is 24 per cent (Kiehl et al., 1992).

soils, these variations are relatively small per cm depth, but with soilless substrates the moisture tension changes dramatically with depth.

With the moisture release curve (Fig. 4.1), it is possible to estimate the water content of such a containerized root zone immediately after an irrigation event. Taking into account the gravimetric potential, when a steady state is reached at the bottom of the root zone of a container with a depth of 20 cm, it will have a moisture tension of 0 kPa while the top of the root zone will be at 2 kPa (= 20 cm depth). Reading off the curve shows that moisture content at the bottom will be 75 per cent, while the substrate at the top will be at about 50 per cent. The differences are even greater as the dry-down progresses. Many growers typically think that the entire root zone is the same, but that is clearly not the case and it can be deceptive since the top layer (which the grower can readily observe) is always drier than the bottom layer.

The moisture release curve also illustrates some general points about dynamics during an irrigation. The wettest condition (i.e. after a thorough irrigation) is represented on the curve at lower tensions. The general shape for most potting substrates is similar, but they do vary enough to affect final set-point recommendations for use in making irrigation decisions. As water is removed from the substrate by the plant (and by evaporation), the status progresses along the curve to higher tensions (lower water content). The pattern is approximately asymptotic to some volumetric water content with increasing high tensions. Most plants will show wilting when they are subjected to moisture content that approaches the levels represented by this line. Thus, in the literature, the water that remains in the substrate at this point is termed 'unavailable water', while the difference between container capacity and this level is called 'available water'. Also, the term 'easily available water' is generally used to refer to water released between 1 and 5 kPa, while the range 5–10 kPa is generally referred to as 'water buffering capacity'. During rewetting, typically a much faster process than dry-down in soilless substrates, the pattern is always similar, even if slightly different (a phenomenon called 'hysteresis'). However, with regard to the precision needed to make irrigation decisions, this aspect can be ignored.

4.1.3 THE ROOT ZONE

Within this book, we use the term 'root zone' to represent the space (generally filled with substrate) which has been made available in a production system to be occupied by plant roots. There may be areas within the root zone where few roots are present because of adverse conditions at various times or because the plant has not yet filled the root zone with roots. In root zones that are designed and assembled well, such 'voids' are present only in early phases of plant production; in poorly designed systems and in systems where irrigation is not controlled well, there may be many such pockets. But even in the most uniform systems, there will be areas with fewer roots. This heterogeneity in root distribution can cause problems.

One factor contributing to this heterogeneity is that root growth is typically geotropic, meaning that they grow downwards. In constrained root systems where the bottom is an impervious layer with access to oxygen through drainage holes, there will generally be more roots there. In the areas in the root zone where many roots are present, water can be removed at a more rapid rate than in areas where fewer roots are present. At the same time, water is removed through evaporation from the parts of the root zone that are exposed directly to the atmosphere. The result is that the removal of water is not uniform throughout the entire root zone.

One important facet in soilless production that directly affects irrigation is a substrate characteristic known as 'hydraulic conductivity'. This refers to the rate at which water can move within the root zone. As discussed elsewhere in this book (Chap. 3), hydraulic conductivity is generally much higher in soilless media than soil-based media. For all substrates, there is a huge difference between the hydraulic conductivity when the substrate is moist, compared to when it is dry. With regard to irrigation schemes, it is the unsaturated hydraulic conductivity (UHC, i.e. the value when moisture is present, but not saturating) that is of particular relevance. The higher the UHC, the faster water will move laterally during, and immediately after, an irrigation event. With drip irrigation systems, this rate dictates the flow rate of the emitters that should be used (if UHC is low, then emitters must be used that have a low flow rate).

It should be noted that if the rate at which water moves in the root zone is slow, then it may be possible for the plants to extract some of the dissolved elements at a rate higher than can be replenished with future waterings. This can result in problems with nutrient deficiency as well as oxygen deficiency. For this reason, soilless production systems always consist of substrates that have higher unsaturated hydraulic conductivity than soil.

In intermittent irrigation approaches there may be the risk of drying out portions of the root zone to the extent that the hydraulic conductivity is so poor that water being applied to the top or bottom of the root zone cannot distribute itself throughout the root zone during the interval between irrigation events. The regions in the root zone that are most prone to drying out are those that have a lot of roots. Thus, when this happens, it becomes very difficult to rewet the entire root zone uniformly and especially those areas where the roots are most prevalent. This can lead to situations where the plant appears to be water stressed despite the fact that the container is heavy with water. In such instances, the plant is responding to the fact that there is little water around the roots and as this is further depleted, the low hydraulic conductivity makes replenishing the substrate surrounding the roots a slow process. This can be particularly problematic if the plants are exposed to high temperature, high light levels, and dry winds, representing a need for urgent action. The best approach to dealing with such an emergency is to apply water to the plant directly to provide water for cooling and increase the humidity to reduce the water loss rate. At the same time, the root zone may need to be irrigated several times within a few hours so as to attempt to elevate the hydraulic conductivity throughout the root zone and to replenish oxygen through bulk flow.

If pockets of the root zone are allowed to dry out, then this can result in root death in those pockets. Also, as the water is removed preferentially from such areas, the concentration of solutes increases to the point where some of them may be toxic to the plant. Historically, the main technique for preventing such problems has been to provide an excess amount of water (i.e. more water than the root zone can hold) and to apply the water slowly so that it has time to distribute itself laterally in the root zone even as it slowly leaches from the bottom of the root zone.

There is a facet to irrigation that is particularly counter-intuitive. There is a widespread knowledge that oxygen starvation can occur and that flooding can cause this, so that some practitioners associate the application of excessive amounts of water with oxygen starvation. Many irrigators are careful to not irrigate too often, but do try to keep the root zone as moist as possible; and it is exactly this scenario that is likely to result in oxygen deficiency. When new water is applied, it carries into the root zone dissolved oxygen and this mass flow of oxygen can resolve the problem very quickly. But if that oxygen is consumed by the plant at a faster rate than water use and this stagnant water is not replaced, then oxygen starvation can occur, leading to poor growth and plant diseases.

4.1.4 WATER QUALITY

As mentioned above, water quality is an important consideration in irrigation, and especially with regard to recirculation. Since this can have a dramatic effect on irrigation practices, it is important to explain some of the details that are particularly important with regard to irrigation.

Pure water consists of just H_2O. There are no water sources in commercial soilless production that are pure. All have some minerals and/or salts dissolved in the water. For the practitioner, it is important to know that materials dissolved in the water do not change the clarity of the water although the colour may change. To the naked eye, salty water appears just as clear as pure water. If the water appears cloudy, then it is said to be turbid and this is due to materials that are suspended but not dissolved in the water. It is very important to understand the difference here, because the particles that cause turbidity can typically be filtered out of the water, while dissolved materials will pass through fine filters unimpeded. Extraction of dissolved materials is accomplished with membranes (which might also be called 'filters' by some).

Water is typically evaluated for EC, pH and alkalinity. EC, or electrical conductivity, is a measure of the amount of salt that is dissolved in the water. There are various salts that could be involved – some are harmful; some may not be toxic, but are undesirable; while others are nutrients that the plant uses as part of its metabolism. Virtually all water sources will contain some of each of these types of materials. As with virtually all environmental variables, there is a response curve for each such material for any plant.

Such a curve generally has a range where the plant will have inadequate amounts (deficiencies), a range where the plant can tolerate the material, and a range of concentrations where the plant will respond in a negative way, possibly showing foliar damage (Fig. 4.2).

EC, generally measured in units of $dS\ m^{-1}$, is typically seen in soilless production from 0 to $5\ dS\ m^{-1}$. Historically, growers have been advised to keep the EC below $3\ dS\ m^{-1}$ to assure rapid growth of plants. Clearly this is impossible if the source water

4.1 INTRODUCTION

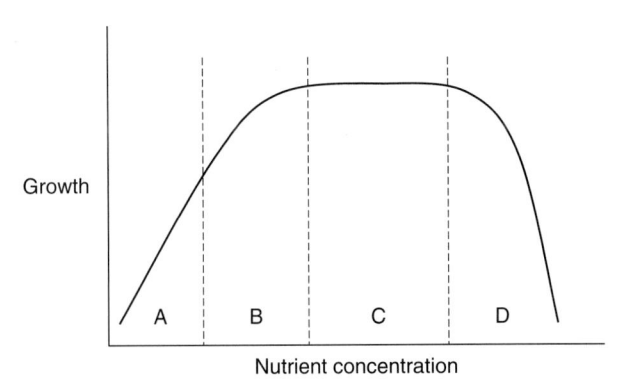

FIGURE 4.2 A nutrient response curve. At low concentrations, small increases in availability results in large changes in growth (A). Further increases in nutrient concentrations have smaller affects as nutrient levels approach optimal levels (B). At some point, additional amounts do not increase growth. This is the range of luxury consumption (C). At high levels, toxicity is reached and growth diminishes. Eventually, plant death occurs (D) (adapted from Nelson (2003). *Greenhouse Operation and Management* (6th edn). Upper Saddle River, NJ: Prentice Hall.

already is high in dissolved salts and the addition of nutrients will raise the EC to greater than $3\,dS\,m^{-1}$. If the source water has too high an EC, then the only options the grower has is to remove some of the salts, blend with low-EC water such as stored rain water, or to grow plants that can tolerate the elevated salt content. It should be noted that with development of complex hydroponic systems that include recirculation, it has become apparent that higher levels of EC might be feasible (see Chap. 9).

pH is a measure of the hydrogen ion concentration, describing the extent that the fluid is an acid or a base. Water is said to be 'neutral' if it has a pH of 7. For soilless production, however, this term is deceptive because for virtually all crops in such production systems, water that has a pH significantly lower than 7.0 is necessary. In soilless production, a range of irrigation solution pH of between 5.6 and 6.8 is required. For some crops, even lower pH levels have been found to be ideal. In general, pH levels of the solution around the roots should not be below 5 or above 7.

One of the most important aspects of water quality is that water containing extensive dissolved minerals that may build up in the root zone. This is due to the fact that plants can preferentially remove water over dissolved minerals. Thus as the water is extracted from the root zone, they become more concentrated as water is depleted. At some point, the concentration may reach a point where the material comes out of solution as mineral deposits. This condition should be avoided.

Alkalinity is a measure of the irrigation water to raise the pH of the root zone solution. This is measured in units of milliequivalents (meq) or ppm of $CaCO_3$. In soilless production, low alkalinity (<3 meq) is generally acceptable, while a level greater than 8 meq is so problematic for growers that such water is not suitable for crop production. When the source water has alkalinity below 8 meq, it can be used if treated with acid injection.

4.2 ROOT ZONE MOISTURE DYNAMICS

The moisture dynamics in the root zone during irrigation are typically much faster than during the time between irrigation events. There are significant complexities related to this that affect irrigation decisions.

4.2.1 DURING AN IRRIGATION EVENT

The irrigation event is generally triggered by a human or automated (pre-programmed) decision-making process. The onset of the event is characterized by the opening of a valve which leads to water or dilute nutrient solution, flowing under pressure. An irrigation event lasts a particular period of time (typically minutes) and the end of the event is specifically controlled to achieve a desired effect. Usually the objective is to bring the root zone up to full capacity, although deviations from this mission are possible and discussed below as part of irrigation decisions.

While the focus during irrigation is generally on the delivery of water, dissolved substances and gasses are also provided to the plant. These other substances may be present because they were part of the source water or because they were purposely added to the irrigation water as part of a crop production strategy.

It is also noteworthy that during the time of an irrigation event, the spatial distribution of water in the root zone is heterogeneous and dynamic (changing from second to second). Water moves downwards through the substrate fast (in seconds) and slow laterally (in minutes or hours). As the water enters the area surrounding the roots, it carries with it substances that were dissolved in the irrigation solution and substances that were already in the root zone. Thus the bulk movement of water moves materials to the root surface.

4.2.2 BETWEEN IRRIGATION EVENTS

Immediately after an irrigation event where the volume of water applied exceeds the amount needed to achieve container capacity, there is a period of time where water continues to leach from the root zone. The leachate has to be replaced with air, which is pulled into the root zone at any point where the substrate is exposed to air. This action is generally overlooked, but it should be noted that air entering this way consists of over 21 per cent oxygen, so that this can account for a substantial amount of oxygen.

Once the leaching has stopped, there is a period of gradual decline in water content due to evapotranspiration: water being removed by the plant (transpiration) plus evaporation of water from the substrate directly to the air. It should be noted that water moves down an energy gradient. A number of variables affect this gradient between the leaf and the atmosphere and, consequently, the rate at which this removal occurs.

Nutrient uptake by the roots may be at rates that are not the same as the rate of water uptake. Thus nutrients could be depleted more rapidly or less rapidly. In the latter case, this could lead to a concentrating effect which can manifest itself as an increase in salinity. An important part of irrigation management is salinity control, and between irrigations it is very likely that salinity is increasing in the liquid that surrounds the roots. There may well be times when an irrigation event is needed to flush out salts that have accumulated in this way.

Sensors that monitor soil moisture may also be able to monitor EC. Time domain reflectometry (TDR) may be able to measure EC in substrates, but these sensors are affected by organic matter content. In addition, when EC is high, the TDR signal is also affected so that these measurements are difficult to make. Decagon Devices (Pullman, Washington) released a sensor in 2006 that measures substrate moisture, temperature, and EC. Oki and Lieth (2004) used a suction lysimeter with an embedded EC cell to measure substrate solution EC.

4.2.3 PRIOR TO AN IRRIGATION EVENT

One of the most notable characteristics evidenced by moisture release curves of soilless substrates is that the dry-down process is not linear with regard to moisture tension and moisture content. Thus as water is depleted past some threshold (5 kPa for UC mix, Fig. 4.1), the tensions increase more and more rapidly with each unit of water depleted. This acceleration forces irrigation managers to schedule irrigations well in advance of critical high-tension conditions.

If the plant is not able to adequately control water loss as the moisture tension rises towards a critical point, then the plant will begin to dehydrate. This typically results in a loss of turgor pressure in the leaves, resulting in the visible symptoms of wilting leaves and flagging stem tips. Within a very short time, the leaves loose the ability to cool themselves since the amount of water evaporating from them is inadequate to meet the cooling demand of the tissue. If this persists, then the temperature of the leaves will rise in the presence of high light, possibly resulting in temperatures high enough to cause foliage damage. Plants have the capacity to adapt somewhat to this. If a plant experiences such high tensions frequently, then it adapts to close stomata at lower moisture tensions so as to avoid wilting.

Effective irrigation practices are needed to prevent this from happening. For CO_2 to enter the leaves, the stomata must be open. As a consequence, however, water also passes through these openings. So, irrigation management that keeps stomata open will also maximize productivity. Closed stomata lead to reduced photosynthesis and thus to lower biomass production.

It is important to note that other deleterious processes are occurring before wilting is observed. Leaf and stem expansion are among the processes most sensitive to water stress (Hsiao, 1990). Maximizing turgor pressure within young cells is necessary to maximize cell enlargement and stem and leaf growth. Water stress at very low levels can reduce plant growth. And with less water moving in the transpiration stream of the plant, there are also fewer mineral nutrients moving within the plant to the sites where they are needed. In addition, there is less water available for the essential metabolic processes within the plant, resulting in suboptimal growth.

While the same phenomena and processes occur in soil-based and soilless systems, there are some significant differences. In general, plants growing in soil-based systems are subjected to slower movement of water in the root zone, so that the plant has to grow much larger root systems to mine a larger soil volume for water in order to meet its transpiration demand. The smaller root systems in soilless systems require greater attention by the irrigation manager as the plant has less protection from catastrophic irrigation system failure. Also, with smaller root systems, the plant may be less capable of responding to climatic extremes that increase water demand.

4.3 IRRIGATION OBJECTIVES AND DESIGN CHARACTERISTICS

An optimally designed irrigation system will deliver water to the plants to maximize water use efficiency (WUE). This means that water is provided in time just as the plants require it and in a manner so that all the delivered water is utilized by the plants and none is wasted. This requires that the delivery and control systems are properly designed for the application that they are to service.

Irrigation delivery systems are always designed to optimize various specific characteristics. This includes system capacity as well as system uniformity. The degree to which these characteristics are optimized has a greater impact on how irrigation is controlled rather than on the water utilization of the crop.

4.3.1 CAPACITY

The capacity of an irrigation system relates to the number of plants that can be irrigated at any one point in time. This is governed by the maximum rate that water can be extracted from its source. In some cases, production facilities have to install tanks to store water, and the capacity of these tanks and the dynamics of how they are filled and emptied governs the maximum rate of water use during production. For example, a grower needs 16 000 L/day to maintain the crop. The well that is available can produce only 1000 L/hr. If an 8000 L water tank is available, it can be filled when there is no water demand, for example at night. Both the well and the storage tank will be accessed to provide water for the crop during the day.

Irrigation systems should group plants so that this source capacity is distributed over time. As such, each irrigation circuit is separate and needs to have its own decision-making. Most large soilless production operations have considerably more plants than could be watered with new source water all at once. Each circuit must be queued (scheduled) for irrigation and care must be taken to only irrigate as many circuits at any one time as can be supplied by the fertigation equipment (pumps and injectors).

4.3.2 UNIFORMITY

The grouping of plants is generally done so as to combine plants of similar water and nutrient use patterns into the same irrigation circuit. Typically there is an expectation by the grower that the entire crop stays uniform in water and nutrient use and that each plant reaches the production goal at the same date. Without this temporal and spatial uniformity, the grower may face economic problems. Since the availability of water and nutrients directly impacts plant growth, the need for this uniformity dictates that the irrigation system delivers water in as uniform a manner as possible. This means that within an irrigation circuit, water is delivered to each plant at nearly the same rate to provide nearly the same amount of water and nutrients as possible.

Absolute uniformity is generally not economically feasible for any crop (Seginer, 1987). Crop response to water, cost of the irrigation solution, type of irrigation system, all affect the level of practicability of taking special measures to achieve greater uniformity (e.g. pressure compensation at each emitter, re-spacing plants as they grow, etc.). For any situation, there is a level of uniformity that will result in the greatest level of economic feasibility (Chen and Wallender, 1984; Seginer, 1987).

Precise evaluation of irrigation system uniformity is difficult. There are a number of sources of variation that can be considered and measured. For any one irrigation system, the distribution of water from an irrigation system after the system is fully pressurized is typically different from the distribution during the time that it is pressurizing. While many practitioners consider the latter to be insignificantly short, in practice, it is almost always a significant percentage of the duration of the irrigation event.

Uniformity in an irrigation system is generally measured by sampling within the system. This can be done by placing a container at each emitter to collect the water that is delivered for the duration of an irrigation event. The water collected can be measured to determine the amount of water delivered and delivery rate to each emitter. Uniformity has been quantified in two differing approaches: (1) by summing the absolute difference of each rate from the mean and (2) by calculating variances or standard deviations. Both methods are equally effective for optimization purposes (Seginer, 1987), although the first of these is somewhat easier for practitioners, while the second is easier to implement in mathematical treatments of the problem. Also, if the uniformity measure is used within economic analyses, then it is important to note that the use of variance tends to give disproportionately greater weight to larger deviations.

Thus for a specific set of data collected as above, a 'Coefficient of Uniformity' (U) can be calculated as a percentage value:

$$U = 100 \times [1.0 - \text{sum}(|x_{mean} - x_i|)/(n \times x_{mean})]$$

where x_i is the individual amount of water collected at a particular location, x_{mean} is the mean of these values, and n is the total number of these samples (Lieth and Burger, 1993). The alternative would be

$$U = 100 \times \{1.0 - [\text{sum}((x_{mean} - x_i)^2)]^{1/2}/(n \times x_{mean})\}$$

Other methods have also been presented, but, as Seginer (1987) points out, the high degree of correlation between the methods suggests that the choice of method is not particularly important. Seginer concludes that while both approaches are useful under various conditions, the second of these two methods should generally be used.

The duration of a uniformity test can have an effect on the outcome. Water delivery to the plants close to the irrigation valve begins very quickly after the valve has opened, while the emitter farthest away from the valve begins delivering water later and at a lower rate due to pressure loss over the length of the flow of water. Thus, while conducting such tests, it is important to operate the system for durations that are similar to the durations that will be used in practice. Some emitters are pressure compensated and will minimize or eliminate fluctuations in flow rate due to pressure variations.

It should be noted that subirrigation systems are difficult, if not impossible, to assess for uniformity. Some practitioners have the notion that subirrigation systems are highly uniform, but this is not necessarily true. Subirrigation systems must be completely level to be absolutely uniform. At the same time, the ability to drain the system in a timely manner is very important to prevent the stagnation problems described above. Ebb-and-flood tables and trays with channels moulded into the bottoms to facilitate water flow and distribution can achieve this in a way that still allows the system to be filled uniformly and drained fairly rapidly. Other subirrigation systems (troughs, flooded floors, and capillary mats) typically are not uniform since the bases of the plants sit at different levels along a drainage gradient. The troughs and flooded floors need to be sloped to provide drainage. Capillary mats installed on long, sloping benches can be problematic. If the height difference from one end of the bench to the other is large, it may create a pressure head in the mat adequate to suck nearly all water out of the root zones of plants sitting at the highest part of the bench.

The only type of system that has the potential to be completely uniform is liquid culture where the roots are always submerged or wetted and have access to water at very low tensions at all times.

4.4 IRRIGATION DELIVERY SYSTEMS

All irrigation systems consist of tubing and/or pipe to transport the irrigation solution from the source to the individual plants or to a group of plants. Generally the irrigation solution is prepared from water which is under pressure, and injectors are used to inject various soluble fertilizers into the water. Mixing or blending tanks assure that the dissolved materials are distributed uniformly within the water and filters remove any insoluble materials. Pumps and pressure tanks may be used to increase or stabilize the pressure at which the solution arrives at the valve to the irrigation circuit. More detail on such systems and their design can be found in Chap. 5.

In all irrigation systems, the flow of water requires some form of pressurization to facilitate moving the water. The higher the pressure, the more rapidly the liquid can flow in the system. Friction within the pipes imposes resistance to flow, causing the pressure to drop over the length of the pipe. The diameter of the pipe has a significant effect on this phenomenon, with pipes of smaller diameter having greater pressure drop along the length of pipe. Tables are available that indicate the pressure drop for various lengths of pipes of various diameters and materials. It is important to properly design irrigation systems so that all of the parts (main lines, valves, pressure regulators,

4.4 IRRIGATION DELIVERY SYSTEMS

laterals, and emitters) are correctly sized to deliver water at the desired pressure and flow rate.

When a valve is opened in a system where the irrigation water is under pressure, this allows the irrigation solution to flow. While there is active flow in the system, there will be a pressure gradient with the highest pressure at the source and the lowest pressure values at the emitters. Any one irrigation circuit must be designed so that this pressure at each emitter is above the minimum specified value for that emitter. If too many emitters are included, then some emitters experience inadequate pressure and consequently deliver little or no water, while other emitters (closer to the valve) experience higher pressure and thus deliver water at a higher rate. If the total potential rate of flow out of all emitters is less than the capacity of the system, pressure regulators, or pressure compensating emitters, can be used to assure that each plant receives water as designed. But if the total rate of water delivery through all emitters in the circuit is greater than that the source can provide, then there is no way that this problem can be solved without increasing the source pressure or installing pipes of larger diameter to increase the flow rate within the system. Under such a circumstance, replacing all of the emitters with ones that have a lower delivery rate (e.g. slower drip) would be a solution. Obviously, the duration of operation of the irrigation circuit would have to increase to assure that each plant receives the appropriate amount of irrigation solution. Another solution would be to reduce the number of emitters serviced by a valve.

In commercial soilless production, one of the most important crop production factors is uniformity. This is particularly important if the entire crop needs to be shipped (harvested) at once. Since the availability of water and fertilizer is a major factor for growth, variations in available water and nutrients will result in variation in the size and quality of the plants. For this reason, the uniform delivery of irrigation water is important in many such systems. Minimizing the pressure drop within the system, and especially the variation in pressure among the emitters, maximizes irrigation uniformity.

The most common errors in commercial production consist of having irrigation circuits too large (attempting to irrigate too many plants through one control valve) and minimizing the plumbing costs by using the smallest diameter pipe that is feasible. Generally, this is an economic decision designed to minimize installation costs. However, a price will be paid during the years of operation. If each circuit is large, then it may be impossible to irrigate all plants during a hot dry day because of inadequate flow rates. By having many smaller circuits, it is possible to make better use of the available pumping capacity, while providing better uniformity.

Various types of irrigation systems are in use in soilless production in greenhouses or nurseries. These include various approaches to delivering the water to the plant: overhead systems (which apply irrigation solution 'over the top' of the foliage), drip or microspray systems that apply water to the substrate surface (which keep the foliage from getting wet), and subsurface systems (which apply water to the base of the root zone and rely on capillary action or flooding to bring water into the root zone from below).

4.4.1 OVERHEAD SYSTEMS

Overhead irrigation systems (Fig. 4.3) deliver water to the space above and surrounding the foliage of the plants with the expectation that this water will disperse through the canopy of the plant to the top of the root zone. While some water will evaporate in the process, effecting some cooling, much will penetrate the canopy and end up at the top of the root zone. If the plants are grown in pots or other containers spaced on the ground or bench, these systems are very inefficient and have a low WUE. Most of the water applied will not reach the substrate surface, wetting the bench, floor, or ground, and will become unused run-off.

Overhead irrigation systems are widely used in commercial nurseries because they tend to be relatively inexpensive to install and can irrigate large areas. However, if materials are dissolved in the water (fertilizer), then this method can result in unsightly residues on flowers and leaves. This is generally only a problem for ornamental plants and many times such residues can be washed off with clear water irrigation or with rain. Water on the foliage may also cause disease problems if water droplets stay long enough to allow germination and growth of fungi and bacteria. Such situations are

FIGURE 4.3 Overhead irrigation systems: (A) Hand watering, (B) Boom system, (C) Mist system, (D) Sprinkler system in an outdoor nursery (see also Plate 4).

most likely to occur when relative humidity is high (e.g. at night or during rainy or foggy weather).

While it is frequently undesirable to wet the foliage, overhead irrigation can be used to cool plants directly on hot days by wetting the plants and allowing the water to evaporate. This type of water application is called 'syringing' and is generally done with clear water and scheduled to allow the foliage to dry completely before the night so as to avoid disease problems.

There are several specific types of overhead irrigation systems that are of use in soilless plant production.

4.4.1.1 Sprinkler Systems

Overhead sprinkler irrigation (Fig. 4.3D) is one of the least expensive water distribution method since each emitter serves many plants. With potted plants, the major drawback to this type of irrigation is the large quantity of wasted irrigation solution due to water that misses the root zone. Also, making such systems uniform requires that the circular distribution patterns of each sprinkler head be overlapped extensively; even with that precaution in the design, it is very difficult to create a uniform distribution pattern. Thus in such systems, extensive over-irrigation of many plants is needed to assure that those plants in the locations receiving the least amount of water get an adequate amount of irrigation solution.

Sprinkler heads may throw a stream of water while rotating in a circular pattern (e.g. gear driven, or impulse sprinklers) or spray water in all directions simultaneously (spray heads) through a particular nozzle design. They are available in full-circle or partial-circle patterns and at various rates of water delivery. Sprinkler systems lend themselves to automated control of irrigation in potted outdoor nursery production.

4.4.1.2 Mist and Fog Systems

While mist and fog systems also dispense water into the air surrounding the plants, these systems are primarily cooling or humidification systems and are generally not designed to deliver water to the root zone. On propagation benches where cuttings are rooting, mist systems can provide adequate water to meet the plants' irrigation needs, but this is largely due to the fact that the transpiration rate is significantly reduced and care needs to be taken to supplement the water to the newly forming roots if the mist system does not deliver enough water to wet the root zone. Mist systems (Fig. 4.3C) generally produce smaller water droplets than sprinkler systems. Their objective is to wet the foliage, either for cooling the crop or to prevent or reduce transpiration (as in the rooting of cuttings). Fog systems are also not irrigation systems. The water droplets are even smaller than with mist and the objective of a fog system is to be a cooling or humidification system by evaporating water droplets in the air before they reach the plants. Both fog and mist systems must be controlled very precisely to avoid disease and algae problems. They must be used in concert with ventilation to remove the moist air before more water can be dispensed.

Controlling irrigation in greenhouses with mist systems requires more care since it is possible for the grower or installed sensors to be deceived by water in the air, on the foliage, and on the root zone surface. In this setting, sensor-based irrigation that measures the moisture in the root zone is very useful to determine when the root zone is becoming too dry since this can easily happen when everything that can be seen is completely wet. It should be noted that sensor inputs into model-based irrigation control systems may also be deceived by the plethora of water surrounding the plants. Properly managed fog systems should not result in wet surfaces.

4.4.1.3 Boom Systems

A boom system (Fig. 4.3B) consists of a machine which travels above the plants on rails, dispensing irrigation solution to the plants below through nozzles that are mounted on a boom suspended over the plants. In protected cultivation, the travel of such a device is always linear. In some cases, the nozzles generate a spray similar to a sprinkler system, while in other cases, water is allowed to flow in a stream to plants below. In the latter case, the plants need to be positioned very precisely under the path of the boom and the substrate needs to have adequate bulk density and consistency so that the water stream does not wash it out of the pot or knock plants over.

Some boom systems can be programmed to modify the delivery rate of the irrigation solution along the path and to even skip over sections. In many nurseries, boom systems are installed so that each section of plants, or perhaps an entire greenhouse consisting of many benches of plants, has a dedicated boom. Such systems can operate fully automated.

Control of such systems commonly is through manual operation, especially if intervention is needed to start and stop the irrigation. The duration of the irrigation event for each plant is dictated by the travel speed of the boom.

Boom systems allow for uniform overhead irrigation and customized delivery to each plant. The main drawback is the installation expense and (perhaps) the need for manual intervention (labour costs). In highly automated (and capital intensive) moving-tray systems, it is possible to mount one boom permanently and have the trays with the plants move underneath.

4.4.2 SURFACE SYSTEMS

Surface systems (Fig. 4.4) come in a wide range of schemes, delivering water directly to the top surface of the root zone with minimal wetting of the foliage.

4.4.2.1 Hand-Watering

Irrigating by hand with a hose and wand affords the possibility of customizing the delivery of water to the top of the plant (Fig. 4.3A) or to just the root zone of each plant. Some sort of diffuser is generally used to prevent washing substrate out of the pot and to disperse the water over the entire top surface of the root zone. Potted plants should have adequate headspace at the top of the root zone since enough water needs to be placed there quickly to allow replenishment of a considerable amount of readily available water (perhaps as much as 50 per cent of the container volume).

FIGURE 4.4 Surface irrigation systems: (A) Spaghetti system, (B) Micro-spray emitter, (C) Drip irrigation monitoring, (D) Perimeter micro-spray irrigation, (E) Drip tape and (F) Custom-spaced in-line emitters (see also Plate 5).

Hand watering also dictates that the infiltration rate must be fast enough to accept water quickly while the pot is being watered, but not so fast as to have the applied water run through the pot without being retained.

Hand-watered plants should always be watered to bring each pot to container capacity. Partial irrigation can cause a portion of the root zone to dry out. With most container mixes, it is difficult to know whether the substrate has become completely rewetted. Water draining out of the bottom of a pot is not necessarily a good indicator as to whether the substrate has been uniformly rewetted to container capacity, since in porous substrates, water can rush (channel) through the root-zone and not be uniformly adsorbed by the substrate.

Although hand watering can be very effective, this is entirely dependent on the competency of the irrigator and very experienced and knowledgeable ones are rare. More typically hand watering is the most inefficient method of irrigation. The applications are usually made based on the schedule of the irrigator and not necessarily when it is best for the plants. The volumes of water applied and application rates are usually much greater and faster than the substrate can accept the water. Water continues to flow between containers, so it is applied where it is not effective. This results in a very low WUE and a significant amount of runoff containing nutrients.

4.4.2.2 Drip Irrigation Systems

Drip irrigation systems (Fig. 4.4A) deliver a spray, drip, or slow flow of irrigation solution directly to the top of the root zone. While some of these types of systems do not technically 'drip', they all have water flowing from an emitter at a relatively slow rate.

Drip systems deliver water directly to the base of each individual plant so slowly as to allow water to move laterally in the root zone before water starts coming out the bottom of the root zone. Water that is applied too fast will channel straight through the substrate and out the drainage holes, where it is lost from the root zone. Other ways to facilitate getting water to all parts of the root zone include the use of more emitters per plant or pot or to use emitters which deliver a spray of water (Fig. 4.4B).

Drip systems should always be equipped with filters to prevent clogging and pressure-regulators to assure proper pressure (25–40 psi). Fluctuations in pressure can result in erratic delivery volumes and poor system uniformity. Also, excessive pressure (e.g. during surges when valves are opened or closed or due to water hammer at the end of an irrigation cycle) can blow out or destroy emitters.

Since this type of irrigation places an emitter at each plant, considerably more supply pipe and/or tubing is needed than with other irrigation systems. Near the plants, this pipe is generally made of a soft plastic (typically polyethylene) to allow emitters to be attached directly to them or with drip tubes.

Tube or 'spaghetti' systems (Fig. 4.4A) are those which use smaller diameter tubes that are connected to laterals to deliver water to each plant. Spaghetti systems can be used with or without emitters. In fact, some emitters actually function only as weights attached at the end of a tube. The tube itself has a small inner diameter, slowing the flow of water. Each tube must be exactly the same length since the flow rate from the tube varies with its length. Another function of the weight is to diffuse the water emerging from the tube. Some such emitters have also been developed which allow the tube to be shut off when not in use.

In a spaghetti system, the rate of water delivered to each pot will depend on the distance of the end of each tube from the main valve with the further away from the valve, the lower the delivery rate at the emitter. Furthermore, this rate will vary with pressure fluctuation. Pressure-compensating emitters or emitters with small diameter orifices can be used to make such systems more uniform. However, the use of pressure-compensating emitters is necessary to maximize uniformity and WUE.

Drip systems allow the highest degree of precision and uniformity but the substantial acquisition costs, installation labour, and maintenance costs mean that they may not be feasible for smaller pots. Drip systems can easily be monitored by measuring the performance of a particular emitter at a representative position within the irrigation circuit (Fig. 4.4C)

In-line systems have emitters embedded at regular distances directly in tubing. These can either be in the form of drip tape (Fig. 4.4E), tubing that has emitters embedded at regular intervals, or tubing to which emitters are attached directly (Fig. 4.4F). Such a system is precise in how it places water near the plants, but it reduces the flexibility a grower has in spacing options. For example, pipe which has drippers built in at specific intervals dictates the spacing for the crop.

Drip irrigation systems are also used in conjunction with bag or slab crop production systems using various materials such as stone wool. In such production systems, the irrigation system delivers a slow drip to the substrate and this water moves relatively quickly through the substrate to the base of the bag, pot, or tray. Holes or slits cut into the base of the container allow water to drain, but some puddling inside the container is unavoidable (and by design) assuring that the substrate does not dry out as long as irrigation occurs frequently. Irrigation control in such systems is typically by timer, but it is also possible to have one bag mounted in a tray that is designed to trigger irrigation whenever the water content in this tray drops below a certain level.

4.4.2.3 Perimeter and Microspray

Similar to overhead sprinkler systems, irrigation systems can also be constructed to deliver a spray directly to the base of the plants (Fig. 4.4D). Such systems are sometimes characterized as drip system due to the relatively slow rate of deliver of water. In cut flower production, this type of system generally is installed around the perimeter of a bench of plants growing in a raised bed of substrate or amended soil.

4.4.3 SUBSURFACE

Subsurface systems (Fig. 4.5) bring irrigation solution into the root zone from below. These systems include capillary mats (Fig. 4.5A), troughs (Fig. 4.5B), flooded trays, benches (Fig. 4.5C), and floors (Fig. 4.5D). When irrigating with these, the bottom of the substrate of the root zone is put in contact with water, and capillary action in the growing medium carries the water into and throughout the root-zone. Substrates that have poor unsaturated hydraulic conductivity will not distribute water adequately above the level to which the base of the root zone is flooded. Thus with subirrigation, it is imperative that (1) root zones cannot be allowed to dry out to any extent and (2) only shallow root zones (<10 cm) can be used unless the substrate is specifically designed to have high water-holding capacity at the top of pots of greater depth.

While there are advantages to using subirrigation, all such methods share a drawback: there is a tendency for salts to build up in the upper portion of the root zone. This occurs because irrigation solution (containing water and fertilizer salts) enters at the bottom. While some is taken up by the plant, water evaporating from the soil surface will leave salts behind, concentrating them at that surface. Monitoring of salt concentrations and leaching by occasional irrigation from above is essential with subsurface irrigation, unless the source water has very low EC and the crop is grown for a relatively short time.

4.4.3.1 Capillary Mats

Capillary action can be used by setting the pots on a wet fibrous mat and allowing irrigation solution applied to the mat to be absorbed into the bottom of the pot

FIGURE 4.5 Subirrigation Systems: (A) Capillary mat, (B) Trough system, (C) Flooded tray or table, (D) Flooded floor system (see also Plate 6).

(Fig. 4.5A). The mat must be on a level, waterproof surface and the base of the pots must have holes so that the potting medium can come into contact with the mat. The basic idea is to saturate the mat with irrigation solution and to keep the rooting medium in every pot at the same moisture tension (and thus the same moisture content).

Improper levelling can result in problem with capillary mat systems. If the bench has just a 10 cm drop over the length, then the plants at the lower level will have moisture tensions 1 kPa lower than those of higher levels. For typical container media (e.g. represented in Fig. 4.1), this can result in substantial differences in water content along the length of the bench, resulting in lack of uniformity.

Mats are available with a layer of perforated plastic film which is designed to inhibit evaporation and algae growth and to reduce the incidence of roots growing into the mat.

The edges of the mat should not be draped over the edge of the bench, since the wicking action will cause water to be drawn off the mat and lost as run-off.

Three sides of the perimeter of the bench can have the edge of the mat and water-tight plastic under the mat slightly elevated (1/2 to 1″), so that all run-off occurs at the end opposite to the water supply (Fig. 4.5A). If water is applied to the mat slowly, then the water movement across the bench will be through the mat and run-off from the bench will be slow until the mat is saturated. If water is applied rapidly, then water will travel through, as well as on top of, the mat. In closed systems, the drained water is recycled and hence can be applied very frequently. In this case, fast application is possible. If the edges of the mat are not elevated slightly, then water may run off the bench, near the source, long before the remainder of the mat is saturated. Some growers also elevate the edge on the fourth side during the irrigation, and then raise the water level above the surface of the mat. Once all plants have absorbed water, the excess water is allowed to drain off by lowering one edge or folding down a flap of the mat.

A manufactured mat product is available that integrates tubing within the fibrous portion between two plastic layers. The tubing is connected to the laterals of an irrigation system and facilitates the distribution of water throughout the mat.

4.4.3.2 Troughs

Irrigation troughs consist of flat-bottomed metal or plastic channels, mounted at a slight slope. Pots are set into such channels or the troughs can be filled with substrate (Fig. 4.5B). Irrigation solution is applied at the elevated end and allowed to run slowly down the length of the trough. Warping or twisting of the trough results in poor distribution causing water to miss pots. While it may be tempting to line the trough with matting material, this will essentially convert the trough to a capillary mat system. The configuration of the pots within the trough, particularly the location of holes in the base of the pot, is very important since water has to migrate to these holes. Water which reaches these holes is drawn into the pots via capillary action. During an irrigation event, water runs in the channel continuously, much of it may miss pots entirely. Unless this water is recycled, it represents a large amount of wasted water.

4.4.3.3 Flooded Trays or Benches

Flooded trays or benches (also called 'ebb-and-flood', 'ebb-and-flow', or 'flood-flow' systems) are designed to be flooded with irrigation solution for 5–20 min, submerging the bottom of the pots. Each of the pots should be submerged to the same depth and for the same length of time to achieve maximum uniformity. Modern trays are equipped with grooves (Fig. 4.5C) to facilitate drainage and to move water rapidly to all parts of the tray at the start of an irrigation event.

A typical irrigation cycle starts by flooding the bench or tray. The duration of this phase is dependent on the rate at which water comes onto the tray and the size of the tray. Care should be taken that water coming onto the tray does not push pots over. The water level in the tray is raised so that the bottom 2–5 cm of each pot is submerged. The optimal level would be one where the holes in base of all of the pots are submerged simultaneously at the start and the water level is maintained to keep the pots in water throughout the flooding period. The depth to which the water must be raised is also dictated by the hydraulic conductivity of the substrate. The lower the unsaturated hydraulic conductivity, the deeper the water must be and the longer the duration that the plant must be allowed to stand in this water. The duration needs to be long enough to allow water to be drawn up into the top portions of the root zone by capillary action. It should be noted that during this time a significant portion of the root zone is above container capacity. As the tray is drained, the excess water in the pots also leaves the pots. This can have significant consequences if some pots are infected with water-borne pathogens since these pathogens are then distributed to the table or tray and subsequently delivered to all nearby pots with the next irrigation. The duration should be minimized to avoid predisposing the roots to disease infection.

Usually there will be a lot of water which needs to be drained from such a system. This water can be either recycled in a separate recycling system or pumped directly onto the next bench. There is, however, a risk that pathogens from diseased plants, or from plant material lying on the bench, may be transmitted from bench to bench if the irrigation solution is not disinfected between applications. This also means that growers must typically be more diligent about roguing infected plants and proper sanitation of benches.

4.4.3.4 Flooded Floors

Subirrigation can be carried out on an even larger scale by lining large sections of the floor with concrete or plastic, contoured so as to allow all water to drain into a central drain (Fig. 4.5D). Water supply lines and drainage pipe are incorporated directly into the concrete. Low walls (curbing or rubber bumpers) are used to section the system into functional circuits. Operation is the same as with flooded tables.

Proper design and construction is critical since even small improper variations in grade can have extensive consequences that are virtually impossible to repair. Flooded floors are on a slope to allow drainage. Incorporation of sloped channels into a perfectly level pad (i.e. akin to trays) would be prohibitively expensive and/or weaken the concrete pad to where it could easily crack. Also, since the surface is shaped by human hands, the concrete is likely to have some areas where shallow will form. Drainage grooves need to be cut to drain these. Otherwise the plants in these areas will be at saturation much longer than others and the continual moisture can cause disease problems.

Another potential problem is that on most soils it is very difficult to pour a large pad of concrete without some cracking occurring in the concrete some time after it has been put into use. It is extremely important during the design phase to verify how large a pad can be poured so that the floor will not crack later. Once a crack has occurred, allowing water to seep through it will exacerbate the situation. Driving forklifts or other heavy equipment onto the pad may also contribute to cracking. Obviously, flooded floors should be designed by an experienced engineer and installed by an experienced concrete contractor, only after the soil has been tested and properly prepared, so that the concrete pad will not be susceptible to cracking.

Sanitation on flooded floors is very important, since workers can carry pathogens on the soles of their shoes directly into the irrigation system. Once disease symptoms appear in one area, it is likely that workers walking on the floor will already have transmitted the problem throughout the nursery.

4.5 IRRIGATION SYSTEM CONTROL METHODS

For all irrigation systems, decisions must be made when to irrigate and for how long. With regard to irrigation, we refer to irrigation 'events' and 'schemes'. An irrigation event is a single application of water or nutrient solution. With soilless growing systems, an irrigation event lasts a period of time, ranging from a few seconds to several minutes (rarely longer than one hour unless the irrigation scheme is continuous). Irrigation schemes are management strategies developed to attain specific crop production goals utilizing various delivery and monitoring methods and refer to the overall plan of managing the irrigation water for the duration of the crop.

4.5.1 OCCASIONAL IRRIGATION

Conventional irrigation schemes can best be termed 'occasional' irrigation. Many times in the horticulture literature, irrigation is described as being on an 'as needed' basis. This vague terminology refers to the use of occasional irrigation such that the root zone does not dry out to an extent that might result in 'water stress'. In this approach, the objective is to deliver water at intervals dictated by the plants' removal of water from the root zone, but applications usually occur based on the timing of the observations of the irrigation manager and secondarily on the plant's need.

In general, this involves an interval of time between irrigation events that ranges from a few hours to several days. This irrigation scheme generally involves bringing the root zone to full capacity of water (i.e. the 'water holding capacity') and then providing additional water which will result in a percentage of the applied water leaching from the root zone.

4.5.2 PULSE IRRIGATION

Typical overhead irrigation systems apply water at rates greater than can be absorbed and held by the substrate. This can generate substantial amounts of run-off. 'Pulse Irrigation' is an approach that utilizes more frequent irrigations with shorter durations. The objective of 'pulse' irrigation is to match the application rate of the irrigation system with the absorption rates of the substrate (Zur, 1976). By applying the water in cycles of a short duration irrigation followed by a 'rest' period, the substrate is allowed to absorb water that is applied before additional water is added (Lamack and Niemiera, 1993).

Irrigations that are normally applied in a single relatively long duration application are divided into several shorter intervals or 'pulses.' For example, an irrigation duration of 40 min may be divided into four individual 10-min intervals separated by a period of 50 min. Pulsing may result in a shorter total irrigation duration that may translate into water savings and reductions in run-off. Using the previous example, the pulses might be reduced to 8 min and still provide adequate water to the plants (Levin et al.,

1979; Fare et al., 1994). On the other hand, dividing water applications over time lengthens the periods of high water availability. This usually results with increased transpiration rate but also with increased water use efficiency of the marketable part of the crop (Fernández et al., 2005; Katsoulas et al., 2006).

With pulse irrigation, there is a danger of pockets within the root zone drying out. For instance, if the roots within the root zone are much denser in a particular layer within the root zone, then this layer will dry out more rapidly. If the pulse is too short to replenish this zone completely, then over time this zone will become hydrophobic and prevent water movement into this substrate. Furthermore, this barrier to flow may prevent water from flowing to other areas of the root zone, which then also dry out. Generally these pockets are not visible to the grower, so that the grower will not recognize this until the killed roots allow entry of pathogens into the plants, causing plant diseases to appear. It is generally best to avoid pulse irrigation unless (1) the roots are distributed uniformly throughout the root zone and (2) the unsaturated hydraulic conductivity of the substrate is such that water still moves rapidly within the root zone when it is somewhat dry.

4.5.3 HIGH FREQUENCY IRRIGATION

In some types of soilless production systems, irrigation events may occur several times per day or hour and each (or at least some) of the irrigations result in bringing the root zone to its full water-holding capacity. In such situations, the system is dependant on the bulk flow of water, nutrients, and oxygen. This approach can only be used in systems where the substrate allows complete drainage of all excess water between irrigation events.

4.5.4 CONTINUOUS IRRIGATION

A number of soilless production systems are best referred to as 'continuously irrigated' systems. In such systems, there is no irrigation *per se*, but rather the root zone is managed in a manner that the roots are constantly submerged and bulk flow is imposed on a continual basis either by moving the water with pumps or by bubbling air into the water to induce water currents. This irrigation approach is appropriate in systems where the root zone consists of substrates that have little or no water-holding capacity, or possibly have no substrate at all. NFT (Nutrient Film Technique) (Hurd, 1978) and other 'water culture' systems are of this type. Another system of this type is aeroponics where the irrigation solution is continually sprayed onto the roots (Zobel et al., 1976; Soffer, 1986; Soffer et al., 1991).

Continuous irrigation can also be applied in the form of micro-irrigation. In this case, water is applied at a rate that coincides with the transpiration, so roots are never submerged in water. In this case, an additional amount of water is applied periodically so as to discharge accumulated salts.

4.6 IRRIGATION DECISIONS

Irrigation control involves two facets: making the decision to irrigate and implementing the decision through a particular irrigation-control approach. The main irrigation decisions are (1) when to initiate the irrigation event (how frequently to open a valve) and (2) how much irrigation solution to deliver during the irrigation event (duration of irrigation events).

4.6.1 IRRIGATION FREQUENCY

To optimize productivity, plants must never be subjected to conditions that cause stress and reduce plant growth. Plants should never be allowed to run out of readily available water since this may delay the crop, cause death of root tissue, or even the entire plant. Also, many substrates, particularly those which include peat, are difficult to re-wet once they have dried out.

The decision to irrigate is linked to the water use of the plant as well as various other circumstances. Irrigations that substantially increase the humidity should be done so that the relative humidity does not stay near 100 per cent for more than 4 h. In situations where the irrigation is likely to be followed by a period where there is no ventilation (e.g. late afternoon, evening, or cloudy conditions), relative humidity can rise to levels where condensation will occur. Since the leaves and greenhouse glazing are generally slightly cooler than the air in the greenhouse, those are the surfaces where condensation will occur first. For many plant species, liquid water on the foliage should not be allowed to remain for any extended period of time since this is likely to allow disease organisms to proliferate, infect the plants, and result in disease problems.

Another consideration is the availability of water pressure in the supply lines. In most production systems, the water supply capacity is limited so that not all irrigation circuits can be irrigated at the same time. If all plants need to be irrigated during a short period of time (as may happen on hot dry summer days), it may be necessary to use pulse irrigation during the day to minimize water stress effects, followed by irrigation events later in the day that are more complete.

In automated irrigation control systems, two parameters need to be explicitly set to prevent the system from irrigating too frequently or too infrequently:

Maximum duration between irrigation events – this parameter identifies the longest tolerable interval between irrigations (in hours or days). It is needed to prevent damage to plants due to failure of other control decisions or equipment. System failures can happen in various ways: the solenoid valve leaks and allows enough water to get to the pot that is being monitored (the representative plant) so that it never dries out. This flow rate would be inadequate for the entire crop; so, many plants would die without use of this parameter. Failure can also occur if water gets into the electronics associated with sensors causing faulty measurements.

Minimum duration between irrigation events – this parameter sets the shortest allowed interval between irrigations (in minutes, hours, or days). This parameter is needed to prevent crop damage in case the sensor fails so that the controller continually sees a signal that represents a dry condition that does not change. This could occur if the emitter to the representative plant has been inadvertently pulled out of the pot. It could also be caused by a signal problem with the sensor or controller or a break in the cable between them.

4.6.2 DURATION OF IRRIGATION EVENT

Deciding on the duration of an irrigation event is not always straightforward since it is a factor of the uniformity of the crop, the uniformity of the distribution system, the capacity of each root zone, and the expected level of depletion at the start of an irrigation event. Typically within a crop, plants that are exposed to more sunlight or air movement or to low VPD deplete more water. Also, the plants that are closest to the irrigation valve may receive the highest rate of water application if the irrigation system has low distribution uniformity. The degree of uniformity of the irrigation system will influence the length of time between the opening of the irrigation valve and the time when the last plant to get water begins to receive it. In large irrigation circuits, this can take tens of seconds or longer. Typically, the plants furthest from the irrigation valve will see the slowest water application. Plants in this area that are large and have plenty of light and air movement will require the longest time before the substrate reaches container capacity.

The duration of an irrigation event is mainly dictated by the amount of water that needs to be supplied to the plant, plus any amount of leachate that is needed. The percentage of leachate in relation to the total amount of water applied is called the 'leaching fraction'. In conventional irrigation approaches, the leaching fraction can range from 0 to 20 per cent under well-controlled conditions. In rare instances, greater percentages are targeted. Leaching fraction can be measured in real-time using, for example an electronic tipping bucket located beneath the drainage holes of a representative plant.

The leaching fraction is typically a function of the degree to which salt-build-up is likely to be a problem. If the source water is somewhat saline and fertilizer salts are added to this water, then the salinity may well rise to a level where it is problematic to the plants. If this is the case, then the amount of leaching should be relatively high. Leaching may be desirable in cases where the water used for irrigation is low in quality, as it will ensure that salts that may accumulate are moved out of the root zone. If the water is of high quality and fertigation results in little build-up of salts in the root zone, then the leachate volume can be minimized. The leaching fraction can be calculated by dividing the EC of the irrigation water by the maximum tolerable EC of the leachate (Marshall et al., 1996). That is, if the EC of the irrigation water is one-fifth of the EC of the maximum desired EC of the leachate, then the leaching fraction should be one-fifth of the volume applied.

The plants that are the last to receive water are typically also the last to reach container capacity. It is these plants that should also be used to determine the desired leaching fraction. By measuring the leachate volume and timing the duration required to achieve the minimal amount of leaching, it is possible to identify the length of time to add to the irrigation duration to achieve a minimum leaching fraction. It should be noted that the volume of the leaching fraction from each plant will be highly variable due to factors such as the nonuniformity of the delivery system and various differences among the plants.

4.7 APPROACHES TO MAKING IRRIGATION DECISIONS

There are several approaches to making irrigation decisions in the nursery: (1) look-and-feel, (2) gravimetric, (3) timer-based, (4) sensor-based, and (5) model-based methods. Some of these methods involve explicit considerations of the plants while others may not and are based on approximations of information about the plants.

4.7.1 'LOOK AND FEEL' METHOD

The look-and-feel method involves close inspection of the plant, paying particular attention to the colour of the substrate surface, the occurrence of flagging foliage, or slight colour changes in the foliage which occur in some crops just prior to wilting. In containerized production, this might be accompanied with lifting up one or two pots to gauge their weight or sticking a finger into the growing medium to attempt to feel whether a significant amount of water has been removed from the pot. Doing this accurately requires a substantial amount of experience of relating these 'human sensor readings' to moisture content and moisture tension, and is very difficult to do reliably and consistently.

While this approach can be fine-tuned to be superior to a time-based approach, there are numerous problems. One is that human touch is not a particularly good measure of moisture content. As long as there is any amount of readily available water, moisture can be felt by the skin surface. The absence of moisture is generally felt only at fairly dry conditions that are suboptimal for plant growth. In general, moisture levels much greater than those coinciding with water stress should be maintained for optimal growth. Thus, if irrigation is based solely on the appearance of the plant or the feel of the soil, the plants will be consistently subject to stress. As the plants adapt to this stress, it will become less noticeable while still being suboptimal for growth.

Judging the weight of pots by lifting a few representative pots is a better approach than looking at or feeling the substrate. However, it is subject to problems if the grower is not aware of the moisture release characteristics of the medium and has not learned how various water contents relate to how heavy the pot feels. Also, when doing this, the grower has to somehow subtract the weight of the plant and substrate. The fresh weight of the plant also varies dramatically with its water status and as it grows (A quick check can be done by irrigating a pot to container capacity. Then the weight of that pot can be compared to other pots. The difference would be the amount of water that had been retained in the irrigated pot.)

Many practitioners (and some scientists) consider the concept of 'water stress' to be an important consideration in irrigation management. However, it should be noted that this term has no specific, quantitative definition and thus means different things to different persons. It is purely a qualitative term. Many persons feel that 'water stress' represents a condition in the plants that results in the appearance of observable symptoms (e.g. wilting). However, it has been shown that such conditions always lead to reductions in plant growth. Even cacti, which are known to be extremely resistant to water stress, only grow at appreciable rates if ample water is present. In fact, the reduction in productivity can be observed even under conditions that might not qualify as 'water stress'. Thus the term 'water stress' is relatively useless in irrigation management, except to say that if water stress is occurring, then there has been a failure in either the irrigation system or the irrigation scheme. In general, 'water stress' should not be the deciding factor for starting an irrigation event; doing so assures that the irrigation event will always come too late.

4.7.2 GRAVIMETRIC METHOD

It is possible to use weighing as a tool in irrigation scheduling. Generally, weight changes due to plant growth are much smaller and less rapid than weight changes due to loss of water from the pot. Recording the weight of the container after the previous irrigation can enable the irrigation control system to determine the amount of water lost since that irrigation. Thus, it is possible to weigh the pot day-to-day and determine the cumulative amount of water that has been lost to ET. Since the container volume and water-holding capacity of media are known, it is possible to approximately track water use. There are, however, possibilities for inaccuracies. For example, as the roots fill in the pot, they take over some of the space which would otherwise be occupied by water. Thus, the volume of readily available water decreases as the plant grows. In root-bound plants (where the root mass has filled most of the root zone) this loss can be substantial.

Devices specifically designed for use with irrigation control are available. These are, however, not commonly used as they require frequent adjustments and tend to be inaccurate for indicating how much readily available water remains. Most growers find the labour needed for this approach not to be cost-effective. Still, with proper care, a load cell is a very good research tool (Graaf et al., 2004; Wallach and Raviv, 2005).

4.7.3 TIME-BASED METHOD

The simplest and least costly approach to automation of irrigation is to use timers to turn irrigation systems on and off. This approach is feasible since soilless media, although having high water holding capacity, typically have high infiltration rates and high porosity allowing excess water to readily drain away. Under these circumstances, it is possible to apply irrigation solution on a fixed schedule, so watering always occurs sooner and for longer than is presumably needed by the crop. While this generates wasted irrigation solution, many growers currently consider this a small price to pay for the insurance that the plants will always be watered and fertilized adequately. This price, however, is likely to rise steeply as economics, environmental considerations, government regulations, and public pressure force growers to eliminate this waste. The combination of time-based irrigation and a closed irrigation system is, therefore, a natural solution.

A purely time-based approach can also lead to suboptimal moisture conditions in the root zone if the substrate cannot provide adequate oxygen for the roots under saturated conditions. This can easily happen especially if the substrate depth is shallow and the plants use water relatively slowly.

Timed irrigation is easily done with the use of electrical or electronic timers specifically designed for controlling irrigation valves. Adjustments to irrigation timing should be made to reflect changes in climate or season or other factors that would affect plant water use. Timers can also be used where the decision to irrigate is made by other methods (e.g. look-and-feel or sensors) and a manually started timer is then used to control the duration of the irrigation.

4.7.4 SENSOR-BASED METHODS

Sensors can be used to register some facet of the environment surrounding the crop and to use the resulting signal as part of irrigation decisions. 'Sensor-based' irrigation control involves a sensor to directly measure some aspect of the moisture content of the root zone or water demand of the plant.

It is possible to use sensors to measure the moisture content (or related characteristics) in the root zone and then use this information to control moisture levels there. A number of different sensor types exist. One feasible sensor for this is the tensiometer, due to its relatively low cost among the sensors which are not affected by salts (fertilizer) in the irrigation solution. Generally, the sensors that are used in this type of irrigation attempt to measure moisture content of the substrate of the root zone.

4.7.4.1 Tensiometers

A tensiometer is a device which measures moisture tension or matric potential. It consists of a tube fitted with a porous ceramic tip on one end and a pressure/suction gauge on the other end. In automated systems, the gauge is supplemented with or replaced by a transducer to convert the tension (suction) to an electrical signal that can be sensed by a computer or electronic controller. The tube is filled with water and the device is sealed and inserted in the substrate so that the ceramic tip is in the middle of the root zone. Tensiometers may be installed oriented vertically, horizontally, or to any other angle as long as the ceramic tip is at the lowest point. Considerations in determining the angle include the effect of the water column within the tensiometer on the pressure measurement, the ability of applied water to channel along the tensiometer, and the size of the container or pot. Tensiometers with 'high-flow' ceramic tips respond rapidly to changes in moisture and are thus better suited for use in soilless substrates than tensiometers designed for use in soils.

The basic operation is to have one tensiometer coincide with each irrigation valve. This sensor needs to be in the root zone of a plant in a location that is representative of the whole crop. The irrigation system should be designed and operated to apply irrigation water as uniformly as feasible. One approach to using tensiometers in irrigation control is to use the signal to merely override irrigation systems (using the rain detection cut-out circuit) to prevent timer-controlled solenoid valves from coming on, unless a specific level of dryness has been reached. Another approach involves two set-points: (1) a high-tension set-point, representing the level of dryness (e.g. where 80 per cent of the readily available water has been removed) when irrigation is initiated, and (2) a low-tension set-point representing the set-point at which an ongoing irrigation event will stop. These set-points are used as follows.

As the tension in the root zone rises with water depletion, the tensiometer is monitored. Once the high-tension set-point is reached, the irrigation is initiated (in large-scale operations, this will mean that this particular irrigation circuit is scheduled for irrigation by placing it in an irrigation queue). Once the irrigation is in progress, the tensions will drop (usually over seconds or minutes) as water is applied. Hopefully the hydraulic conductivity and application method adequately disperses water laterally in the root zone. The rate of water application should be slow enough to allow this to occur and also to allow the tensiometer to follow the changes in moisture condition. Applications may be applied in pulses to match substrate infiltration and water application rates. When the low-tension set-point is reached, the irrigation is stopped (or allowed to continue for a specific length of time to obtain the desired leaching fraction).

Kiehl et al. (1992) investigated the use of tensiometers in UC mix and found that in potted plant production in this substrate, much of the readily available water in the substrate is exhausted by the time the dry-down has reached 7 kPa. At this point, extraction of a little more water sends the tension to over 10 kPa. They observed that at tensions over 10 kPa, a plant that is accustomed to fairly moist conditions will start showing signs of wilting. Unless irrigation occurs at that point, the plant will be exposed to damaging conditions as the substrate dries out further. Thus it is generally wise to irrigate when the tension is around 5 kPa; after that the urgency increases with increasing tensions. They also investigated a number of scenarios involving lower levels of high tension set-points and found that the an irrigation regime that involves a significant draw-down to about 5 kPa is best when working with UC Mix. In later work, Raviv et al. (2001) found that UC Mix, despite having relatively good aeration, does subject the plants to oxygen deficiency if the tension is not allowed to rise so as to deplete some of the water in the root zone.

Kiehl et al. (1992) also investigated the suitability of various low-tension set-points. They found that a low-tension set-point fairly close to the high-tension set-point (attempting to maintain relatively constant non-saturating moisture conditions at all times) was not conducive to optimal growth, even if such an approach allowed them to nearly eliminate all run-off from the crop. They also found that a low leaching percentage (close to 0 per cent) could be achieved with much lower low-tension set-points of 0.5–1 kPa. While low- and high-tension set-points of 1 and 5 kPa, respectively, appear to be suitable for UC mix or other similar potting mixes, other substrates (e.g. peat, coir, stone wool and perlite) would require lower set-points. It has not been shown whether tensiometer-based irrigation control is feasible with these systems.

The system parameters described earlier in the chapter for irrigation control are all pertinent when using tensiometers to control irrigation. As described above, high- and low-tension set-points can be used to trigger irrigation and to determine when to stop an irrigation. As described earlier, various scheduling overrides based on time or other factors should be included as safety features. Such systems can also include alarms that alert growers that something is potentially wrong and that human intervention is necessary to resolve a conflict. For example, with tensiometers in soilless substrates, a high-tension alarm should be implemented so that the operator can be made aware of tensions that have risen to levels that represent dangerous conditions. This can happen if the irrigation system fails (due to, e.g. defective pump, solenoid valve, or emitter in representative pot). It could also signal defective signal wiring (depending on design of signal wiring/processing). It might also be an indicator that the tensiometer has been removed from the root zone and is drying out.

There may be other reasons for overriding sensor-based irrigation control. The grower may have other motives for the crop that would suggest more or less frequent irrigation. For instance, if plants are relatively small and are not able to remove water from the entire root zone, then the substrate solution immediately around the roots may become depleted of nutrients and oxygen without becoming depleted of water. Under these conditions, the grower should schedule extra irrigations so as to provide the plants with needed oxygen or fertilizers.

4.7.4.2 Electrical Conductance Sensors

In field soils, a common type of moisture sensor is one that has electrodes separated by a few centimeters; the conductance between the electrodes is related to the moisture content. The main problem with these types of sensors is that the conductance of pure water is quite different from water that has salts dissolved in it. Commercial plant production implies explicitly that there will be fertilizer salts in the root zone. If the salinity is relatively constant, then these types of sensors may be feasible as part of sensor-based irrigation control. However, if there will be times where irrigation is with just clear water or when salinity of the root zone solution fluctuates separate from moisture status, then this type of sensor is not ideal for controlling irrigation.

The electrodes can be imbedded in a material such as ceramic, gypsum, or other porous material to minimize problems associated with salinity. In gypsum block sensors, for example, the water within the block is saturated with Ca^{2+} and SO_4^{2-} ions making it less sensitive to changes in the salinity of the water in the substrate. However, this solubility means that the blocks slowly dissolve and have a finite life. Soil Moisture Equipment Corp. (Santa Barbara, CA), for example, rates the life of their gypsum block at 2–5 years.

The sensor needs to be in good contact with the substrate to facilitate water movement between the substrate and the gypsum as moisture conditions change. If the structure of the substrate is inconsistent, as moisture content changes, the contact with the substrate may be affected. Imbedding the gypsum blocks into a granular matrix (e.g. Watermark, Irrometer Co., Riverside, CA) reduces this effect and improves precision, although not to the level of other sensors such as tensiometers.

Since the movement between the gypsum block and the substrate is usually very slow, so is the response of these types of sensors to changes in moisture conditions (Munoz-Carpena et al., 2005). For example, after a rapid change in moisture content due to irrigation, a gypsum block may take up to an hour to equilibrate to the change.

Implementation of this type of sensor (specifically granular matrix sensors) can be to sense an alarm or a high set-point. For example, a system can be set to initiate an irrigation event upon reaching a high set-point (dry) condition and an application of a set duration would deliver water to the crop. Since these sensors are slow to respond to changes, a timed application is required.

It should be noted that electrical resistance sensors can be implemented very inexpensively. Thus if precision, salinity, or rapid responsiveness are not issues, then this type of sensor can work well.

4.7.4.3 Dielectric Capacitance (TDR, FD)

Methods to measure the moisture content of substrate by measuring dielectric capacitance or dielectric permittivity of substrates have recently become more prevalent, since they are fast (fractions of seconds) and some methods also have the capability to measure salinity (EC) of the substrate (Dalton et al., 1984). Time domain reflectometry (TDR) utilizes parallel metal rods (wave guides) to carry an electromagnetic pulse into the substrate. An oscilloscope is used to analyse the electrical pulse as it enters the substrate, travels the length of the wave guides, and then returns. The shape of the waveform can be analysed to determine the moisture content and salinity of the substrate. The time that it takes for the pulse to travel down the waveguides is related to water content, and the dissipation of the signal is indicative of the electrical conductivity of the substrate (Dalton and Poss, 1990).

The TDR probes have either two or three parallel bare metal rods ranging in length from 10 to 50 cm, are relatively inexpensive ($\sim$\$50–\$100), and are very robust due to their simplicity. The cost of the electronics to generate the electromagnetic pulse and to detect and analyse the signal, however, can cost $\sim$\$5000 for research grade equipment. Variations in sensors include coiling the waveguides to reduce the overall length of the probes (Nissen et al., 1998) and embedding in gypsum to measure matric potential (Persson et al., 2006).

Frequency domain (FD) reflectometry is similar to TDR. However, as TDR measures changes in time characteristics of the electromagnetic pulse as it travels down the waveguides, FD analyses changes in the frequency characteristics of the pulse (Lin, 2003). The advantage over TDR is the ability to use much shorter (<10 cm) waveguides and is much simpler in implementation as it does not require a complex set of electronics. The sensors can be connected directly to a datalogger without the need for the pulse generator as a TDR probe requires. A sensor variant imbeds the waveguide in a circuit board (fibre reinforced epoxy) material (ECH$_2$O, Decagon Devices, Inc., Pullman, WA) that makes them very easy to install and even more durable. As with TDR, these sensors require no regular maintenance.

Although there are advantages of TDR and FD such as sensor durability and rapid measurement, there are drawbacks. These measurement systems need to be calibrated for each substrate type, as texture, structure, and organic matter content affect the accuracy of measurements (Baas and Straver, 2001; Morel and Michel, 2004). They are also affected by high ECs ($\sim$5 dS m^{-1}), as the electromagnetic pulse is unable to be reflected at the end of the waveguides as the conductance of the substrate increases.

The use of TDR (Murray et al., 2004) and FD systems can be similar to a system using tensiometers: they can sense when a dry set-point is reached. Since they are very rapid, they can be use to monitor soil moisture changes, as water is applied to terminate the irrigation event. A major difference from tensiometers is what is being measured. Tensiometers, as their name reflects, measure matric tension or potential, the x-axis on a soil moisture curve (Fig. 4.1). Capacitance sensors, TDR, and FD measure volumetric water content, which would be the y-axis of the curve.

Various plant-based sensors such as stem flow gauges, dendrometers, leaf thickness sensors, stem dielectric sensor are used in research related to plant water status, but are generally not used in commercial production settings or urban horticulture.

4.7.4.4 Pros and Cons of Various Types of Sensor-Based Control

Tensiometers tend to be immune to issues regarding salinity. They respond purely to moisture tension and thus measure the force that the plants have to overcome to extract water from the root zone. However, the tensiometer contains water within the tube which, under high tension conditions, can be pulled out. Thus when high tensions are allowed to occur (depending on the characteristics of the ceramic tip), the sensor will require service to replace water lost from the tensiometer tube.

Electrical resistance sensors are generally less expensive than tensiometers and require less service while in use. However, since the sensor reading is affected by salt content even at normal fertilizer concentrations, such sensors are not suitable for use in situations where precision is desired.

Dielectric capacitance sensors generally require considerably more electronics to be feasible as sensors in irrigation control. This means that they cost considerably more to use than the other sensors. However, with miniaturization and electronic customization, this is changing and these types of sensors are becoming more cost-effective in commercial production settings.

4.7.5 MODEL-BASED IRRIGATION

It is possible, through the use of mathematical models, to compute how much water a crop has used and to then irrigate so as to replace this water. One way to accomplish this in the greenhouse is to set out a pan of water and track the water level in it as it evaporates. The evaporation of water from this pan can be related to the evapotranspiration (ET) of the crop and thus irrigation schemes can be used that are based on this measurement. Evaporation pans can be obtained for precisely this use. A device has also been developed that uses this same principle to estimate ET without using an open pan. It protects that water and automates the measurement process so that a data signal is generated which can be used as part of input to a control device.

These type of devices are frequently used in outdoors production of field crops and orchards.

Another similar approach is to use a reference plant with known water-use characteristics, measure its ET rate by weighing it and tracking its water loss. This is basically identical with the gravimetric approach, described previously.

In all these cases, calculations have to be made to relate the measured evaporation rate to the ET of the crop and to also integrate this rate over time so as to determine a cumulative amount of water that has been lost from the root zone of the plants. Such calculations are made with mathematical models and methods based on such calculations are known as model-based irrigation control.

By relating the total volume of water removed from the root zone to the volume of the root zone, it is possible to calculate when a particular level of depletion has occurred. As discussed previously, the moisture release curve for the substrate can be used to identify the level of depletion that should be used to trigger or schedule an irrigation event, or to elevate the priority of a previously scheduled event.

In general, all model-based irrigation control methods involve the use of inputs from sensors in the environment. Calculations are typically made by computers, but it is feasible to program controllers to do this as well, although controllers are generally limited in the level of complexity that can be implemented.

Greenhouse environmental control computers are used to control many characteristics of the crop environment. Many schemes for irrigating crops have been devised to use data being recorded by these computers. As scientists develop models for water-use in crops, the various manufacturers of these computer systems implement these models in their systems, providing growers with a continually wider array of choices of model-based irrigation strategies. These methods attempt to relate the amount of water which has been lost by the plant and from the root zone, to one or more environmental variable (light, temperature, VPD, and wind speed), so that irrigations can be scheduled to replace this lost water.

Various methods exist for approximating how much ET has occurred since the last irrigation. Implementation of these methods in irrigation control in soilless production is only feasible with computerized irrigation systems, since this estimation requires extensive computation using sensor data. The most common method for estimating ET is the Penman–Monteith model (Penman, 1948; Monteith, 1965) although other methods can also be found in the scientific literature (Sharma, 1985). These approaches generally involved the above-mentioned four environmental variables as well as a variable related to the size of the plant's foliage surface area. It should be noted that these models were developed for plants growing in an outdoor environment, so that certain considerations are needed when dealing with protected cultivation. In virtually all cases, implementation of such models involves simplification so as to remove variables and parameters that typically are not easy to monitor or evaluate in a practical situation. Estimations of plant water use are available to assist in irrigation management from on-line computer programs such as the California Irrigation Management Information System (CIMIS, wwwcimis.water.ca.gov). The worldwide standards for measuring reference ET are detailed in Allen et al. (1998) and are available on the web at: http://www.fao.org/docrep/X0490E/X0490E00.htm.

One irrigation control method that focuses on a single environmental parameter sums light measurements at regular intervals. Once a specified light-sum level is reached, an irrigation is initiated. The light-sum-levels set-points are refined by the individual grower for each crop. Another method estimates the vapour pressure deficit (VPD) of the air and relates this to the rate of ET. VPD essentially measures how much water can be taken up by the air and, in conjunction with other variables, can be used to compute the amount of ET over time. However, when implemented without the use of light, temperature, wind speed, and leaf area, the calculations can only serve as a rough approximation requiring frequent set-point modifications.

When using a particular method, it is important to understand how calculations are being made and which variables are being used in the computations. Using a method that makes ET calculations based on measurements of temperature, light, relative humidity, and wind speed, such as the Penman–Monteith model, has the greatest potential for success, but still requires frequent grower intervention as the crop grows, is harvested, affected by insects and diseases, and so on. As models for plant growth become available, this need for grower intervention may well decrease.

There are generally circumstances under which any of the irrigation systems and control strategies are appropriate. However, for any one crop and cropping system, there is generally one system which is best. The main variables determining which is best include labour and installation costs as well as the degree to which each is able to keep the crop at optimal moisture conditions.

4.8 FUTURE RESEARCH DIRECTIONS

In soilless production, growers typically pay much closer individualized attention to the plants than in field production. Current research into methods that are broadly termed 'precision agriculture' focus on striving towards greater individualized attention to specific plants or groups of plants. As such, soilless crop production leads the way, with various implementations already in practice or close to adoption. For example, wireless irrigation control has already been developed so as to eliminate wiring associated with sensors and valves. Some companies are currently developing commercial products based on this concept. Further innovations are certain to come first to soilless production agriculture, striving for specific identification of the irrigation and fertility needs of particular plants to groups of plants using sensor technologies and controlling delivery of water to each plant on an individualized basis. Further advances can be anticipated to also control other root zone variable in this way.

Another area where more research and development are needed is in the integration of more sensors into irrigation control to dynamically account for more variables that affect plant growth. As such, soilless production could benefit significantly from a control strategy to simultaneously optimize root zone temperature, oxygen and nutrient concentrations, salinity, pH, as well as water content. Currently the most advanced control strategies use a few environmental variables (estimated integrated VPD or light) and water content. Ultimately it may be possible to control fertigation based on all measurable root zone variables. It may, in fact be possible to involve measurements of non-root-zone variables. For example, one could even integrate irrigation system hydraulic data (e.g. flow rates or water pressure) so as to optimize flow characteristics and optimize water pressure at each emitter. Ultimately it may also be feasible to involve sensor data for various plant processes (e.g. transpiration rate), as this information can be useful in diagnosing unexpected conditions and plant problems. For example, if the root zone sensors indicate that there is adequate water, but the transpiration rate and environmental data suggest that the plant is not using enough water to compensate for the atmospheric demand, then this will indicate particular conditions that may well require a change in the irrigation control or another intervention by the grower.

Another area where research is needed in the future is in queuing irrigation events subject to sensor input or model calculations. For example, if irrigation for a particular irrigation circuit is planned when the representative plant reaches 5 kPa of moisture tension, then in most commercial nurseries it is not feasible to turn on the corresponding irrigation valve if other plants are already being irrigated, due to inadequate capacity. In general, this is handled through irrigation scheduling. On days when the plants consume a lot of water, it needs to be anticipated that many plants will be scheduled for irrigation. As the plants wait in the queue, they continue to consume water so that the tension continues to rise. Clearly some optimization strategy and research are needed to identify how to best juggle the rising moisture tension, irrigation capacity, and risk assessment to minimize financial losses in the event that some plant may end up seeing water stress.

In the same way, modelling approaches can also be integrated with queuing methods to provide anticipatory control. Using all the root zone variables, as well as plant and environment measurements, it should be feasible to forecast when harmful conditions can be anticipated. With such an approach, models would be used to forecast when an irrigation will be needed for each group of plants, by accounting for the aforementioned queuing approach and using simulation modelling to calculate whether the currently selected high-tension set-point will result in a situation where not all plants can be irrigated as needed. If that situation occurs, then the control algorithm would either lower this set-point or reduce the irrigation duration so as to eliminate leaching and to perhaps do partial irrigations throughout the day to avoid financial losses due to wilting of plants. Research is needed to develop such models, sensors, and integration strategies. Engineering research is needed to develop decision-support tools and control algorithms that encompass these concepts.

Emitters could be developed to do more than just deliver water in a particular fashion. In addition to pressure regulation, which is already common, emitters could be developed to include sensor technologies so as to dynamically adjust flow through the emitter based on sensed root zone conditions. In particular, if a sensor could be developed to assess hydraulic conductivity of the substrate, then the emitter flow rate could be dynamically adjusted to match the ability of the substrate to carry irrigation water to the roots of the plants.

Perhaps the most important facet of future developments in irrigation in soilless production is to recognize that current approaches are quite primitive in comparison with what they could be. We know most of the variables that are important as part of optimal root zone moisture management. Yet the vast majority of commercial soilless plant production uses nearly none of these tools, relying instead on substrates that can be over-irrigated without a significant penalty and plants that can tolerate the resulting conditions. Clearly there is a lot of room for advancement towards optimization, especially as the cost of electronics and sensors continues to decline.

REFERENCES

Allen, R.K., Pereira, L.S., Raes, D. and Smith, M. (1998). *Crop Evapotranspiration: Guideline for Computing Crop Water Requirements*. FAO Irrigation and Drainage Paper No. 56. Rome: United Nations Food and Agricultural Organization.

Baas, R. and Straver, N.A. (2001). In situ monitoring water content and electrical conductivity in soilless media using a frequency-domain sensor. *Acta Hort.* (ISHS), **562**, 295–303.

Chen, D. and Wallender, W.W. (1984). Economic sprinkler selection, spacing and orientation. *Trans. ASAE*, **27**, 737–743.

Dalton, F.N. and Poss, J.A. (1990). Soil water content and salinity assessment for irrigation scheduling using time-domain reflectometry: Principles and applications. *Acta Hort.* (ISHS), **278**, 381–393.

Dalton, F.N., Herkelrath, W.N., Rawlins, D.S. and Rhoades, J.D. (1984). Time-domain reflectometry: Simultaneous measurement of soil water content and electrical conductivity with a single probe. *Science*, **224**, 989–990.

Fare, D.C., Gilliam, C.H., Keever, G.J. and Olive, J.W. (1994). Cyclic irrigation reduces container leachate nitrate-nitrogen concentration. *Hort. Science*, **29**, 1514–1517.

Fernández, M.D., Gallardo, M., Bonachela, S., et al. (2005). Water use and production of a greenhouse pepper crop under optimum and limited water supply. *J. Hortic. Sci. Biotechnol.*, **80**, 87–96.

Graaf, R. de, Gelder, A. de. and Blok, C. (2004). Advanced weighing equipment for water, crop growth and climate control management. *Acta Hort.* (ISHS), **664**, 163–167.

Hsiao, T.C. (1990). Measurements of plant water status. In *Irrigation of Agricultural Crops – Agronomy Monograph no. 30* (B.A. Stewart and D.R. Nielsen, eds). Madison, WI: Amer. Soc. Agron.; Crop Sci. Soc. Amer.; Soil Sci. Soc. Amer., pp. 243–279.

Hurd, R.G. (1978). The root and its environment in the nutrient film technique of water culture. *Acta Hort.* (ISHS), **82**, 87–97.

Katsoulas, N., Kittas, C., Dimokas, G. and Lykas, C. (2006). Effect of irrigation frequency on rose flower production and quality. *Biosyst. Eng.*, **93**, 237–244.

Kiehl, P.A., Lieth, J.H. and Burger, D.W. (1992). Growth response of chrysanthemum to various container medium moisture tension levels. *J. Am. Soc. Hortic. Sci.*, **117**(2), 224–229.

Lamack, W.F. and Niemiera, A.X. (1993). Application method affects water application efficiency of spray stake-irrigated containers. *Hort. Science*, **28**, 625–627.

Levin, I., Van Rooyen, P.C. and Van Rooyen, F.C. (1979). The effect of discharge rate and intermittent water application by point source irrigation on the soil moisture distribution pattern. *Soil Sci. Soc. Am. J.*, **43**, 8–16.

Lieth, J.H. and Burger, D.W. (1989). Growth of chrysanthemum using an irrigation system controlled by soil moisture tension. *J. Am. Soc. Hortic. Sci.*, **114**(3), 387–392.

Lieth, J.H. and Burger, D.W. (1993). Irrigation. In *Geraniums IV The Grower's Manual* (J. White, ed.). Geneva, Illinois: Ball Publishing, pp. 31–38.

Lin, C.-P. (2003). Frequency domain versus travel time analyses of TDR waveforms for soil moisture measurements. *Soil Sci. Soc. Am. J.,* **67**, 720–729.

Marshall, T.J., Holmes, J.W. and Rose. C.W. (1996). *Soil Physics* (3rd edn). Cambridge, UK: Cambridge University Press.

Monteith, J.L. (1965). Evaporation and environment. *Symp. Soc. Exp. Biol.,* **19**, 205–234.

Morel, P. and Michel, J.-C. (2004). Control of the moisture content of growing media by time domain reflectometry (TDR). *Agronomie (Paris),* **24**, 275–279.

Munoz-Carpena, R., Dukes, M.D., Li, Y.C. and Klassen, W. (2005). Field comparison of tensiometer and granular matrix sensor automatic drip irrigation on tomato. *HortTechnology,* **15**, 584–590.

Murray, J.D., Lea-Cox, J.D. and Ross, D. (2004). Time domain reflectometry accurately monitors and controls irrigation water applications in soilless substrates. *Acta Hort.* (ISHS), **633**, 75–82.

Nissen, H.H., Moldrup, P. and Henriksen, K. (1998). High-resolution time domain reflectometry coil probe for measuring soil water content. *Soil Sci. Soc. Am. J.,* **62**, 1203–1211.

Oki, L.R. and Lieth, J.H. (2004). Effect of changes in substrate salinity on the elongation of Rosa hybrida L. 'Kardinal' stems. *Sci. Hortic.,* **101**, 103–119.

Penman, H.L. (1948). Neutral evaporation from open water, bare soil and grass. *Proc. R. Soc. Lond. Ser. A,* **193**, 120–146.

Persson, M., Wraith, J.M. and Dahlin, T. (2006). A small-scale matric potential sensor based on time domain reflectometry. *Soil Sci. Soc. Am. J.,* **70**, 533–536.

Raviv, M. and Blom, T.J. (2001). The effect of water availability and quality on photosynthesis and productivity of soilless-grown cut roses. *Sci. Hortic. (Amsterdam),* **88**, 257–276.

Raviv, M., Lieth, J.H., Burger, D.W. and Wallcah, R. (2001). Optimization of transpiration and potential growth rates of 'Kardinal' rose with respect to root-zone physical properties. *J. Am. Soc. Hort. Sci.,* **126**, 638–643.

Richards, L.A. (ed.) (1954). *Diagnosis and Improvement of Saline and Alkali Soils.* Agriculture Handbook No. 60. Washington, DC: Soil and Water Conservation Branch, Agricultural Research Service, US Department of Agriculture.

Seginer, I. (1987). Spatial water distribution in sprinkler irrigation. In *Advances in Irrigation,* Vol. 4, (D. Hillel, ed.). New York, Academic Press, pp. 119–168.

Sharma, M.L. (1985). Estimating evapotranspiration. In *Advances in Irrigation,* Vol. 3, (D. Hillel, ed.). New York, Academic Press, pp. 213–281.

Soffer, H. (1986). A device for growing plants aero-hydroponically: The E.G.S. (Ein Gedi System) miniunit. *Soilless Cult.,* **2**, 45–52.

Soffer, H., Burger, D.W. and Lieth, J.H. (1991). Plant growth and development of Chrysanthemum and Ficus in aero-hydroponics: Response to low dissolved oxygen concentrations. *Sci. Hortic.,* **45**, 287–294.

Wallach, R. and Raviv, M. (2005). The dependence of moisture-tension relationships and water availability on irrigation frequency in containerized growing medium. *Acta Hort.* (ISHS), **697**, 293–300.

Zobel, R.W., Del Tredici, P. and Torrey, J.G. (1976). Method for growing plants aeroponically. *Plant Physiol. (Rockville),* **57**, 344–346.

Zur, B. (1976). The pulsed irrigation principle for controlled soil wetting. *Soil Sci.,* **122**, 282–291.

5

TECHNICAL EQUIPMENT IN SOILLESS PRODUCTION SYSTEMS

ERIK VAN OS, THEO H. GIELING
AND J. HEINRICH LIETH

5.1 Introduction
5.2 Water and Irrigation
5.3 Production Systems
5.4 Examples of Specific Soilless Crop Production systems
5.5 Discussion and Conclusion
References

5.1 INTRODUCTION

The introduction of soilless culture on a commercial scale (Steiner, 1967) was motivated by a potential for increased crop productivity and efficiency. As part of this development, technical developments were made related to problems with root diseases, root zone oxygen deficiency, fertility control and increased complexity in irrigation strategy. Technical solutions to these problems and opportunities resulted in widespread adoption of soilless container plant production in outdoor nurseries in the 1950s and 60s. In the early 1970s, production of greenhouse crops in stone wool dramatically expanded commercially viable soilless crop production (Verwer, 1978; Cooper, 1979). Technical innovations in fertilization and irrigation resulted in adoption of fertigation technologies wherein completely soluble fertilizers are dissolved in irrigation water so as to deliver to plants the nutrients they need for optimal growth.

In all modern production systems, fertilization and irrigation have been integrated into a system that the grower seeks to optimize. Traditionally growers applied dry fertilizer products to the top of the root zone, resulting in fertilizer nutrients being carried to the root surfaces with the applied irrigation water. Once it became evident that all essential fertilizers could be supplied through completely soluble fertilizer salts, systems were developed where such salts are dissolved at relatively high concentrations in special stock solutions. By using one or more injectors, such concentrated solutions could be injected into the irrigation water (fertigation). This chapter focuses on the technical equipment associated with fertigation and irrigation in greenhouse production. In most such soilless production systems, fertilization is accomplished through injection of soluble fertilizer into irrigation water. We also discuss which system is best suited to which crop and where continued innovation is likely to result in future technical advances. Irrigation control methods are the subject of the previous chapter.

5.2 WATER AND IRRIGATION

5.2.1 WATER SUPPLY

An adequate supply of high-quality water is essential in soilless crop production, regardless of whether in outdoor nursery production or in greenhouse crop production. Potential water sources include rainwater, surface water and ground water. The latter two can be secured either directly or though municipalities as part of their drinking water supply.

The quality of irrigation water is typically evaluated through consideration of the dissolved minerals and salts in the water. Salinity is typically measured as electrical conductivity (EC) and it is known that water with high salinity (EC>2 mS cm^{-1}) can result in growth suppression (and perhaps also other problematic manifestations) for many plants. In some areas of the world, the available supply of water has EC nearly that high (or perhaps higher), so that this presents the grower with serious challenges.

The EC of the irrigation water results from a combination of the dissolved materials in the supply water plus any fertilizers that are dissolved in the water. It is desirable for the supply water to have an EC of less than 0.5 mS cm^{-1} and the sodium concentration less than 0.5 mmol l^{-1} (Sonneveld, 2000). This then allows for addition of ample fertilizer ions, so that the irrigation solution seen by the plants is below a level where problems might result. In recirculating irrigation systems, where some soluble materials will build up over time, EC management can be particularly challenging if the EC of the source water is higher than 1 mS cm^{-1}. Management of this type of situation is discussed in Chap. 9 and is an area of continued innovation in soilless production because water quality in many areas of the world is low, containing higher levels of sodium (Na$^+$) and other problematic elements such as SO$_4^{2-}$, Fe^{2+} or Mg^{2+}.

In addition to causing problems for plants, it is also possible for materials dissolved in the water to cause problems to the irrigation system. For example, iron and carbonates can deposit in the pipes causing blockages in pipes and filters.

5.2.1.1 Rainwater

Since the advent of agriculture, rain has always played a role in crop production. This source of water has the advantage that it is generally very clean (with low EC), but has the disadvantage that its availability is generally sporadic. Many soilless production systems include full protection of the plants from precipitation because of the uncertainties and potential production problems associated with rain events imposing uncertainty on the nutritional status of the root zone. Thus, the use of rainwater in soilless production involves capture of precipitation, storage in reservoirs and pumps (along with filters) to put the water under pressure for use in the nursery.

In general, rainwater has very low EC levels relative to the needs and limitations discussed above. Sometimes Na^+ concentration in rainfall is increased by rain and wind near oceans. Rainwater can be collected from the roofs of the greenhouses. In the Netherlands, where rain throughout the year is plentiful, growers are required to have a storage capacity of at least 500 m^3 of rainwater per hectare of production facility. Rainfall averages may range from 50 to 90 mm per month, so that with a storage capacity of 500 m^3 per hectare can provide for about 60 per cent of the irrigation water needs (Van Os and Stanghellini, 2002); if the storage capacity is increased to 1500 or 4000 m^3 ha^{-1} then 75 or 95 per cent, respectively, of the necessary water can be secured this way. This rainwater is stored in basins or tanks depending on the desired capacity. Earthen basins are used for capacities of more than 1500 m^3, while tanks (Fig. 5.1) are used for smaller volumes, since this is less expensive and less production space is needed to house such tanks.

Collection of rainwater is not used in some production areas for various reasons. If the quality of other water sources is high and it can be obtained at low cost, then it may not be economically feasible to make the investment in collection equipment and water storage, since the space lost to this equipment can remove a substantial amount of production space from the nursery. Also, in areas where low-cost greenhouse structures are in use, rainwater collection may not be feasible if the greenhouses are not designed with this in mind (lacking gutters or drainage plumbing). Also, in many areas (e.g. Northern California, Mediterranean countries) precipitation patterns are more uneven, requiring that the collection reservoirs be substantially larger than those identified above for Northwest Europe.

In some regions, rainwater can be stored underground in aquifers. This saves using expensive space, and water quality is maintained. In fact, many ground water sources are such sources, although on a much larger scale.

One salient issue related to the use of collected rainwater is that it is never adequate for 100 per cent of the needs of most nurseries. Thus irrigation systems must always include another water supply. This must be considered as part of the investment cost.

FIGURE 5.1 Rainwater storage in metal tanks (see also Plate 7).

5.2.1.2 Municipal Tap Water

Tap water may be unsuitable for use in closed soilless systems if it has been treated for human consumption to improve its flavour and kill bacteria. If calcium and chlorine have been added during processing, then it is particularly important that chlorides be present at less than $1.5 \, \text{mmol} \, l^{-1}$. In arid and semi-arid regions of the world, chloride levels are typically higher than this so that growers have to manage this by using salinity-resistant cultivars and other means.

In many areas of the world, municipal drinking water is relatively cheap compared to creating such water quality and capacity on site. Where this is not the case, the greenhouse operation must use ground water or surface water, perhaps in combination with stored rainwater and captured irrigation run-off.

5.2.1.3 Surface Water

Water from surface sources (creeks, rivers and lakes) is sometimes available in large quantities at a low price. While such water can be of high quality if relatively remote from human impact, the quality can also be low, especially if the water has been inadequately treated after municipal use or due to other human environmental impacts (e.g. canal water). The water may be polluted by salts, agricultural chemicals, inadequately treated sewage or storm water run-off from nearby urban areas.

Even when surface water is generally of good quality, it should be noted that water quality might not be uniform throughout the year, requiring frequent testing by the grower to assure its continued suitability for crop production. If such water quality deteriorates at particular times of the year, then the grower will need water treatment equipment to improve the water quality. Alternately, a second source of high-quality water can be used to blend with poorer quality water, as is done by many growers in the Netherlands with captured rainwater. Reverse osmosis is a commonly used treatment approach; while this process is effective, it increases the net price of supply water dramatically.

Desalination of sea water has been tested in an attempt to develop water resources for greenhouse horticulture. Unfortunately, the price of fresh water resulting from this is so high that it is not economically feasible without government subsidies.

5.2.1.4 Groundwater

In regions of the world with aquifers, groundwater can be an excellent source of water. Typically such aquifers are large and well buffered so that the supply quality is relatively stable. This is important for growers since it means that the fertigation systems design does not need to change over time. Since variations in water quality can, nonetheless occur, it is wise for the grower to have the source water tested every 6–12 month to monitor water quality. Many domestic groundwater sources have been unchanged for centuries but are now experiencing changes as run-off water from agricultural operations reaches the aquifer (a process that can take decades).

Also, in many areas where intensive agriculture is practised, groundwater depletion is a problem. Along coastal areas, where agricultural production is frequently intense due to the milder climates, the salt water intrusion into the aquifer can result in groundwater too high in salt for use in plant production (e.g. EC 10–15 mS cm^{-1}). In the region of Spain around Almería, groundwater, which mainly originates from the nearby mountains, is often used (EC ranges between 0.4 and 3.5 mS cm^{-1}). Water is pumped from wells between 150 and 600 m deep and, consequently, the groundwater table is decreasing, raising environmental and political concerns (Heuvelink and da Costa, 2000). The same phenomenon has been observed in the Central Coast area of California, particularly around Watsonville and Salinas where salt water intrusion has forced some growers to find alternate sources of water or close down their cropping system.

5.2.2 IRRIGATION APPROACHES

The equipment used to convert the raw supply water into water that is suitable for use in soilless cultivation should focus on improving water quality and to put the water under pressure so that it will flow through the plumbing. Sand filtration is frequently used to clean up water to make it suitable for use. In addition to filters, pumps are needed to create water pressure. All soilless production systems require water to be pressurized so as to make it feasible to have uniform irrigation. Water delivery systems typically involve the use of pipes that are of largest inner diameter near the pumps. The further the water in the pipe is from the source, the lower the pressure. This problem can partly be mitigated by decreasing the pipe cross section as the water travels through the pipe. But ultimately the irrigation water supply system

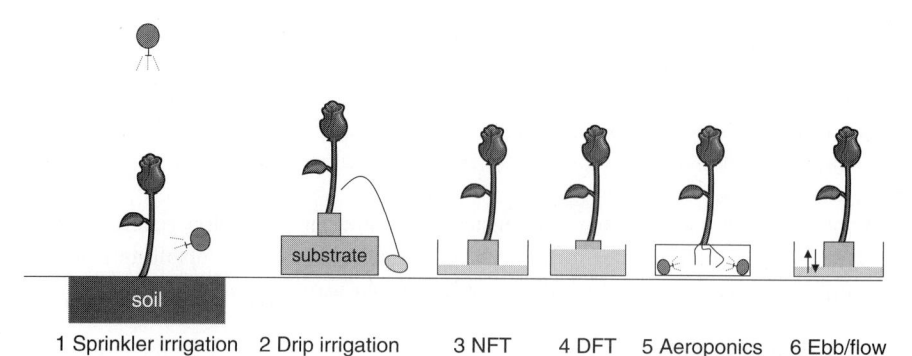

FIGURE 5.2 Various types of watering methods (see also Plate 8).

1 Sprinkler irrigation 2 Drip irrigation 3 NFT 4 DFT 5 Aeroponics 6 Ebb/flow

has to be divided into circuits in such a way that the pressure in each circuit is adequate to assure that the plants are irrigated uniformly.

As described in the previous chapter, irrigation can be from above the plant canopy, to the top of the root zone, or from below the root zone. Each of these irrigation approaches requires different technical equipment directly at the plant. The irrigation equipment needed in soilless production will be described below; a schematic overview is illustrated in Fig. 5.2. Generally in greenhouse soilless production, irrigation and fertilization methods are integrated.

5.2.2.1 Sprinkler Irrigation

Sprinkler irrigation systems apply irrigation water to the plants from above (Fig. 5.2). In some greenhouse and shade structures, plumbing is mounted overhead. Overhead installation does have the advantage that the plumbing is protected from mechanical damage by vehicles and persons. Installations can also have supply lines buried in the soil below the crop or mounted to the infrastructure. Movable sprinkler systems are also in use in various parts of the world. Crops for which it is undesirable to wet the foliage (e.g. some ornamental crops) can be irrigated with micro-sprinklers to the base of the plants. One positive feature of sprinkler irrigation from above is that it supplies water to most of the top of the root zone with relatively low investment and low maintenance costs (Heemskerk et al., 1997). Each sprinkler head type has a particular circular water distribution pattern; multiple sprinkler heads with short distances between emitters can be used to create a more uniform distribution pattern, but inherently sprinkler systems have uneven water distribution. For some crops, wetting of the foliage introduces a higher risk of plant disease development.

Micro-sprinkler irrigation systems, which deliver water to the base of the plants, overcome some of the drawbacks of overhead supply of water: there is no light interception due to overhead pipes and the crop becomes only partly wet. However, the plants themselves interfere with the even distribution of the water and prevent irrigation of large surface areas so that many more sprinkler heads are required, each delivering smaller amounts of water, to provide water to all plants.

Modern sprinkler installations are typically designed for the type of cropping system and the specific demands of the grower. In outdoor nursery production, sprinkler emitters may include rotating impact sprinklers and gear-driven ones which throw a stream of water 2–20 m to ones with no moving parts which provide a circular spray pattern with a radius of 1–5 m. In greenhouse production, sprinkler systems are less common and emitters are frequently micro-sprinklers which deliver water at lower rates with greater precision to the base of the plants. Emitters have also been designed with other features that improve performance; pressure compensation assures that emitters deliver water only while the pressure is between specified minimum and maximum levels to avoid leaking of the emitters.

Due to the low expense of sprinkler systems relative to other systems, it is the most widely used system in outdoor container production. But even in these types of operations, growers typically use sprinkler irrigation only with tightly spaced plants. Plants which are spaced far apart are typically irrigated with drip or micro-sprinkler systems to reduce the amount of irrigation that misses the plants entirely. In greenhouse production, sprinkler irrigation is less commonly used due to the drawbacks mentioned above.

5.2.2.2 Drip Irrigation

Drip irrigation is currently the most common irrigation approach in soilless culture in greenhouses. Drip systems are of two types: microtube systems or in-line systems. The former generally involves many emitters and supply tubing to allow an emitter to place an irrigation solution at a specific location, generally near the base of a plant. The latter consists of supply tubing with emitters attached directly to, or embedded in, the supply pipe.

Drip irrigation with microtubing (capillaries) and drip emitters (Fig. 5.2) allow the system to operate with relatively high pressure (at the pump or source), reduced to uniform lower pressure at the emitters. This approach allows the grower to maximize uniformity of delivery of irrigation solution to every plant. This is important for crops with a low planting density or where each plant is of high value, making it financially feasible to optimize fertigation of each plant. Each emitter is typically connected either to a small stake (Chap. 4, Fig. 4.4A) or to a weight (Chap. 4, Fig. 4.4B) to assure that they release the irrigation solution at the desired spot. Emitters or nozzles with various flow rates and working pressures are available and chosen according to the requirements of the crop and substrate. Emitters are typically matched to particular types of substrates so that emitters with low flow rates are typically used with substrates that have high infiltration and drainage rates. Pressure-compensated emitters can be used to obtain even water distribution. Leakage-protected emitters assure that water is only supplied during irrigation events (Fig. 5.3) and prevent dripping at various points in the system; by keeping the irrigation system from draining between irrigation events, the system also starts up faster during an irrigation, leading to greater irrigation uniformity especially for short-duration irrigation events.

When applied to crops in beds where plants are growing at a specified spacing, in-line drip irrigation can be used by matching the planting density to the fixed in-line emitter spacing. In-line emitters can have the same properties as the standard emitters.

FIGURE 5.3 Drip irrigation using pressure-compensated emitters to avoid leaking (see also Plate 9).

5.2.2.3 Nutrient Film Technique

Nutrient Film Technique (NFT) (Fig. 5.2) involves growing plants by maintaining a thin layer of nutrient solution around the roots, without the use of a substrate. When NFT first appeared, it seemed to be an ideal growing system because it seemed to offer optimal control over the watering of the roots without the expense of a substrate (Cooper, 1979; Graves, 1983). Today, however, NFT is used only for a few specific crops because of the expense and difficulty of solving a variety of technical issues related to the lack of a buffer and potential for outbreak of plant disease. Technically most crops could be grown in an NFT system (examples amongst others: Morgan and Tan, 1983; Lataster et al., 1993; Ito, 1994; Hortiplan, 2005) but widespread adoption has not occurred probably because such systems lack the ability to buffer even the slightest interruption in water and nutrient supply and there is a considerable risk of spreading root-borne diseases.

The system consists of a trough on a slope of 0.3–2 per cent; the roots of the plant lie inside the bottom of this trough. Nutrient solution is continually applied at the elevated end, so that the solution flows down through the trough at exactly the rate required to keep the roots completely wet. At the bottom end of the trough, the solution is allowed to drain. The nutrient solution layer should be as thin as possible, almost as a film. The width of the trough varies according to the crop; troughs of 4–8 cm are sufficient for crops such as lettuce and chrysanthemums, while for tomato and sweet pepper a trough of 15 cm is needed. The length of the trough varies from 1 to 20 m. Depending on the crop and the sizes of the troughs, various types of materials have been used: polyethylene liner, polyvinylchloride (PVC), polypropylene and coated metal. Ideal water flow rates have been identified to be between 3 and 8 $l\,m^{-2}\,h^{-1}$ for crops such as chrysanthemums and lettuce (Benoit and Ceustermans, 1989b; Ruijs et al., 1990a,b; Benoit and Ceustermans, 1994). Fruiting vegetables benefit from a faster flow rate. The slowest flow rate that is adequate to keep the roots coated with water may not be adequate in an NFT system; if the flow rate is too low, the problem is not lack of water, but lack of nutrients, especially for plants whose roots are downstream in the trough and are exposed to water from which many other plants have already extracted some nutrients. The last plants in the row get the least nutrients, especially potassium. Sometimes a distinction is made in flow rates needed for a young crop ($2l\,m^{-2}\,h^{-1}$) versus a mature crop ($5l\,m^{-2}\,h^{-1}$).

Plants destined for use in NFT systems are raised in small pots which are placed in the trough when a substantial root system has formed. In situations where the flow of water meanders in the trough, thus bypassing some plants, a lining of tissue in the bottom of the trough can be used to minimize this problem (Van Os and Kuiken, 1984). Shaped troughs (e.g. V-shaped) also prevent this problem (Formflex, 2005).

5.2.2.4 Deep Flow Technique

Deep Flow Technique (DFT) refers to another method that attempts to keep the roots continuously exposed to moving water and nutrients. While with NFT the water film is as thin as possible, with DFT the continuously flowing nutrient solution has a depth of about 5–15 cm (Fig. 5.2). The large buffer of water and nutrients make it considerably simpler to control the nutrient solution. Only a relatively small fraction of the water and nutrients are actually taken up by the plants. The large volume of water also buffers the temperature, making the system practical in regions where nutrient solutions temperature fluctuations can be a problem (Ikeda, 1985; Ito, 1994; Park et al., 2001; Both, 2005). The width of the troughs in a DFT system are typically about 100–130 cm. Plants are secured in holes on polystyrene panels by means of a polyurethane foam; the panels float on the water or rest on the troughs sidewalls. The system is often installed at working height so that crops such as the lettuce or herbs can be easily planted and harvested.

5.2.2.5 Aeroponics and Aerohydroponics

In aeroponic growing systems the roots of plants are suspended in a volume where emitters continually spray the roots with nutrient solution. The construction is similar to DFT (closed square box of about 1.2 m wide and 5–10 m long) but there is no water layer; rather there is a constant misting of the roots. A similar enclosed space

FIGURE 5.4 Aeroponics, nutrient solution is divided by sprinklers (see also Plate 10).

is created in a triangle construct of polystyrene boards, as in a Japanese commercial design (Panel, 1988) and in a research arrangement (Leoni et al., 1994). In the triangle construct, the relative humidity is 100 per cent and oxygen availability is increased around the roots (Fig. 5.4). However, care should be taken that the upper plants receive enough water. Often a thin layer of water is formed on the bottom, which acts as a buffer to the plants. Kratky (2005) showed the usefulness of the additional oxygen available in these systems by measuring higher lettuce yields in comparison to a continuous water layer around the roots. The amount of water supplied to the plants is not mentioned in the literature; often a timer is used, but the release rate of the nozzle is not given. Ruijs et al. (1990b) suggest a flow rate of $2 l m^{-2} h^{-1}$ for chrysanthemum crops.

The NFT and aeroponic systems share the disadvantages due to lack of a buffer around the roots for water, nutrients and heat. A hybrid system was devised by Soffer and Levinger (1980) by combining aeroponics with DFT, which they called 'The Ein Gedi System' and coined the term 'aerohydroponics' for it. Here the roots of the plants extend through a panel, dangling in an airspace as with aeroponics, but into a substantial quantity of flowing nutrient solution, rather than just a thin film. One facet of this system is that the nutrient solution is continuously aerated and the rate at which the bulk flow of water and nutrients bathes the roots exceeds that of any other soilless production system.

With all variations of aeroponics, electricity is used to achieve all the bulk movement of water and this movement is critical to the survival of the plants. Evaluation of the commercial feasibility of such system requires balancing this energy cost with the enhancements in nutrient uptake and oxygen availability.

FIGURE 5.5 Ebb and flow on a concrete floor for foliage plants (see also Plate 11).

5.2.2.6 Ebb and Flow/Ebb and Flood

In 'Ebb and Flow' or 'Ebb and Flood' systems (EF), container-grown plants are flood-irrigated on a water-tight tray or floor. During irrigation, water flows onto the tray or floor so that the base of each pot is submerged as the tray or floor is flooded (Fig. 5.5). The required duration of the flooding depends on the hydraulic conductivity of the substrate in the containers. The duration should be adequate to allow the water within the root zone to be wicked to the top of the root zone: generally 10–30 min if the substrate has high unsaturated hydraulic conductivity (longer if not). Then the tray or floor is drained which also allows the substrate to drain. Sloping surfaces and drainage grooves or channels are engineered into these systems to maximize the uniformity in water contents across a floor or tray between all pot/containers while facilitating rapid drainage. It is particularly important that the water be allowed to drain away completely so as to prevent root diseases. If wet spots remain around the plant, algae growth, root diseases and uneven watering will occur.

5.2.3 FERTIGATION HARDWARE

Water serves two important functions in plant production: it provides a vital resource for growth and also acts as a transport system for nutrients. Irrigation practices where both of these functions are actively combined in one system through the use of

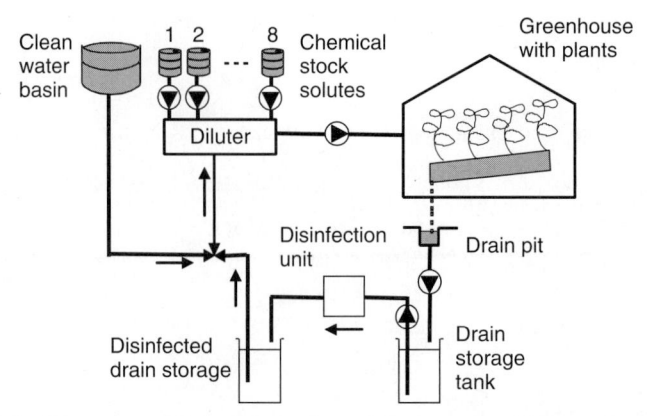

FIGURE 5.6 Schematic overview of the nutrient solutions way in a closed production system. The diluter/dispenser unit sees to it that concentrated fertilizers are diluted to a nutrient solution to be taken up by the plant.

completely soluble fertilizers are called fertigation. In this section, we focus on the dynamic control of the supply of nutrients in fertigation systems.

A diluter/dispenser unit combines nutrients as concentrated chemical solutes from two to eight stock tanks (Fig. 5.6) with irrigation water. A system of pumps, valves and irrigation capillaries delivers the resulting nutrient solution to each individual plant.

The amount of irrigation is often controlled by time-duration operation of the supply or by a sequence of a controlled number of equal volumetric additions. Valves connect the nutrient diluter/dispenser unit to the various irrigation sections of the greenhouse. Approximately, 50 section valves can be allocated flexibly; each section is equipped with instruments to measure and log the amount of supply and drainage and daily averages of EC, pH and water usage.

Supporting programs manage the rainwater stock storage tanks or basins, the drainage storage and the storage for cleaned drain water. When additional water is needed in the system, selection criteria determine the source of the clean water. Since clean water is a highly valuable resource, water from different sources should not be mixed beforehand.

In closed growing systems the drainage water is recycled. Precipitation of crystallized salts may occur when highly concentrated stock fertilizer are directly mixed with the water from the drainage tanks. The recycled drainage water will be mixed with fresh water in a proportion selected to achieve a pre-defined EC value of the input water to the diluter/dispenser unit. An accurate control of the EC and pH of this input water should be incorporated into the unit's controller, in combination with procedures for automatic cleansing of filters in the recirculation circuit. Comprehensive and detailed fertigation management should include these techniques to prevent failures due to unexpected EC and pH values or blockages in the system due to precipitation. See chapter 9 for more information on recycling if irrigation water in soilless production.

5.2.3.1 Sensors and Measurement

Electrical conductivity and pH

The EC and pH sensors are the simplest form of direct measurement in electrolyte solutions and are basic instruments for any grower using soilless production systems.

The EC sensors that are used in equipment for the processing of fertilizers measure using three equidistant ring-shaped electrodes which are mounted inside the water transport pipe at equal distances. The two end-electrodes are connected to each other and to ground. The temperature of the fluid is measured and is used to modify the value of the alternating current (AC) voltage applied between the central electrode and the ground electrodes as a means of temperature compensation. This AC voltage is typically around 1 V. The AC frequency that may range from 400 Hz to 50 kHz AC voltage is used to avoid polarization of the electrodes. The EC is determined by dividing the AC voltage by the electric current measured between the central electrode and the two end-electrodes. The current ranges from 0.1 to 10 mA. Both end-electrodes are connected to each other and to the electrical ground terminal to allow the serial or parallel connection of several EC electrodes in one water supply system. In horticultural practice, two distinct EC electrodes are used in parallel to provide a check on the functioning of both EC sensors against each other. The measuring range of an EC sensor is between 0 and 10 dS m^{-1}.

The pH of a solution indicates how acidic or basic (alkaline) it is. The pH sensor measures the potential across a thin glass bulb or membrane caused by the difference in activity of H_3O^+ ions (protons) in the electrolyte on one side of the membrane and the measurant on the other side. In fertigation equipment, pH is measured with a standard combination type of sensor and includes a measuring electrode and a reference electrode in the same sensor body. A gel is used as an electrolyte, which means that the electrolyte needs no further replenishing. In this way, the slow deterioration of the pH sensor is avoided. The lifetime expectancy of these pH sensors is about one year. Again, two sensors can be used, so that one sensor is checked against the other. The values obtained with the EC and pH sensors are compared with the results of a bi-weekly laboratory analysis as an additional check. The laboratory analysis is also used for determination of the composition of the nutrient solution.

Sensors for individual ions

An Ion-selective Electrode (ISE) is a sensor which converts the activity of a specific ion dissolved in a solution into an electrical potential which can be measured by a voltmeter (Fig. 5.7). The voltage is theoretically dependent on the logarithm of the ionic activity, according to the Nernst equation (Chang, 1990). The sensing part of the electrode is usually made as an ion-specific membrane, along with a reference electrode. ISEs are used to measure the activity of cations and anions in the root environment. ISE sensors are available for most macro-nutrients, like K^+, Ca^{2+}, NO_3^-, SO_4^{2-}, NH_4^+ and for Na^+ and Cl^-.

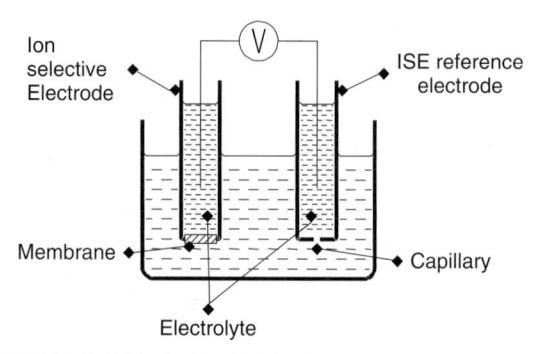

FIGURE 5.7 Basic principle of an Ion-selective Electrode.

Substrate moisture

Close monitoring of soil or substrate moisture may be useful to maintain an efficient growth and to protect the environment. In open growing systems, waste of valuable nutrients and pollution of the surrounding environment are caused by the practice of providing excess nutrient solution at each irrigation (10–50 per cent) and letting the excess drain away. Closed growing systems solve this problem by catching and recycling the drainage water.

The moisture content of the growing medium is an important variable in the uptake of water. Sensors to measure water content determine the dielectric properties or the hydraulic properties of the growing medium, or the change of weight of the growing medium is established.

Dielectric sensors

Dielectric methods to determine water content from the dielectric constant of the substrate are becoming more common. The relationship between the dielectric constant for each specific growing medium type and its water content has to be established in a calibration procedure.

In time domain reflectometry (TDR), a short electrical pulse is sent into a pair of electrodes and the time-dependent reflected signal is a measure of the water content of the medium between the electrodes. TDR equipment is commercially available. The basic principle of dielectric sensor operation is described in detail in Chap. 3.

In an alternative approach, impedance between two electrode needles is measured at one carefully chosen (high) frequency (Hilhorst et al., 1992). This Frequency Domain (FD) Method allows calculation of both water content and EC and is easier to automate and miniaturize than TDR. A commercial version is available as a sensor for *W*ater content of the substrate, *EC* of the substrate and *T*emperature of the substrate (the so-called WET sensor, Balendonck et al., 1998; Delta-T, 2006). Special designs of the two electrodes of the FD sensor offer opportunities to build sensors with one electrode rod (Balendonck et al., 1998; Delta-T, 2006). FD sensors are commercially available.

Hydraulic tensiometer

A porous cup at the end of a sealed tube filled with distilled water can be used to measure water potential under relatively moist conditions. When the cup is in contact with the growing medium, equilibrium will develop in which the pressure inside the cup equals the suction of the soil or the substrate. The pressure inside the tensiometer can be measured with a pressure transducer. The basic principle of tensiometer operation is described in detail in Chap. 3.

The method works for suction pressures as low as -80 kPa. At lower pressures, there is a risk of air entering the cup. Errors will arise when contact between cup and substrate is lost. Hydraulic tensiometers are commercially available. For more information on the use of tensiometers in soilless production, see Chap. 4.

Gravimetric sensing

Changes in plant weight can also be sensed so that this input can be used as part of irrigation control. One example is shown in Fig. 5.8 where metal trough with approximately 16 plants (to give a representative sampling) is suspended from the greenhouse frame by wires. The wires are connected to load cells. An extra set of load cells measures the change of the fresh weight of the plants (Fig. 5.8). The weight changes are due to growth, supply of nutrient solution, run-off of drainage water, evaporation of water and the transpiration of the plants. The change of weight due to growth, evaporation or transpiration is slow and over short time periods it shows itself as an almost steady signal.

The amount of supplied nutrient solution is determined by detecting the sudden change in weight when the supply is started. The exact increase in weight from that moment on till the end of supply is the accurate measuring signal for the supplied amount.

The drainage water is collected in a small break-out tank ($\sim$50 mL), which is fixed to the trough. The tank discharges when a fixed level is reached in the break-out tank. The sudden discharge of the collected drainage by the break-out tank is detected by the sudden change in the overall weight signal of the load cells. The exact decrease in weight from that moment till the end of the discharge is the accurate measure for the amount of drainage water.

On the basis of this precise information, the supply of nutrient solution can be controlled with high accuracy. The change of the fresh weight of the plants and the value of the transpiration are used to inform the grower about the status of the crop. The basic principles of gravimetric sensing, and examples for its use, are described in detail in Chap. 3.

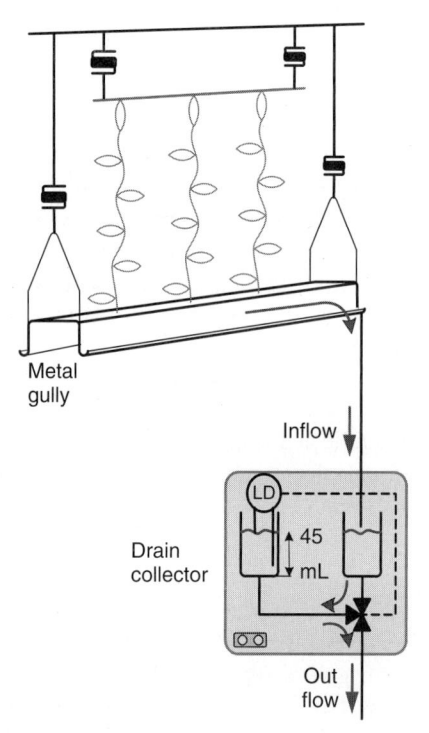

FIGURE 5.8 Measuring trough: by measuring the weight of the plants and the trough and the incoming water supply and outgoing leaching part, there is exact knowledge about the quantity of nutrient solution to be dosed to the plant at exactly the right moment.

5.2.3.2 Stock Solutions: A and B Solutions Versus Individual Liquid Fertilizers

A specific recipe of nutrients can be made for the particular crop in relation to the chemical composition of the source water. In 1970, when hydroponic production was introduced in The Netherlands, a system was developed by Verwer (1978) with one central tank to hold irrigation solution created by combining water with two stock solutions, called A and B solutions.

This type of system is still in widespread use today. The two stock solutions are prepared at concentrations 100 or 200 times higher than the average concentration of the solution to be supplied to the plants. Tank A contains all calcium compounds. In container B, all sulphates and phosphates are dissolved (Fig. 5.9). If the fertilizers in the two stock solutions were to be mixed at these concentrations, precipitation would occur in the pipes, valves and emitters, with disastrous results. For these reasons, solutes from the A and B tanks are diluted in fresh water in the mixing tank to achieve the right concentration and the appropriate EC level of the nutrient solution. An extra amount of an acid or base solute may be used to control pH of the batch of nutrient solution in the mixing tank.

PLATE 1 The orangery at Pillnitz Palace near Dresden, Germany (see also Figure 1.2, p. 3).

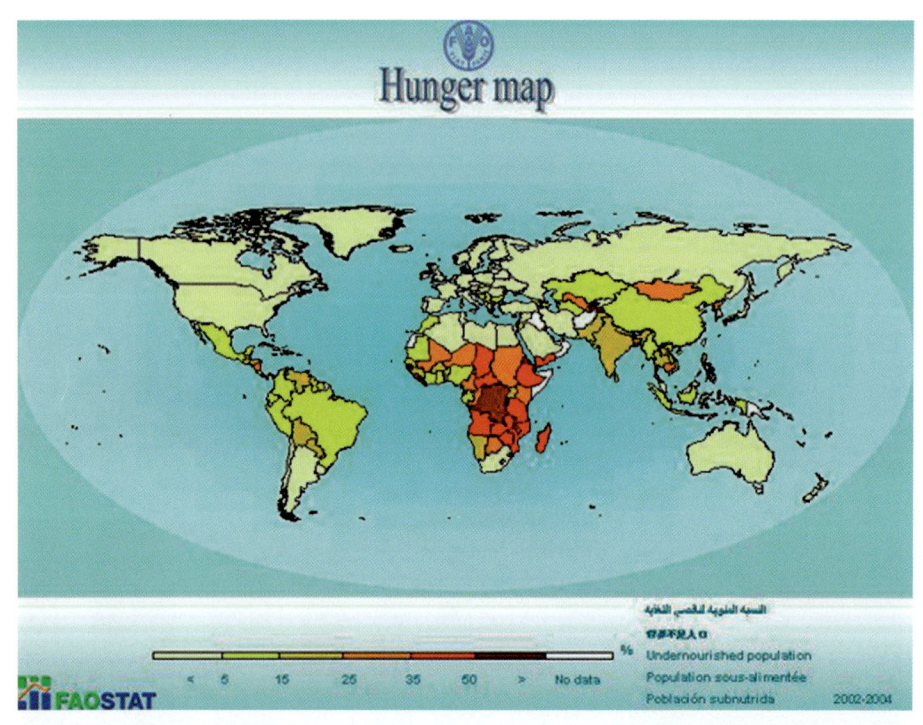

PLATE 2 Percentage of undernourished population around the globe (see also Figure 1.4, p. 9; with kind permission of the FAO).

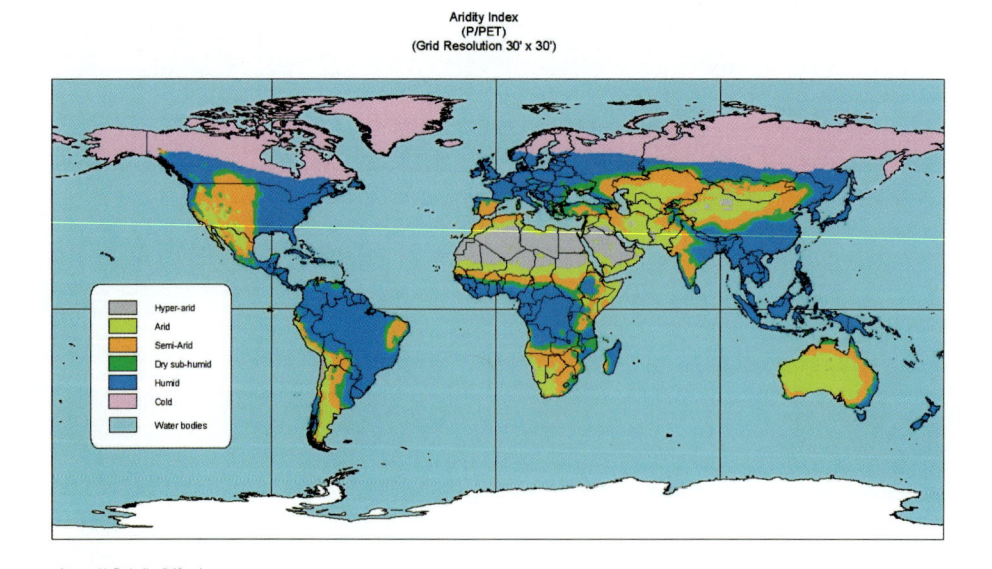

PLATE 3 Aridity index around the globe (see also Figure 1.5, p. 9; with kind permission of the FAO).

PLATE 4 Overhead irrigation systems: (A) Hand watering, (B) Boom system, (C) Mist system, (D) Sprinkler system in an outdoor nursery (see also Figure 4.3, p. 132).

PLATE 5 Surface irrigation systems: (A) Spaghetti system, (B) Micro-spray emitter, (C) Drip irrigation monitoring, (D) Perimeter micro-spray irrigation, (E) Drip tape and (F) Custom-spaced in-line emitters (see also Figure 4.4, p. 135).

PLATE 6 Subirrigation Systems: (A) Capillary mat, (B) Trough system, (C) Flooded tray or table, (D) Flooded floor system (see also Figure 4.5, p. 138).

PLATE 7 Rainwater storage in metal tanks (see also Figure 5.1, p. 160).

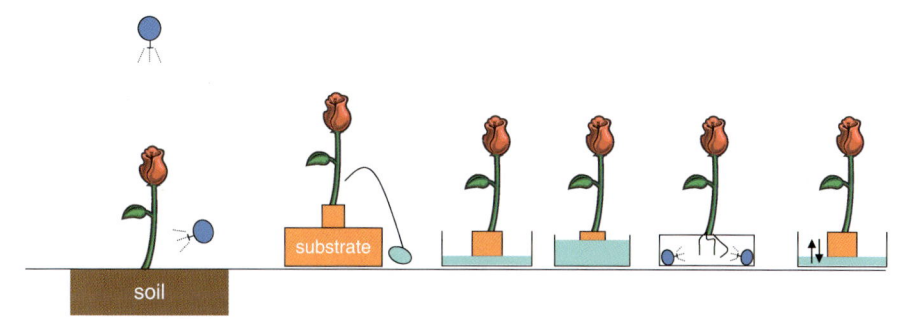

1 Sprinkler irrigation   2 Drip irrigation   3 NFT   4 DFT   5 Aeroponics   6 Ebb/flow

PLATE 8 Various types of watering methods (see also Figure 5.2, p. 162).

PLATE 9 Drip irrigation using pressure-compensated emitters to avoid leaking (see also Figure 5.3, p. 164).

PLATE 10 Aeroponics, nutrient solution is divided by sprinklers (see also Figure 5.4, p. 166).

PLATE 11 Ebb and flow on a concrete floor for foliage plants (see also Figure 5.5, p. 167).

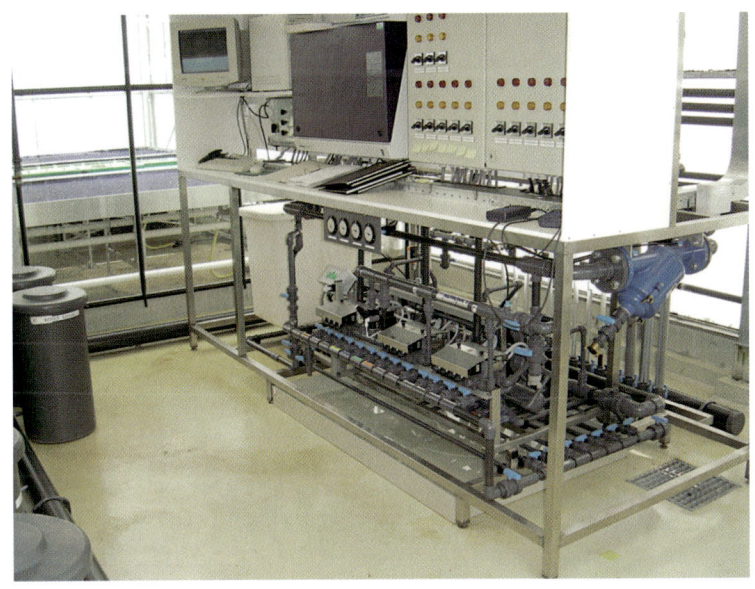

PLATE 12 More sophisticated direct injection system for single element liquid fertilizers (see also Figure 5.12, p. 177; courtesy Priva bv, De Lier, Netherlands).

PLATE 13 Suspended troughs for uniform water contents of the substrate and drainage of the nutrient solution (see also Figure 5.15, p. 182).

PLATE 14   Belgian endive in stacked 120 × 120 cm containers (see also Figure 5.17, p. 184).

PLATE 15 Movable benches can be transported out of the greenhouse bay to another place. Mostly the pipe-rail system is used; In some applications a robotic lift is used to pull out an entire individual bench for rapid movement to another location (see also Figure 5.22, p. 190).

PLATE 16 Movable row system for rose (see also Figure 5.23, p. 191).

PLATE 17   Walking plant: an individual approach (see also Figure 5.24, p. 192).

PLATE 18 Strawberry in suspended troughs: optimized utilization of space (see also Figure 5.25, p. 196).

PLATE 19 The lower test cylinder extended to double its height by putting a second test cylinder on top, held in place by using a clamp and a rubber. The test cylinder and rim (second holder) are then filled with a few gentle taps to avoid large voids and create an even filling. The sample is brought to a specific suction force and finally the rim or top cylinder is removed and the excess material is wiped or scraped off with a rigid blade. After weighing, drying and weighing, the bulk density is calculated (see also Figure 7.1, p. 251).

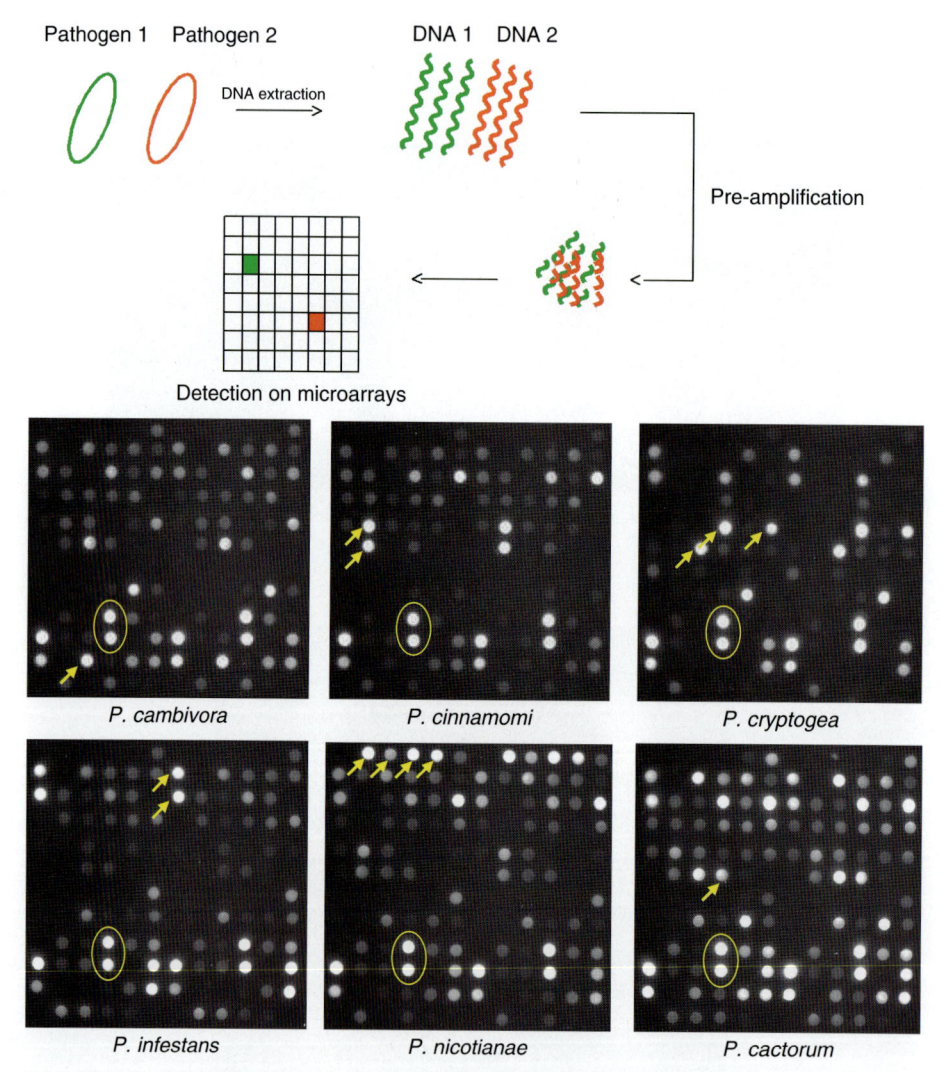

PLATE 20 Schematic presentation of the micro-array-based multiplex detection of several pathogens on a single chip (upper part). Below part: Hybridization patterns for six different *Phytophthora* species on a micro-array on which species-specific probes were spotted. Examples of common *Phytophthora* DNA sequences (circle) and species-specific sequences (arrows) are indicated (see also Figure 10.2, p. 434).

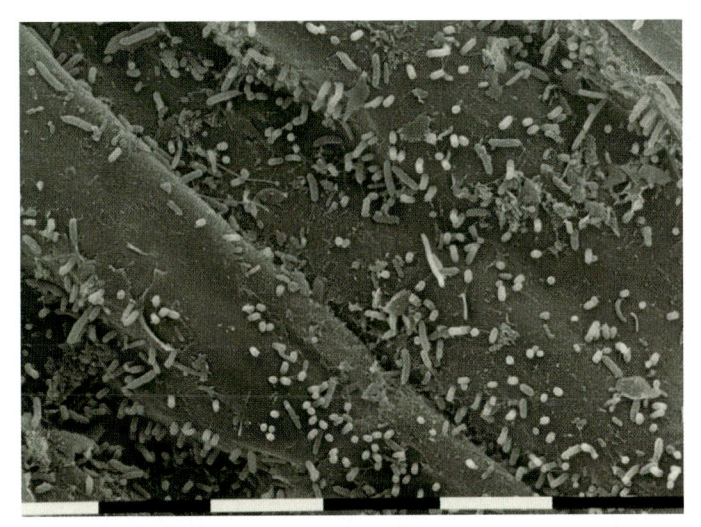

PLATE 21 Bacterial population on a three-week-old cucumber root grown in stone wool visualized by scanning electron microscopy (SEM) (bar = 10 μm) (photo by Anke Clerkx, Plant Research International) (see also Figure 10.4, p. 437).

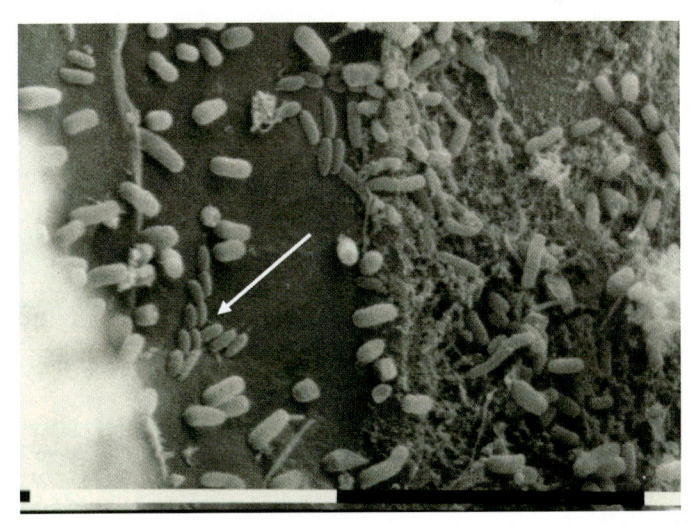

PLATE 22 Bacterial populations on a cucumber root grown in nutrient solution (bar = 10 μm). Arrow points at a micro-colony (photo by Anke Clerkx, Plant Research International) (see also Figure 10.5, p. 438).

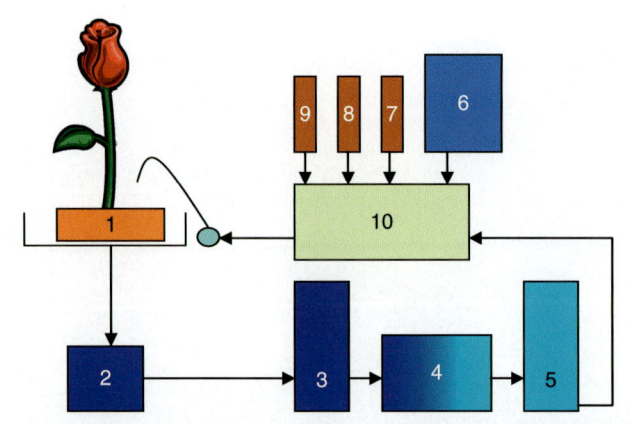

PLATE 23 Closed system scheme focusing on disinfestation. Surplus water of the plants (1) flow by gravity to a central recatchment tank (2), from where it is irregularly pumped to a day storage tank (3). Twenty-four h continuous disinfestation (4) takes the nutrient solution from (3) to a tank (5), where it is stored to be used for watering. In the mixing container (10) disinfected drain water (5) is mixed with fresh water (6) and A and B nutrients (7, 8) and if needed acid or lye are added (9). After controlling EC and pH, the water is supplied to the plants (see also Figure 10.6, p. 443).

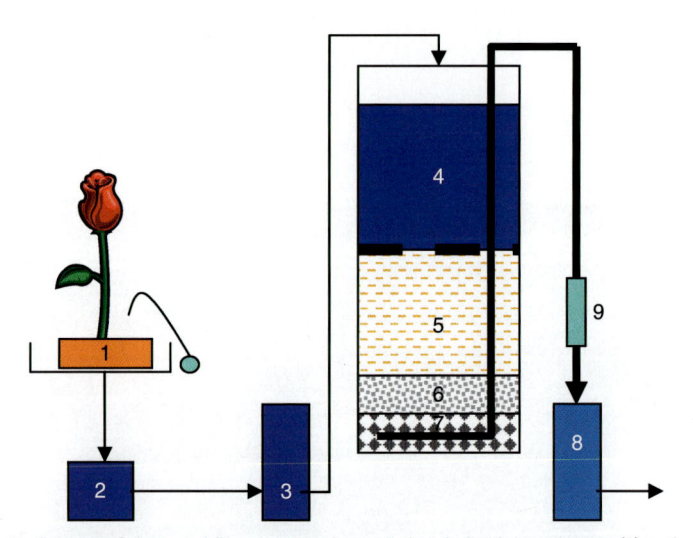

PLATE 24 Scheme of slow sand filtration. Nutrient solution drains from substrate (1) to the recatchment tank (2). From there it is pumped to the day storage tank (3) and into the top of a large container or metal silo (4), from which it drips into a sand layer of 1 m thickness (5). The layer between (4) and (5) is called the Schmutz-decke or filter skin. (6) and (7) are a 10 cm fine and a 15 cm coarse gravel layer, respectively. The filtrate is pumped out of the gravel layer to container (8). In a metal silo it is done via the top, in a synthetic filter it is possible to drain via the bottom of the filter. For initial filling of the filter water is pumped from (8) into the gravel layers (7) and (6) and to above the sand layer. Flow meter (9) controls the filtration rate. From container (8) the filtrate will be mixed with fresh water to a new nutrient solution for the plants (see also Figure 10.7, p. 445).

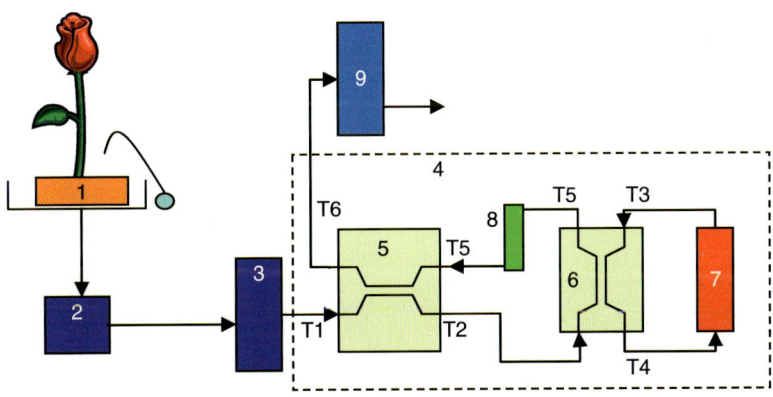

PLATE 25 Scheme of heat treatment. Nutrient solution drains from substrate (1) to the recatchment tank (2). From there it is pumped to the day storage tank (3) and into heat treatment unit (4). The solution is pumped with temperature T1 (about 17–25°C) into heat exchanger (5) and preheated to T2 (70–80°C). In heat exchanger (6), it is further heated to T5 (85–97°C) by an external heat source (boiler, 7) which has an incoming temperature of T3 (95–105°C) and an outcoming temperature T4 (85–95°C). The water is kept on temperature T5 in unit (8) for an exposure time of about 30s and cooled down to temperature T6 (22–30°C) and stored in container (9) to be mixed with fresh water and nutrients for watering the plants (see also Figure 10.8, p. 446).

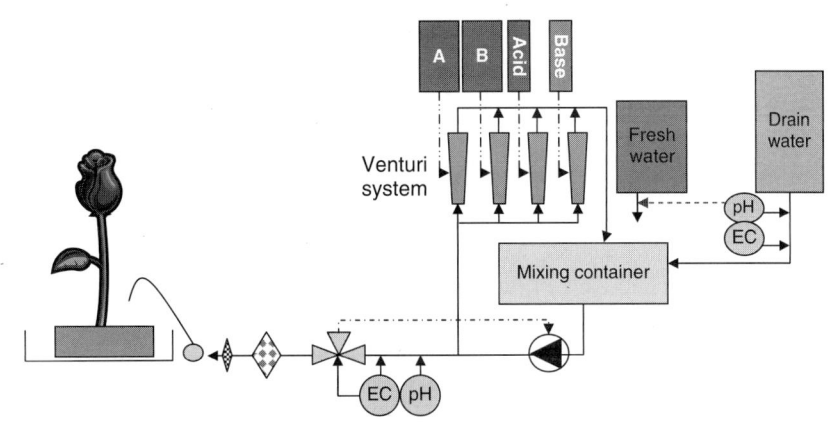

FIGURE 5.9 Example of a fertilizer diluter/dispenser unit where a Venturi fitting in the main supply pipe creates a pressure drop to transport concentrated fertilizers.

The composition of the A and B mixtures is selected according to the requirements of the crop to be fertigated and is based on a laboratory analysis performed every two weeks. Fertilizers are usually added to the A and B tanks by hand. This is heavy and unpleasant work, especially when in large commercial production green houses.

One option for dealing with this difficulty is to obtain premixed individual liquid fertilizers by tanker from a supplier. In general, depending on the supplying company, 6–10 liquid fertilizers are combined in a standard recipe for most commercial greenhouse crops. The tanker is equipped with separated segments, one segment for each concentrated fertilizer. A standardized discharge procedure ensures that each chemical fluid ends up in the right container, thus minimizing the chance of accidents. The need for additional manual labour is eliminated. However, a significant disadvantages of this method is the cost for the transportation, since a large quantity of water is transported. The tanker method is only of particular interest in areas of greenhouse concentration, such as the Westland area in The Netherlands, the Almería region in Spain or the Brittany region in France.

A second option is to manually mix solid fertilizers in the A and B stock tanks from 25 kg bags with standard fixed mixtures. The amount of labour can be decreased by the use of large plastic bags of 1 or 2 m^3, where the solid fertilizers are pre-mixed by the supplier in an A bag and a B bag, in accordance to the latest laboratory analysis. The bags are delivered on standard pallets, placed on a platform above the A and B tanks by a fork lift truck and easily emptied through an opening at the bottom of the bag. The whole content of the big bags is used at once.

New recipes of the A and B stock solution are prepared sequentially, but the release of the new recipe in both stock tanks is done at the same time to ensure that changes in the composition of the A and B solutions happen at the same time.

The recipe of the nutrient solution must be adapted to the needs of the crop during the development of the plant from a seedling stage to a reproductive stage. Cultivation starts with a species-specific standard recipe. This recipe is then based on advice from a specialized laboratory after a test performed every two weeks on a sample taken from the root environment. The advice takes into account the composition of the different stock nutrients, whether they are solid or liquid stock fertilizers and information about the status of the crop. The appropriate adjustments to the recipe are calculated on the basis of this information. Not all the possible outcomes of these calculations result in recipes that the diluter/dispenser unit is capable of producing. During recent years, fertilizers have evolved from simple single- or multi element fertilizers to more sophisticated combinations of elements, fine-tuned for a specific growth strategy. In liquid fertilizers, the option of adjusting pH levels has also been introduced (Fig. 5.10) by adding acid or basic fertilizer solutions. Manual calculation of the recipe has been replaced by computerized calculation, which has made it possible to take all kinds of constraints into account. Nowadays these algorithms are also directly available to the grower as a spreadsheet or as a specialized calculation program related to a particular brand of fertilizer. Certain mixtures may lead to precipitation or to combinations of ions that cannot be produced from the available fertilizer stock solids or stock solutes.

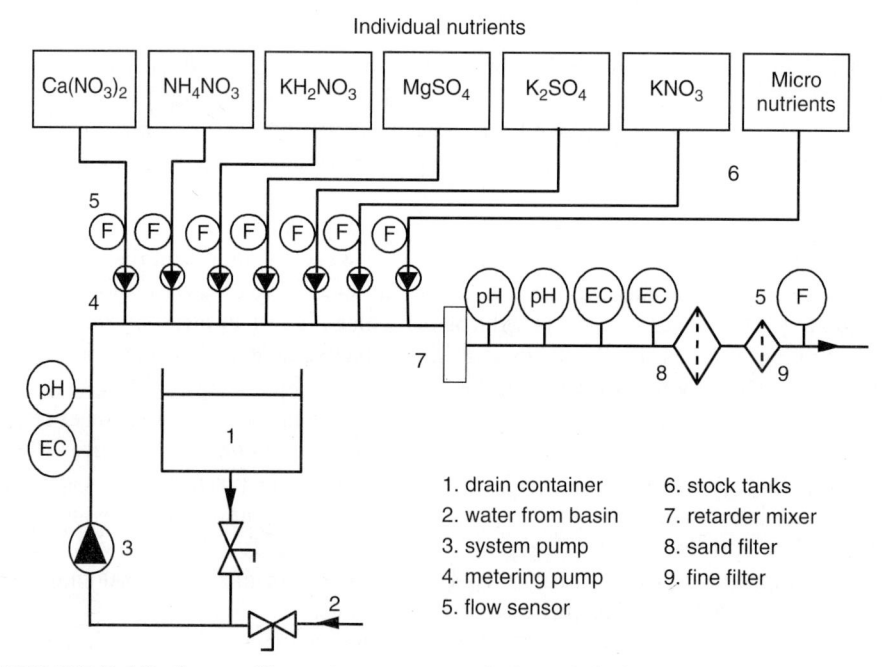

FIGURE 5.10 System with metering pumps to supply the stock liquids directly into the main supply line or into an open non-pressurized mixing tank.

If the dilutor/dispenser units were forced to proceed with an 'impossible recipe', it could cause precipitations, blockages in the transport conduits or even explosions (Schrevens and Cornel, 1990). Additionally, optimization algorithms may on occasion lead to a more cost-effective choice of similar fertilizers.

5.2.3.3 Commercial Injector Systems for the Dilution of Concentrated Fertilizers

The mixing tank system

The operating principles of mixing tank diluters are as follows: A mixing tank is connected to the various fertilizer stock tanks (Fig. 5.9). In these storage tanks, solid fertilizers are dissolved in water and diluted at a standard high concentration. Quantities of the highly concentrated fertilizer solutions are fed into the tank of the nutrient dispenser system. They are mixed in accordance with a recipe, diluted with fresh water to a suitable concentration for the plants and finally pumped to the crop in the greenhouse.

There are several ways of feeding the highly concentrated stock mixtures into the mixing tank:

- Metering pumps supply the stock liquids into an open non-pressurized mixing tank or directly into the main supply line. One batch is prepared at a time in the mixing tank and then pumped to the greenhouse (Figs. 5.9 and 5.10). Examples of these systems are found in the early designs of computerized nutrient fertilizer systems (Bauerle et al., 1988; Papadopoulos and Liburdi, 1989).
- A Venturi fitting inside the main supply line creates a pressure drop which forces the stock fertilizers (Fig. 5.9) into the mixing tank or into pre-mixing chambers.
- Small container tubes (~ 5 L) are placed just above a mixing tank, one tube for each concentrated stock solution (Fig. 5.11). A valve at the bottom of the tube releases a precise quantity of stock solution into the mixing tank. A pressure sensor in the bottom of the tube measures the amount of stock solution released. The container tubes are filled by means of simple pumps or gravity (the stock tanks are in an elevated position).

The EC of the supply solution is controlled by altering the extent of the dilution of nutrients and water. The simplest systems are laid out to mix a small number of stock solutions, such as systems where four nutrient stock fertilizers are used with an extra acid or base solution for pH correction. EC- and pH-sensors are installed in duplicate to ensure safe operation. An embedded computer system controls the mixing process, the alarms and operator communication.

Mixing tank dispensers are used in systems with overhead sprinklers, drip irrigation systems or in ebb-flow systems. They are configurable for both in-line as well as in batched by-pass mode. The systems are available for various output capacities in the range of $1 \, m^3 \, h^{-1}$ to more than $40 \, m^3 \, h^{-1}$. A small number (4–6) of different nutrient recipes is possible. The EC set value is dependent on changes in global radiation.

FIGURE 5.11 Small container tubes (~5 L) are placed just above a mixing tank to precisely release a quantity of stock solution into the mixing tank.

Direct injection system

The injection of concentrated fertilizers by means of a pressure drop in the main supply line is a modern and attractive alternative to the use of metering pumps. The pressure drop is created across a Venturi (Fig. 5.9), sucking the concentrated fertilizer fluids into the flowing liquid. This means of fertilizer transport uses fewer devices and moving parts that require maintenance, than a system with metering pumps. Direct injection diluter/dispenser units are available in a range of sizes and capacities.

Simple direct injection systems use a mixing tank to mix and dilute the stock fertilizers from the A and B tanks by using separate pre-mixing chambers into which the highly concentrated solutions of stock fertilizers are combined with supply water. This solution is then injected into the main supply line, where it is diluted to the concentration that is needed for the plants.

Systems that blend and dilute highly concentrated stock solutions of individual liquid fertilizers are more complex. The dispenser unit uses a fertilizer channel for each stock tank. A Venturi pressure drop in the main supply line is again the driving force behind the movement of the concentrated fertilizer fluids. The channel is closed off with a small ceramic insert with an orifice. The orifice ensures a fixed and limited flow rate. Ceramic inserts with different orifice sizes are used for the different channels. A flow meter and a fast-switching valve control the flow. Four channels are connected to the mixing chamber. The amount of fertilizer transported into the mixing chamber is determined by the pulse activating the valves. A static retarder in the main supply line

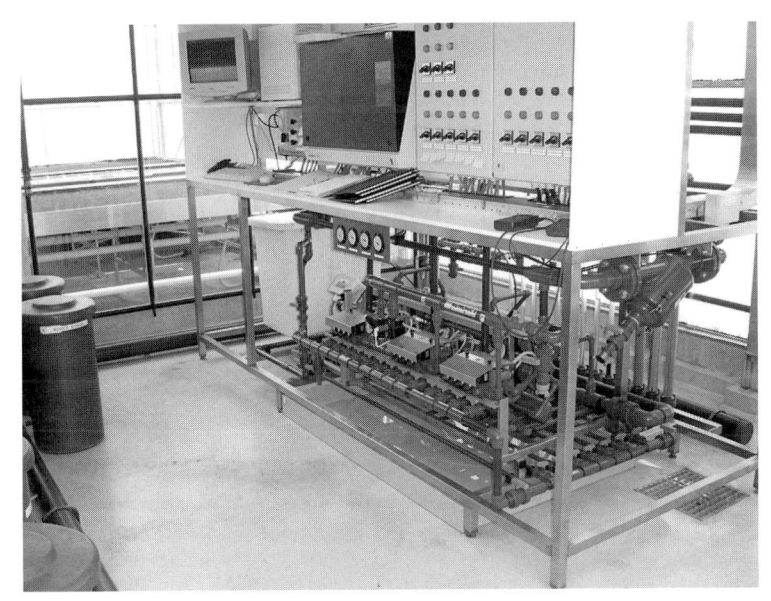

FIGURE 5.12 More sophisticated direct injection system for single element liquid fertilizers (see also Plate 12; courtesy Priva bv, De Lier, Netherlands).

mixes the highly concentrated fertilizers with the supply water (Fig. 5.12) and prevents the contact of hazardous combinations of highly concentrated stock fertilizers with residues of nutrient ions in the supply water (recirculated in case of closed systems).

The process of mixing and diluting the various individual fertilizers is controlled by feeding back the measured EC value of the supply water into a computer controller. The amount of each individual fertilizer is calculated and corrected by the controller as a function of the measured water supply flow rate and the volumetric relation between the various fertilizers, as prescribed in the recipe selected by the grower. The intended pH and EC will be realized in the solution when the mixing algorithm is correctly applied. However, other factors, such as small changes in the composition of the clean input water of the system or the use of recirculated drainage water, introduce the need for fine-tuning the pH and EC value. This is carried out automatically by adjusting the output of the appropriate individual fertilizer channel in relation to its recipe value. An excess of a particular nutrient ion can be corrected in the next fertilizer supply cycle.

The concentration of the individual stock solutions is used as an input parameter in the calculation of the fertilizer recipes. The system is thus easily adjusted to accommodate all available concentrations of stock solution, such as standard industrial solutions or special stock solutions prepared for horticulture. Durable materials and a small number of moving parts ensure intrinsic reliability.

In general, eight liquid fertilizer fluids, including one channel for micro-nutrients, are sufficient to realize every possible recipe. Injection dispensers are available at a capacity in the range of $100 \, m^3 \, h^{-1}$.

Safety and protection

Mixing highly concentrated ionic nutrient liquids can be hazardous. Extra precautions must be built into the system to protect the grower and his staff from hazardous situations. For this purpose, extra sensors should check the process of dilution, mixing and dispensing for potential hazards, such as

- excessive or insufficient flow in the individual fertilizer channels;
- liquid fertilizer leakages;
- intake of air into the fertilizer mixing channels;
- unexpected drop in capacity;
- jamming of the channel.

General checks on the following should also be made:

- sudden or large deviations from the expected EC and pH values;
- leakage from the stock tanks (also when the equipment is switched off);
- the stock tanks should be stored in a containment pool with the capacity of at least one stock tank.

5.3 PRODUCTION SYSTEMS

Since the advent of the use of soilless culture, various types of production systems have been developed to produce specific crops. In general, crops grown this way are 'high-value' crops that have a high rate of economic return, thus ensuring that the deployed inputs-intensive system is profitable for the grower. Initial phases of this industry involved production in the ground in highly amended soils with above-ground soilless systems being a more recent development. In soilless production today, some systems sit directly on the ground, while others are raised above ground, allowing for no connection between plants and soil.

For each of these types of production systems, we will discuss the advantages and disadvantages, and the suitability for specific groups of crops (crops in rows, crops in beds). Substrates will not be discussed in this section, although a distinction will be made between solid substrates (such as stone wool or polyurethane slabs), granulated substrates (such as perlite, pumice, coir or peat) and no substrate (as in NFT and aeroponics). For a complete discussion on substrates see chapters 3, 6, 11 and 12. Most systems can be either open (all leachate flows into the subsoil without recirculation) or closed (leachate is reused).

5.3.1 SYSTEMS ON THE GROUND

5.3.1.1 Ground Preparation

Ground preparation is extremely important when installing a production system that will be placed directly on the ground, so as to minimize the incidence of pathogens and weeds. Steam sterilization, a chemical treatment or a solarization treatment can be used to ensure a disease-free start and to prevent the germination of weeds. Treatment of the top 10 cm of soil is generally sufficient. Production systems which sit directly on the ground also require that the ground be prepared to accommodate the hydraulic needs of the system so that the drainage water from the production system is managed correctly. Wet and dry areas will develop if the root zone is not perfectly horizontal or on an even incline. Consequently, water availability for the plants will not be uniform resulting in oxygen deficiency in some pockets, leading to root rot. Deviations from this may also lead to variations in EC levels, leading to reduced productivity and quality.

Levelling is mostly done at a slope of 0.3 to 0.5 per cent along the length of the greenhouse span. The overall length is subdivided into sections of 20–25 m to avoid excessive differences in height within the greenhouse. A moulding machine can be used to adequately shape the subsoil after levelling. The substrate bed that will make up the root zone may be lower or higher than the aisle and small canals are dug for drain water collecting tubes. Shaping by hand is impractical for larger areas, and the accuracy of the work is likely to be inadequate. Compaction of the soil may also be needed in cases where the production system is heavy.

A plastic barrier between the production system and the ground is generally used to manage weeds and excess moisture. This can take the form of a plastic film or durable woven material with a tight mesh. Selection of a liner made of a light-coloured material provides the feature that it reflects light back up towards the plants. This is especially important in latitudes with prolonged dark periods and short daylight periods. A disadvantage of a light-coloured (white) material is that it generally allows enough light to penetrate into and through the material to allow for some weed germination and UV degradation of the plastic. A black liner may be more advantageous than a white liner in areas where it is of greater value to have the ground below the liner attain slightly higher temperatures (cold greenhouses without possibilities to heat during the night) since it will store more heat during the day and release it during the night. A dark material is also generally more resistant to the degrading effects of UV radiation. Additionally, a liner prevents the movement of plant diseases from the ground into the production system.

Woven material, rather than impervious film, should be used in cases where water needs to seep into the ground below the aisle and production system. In such cases, the ground preparation should be such as to avoid excessive compaction of the ground where the aisle will be located.

There is a significant difference in the type of layout of plants within ground-based systems. Some production systems consist of long rows of individual plants spaced at some particular spacing (i.e. 'single rows'). Other systems consist of rows of plants with several plants side by side (i.e. a 'bed' of plants). In this chapter, we define 'single row' to be a row of individual plants in a substrate and/or trough forming a root zone that is 4–25 cm wide. On the other hand, a 'bed' is defined as consisting of several plants growing side by side in wider arrangement (>1 m) with the root zone completely filled with substrate.

5.3.1.2 Crops in Single Rows

Single row systems are in use for crops such as roses, cucumber, tomato, pepper, strawberry and lettuce (Benoit and Ceustermans, 1989a). For crops in narrow rows, small quantities of substrates are normally used during propagation in the form of pressed peat pots or plugs. Propagules are started in these, from either seed or cuttings, and later transferred to a production system. Many times, single row systems are placed so that two single rows are adjacent (forming pairs of rows); this is done when the width of a single row is such that harvesting and plant management can be accomplished from one side of the row. This effectively eliminates a large percentage of aisle space, resulting in greater overall production efficiency.

Generally, the soilless substrate that makes up the root zone of plants grown in single rows is enveloped in plastic film so as to contain the irrigation water. Preformed materials, such as stone wool slabs, as well as some granulated substrates are generally enclosed in bags which have adequate support to allow them to be placed directly on top of the liner without the need for a reinforcing container. Substrate products which are not in these forms require some physical containment to hold the root zone in its desired form. This is either in the form of nursery containers, buckets or troughs.

Production systems which are placed directly on the ground are generally more difficult to configure as closed systems. In an open system, drainage water seeps directly from the root zone through the liner into the ground below the crop (Fig. 5.13). This traditional approach has the benefit that the required investment in materials is low, but a lot of water and fertilizer is wasted while also polluting the environment. Governmental regulations in most of Europe and North America now control the permitted amount of nitrate leaching (Agricultural Structure Memorandum, 1989; National Environmental Policy Plan, 1989) so that open systems are being phased out in many such areas. This has accelerated the development towards closed systems.

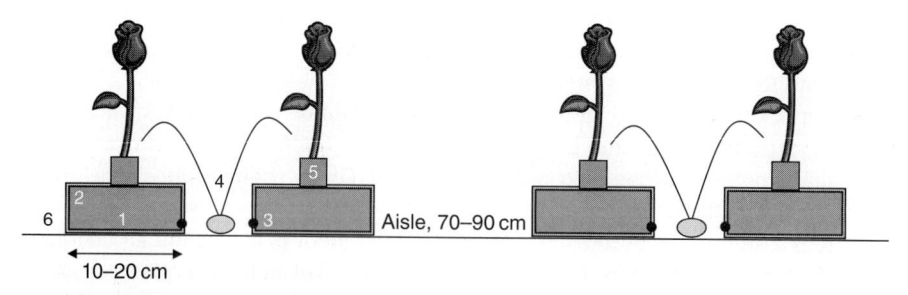

1: substrate slab or granulated substrate
2: foil around the substrate
3: drain holes are made 1–2 cm above ground level and at the non-aisle side
4: drip irrigation at the non-aisle side
5: pots for plant raising, using the same substrate
6: plastic liner covering the soil

FIGURE 5.13 Open systems for single row crops.

5.3 PRODUCTION SYSTEMS

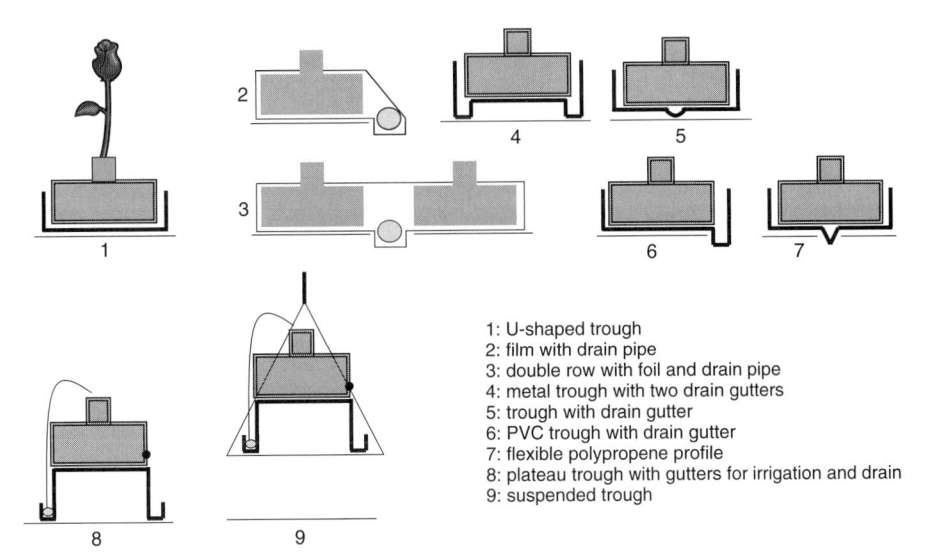

FIGURE 5.14 Troughs and profiles for closed single row systems.

1: U-shaped trough
2: film with drain pipe
3: double row with foil and drain pipe
4: metal trough with two drain gutters
5: trough with drain gutter
6: PVC trough with drain gutter
7: flexible polypropene profile
8: plateau trough with gutters for irrigation and drain
9: suspended trough

A variety of materials have been used to create the troughs that collects the drainage water from slabs and bags (Lataster et al., 1993): plastic film, with or without integrated drain pipe, U-shaped polypropylene troughs, V-profiles and PVC profiles, metal troughs (Fig. 5.14).

One of the least expensive closed systems consists of a plastic bag or film wrapped around a substrate slab with collection pockets on one side. Often a drain pipe is laid into the channel along the length of the row to keep the drainage channel open.

Plastic sheeting material consisting of polypropylene or polyethylene, 1–2 mm thick, delivered in rolls, can be used to create a system of drainage troughs. These are typically 15–25 cm wide and have vertical edges 5 cm high. The bottom can be flat or V-shaped. The system requires careful levelling and compaction of the ground to prevent depressions where water can stagnate. Recirculation of the nutrient solution to the water treatment unit is easy because special fittings interconnect the troughs and the main pipeline.

Troughs have also been made of polyvinyl chloride (PVC), but such systems are much less flexible. This material is brought in its final shape from the factory and, consequently, cannot be delivered in roll form to the greenhouse for production of the trough on the spot. Lengths of 6–12 m have to be assembled at the greenhouse into the desired configuration. There is a risk of breakage.

Metal troughs can also be created either through shaping on-site off of large rolls of sheet metal (with a machine) or from preshaped trough segments. While expense may be a deciding factor as to whether plastic of metal needs to be used, it is important to note that some subsoil may not have adequate support for a plastic system, so that the stronger metal systems are needed so as to prevent breakage and reduce sagging.

FIGURE 5.15 Suspended troughs for uniform water contents of the substrate and drainage of the nutrient solution (see also Plate 13).

One consideration when selecting a material for drainage troughs is the need to disinfect the troughs between crops. If steam or hot water is applied to a trough made of thin plastic, then the trough is likely to deform. Solarization of these materials is possible, but temperatures must reach at least 70°C, and this is obviously dependent on the prevailing weather conditions. PVC has the added disadvantage that burning during disposal releases chlorine into the atmosphere and is therefore prohibited in many countries, while polyethylene and polypropylene can be burned completely without introducing too much pollution.

A system consisting of both a metal trough and a plastic film may be needed in situations where the structure of the soil of the ground is such that it results in sagging of a production system installed directly on the ground. Coated metal troughs are available in many widths and shapes (Formflex, 2005, Fig. 5.15) and can be placed directly on the ground or above the ground and even as movable systems (as described below). A metal trough system is typically more expensive than a plastic trough, but it is less susceptible to settling and sinking of the ground, and may be more durable, lasting more than 10 years.

5.3.1.3 Vertical Systems

Vertical systems are ones where some plants are on the ground while others are stacked in rows vertically. Such systems have been tested for crops with small plants like strawberry and lettuce with limited results (Fig. 5.16). The nutrient solution is

5.3 PRODUCTION SYSTEMS

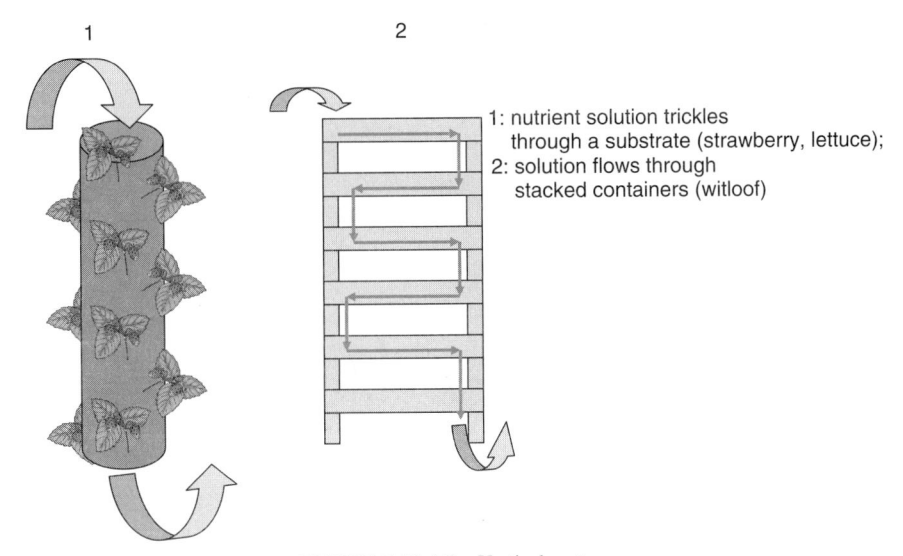

FIGURE 5.16 Vertical systems.

1: nutrient solution trickles through a substrate (strawberry, lettuce);
2: solution flows through stacked containers (witloof)

applied at the top and drips through a bag filled with substrate (Massantini, 1977). Drain water is collected at the bottom. Similar systems are in use as aeroponics or as stacked containers (Liu, et al., 2005) or containers fixed to a pillar (Vincenzoni, 1982). The main reason for using such an approach is to make optimal use of available space in an attempt to maximize the yield per square meter. However, the available light intensity is distributed over more plants making less light available to those at the bottom than those at the top. The consequences of this on year-round cropping should not be neglected: if harvesting of all plants begins at the moment the upper ones are ready, then the top layer crop is of good quality, but the yield from the lower layers decreases; if, on the other hand, replanting is only begun after the bottom plant is ready, each layer will yield a high-quality crop, but it will take longer before all plants are harvested and the system can be replanted.

Vertical systems are not widely used commercially. A similar approach to seek increases in yield per square meter greenhouse area using A-frames is described below.

Vertical systems for production of Belgian Endive (*Chicorium endivia* L. witloof, Fig. 5.17) is feasible despite this problem (Van Kruistum, 1997). Plants are placed in containers outside the growing area where they are grown to until they reach a specified size. These are later installed in 4–8 tiers of containers of 100×120 cm. Nutrient solution can flow through the containers in a zigzag pattern from top to bottom; the solution is then recycled. The plants are watered, kept at a temperature of approximately 22°C and are protected from light. White heads of Belgian Endive can be harvested after a forcing period of 3–4 weeks. A similar approach has been tried for asparagus (Van Os and Simonse, 1988).

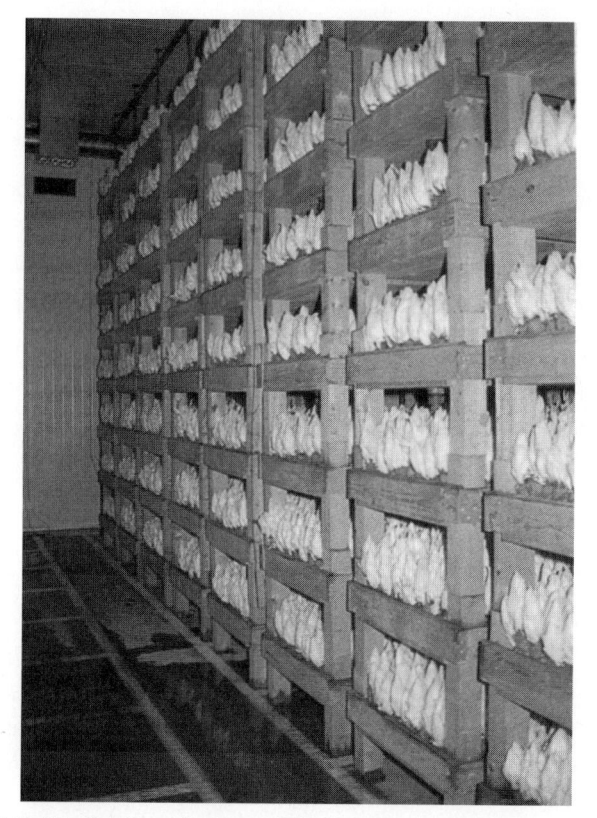

FIGURE 5.17 Belgian endive in stacked 120 × 120 cm containers (see also Plate 14).

5.3.1.4 Crops in Beds

Several flower crops, such as chrysanthemums, carnation, freesia (*Freesia* spp.) or alstroemeria (*Alstroemeria aurantiaca* L.), are grown in beds at a density of 20–60 plants per square meter (Fig. 5.18).

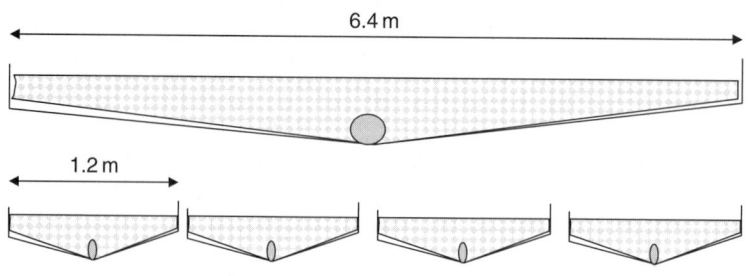

FIGURE 5.18 Systems for crops in beds.

In this case, row systems as described above with slabs or substrates in bags are not used because the return on investment is not adequate to support such a system for these crops.

Polypropylene liner and substrate

The most widely used system for crops grown in beds consists of a 110–140 cm wide bed, with a 1–1.5 mm thick polypropylene liner between the subsoil and a layer of 5–10 cm of granulated substrate. Sidewalls are needed when the system is above the subsoil. If the aisles are raised, they can serve as sidewalls. A drainage pipe is laid at the lowest point of a bed. Depending on the type of crop, either heating or cooling pipes can be embedded. This system is especially useful for crops such as bulbs, tubers and rootstocks [lily (*Lilium* ssp.), freesia and alstroemeria] which have to be removed entirely between cropping cycles.

Polystyrene panels

Sometimes crops like lettuce or chrysanthemums are grown in beds without substrate, but with an NFT or aeroponic system (Ikeda, 1985; Ruijs et al., 1990a,b; Ito, 1994; Both et al., 1999; Kratky, 2005; Both, 2005). In this case, plugs are firmly secured by means of a sponge substrate (polyurethane foam) in polystyrene panels. Sprinklers moisten the roots from below (aeroponics, Figs. 5.2 and 5.4) or there is a thin (NFT, < 2 cm) or thick (deep flow technique, DFT; 5–10 cm) water layer.

Plant plane hydroponics

The 'Plant Plane Hydroponic' (PPH) system (Schröder, 1994) consists of a polyethylene or polypropylene liner below a layer of fleece material (plastic or cellulose), covered by a second liner to prevent evaporation and algae growth. The whole system is placed on a slight slope within the width of the bed of about 0.2 per cent. Water flows down the slope through the fleece to a drainage line in the centre or at the side of the bed. In large installations, the uniformity of the slope is very important and, consequently, subsidence of the ground beneath can be a problem.

Eco-organic soilless system

It is also worth mentioning a cheap soilless bed production system developed for Chinese growers (Jiang, et al., 2004). In a traditional Chinese solar lean-to greenhouse, beds of 48 cm wide, 15 cm high and 5–7 m long are walled with bricks and lined with a 0.1 mm polyethylene liner. The beds are filled with locally available substrates (rice husks, sunflower or maize stems, coal cinder, sawdust, sand). Solid chicken manure is an essential part of the substrate which means there is no need for liquid fertilizers and the plants receive only plain water.

Span wide

Beds have also been installed with much greater widths than those mentioned above. Ruijs et al. (1990b) described a system for chrysanthemums, with a 0.5–1.5 mm polypropylene film below a 10 cm soil layer. In the middle of the span, at the lowest point, there is a drain pipe to collect the drain water. In practice, a thicker layer of soil proved necessary because a thin layer stayed too wet. Fair results were achieved by Bleijaert et al. (2004) when cultivating lettuce and cucumber in a 40 cm soil layer lying upon a plastic liner. Watering should be adapted to the substrate, which should be rather coarse. The system is not widely used on a commercial basis because the investment required is too high compared to the profit potential, particularly as far as leaf vegetables are concerned. Many times growers avoid wide beds as it makes pest management difficult since it is impossible for workers to inspect the plants in the centre of the bed. Wide beds are only feasible where all plants are harvested at the same time.

Concrete floor

One variant of the wide bed approach incorporates a flooded concrete floor (Fig. 5.5) or sprinkler irrigation. This type of system is used mainly for labour-extensive pot and bedding plants, raising young plants for fruiting vegetables and rooting cuttings in crates (60 × 40 cm). The concrete floor sometimes has embedded heating tubes. The floor is divided into watering sections; irrigation water enters the section and through gutters which are also used for drainage. The floor has a slope of 0.2–0.4 per cent to facilitate drainage. If sections are too wide or if the slope is too steep, the difference in water height becomes too great to realize a uniform water level for all containers on the floor. The depth of water during irrigation can be as much as 5 cm. Containers placed on an ebb and flow floor must have a special elevation of 0.3–0.5 cm between the floor and the substrate to facilitate drainage. This type of growing system is especially suited to highly mechanized methods where containers are placed or collected with specialized machines (automaton/robotics). The expected lifetime of a concrete floor is the same as for a greenhouse (around 15 years). It should be considered that removing the concrete floor is just as costly as its original construction, and it is therefore hard to conclude whether the use of such a floor is economically feasible.

A floor should be designed with flexibility in mind. Birch (1979) designed a concrete floor system with a 10 cm trough-profile for year-round lettuce production. Transport equipment was incorporated into the design. Unfortunately, such a system excludes the possibility of changing to any other crop, forcing the grower to achieve the payback on the investment without any opportunities for changing the cropping system over the years.

5.3.2 ABOVE-GROUND PRODUCTION SYSTEMS

It has also been found that elevating the production system above the ground can have significant benefits due to better air circulation around the plants, reducing the incidence of plant diseases. Some plants also benefit from improved control over the temperature of the root zone. Another reason for elevating crops above ground level is to improve working conditions in the greenhouse by allowing workers to reach the plants or fruits to be harvested more easily. Raised systems can be either fixed or movable and are used for crops such as pot plants, strawberry, lettuce, rose, tomato, pepper, cucumber and gerbera. As with production systems which are placed on the ground, crops can be in single rows or in beds. Above-ground beds are typically called 'benches' and represent the majority of container-grown greenhouse potted plant production.

5.3.2.1 Single Rows

Racks constructed of metal or wood are installed on stone blocks or risers made of metal or treated lumber, so as to form a single or double row of container-grown plants from which harvests are taken on an ongoing basis. The root zone can either be formed by a set of pots or by elongated pans or troughs. Examples of crops which have been grown in such systems are roses, gerbera and strawberry (Fig. 5.19) and various other crops are being tested by growers. The racks are placed on the ground or secured partially in the ground depending on degree to which the ground is able to support the system without subsiding.

One example of this type of system consists of 15–25 cm diameter containers placed in a metal rack, where a trough is placed just below the containers to carry away drainage water (Worm, 2005). This trough can be made of polypropylene, PVC or coated metal. The system can be tailor-made on site by the grower.

Another example uses long trays (100–133 cm long, 15–20 cm wide and 7.5–10 cm high) to hold bags or slabs. Each tray is supported by a rack at the point where two trays meet (every 100–133 cm). More sophisticated systems consist of an aluminum frame into which containers are hung. Two frames are placed upon a rack, the height of which can be adjusted.

Outdoor crops can be grown in a one-row system, where the supporting frame is rigidly placed in the subsoil. Special accessories are also available to fit a rain shelter over the one-row system.

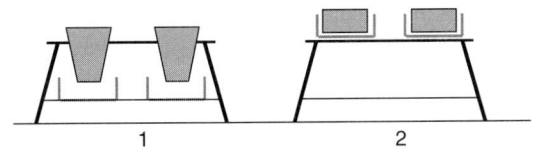

1: fixed racks for containers (gerbera, strawberry)
2: troughs or long-shaped containers on racks

FIGURE 5.19 Systems above the ground.

5.3.2.2 Suspended Troughs

Some systems have also been developed where the entire production system is suspended from the greenhouse frame by steel cables (Fig. 5.15). In this situation, there is no need for racks to support the system from below. Clearly such an installation is only feasible if the greenhouse is engineered to support this significant extra weight. Cables attached to the structural supports of the greenhouse at 2.5–4 m intervals are installed so as to give the suspended rows a smooth slope of 0.2–0.5 per cent. Troughs of various cross-sectional shapes are available for such application. A commonly used design is called 'the plateau trough' (10–30 cm wide) where slabs or bags lie on an almost flat plateau, with small channels (4.5 cm) on both sides (about 10 cm lower) for drainage water collection (Fig. 5.15). The height difference is especially useful for avoiding root growth in the drain water gutter, thus minimizing the risk of dispersal of pathogens among plants in the row.

5.3.2.3 A-frames

The use of A-frames or arches to increase the efficiency of space utilization has been investigated. Morgan and Tan (1983) described the principle for a trough system with 12 tiers; Leoni et al. (1994) described the use of an A-frame for the growth of tomato and lettuce with aeroponic watering (Fig. 5.20). The system by Morgan and Tan was 240 cm wide at its base, 40 cm wide at the top and 230 cm high. It accommodated 12 troughs with an NFT system for lettuce, and was designed for a 3.2 m span of greenhouse space. This configuration doubled the cropping surface (up to 40 lettuce plants per square meter) with 8–9 crops annually (Morgan and Tan, 1983). The system of Leoni et al. consisted of two polystyrene panels of 100 cm wide. The two panels formed a triangular A-frame shape with 120 cm at its base. The system included two A-frames in a 3.2 m span.

In such A-frame systems, the lower tiers of plants receive less light than the higher ones resulting in lower productivity for the lower plants. In the format of the system by Leoni et al., there is also a risk that plants such as head lettuce will grow in a crooked shape.

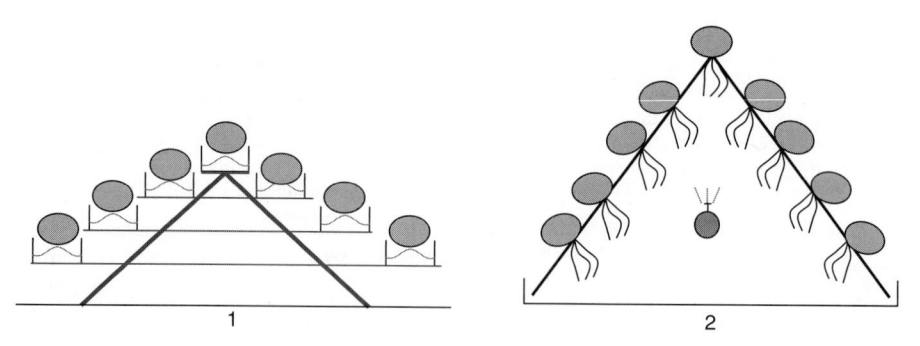

1: troughs with NFT system
2: polystyrene paels with aeroponics

FIGURE 5.20 Archway or panel system to increase the utilization of space.

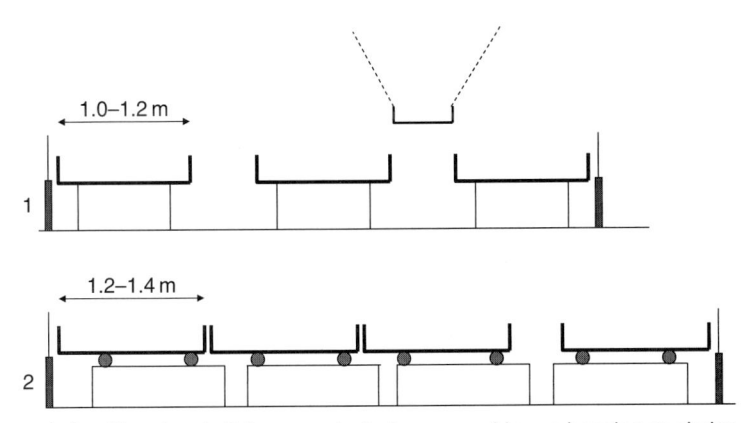

1: fixed benches in 3.2 m span, including a movable one hanging on chains
2: rolling benches in a 6.4 m span with one movable aisle

FIGURE 5.21 Fixed and rolling benches.

5.3.2.4 Benches

Benches are generally used to hold a large number of plants side by side so as to minimize the aisle space in the production area. The bench width is dictated by the distance that workers can reach into the crop to manipulate the plants. Benches can consist of fixed (Fig. 5.21), rolling benches or movable benches. Fixed benches cannot move and always have the same position in the greenhouse. They are mostly used for the cultivation of plants where aisles are needed for crop maintenance. Space utilization is relatively low (65–75 per cent).

Rolling benches are made of lighter-weight reinforced materials installed so that they sit on two long metal pipes which extend the full length of the bench. Cranks on the ends of the pipe allow the entire bench to be rolled side-to-side. The distance that a bench can roll is governed carefully with stops to prevent the bench from rolling off its supports. In such systems, only one aisle is needed for all the benches since every bench can be rolled so that the utilization of space is increased to 85–93 per cent.

Another step in the development of rolling benches was the introduction of aluminum benches, with water-tight plastic liner.

There has also been a trend towards wider benches, some as wide as 1.5 m. This, however, appears to be too wide for rolling benches with labour-intensive crops. For ergonomic reasons, a bench width of less than 1.4 m is recommended, because the reach of an average Northern-European worker is about 65–70 cm. If workers are of a shorter average height, then the width of the bench would need to be decreased.

Movable benches

Movable bench systems were developed in regions where labour and land costs are high, for use with plants that require a lot of manipulation (spacing). Special transport

FIGURE 5.22 Movable benches can be transported out of the greenhouse bay to another place. Mostly the pipe-rail system is used; In some applications a robotic lift is used to pull out an entire individual bench for rapid movement to another location (see also Plate 15).

systems are designed to allow an entire tray of plants to be moved at one time by rolling it to a different location in the transport system (Fig. 5.22). This allows the plants to be easily moved in large batches to the worker, instead of having the workers spend time going to the plants. Such systems also allow for minimization (sometimes elimination) of aisle space.

Movable benches are manufactured to standard sizes with length of 3, 4 or 6 m. The width of the bench varies between 120 and 150 cm. This results in a greenhouse space utilization of about 90–95 per cent. Many such movable benches are also equipped with a water tight bottom (PVC) or liner so that irrigation can take place via ebb and flow or capillary mats.

The concept of bringing the plants to the workers is also currently being tested for cut-flower production. For rose and gerbera, labour-intensive crops with a daily harvest from the same each plant, transporting the single rows of plants from the greenhouse to a separate harvesting area is being tested (Van Weel, 2005; Van Henten et al., 2005; Van Tuijl et al., 2005; Fig. 5.23). When set up in this way, all the plants in the row can be reached easily for maintenance or harvesting. Two problems have yet to be overcome with this type of system: one is the development of an optimal drip irrigation systems that can easily (and in an automated fashion) be disconnected and reconnected; this is a technical challenge that can probably be overcome. The other problem is related to the frequency with which such small rows of plants need to be

FIGURE 5.23 Movable row system for rose (see also Plate 16).

moved and the rate at which they can be transported. The rows need to be transported at an interval of 4–8 h. Rows in one span come to the main path, where maintenance or harvesting takes place; it then returns to is normal location via a separate aisle. Managing the cycles of harvests for each plant is difficult.

Individual plant

In intensive greenhouse production, plants in crops are being treated in an ever more individualized fashion. In the case of rolling benches, the smallest movable unit is the 1.5 m wide bench with its whole span length of 20–50 m. In case of transporting benches, the smallest movable unit is the bench of 1.5 m by 3–8 m. In case of the movable troughs, the smallest movable unit is one row of plants with a size of 0.8 by 3–8 m. Ultimately, it may happen that individual plants will be moved and treated individually.

The 'Walking Plant System' offers the most individualized treatment (Formflex, 2005; Fig. 5.24). The potted plants are transported on a moving belt. A machine places the plants onto a movable belt in the bottom of a trough. This system is programmed by the grower to place the plants into a certain row at a particular spacing.

Other robotic systems are used to space the plants within the trough by using one automaton or machine to pull the plants out of the trough and place them onto a main transport belt; a second automaton sets the plants into a trough at a specified spacing.

FIGURE 5.24 Walking plant: an individual approach (see also Plate 17).

5.4 EXAMPLES OF SPECIFIC SOILLESS CROP PRODUCTION SYSTEMS

The list of soilless growing systems is continually expanding. Many different crops have been tried and a considerable number of them are in use in commercial production. In this section, a variety of such production systems will be described and some advantages and disadvantages will be discussed (Table 5.1).

5.4.1 FRUITING VEGETABLES

One of the largest segments of soilless production is the production of fruiting vegetables in greenhouses. These crops generally have long cropping period (4–12 months), 1–3 plantings per year, 2–5 plants per square meter, rows separated by 0.8–1.2 m and within the row a plant distance of about 0.3–0.6 m.

Along the length of the row, substrate slabs (stone wool) or bags are laid in different types of trough systems or directly on the ground. Plants are propagated in the same substrate to ensure the correct rate of drainage when planted on the slab or in the bag.

TABLE 5.1 Overview of Crop Groups

No	Group name	Crops
1	Fruiting vegetables	Tomato (truss, cherry, beef, round), sweet pepper, cucumber, melon, aubergine, courgette, French beans, hot pepper
2	Single-harvest leaf vegetables	Butterhead lettuce, iceberg lettuce, kohlrabi, endive, Chinese cabbage, other salads
3	Single-harvest vegetables, drilling	Radish, spinach, (medicinal) herbs
4	Other vegetable crops	Strawberry, witloof, asparagus
5	Cut flowers	Rose, carnation, gypsophila, bouvardia
6	Single-harvest cut flowers	Chrysanthemum, aster, lisianthus
7	Flowering bulb, tuber and rootstocks	Freesia, amaryllis, alstroemeria, lily, tulip, iris, hyacinth
8	Other cut-flower crops	Anthurium, gerbera, cymbidium
9	Flowering potted plants	Cyclamen, begonia, saintpaulia, pot chrysanthemum, fuchsia, kalanchoe
10	Foliage potted plants	Ficus, dracaena, monstera, schefflera, ferns

Drip irrigation is used to fertigate the plants. Nearly all fruiting vegetable crops in the United States, Canada and Europe are grown in this type of system. There are few other types of systems such as NFT (Cooper, 1979) or ebb and flow. This is mainly because of the potential for the dispersal of soil-borne pathogens via the recirculated water, the enormous quantity of circulating nutrient solution and, in places, the relatively poor water quality which can lead to sodium accumulation.

Currently in the Netherlands, all new greenhouses are installed with a suspended trough system, developed for tomato, cucumber and sweet pepper (Fig. 5.12). The big crop advantage is the uniform water content of the substrate and, after a few weeks, the uniform height at which the fruits hang to be harvested and, consequently, more efficient use of labour.

Movable systems per row have been tested (Van Os et al., 1993). In this design, the plants move in the length of the row, hanging on a rail, they return via the next row at 0.8 m distance. Continuous moving damages the leaves extensively because they rub against each other. Irrigation is carried out in one place along the path of travel. Further developments are focusing on making the plants movable so as to travel to stations where harvesting can occur. Here, one unit consists of two rows of plants, trained to a high wire system; the unit moves along one bay at a time, until it reaches the harvesting point at the main path. Here harvesting takes place while the unit moves sideways. After harvesting, the unit moves into the next bay. Watering takes place via drip irrigation, but connection to, and disconnection from, the main water pipe line is optimized. A mobile system should not be considered if the utilization of space cannot be increased or a saving of labour input cannot be realized, since the system is very expensive in terms of both the initial investment required and the annual operating and maintenance costs. For sweet pepper, a much slower growing crop, a system of movable containers is under development in which utilization of space in the beginning of the crop is of importance. Here the walking-plant-system is being used (Formflex, 2005; see Sect. 5.3.2.4, Fig. 5.24).

5.4.2 SINGLE-HARVEST LEAF VEGETABLES

Many leafy vegetables are grown to size and then harvested when each plant has reached some specified size. Row spacing is generally 10–30 cm depending on plant size, with similar distances between plants along the length of the row. For these mostly short-term (4–12 weeks) crops, it is generally not feasible to use drip irrigation nor substrate slabs or bags. Most systems used are based on troughs with NFT or beds with DFT or aeroponics.

In the United Kingdom in the 1970s, commercial hydroponic lettuces production (2 ha) was set up using 10 cm wide concrete troughs (Birch, 1979). In the Netherlands in the 1980s, a commercial system (Van Os and Kuiken, 1984) was introduced to facilitate the seasonal switch between lettuce and tomato to address the issue that economic feasibility is generally related to competition with field-grown vegetables where such systems are profitable during the winter, but not during the summer. Seven troughs of 25 cm with two rows of lettuce per trough in winter are combined with only four troughs with slabs for tomato in summer. However, Dutch growers who invested in soilless tomato systems and butterhead lettuce have disappeared from the greenhouse market. Many other systems were developed for research or on a semi-commercial level (Ikeda, 1985; Morgan, 1985; Panel, 1988; Ruijs et al., 1990a; Benoit and Ceustermans, 1994; Ito, 1994; Leoni, et al., 1994). In north-western Europe, lettuce is not widely grown in soilless systems because market prices are often too low to compensate for the high annual costs.

On a large commercial scale, the movable gully system of Hortiplan, based on a Swedish design (Hortiplan, 2005), appears to be economically feasible. However, the level of investment required is also high. It is a system in which troughs are spaced during cropping; planting and harvesting take place at the same side of the greenhouse. Another spacing system was developed in the United States by Both et al. (1999) and Both (2005) where lettuce is produced on floating panels or floating rafts. The panels have a fixed planting density and various densities are used throughout the year. A similar large-scale system is in use in Canada for year-round production of lettuce and herbs. In Australia and the United States, variants of trough systems are in use, which are more suitable for small-scale production. Special systems have been developed for special crops such as lamb's lettuce (Benoit and Ceustermans, 1989a), cabbage (Anderson, 2004) and medicinal herbs (Dorais et al., 2001). Other crops use systems developed for butterhead lettuce with some adaptations, but these systems are only economically viable under certain special circumstances.

The archway concept for lettuce production (Sect. 5.3.2.3) has progressed no further than use as a research tool. The practical problem of growing differences within one panel reduces the economic feasibility of the system. Lettuce is often used as a pilot crop in plant factories. In a plant factory, mostly of Japanese or Korean origin, lettuce plants are placed in a huge growing chamber with artificial lighting from which sunlight is excluded. Benches with a water buffer and without substrate are moved to a central place to be harvested. However, economic feasibility proved poor.

5.4.3 SINGLE-HARVEST SOWN VEGETABLES

Crops such as radish or spinach are grown in soil in a system covering the entire area in rows 5–15 cm apart from each other and a spacing of 2–5 cm within the rows. For soilless systems, this means that the entire greenhouse area has to be covered with substrate, which is generally not economical. Alternately, a NFT or DFT system has to be used. In this case, a separate system must first be used to germinate the seeds, grow the seedlings and then transplant them into the soilless system where they will grow to maturity. Both et al. (1996), Ikeda et al. (1995) and Maruo et al. (2001) describe such a system for spinach. In Japan, spinach is commercially grown in soilless systems, while in north-western Europe, economic feasibility is too low because of competition from outside production. Ruijs et al. (1990a) also describe several systems for radish, but here again it appears that economic feasibility is poor.

5.4.4 OTHER SPECIALITY CROPS

5.4.4.1 Strawberry

Problems associated with soil-based production of strawberry – such as the low utilization of space, loss of yield through diseases, and limited potential for crop rotation – have prompted growers to seek soilless cultivation methods. In the 1980s, growers tried to increase the utilization of space by putting plants in large containers (25 cm diameter buckets) so that the ripening fruit were hanging along the container, resulting in less botrytis and easier harvesting. In a movable suspended trough system, every second trough was hanging near the top of the greenhouse creating a path to pick the lower troughs, hanging at easy picking height (Van Os, 1986). This system proved too expensive and was replaced by a system where the containers were mounted higher in the greenhouse without any aisle space. Here the fruit was hanging above the harvesters, making the harvesting of fruit uncomfortable. Many other systems have been tested for strawberry production such as a vertical wall with plant holes to collect solar heat for the night; archway concepts; DFT systems with floating panels; vertical bags filled with substrate. Nowadays, many growers use a system of metal, plastic or polystyrene troughs with a special profile into which bags or slabs can be laid. The strawberries to be harvested hang at eye level. A cheaper alternative is the use of containers filled with peat or coir, mostly placed in round pipes at a similar height. The 20 cm diameter white PVC pipe is a drain water collector and a container support at the same time. In tunnels and shelters, strawberries are also grown in a soilless

FIGURE 5.25 Strawberry in suspended troughs: optimized utilization of space (see also Plate 18).

system. These systems are less sophisticated but work well technically. Containers are often placed at a height of between 50 and 100 cm to make harvesting easier and prevent the hanging fruits coming into contact with the wet soil. Alternatively, metal troughs are placed at a height of 1.30 m, into which substrate bags are laid; the crop is watered via drip irrigation (Fig. 5.25).

5.4.4.2 Belgian Endive (witloof)

In France, Belgium and The Netherlands, Belgian Endive (witloof) is an important hydroponically grown crop. The first stage of cultivation takes place in the open field where a root is grown from seed. Before winter, the roots are lifted and stored. Over the course of the winter, the roots are taken in batches and forced in dark temperature-controlled chambers, where around eight containers are placed on top of each other. A 3–5 cm water layer flows from the upper to the lower container over the roots. After about three weeks, a yellow-white head can be cut from the roots (Fig. 5.17). This has become a year-round forcing method which has completely replaced soil forcing.

5.4.4.3 Asparagus

In a similar system to that used for witloof, asparagus plants can be forced for white spears (van Os and Simonse, 1988). Growers became aware that asparagus plants could also produce spears under temperature-controlled conditions during the winter, but no way has been found to exploit this commercially. The supply season was considerably extended by the use of greenhouses, soil heating and other planting methods (ridges of 1 m high).

5.4.5 CUT FLOWERS

One type of cut-flower production involves the use of the same plant for several months or years with daily harvests. The crop stays in the same substrate at a spacing that is dictated by harvesting needs. Crops such as rose are grown in single or double rows, while crops such as carnation, alstroemeria and bouvardia were grown in wider beds of 1.30–1.45 m. A granulated or preformed slab substrate is used. The feasibility of producing cut flowers in soilless production is linked to the profitability of the crop. Carnation, for instance, has proved uneconomical in soilless production in northern Europe and North America. This crop is still grown in tuff beds in Israel and in both soil and soilless substrate in South America. Rose production as a whole changed dramatically to where nearly all is now in soilless production. The change to soilless cultivation resulted in an increase of the production level and fewer diseased or dead plants. This was the main reason why the change to soilless systems for roses happened so rapidly. In other parts of the world, soil dispersion due to continuous application of monovalent ions (K^+ and NH_4^+) was another cause for shifting to soilless culture. Later, rose bushes were placed in single rows with a stone wool slab or a bag with granulated substrate. This made the system cheaper, as less substrate per square meter was needed. The introduction of the bent cane system in rose production became feasible when production was shifted to above-ground systems and this approach has become standard practice to control the quality of rose cut flowers. The introduction of the movable trough may be seen as a system for the future (Van Weel, 2005). The trough is at a slight incline to collect the drain water, and watering takes place by drip irrigation. Connecting the rigid water network to the movable drip irrigation system was a problem for a long time but has been solved (Van Zaal, 2005). During harvesting and crop maintenance, the troughs are disconnected from the watering system and pushed to the main path to perform the necessary tasks. After that, the trough moves sideways over a conveyor belt and into the following span. Each rotating unit consists of two spans 12–16 m wide and 50–100 m long. The advantages of the mobile trough system are the reduced labour requirements (the plants come to the worker), and the improvement of the quality of the labour (the roses to be harvested can be seen better; the worker does not need to reach to the middle of a bed). For the future, this mobile trough system can be used for a stage-related rose production giving each stage its own optimized climate conditions and leading to a synchronized crop. For this, the whole greenhouse area will be divided into three sections: one for sprouting of the buds to a stem length of about 5 cm; a second for the growth to a visible flower bud; and a third for development into a harvestable flower. In the last stage, all flowers must be harvested at one time. After harvesting, the bent branches are still on the plant and allow the plant to sprout again. First experiments are under way and look promising (Van Weel, 2005). The main problem to address has been the deviation in the harvesting time between the plants; over a course of six weeks, the harvesting period has been reduced from 14 to 5 days within one group of plants. The challenge is to reduce the harvesting period even more, if it can be done in one day, mechanical harvesting becomes possible. In addition to this, each individual plant must be of a similar perfect quality to flower at exactly the same time as the adjacent plant. After harvesting, the plants move to another section in the greenhouse to obtain an optimal climate to sprout.

Anthurium, gerbera and cymbidium cut-flowers can also be produced in a soilless production using containers, slabs or bags. For cymbidium as cut flower and several other orchids, the substrate is only needed for physical support. A large container filled with substrate is sufficient.

Over the course of a number of years, Anthurium grows into a large plant. A heavy substrate, such as clay or lava granules, is often chosen to fill large containers. Sometimes containers are placed in a bed of an ebb and flow system, otherwise drip irrigation is used. Less frequently, beds filled with granules (Fig. 5.18) are used; in this case, too much substrate is needed to fill the bed, and the plant needs support in height to avoid falling by imbalances.

Gerbera flower production is currently entirely done in soilless production, frequently in containers, but also in bags and slabs (Worm, 2005) which are placed in racks or in long-shaped containers filled with substrate. These rack systems are designed to have the plant at a height suitable for easy harvesting. As most flowers are pulled from the plant, it is important that the plant is fixed well in the substrate. Mobile trough systems are under development, similar to those used for rose (Van Tuijl et al., 2005), which involve harvesting in a central place.

5.4.5.1 Single-harvest Cut Flowers

Crops such as chrysanthemum, lisianthus and aster are grown from cuttings or plugs in beds 1.0–1.2 m wide and at a density of 30–70 plants per square meter with the expectation that all flowers will be ready for harvest at the same time so that the entire bed can be harvested at once. This means that the total area of the greenhouse is covered with plants, so the introduction of a soilless system means that the whole greenhouse area must be covered with substrate, which is economically unattractive. In the 1980s, NFT systems were introduced for research purposes (Van Os, 1980; Morgan and Tan, 1983). Flat-bottomed troughs 3–5 cm wider than the stem or the rooting medium block appear to work best. In that case, the flowing nutrient solution easily finds its way past the plants and blocks, even if the root mat grows to fill the entire trough at later stages. In the 1990s, several studies were undertaken to investigate if commercial chrysanthemum production is feasible in soilless culture (Ruijs et al., 1990b; Lataster et al., 1993). Production in soilless systems resulted in only a small increase in yield, while investment and running costs are much higher, resulting in reduced economic feasibility. Systems using beds filled with substrate, aeroponics or movable benches appear more expensive than NFT, and even NFT is no more financially viable than soil growing (Ruijs et al., 1990b). In 2004, a new NFT system was developed using troughs: the plants are gradually spaced further and further apart during the long-day period; by the short-day period, they have reached their final spacing distance. Optimum light, temperature and nutrient solutions are supplied during all the stages of growth (Van Henten et al. 2005; Van Os, 2006). Calculations made clear that up to 30 per cent extra yield per square meter is possible, which must be used to pay the system costs.

5.4.5.2 Flowering Bulbs and Tubers

Flowering bulbs, tubers and rootstocks must be discussed in two separate groups: those which are grown in beds to produce flowers; and those which are first grown in open fields to a marketable bulb size to be forced in soilless production during winter and spring. These production systems are also single-harvest systems with the planted plant material being replaced after harvest.

Freesia, amaryllis, alstroemeria, lily

Freesia and some varieties of alstroemeria need root cooling to obtain a perfect year-round crop. These crops need a granulated or loose substrate to facilitate the planting and digging of the bulbs and tubers. Rolling benches (four benches of 1.45 m per 6.4 m span) appear to be the most economical soilless system to use (Hendrix and Ruijs, 1992), followed by 60 × 40 cm trays/boxes placed upon a plastic liner in which a drain pipe has been laid (Fig. 5.18). In both systems, combined heating/cooling pipes are used, while watering takes place by drip irrigation. The increase in the utilization of space leads to an improvement in efficiency. Alstroemeria is currently being tested in a commercial greenhouse in which 1.2 m beds are used, made of a plastic film and filled with a loose substrate.

Tulip, iris, hyacinth

In The Netherlands, around 2 billion tulip bulbs are produced per year in open fields. A small fraction of these bulbs are not exported but forced hydroponically for cut flowers. Here, a forcing method is employed that uses clear water, relying on the fact that the bulbs already contain the nutrients they need to grow. This fraction represents an area of around 500 ha. The main problem is keeping the bulbs in the correct position, with their tip pointing upwards, and preventing the stem from dropping during the growth. After many years of trial and error, two systems are currently in use: (1) containers with an aluminium device (90 × 120 cm, like those used in witloof forcing) with holes in which to place the bulbs and to allow them root into the water layer beneath the device; (2) bulbs are impaled on pins in the base of the case to keep them in the right position (Armstrong, 2002a). With either method, the vernalized bulbs are placed in the greenhouse at a temperature between 16 and 21°C. Water is added via in-line drip irrigation. This approach saves on labour and avoids the expense of a substrate. Similar systems are being developed for forcing hyacinths (Armstrong, 2002b). Recent experiences show that the best results are achieved with the NFT and ebb and flow systems. Stagnant water decreases yield, probably because of the low quantity of dissolved oxygen.

5.4.6 POTTED PLANTS

Flowering potted plants are relatively labour-intensive compared with cut-flower production and are grown on benches (orchids, cyclamen, begonia, saintpaulia, pot chrysanthemum, fuchsia, kalanchoe and a host of others (Fig. 5.22). Many are forced to flower on particular dates (e.g. Mothers Day, Christmas), requiring manipulation and grading. Green foliage plants typically have less labour input and are not scheduled as precisely. They are grown to reach a particular size, rather than a stage of flowering. A wide range of production systems are in use, ranging from gravel beds and concrete floors to fixed or moving benches or trays. Potted plant lends itself to extensive automation since the plants are individually grown.

5.4.6.1 Flowering Potted Plants

Introducing movable benches is the first step to the automation of the potted plant production process. The increased utilization of space translates into an increase in yield, which is generally enough to achieve a return on the investment. High labour input and increasing labour costs are driving investment in automation. Movable trays and benches allow work to be performed in a central location: the plants come to the worker rather than the worker going to the plants. Labour conditions can be improved in the work area and the less-favourable working conditions in the greenhouse are avoided. Nearly all facets of potted plant production lend themselves to automation so that it can be expected that ultimately all tasks from the filling of pots to planting, grading and packing of the plants can be automated. All this automation comes at a price; consequently it is not suitable for all types of potted plants. Generally flowering pot plants require the most labour so that it is the cultivation of these plants where the greatest savings can be derived from automated system.

Potted flowering orchids, such a phalenopsis, are also grown with such a system and this type of automation is being used by several growers. This crop takes a relatively long time to produce and benefits from various movements of plants through various temperature regimes and pot spacing as the plants grow.

Another automation system that is seeing continued development is the process of colour automated grading. Saintpaulia (African Violet) is grown in production systems in Europe, where the plants pass through an image analysis system which identifies how much colour the plant is showing. This can be used to assess whether the plant is ready to be marketed, and in conjunction with spacing machines can be used to sort plant onto several trays depending on how far from harvest each plant is. The various trays are transported to differently controlled greenhouse compartments to adjust the rate of development and force the plants to be ready on desired target dates. Similar image analysis systems are feasible for many other high-value potted flowering plant crops.

5.4.6.2 Foliage Plants and Outdoor Container Production

Large foliage plants (Ficus species, Monstera, Dracaena species, palms, ferns) spend longer periods in the greenhouse without little need for maintenance except for watering and fertilizing. For this type of plants, movable systems are often too expensive and the plants are also bigger, which make transporting more difficult. Here, the process of collecting and placing of larger containers has been automated. If watering takes place by ebb and flow upon a concrete floor, special adaptations to the pot and floor need to be made.

The outdoor production of container plants is comparable with the foliage plants mentioned above. The growing period is relatively long, while labour input per plant is lower than that for other potted plant production. Overhead sprinkler irrigation is generally used on gravel beds covered with a weed barrier (Bailey et al., 1999). Robotic automation is also being tested for outdoor nursery crops with considerable success. Such equipment includes automated pot filling, transplanting, placement of plants onto carts and spacing of plants into bed using robotic fork-lifts.

5.5 DISCUSSION AND CONCLUSION

Irrigation and fertigation systems, along with all the technology related to them, are undergoing continual innovative transformations leading to widespread use of automation and new developments in many areas. Irrigation approaches, in particular, have undergone considerable changes. Irrigation emitters improved with the introduction of pressure-compensated emitters, for both drip and sprinkler irrigation. In areas where water is becoming more expensive and where pollution due to excess fertigation is being regulated, sprinkler irrigation is being replaced by drip irrigation with more precise emitters and control strategies. Improvements in filtration equipment are responsible for major advancements in soilless production technology.

There is a small niche for NFT in crops such as chrysanthemums and lettuce in narrow troughs. DFT is in use to avoid excessive increase of the temperature of the recirculating nutrient solution and where control options are minimal. Ebb and flow is mainly in use for pot plants on benches or concrete floors. Aeroponics is rarely used on a commercial basis but could see more widespread use if the technological hurdle of lack of buffering can be resolved. Good production is possible with all these methods, but each method requires a different strategy to achieve optimal growth conditions and economic viability.

In the 1980s, the first soilless systems were open or run-to-waste systems. The excess of nutrient solution flows into the soil or to surface or groundwater was polluting the environment. Research projects showed that savings in water consumption and a decrease in the eutrophication of surface water by greenhouse crop production could be accomplished through recirculation of irrigation water (Van den Boogaard et al., 2003; Lecomte et al., 2004). In many countries and states this has led policy makers to force growers to switch to closed systems which involved reusing the nutrient solution. The basic logic behind this approach is well understood by growers: it saves water and fertilizers. But the resulting systems are much more expensive and need to incorporate the disinfection of the solution. Again, this is technically feasible, but not economically viable unless all competing growers are forced to adopt the same expenses. With a substantial increase in government regulation in Europe and North America since 2000, many growers are now seeing the need to modify their systems into closed or semi-closed systems to avoid fines. At the same time, this area is developing rapidly as new components and technologies related to recirculation are coming on the market. Researchers are also now developing information that helps this field evolve. Stanghellini et al. (2003) calculated the water-use efficiency of tomato grown in a closed climatized greenhouse to be 15–25 l of fresh water for one kilogram of tomato, as compared to 25–40 l for a soil-grown, non-climatized greenhouse and 40–60 l for outdoor production. The variation between 15 and 25 l water per kilogram is accounted for by the difference between an open soilless system (25 l) and a closed system (15 l) in a climatized greenhouse. This means that greenhouse production in dryer or hotter climates may be able to use less high-quality water, compared to outdoor production, but also that still more water (and expensive fertilizers) can be saved if a closed soilless system is used rather than an open system.

The commercial acceptance of soilless-grown crops in the 1980s and early 1990s raised expectation that many more crops would soon change over to soilless systems. Indeed, when one looks at all the systems developed for various crops, there are few technical limits to soilless cultivation: any crop can be grown perfectly well in soilless systems. However, to date not all crops have been changed over to soilless systems apparently, in many cases, due to the large economic investment. The required investment and annual costs in particular are too high. Insufficient increases in yield and quality are realized, while the price per unit of production area is declining. Those crops where the transition was made where commercial soilless systems is now the norm achieved this because of at least two of the following benefits to the grower: an increase in production at the same plant density, an increase of the utilization of space, or a lower planting density. Upon transition from soil- to soilless-production crops such as tomato, cucumber and sweet pepper, that are characterized by a relatively low plant density (2–5 plants per square meter), showed an increase in yield. When grown in soil, these crops were subject to many soil-borne diseases which could not be eliminated. At the same time, they changed from a rather low-quality water supply to an excellent quality rainwater supply. All of this has led to increased production. Almost all of these crops grown in The Netherlands now use a soilless system and are the major crops grown in soilless systems in other countries too.

On the other hand, leaf vegetables are grown in the ground at many plants per square meter (about 10–20), completely covering the ground. Changing to a soilless system results in relatively little increase in yield compared to soil production. The utilization of space can only be increased in some special cases. In such systems, increases in productivity would have to be realized through more-rapid accumulation of biomass, through shorter intervals between harvests. Consequently, soilless systems for these crops are feasible where this increase in production per unit greenhouse space per unit time is adequate to compensate for the large investment that is required. The costs in such systems may be due to the need for the entire greenhouse area to be covered with substrate or due to the need for an extensive and expensive NFT or DFT system. At the same time, competition from open-field production or from other countries is keeping the prices low, limiting the extent to which growers can recoup investment cost. Similar results can be seen in floriculture. Crops such as chrysanthemum, freesia and alstroemeria have been widely tried in soilless systems. The technical system and the cultivation method are not the problem, but the fact remains that for a commercial grower, there is no reason to change if his crop has a high plant density and there is insufficient improvement in yield or in the utilization of greenhouse space. With rose and gerbera, on the other hand, the plant density is low and the utilization of space increases in a soilless system; consequently, it is financially beneficial for the grower to change to a soilless system. Additionally, these crops suffer from soil-borne diseases, which can be avoided in soilless cultivation and, consequently, direct yield also increases.

Fertigation control involves precise control over the various fertilizer ions that are needed for plant nutrition. In the past, this has been done mostly by controlling the stock solutions and precise combination into irrigation solution. With the advent of recirculation, this has become more difficult since augmentation of nutrients to blended irrigation water requires knowledge of concentrations of the various ions. Currently many growers collect run-off in large storage basins and then test this water for available nutrients. This water is then used by blending with fresh water at a known rate and injecting fertilizers at a modified rate to achieve the desired recipes. While ISE can theoretically be used to dynamically monitor recirculating water, this technology is not yet stable and thus not feasible. Further research and development are needed to improve this.

Mathematical models which describe plant nutrition-related phenomena are widely used in research (Heinen, 1997; Gieling et al., 1998b,c) and can also be adapted for use in fertigation management through implementation in computer-controlled systems. Such models can replace direct measurements with predictions based on models (Le Bot et al., 1998). In some cases, this has been shown to be successful, but more research is needed.

Alternately if ion-specific electrodes can be developed to be as stable as current in-line pH and EC sensors, then such model-based ion concentration predictions may not be needed. Ion-specific measuring principles promise much greater degree of accuracy in the supply of nutrients to the plants than is possible with existing methods. The ratio of the uptake by the plant of various ions, and also the uptake-rate itself, may change over time, due to variations in the plant's needs. Using ion-specific control allows faster compensation for these fluctuations by adjusting the ion content of the incoming water and opens the possibility for feedback on the actual uptake of these ions (Gieling, 2001; Gieling et al., 2005). One of the biggest advantages would be the fact that water could be recycled dynamically rather than first accumulating batches.

Application of this innovative control technology would ultimately include the following:

- ion-specific monitoring of the nutrient solution in a closed growing system;
- monitoring of plant-related variables like photosynthesis, leaf area index (LAI), water uptake, leachate flow;
- control of fertigation based on feedback on the ion content of the drainage water;
- prediction of the uptake of water and nutrients using a model-based soft sensor or observer;
- decision-support systems for feed-forward control of the diurnal nutrient supply, using a plant-substrate model;
- use of a plant-substrate model in optimal model-based control of water and nutrient supply.

The idea of controlling the supply of water and nutrient ions by means of feedback measurements is the subject of considerable research and development into new control technologies. Such controllers have been shown to operate consistent with expectations (van Straten et al., 2006). This and other research projects exemplify the role that model-based optimal control currently plays and will play in future issues related to systems engineering and automatic control of plants in protected cultivation (Tap et al., 1993; Van Henten, 1994; Van Henten and Bontsema, 1996; Van Henten et al., 1997).

REFERENCES

Agricultural Structure Memorandum (1989). *Ministry of Agriculture*. The Hague: Nature Management and Fisheries, 174pp.
Anderson, R.G. (2004). Production and yield of selected edible greens in hydroponic ponds (float beds) in a greenhouse. University of Kentucky, Floriculture Research Report 16-04, 4p.
Armstrong, H. (2002a). Hydroponic tulips succeed at second attempt. *FlowerTech.*, **5**(1), 8–10.
Armstrong, H. (2002b). Forcing hyacinths for winter cut flowers. *FlowerTech.*, **5**(1), 10–12.
Bailey, D., Bilderback, T. and Bir, D. (1999). *Water Considerations for Container Production of Plants*. North Carolina Cooperative extension service, Horticulture information leaflet, p. 557.
Balendonck, J., Kuyper, M.C. and Hilhorst, M.A. (1998). A water management method for non-closed crop production. In: Proceedings of the AgEng 98 Conference, Oslo Norway.
Bauerle, W., Short, T., Mora, E., et al. (1988). Computerized individual nutrient fertilizer injector: The system. *Hort. Science*, **23**(5), 910.
Benoit, F. and Ceustermans, N. (1989a). Growing lamb's lettuce (Valerianella olitoria L.) on recycled polyurethane (PUR) hydroponic mats. *Acta Hort.* (ISHS), **242**, 297–304.
Benoit, F. and Ceustermans, N. (1989b). Verworvenheden van het toegepast wetenschappelijk onderzoek in de NFT-kropslateelt. Landbouwtijdschrift, **42**(1), 29–37 (in Dutch).
Benoit, F. and Ceustermans, N. (1994). Eerste bevindingen met het continu productiesysteem van nft-kropsla in plastieken potjes, St. Katelijne Waver, 7p. (in Dutch).
Birch, P. (1979). Concrete costs repaid by easier lettuce management. *Grower*, May **3**, 31–35.
Bleijaert, P., Pollet, S. and Lemeur, R. (2004). Possibilities of growing glasshouse vegetables in a closed system using common soil as a substrate. *Acta Hort.* (ISHS), **633**, 213–221.
Both, A.J. (2005). *Lettuce Production System in DFT*. http://aesop.rutgers.edu/~horteng/
Both, A.J., Albright, L.D. and Langhans, R.W. (1999). Design of a demonstration greenhouse operation for commercial hydroponic lettuce production. ASAE paper No. 994123. ASAE, 2950 Niles Road, St. Joseph, MI 49085-9659, USA. 12pp.
Both, A.J., Leed, A.R., Goto, E., et al. (1996). Greenhouse spinach production in a NFT system. *Acta Hort.* (ISHS), **440**, 187–192.
Chang, R. (1990). *Physical Chemistry with Applications to Biological Systems*. Maxwell Macmillan International Editions (Second edn). New York: Macmillan.
Cooper, A. (1979). *The ABC of NFT*. London, UK: Grower Books.
Delta-T (2006). *WET Sensor*. http://www.delta-t.co.uk.
Dorais, M., Papadopoulos, A.P., Luo, X., et al. (2001). Soilless greenhouse production of medicinal plants in North Eastern Canada. *Acta Hort.* (ISHS), **554**, 297–303.
Formflex (2005). *Coated Metal Troughs*. www.formflex.nl
Gieling, Th.H. (2001). Control of water supply and specific nutrient application in closed growing systems. PhD Thesis Wageningen University, ISBN 90-5808-525-2, 2001.
Gieling, Th.H., Corver, F.J.M., Janssen, H.J.J., et al. (2005). Hydrion-line, towards a closed system for water and nutrients: Feedback of water and nutrients in the drain. *Acta Hort.* (ISHS), **691**, 259–266.
Gieling, Th.H., Jansen, H.J.J., Vree, H. de. and Loef, P. (1998b). Feedback control of water supply in an NFT growing system. In: Proceedings 3d workshop on 'Sensors in Horticulture', Tiberias, Israel.

Gieling, Th.H., Van Straten, G., Janssen, H.J.J. and Berge, A. (1998c). Control of water supply in closed growing systems in a greenhouse. In: Proceedings of the IFAC/CGIR/ASAE Workshop on Control and Ergonomics in Agriculture, pre-prints, pp. 167–170.

Graves, Ch. (1983). The nutrient film technique. *Hortic. Rev.*, **5**, 1–44.

Heemskerk, M.J., Van Os, E.A., Ruijs, M.N.A. and Schotman, R.W. (1997). Verbeteren watergeefsystemen voor grondgebonden teelten. PBG rapport 84, 63p. (in Dutch).

Heinen, M. (1997). Dynamics of water and nutrients in closed, recirculating cropping systems in glasshouse horticulture – with special attention to lettuce grown in irrigated sand beds. PhD Thesis Wageningen University, 270pp.

Hendrix, A.T.M. and Ruijs, M.N.A. (1992). Simulatie van milieuvriendelijkere bedrijfssystemen in de glastuinbouw; Gewasgroep 'bloeiende bol- en knolgewassen' (freesia), verslag nr 16, Naaldwijk, 34p.

Heuvelink, E. and da Costa, J. (2000). *The Almería Greenhouse District: Strong and Weak Points, Opportunities and Threats*, Wageningen: Horticultural Production Chains Group.

Hilhorst, M.A, Groenwold, J. and de Groot, J.F. (1992). Water contents measurement in soil and rockwool substrates: Dielectric sensors for automatic in situ measurements. *Sensors in Horticulture, Acta Hort.* (ISHS), **304**, 209–218.

Hortiplan (2005). www.hortiplan.com/MGS.

Ikeda, H. (1985). Soilless culture in Japan. *Farming Japan*, **19**(6), 35–45.

Ikeda, H., Wada, T., Okabe, K., et al. (1995). Year-round production of spinach by NFT and DFT in the greenhouse. *Acta Hort.* (ISHS), **396**, 257–264.

Ito, T. (1994). Hydroponics. In *Horticulture in Japan* (K. Konishi, S. Iwahori, H. Kitagawa, T. Yakuwa, eds). Tokyo: Asakura Publishing Co., pp. 124–131.

Jiang, W., Qu, D. and Mu, D. (2004). Protected cultivation of horticultural crops in China. *Hortic. Rev.*, **30**, 115–162.

Kratky, B.A. (2005). Growing lettuce in non-aerated, non-circulated hydroponic systems. *Journal of Vegetable Science*, **11**(2), 35–42.

Lataster, J.M.J, Van Os, E.A., Ruijs, M.N.A., et al. (1993). Teeltsystemen los van de ondergrond, groenten en snijbloemen onder glas. IKC Information and Knowledge Centre greenhouse vegetables and floriculture, Naaldwijk/Aalsmeer, 104pp. (in Dutch).

Le Bot, J., Adamowicz, S. and Robin, P. (1998). Modelling plant nutrition of horticultural crops: A review. *Sci. Hortic.*, **74**, 47–82.

Lecomte, A., Flaus, J.M., Brajeul, E., et al. (2004). Multivariable greenhouse control: Applications to fertigation and climate management. *Acta Hort.* (ISHS), **691**, 249–258.

Leoni, S., Pisanu, B. and Grudina, R. (1994). A new hydroponic system tomato greenhouse cultivation: High density aeroponics system (hdas). *Acta Hort.* (ISHS), **361**, 210–217.

Liu, W., Chen, D.K. and Liu, Z.X. (2005). High efficiency column culture system in China. *Acta Hort.* (ISHS), **691**, 495–500.

Maruo, T.H., Hoshi, M., Hohjo, Y., et al. (2001). Quantitative nutrient management at low concentration condition in NFT spinach culture. *Acta Hort.* (ISHS), **548**, 133–140.

Massantini, F. (1977). Impianto aeroponico sperimentale a superfici di coltura verticale. *Colture Protette*, (8–9), 29–34.

Morgan, J.V. and Tan, A.L. (1983). Greenhouse lettuce production at high densities in hydroponics. *Acta Hort.* (ISHS), **133**, 39–46.

National Environmental Policy Plan (1989). *Ministry of Housing, Planning and the Environment*. The Hague, 258pp.

Panel Design Doubles Growing Area (1988). *Greenhouse Manager*. November 1988, http://www.drtomorrow.com/lessons/lessons7/18.html; http://www.kewpie.co.jp/english/veg.html.

Papadopoulos, A.P. and Liburdi, N. (1989). The Harrow fertigation manager, a computerized multifertilizer injector. *Acta Hort.* (ISHS), **260**, 255–266.

Park, K.W., Kim, Y.S. and Lee, Y.B. (2001). Status of the greenhouse vegetable industry and hydroponics in Korea. *Acta Hort.* (ISHS), **548**, 65–70.

Ruijs, M.N.A., Van Os, E.A., Hendrix, A.T.M., et al. (1990a). Simulatie van milieuvriendelijkere bedrijfssystemen voor de gewasgroep 'eenmalig oogstbare groenten' (sla, radijs), verslag 4, Naaldwijk, 34p. (in Dutch).

Ruijs, M.N.A., Van Os, E.A., Hendrix, A.T.M., et al. (1990b). Simulatie van milieuvriendelijkere bedrijfssystemen voor de gewasgroep 'eenmalig oogstbare snijbloemen' (chrysant), Verslag 2, Naaldwijk, 45p. (in Dutch).

Schrevens, E. and Cornel, J.A. (1990). Design and analysis of mixture systems: Applications in hydroponic plant nutritional research. *Plant and Soil*, **154**, 45–52.

Schröder, F.G. (1994). Technological development, plant growth and root environment of the plant plane hydroponic system. *Acta Hort.* (ISHS), **361**, 201–209.

Soffer, H. and Levinger, D. (1980). The Ein Gedi System – Research and development of a Hydroponic System, ISOSC Proceedings 5th Int. Congress on Soilless Culture, pp. 241–253.

Sonneveld, C. (2000). Effects of salinity on substrate grown vegetables and ornamentals in greenhouse horticulture. PhD Thesis, 151p.

Stanghellini, C., Kempkes, F.L.K. and Knies, P. (2003). Enhancing environmental quality in agricultural systems. *Acta Hort.* (ISHS), **609**, 277–284.

Steiner, A.A. (1967). Achttien jaar onderzoek inzake plantenteelt zonder aarde in Nederland. TNO-nieuws, **22**, 33–37 (in Dutch).

Tap, R.F., Van Willigenburg, L.G., Van Straten, G. and Van Henten, E.J. (1993). Optimal control of greenhouse climate: Computation of the influence of fast and slow dynamics. In: (G.C. Goodwin, R.J. Evans, J.B. Cruz Jr, U. Jaaksoo, eds). Proceedings of the 12th IFAC World Congress, Sydney, pp. 321–324.

Van den Boogaard, H.A.G.M., Marcelis, L.F.M. and Gieling, Th.H. (2003). Hydrionline III: On-line monitoring and control system for process water in closed growing systems in greenhouse industry. (Min. Economic Affairs).

Van Henten, E.J. (1994). Greenhouse climate management: An optimal control approach. PhD Thesis Wageningen Agricultural University, Wageningen, The Netherlands, 329p.

Van Henten, E.J. and Bontsema, J. (1996). Greenhouse climate control: A two time-scale approach. *Acta Hort.* (ISHS), **406**, 213–219.

Van Henten, E.J., Bontsema, J. and Van Straten, G. (1997). Improving the efficiency of greenhouse climate control: An optimal control approach. *NJAS Wageningen Journal of Life Sciences*, **45**(1).

Van Henten, E.J., Van Os, E.A., Van Tuijl, B. and Van Weel, P. (2005). Mobiles Anbauverfahren für Schnittblumen. *Das Magazin für Zierpflanzenbau*, **10**, 26–28.

Van Kruistum, G. (ed.) (1997). Productie van witlof en roodlof, teelthandleiding, **79**, 226p. (in Dutch).

Van Os, E.A. (1980). Complete mechanization of the growing of cut chrysanthemums in nutrient film. ISOSC Proceedings 5th Congress, pp. 187–196.

Van Os, E.A. (1986). Technical and economical consequences and mechanization aspects of soilless growing systems. *Acta Hort.* (ISHS), **178**, 85–92.

Van Os, E.A. (1994). Closed growing systems for more efficient and environmental friendly production. *Acta Hort.* (ISHS), **361**, 194–200.

Van Os, E.A. (2006). Mobysant: Meer dan 28% energiebesparing per tak is mogelijk Onder Glas, **3**(1), 44–45. (in Dutch).

Van Os, E.A. and Kuiken, J.C.J. (1984). Mechanization of lettuce growing in nutrient film technique. ISOSC Proceedings 6th Congress, pp. 483–492.

Van Os, E.A. and Simonse, L. (1988). Forcing asparagus in water. *Acta Hort.* (ISHS), **221**, 335–346.

Van Os, E.A. and Stanghellini, C. (2002). Water reuse in greenhouse horticulture. In *Water Recycling and Resource Recovery in Industry: Analysis, Technologies and Implementation* (P. Lens, et al., eds). London, UK: IWA Publishing, pp. 655–663.

Van Os, E.A., Van Zuijdam, R.P., Hendrix, A.T.M. and Koch, V.J.M. (1993). A moving fruit vegetable crop. *Acta Hort.* (ISHS), **342**, 69–76.

Van Straten, G., Vanthoor, B., Van Willigenburg, L.G. and Elings, A. (2006). A 'big leaf, big fruit, big substrate' model for experiments on receding horizon optimal control of nutrient supply to greenhouse tomato. *Acta Hort.* (ISHS), **718**, 147–155.

Van Tuijl, B.A.J., Van Henten, E.J. and Van Os, E.A. (2005). Mobiele gerberateelt en toekomstige ontwikkelingen. *A&F Rapport*, **330**, 42p.

Van Weel, P.A. (2005). Movable gutters role in rose production. *FlowerTech.*, **8**(7), 30–32.

Van Zaal, F. (2005). *Mobile Flower Systems*. www.fransvanzaal.nl

Verwer, F.L.J.A. (1978). Research and results with horticultural crops grown in rockwool and nutrient film. *Acta Hort.* (ISHS), **82**, 141–148.

Vincenzoni, A. (1982). La colonna di coltura per la produzione di fragole. *Colture Protette*, **11**(1), 51–55.

Worm (2005). *Racks for Cutflowers*. www.wormlathum.nl

6

CHEMICAL CHARACTERISTICS OF SOILLESS MEDIA

AVNER SILBER

6.1 Charge Characteristics
6.2 Specific Adsorption and Interactions between Cations/Anions and Substrate Solids
6.3 Plant-induced Changes in the Rhizosphere
6.4 Nutrient Release from Inorganic and Organic Substrates
References

ABBREVIATIONS

AEC – anion exchange capacity;
CEC – cation exchange capacity;
HA – [$Ca_5(PO_4)_3OH$];
OCP – [$Ca_4H(PO_4)_3 2.5(H_2O)$];
P_W – solution P concentrations;
PZNC – point of zero net charge;
PZSE – point of zero salt effect;
SAI – specifically adsorbed ion;
βTCP – [$Ca_3(PO_4)_2$];
ZPT – suspension pH prior to the addition of protons or hydroxyls.

The surfaces of particles and other solids in horticultural substrates bear permanent and/or variable electrical charges. The surface charge properties and the dissolution characteristics of the substrate solids are important, since they affect the ionic composition of nutrient solutions. Ordinarily, the chemical properties of substrates are established prior to their use for growing plants; therefore undisrupted, homogeneous solid phases are addressed. However, throughout the growth period organic compounds excreted from plant roots or resulting from decomposition processes accumulate in the substrate (Tate and Theng, 1980; Huang and Violante, 1986; Tan, 1986; Silber and Raviv, 1996). Thus, the surfaces of the newly formed solids become heterogeneous and the chemical properties of the mixture may be significantly different from those of the single, well-defined component that was characterized before use (Silber and Raviv, 1996). Plants growing in soilless culture typically have smaller root-system volume than plants growing in soil culture. Therefore, the root density of soilless-grown plants is higher so that such plants have a greater impact on the rhizosphere than soil-grown plants. Despite extensive research relating to irrigation and fertilization management or plant growth in soilless culture, the literature regarding the chemical properties of soilless media is scarce. Although peat, perlite, stone wool and mixtures of organic with inorganic materials are abundant in the horticulture industry, the literature regarding the chemical properties of these media is insufficient. Tuff is a volcanic material used as a substrate for horticultural crops in Italy, Spain, France, Turkey and Israel, and had been the subject of considerable research with respect to chemical properties as a horticultural substrate. Thus tuff is used in this chapter as a model for illustration and elucidation of the chemical processes taking place in soilless media systems.

6.1 CHARGE CHARACTERISTICS

The surfaces of solids in horticultural substrates of volcanic or organic origin carry permanent and/or variable positive and negative electrical charges. Permanent negative charge results from isomorphous substitutions (Gast, 1977; McBride, 1989, 2000; Sumner and Miller, 1996; Sposito, 1989, 2000), that is substitution of structural cation by lower valency cation that has the same coordination number and size in layer-silicates (common examples: Si^{4+} by Al^{3+} or Al^{3+} by Mg^{2+}). The extent of cation adsorption to the surfaces is referred to as the *cation exchange capacity* (CEC; $cmol\,kg^{-1}$) and is used to characterize the cation exchange properties of the medium (Gast, 1977; Sposito, 1989).

The CEC of several inorganic and organic horticultural substrates are presented in Table 6.1. These values represent CEC measurement of intact materials prior to use, and it is important to note that plant growth may affect the chemical properties of the substrate solids. CEC values of soil components are usually referred by weight (i.e., $cmol\,kg^{-1}$). However, soilless production system involve a relative small fixed root zone volume and therefore it is more practical to relate substrates CEC to volume unit ($cmol\,L^{-1}$). By using volume-based comparison it is possible to properly compare both 'heavy' and 'light' media.

TABLE 6.1 Cation Exchange Capacity (CEC) of Several Horticultural Substrates

Substrate	CEC cmol kg^{-1}	CEC cmol L^{-1}	Reference
Inorganic Substrates			
Perlite	25–35	2–4	Doğan and Alkan (2004)
Stone wool	34	5	Argo and Biernbaum (1997)
Tuff	10–60[a]	10–60	Silber et al. (1994)
Clinoptilolite (zeolite)	200–400	400–800	Mumpton (1999)
Organic Substrates			
Coconut coir	39–60	2–4	Evans et al. (1996)
Peat	90–140	7–11	Puustjarvi and Robertson (1975)
Pine bark	98[b]	10	Daniels and Wright (1988)
Compost	160–180	15–20	Inbar (1989)

[a] CEC in pH 7, tuff has variable charge surfaces as shown in Fig. 6.2.
[b] CEC in pH 7, pine bark has variable charge surfaces as shown in Fig. 6.4B.

The variable charge (both negative and positive, depending on the solution pH) is generated mainly from the adsorption of H$^+$ and OH$^-$ on solid surfaces such as metal oxides, hydroxides, microcrystalline silicates (allophane and imogolite), or on functional organic groups (Stevenson, 1994). The effects of solution pH on the magnitude of the surface charge may be determined experimentally by measuring the CEC and the *anion exchange capacity* (AEC) over a range of pH values. The pH at which the AEC is equal to the CEC is referred to as the *point of zero net charge* (PZNC) (Parker et al., 1979; Sposito, 1981, 1984, 2000) and is frequently used to characterize the charge properties of the medium. Note that in all cases, pH is referred to the solutions. An alternative experimental method is to conduct a pH titration curve using an indifferent electrolyte (Gast, 1977; McBride, 1989, 2000; Sparks et al., 1996). A typical potentiometric titration for woody peat is presented in Fig. 6.1 (Bloom, 1979).

The pH value at which a series of titration curves for various electrolyte concentrations intersect is referred to as the *point of zero salt effect* (PZSE; Parker et al., 1979; Sposito, 1981, 1984, 2000). However, common substrates (compost, peat moss, tuff, stone wool) rarely consist of a single, well-defined component. They typically contain a mixture that presents both permanent and variable surface charges. The interpretation of the experimental CEC, AEC or pH titration data obtained from these mixtures is, therefore, complicated. Moreover, the existence of additional sources or sinks for H$^+$/OH$^-$ derived from the dissolution/precipitation of minerals during the CEC, AEC or pH titration analyses makes direct interpretation of analytical results even more ambiguous.

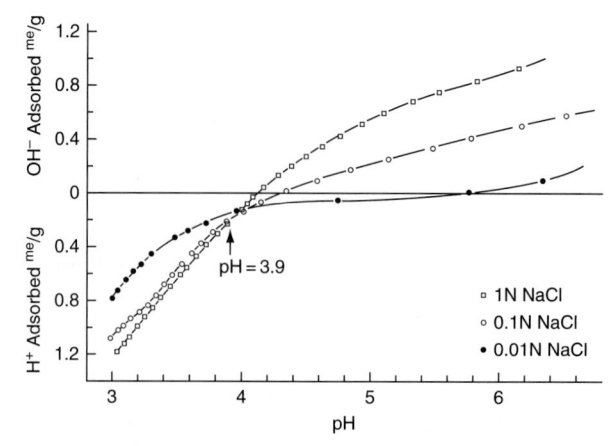

FIGURE 6.1 Acid–base potentiometric titration curves of woody peat. Reprinted from Bloom (1979), with kind permission from the Soil Science Society of America Journal.

The difficulties involved in the determination of the actual surface charge of a heterogeneous material that contains both pH-dependent and permanently charged surfaces, together with primary and secondary minerals, can be seen in the case of tuff (Figs. 6.2 and 6.3). Yellow, Red and Black tuff, commonly used in greenhouse production, are characterized by the degree of weathering, with Yellow tuff being most weathered and Black tuff being the least weathered. The CECs of Yellow, Red and Black tuff were found to be pH dependent, with the CEC increasing by 6.6, 4.2 and 1.9 cmol kg^{-1}, respectively, for every unit increase of pH (Silber et al., 1994).

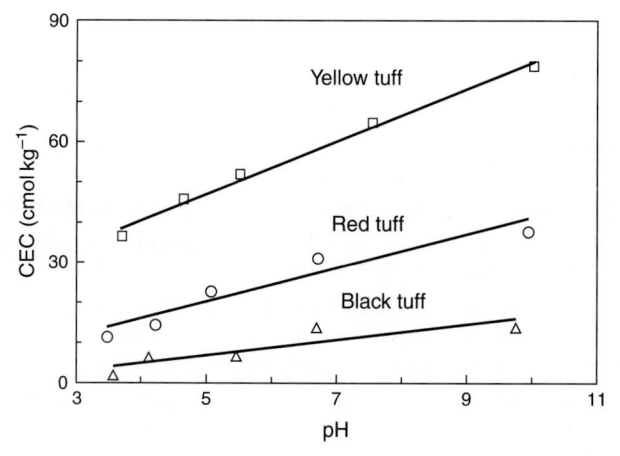

FIGURE 6.2 Cation exchange capacity *versus* pH for black, red and yellow tuffs from northern Israel. Reprinted from Silber et al. (1994), with kind permission from Elsevier Science.

6.1 CHARGE CHARACTERISTICS

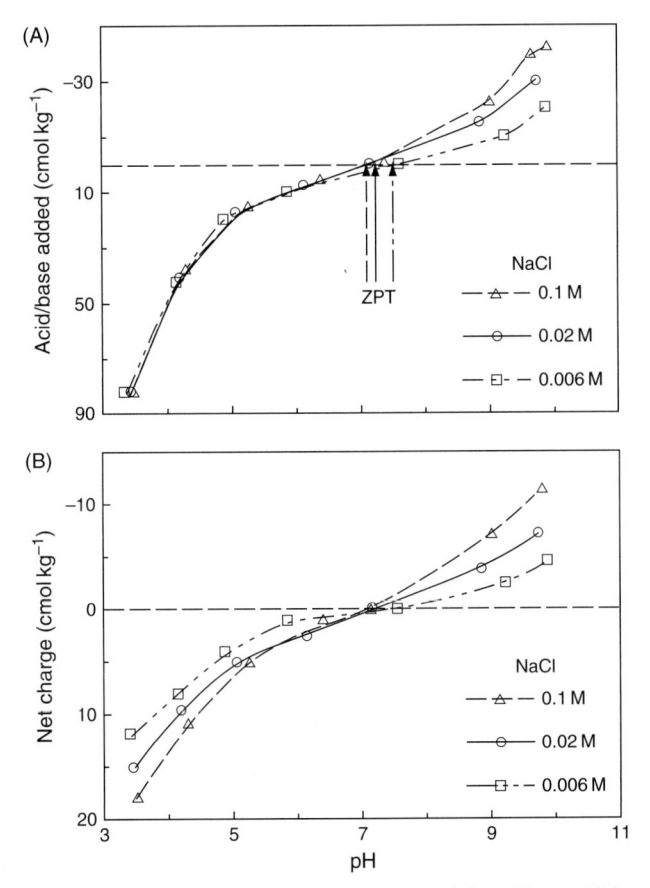

FIGURE 6.3 Potentiometric titration curves of yellow tuff: (A) acid/base addition (0.1 N HCl and NaOH, respectively) as functions of pH and NaCl concentrations, tuff/solution ratio of 1/40, equilibrium period of 1 week. ZPT is the suspension pH prior to the addition of acid or base; and (B) net surface charge calculated with Eq. (1). Based on Silber (1991).

The differing values of CEC slope among the three types of tuff reflect differences in their contents of amorphous materials with variable-charged surfaces. Silber et al. (1994) found, however, that it was not possible to determine the AEC of these tuffs because phosphorus released from indigenous Ca–P minerals during their analysis was re-adsorbed on the tuff surfaces and thus interfered with the AEC determination. The CEC of tuff at pH 3.5, especially that of the weathered Yellow tuff, was relatively high (35 cmol kg^{-1}) and may indicate the presence of components having a permanent charge. The pH titration curves of the tuff types differed significantly from those of homogenous variable-charged materials. In the cases of the less-weathered tuffs (Red tuff and Black tuff), the pH titration lines obtained with three different NaCl concentrations overlapped (data not presented). Distinction of an ionic strength effect was possible only for the Yellow tuff at a pH above 7 (Fig. 6.3A).

The overlap of potentiometric titration curves at low pHs in volcanic material from Chile was attributed to the exchange of added H^+ ions with cations associated with a permanent negative charge (Espinoza et al., 1975; Gast, 1977). Wann and Uehara (1978) reported that specific adsorption of P caused overlapping of potentiometric titration curves of soil; therefore, re-adsorption of P may be partially responsible for the overlap of the pH titration curves in the case of tuff. Nevertheless, on the assumption that H^+/OH^- consumption occurs in cation/anion exchange, the net charge at each pH value can be evaluated by subtracting the net quantity of cations/anions accumulated in the liquid solution of the substrate (Δcharge) from the quantity of acid/base added at the pertinent pH. Hence, Δcharge (cmol kg^{-1}) is defined as

$$\Delta\text{charge} = \left(\sum\text{cat} - \sum\text{anion}\right)\text{pH} - \left(\sum\text{cat} - \sum\text{anion}\right)\text{ZPT} \quad (1)$$

where ZPT is the suspension pH prior to the addition of protons or hydroxyls (in Fig. 6.3a: 7.1, 7.2 and 7.6 in 0.1, 0.02 and 0.006 M, respectively) and $\sum$cat and $\sum$anion include all the cationic and anionic species in the solution, according to speciation calculations. The differences between the quantities of acid/base added to tuff suspensions, and the net surface charge calculated with Eq. (1) (Fig. 6.3a and 6.3b, respectively), demonstrate that the major quantities of acid/base added to tuff solutions were exchanged with adsorbed cations/anions or consumed in the dissolution of indigenous Ca–P minerals and of very fine amorphous particles (Silber et al., 1999).

The CEC of common organic substrates such as peat, pine bark or composts is generally high (80–160 cmol kg^{-1}) and is pH dependent (Brown and Pokorny, 1975; Puustjarvi, 1977; Ogden et al., 1987; Daniels and Wright, 1988). The charge is derived mainly from ionization of COOH groups and, to a lesser extent, from phenolic OH (Stevenson, 1994). The pH effect on CEC of organic material was found to be more pronounced than that of soil inorganic materials (Hallsworth and Wilkinson, 1958; Helling et al., 1964; Ogden et al., 1987; Daniels and Wright, 1988; Stevenson, 1994). The contribution of a unit pH increase to the CEC of soil organic material was found to be 51 cmol kg^{-1} of organic C (Fig. 6.4A).

Assuming that organic C constituted 58 per cent of the organic matter (Stevenson, 1994), then each unit increase in pH raised the CEC of soil organic matter by 29.6 cmol kg^{-1}. Cation exchange properties of pine bark growing media increased linearly between pH 4 and 7 and the contribution of a unit pH increase to the CEC was 23.1 cmol kg^{-1} for particles of size smaller than 0.05 mm (Fig. 6.4B). Note that the increase in measured CEC slope of the fine pine bark was close to the calculated value of soil organic matter, and that these were an order of magnitude higher than the aforementioned slopes of 1.9, 4.2 and 6.5 cmol kg^{-1} per pH unit measured for Black, Red and Yellow tuff, respectively (Fig. 6.2).

The charge characteristics of composts were found to be dependent on composting time (Harada and Inoko, 1980; Inbar, 1989; Inbar et al., 1989, 1991; Iglesias-Jimenez and Perez-Garcia, 1992; Saharinen, 1996; Jokova et al., 1997), mainly because of transformations of organic constituents such as C/N ratio, humic material and lignin

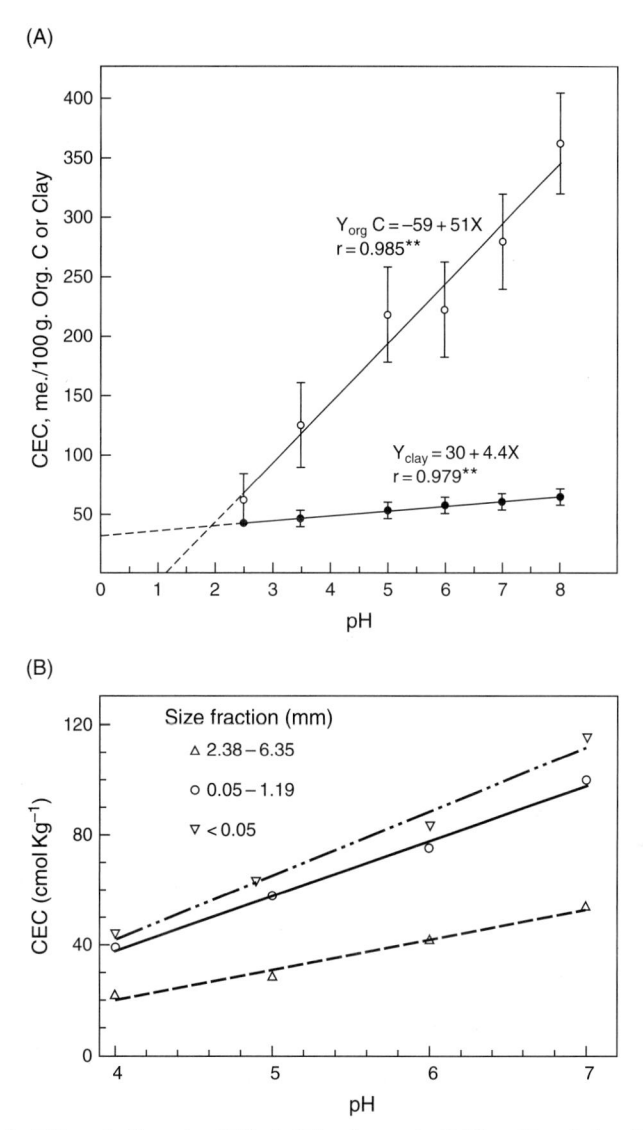

FIGURE 6.4 Effect of pH on the CEC of: (A) soil organic C (Yorg C) and clay (Yclay). Vertical lines show standard errors of the individual values. From Helling et al. (1964), with kind permission from the Soil Science Society of America Journal; (B) several particle size fraction of pine bark growing media. Based on Daniels and Wright, (1988).

contents during the composting process (Inbar et al., 1989, 1991; Saharinen, 1996). The C/N ratio and the lignin concentration correlated with the composting time (Raviv et al., 1987; Tarre et al., 1987; Inbar, 1989; Inbar et al., 1989, 1991), and could be used in estimating the CEC (Figs. 6.5A–6.5C) and maturity indices of compost quality.

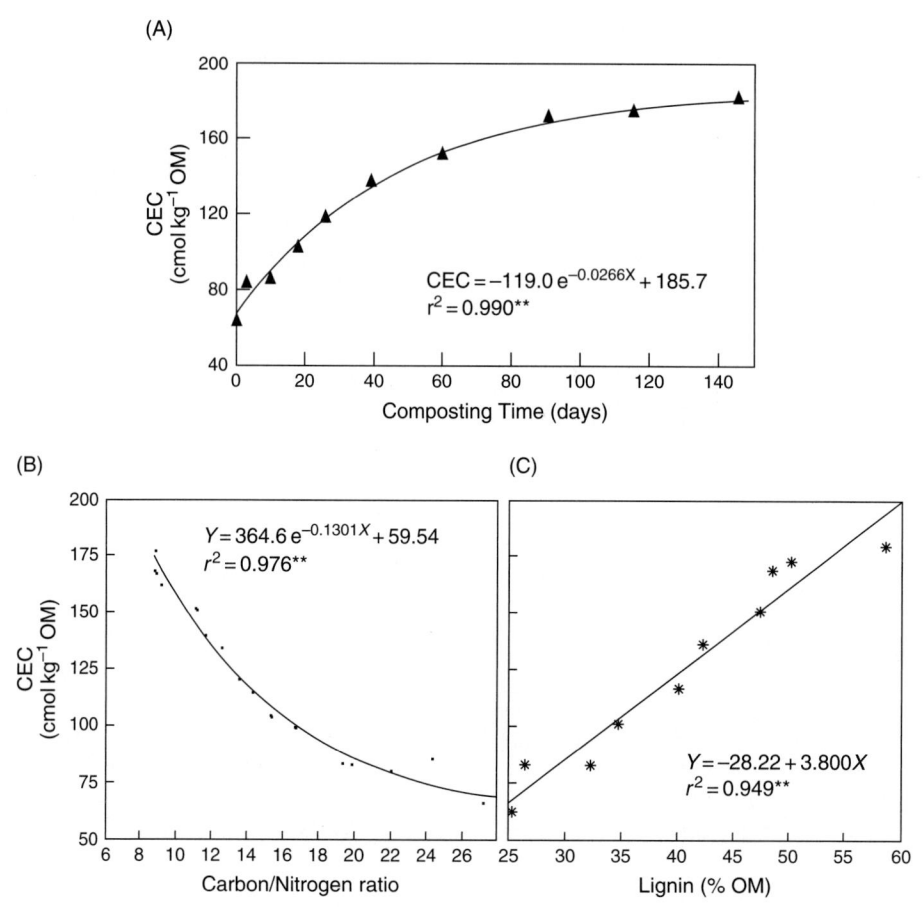

FIGURE 6.5 Cation exchange capacity of compost during the course of the composting process as functions of: (A) composting time; (B) carbon/nitrogen ratio; and (C) lignin content. Based on Inbar (1989).

6.1.1 ADSORPTION OF NUTRITIONAL ELEMENTS TO EXCHANGE SITES

The affinity of cations for negatively charged surfaces under equal concentration in the aqueous portion of the substrate is affected by ion characteristics such as valence, size and hydration status. The affinity of divalent cations are higher than that of monovalent cations because of their greater charge, and usually, the relative adsorption strength follows the order of hydration, that is, K≈NH_4 >Na and Ca > Mg (Barber, 1995). However, concentrations of each of the cations in a typical irrigation solution for greenhouses crops are not equivalent, with the K concentration commonly exceeding the concentration of Ca, NH_4 or Mg (Adams, 2002, Sonneveld, 2002). As an example, recommendations for K, NH_4, Ca and Mg concentrations in the nutrient solution for soilless production of ornamental and vegetable crops in The Netherlands (Naaldwijk) consist of 5.0, 1.0, 4.5 and 2.5 mmol L^{-1}, respectively (Sonneveld, 2002). Thus, it is expected that exchange reaction, that is displacement of cations from the substrate solids by K$^+$, will reduce its concentration and subsequently will increase the concentration of other cations in the liquids within the substrate. The exchange reaction may be horticulturally beneficial in cases where: (i) concentration of a toxic or detrimental cation such as heavy metals or Na$^+$ in the liquid solution of the growing medium decline as a result of adsorption or (ii) concentration of a beneficial cation in that liquid increases as a result desorption. Alternatively, the opposite could occur, with expected detrimental effects, of cation exchange including: (i) increased concentrations of a toxic or detrimental cation in the solution followings desorption or (ii) decreased concentrations to a deficiency level of beneficial cation followings its adsorption to substrate solids.

6.2 SPECIFIC ADSORPTION AND INTERACTIONS BETWEEN CATIONS/ANIONS AND SUBSTRATE SOLIDS

The solubility and the interactions of nutritional elements with the substrate solids are associated with ion characteristics such as valence, size and hydration status as detailed above. The solubility of NO_3^-, the major N source for plants grown in soilless culture (Chap. 8), is very high, its affinity to positive charged surfaces is very low compared with that of $H_2PO_4^-$ or SO_4^{2-}, and therefore, the concentrations of NO_3^- in the substrate solution is nearly unaffected by precipitation and/or adsorption reactions. The effects of microbes on NO_3^- availability in the rhizosphere are detailed elsewhere (Chap. 8). NH_4^+ is also an N source for plants in soilless production and, similar to NO_3^-, generally does not precipitate. Unlike NO_3^-, the high affinity of NH_4^+ to negatively charged surfaces affects its availability to plant.

Nitrification/denitrification reaction and the effects of NH_4^+ nutrition on rhizosphere pH and the availability of other nutritional elements are discussed in Chap. 8. Potassium is an important nutritional element and is also not involved in precipitation reactions to an appreciable extent. However, as a charged cation, K$^+$ may be adsorbed by negatively charged surfaces, and this may reduce its availability to plants. Note that the partitioning of K$^+$ as well as other cations between the aqueous and the solid parts of the substrate depends on its relative concentration in the solution and composition of the exchange sites. The affinity of the divalent cations Ca^{2+} and Mg^{2+} to negatively charged surfaces is very high (Barber, 1995) and they are highly involved in precipitation reactions, especially with $H_2PO_4^-$ and SO_4^{2-}. Actually, it is well accepted that P availability in the rhizosphere is governed by Ca^{2+} and Mg^{2+} activities (Lindsay, 1979). Metal micronutrients such as Cu, Fe, Mn and Zn are considered as specifically absorbed ion (SAI), their affinity to charged surfaces is very high and considerably higher than other cations. These metals are also highly involved in precipitation reactions so that their availability to the plant is very low. Fertigation managements used in the horticulture industry for guaranteeing the availability of nutritional elements for plants, including that of micronutrients, are discussed in Chap. 8.

Phosphorus and zinc were selected below for detailed discussion to illustrate the interactions of cations and anions with the solid constituents of substrate because: (i) both are important nutritional elements for plants; (ii) under common agricultural practices, low availability in the vicinity of the roots of either of these elements often limits crop growth and productivity, even if P and Zn concentrations in the irrigation solutions are adequate (Marschner, 1995); (iii) ions of both elements are classified as SAIs because of their high affinity to charged surfaces; and (iv) both elements easily form new solid compounds as a result of precipitation reactions.

6.2.1 PHOSPHORUS

Continuous decline of solution P concentration (P_W) following fertilizer application is a widespread phenomenon, and temporary P deficiency often restricts crop productivity. P_W decreases through two mechanisms: (i) very fast (on a scale of seconds or minutes) electrostatic reactions of adsorption onto the substrate, as a result of the high affinity of P ions to charged surfaces; and (ii) slow (on a scale of hours or days) formation of new solid metal-P compounds (with Al, Fe and Mn under acidic pHs and with Ca and Mg under neutral or basic pHs; Lindsay, 1979).

6.2.1.1 Effect of pH on P Partitioning between Substrate Solids and Liquids in Tuff Suspensions

Solution pH plays the dominant role on P partitioning between substrate solids and liquids as presented in Fig. 6.6A for Yellow tuff. Negative values for P retention indicate that the quantities of soluble P in tuff suspensions were greater than the added quantities, as a result of dissolution of indigenous hydroxyapatite during the analysis procedure (Silber et al., 1994, 1999). Retention of P by Yellow tuff at acidic pHs could be fitted to a pH-specific Langmuir isotherm (Fig. 6.6A). The activity products of Al^{3+}, Fe^{3+} and PO_4^{3-} in acidic tuff suspensions (pH 4–7) were below the Ksp of variscite ($AlPO_4 \cdot 2H_2O$) and strengite ($FePO_4 \cdot 2H_2O$), respectively, therefore it is logical to deduce that adsorption reactions governed P solubility in acidic tuff suspensions. This conclusion is in accord with the high content of amorphous materials in the Yellow tuff (51.5 g per 100 g of Halloysite-like allophane) having variable charged surfaces with high P-binding capacity (Silber et al., 1994). However, the formation of an X-ray-amorphous analogue of variscite or $AlOHNaPO_4$ (Veith and Sposito, 1977) is also possible. At higher pH values, the Langmuir rule of declining slope with increased P_W was not maintained: at pH 8.3 and 9.3 and at P_W above 0.46 and 0.31 mM, respectively, an increase in the slope of P-sorption isotherm was observed (see arrows in Fig. 6.6A). The increased slope indicates that a change in the phase-controlling P partitioning took place: increasing P addition rates enhanced the feasibility of Ca–P mineral precipitation because above pH 7, all the activity product of Ca^{2+} and PO_4^{3-} in tuff solutions exceeded the Ksp of hydroxyapatite (HA; $Ca_5(PO_4)_3OH$) and shifted towards the solubility line of βTCP ($Ca_3(PO_4)_2$) or

6.2 SPECIFIC ADSORPTION AND INTERACTIONS

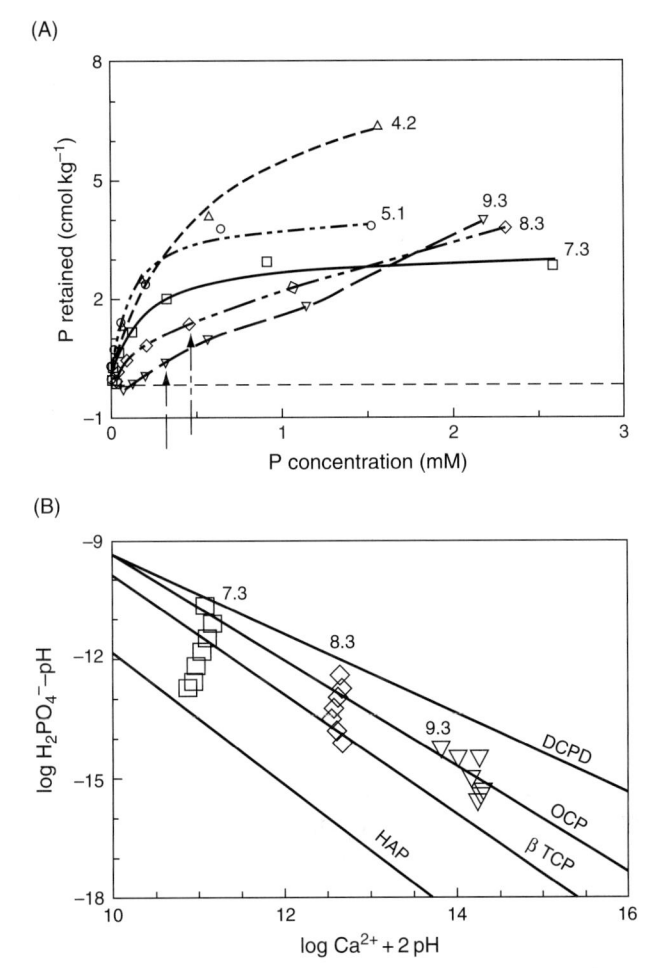

FIGURE 6.6 (A) pH-dependent adsorption of phosphate onto yellow tuff (pH values are shown above each isotherm), constant ionic strength (0.01 M NaCl), tuff/solution ratio of 1/40, 24 h of reaction. Arrows indicate experimental points above the Ksp of OCP; (B) experimental activities in tuff extracts at pH 7.3, 8.3 and 9.3 in relation to the theoretical solubility lines for calcium-phosphate minerals (HAP: $Ca_5(PO_4)_3OH$, βTCP: $Ca_3(PO_4)_2$, OCP: $Ca_4H(PO_4)_3 \cdot 2.5(H_2O)$ and DCPD: $CaHPO_4 \cdot 2H_2O$). Based on Silber (1991).

OCP ($Ca_4H(PO_4)_3 \cdot 2.5(H_2O)$) (Fig. 6.6B). HA is the most thermodynamically stable Ca–P mineral, but more soluble compounds such as βTCP or even OCP temporarily controlled P in tuff solutions. This conclusion is supported by the data of Imas et al. (1996) and Salinger et al. (1993), who found that on a time scale of a few days, P solubility in the $Ca\text{-}CO_3\text{-}PO_4$ system was controlled by a metastable Ca phosphocarbonate phase. Adopting the conclusion of Sanyal and De Datta (1991) that βTCP governs P dissolution in neutral and basic soil solutions, adsorption reactions may govern P solubility in tuff suspensions at pH 7.3, as long as P_W does not exceed 0.9 mM at the pertinent Ca concentration (0.4 mM). At pH 8.3, the P_W value was below the βTCP line, at 0.09 mM, whereas at pH 9.3 P_W was above βTCP solubility even at zero P addition (Fig. 6.6B). If the solution concentration of Ca and Mg are assumed to be 2 mM, which is typically required for plant growth, then the calculated P_W concentration at pH 7.3, according to βTCP solubility, will not exceed 2.5 µM; its actual value will depend on the CO_2 and SO_4 concentrations as well. It is possible to assume, therefore, that adsorption reactions govern P_W in tuff substrate at basic pHs for a short time (minutes or hours), but as time proceeds, P_W is expected to decline until the activity product of Ca^{2+}, Mg^{+2} and PO_4^{3-} attains the Ksp of a thermodynamically stable compound. Thus, the reaction kinetics must be taken into account when evaluating the relative importance of adsorption and precipitation reactions in tuff solutions.

6.2.1.2 Effect of solution ionic strength on P solubility

The solution ionic strength (I) may affect P solubility via three mechanisms:

1. Alteration of the ion activity coefficient: increase of I decreases the ion activity coefficient, in turn increasing the possible concentration of P in the substrate solution.
2. Variation of charge density: the solution ionic strength affects the charge density in the diffuse double layer; therefore, the effect of I on P adsorption will be pH-dependent. Similarly to the effect of I on the surface charge characteristics presented in Figs. 6.1 and 6.3, below the PZSE of the sorbent, increase of I causes a reduction of anion adsorption, whereas above the PZSE the opposite trend is expected, and increase of I causes increased anion adsorption (Sposito, 1984).
3. Variation in anion/cation concentrations in the substrate solution: increases of I may affect the release of anions and cations from the solid phases, on which they might have been adsorbed or of which they might have been an indigenous part.

The effect of I on P_W in Yellow tuff solutions can be used to illustrate these mechanisms (Fig. 6.7). The activity products of Ca^{2+} and PO_4^{3-} with no acid/base addition (ZPT) were above the HA solubility line but below that of βTCP (Fig. 6.7A). P_W in Yellow tuff suspensions at the ZPT decreased as a result of increases in I (Fig. 6.7B). In the light of the conclusions of Sanyal and De Datta (1991) that βTCP controls P_W dissolution/precipitation reactions, it may be deduced that adsorption governed P_W in this pH range. The PZSE of the Yellow tuff was estimated to be 6.2 (Silber, 1991), therefore, the increase of P_W with decreasing I was consistent with mechanism (ii), described above.

At higher pH the electrostatic potential of the solid surfaces became less negative (Figs. 6.1 and 6.3) and P_W increased significantly with decreasing values of I (Fig. 6.7B). An increase in I affected metal concentrations as well (mechanism (iii),

6.2 SPECIFIC ADSORPTION AND INTERACTIONS

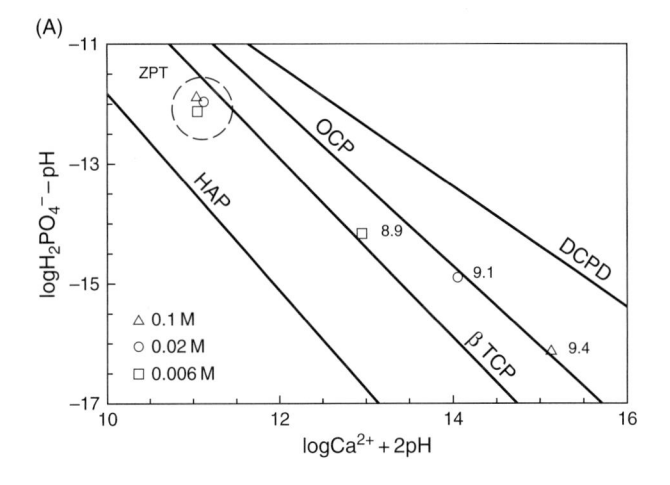

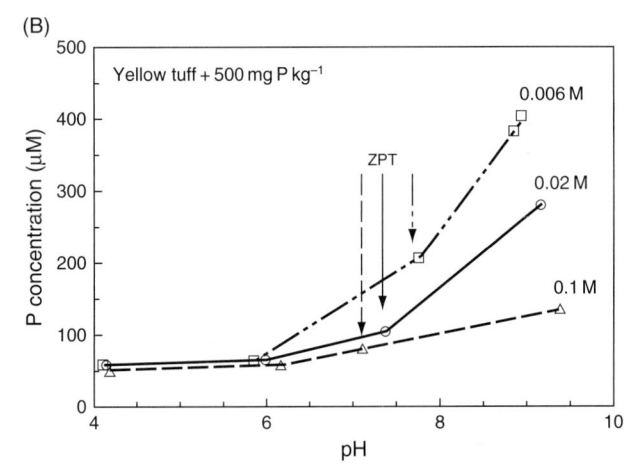

FIGURE 6.7 (A) pH-dependent ionic strength (NaCl) effect on solution-P concentration in tuff extracts after addition of P at 500 mg kg^{-1}, tuff/solution ratio of 1/40, 24 h of reaction, ZPT is the suspension pH prior to the addition of acid or base; (B) experimental activities in tuff extracts at pH 8.9, 9.1 and 9.4 in relation to the theoretical solubility lines for calcium-phosphate minerals. Based on Silber (1991).

above), and the Ca and Mg concentrations in tuff suspensions increased (data not presented). The activity products of Ca^{2+} and PO$_4^{3-}$ in tuff solutions were between the solubility lines of βTCP and OCP (Fig. 6.7A). Adsorption reactions probably controlled P$_W$ at high pH but it would be expected that at equilibrium, the P$_W$ in 0.1 and 0.02 M solutions would decline further, until the activity products of Ca^{2+} and PO$_4^{3-}$ attained the Ksp of βTCP. In acidic tuff solutions the effect of I on the P$_W$ was insignificant (Fig. 6.7B), probably because of P release from indigenous Ca–P minerals (Silber, 1991, Silber et al., 1994, 1999) during the analyses.

6.2.1.3 Horticultural Implications

Phosphorus retention by substrates has an important effect on P fertilization management. For example, a typical substrate quantity used for vegetables or flowers consist of 50–60 kg of tuff m^{-2}. The daily evapotranspiration rate during the growth period of roses (*Rosa* × *hybrida*) in a semi-arid climate is usually 5–10 L m^{-2}; therefore, the daily irrigation supply is 0.1–0.2 L kg^{-1} of tuff. The recommended P concentration in the irrigation water for cut-rose culture is usually 30–45 mg L^{-1} (Sonneveld and de Kreij, 1987; Jones, 1997; Sonneveld, 2002), therefore the daily input of P is 3–9 mg kg^{-1}. Under this daily P rate and ignoring leaching, the possible P_W at pH 7.3 according to the P retention curve will be lower than 0.05 mmol L^{-1} (Fig. 6.6A). In the light of the above data and on the assumption that adsorption reactions control P availability in tuff suspensions, it is expected that the decline of P solubility in the solution will be very fast, so that shortly after irrigation, the P_W will be deficient, and insufficient for rose production. Johanson (1978) found that the yield of cut roses fell to 40 per cent following low P application (0.16 mmol L^{-1}), compared with that under an adequate P management (1.3 mmol L^{-1}). Thus, low P concentrations in tuff substrates may restrict plant production. If a P concentration of 0.5 mmol L^{-1} is taken as the minimum for adequate growth of cut roses, the amounts of P adsorbed by Yellow tuff at pH 6 and 7.3 should be 840 and 600 mg kg^{-1}, respectively (Silber, 1991). If the quantity of P added daily via fertilization is taken to be 4.5 mg kg^{-1} (0.15 L kg^{-1} at a P concentration of 30 mg L^{-1}), then with solution pH of 6 or 7.3, at least 187 or 133 days would be required, respectively, to fill all the available sites for P adsorption on the Yellow tuff surface and to attain the desired P_W. During this period, cut-rose development would certainly be impaired. With a solution pH above 7 and Ca and Mg concentrations assumed to be 1.5–3 mmol L^{-1}, as required for optimal plant growth, P deficiency will restrict plant growth because the activity products of Ca^{2+}, Mg^{2+} and PO_4^{3-} are far above the Ksp values of known Ca–P and Mg–P minerals (Lindsay, 1979).

The direct outcome of the adsorption and the precipitation reactions detailed above is that, under equilibrium conditions, P_W may be lower than plant requirements throughout the pH range of plant growth in tuff substrates. It is important to note that conditions of P deficiency are not exclusive to tuff media. Precipitation of P with Ca and Mg at neutral and high pH is expected to be the general phenomenon in all substrates, as long as nutrient concentrations in the irrigation water are kept at adequate level for plant growth. Considering the effect of pH on adsorption and precipitation reactions, it is reasonable to expect that optimal P availability should be obtained at pH 6.0–6.5.

P adsorption by organic compounds is well known. High P retention by pine bark media have been reported (Cotter, 1979; Yeager and Wright, 1982; Ogden et al., 1987), and low P availability inhibited plant growth (Cotter, 1979). Shortening the time between consecutive P additions, that is increasing the frequency of fertigation, and inducing non-equilibrium conditions between successive fertigation events may be a preferable management means to enhance P availability in the rhizosphere and to improve P uptake by plants (Silber et al., 2003, 2005; Xu et al., 2004). The beneficial

6.2 SPECIFIC ADSORPTION AND INTERACTIONS

effects of high irrigation frequency on P availability to plant roots are detailed in Chap. 8. Alternatively, addition of organic components, such as mature compost, which contain humic and fulvic substances, to tuff substrates might significantly reduce P retention by the substrate (Gottesman, 1989).

6.2.2 ZINC

Zn ions are SAIs that have affinity to charged surfaces. Zn adsorption increases as pH increases, reflecting typical cation behaviour. Zn solubility may be controlled by hopeite [$Zn_3(PO_4)_2 \cdot 4H_2O$] solubility (Lindsay, 1979). Similar to P, Zn concentration in the rhizosphere may be lower than plant requirements throughout the pH range of plant growth in many substrates. The effect of solution-pH on Zn retention in Yellow tuff suspensions is presented in Fig. 6.8.

All the activity products of Zn^{2+} and PO_4^{3-} in the tuff solutions were below the Ksp of hopeite or of any Zn-oxide mineral, therefore the retention of Zn on Yellow tuff surfaces was related solely to electrostatic adsorption. Addition of Zn to Yellow tuff suspensions decreased the P_W at all the pHs examined (Fig. 6.9). Since the possibility of Zn precipitation with P was ruled out, it is most likely that the P_W decreased as a result of increased P adsorption, promoted by an increase in the positive charge on tuff surfaces following Zn adsorption. However, under common agricultural practices P_W is higher by at least three orders of magnitude than the Zn concentration (Sonneveld, 2002), therefore, the opposite trend is to be expected, and usually Zn availability is restricted through hopeite solubility (Lindsay, 1979) and/or the increase of negative charge that follow P adsorption on the solid phase surfaces (Diaz-Barrientos et al., 1990; Pardo, 1999).

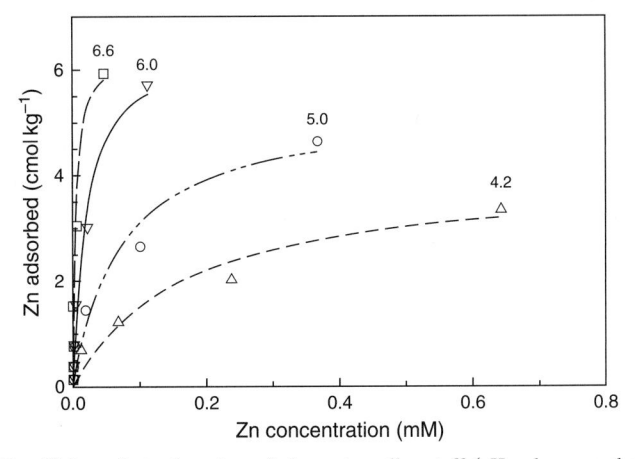

FIGURE 6.8 pH-dependent adsorption of zinc onto yellow tuff (pH values are shown above each isotherm), constant ionic strength (0.01 M $NaClO_4$), tuff/solution ratio of 1/40, 24 h of reaction. Based on Silber (1991).

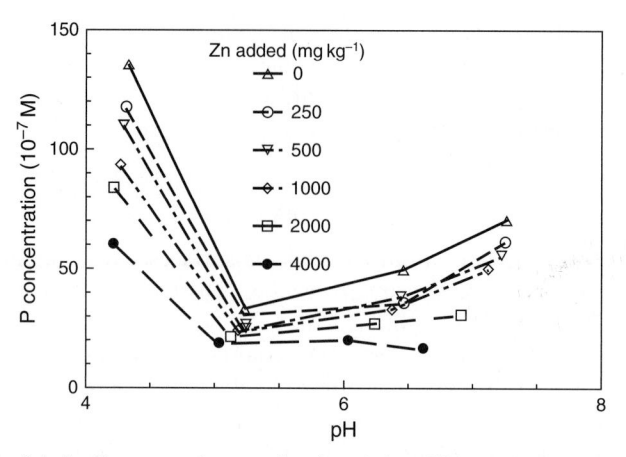

FIGURE 6.9 Solution-P concentrations as a function of zinc addition and pH in tuff extracts. Constant ionic strength (0.01 M NaClO$_4$), tuff/solution ratio of 1/40, 24 h of reaction. Based on Silber (1991).

Zn adsorption by substrates is expected to follow the CEC values presented in Table 6.1, and therefore some substrates commonly used in soilless culture (tuff, zeolite, compost) exhibit high adsorption capacity for Zn and other micronutrients such as Fe, Mn and Cu. Zn adsorption to more inert substrates such as perlite and stone wool is expected to be insignificant (on volume basis); however, plant growth may alter the surface properties of their solid phases, and subsequently, Zn adsorption may be significant even in 'inert' substrate as a result of adsorption to dead or inactive roots or to new organic material in the substrate from root decomposition. Thus, under the solution-pH conditions commonly used in soilless production (pH of 5.5–7.5), adsorption reactions will reduce the concentrations of metal micro-nutritional elements to deficiency levels. Bearing in mind the effect of pH on adsorption and precipitation reactions, it is reasonable to expect that optimal Zn availability should be increased at acidic pHs. Increasing the solution-metal concentration is useless for the same reasons as detailed above for phosphorus. Chelating agents such as EDTA (ethylene-diamine tetracetic acid), DTPA (diethylenetriamine penta-acetic acid) or EDDHA (ethylenediamine di(o-hydroxyphenylacetic acid)) are commonly used to maintain adequate availability of micronutrients (Parker et al., 1995). In addition, as in the case of P, increasing the fertigation frequency was found to be an effective means of enhancing metal uptake by plants (Silber et al., 2003, 2005).

6.2.3 EFFECTS OF P AND ZN ADDITION ON SOLUTION SI CONCENTRATION

Addition of SAI anion or cation, respectively, enhances the negative or positive surface charge (Sposito, 1984). Silicon is an important constituent of substrate solids and Si ions have pH-dependent adsorption properties (Hingston and Raupach, 1967; Goldberg and Glaubig, 1988; Hansen et al., 1994; Nartey et al., 2000). Therefore, the

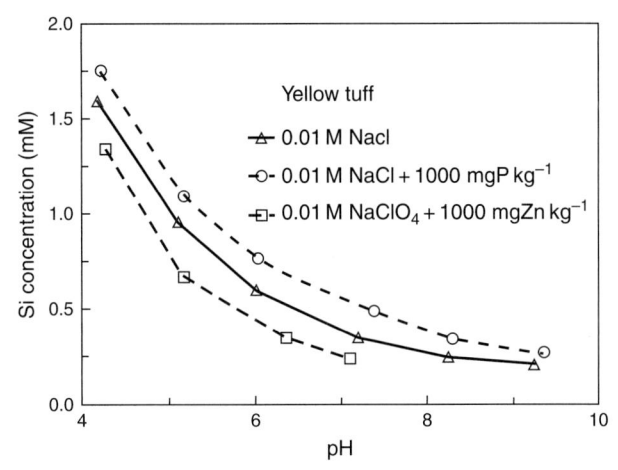

FIGURE 6.10 Solution-Si concentrations as a function of phosphorus and zinc addition to tuff suspensions. Ionic strength and electrolytes used are indicated inside the Figure, tuff/solution ratio of 1/40, 24 h of reaction. Based on Silber (1991).

effects of the addition of SAI anion or cation on solution Si concentration is of interest; Si concentrations increased with decreasing pH in all tuff suspensions (Fig. 6.10) because of the dissolution of volcanic glass and halloysite-like allophane (Silber et al., 1999).

Addition of P to tuff solutions induced an increase of Si concentrations, in accordance with the findings of Rajan (1975), Rajan and Perrott (1975), Rajan and Fox (1975), Veith and Sposito (1977), Pardo and Guadalix (1990) that P retention was followed by displacement of SiO_2 from aluminosilicates, either because of displacement from the surface or because of partial disruption of the structure of the substrate solids. Contrary to the effect of anion addition, Zn addition to Yellow tuff suspensions decreased Si concentrations at all the pHs examined (Fig. 6.10). While silicon is generally not considered to be an essential nutrient, it does play a significant role in plant resistance to stress conditions (Epstein, 1999). Therefore, the effects of P and Zn addition on Si concentrations may be of interest.

6.3 PLANT-INDUCED CHANGES IN THE RHIZOSPHERE

6.3.1 EFFECTS ON CHEMICAL PROPERTIES OF SURFACES OF SUBSTRATE SOLIDS

Plant growth may affect the chemical properties of the substrate solids through three main mechanisms:

1. changes of the surface charge and chemical properties of the solids through the addition of specifically adsorbed ions such as orthophosphates in the fertigation nutrients

2. addition of new solid materials, that is organic compounds, mainly because of root growth
3. accumulation of root exudates and decomposition products, especially humic and fulvic acids that may cause decomposition of substrate solids (Tan, 1986) and/or modification of the surface charge (Tate and Theng, 1980; Huang and Violante, 1986).

The effect of SAI addition on the chemical properties of substrate solids (mechanism i) is well understood (Sposito, 1984), and as long as the solution-pH is not extreme (below pH 4) and does not cause substantial dissolution of the solids, this effect is expected to be reversible. The direct effect of plant growth on the chemical properties of the substrate, that is alteration of the chemical characteristics of the media resulted from formation of new organic material in the form of roots, is less understood. Generally, the CEC of root dry matter has the same order of magnitude as soil constituents (10–100 cmol kg^{-1}) and is higher in dicotyledonous species than in monocotyledonous ones (Blamey et al., 1992; Marschner, 1995). The root CEC was found to be pH-dependent, and adsorption of protons and copper to root cell walls could be evaluated by the same surface complexation model (single capacitance model) used for adsorption to soil (Allan and Jarrell, 1989), suggesting that the same mechanism of adsorption are similar to that of soil. It is, however, doubtful that intact roots contribute to the charge characteristics of a medium, because the high CEC values reported in the literature were related to root components such as cell walls. Therefore, although the high CEC values may reflect the 'potential' CEC, the 'actual' CEC of intact plant root *in vivo* is probably lower because of the low accessibility of ions from the external solution to the internal apoplast (Marschner, 1995).

The effects of root excretions and root-induced pH changes or modifications of nutrient availability in the rhizosphere were reviewed by Marschner and Römheld (1996), Bar-Yosef (1996), Jones (1998) and Hinsinger et al. (2003). The effects of plant growth on the chemical characteristics of the substrate have been studied by Silber and Raviv (1996). Surface characteristics of Red tuff were evaluated prior (T0) to using it for rose culture as well as after it had been exposed to fertigation solution in the absence and presence of plants (Tf and Tp, respectively). Extractions were carried out at various pH levels with solutions differing in ionic strength. The ionic strength and pH were selected to reflect the relevant conditions at the rhizosphere (Silber and Raviv, 1996). The pH titration lines of intact unused Red tuff (T0) overlapped, and distinction of an ionic strength effect was not possible (Fig. 6.11A).

Red tuff consists mainly of primary minerals and un-weathered volcanic glass. It contains only 19 per cent of amorphous material that may contribute to the pH-dependent charge (Silber et al., 1994). Therefore, major proportions of added acids/bases were probably consumed either through exchange with cations/anions adsorbed on permanently charged sites, through the dissolution of indigenous Ca–P minerals and very fine amorphous particles, or both, as suggested by Silber et al. (1999). As a consequence, these processes masked the effect of solution ionic strength on variable-charged surfaces (Fig. 6.11A).

6.3 PLANT-INDUCED CHANGES IN THE RHIZOSPHERE

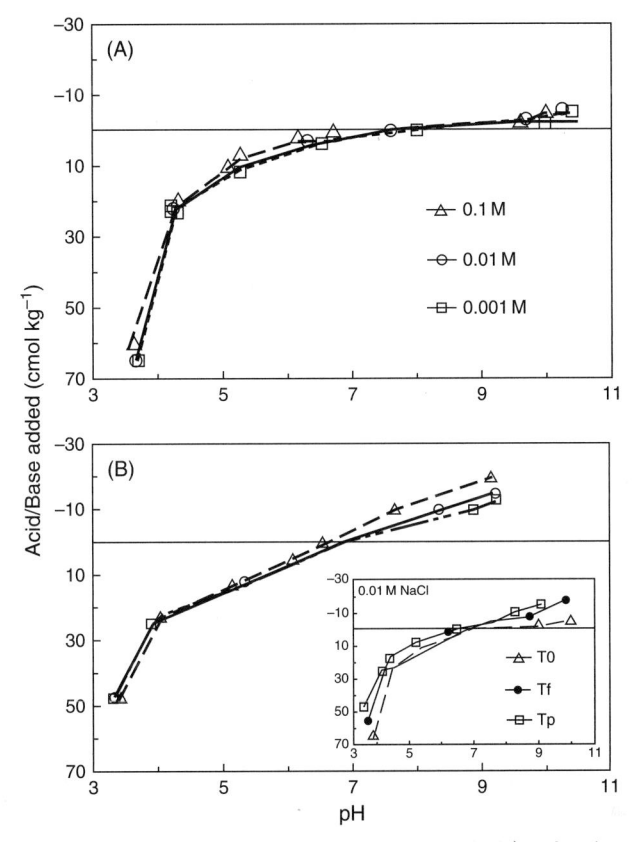

FIGURE 6.11 Acid/base addition (0.1 N HCl and NaOH, respectively) as functions of pH and NaCl concentrations, tuff/solution ratio of 1/40, equilibrium period of 1 week. (A) intact unused red tuff; and (B) red tuff after two years of rose growing. The inset curves show a comparison of acid base addition at constant ionic strength of 0.01 M NaCl to: unused tuff (T0), and tuff irrigated with fertilization solution in the absence and presence of plants (Tf and Tp, respectively). Based on Silber and Raviv, (1996).

Plant growth induces an increase in the negatively charged surfaces (Fig. 6.11B) analogous to the trend observed after the addition of citric acid to goethite and clay minerals (Lackovic et al., 2003). In addition, the effect of solution ionic strength at pH values above 7 was apparent (Fig. 6.11B), and was similar to that observed for Yellow tuff (Fig. 6.3). The increase of negatively charged surfaces of Red tuff in the presence of plants (Tp) was probably due to the specific adsorption of root-excreted organic compounds. Two years of irrigating tuff substrate with fertilizer solution in the absence of plants (Tf), also induced an increase of positive charge on Red tuff particle surfaces, but this effect was less significant than under plant growth (inset in Fig. 6.11B). Specific adsorption of orthophosphate ions added via fertigation and the accumulation of soluble low-molecular-weight compounds exuded from roots or released during organic matter decomposition in the tuff substrate induced pH decrease in Tf and Tp solutions, respectively (Silber and Raviv, 1996). In almost all cases, addition of equal quantities of acid or base lowered the pH in tuff solutions, with intact unused tuff being most affected, and tuff exposed to fertigation solutions in the presence of plants being the least (inset in Fig. 6.11). It is important to note that all the root-mediated changes in the chemical properties of substrate solids are affected by the solid characteristics, the plant type and growth stage, and the prevailing agricultural management (rate of fertilization, proportions of fertilizer components, such as the NH_4:NO_3 ratio, fertigation frequency and the volume of medium available for root growth). Morel and Hinsinger (1999) discussed the effect of root-induced modifications of P partitioning between the substrate solid and the substrate solution and concluded that it was notably influenced by root density. They also found that growth of crops with low root density had little or no effect on soil properties over a long period of 80 years (Morel et al., 1994), while one week of intensive plant growth with 100-fold increase in root density significantly affected soil properties (Morel and Hinsinger, 1999). Soilless culture is commonly characterized by root-system constraints that differ from those found in conventional soil culture, and that lead to higher root density of plants in soilless production. The conclusion by Morel and Hinsinger (1999) suggest that in soilless production, plants have a significantly greater impact on medium properties than plants growing in soil.

The effects of root-mediated pH changes in the rhizosphere are well documented (for review see Nye, 1986 and Hinsinger et al., 2003); they are caused by excretion of ions (mainly H^+, OH^- and organic bases or acids) which balance the electric charge, following cation or anion uptake by roots, root respiration, carbon exudation or redox processes. All these processes are strongly dependent on plant type and nutritional status (Marschner et al., 1986; Nye, 1986; Neumann and Römheld, 1999; Abadia et al., 2002; Hinsinger et al., 2003). The effect of plants on the rhizosphere pH is largely related to the chemical properties of the solid phases. Accumulation of soluble organic acid excreted from roots in the presence of an inert medium may induce a pH decrease because the adsorption reactions are negligible and the ligand acts as weak acid, whereas, in contrast, the quantities of OH^- released following ligand adsorption onto a substrate with a high surface charge may be significant and much higher than that of H^+ added via ligand dissolution. The effects of tomato (*Lycopersicon esculentum* Mill.) plants on the pH in leachate from pots filled with various substrates are presented in Fig. 6.12A.

To simulate the effect of roots on the pH of the medium, citric acid, which is an important organic acid excreted from the roots, was added to tuff suspensions equilibrated at three pH levels (3.3, 4.8 and 6.5) which approximately fit the log K^0 of citric acid (3.1, 4.8 and 6.3) (Martell and Smith, 1976–1989; Smith and Martell, 1989). At the three pH values examined, the pH rose as a result of increasing citric acid addition, but the size of the change in pH resulting from the citric acid addition (ΔpH) varied (Fig. 6.12B). The predominant citrate species at pH 3.3 were H_3Cit^0 and H_2Cit^- (38 and 60 per cent, respectively, of the total citrate ions in the substrate solution) whereas the monovalent $H_2PO_4^-$ was the main P species (Lindsay, 1979). Thus, the competing effect of citrate ions was relatively small and, therefore, only

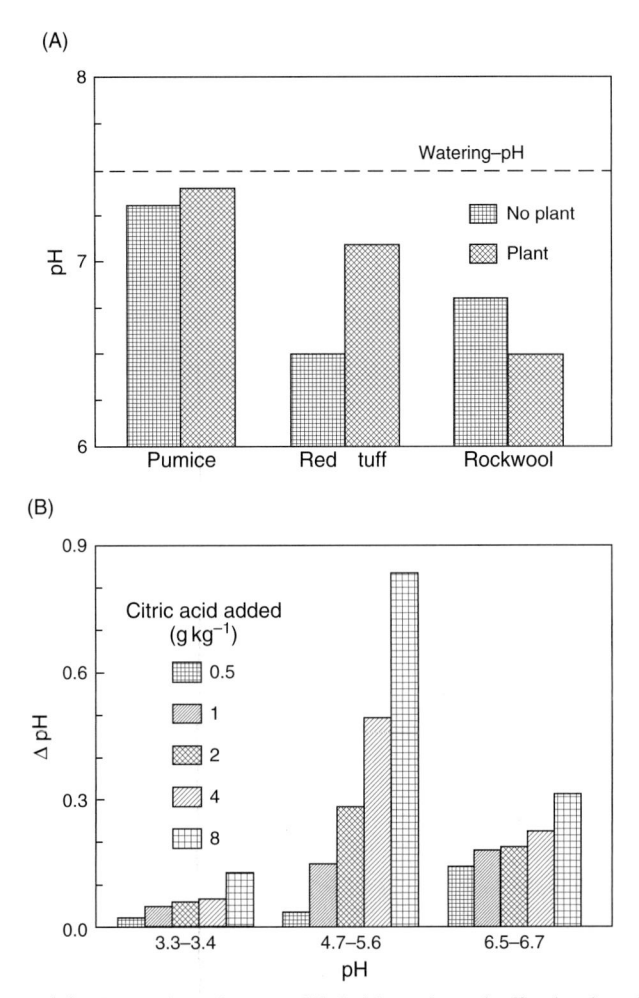

FIGURE 6.12 (A) pH in leachates from pots filled with pumice, red tuff and rockwool in the presence and absence of plants. The solution NH_4-N/NO_3-N ratio applied was 1:3 and the solution-pH is indicated by the horizontal line. Based on Bar-Yosef et al. (1997); (B) The effect of citric acid addition to tuff suspensions equilibrated at three pHs (3.3, 4.8 and 6.5), ΔpH values represents the difference between actual pH and the initial pH, i.e., before citric acid addition. Based on Silber et al. (1998).

small quantities of OH^- were released from tuff surfaces after citric acid addition. In addition, dissolution of indigenous Ca–P minerals during the pH equilibration process and the partial dissolution of tuff surfaces (Silber et al., 1999) probably added competing anions. Evaluation of the impact of citric acid on pH changes at each pH should be based on the quantities of OH^- released. The largest quantity of citric acid added to tuff ($8\,g\,kg^{-1}$) at initial solution-pH values of 3.3, 4.8 and 6.5 induced increases of 3.1×10^{-10}, 1.4×10^{-7} and 1.6×10^{-6} moles, respectively, of

OH⁻ per kilogram of tuff (calculated from solution-pH changes in Fig. 6.12B). In general, in a multifaceted system consisting of substrate, solution and plant, numerous factors may have significant impacts on the direction and extent of ΔpH. Such factors include organic ions excreted from roots or inorganic specifically adsorbed anions (mainly P) added via the irrigation solution, biological processes in the rhizosphere (nitrification/denitrification), surface charge characteristics of the solid phase, root density and, especially, the nature of the plant and its growth conditions.

6.3.2 EFFECTS ON NUTRIENTS AVAILABILITY

The effects of plant growth on the solubility of nutrient elements in the rhizosphere are specific and primarily related to the formation constants between the inorganic ions and root-excreted ligands. For example, the formation constants (log K^0) of citric and oxalic acids with K are 1.2 and 1.1, respectively, as compared with 4.8 and 3.8, 4.7 and 3.4, and 9.9 and 13.0, for the same two acids with Ca, Mg and Al, respectively (Martell and Smith, 1976–1989). Therefore, in the presence of plants and with the consequent build-up of root-excreted organic acids in the rhizosphere, the soluble complexes of Ca-, Mg- and Al-organic ligands will be predominant while the proportions of K-ligands will be minor. These differences may account for the inconsistency between the effects of plants on solution-K and on solution-Ca in tuff suspensions (Fig. 6.13): solution-K concentrations decreased while those of solution-Ca and solution-Mg (data for Mg are not shown because of their similarity with those for Ca) increased in the presence of plants compared with their concentrations in the same tuff leached with fertilizer solution (Figs. 6.13A and 6.13B, respectively).

The solution concentrations of Ca and Mg increased as a result of two consecutive processes: increase in adsorption of Ca and Mg by tuff surfaces during the growth period and subsequent displacement by electrolyte ions during the titration procedure. The Ca and Mg adsorption increased because of two independent, yet simultaneously acting mechanisms: (i) augmentation of the negatively charged surface area (Fig. 6.11) which resulted in increased adsorption of Ca and Mg to inorganic particles and (ii) augmentation of negatively charged soluble complexes (e.g., Ca-citrate) that adsorbed onto positively charged surfaces.

Enhancement of P availability may be the most important outcome of plant growth, because solution-P concentrations are governed by adsorption/desorption and precipitation/dissolution reactions; they frequently fall to deficiency levels, as detailed above. Thus, any modification of the surface properties that facilitates an enhancement of P availability may be valuable. Numerous studies have found significant enhancement of P mobilization towards roots as a result of rhizosphere modification by organic acids (mainly citric) exuded by plant roots (Gardner et al., 1983; Dinkelaker et al., 1989; Dinkelaker and Marschner, 1992; Hoffland, 1992; Gerke et al., 1994; Bar-Yosef, 1996; Jones, 1998; Jones and Brassington, 1998; Geelhoed et al., 1998, 1999; Neumann and Römheld, 1999, 2001; Bertin et al., 2003), and recent up-to-date reviews have been collected by Lambers and Poot (2003). The high root density of plants in soilless production probably accounts for a greater effect on P availability than that of soil-grown

6.3 PLANT-INDUCED CHANGES IN THE RHIZOSPHERE

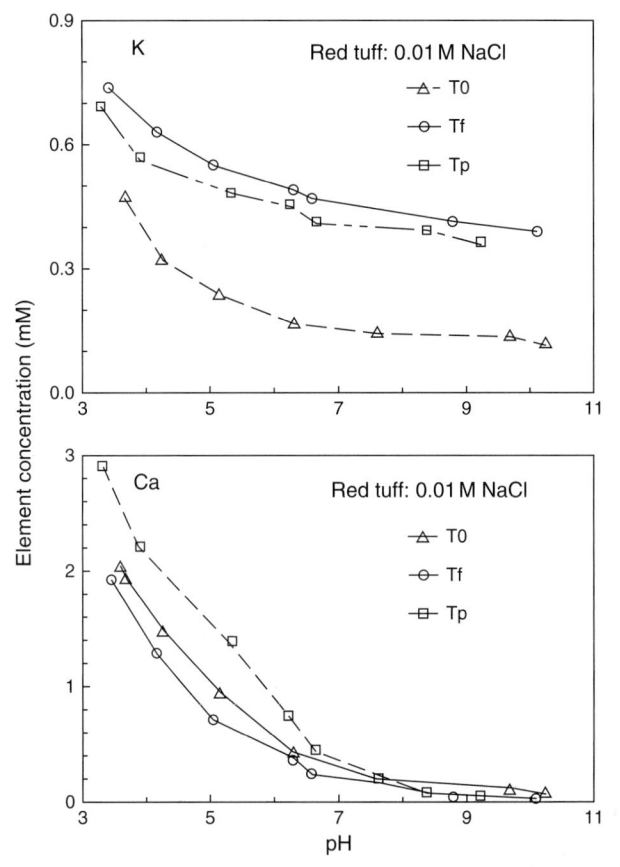

FIGURE 6.13 Potassium and Ca concentrations in three tuff extracts (unused tuff (T0), and tuff irrigated with fertilization solution in the absence and presence of plants (Tf and Tp, respectively) as functions of pH at constant ionic strength (0.01 M NaCl). Based on Silber and Raviv, (1996).

plants. Rose growth in tuff substrate induced a considerable increase in solution-P concentrations compared with that in tuff without plants, irrigated with identical fertilizer solution (Fig. 6.14). Surface reactions probably governed P solubility in the absence of plants, since the activity products of Al^{3+}, Fe^{3+} and PO_4^{3-} at pH 4–7 were below the Ksp of variscite and strengite, respectively, and that of Ca^{2+} and PO_4^{3-} at higher pHs were between the solubility lines of βTCP and OCP (only data of Ca–P solubility lines are presented, inset in Fig. 6.14A). The P partitioning between the solid and the aqueous phases was typical to specifically anions, that is, the concentration increased with increasing pH, and there was a clear effect of ionic strength (I). Below the zero point of salt effect (ZPSE) an increase of I increased the positive surface charge of the sorbent, whereas above the ZPSE it decreased the positive surface charge, as illustrated in Fig. 6.1. Hence, if P solubility were governed solely by surface reactions, then P adsorption would be expected to increase or decrease with increasing I below or above

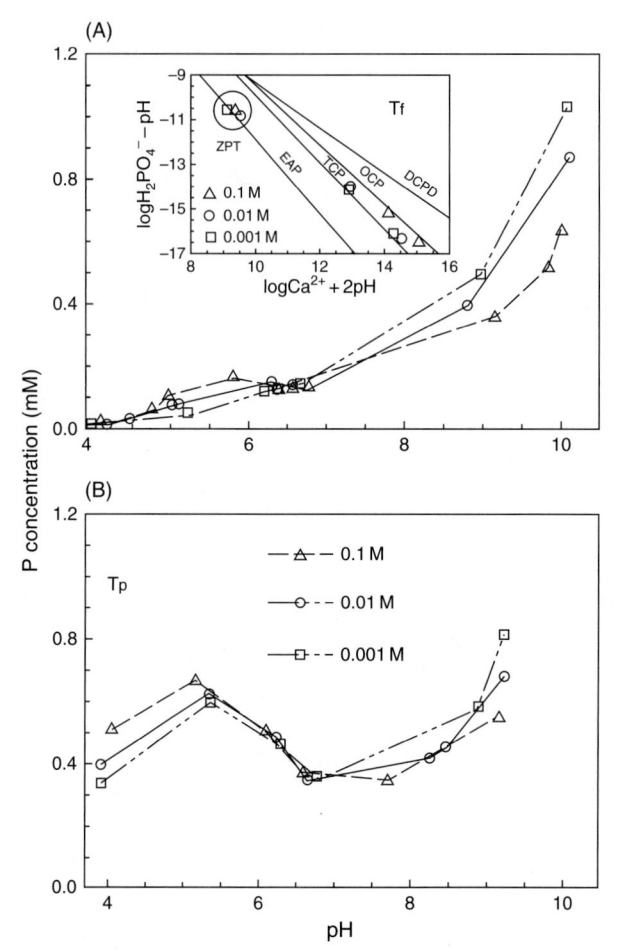

FIGURE 6.14 Phosphorus concentration as a function of pH at three initial ionic strength (NaCl): (A) tuff irrigated with fertilized solution in the absence of plants (Tf), the inset present the experimental activities in tuff extracts in relation to the theoretical solubility lines for calcium-phosphate minerals, ZPT is the suspension pH prior to the addition of acid or base; (B) tuff irrigated with fertilized solution in the presence of plants (Tp). Based on Silber and Raviv, (1996).

the ZPSE, respectively, as indeed presented in Fig. 6.14A. The overlap around pH 4 probably resulted from fast dissolution of tuff surfaces. The clear intersection around pH 6.5 (Fig. 6.14A) is consistent with previous results that indicated pH 6.5 to be the ZPSE of the Red tuff (Silber, 1991).

In contrast to intuitive thought, plant growth significantly improved P availability in the rhizosphere. Despite consumption by plants, P_W was higher in Tp than in Tf extracts at all ionic strengths, especially at acid pHs (Figs. 6.14A and 6.14B). The higher Ca and Mg concentrations in Tp extracts than in Tf ones (Fig. 6.13) did not prevent the P_W from being higher as well, probably because the Ca^{2+} and Mg^{2+} activities were low because of organic complexing agents excreted by the roots (Inskeep and Silvertooth, 1988; Grossl and Inskeep, 1991). The effect of I on P_W in Tp extracts was almost negligible compared with that in Tf extracts, and no clear intersection could be discerned (Fig. 6.14B). The increase of negative-charged surface induced by plant growth (Fig. 6.11), which were probably caused by specific adsorption of carboxylic acids excreted from roots, also reduced P adsorption on tuff surfaces, in accord with known mechanisms reported in the literature (Nagarajah et al., 1970; Earl et al., 1979; Lopez-Hernandez et al., 1979; Traina et al., 1986a,b, 1987; Kafkafi et al., 1988; Geelhoed et al., 1998, 1999; Jones, 1998; Jones and Brassington, 1998; Morel and Hinsinger, 1999; Lackovic et al., 2003).

6.3.3 ASSESSING THE IMPACT OF PLANTS: THE EFFECT OF CITRIC ACID ADDITION ON P AVAILABILITY

By adding citric acid to tuff suspensions it is possible to assess the role of low-molecular-weight organic acids excreted from roots in P partitioning between the aqueous and the solid parts of the substrate. This also simulates the beneficial effect of plant growth on P availability that was detailed above as shown by Silber and Raviv (1996) (Fig. 6.14). They also found that the isotherms of citric acid adsorption by Yellow tuff fell as pH increased from 3.5 to 4.9 and to 6.4 (Fig. 6.15) and that these curves were pH-specific Langmuir curves. This was consistent with previous findings of specific adsorption of citrate by variable-charged surfaces (Bowden et al., 1980; Bar-Yosef, 1996; Jones and Brassington, 1998; Geelhoed et al., 1999; Lackovic et al., 2003). On the other hand, it is inconsistent with citric acid isotherms that the quantities of P released is greatest for pH 3.5, intermediate for pH 6.4 and lowest for pH 4.9 (Fig. 6.16).

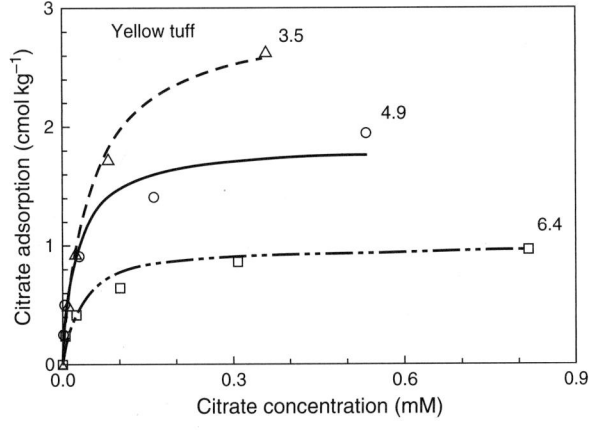

FIGURE 6.15 pH-dependent adsorption of citric acid onto yellow tuff (pH values are shown above each isotherm), constant ionic strength (0.01 M $NaClO_4$), tuff/solution ratio of 1/40, 24 h of reaction.

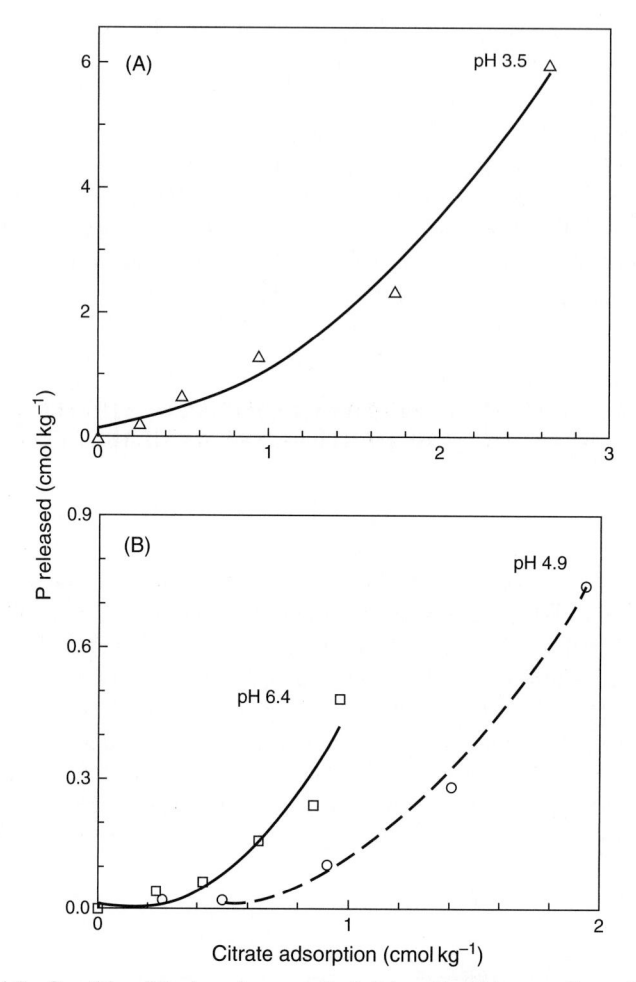

FIGURE 6.16 Quantities of P released as a result of citric acid addition to yellow tuff, constant ionic strength (0.01 M NaClO$_4$), tuff/solution ratio of 1/40, 24 h of reaction. (A) initial pH 3.4; (B) initial pH 4.9 and 6.4.

Citrate adsorption onto an iron oxide (goethite) surface has been suggested to be as an inner-sphere complex (no molecule of the bathing solvent is interposed between the surface functional group and the molecule unit it bounds (Zelancy et al., 1996)) at pH < 7 (Geelhoed et al., 1998, 1999; Harsh, 2000; Lackovic et al., 2003), or as outer-sphere adsorption (non-specific adsorption by electrostatic attraction) at pH 8.8 (Lackovic et al., 2003). However, the inconsistencies in the order and slope of the curves of P release vs. citric acid adsorbed (Fig. 6.16) suggest that an additional process may be involved in the release of P into the substrate solution. It is plausible that, in addition to the mechanism of citrate-P competition for the adsorption sites, citrate addition accelerated the dissolution of indigenous P through the formation of soluble metal-citrate complexes. Assessment of the effects of low-molecular-weight organic acids, such as citric acid excreted from roots, on rhizosphere-P solubility should take into account miscellaneous factors such as rate of excretion by plant roots, the involvement of microflora that may secrete or consume organic acids, fertilization management, content in the system of metal-P compounds that might be prone to dissolution and properties of the substrate solids. The relation between plant-P status and the rate of root excretion is well established, and usually solution-P concentrations and the amounts of organic acids excreted exhibited opposing trends.

For example, root excretion significantly decreases as solution-P increases (Gardner et al., 1982, 1983; Lipton et al., 1987; Gerke et al., 1994; Johnson et al. 1994, 1996a,b; Imas et al., 1997; Keerthisinghe et al., 1998). Thus, with the solution-P concentrations normally used in soilless culture (0.5–1.5 mM), the enhancement of P availability by organic acids is expected to be low. In addition, rapid utilization and degradation by micro-organisms may diminish the content of organic acid in the rhizosphere (Jones, 1998). Therefore, even if adsorption by solids reduced the rate of degradation (Jones and Edwards, 1998), the actual contribution of organic acids to nutrient acquisition by plants grown under proper fertilization regime would be minor (Jones et al., 2003).

The diffusion-coefficients of organic acids in the rhizosphere are very low (Darrah, 1991) and therefore the concentration of organic acid is estimated to be heterogeneous in the root zone (Jones et al., 2003). Models can provide insight into the complex processes occurring in the rhizosphere, specifically with regard to examination of the effects of root excretions on nutrient availability under diverse growth conditions as illustrated by Geelhoed et al. (1999) in Figs. 6.17A–D. In particular, solution-citrate concentrations in the substrate solution increase with citric acids excretion from roots and decrease through degradation. According to model simulations, the profiles of citrate concentration are expected to decrease sharply with increasing distance from the root surface over short time of few hours, so that the citrate concentration will fall almost to zero at a distance of few millimetres from the root (Fig. 6.17A). This phenomenon is valid as long as: (i) electrostatic adsorption reactions are instantaneous; (ii) no degradation processes are taking place; (iii) only competition between citrate and P on the solid phase surface sites is involved (i.e., no involvement of chelating effect of citrate on indigenous metal-P compounds). However, over time and in the absence of degradation, citrate is diffused to few millimetres from the root surfaces. As citrate accumulates in the boundary layer around the roots, adsorbed phosphate is released from the surfaces of substrate solids into the substrate solution and P is mobilized (Fig. 6.17B). The presence of a microflora population and its utilization of citrate significantly decreases citrate quantities in both the solid and the aqueous parts of the substrate (Fig. 6.17C), thus reducing the effect of root-excreted citrate on P partitioning (Fig. 6.17D).

By definition, the patterns resulting from this model simulation (Fig. 6.17) are valid only under the assumptions of the model and for the parameters used. The results obtained may be affected significantly by the presence of different amounts of P and citrate in the system or different values for the model's parameters such as P and

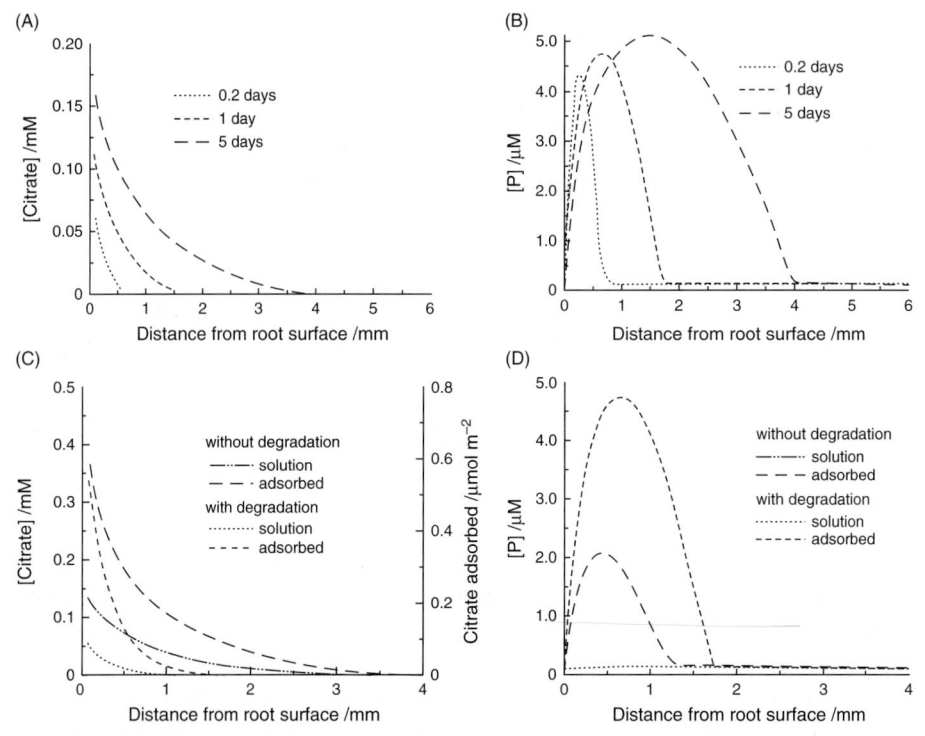

FIGURE 6.17 (A) Solution-citrate concentration as a function of distance from root surface in the absence of degradation processes after 0.2, 1 and 5 days of excretion; (B) Solution-P concentration as a function of distance from root surface in the absence of degradation processes after 0.2, 1 and 5 days of excretion; (C) predicted concentration profiles of citrate in solution and adsorbed onto goethite with and without degradation; (D) effect of citrate exudation, with and without degradation, on the predicted P concentration in solution. Reprinted from Geelhoed et al. (1999), with kind permission from Blackwell Science, Ltd.

citrate adsorption rates and the function relating citrate degradation rates to pH. The model result may also differ with alternate submodels of related processes such as the rate dependence that pH has on the activities of the microflora population. Also, other factors used in the model of Geelhoed et al. (1999) such as root density, diffusion coefficients, tortuosity factor or volumetric water content may significantly affect the model predictions.

6.4 NUTRIENT RELEASE FROM INORGANIC AND ORGANIC SUBSTRATES

Reducing the pollution hazards posed to the environment by excess fertilization in soilless production has led to efforts to regulate nutrient availability in the rhizosphere and to develop substrates that might contribute nutrients during the growth period of horticultural crops. The goal of reducing excess fertilization may be achievable through two approaches: (i) pre-charging the substrate with nutrients that are subsequently released during the growth period, that is, the substrate functions as slow-release fertilizer and (ii) release of indigenous nutrients from the substrate as a result of chemical/biological processes such as N mineralization in organic media or dissolution of indigenous nutrients from inorganic components.

Pre-charging of substrates with nutrients requires that the material have high CEC or AEC values. The low values of these characteristics of most inorganic substrates, such as stone wool, perlite and pumice, suggest that these widely used horticultural materials may not be suitable for this. Zeolites, however, possess extremely high CEC values of 200–400 cmol kg^{-1} (Mumpton, 1999). In fact, zeolites are used as nutrient sources in infertile soils and substrates because of their high affinity for K or NH$_4$ ions (Hershey et al., 1980; Chen and Gabelman, 1990; Notario del Pino et al., 1994; Williams and Nelson, 1997; Dwairi, 1998; Mohammad et al., 2004). Although the use of zeolites as a single-component growing substrate is not recommended because of the high bulk density (1.9–2.3 Mg m^{-3}, Ming and Mumpton, 1989), mixed substrates which may include organic or inorganic materials as well as zeolites are used worldwide in the production of flowers and vegetables. Theoretically, if all the exchange sites of a zeolite substrate were pre-charged with K ions, the total potential pool for K would be 80–160 g kg^{-1}. Since the total K acquisition by tomato or pepper (*Capsicum annum* L.) crops under intensive conditions is 15–20 g per plant (Bar-Yosef, 1999), approximately 200 g of zeolite are required to satisfy the total plant demand for K throughout a growing season. Thus, from the quantitative point of view, the pre-charging approach can be regarded as potentially successful; however, the release rate may present problems. Basically, the optimal rates of nutrient release must match the plant demand, which is obviously low at the beginning of the growing period and increases gradually during the various growth stages. However, the pre-charged nutrients are released from the surface sites by exchange reactions with competing ions (in the case of K$^+$ these are NH$_4^+$, Na$^+$, Ca^{2+} and Mg^{2+}). Thus, factors such as concentrations of competing ions in the substrate solution, irrigation doses and plant activity all affect the release rate; therefore, nutrient availability is not actually regulated and may be different from the plant's demands.

A combination of approaches (i) and (ii) to reducing excess nutrient run-off was implemented by mixing phosphate rock with zeolite pre-charged with NH$_4^+$ ions (Allen et al., 1995; Williams and Nelson, 1997; Pickering et al., 2002). While the NH$_4^+$ ions served as an N source for plants during the growth period, the main purpose of this process was to induce phosphate rock dissolution, either through acidification or through exchange with Ca ions (Allen et al., 1993; Notario del Pino et al., 1994) so as to slowly release phosphate over time from the zeolite. Zeolites can also be used as slow-release sources of micronutrients and K (Opena and Williams, 2003; Puschenreiter and Horak, 2003).

The contribution of indigenous nutrients released from inorganic substrates is usually very small and, except in a few cases of utilisation of phosphate rock, the release of indigenous nutrients is practically negligible. Organic media, on the other hand, and especially composts, contain large quantities of valuable N resources and, therefore, the production of composts that may contribute N during the growth period (approach (ii)), especially for organically grown crops, may be of interest. Generally part of the initial N content of organic constituents is lost during the composting process. In fact, high-quality mature composts are characterized by their low mineralization rate and low N availability (Hadas and Portnoy, 1994, 1997). It is, however, possible to reduce the N loss during the composting process by using a high C/N ratio, lowering the pH and manipulating the air flow and the composting temperature (Raviv, 2005). Raviv found that composts enriched with wheat straw or orange peels (imparting a high initial C/N ratio and a low pH, respectively) satisfied the N requirement of tomato plants for several months under conditions of minimal leaching fraction (Raviv et al., 2005) but they were, nevertheless, not able to match the rate of N release to the plant requirement. The rate of N release from a successful N-conservation compost consisting of orange peels and separated cattle manure (OP-SCM) (Raviv et al., 2005) was very high during the first days. The rate far exceeded plant demand for a crop such as, for example, lettuce (Fig. 6.18). However, the release rate declined very rapidly so that after 22 days it was lower than the N requirement of the lettuce plants, and the discrepancy between the two rates increased with time. Nitrogen release from composts is usually described by first-order kinetic equations (Hadas and Portnoy, 1994, 1997) and is influenced by temperature, water content and other environmental factors that affect biological and chemical processes. If the amount of N required daily by mature plants grown under intensive growth condition in greenhouses is taken to be 120–170 mg (Bar-Yosef, 1999), it is not likely that organic composts could contribute sufficient N, even under less favourable conditions for N degradation as presented in Fig. 6.18.

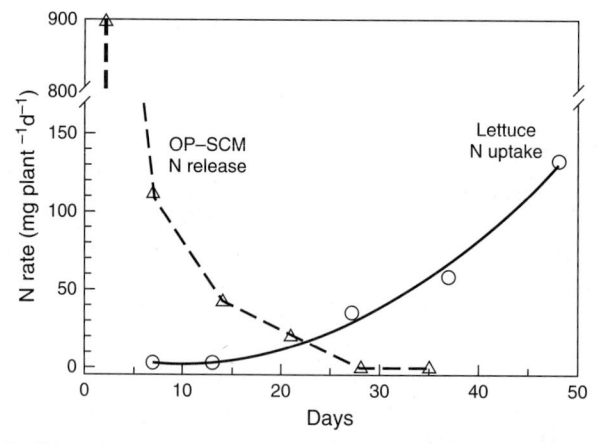

FIGURE 6.18 Comparison between the rate of N release from compost containing orange-peel amendment to separated cattle manure (OP-SCM) and the N requirements of lettuce. The N release data were computed from Fig. 6.2 of Raviv et al. (2005), assuming application of 3 l of compost per plant. The lettuce N uptake data were taken from Silber et al. (2003).

Nonetheless, inclusion of high-N composts in substrates may allow for reduced N application rates during the initial growth period. This is particularly important under organic agriculture management and in situations where reduction of pollution hazards is important. This finding may be particularly applicable in short-term crops such as organic transplant production which usually takes up to 35 days (Raviv et al., 1998).

REFERENCES

Abadía, J., López-Millán, A.F., Rombolà, A. and Abadía, A. (2002). Organic acids and Fe deficiency: A review. *Plant Soil*, **241**, 75–86.

Adams, P. (2002). Nutritional control in hydroponics. In *Hydroponic Production of Vegetables and Ornamentals* (D. Savvas and H.C. Passam, eds). Athens: Embryo Publications, pp. 211–262.

Allan, D.A. and Jarrell, W.M. (1989). Proton and copper adsorption to maize and soybean root cell walls. *Plant Physiol.*, **89**, 823–832.

Allen, E.R., Hossner, L.R., Ming, D.W. and Henninger, D.L. (1993). Solubility and cation exchange in phosphate rock and ammonium and potassium-saturated clinoptilolite mixtures. *Agron. J.* **57**, 1368–1374.

Allen, E.R., Ming, D.W., Hossner, L.R., et al. (1995). Growth and nutrient uptake of wheat in clinoptilolite-phosphate rock substrates. *Agron. J.*, **87**, 1052–1059.

Argo, W.R. and Biernbaum, J.A. (1997). The effect of root media on root-zone pH, calcium and magnesium management in containers with impatiens. *J. Am. Soc. Hortic. Sci.*, **122**, 275–284.

Barber, S.A. (1995). *Soil Nutrient Bioavailability: A Mechanistic Approach* (2nd edn). New York: Wiley.

Bar-Yosef, B. (1996). Root excretion and their environmental effects: Influence on availability of phosphorus. In *Plant Roots, The Hidden Half*, 2nd edn (Y. Waisel, A. Eshel and U. Kafkafi, eds). New York: Marcel Dekker, pp. 581–605.

Bar-Yosef, B. (1999). Advances in fertigation. *Adv. Agron.*, **65**, 1–77.

Bar-Yosef, B., Silber, A., Markovitch, et al. (1997). The effect of substrate volume and type and fertilizer level on greenhouse tomato growth. Report submitted to the Chief Scientist of the Israeli Ministry of Agriculture and Rural Development (in Hebrew), p. 33.

Bertin, C., Yang, X. and Weston. L.A. (2003). The role of root exudates and allelochemicals in the rhizosphere. *Plant Soil*, **256**, 78–83.

Blamey, F.P.C., Robinson, N.J. and Asher, C.J. (1992). Interspecific differences in aluminium tolerance in relation to root cation-exchange capacity. *Plant Soil*, **146**, 77–82.

Bloom, P.R. (1979). Titration behavior of aluminum organic matter. *Soil Sci. Soc. Am. J.*, **43**, 815–817.

Bowden, J.W., Nagarajah, S., Barrow, N.J., et al. (1980). Describing the adsorption of phosphate, citrate and selenite on a variable-charge mineral surface. *Aust. J. Soil Res.*, **18**, 49–60.

Brown, E.F. and Pokorny, F.A. (1975) Physical and chemical properties of media composed of milled pine bark and sand. *J. Amer. Soc. Hortic. Sci.*, **100**, 119–121.

Chen, J.L. and Gabelman, W.H. (1990). A sand-zeolite culture system for simulating plant acquisition of potassium from soils. *Plant Soil*, **26**, 169–176.

Cotter, D.J. (1979). Effect of fresh bark media on phosphorus status of tomato plants. *Commun. Soil Sci. Plant Anal.*, **10**, 1381–1384.

Daniels, W.L. and Wright, R.D. (1988). Cation exchange properties of pine bark growing media as influenced by pH, particle size, and cation species. *J Amer. Soc. Hort. Sci.*, **113**, 557–560.

Darrah, P.R. (1991). Measuring the diffusion-coefficients of rhizosphere exudates in soil. 2. The diffusion of sorbing compounds. *J. Soil Sci.*, **42**, 421–434.

Diaz-Barrientos, E., Madrid, L., Contreras, M.C. and Morillo, E. (1990). Simultaneous adsorption of zinc and phosphate on synthetic lepidocrocite. *Aust. J. Soil Res.*, **28**, 549–557.

Dinkelaker, B. and Marschner, H. (1992). In vivo demonstration of acid phosphatase activity in the rhizosphere of soil-grown plants. *Plant Soil*, **144**, 199–205.

Dinkelaker, B., Römheld, V. and Marschner, H. (1989). Citric acid excretion and precipitation of calcium citrate in the rhizosphere of white lupin (*Lupinus albus* L.). *Plant Cell Env.*, **12**, 285–292.

Doğan, M. and Alkan, M. (2004). Some physiochemical properties of perlite as an adsorbent. *Fresenius Environ. Bull.*, **13**, 252–257.

Dwairi, I.M. (1998). Evaluation of Jordanian zeolite tuff as a controlled slow-release fertilizer for NH_4^+. *Environ. Geol.*, **34**, 1–4.

Earl, K.D., Syers, J.K. and Mclaghlin, J.R. (1979). Origin of the effects of citrate, tartarate and acetate on phosphate sorption by soils and synthetic gels. *Soil Sci. Soc. Am. J.*, **45**, 674–678.

Epstein, E. (1999). Silicon. *Annu. Rev. Plant Physiol. Mol. Biol.*, **50**, 641–664.

Espinoza, W., Gast, R.G. and Adams, R.S. (1975). Charge characteristics and nitrate retention by two andepts from south-central Chile. *Soil Sci. Soc. Am. J.*, **39**, 842–846.

Evans, M.R., Konduru, S. and Stamps, R.H. (1996). Source variation in physical and chemical properties of coconut coir dust. *Hort. Science*, **31**, 965–967.

Gardner, W.K., Barbery, D.G. and Barber, D.A. (1982). The acquisition of phosphorus by *Lupinus albus* L. II The effect of varying phosphorus supply and soil type on some characteristics of soil/root interface. *Plant Soil.*, **68**, 33–41.

Gardner, W.K., Barbery, D.G. and Barber, D.A. (1983). The acquisition of phosphorus by Lupinus albus L. III. The probable mechanism by which phosphorus movement in the soil/root interface is enhanced. *Plant Soil*, **70**, 107–124.

Gast, R.G. (1977). Surface and colloid chemistry. In *Minerals in Soil Environments*, 1st edn (J.B. Dixon, S.B. Weed, J.A. Kittrick, et al., eds). SSSA Book Series No. 1. Madison: Soil Sci. Soc. Amer., pp. 27–73.

Geelhoed J.S., Hiemstra, T. and Van Riemsdijk, W.H. (1998). Competitive phosphate between phosphate and citrate on goethite. *Environ. Sci. Technol.*, **32**, 2119–2123.

Geelhoed, J.S., Van Riemsdijk, W.H. and Findenegg, G.R. (1999). Simulation of the effect of citrate exudation from roots on the plant availability of phosphate adsorbed on goethite. *Eur. J. Soil Sci.*, **50**, 379–390.

Gerke, J., Römer, W. and Jungk, A. (1994). The excretion of citric and malic acid by proteoid roots of *Lupinus albus* L.; effects on soil solution concentrations of phosphate, iron, and aluminum in the proteoid rhizosphere in samples of an oxisol and a luvisol. *Z. Pfanzenernahr. Bodenk.*, **157**, 289–294.

Goldberg, S. and Glaubig, R.A. (1988). Boron and silicon adsorption on an aluminum oxide. *Soil Sci. Soc. Am. J.*, **52**, 87–91.

Gottesman, A. (1989). Effects of compost amendment on protected vegetable production in container media. MSc Thesis, Hebrew Univ. of Jerusalem (in Hebrew with English abstract), p. 144.

Grossl, P.R. and Inskeep, W.P. (1991). Precipitation of dicalcium phosphate dihydrate in the presence of organic acids. *Soil Sci. Soc. Am. J.*, **55**, 670–675.

Hadas, A. and Portnoy, R. (1994). Nitrogen and carbon mineralization rates of composted manures incubated in soil. *J. Environ. Qual.*, **23**, 1184–1189.

Hadas, A. and Portnoy, R. (1997). Rates of decomposition in soil and release of available nitrogen from cattle manure and municipal solid waste composts. *Compost Sci. Util.*, **5**, 48–54.

Hallsworth, E.C. and Wilkinson, G.K. (1958). The contribution of clay and organic matter to the cation exchange capacity of the soil. *J. Agr. Sci.*, **51**, 1–3.

Hansen, H.C.B., Raben-Lange, B., Raulund-Rasmussen, K. and borggaard, O.K. (1994). Monosilicate adsorption by ferrihydrite and goethite at pH 3–6. *Soil Sci.*, **158**, 41–46.

Harada, Y. and Inoko, A. (1980). The measurement of cation exchange capacity of composts for the estimation of the degree of maturity. *Soil Sci. Plant. Nutr.*, **26**, 127–134.

Harsh, J. (2000). Poorly crystalline aluminosilicate clays. In *Handbook of Soil Science* (M.E. Summer, ed.). Boca Raton: CRC Press, pp. F169–F182.

Helling, C., Chesters, G. and Corey, R.B. (1964). Contribution of organic matter and clay to soil cation-exchange capacity as affected by the pH of the saturating solution. *Soil Sci. Soc. Am. J.*, **28**, 517–520.

Hershey, D.R., Paul, J.L. and Carlson, R.M. (1980). Evaluation of potassium-enriched clinoptilolite as a potassium source for potting media. *Hort. Science*, **15**, 87–89.

Hingston, F.J. and Raupach, M. (1967). The reaction between monosilicic acid and aluminium hydroxide. I. Kinetics of adsorption of silicic acid by aluminium hydroxide. *Aust. J. Soil Res.*, **5**, 295–309.

REFERENCES

Hinsinger, P., Plassard, C., Tang, C. and Jaillard, B. (2003). Origins of root-mediated pH changes in the rhizosphere and their responses to environmental constraints: A review. *Plant Soil*, **248**, 43–59.

Hoffland, E. (1992). Quantitative evolution of the role of organic-acid exudation in the mobilization of rock phosphate by rape. *Plant Soil*, **140**, 279–289.

Huang, P.M. and Violante, A. (1986). Influence of organic acids on crystallization of precipitation products of aluminium. In *Interactions of Soil Minerals with Natural Organics and Microbes* (P.M. Huang and M. Schnitzer, eds). SSSA, Spec. Pub. No. 17. Madison: Soil Science Society of America, pp. 159–222.

Iglesias-Jimenez, E. and Perez-Garcia, V. (1992). Determination of maturity indices for city reuse composts. *Agr. Ecosyst. Environ.*, **38**, 331–343.

Imas, P., Bar-Yosef, B., Kafkafi, U. and Ganmore-Neumann, R. (1997). Phosphate induced carboxylate and proton release by tomato plant. *Plant Soil*, **191**, 35–39.

Imas, P., Bar-Yosef, B., Levkovitch, I. and Keinan, M. (1996). Orthophosphate solubility in waters of different ionic composition. *Fertilizer Res.*, **44**, 73–78.

Inbar, Y. (1989). Formation of humic substances during the composting of agricultural wastes and characterization of their physico-chemical properties. PhD Thesis, Hebrew University of Jerusalem.

Inbar, Y., Chen, I. and Hadar, I. (1989). Solid-state carbon-13 nuclear magnetic resonance and infrared spectroscopy of composted organic matter. *Soil Sci. Soc. Am. J.*, **53**, 1695–1701.

Inbar, Y., Chen, I. and Hadar, I. (1991). Carbon-13 CPMAS NMR and FTIR spectroscopic analysis of organic matter transformation during composting of solid wastes from wineries. *Soil Sci.*, **152**, 272–282.

Inskeep, W.P. and Silvertooth, J.C. (1988). Inhibition of hydroxyapatite precipitation in the presence of fulvic, humic and tannic acids. *Soil Sci. Soc. Am. J.*, **52**, 941–846.

Johanson, J. (1978). Effect of nutrient levels on growth, flowering and leaf nutrient content of greenhouse roses. *Acta Agric. Scand.*, **28**, 363–386.

Johnson, J.F., Allan, D.L. and Vance, C.P. (1994). Phosphorus stress-induced proteoid roots show altered metabolism in *Lupinus albus*. *Plant Physiol.*, **104**, 637–645.

Johnson, J.F., Allan, D.L., Vance, C.P. and Weiblen, G. (1996a). Root carbon dioxide fixation by phosphorus-deficient *Lupinus albus*. *Plant Physiol.*, **112**, 19–30.

Johnson, J.F., Allan, D.L. and Vance, C.P. (1996b). Phosphorus deficiency in *Lupinus albus*. *Plant Physiol.*, **112**, 31–41.

Jokova, M., Kostov, O. and van Cleemput, O. (1997). Cation exchange and reducing capacities as criteria for compost quality. *Biol. Agric. Hort.*, **1493**, 187–197.

Jones, B., Jr. (1997). *Hydroponics: A Practical Guide for the Soilless Grower*. Boca Raton: St.-Lucie Press.

Jones, D.L. (1998). Organic acids in the rhizosphere — a critical review. *Plant Soil*, **205**, 25–44.

Jones, D.L. and Brassington, D.S. (1998). Sorption of organic acids in acid soils and its implications in the rhizosphere. *Eur. J. Soil Sci.*, **49**, 447–455.

Jones, D.L. and Edwards, A.C. (1998). Influence of sorption on the biological utilisation of two simple carbon substrates. *Soil Biol. Biochem.*, **30**, 1895–1902.

Jones, D.L., Dennis, P.G., Owen, A.G. and van Hees, P.A.W. (2003). Organic acid behaviour in soils – misconceptions and knowledge gaps. *Plant Soil*, **248**, 31–41.

Kafkafi, U., Bar-Yosef, B., Rosenberg, R. and Sposito, G. (1988). Phosphorus adsorption by kaolinite and montmorillonite. II. Organic anion competition. *Soil Sci. Soc. Am. J.*, **52**, 1585–1589.

Keerthisinghe, G., Hocking, P.J., Ryan, P.R. and Delhaize, E. (1998). Effect of phosphorus supply on the formation and function of proteoid roots of white lupin (*Lupinus albus* L.). *Plant Cell Environ.*, **21**, 467–478.

Lackovic, K., Johnson, B.B., Angove, M.J. and Wells, J.D. (2003). Modeling the adsorption of citric acid onto Multoorina illite and related clay minerals. *J. Colloid Interface Sci.*, **267**, 49–59.

Lambers, H. and Poot, P. (2003). *Structure and Functioning of Cluster Roots and Plant Response to Phosphate Deficiency*. Dordrecht: Kluwer Academic Publishers.

Lindsay, W.L. (1979). Chemical Equilibrium in Soils. New York: Wiley.

Lipton, D.S., Blanchar, R.W. and Blevins, D.G. (1987). Citrate, malate, and succinate concentration in exudates from P-sufficient and P-stressed *Medicago sativa* L. seedlings. *Plant Physiol.*, **85**, 315–317.

Lopez-Hernandes, D., Flores, D., Siegert, G. and Rodriguez, J.V. (1979). The effect of some organic anions on phosphate removal from acid and calcareous soils. *Soil Sci.*, **128**, 321–326.

Marschner, H. (1995). *Mineral Nutrition of Higher Plants* (2nd edn). New York: Academic Press.

Marschner, H. and Römheld, V. (1996). Root-induced changes in the availability of micronutrients in the rhizosphere. In *Plant Roots, the Hidden Half*, 2nd edn (Y. Waisel, A. Eshel, and U. Kafkafi, eds). New York: Marcel Dekker, pp. 557–579.

Marschner, H., Römheld, V., Horst, W.J. and Martin, P. (1986). Root-induced changes in the rhizosphere: Importance for the mineral nutrition of plants. *Z. Pfanzenennaehr Bodenk.*, **149**, 441–459.

Martell, A.E. and Smith, R.M. (1976–1989). *Critical Stability Constants*. Vol. 6, New York: Plenum Press.

McBride, M.B. (1989). Surface chemistry of soil minerals. In *Minerals in Soil Environments* 2nd edn (J.B. Dixon and S.B. Weed, eds). SSSA Book Series No. 1. Madison: Soil Science Society of America, pp. 35–88.

McBride, M.B. (2000). Chemisorption and precipitation reactions. In *Handbook of Soil Science* (M.E. Summer, ed.). Boca Raton: CRC Press, B264–B302.

Ming, D.W. and Mumpton, F.A. (1989). Zeolites in soils. In *Minerals in Soil Environments*, 2nd edn (J.B. Dixon and S.B. Weed, eds). SSSA Book Series No. 1. Madison: Soil Science Society of America, pp. 873–909.

Mohammad, M.J., Karam, N.S and Al-Lataifeh, N.K. (2004). Response of cotton grown in a zeolite-containing substrate to different concentrations of fertilizer solution. *Commun. Soil Sci. Plant Anal.*, **35**, 2283–2297.

Morel, C. and Hinsinger, P. (1999). Root-induced modifications of the exchange of phosphate ion between soil solution and soil solid phase. *Plant Soil*, **211**, 103–110.

Morel, C., Tiessen, H., Moir, J. and Stewart, J.W. (1994). Phosphorus transformations and availability due to crop rotations and mineral fertilisation assessed by an isotopic exchange method. *Soil Sci. Soc. Amer. J.*, **58**, 1439–1445.

Mumpton, F.A. (1999). La roca magica: Uses of natural zeolites in agriculture and industry. *Proc. Nat. Acad. Sci. U.S.A.*, **96**, 3463–3471.

Nagarajah, S., Posner, A.M. and Quirk, J.P. (1970). Competitive adsorption of phosphate with polygalacturonate and other organic anions on kaolinite and oxide surfaces. *Nature*, **228**, 83–84.

Nartey, E., Matsue, N. and Henmi, T. (2000). Adsorptive mechanisms of orthosilicic acid on nano-ball allophane. *Clay Sci.*, **11**, 125–136.

Neumann, G. and Römheld, V. (1999) Root excretion of carboxylic acids and protons in phosphorus-deficient plants. *Plant Soil*, **211**, 121–130.

Neumann, G. and Römheld, V. (2001). The release of root exudates as affected by the plant's physiological status. In *The Rhizosphere, Biochemistry and Organic Substances at the Soil-plant Interface* (R. Pinton, Z. Varanini and P. Nannpieri, eds). New York: Marcel Dekker,. pp. 41–91.

Notario del Pino, J.S., Artega-Padron, I.J., Gonzales-Martin, M.M. and Garcia-Hernandez, J.E. (1994). Response of alfalfa to a phillipsite-based slow-release fertilizers. *Commun. Soil Sci. Plant. Anal.*, **25**, 2231–2245.

Nye, P.H. (1986). Acid-base changed in the rhizosphere. *Adv. Plant Nutr.*, **2**, 129–163.

Ogden, R.J., Pokorny, F.A., Mills, H.A. and Dunavent, M.G. (1987). Elemental status of pine bark-based potting media. *Hortic. Rev.*, **9**, 103–131.

Opena, G.B. and Williams, K.A. (2003). Use of precharged zeolite to provide aluminum during blue hydrangea production. *J. Plant Nutr.*, **26**, 1825–1840.

Pardo, M.T. (1999). Influence of phosphate on zinc reaction in variable charge soils. *Commun. Soil Sci. Plant Anal.*, **30**, 725–737.

Pardo, M.T. and Guadalix, M.E. (1990). Phosphate sorption in allophanic soils and release of sulphate, silicate and hydroxyl. *J. Soil Sci.*, **41**, 607–612.

Parker, J.C., Zelancy, L.W., Sampath, S. and Harris, W.S. (1979). A critical evaluation of the extension of zero point of charge (ZPC) theory to soil systems. *Soil Sci. Soc. Am. J.*, **43**, 668–573.

Parker, D.R., Chaney, R.L. and Norvell, W.A. (1995). Chemical equilibrium models: Application to plant nutrition research. In *Soil Chemical Equilibrium and reaction Models* (R.H. Loeppert, et al., eds). SSSA, Spec. Pub. No. 42,. Madison: Soil Science Society of America, pp. 163–200.

Pickering, H.W., Menzies, N.W. and Hunter, M.N. (2002). Zeolite/rock phosphate — A novel slow release phosphorus fertiliser for potted plant production. *Sci. Hortic.*, **94**, 333–343.

References

Puschenreiter, M. and Horak, O. (2003). Slow-release zeolite-bound zinc and copper fertilizers affect cadmium concentration in wheat and spinach. *Commun. Soil Sci. Plant. Anal.*, **25**, 2231–2245.

Puustjarvi, V. (1977). Peat and its Use in Horticulture. Turveteollisuusliitto Ry, Publ. 3, Helsinki.

Puustjarvi, V. and Robertson, R.A. (1975). Physical and chemical properties. In *Peat in Horticulture* (D.W. Robinson and J.G.D. Lamb, eds). New York: Academic press, pp. 23–38.

Rajan, S.S.S. (1975). Phosphate adsorption and the displacement of structural silicon in an allophanic clay. *J. Soil Sci.*, **26**, 250–256.

Rajan, S.S.S. and Fox, R.L. (1975). Phosphate adsorption by soils: II. Reactions in tropical acid soils. *Soil Sci. Soc. Amer. Proc.*, **39**, 846–850.

Rajan, S.S.S. and Perrott, K.W. (1975). Phosphate adsorption by synthetic amorphous aluminosilicates. *J. Soil Sci.*, **26**, 458–466.

Raviv, M. (2005). Production of high-quality composts for horticultural purposes — a mini-review. *HortTechnology*, **15**, 52–57.

Raviv, M., Tarre, S., Geler, Z. and Shelef, G. (1987). Changes in some physical and chemical properties of fibrous solids from cow manure and digested cow manure during composting. *Biological Wastes*, **19**, 309–318.

Raviv, M., Zaidman B.Z. and Kapulnik, Y. (1998). The use of compost as a peat substitute for organic vegetable transplants production. *Compost Science and Utilisation*, **6**, 46–52.

Raviv, M., Oka, Y., Katan, J., et al. (2005). High-nitrogen compost as a medium for organic container-grown crops. *Bioresour. Technol.*, **96**, 419–427.

Saharinen, M.H. (1996). Cation exchange capacity of manure-straw compost — does sample preparation modify the results? In *The Science of Composting; part 2* (M. de Bertoldi, P. Sequi, B. Lemmes, T. Papi, eds), Glasgow: Blackie Academic and Professional, pp. 1309–1311.

Salinger, Y., Geifman, Y. and Aronowitch, M. (1993). Orthophosphate and calcium carbonate solubilities in the Upper Jordan watershed basin. *J. Environ. Qual.*, **22**, 672–677.

Sanyal, S.K. and De Datta, S.K. (1991). Chemistry of phosphorus transformations in soil. *Adv. Soil Sci.*, **16**, 1–120.

Silber, A. (1991). Chemical properties and surface reactions of pyroclastic materials from Mt. Peres, the Golan Heights. PhD Thesis, Hebrew Univ. of Jerusalem (in Hebrew with English abstract). p. 171.

Silber, A. and Raviv, M. (1996). Effects on chemical surface properties of tuff by growing rose plants. *Plant Soil*, **186**, 353–360.

Silber, A., Bar-Yosef, B., Singer, A. and Chen, Y. (1994). Mineralogical and chemical composition of three tuffs from northern Israel. *Geoderma*, **63**, 123–144.

Silber, A., Ganmore-Neumann, R. and Ben-Jaacov, J. (1998). Effects of nutrient addition on growth and rhizosphere pH of *Leucadendron* "Safari Sunset". *Plant Soil*, **199**, 205–211.

Silber, A., Bar-Yosef, B. and Chen, Y. (1999). pH-dependent kinetics of tuff dissolution. *Geoderma* **93**, 125–140.

Silber, A., Xu, G., Levkovitch, I., et al. (2003). High fertigation frequency: The effect on uptake of nutrients, water and plant growth. *Plant Soil*, **253**, 467–477.

Silber, A., Bruner, M., Kenig, E., et al. (2005). High fertigation frequency and phosphorus level: effects on summer-grown bell pepper growth and blossom-end rot incidence. *Plant Soil*, **270**, 135–143.

Smith, R.M. and Martell, A.E. (1989). Critical Stability Constants. Vol. 6, 2nd Supplement, New York: Plenum Press.

Sonneveld, C. (2002). Composition of nutrient solutions. In *Hydroponic Production of Vegetables and Ornamentals* (D. Savvas and H.C. Passam, eds.), Athens, Greece: Embryo Publications, 179–210.

Sonneveld, C. and de Kreij, C. (1987). Nutrient solutions for vegetables and flowers grown in water or substrates. Serie: Voedingsoplossingen glastuinbouw, 8, (6th edn). Naaldwijk.

Sparks, D.L., Page, A.L., Helmke, P.A., et al. (1996). Methods of Soil Analysis: Part 3 — Chemical Methods. Number 5 in the Soil Science Society of America Book Series. Madison: Soil Science Society of America and American Society for Agronomy, 1358 pp.

Sposito, G. (1981). The operational definition of the zero point of charge in soils. *Soil Sci. Soc. Am. J.*, **45**, 292–297.

Sposito, G. (1984). *The Surface Chemistry of Soils*. Oxford University Press, Oxford.

Sposito, G. (1989). *The Chemistry of Soils.* Oxford University Press, Oxford.

Sposito, G. (2000). Ion exchange phenomena. In *Handbook of Soil Science* (M.E. Summer, ed.). Boca Raton: CRC Press, pp. B241–B263.

Stevenson, F.J. (1994). *Humus Chemistry, Genesis, Composition, Reactions* (2nd edn). New York: John Wiley and Sons.

Summer, M.E. and Miller, W.P. (1996). Cation exchange capacity and exchange coefficients. In *Methods of Soil Analysis Part 3: Chemical Methods* (D.L. Sparks, et al., eds). No 5 in the Soil Science Society of America book, Madison: Soil Science Society of America, Inc., pp. 1201–1229.

Tan, H.K. (1986). Degradation of soil minerals by organic acids. In *Interactions of Soil Minerals with Natural Organics and Microbes* (P.M. Huang, M. Schnitzer, eds). SSSA, Spec. Pub. No. 17, Madison: Soil Science Society of America, pp. 1–27.

Tarre, S., Raviv, M. and Shelef, G. (1987). Composting of fibrous solids from cow manure and anaerobically digested manure. *Biological Wastes*, **19**, 299–308.

Tate, K.R. and Theng, B.K.G. (1980). Organic matter and its interactions with inorganic soil constituents. In *Soils with Variable Charge* (B.K.G. Theng, ed.). New Zealand: Lower Hutt, pp. 225–249.

Traina, S.J., Sposito, G., Hesterberg, D. and Kafkafi, U. (1986a). Effects of pH and organic acids on orthophosphate solubility in an acidic, montmorillonitic soil. *Soil Sci. Soc. Am. J.*, **50**, 45–52.

Traina, S.J., Sposito, G., Hesterberg, D. and Kafkafi, U. (1986b). Effects of ionic strength, calcium, and citrate on orthophosphate solubility in an acidic, montmorillonitic soil. *Soil Sci. Soc. Am. J.*, **50**, 623–627.

Traina, S.J., Sposito, G., Bradford, G.R. and Kafkafi, U. (1987). Kinetic study of citrate effects on orthophosphate solubility in an acidic, montmorillonitic soil. *Soil Sci. Soc. Am. J.*, **51**, 1483–1487.

Veith, J.A. and Sposito, G. (1977). Reactions of aluminosilicates, hydrous oxides, and aluminum oxide with O-phosphate: The formation of X-ray amorphous analogs of variscite and montebrasite. *Soil Sci. Soc. Am. J.*, **41**, 870–876.

Wann, S.S. and Uehara, G. (1978). Surface charge manipulation in constant surface potential soil colloids.: I. Relation to sorbed phosphorus. *Soil Sci. Soc. Am. J.*, **42**, 565–570.

Williams, K.A. and Nelson, P.V. (1997). Using precharged zeolite as a source of potassium and phosphate in a soilless container medium during potted chrysanthemum production. *J. Amer. Soc. Hort. Sci.*, **122**, 703–708.

Xu, G., Levkovitch, I., Soriano, S., et al. (2004). Integrated effect of irrigation frequency and phosphorus level on lettuce: P uptake, root growth and yield. *Plant Soil*, **263**, 293–304.

Yeager, T.H. and Wright, R.D. (1982). Pine bark – phosphorus relationships. *Commun. Soil Sci. Plant Anal.*, **13**, 57–66.

Zelancy, L.W., Liming, H. and Vanwormhoudt, A.M. (1996). Charge analysis of soils and anion exchange. In *Methods of Soil Analysis Part 3: Chemical Methods* (D.L. Sparks, et al., eds), No 5 in the Soil Science Society of America book. Madison: Soil Science Society of America, Inc., pp. 1231–1253.

7

ANALYTICAL METHODS USED IN SOILLESS CULTIVATION

Chris Blok, Cees De Kreij, Rob Baas
and Gerrit Wever

7.1 Introduction
7.2 Physical Analysis
7.3 Methods
7.4 Chemical Analysis
7.5 Biological Analysis
    References

7.1 INTRODUCTION

7.1.1 WHY TO ANALYSE GROWING MEDIA?

In the last decades numerous new raw or processed products have been offered as potential growing media or media constituents. The products can be as diverse as non-processed raw materials (e.g. peat, pumice, tuff), waste products from agricultural, forest and food industries (e.g. bark, rice hulls, coir), recycled materials (e.g. from mattresses, tires, paper waste), up to processed materials (e.g. poly phenol foam, urea formaldehyde foam, stone wool) or composted materials. The number of possible growing media is further enhanced by mixing these diverse materials. Ideally each potential growing medium should be tested under commercial growing conditions in field trials in order to assert its performance in crop production. However, besides the costs and duration of these 'biotests', even these tests can give no 100 per cent guarantee for optimal performance, since commercial growing conditions such as temperature or fertigation are never identical to the experimental conditions. Moreover, the performance of a product may be crop-specific, which would considerably increase the number of required field trials. Manufacturers of new growing media sometimes do this extensive testing of their products under commercial conditions. However, laboratory tests which should give a reliable indication of the performance of the medium with respect to, for example, water, nutrient and oxygen availabilities on the short and long term are normally more cost-effective, and can be performed under standardized conditions with reference samples. It is therefore logical that analytical methods – mostly adopted from soil science analytical methods – have become available as part of **the selection process of growing media** and for setting soilless growing media standards.

Another reason for the use of reliable analytical methods is that producers of growing media and/or laboratories should be able to use these methods for **quality control** by themselves or by independent organizations for production of described quality in order to provide growers or retailers with media with prescribed physical and/or chemical characteristics. Among the most prominent of these organizations are ISO (International Organization for Standardization), ASTM (American Society for Testing and Materials, www.astm.org), CEN (European Committee for Standardization, www.cenorm.eu), DIN (Deutsches Institut für Normung e.V., www.din.de), VDLUFA (Verband Deutscher Landwirtschaftlicher Untersuchungs- und Forschungsanstalten www.vdlufa.de), AFNOR (Association Française de normalisation, www.afnor.org) and RHP (Regeling Handelspotgronden, www.rhp.nl).

The next, difficult, step for analytical methods to be really useful for end-users such as growers is a **recommendation of growing media** based on the physical, chemical and biological characteristics of the media, in combination with the requirements. These recommendations will vary according to the crop, the growing system (slabs, containers), the water supply system (e.g. ebb-flood, overhead irrigation, drippers), water quality and the growing season and duration. Each combination of crop, growing system and water supply system thus may require specific characteristics of a growing medium. In the International Substrate Manual (Kipp et al., 2001), an effort was made to set up such a system based on salinity, oxygen and drought sensitivity of crops. Although this system can not be totally justified by scientific results, the philosophy is that it is 'better than nothing', and that this system can be improved according to ongoing experience and research results.

Finally, analysis of growing media is done routinely for **advice on fertilization**. Systems for fertilization advice based on various nutrient analysis techniques have been in use for many years, and are indispensable in hydroponics. Frequent (weekly) sampling with analysis and advice within two days has become a standard procedure for vegetables and cut flowers in hydroponics.

Obviously, not all parameters which are determined by means of physical, chemical and biological methods are equally important for the above-mentioned applications of the analysis methods. The overview (Table 7.1) shows that some of the parameters such as directly available nutrients and EC are rather important and frequently used for all applications. For physical analysis, water retention characteristics including rewetting

TABLE 7.1 Overview of Parameters from Physical, Chemical and Biological Analyses Used for Growing Media and their Relevance and Use for the Various Applications

	Application of growing medium analysis				
Parameters	Selection	Quality control	Fertilization advice	Recommendation	Remarks
Physical analysis					
Bulk density	+	++	−/+	++	
Porosity	++	++	−	++	
Water retention including rewetting	+++	++	−	+++	
Particle size	+	+	−	+	
Shrinkage	++	+	−	++	
Hardness, Stickiness	+	−	−	+	
Penetrability	+	−	−	+	
Hydrophobicity	+	−	−	+	
Hydraulic conductivity	+	−	−	+	Little experience
Oxygen diffusion	+	−	−	+	Little experience
Chemical analysis					
Direct available elements in the rhizosphere,	+	+++	+++	++	
Potentially available elements, EC	++	++	++	++	
Exchangeable ions, CEC, AEC	++	++	++	+	
Total analysis	+	+	+	+	Depending on growing medium
N fixation/P fixation	++	++	++	++	Depending on growing medium
CaCO$_3$	−/++	−	+	+	Depending on growing medium
pH	++	++	++	+	
Biological analysis					
Stability	++	−/+	−	++	Depending on growing medium
Phytotoxicity	+++	−/+	−	++	May depend on crop

Parameters determined in soilless media in relation to their application: selection of growing media, routine quality control, routine fertilization advice, and recommendation for use in growing system (defined by, e.g. height and volume of substrate, irrigation system, crop used). Note: scale increases from − which is equal to parameter not relevant to +++ which is equal to very relevant.

are essential for screening, quality control and recommendations. This does not mean that other analysis methods are less relevant. For example, the $CaCO_3$ content can be very important in screening substrates such as composts. Hydraulic conductivity has been argued to be very essential for horticultural practice (Raviv et al., 1999). However, a routine method for (unsaturated) hydraulic conductivity is not yet available due to technical difficulties (Wever et al., 2004). For an updated discussion of this subject, see Chap. 3. Measurement of some parameters suffer from large variation (e.g. oxygen diffusion), and/or the high costs involved, which makes it less suitable to use it as a method for routine measurements. Table 7.1 gives an overview of the physical, chemical and biological parameters that are measured in growing media and the relevance of their use in specific stages along the decision-making process.

7.1.2 VARIATION

Since there is a large variety of physical and chemical analytical methods used by laboratories, efforts have been made to compare and standardize the laboratory methods. As an example, for physical analysis not all laboratories use the same sample preparation procedures. In the method used in service labs for physical parameter determination in the Netherlands, the potting mix sample is pressed and not moistened, whereas in the European Norm (CEN method) the sample is not pressed and is moistened. Hence, analysis reports can result in confusion to users familiar with their own analytical methods. Naturally, labs and countries are reluctant to change their analytical methods. The ISHS 'Working Group on Substrates in Horticulture Other than Soils in Situ' of the ISHS Commission of Plant Substrates should be mentioned for their efforts to encourage standardization of analytical methods (e.g. Gabriëls et al., 1991; Gabriëls, 1995). Research has shown that in many cases analysis results can be converted from one method into another (e.g. Sonneveld and Van Elderen, 1994; De Kreij et al., 2001; Wever et al., 2005). Moreover, the variation among labs using the same methods on the same samples, both for physical (Gabriëls et al., 1991; Wever and Van Winkel, 2004) and chemical analysis (Baumgarten, 2004a, De Kreij and Wever, 2005), should be decreased in order to increase the reliability.

7.1.3 INTERRELATIONSHIPS

To assess the suitability of a substrate for a well-defined purpose, usually several interrelated measurements should be interpreted. A common example is the relationship between stability at the one hand and water and air transport at the other hand. The underlying mutual cause is the breakdown of structural elements which results in a loss of pore volume. Another example is the relationship between loss of stability and the fixation of nitrogen and phosphorous. Microbiological activity is the mutual cause underlying both. Both examples show the importance of interpretation and the available room for new, more direct measurements.

In this chapter, the physical, chemical and biological parameters (Table 7.1) are described. After each definition and description of the method, common values, special cases and relation to crop growth are given. In case a standardized European method (CEN-method; more information on www.cenorm.eu) is used, this is mentioned with the method number.

7.2 PHYSICAL ANALYSIS

7.2.1 SAMPLE PREPARATION (BULK SAMPLING AND SUB-SAMPLING)

To perform most of the measurements discussed below, a representative sample of the material to be assessed must be prepared. From this bulk sample the final units for individual measurements are prepared usually by sub-sampling into test cylinders such as rings. Materials like pre-shaped plugs for rooting will of course be measured as such, without a test cylinder. Some materials like stone wool, peat boards, coir boards or polyurethane slabs are preformed, that is, they consist of linked or interwoven particles or form a coherent mass by nature, others like peat and perlite are loose, that is, they consist of unlinked particles.

7.2.2 BULK SAMPLING PREFORMED MATERIALS

Preformed materials have a variation in properties within and between the units they are sold in, and within and between the batches they are produced in. Furthermore, most preformed materials are pressed into some shape during production and usually show a density profile in the direction perpendicular to the compression force. The pattern of cutting the material into the final substrate units determines in what way these density differences may manifest themselves in the end product. Finally, preformed fibre materials like stone wool, coir pith and some peats may show differences in fibre orientation in one or two directions. If it is desirable, samples should be taken in different directions of the original product. The sample preparation of pre-formed materials is usually confined to cutting a sample of specific dimensions in such a way that the material is not disturbed. To get a representative sample, two or more samples from different units are necessary (CEN 12579, 1998).

7.2.3 BULK SAMPLING LOOSE MATERIAL

Big differences among samples of the same material may exist among and within heaps or rows, and more importantly among different production dates of the same product. For most products like peat, coir dust and compost, the weather conditions from the time of harvest till the day of transport influence the properties of the end product. Examples are peat excavated during a wet or a dry summer, composts made during a hot or cold period, coir dust from a lot stored for one and up to five years and composts made from summer grown plants or winter grown plants.

Samples are taken from the heap or heaps with a hollow-headed agricultural hand auger. The auger diameter should be larger than three times the maximum particle size of the sampled material. When no auger with a large enough diameter is at hand, a scoop may be used. The samples should be taken from a depth deep enough to avoid any material from the outer layer that has dried or otherwise obviously changed. This is usually deeper than 50 cm. To get a representative sample, 20 or more samples from different places are necessary. From bags (20–100 L), the whole bulk sample is taken from randomly chosen bags. Care should be taken to avoid fractionation of the sample, especially when using a scoop. The total sample size should not exceed what is needed in the laboratory to minimize the need for sub-sampling and the risk of segregation (CEN 12579, 1998).

7.2.4 SUB-SAMPLING PRE-FORMED MATERIALS

The sample preparation is confined to cutting a sample of specific dimensions in such a way that the material is not disturbed. Special attention should be paid to avoiding compression by the knife or saw blade involved. When the material is pre-formed but the form is not stable, for example boards of peat, it can be necessary to put the material in a holder.

7.2.5 SUB-SAMPLING LOOSE MATERIALS

The density of the materials may be stable because of the rigid nature of the individual particles. One might think of clay pellets, pouzolane, perlite or sand. These materials may be poured gently into a test cylinder of the right dimensions (e.g. diameter 5 cm and height 5 cm or diameter 10 cm and height 7.5 cm). No compression by weight or by tapping is applied.

For softer materials, which might be compressed by their own weight and moisture content, a laboratory-compacted bulk density in its 'as received state' has to be estimated. Based on that laboratory-compacted bulk density, the weight of sample to be taken for sub-samples is calculated. Different procedures exist in which the samples are brought to a standard density and/or moisture content. A specific example is given (CEN 13040, 2006). The medium is manually homogenized and passed through a screen beforehand. It is then sieved into a 1 L test cylinder with a removable collar until overflowing. The excess material is wiped off and the sample is then compacted with a plunger of specific weight. After a standard compaction interval, the collar is removed and the sample is wiped off at the level of the test cylinder. The test cylinder is then ready for weighing and the laboratory-compacted bulk density is calculated.

7.3 METHODS

7.3.1 BULK DENSITY

7.3.1.1 Definition and Units

The dry bulk density ($kg\ m^{-3}$) is defined as the ratio of the mass of dry solids to the bulk volume of the substrate (after Blake and Hartge, 1986). Bulk density is defined

7.3 METHODS

as the density of the medium as it occurs. The bulk density is partly dependent on the actual moisture content. Bulk density is of importance for trade and handling.

7.3.1.2 General Principle of Determination

The dry bulk density is measured by determining the dry weight of a known volume of material under specified conditions. Various older norms have been condensed into CEN norms for trade volume, chemical analysis and physical analysis (CEN 13041, 1998; CEN 12580, 1999; CEN 13040, 2006). Usually bulk density, dry bulk density and air and water content at a suction of −10 cm of water column are simultaneously determined (Fig. 7.1).

The bulk density measurement for the determination of quantity (CEN 12580, 1999) is measured using a container with a known capacity of about 20 l and a height to diameter ratio between 0.9:1 and 1:1. A 7.5-cm high collar and a fall controller (a sieve with a defined mesh aperture) with the same diameter as the container are attached. The container and collar are filled to the top after which the collar is removed and the material is levelled off. The container is weighed and the bulk density is then calculated according to Eq. (1).

$$^b\rho = \frac{M}{V} \tag{1}$$

where

- M mass of the material in the container (fresh, kg)
- $^b\rho$ bulk density (fresh, kg m^{-3})
- V volume of the empty container (m^3).

FIGURE 7.1 The lower test cylinder extended to double its height by putting a second test cylinder on top, held in place by using a clamp and a rubber. The test cylinder and rim (second holder) are then filled with a few gentle taps to avoid large voids and create an even filling. The sample is brought to a specific suction force and finally the rim or top cylinder is removed and the excess material is wiped or scraped off with a rigid blade. After weighing, drying and weighing, the bulk density is calculated (see also Plate 19).

7.3.1.3 Special Cases

Non-rigid materials such as peat and composts need to be weighed under carefully defined pressure and filling circumstances as described in 'sampling'. Transport vibrations may alter bulk density considerably.

More than one figure for bulk density is necessary to sufficiently characterize media with a density profile as in some peat products, or with layers of different density as in some types of stone wool (Bullens, 2001).

7.3.1.4 Common Values for Media

The dry bulk densities for most growing media are 3–20 times lower for most soils (soils are about $1500\,kg\,m^{-3}$, for rooting media, see Table 7.2).

7.3.1.5 Influence on Growth

For a given material, an increase in bulk density is associated with decrease in total pore space and thus affects growth mainly through the effects of reduced free pore space. A decrease in total pore space will often decrease oxygen transport and decrease root penetration. A decrease in total pore space may also increase the water retention as pore diameters decrease, which is to say that loss of physical structure often results in an increase in water retention of the remaining material. Interpretations of the influence of bulk density on growth may be improved by focussing on the individual effects of reduced pore space rather than on the broader concept of bulk density.

TABLE 7.2 Bulk Density, Volume Fractions of Solids, Water and Air at a Suction Head of $-10\,cm$ and Total Pore Space (Total Porosity) as Measured in Some Growing Media

	Dry bulk Density Units $kg\,m^{-3}$	Solids % V/V	Water % V/V	Air % V/V	Total pore space % V/V
Glass wool	–	2	59	39	98
Stone wool	49	3	69	28	97
Perlite	105	4	35	61	96
Polyurethane	78	5	18	77	95
Peat	113	9	54	37	91
Pumice	431	17	32	51	83
Clay granules	489	24	21	55	76
Pouzolane	–	47	20	33	53

Source: After Kipp et al. (2001).

7.3.2 POROSITY

7.3.2.1 Definition and Units

The porosity is the total space that is not occupied by solid material, including space where water can not readily enter. The porosity or total pore space (per cent V/V, TPS) is calculated from the total volume minus the volume occupied by the solids. Often, only the maximum volume of water which will enter the material on immersion (effective pore volume) is reported as this is the volume which affects several other properties such as water and air transport. Effective pore volume can be measured more readily.

7.3.2.2 General Principle of Determination

Most siliceous materials have a density of the solid phase (a.k.a. particle density) of about $2650\,g\,L^{-1}$ and most organic materials have a density of the solid phase of about $1550\,g\,L^{-1}$. When the dry bulk density of a sample with single constituent, for example siliceous sand, is known, it is assumed that the density of the solid phase of that sand is $2650\,g\,L^{-1}$. Thus a common dry bulk density of a siliceous material of $1500\,g\,L^{-1}$ would mean that $1500/2650 = 0.57$, that is 57 per cent V/V of the volume is occupied by the sand solids and 43 per cent V/V of the sample is the total pore volume or porosity.

The effective pore volume may be found by measuring the amount of water entering a material upon immersion. Care should be taken to allow the sample to saturate from the bottom up by raising the water table slowly, taking 30 min or more. This is necessary to prevent air from being trapped in the pores.

A more accurate method to find the effective pore volume, using air instead of water, is the pycnometer. In a pycnometer measurement, a sample is put into a vacuum chamber and the volume of air entering the chamber upon breaking the vacuum is measured.

7.3.2.3 Special Cases

Some materials are composed of solids with a different density, for example, some stone wool mineral melts yield a density of about $2900\,g\,L^{-1}$. If the density is not known, matters can become increasingly complicated when mixtures of two or more materials are present.

The porosity calculated from the dry bulk density and the specific weights may have to be corrected for the amount of non-connected pores. These pores are embedded in the material and remain air-filled during immersion as in perlite, volcanic products like tuff, pouzolane and extruded plastics like phenols and polyurethane. There is no fixed method to measure the total porosity but a fair estimate is usually found by grinding the material to a fine, compressed powder. Measuring the volume of the sample before and after grinding yields a volume lost by grinding. Then the effective pore volume of the powder is measured by submersion or preferably a pycnometer. Finally the total pore volume of the original material can be calculated by adding the volume lost upon grinding and the effective pore volume of the powder.

7.3.2.4 Common Values for Media

Total pore space for most growing media is 1.5–2.8 times higher than the values found for common soils (about 35 per cent V/V, Table 7.2).

7.3.2.5 Influence on Growth

An increase in total pore space will often decrease the water retention, increase oxygen transport and increase root penetration. These, in turn, will influence plant growth. The effect of pore space is therefore complex as it affects plants in more than one way.

7.3.3 PARTICLE SIZE

7.3.3.1 Definition and Units

A particle size class is a fraction (per cent W/W) of a material with particles with a diameter larger than a given lower limit and smaller than a given upper limit, for example 2–4 mm.

7.3.3.2 General Principle of Determination

The material is air dried and either gently ground or put directly on a set of sieves with decreasing mesh size. Thus the material is fractionated into classes which pass through the sieve above but not through the sieve underneath. A common sieve set in square mesh has the following sizes: 16.00, 8.00, 4.00, 2.00, 1.00, 0.500, 0.125 and 0.063 mm.

7.3.3.3 Special Cases

Sieve analysis is very sensitive to effects of drying, breaking prior to sieving and the sieving process itself. Therefore, detailed specifications are needed to ensure reproducible results. Samples with a lot of fine organic material, like green composts, may form rigid 'cakes' upon drying. When this happens and the resulting cake does not readily disintegrate, it is better to follow a procedure using lower drying temperatures and periodic gentle agitation.

Breaking prior to sieving is only done to avoid starting with largely oversized clods. There is a maximum amount of material for a given sieve size to prevent clogging the mesh. The amount of energy in moving the sieves and the time of sieving is usually standardized, for example, to 7 min duration with 10 s bursts with a rest interval of 1 s between them and an amplitude of 1 mm.

The fraction passing one sieve size and not another is not a true measure of the particle diameter. It means that there is at least one diameter of the particle which enables it to pass the upper sieve. Particles with largely different dimensions for length, width and height in particular, like reed compost, tend to orientate horizontally and not pass sieves which they can easily pass if handled manually. Finally the sieves may have square or round holes which slightly influence the result (round holes retaining very slightly more material, depending on particle shape).

7.3.3.4 Common Values for Media

Media used in horticulture generally have particles ranging, arbitrarily, from 0.125 to 2 mm to reach an optimum between available air and available water (Abad et al., 2005). Much coarser materials, for example many barks, are in common use to increase aeration for larger plants. Much finer materials, for example clays, are in common use to enhance water retention characteristics. Many materials are offered in different size ranges, for example perlite may be delivered in fractions of 0–1 mm, 0.6–2.5 mm and 1–7 mm. The vast variety of peat products may be offered as a single product, for example fractions of 6–18 mm, but are quite often mixed; a common mix being 70 per cent V/V peat < 2 mm with 30 per cent V/V of fractions 20 mm.

7.3.3.5 Influence on Growth

The particle size distribution (in combination with the total pore volume) indicates the amount of small and large pores. This, in turn, is an important indication of the percentage of water and air at different suction levels and the ease of rewetting. In some cases it gives information on the ease of handling of the material by machinery, for example drilling holes for container plants or filling sowing trays. It is difficult to use the particle size on its own to estimate water characteristics because the different size classes may show interstitial filling, which means a smaller size class can fill the holes among larger size classes thus creating an unexpectedly high bulk density. In addition, water retention is also affected by surface characteristics such as the affinity between water molecules and the surface of the particles.

7.3.4 WATER RETENTION AND AIR CONTENT

7.3.4.1 Definition and Units

The water-holding capacity or retention is a function of the total pore space (per cent V/V) and a suction force applied (either in cm water pressure or kPa). The more suction is applied, the drier the material gets and the higher the air content will be (this is represented by the main drainage curve in Klute, 1986).

7.3.4.2 General Principle of Determination

For media with suctions up to 10 or even 20 kPa suction tables are in common use. The suction is usually applied by placing the samples on a bed of very fine sand or clay in a water-tight container connected to an adjustable overflow (Fig. 7.2).

Water retention measurements are time consuming because equilibrium between the applied suction and the water-filled pore space is essential. Usually each measured point on a curve with 3–10 points requires 1–2 days to reach equilibrium. Frequently, the term 'water retention' is used for the measurement of both the main drainage curve and the main wetting (= rewetting) curve. Common measuring points include the water content at saturation, and at suctions of -2.5 cm, -10 cm, -20 cm, -50 cm, -100 cm and -500 cm suction head. The air content is calculated as the difference between porosity and water content. Usually the main drainage curve and the main wetting curve share the wettest and the driest points but differ in between. This effect is called hysteresis. Its causes are discussed in the section on 'Hydrophobicity'. Hysteresis also influences the **rewetting, rehydration rate** and **unsaturated hydraulic conductivity**. Further discussion appears in Chap. 3.

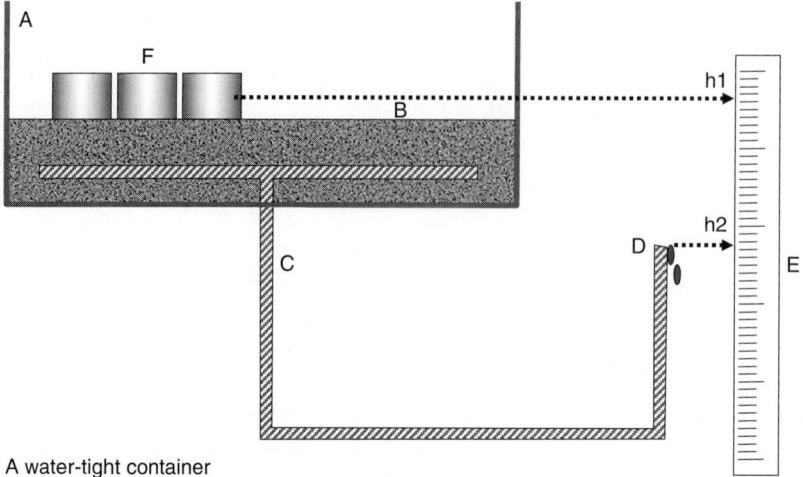

A water-tight container
B fine-graded and packed sand in a flat layer of +/−10 cm
C drainage system with multiple well-distributed entries
D adjustable overflow
E scale in mm to register the suction (h2−h1)
F sample in sample holder

FIGURE 7.2 Schematic suction table made of a moist sand layer in a container, connecting via capillary action the moist samples with an overflow. The suction applied to the samples is usually defined as the difference in centimetres between the overflow and the half height of the samples.

7.3.4.3 Special Cases

Various variants exist, for example using air pressure instead of water columns. Especially at higher suction levels air will enter the sand bed of the suction table. Strict procedures for sample preparation, saturation and measuring time have to be observed. The method is derived from soil science and its application for media creates some interpretation problems. One interpretation problem is related to the sample height of 5 cm, which is treated as a point sample. In fact, for small suction forces applied (from 0 to 15 cm suction force, i.e. 0–1.5 kPa), the sample height does influence the reading to a large degree.

Another interpretation problem concerns the point of 0.0 cm suction, which is defined at a free water table up to half of the sample's height. The point of 0.0 cm suction is thus changing with the sample height. Sample heights other than 5.0 cm are common, for example 3.5 cm for propagation plugs.

A third common interpretation problem is that many substrates used in horticulture are in practice subjected to only a part of the measuring range mentioned. There is no point in measuring substrates such as polyurethane foam at suctions above 20 cm of water column, since at these suctions there is no longer water in interconnected pores.

7.3.4.4 Common Values for Media

Water retention forces in growing media are usually 10–100 times lower than the common values for soil (10–100 kPa). Representative examples are given in Chap. 3.

7.3.4.5 Influence on Growth

The results are indicative of the ease of the uptake of water – and nutrition – by plants as well as the wetness in various growing systems. Growth is highest at low water retention forces, but very low water retention forces are sometimes avoided, for example, when the amount of air-filled pores becomes too low for proper oxygen transport. The air content recommendations for optimal growth are found in kipp et al., 2001. Water retention forces high enough to decrease the fresh weight growth may actually be desirable, for example to create denser, that is better quality pot plants and in transplant production, when hardy plants are preferred by the growers.

7.3.5 REWETTING

7.3.5.1 Definition and Units

The rewetting curve (main wetting curve in Klute, 1986) shows the ability of a material to take up water against gravity from a well-defined point of dryness. The rewetting curve is the water retention curve during rewetting.

7.3.5.2 General Principle of Determination

The rewetting curve is measured with the same apparatus and in the same units as described for measuring the water retention (= water-holding capacity). Common measuring points include -500 cm, -100 cm, -50 cm, -20 cm, -10 cm, -2.5 cm, and saturation.

7.3.5.3 Special Cases

Media are very different in their ability to rewet as well as in the amount of hysteresis they show. After several drying cycles, coir dust is known to rewet to almost the level at saturation but some stone wool types may only rewet to half of the level at saturation. Unfortunately it is still difficult to characterize rewetting ability experimentally because standard methods from soil science require too large pressure heads before rewetting for some growing media. The problem is in choosing the point of defined dryness. The point of dryness has to be related to dry but practical circumstances which are different for different materials. The actual situation during growing is even more complicated as many drying and rewetting cycles are following one another, ending and starting from different points of dryness (Raviv et al., 1999).

Another indication on the rewetting ability of the material is the **rehydration rate**. Some work is being done on a rewetting indicator calculated from the unsaturated hydraulic conductivity.

7.3.5.4 Common Values for Media

Values have to be regarded in relation to the point of dryness used. If stone wool is dried from a water content of 88 per cent V/V (at -0.1 kPa) to 30 per cent V/V (at -2 kPa), rewetting to 60 per cent V/V is normal. Had the point of maximum dryness been -10 kPa, rewetting to 60 per cent V/V would indicate superb rewetting.

7.3.5.5 Influence on Growth

Plants do not react directly to differences in rewetting ability but imperfect rewetting will amplify any unevenness in crop water use of a given area. In practice, this will force growers to use more frequent irrigation cycles.

7.3.6 REHYDRATION RATE

7.3.6.1 Definition and Units

The rehydration rate is defined as the increase in moisture content of a dried sample over a set period of time (per cent V/V over minutes).

7.3.6.2 General Principle of Determination

A substrate sample of standard dimensions (such as a cylinder of 5 or 10 cm diameter and a height of 5 or 7.5 cm) is oven-dried at 105°C for 24 h. After weighing, the sample is placed on a thin layer of water on a coarse mesh. The weight of the sample is measured after 0, 1, 2, 4, 8, 15, 30, 60, 120, 240 and 360 min and one, two and seven days. Special care should be devoted to the drying, as small differences in drying may result in surprisingly large deviations in the rewetting rate. Secondly, care should be devoted to the contact with the free water surface. There should be a permanent full contact of 0–1 mm.

7.3.6.3 Special Cases

Some materials may develop **hydrophobicity**, notably materials containing non-frozen black peat as well as some composts. Hydrophobicity may be caused by specific molecules either formed or deposited. It is difficult to distinguish effects of hydrophobicity from effects of the pore-size distribution or of changes in pore geometry upon shrinking. Rehydration rate is closely related to **unsaturated hydraulic conductivity**. Some work has been devoted to relating rehydration rate and hydraulic conductivity with each other. This line of work can account for pore-size distribution effects and even the effects of shrinkage, but cannot account for hydrophobicity caused by specific molecules.

7.3.6.4 Common Values for Media

Figure 7.3 shows the excellent rewetting rates of coir pith, a sowing soil, peat (H1–3) and a sandy medium used for outdoor ornamentals in large containers. In comparison to the others, coir pith shows the best rewetting rate, reaching over 70 per cent V/V in 15 min.

7.3.6.5 Influence on Growth

A slow rewetting rate, for example less than 40 per cent V/V in 15 min, increases the risk of dry substrate in the top of containers. It also increases the risk on hydrophobicity in places where there is more evapotranspiration such as borders. Some substrates, as the sandy material in Fig. 7.3, rewet too slow to be used in combination with sub-irrigation.

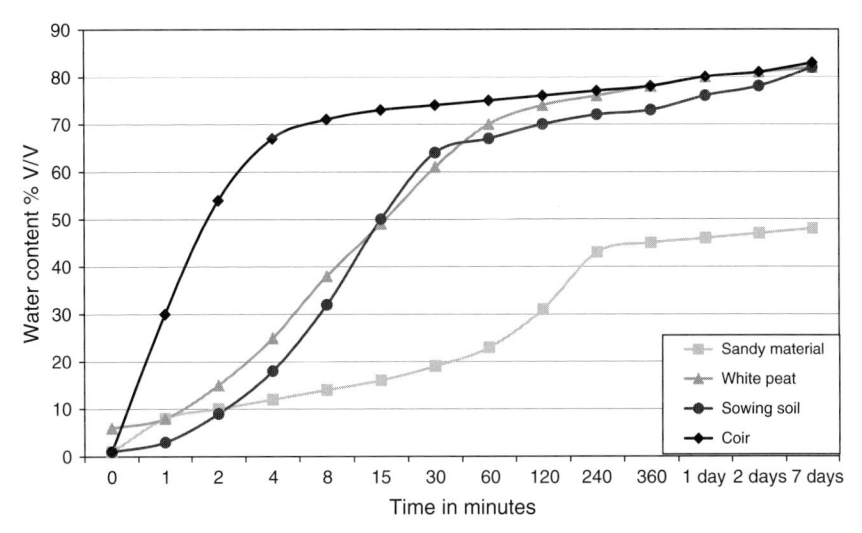

FIGURE 7.3 Rehydration rate of dry samples of four rooting media from a free water surface.

7.3.7 HYDROPHOBICITY (OR WATER REPELLENCY)

7.3.7.1 Definition and Units

Hydrophobicity is the irreversible loss of water retention upon drying caused by the change during drying in molecular organization of organic matter, and, as a consequence, deposition of hydrophobic molecules on pore walls.

7.3.7.2 General Principle of Determination

It can be measured as a change in water contact angle with the pore wall. The direct measurement requires many microscopic observations and is laborious. A method based on capillary rise is more often used to estimate the contact angle (Michel et al., 2001; Goebel et al., 2004).

7.3.7.3 Special Cases

It is still difficult to discern between incomplete rewetting and pore wall hydrophobicity. Incomplete rewetting is also caused by air entrapment and static pore diameter variations (which stands for different diameter sections in one continuous pore). These are the classic explanation of hysteresis, that is the difference between a drainage curve and the subsequent wetting curve. Pore wall hydrophobicity is caused by the deposition of hydrophobic organic substances on soil particles (Ellerbrock et al., 2005). The difference is of practical importance as the **rewetting** method, based on equilibrium between suction and moisture content, shows mainly effects of hysteresis, and the **rehydration rate** shows combined effects of hysteresis and pore wall hydrophobicity. Care should be taken in defining the terms explicitly in publications as there seem to exist several interpretations of hydrophobicity, air entrapment, static pore diameter

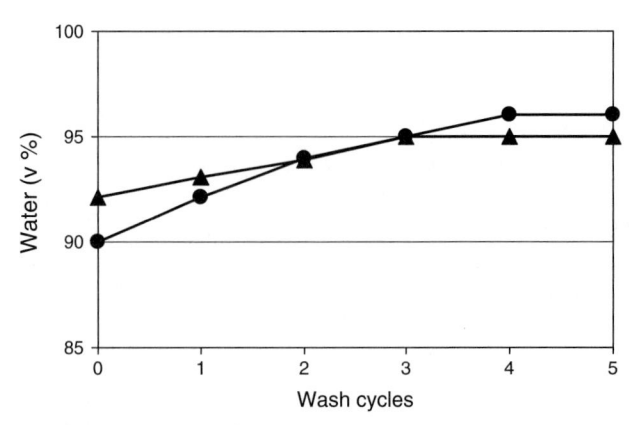

FIGURE 7.4 Volume of water at container capacity after repeatedly 'washing' two brands of stone wool treated with a wetting agent before testing (after Kipp and Wever, 1998).

variations and shrinkage (dynamic pore diameter variations). Wetting agents are sometimes used to facilitate rewetting of hydrophobic growing media like unfrozen black peat, mineral fibres, composts and growing media in which decomposing roots and algae are present (Fig. 7.4). When using wetting agents, the initial maximum water content reached is lower than that after a few rewetting cycles because of the reduced surface tension of the water.

7.3.7.4 Common Values for Media

It is recommended to measure against one or more well-known reference samples, as figures from literature may be difficult to reproduce due to variations caused by temperature, ions from the substrate and wetting agent – medium interactions.

7.3.7.5 Influence on Growth

Hydrophobicity is a common and unwanted property. By nature, hydrophobicity will be more pronounced when the material gets drier. The results are often seen in nurseries, where the boundary rows dry out faster, develop more hydrophobicity and consequently take up less water and get even more hydrophobic. This will reduce yield or quality of the plants from the boundary rows or will cost extra labour to manually rewet the growing medium.

7.3.8 SHRINKAGE

7.3.8.1 Definition and Units

Shrinkage is the volume loss (per cent V/V) of a sample of standard dimensions after drying to 105°C as compared to the volume at standard moisture (usually $-10\,cm$ suction force).

7.3.8.2 General Principle of Determination

The material is moistened to standard moisture and put in a test cylinder of known dimensions in a standardized way. The material is then oven-dried. The dimensions of the remaining samples are measured by hand. The use of image analysis methods to measure shrinkage is emerging (Michel et al., 2004).

7.3.8.3 Special Cases

Some materials stick to the walls of the test cylinder, form cracks and even break in several pieces. The resulting volume may be measured after sealing the sample pieces with wax and measuring the underwater volume of the pieces.

7.3.8.4 Common Values for Media

In laboratories, 5–10 per cent V/V is a common value, some materials like transplant media may shrink over 20 per cent V/V.

7.3.8.5 Influence on Growth

Shrinkage is a problem for outside ornamentals like trees, shrubs, flowering plants which may move in their containers or even be blown by wind. It is often related to hydrophobic effects caused by drying. In these cases, irrigation may be problematic, as any excess water drains very fast through the cracks or the void between container wall and the medium (channelling). Finally the evaporation from the medium is much faster as the surface area to the surrounding air, including the area of the cracks and voids, is much larger than the container surface alone. Shrinkage may be desirable in some transplant mixes used in tray cells as the plugs are to shrink loose from the surrounding tray which facilitates transplanting.

7.3.9 SATURATED HYDRAULIC CONDUCTIVITY

7.3.9.1 Definition and Units

Saturated hydraulic conductivity is the mass of water passing through a unit area in time ($m^3 \, m^{-2} \, s^{-1}$) in a saturated growing medium.

7.3.9.2 General Principle of Determination

The saturated hydraulic conductivity is characterized by Darcy's equation (Eq. [2])

$$Q = K_{sat} \frac{dH}{dx} \qquad (2)$$

where

Q flux ($m \, s^{-1}$)
K_{sat} saturated hydraulic conductivity ($m \, s^{-1}$)
dH head difference (m)
dx distance (m).

According to this equation, it is possible to calculate or model water transport rates for any head difference over any distance in a given growing medium. Apparatus to measure K_{sat} with the constant head method usually consists of a supply container from which water flows to a sample of known dimensions. The saturated sample is kept under a constant head of water. The water passing through the sample is collected in a drain collection container and is measured per unit time. Apparatus for the running head method are similar in construction but the containers are directly connected to the sample and the head is allowed to change during the test. Care must be taken to fully saturate the sample and to prevent air bubbles from appearing in between the substrate particles while running the tests.

7.3.9.3 Special Cases

The application of standard measurement devices from soil science for the measurement of saturated hydraulic conductivity in growing media has proved difficult as the saturated hydraulic conductivities are 100–1500 times larger than that in soils (1–100 cm d^{-1}). At all times it should be checked and reported that the apparatus without sample allows flows of at least one order of magnitude larger than the K_{sat} measured. This check is necessary to rule out the possibility that the hydraulic conductivity of a part of the apparatus such as the sample support or a low diameter valve or bend is limiting, and the measured value reported as if it were that of the test material. Several designs of apparatus have been proposed over the last decade (Da Silva et al., 1995; Wever et al., 2004; Naasz et al., 2005). Consequent validation by other groups has not yet been reported for all of these and many growing media have not yet been measured.

7.3.9.4 Common Values for Media

Un-validated measurements suggest 1 m min^{-1} for stone wool and coarse perlite (1–7.5 mm) and 0.01 m min^{-1} for fine perlite (0–1 mm).

7.3.9.5 Influence on Growth

K_{sat} is usually not of direct importance for plants (but see 'Unsaturated hydraulic conductivity').

7.3.10 UNSATURATED HYDRAULIC CONDUCTIVITY

7.3.10.1 Definition and Units

Unsaturated hydraulic conductivity is the mass of water passing through a unit area in time (m^3 m^{-2} s^{-1}) under unsaturated conditions.

7.3.10.2 General Principle of Determination

The theoretical background of unsaturated hydraulic conductivity is thoroughly discussed in Chap. 3. Clearly, saturated and unsaturated hydraulic conductivity are associated. Unsaturated hydraulic conductivity is, however, more difficult to measure.

The aforementioned formula after Darcy is, with one adapted parameter, used to characterize unsaturated water transport as shown in Eq. (3).

$$Q = K_h \frac{dH}{dx} \qquad (3)$$

where

- Q flux (m s^{-1})
- K_h unsaturated hydraulic conductivity (m s^{-1})
- dH head difference (m)
- dx distance (m).

According to Eq. (3), it is possible to calculate or model water transport rates for any head difference over any distance in a given growing medium. The real difficulty is in finding K_h. K_h may be found with a series of formulae according to the structure in Eq. (4).

$$K_h = K_{sat} \left(\frac{WFP}{TPS}\right)^m \qquad (4)$$

where

- K_h unsaturated hydraulic conductivity (m s^{-1})
- K_{sat} saturated hydraulic conductivity (m s^{-1})
- WFP water filled pore fraction (V/V)
- TPS total pore space (V/V)
- m constant.

The only variable now is the water-filled pore space. The influence of WFP on K_h is very large as can be seen from the exponential function. The relation WFP over TPS is thought to represent the tortuosity of the transport path (Allaire et al., 1996). The measurement of either K_{sat} or K_h requires careful pre-treatment of the samples and the apparatus used (Da Silva et al., 1995; Wever et al., 2004).

7.3.10.3 Special Cases

Several designs of apparatus have been proposed over the last decade (Da Silva et al., 1995; Wever et al., 2004; Naasz et al., 2005). Consequent validation by other groups has not yet been reported for all of these and many growing media have not yet been measured.

When growing media have been properly characterized for K_h, or the relation K_h–K_{sat}, further work may be based upon model calculations for specific growing media dimensions and specific cultivation and supply techniques. It is common to link the water retention characteristics and unsaturated hydraulic conductivity (models using the work of Mualem and the work of van Genuchten). These models are now being expanded to include dynamic rewetting and oxygen diffusion (Caron, 2004).

7.3.10.4 Common Values for Media

The hydraulic conductivities at higher water contents are 100–1500 times higher than the values found for soils (1–100 cm d^{-1}, Fig. 7.5).

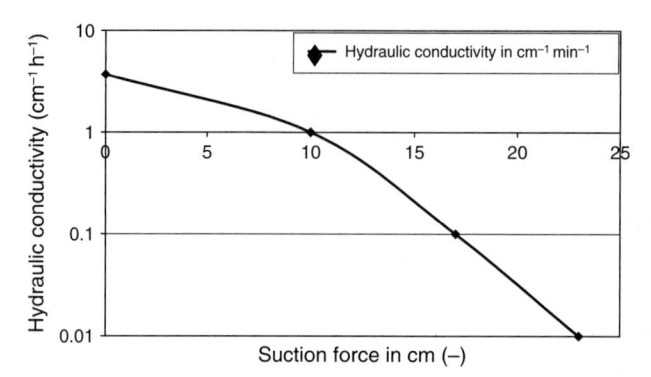

FIGURE 7.5 Unsaturated hydraulic conductivity of stone wool in relation to the suction force applied (after Da Silva et al., 1995).

7.3.10.5 Influence on Growth

Despite the very high initial hydraulic conductivities, water transport rates may be growth limiting. The transport rate drops very rapidly with a decrease in water content, water uptake by horticultural crops is very high and the rooting volume is usually small (Raviv et al., 1999). Growth may also be limited by low water content and transport rates in the rhizosphere, while the overall water content still seems acceptable (Caron and Nkongolo, 1999; Caron et al., 2001).

7.3.11 OXYGEN DIFFUSION

7.3.11.1 Definition and Units

Oxygen diffusion is the mass of oxygen passing through a unit area in time ($g\,m^{-2}\,s^{-1}$) in a growing medium.

7.3.11.2 General Principle of Determination

The oxygen transport by diffusion through air is characterized by Eq. (5).

$$Q = D_h \frac{dC}{dx} \quad (5)$$

where

Q flux ($g\,m^{-2}\,s^{-1}$)
D_h unsaturated oxygen diffusivity ($m^2\,s^{-1}$)
dC concentration difference ($g\,m^{-3}$)
dx distance (m).

According to this equation, it is possible to calculate or model oxygen transport rates for any concentration difference over any distance in a given growing medium. The real difficulty is in finding D_h. D_h may be found with a series of formulas with a structure as in Eq. (6).

$$D_h = D_o (\text{AFP}^a \text{TPS}^b) \quad (6)$$

where

D_o oxygen diffusivity in air
AFP   air-filled pores (V/V)
TPS   total pore space (V/V)
a, b constants.

The only variable is the air-filled pore space. The influence of AFP on D_h is extremely large as indicated by the exponential function (Wever et al., 2001). The measurement of either D_h or D_o requires careful pre-treatment of the samples and the apparatus used (Klute, 1986). It is usually done by measuring the oxygen concentration with an electrode in an open-top chamber initially filled with 0 per cent oxygen. Oxygen can only enter the chamber by diffusion through the substrate which is fixed on top of the chamber.

7.3.11.3 Special Cases

A basic assumption is that diffusion through air-filled space is the main transport mechanism for oxygen to the roots. In some cases, mass transport of oxygen in either air or water will be larger than diffusion. This may be the case when there is a temperature difference in the substrate or when water is frequently or continuously moving as in nutrient flow techniques or aeroponics.

An alternative technique for these measurements is the use of air-tight containers around the roots of a growing plant (Holtman et al., 2005). In such systems with substrate, an oxygen gradient from top to bottom arises which may be used as input for improving models (Blok and Cassamassimo, 2001). The measurement of oxygen in air or water has become much easier with the introduction of fibre optic methods (Blok and Gérard, 2001). The introduction of the sensors in the material still requires great care and a control procedure to be sure that the results are reliable, that is no air leakage is introduced by the measurement. Another approach is the installation of air chambers with local contact to the atmosphere in the growing medium examined (Wever et al., 2001). The oxygen content in the air chambers can be measured either by entering the sensor through a suitable septum or by using an air-tight syringe to take samples for gas chromatography. Thus both oxygen and carbon dioxide may be measured simultaneously, showing a typical rapid decrease respectively increase with depth (Fig. 7.6).

7.3.11.4 Common Values for Media

Actual measurements of diffusivity against air-filled pores are not abundant (Wever et al., 2001). Calculations based on water content measurements prevail (Caron et al., 2002). The results show an exponential decrease of D_h with AFP (Fig. 7.7). An important consequence of the exponential decrease of D_h with AFP is that oxygen diffusion rates in all substrates with interrelated pores may, for lower air-filled porosities, that is under 40 per cent V/V, be simplified to one relation with air-filled porosity without much extra error in the air-filled pore volume found (Wever et al., 2001).

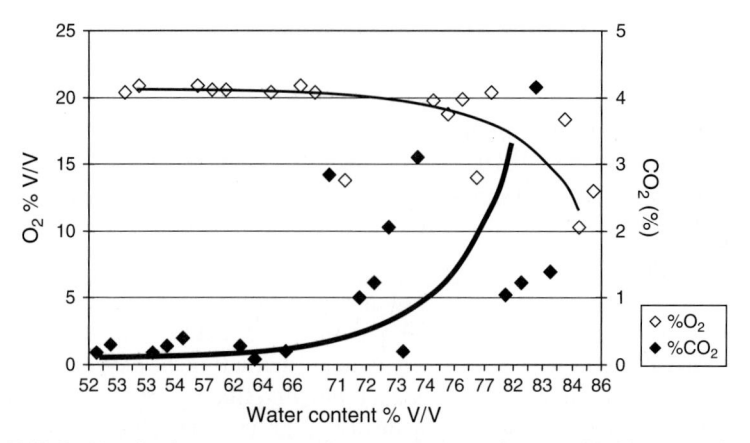

FIGURE 7.6 Very local measurements of oxygen content and carbon dioxide content in the air of pores in stone wool against the local water-filled pore volume in per cent V/V (after Baas, 2001).

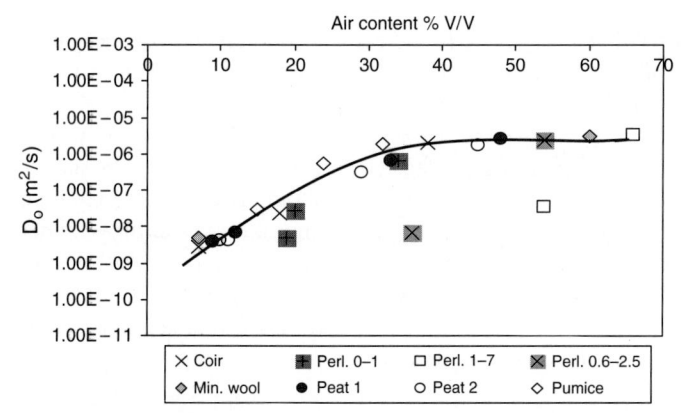

FIGURE 7.7 Measurements of the diffusivity of oxygen in various growing media against the air-filled pore volume in per cent V/V (after Wever et al., 2001).

For horticultural evaluations, flux calculations must be adapted to the dimensions and properties of the system under evaluation. For example, plant containers are open at the top and parts of the bottom, but impermeable at the sides. Furthermore, the shape has to be included to be able to calculate the minimum pathway of oxygen to a particular point, given the changing water content and the oxygen consumption along the path. Based on flux calculations for diffusion only, and with container dimensions of less than 20 cm for both, width and height, oxygen levels in horticultural substrate systems may drop way below 20 per cent in layers with less than 30 per cent V/V air-filled pores (Wever et al., 2001).

7.3.11.5 Influence on Growth

Plant oxygen use may be as high as $0.2\,\text{mg}\,\text{g}^{-1}$ (FW-roots) h^{-1} (Blok and Gérard, 2001). At 20°C, water contains a maximum oxygen content of $8\,\text{mg}\,\text{L}^{-1}$, which allows the conclusion that oxygen supply from the ambient atmosphere is essential. Anoxia may be expected in layers in horticultural substrate systems with less than 10 per cent V/V air-filled pores. The exception in Fig. 7.7 is perlite, which had to be corrected for the amount of closed not interrelated pores before it fitted the common line.

7.3.12 PENETRABILITY

7.3.12.1 Definition and Units

Penetrability is the resistance to rooting (kPa). It can be measured with a cone penetrometer (Fig. 7.8).

7.3.12.2 General Principle of Determination

Penetrometers are devices to measure the force needed to push a metal rod of known diameter into a growing medium. They may be hand-operated and portable, or machine-driven and stationary.

Universal materials testing machine

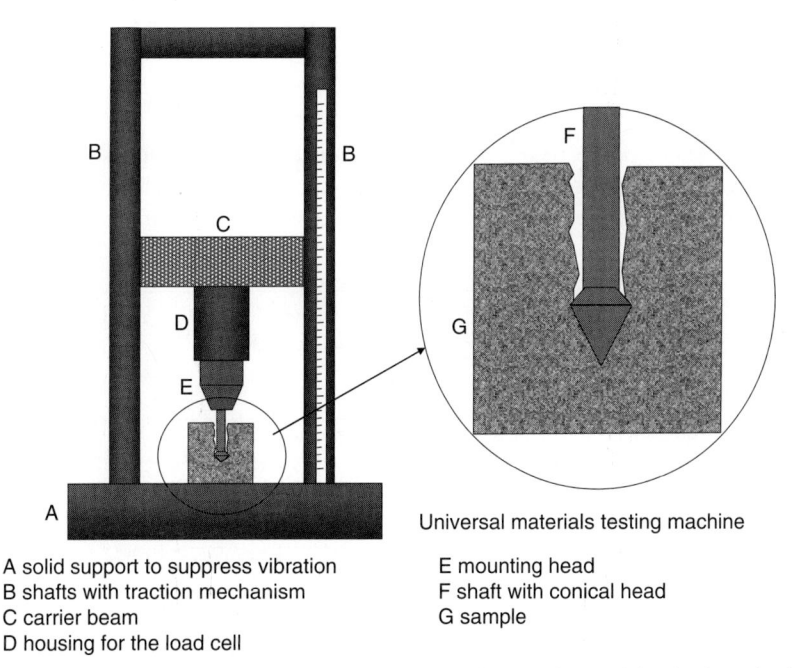

A solid support to suppress vibration
B shafts with traction mechanism
C carrier beam
D housing for the load cell
E mounting head
F shaft with conical head
G sample

FIGURE 7.8 Schematic representation of a universal testing machine used for the determination of mechanical properties as penetrability, hardness and elasticity.

The cone penetrometer is supposed to represent a root growing through the material. The force needed to insert the cone in the material is measured. Ideally, the penetrometer reading is then correlated to measured pressures of growing roots, but more often it is used alone. Before the reading becomes stable, some compression of the material takes place. A larger cone angle and a larger cone diameter increase the distance over which compression takes place before a stable reading is found (Bullens, 2001). To avoid the influence of friction along the shaft, it is necessary that the shaft driving the conical head is smaller in diameter than the head (Fig. 7.10).

7.3.12.3 Special Cases

Fibrous materials like stone wool and coarse peat may be compacted over several centimetres before a stable reading is found unless an adapted cone is used with a maximum diameter of 2 mm and a cone angle of 30°. Rigid materials like stone wool should be measured in three directions as the fibres may be deposited in planes, and roots will prefer certain directions. Rigid materials like stone wool and pressed peat/soil plugs and cubes may also show a marked compaction profile. Material with coarse particles may occasionally show very high readings which are better reported separately. This type of high reading is caused by the probe trying to press such a particle down, whereas roots will grow around it.

7.3.12.4 Common Values for Media

The ease of rooting is determined by the material density and the way the particles are inter-connected. Typical penetrometer values for growing media are 2–20 times lower than the results found for soil (approximately 1200 kPa).

7.3.12.5 Influence on Growth

Root and shoot development react to differences in rooting resistance over the whole range of values found (Fig. 7.9). In comparative research, threshold values for inhibition of fresh weight growth of roots and shoots were different for different

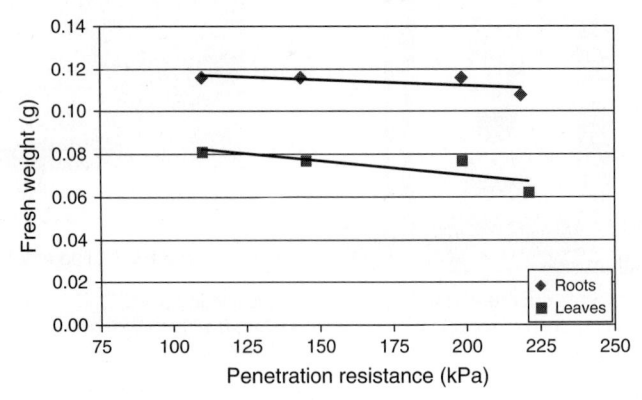

FIGURE 7.9 Decreasing growth with increasing penetration resistance of Irish white peat < 15 mm (after Bullens, 2001).

materials, for example 250 kPa for stone wool and 450 kPa for perlite (Bullens, 2001). This seems to indicate that the method still needs to be improved.

7.3.13 HARDNESS, STICKINESS

7.3.13.1 Definition and Units

Hardness is the maximum resistance value (kPa) measured when pushing a defined shape through a body of rooting medium, just before the body gives way and splits into fragments. Hardness is used to characterize the handling forces that some pre-formed products as plugs can withstand.

Stickiness is the maximum negative force (kPa) measured when lifting a defined surface from a compressed body of rooting medium. Stickiness characterizes the cohesion between particles, a quality which determines whether the material can be used to be pressed into coherent substrate forms like plugs or blocks, mainly for transplanting.

7.3.13.2 General Principle of Determination

Both hardness and stickiness can be measured with a universal material testing machine (Zwick, Instron, many others). These machines measure pressure against replacement and use load cells and displacement meters.

For a specific hardness test, a cylindrical plunger with an accurately measured maximum diameter and a conical head with an angle of 45° are pressed into the material until 75 per cent of the original height of the sample is reached. As with the measurement of penetration resistance, it is important that the shaft driving the conical head is of a markedly smaller diameter than the diameter of the conical head, to be sure that there is no influence of friction of the material along the shaft (Fig. 7.10.). Appropriate values for the material testing machine are a downwards speed of 50 mm min^{-1} for the plunger and a 100 N load cell to measure the force and calculate the pressure in kPa.

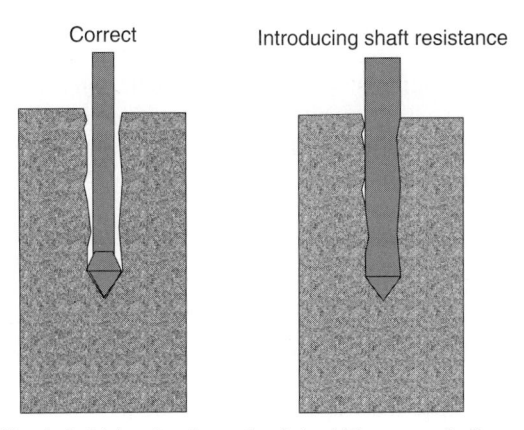

FIGURE 7.10 The shaft driving the plunger head should have a markedly smaller diameter than the plunger head.

The maximum peak load registered during insertion is used to characterize hardness (Wever and Eymar, 1999).

The energy necessary to withdraw a plunger after insertion to 75 per cent of the original height has been used to characterize stickiness or adhesiveness (Wever and Eymar, 1999). For both hardness and stickiness, one has to be aware of the extreme sensitivity of the properties to moisture content. It is therefore necessary to report results together with the moisture contents used. Hardness can be measured at saturation and for oven dry material, but stickiness can only be measured on wet material, for example, subjected to a suction force of -10 cm water column, after saturation.

7.3.13.3 Special Cases

The measurement strongly resembles the penetration resistance measurement. Again fibrous materials like stone wool and coarse peat may be compacted over several centimetres before a stable reading is found, the reason to use an adapted cone with a cone angle of 30°. Rigid materials like stone wool and peat squares should be measured in three directions as the fibres may be deposited differently in some directions and roots will prefer particular directions.

Many other mechanical properties have been measured and procedures have been suggested. Among these are hardness and elasticity (Wever and Van Leeuwen, 1995), cohesiveness (Wever and Eymar, 1999), brittleness, stickiness, and firmness. Many more mechanical properties are known from descriptive work on building materials and food materials. There is, however, no agreement on which properties are essential for substrate evaluation and which methods are appropriate. There is certainly room for a comparative study on this subject!

7.3.13.4 Common Values for Media

Hardness of pressed materials may vary from 400 to 2500 kPa.

7.3.13.5 Influence on Growth

Data from the **penetrability** readings indicate growth reduction above 250–450 kPa. Hardness is, however, often measured to characterize or compare structures like peat fractions which are used to bring stability in an otherwise easily penetrable matrix, or to characterize the suitability for automated handling of blocks and plugs. The readings have therefore usually to be set against a standard acceptable for a specific handling system.

7.4 CHEMICAL ANALYSIS

One single extraction method is not sufficient in substrate analysis due not only to the enormous diversity of materials, but also to the large number of elements and compounds involved. Every extraction method treated below is designed for a specific group of elements/compounds or a particular fraction of such a group. In the first place, the methods are aimed at investigating nutrient element contents or whether the substrate contains high concentrations of elements which are damaging to the plant, for example sodium. Secondly, substrates must not contain high concentrations of elements which may be dangerous for food crops, such as heavy metals. One of the relevant factors is how rapidly or easily the element is released from the substrate. There are, for example, 'mobile compounds' which are water-soluble, so that after analysis an impression may be gained of the readily available amount in the substrate. In addition to this group, there are 'semi-mobile' compounds, which are soluble in an aqueous solution of ammonium acetate or a complex-forming reagent (e.g. DTPA). Mostly, the ammonium acetate extraction is applied for the determination of cations and the DTPA for the determination of trace elements. In addition, there is the 'overall determination' which uses concentrated acids to provide pseudo-total concentrations of inorganic components. With this method, both the organic and the inorganic matrices are accessed. In addition, several other extraction methods are being used for specific aims such as the determination of a single element only.

In the instructions and descriptions given in this chapter, the term 'substrate solution' is used similarly to the term 'soil solution' which is used in soil science. Substrate solution in growing media should represent the solution present in a medium under growing conditions. Thus, the concentrations of dissolved substances in the substrate solution are closely related to their actual concentration in the rhizosphere during most of the growing period. Under growing conditions the moisture content is certainly not constant; consequently, a definition for moisture condition has to be chosen which is achieved artificially, but is sufficiently close to the moisture condition under growing conditions. Fluctuations occurring due to irregularities in the moisture content are obviously not represented in this definition, so that the definition relates to the circumstances in the substrate briefly after irrigation and the termination of free drainage. This moisture condition is called 'container capacity'. However, this is close to saturation. In substrate-grown crops, water is often supplied frequently, so that substantial fluctuations in the moisture condition during the growing period are limited. Under growing conditions in general terms, two situations occur: (a) substrates of which a low negative pressure height is built up, for example peaty materials and coir and (b) substrates which are not suitable for the build-up of any negative pressure height in the matrix and are continuously kept more or less saturated. Examples are stone wool, expanded clay granules and artificial foam. So, in principle, two moisture conditions are defined: (a) for the growing media under a moisture content at a pressure head of -32 cm (pF 1.5) and (b) for the growing media under a moisture content at nearly saturation, called 'container capacity'. The pressure head at which the substrate solution is being defined must be indicated individually for every growing medium and extraction method executed.

For substrate analysis in glasshouse horticulture, aqueous extracts are mostly used. The reason for this is the fact that in substrate analysis for glasshouse horticulture great value is attached to the composition of the substrate solution. The most important parameter of the substrate solution is the osmotic pressure, and estimates of it can only be obtained with the help of aqueous extracts. The osmotic pressure of the substrate solution is an important factor in growth regulation in glasshouse crops. High values cause growth inhibition, while low values may lead to insufficient product quality and particularly under light-deficient conditions to vegetative crop development. Osmotic pressure in substrate solution can be built up by letting salts accumulate or by excessive application of soluble fertilizers. In this way, the osmotic pressure, the salt content and the nutrition are intertwined. It is therefore important that the various parameters for salt and nutrition are determined in the same aqueous extract, in order to be able to regulate them and their interrelations.

The best method to evaluate the ionic composition of the substrate solution is by analysing the substrate solution extracted immediately from the growing medium. In stone wool-grown crops, where the substrate solution can be obtained by means of a simple suction device, this is a standard procedure. With many other growing media, this is more difficult to realize, because the water is bound much stronger to the matrix, and can then only be obtained by way of pressing out or displacement. Such methods are not suitable for routine analysis. For routine analysis in such cases, a known amount of extra water is added to the sample to prepare an extract more easily. On the basis of the concentrations of salt and nutrients found in the extract, the concentrations of salt and nutrient elements in the substrate solution are then calculated. As a rule, the less water is added to the sample, the more accurate is the calculation, since in that case disturbances due to dilution and valence effects are reduced to the minimum.

For peaty substrates, coir dust and comparable media, the 1:1.5 volume extract is used in Denmark, South Africa, Australia and The Netherlands, a method in which the volume is measured out in a standardized way and the moisture content is standardized at a pressure height of -32 cm (pF 1.5). The substrate solution is diluted about three times.

The 1:1.5 volume method, however, is not suitable for some kinds of other growing media, because (a) the volume can not be measured as in the 1:1.5 volume method (substrates are rigid or too compressible), or (b) it is too laborious to estimate the moisture content at the pressure height of -32 cm, or (c) in practice these substrates are used in practical conditions at moisture contents nearby saturation. So, other methods have to be applied. To this purpose, an extraction ratio of 1:3 (medium:water, v:v) has been developed. This ratio was chosen, because this extraction results in the same dilution of the substrate solution as the 1:1.5 extract. For all substrates in this way, an unambiguous interpretation of analytical data can be obtained.

The 1:5 volume method (CEN 13652, 2001) was developed by CEN, because this was necessary for universal application.

7.4.1 WATER-SOLUBLE ELEMENTS

7.4.1.1 The 1:1.5 Volume Method

The 1:1.5 volume extract is meant to be able to estimate the composition of the substrate solution of peaty substrates and comparable media. The substrate sample is mixed manually in such a way that its natural structure is not disturbed. Water is added

7.4 CHEMICAL ANALYSIS

to the 'fresh' ('as received') non-dried material until water can be pinched though one's fingers during gentle pressing of a ball-shaped sample in one's hand. In this way, the water content corresponds reasonably well with a pressure head of −32 cm. Different operators should check this regularly by using a standard sample and determine the water content after adjustment to the desired pressure head. If substrates are wetter than a pressure head of −32 cm, they have to be air-dried a little in advance and the procedure commences with the 'visual' wetting. The material is put into two rings placed one upon the other, the lower having a minimum content of 60 ml at a height of 5 cm. The material is pressed for 10 s with a pressure of 10 kPa. The upper ring is removed by cutting the substrate between the two rings carefully. The content of the lower ring is mixed with 1.5 times its volume of demineralized water. Subsequently, the suspension is shaken firmly for about 15 min, and after a waiting period of 1 h the suspension is filtered using a filter paper which should not contaminate the extract with the elements to be determined. For some elements (mainly trace elements), filtration is done with a 0.45 μm ceramic filter. The method can be applied for all substrates which consist of at least 50 per cent peat or other organic materials such as coir, bark and wood fibre. EC, NH_4, Na, K, Ca, Mg, NO_3, Cl, SO_4, HCO_3, P, Mn, Fe, Zn, B, Cu and Mo are measured in the extract. Element contents are expressed per volume of extract. To convert the contents per volume of extract to contents per volume of substrate, Eq. (7) should be used.

$$C_s = A C_e (1.5 + \theta) \qquad (7)$$

where

C_s element content per volume unit of substrate, mg L^{-1} substrate
C_e element content per volume unit of extract, mmol L^{-1} extract
θ water content on volume basis at pressure head −32 cm, m^3 m^{-3}
A atomic weight, g mol^{-1}.

Typically, peaty substrates have water contents at pressure head of −32 cm of about 0.4 m^3 m^{-3}, which means that (for simplicity) the element contents per volume unit of extract should be multiplied with 1.9 to reach the element content per volume unit of substrate. The pH is measured in the slurry, the semi-solid settled matter below the supernatant after 6 h of waiting time.

For coarse growing media, it is recommended to use a wider ring for measuring out the substrate. As a general rule, it is recommended to use a ring with a diameter of 2.5 times bigger than the substrate's biggest particles. The height of the lower ring, however, remains as 5 cm. The method is described by Sonneveld et al. (1974) and Australian Standard (2003a,b).

Element contents are expressed in mmol L^{-1} per volume unit of extract or could be converted to mg per litre of fresh substrate (see Eq. [7]). Expressing results in moles rather than weight is more rational from both scientific and practical points of view. EC and pH are expressed on extract-basis.

7.4.1.2 The 1:3 Substrate Solution Method

The aim is to determine the composition of the substrate solution for other than peaty substrates which can not easily be hand-pressed. The dilution of the substrate solution is comparable with the 1:1.5 method. In this way, the element contents of the extracts of both methods can be used in the same recommendation system. The moisture content of the fresh substrate is determined. A ring with a diameter of about 10 cm and 5 cm height is filled loosely with fresh substrate with a pressure of 1 kPa on top. Subsequently, it is saturated for 2.5 h and then leached out on a grid for 2.5 h. To a newly prepared volume of substrate, subsequently add sufficient demineralized water, so that it contains three times the amount of moisture at container capacity after leaching out. The measured volume of the substrate chosen should depend on the coarseness of the substrate. For fine substrates with a particle size of < 20 mm, a minimum volume of 100 ml is sufficient, while for substrates with a particle size of > 20 mm, a volume of 400 ml is recommended. The suspension is rotated (using an end-over-end shaker) for 2 h and subsequently remains standing for one night. The rotation rate must be low enough not to damage the texture of the substrate. The suspension is filtered through a filter that does not contaminate the extract with elements that must be determined. For some elements (especially trace elements) a 0.45 μm filter is used. The method is designed for all granular substrates for which the 1:1.5 method is impractical, such as expanded clay granules, perlite, pumice stone, granulated stone wool, wood fibre, rice husk and bark. The following determinations are carried out on the extracts: EC, pH, NH_4, Na, K, Ca, Mg, NO_3, Cl, SO_4, HCO_3, P, F, Mn, Fe, Zn, B, Cu and Mo. Element contents are expressed per volume unit of extract. The method is described by De Kreij et al., 1995 and De Kreij et al., 2001.

7.4.1.3 The 1:5 Method

The 1:5 volume method has been developed by CEN in order to extract growing media and soil improvers with a wide variety of characteristics in an unequivocal way with demineralized water (CEN 13040, 2006; CEN 13652, 2001). The substrate is first measured into a 1 L cylinder having a height of 10 cm and a loose collar extension of 5 cm. The cylinder and collar are filled via a screen which functions as a fall-breaking device (to minimize operator sensitivity), and is subsequently pressed with a pressure of 0.9 kPa. After 3 min the collar is removed and the cylinder is levelled off using a straight edge. The weight of the sample is determined – thereby providing the 'laboratory compacted' bulk density of the sample in $g\,L^{-1}$. For the preparation of the extract, a certain volume necessary for preparing the extract is determined through weighing, based on the calculated bulk density. In materials in which the particles are smaller than 20 mm, a weight equivalent of 60 ml is weighed out. Of the other materials with particles smaller than 40 mm, a weight equivalent of 250 ml is weighed out. The sample is mixed with five times the volume in demineralized water. After shaking for 60 min, the suspension is filtered through filter paper which does not contaminate the extract with the elements to be determined. For some trace elements, filtering is done with a 0.45 μm filter. The method is designed for all granular growing media. In the extracts the following determinations are carried out: EC, NH_4, Na, K, Ca, Mg,

NO_3, Cl, SO_4, HCO_3, P, F, Mn, Fe, Zn, B and Cu. Element contents are expressed per volume unit of substrate. EC is expressed on the basis of the extract in $mS\,m^{-1}$.

The method poses problems when measuring out very coarse and very wet, cohesive materials. Materials which compact very easily also give problems when the weight is mounted. Due to the high dilution of the substrate solution, it gives rise to valance/dilution effects, some elements reach the detection limits of the instruments, and substrates containing gypsum show high EC values due to gypsum dissolution, which in insoluble form do not pose risk for crop growth.

Proficiency tests have been executed by De Kreij and Wever, 2005.

7.4.1.4 The Pour Through Method

The Pour Through (PT) method was developed in the United States for on-farm measurements by Yeager et al., (1983), and is graphically described in http://www2.ncsu.edu:8010/unity/lockers/project/hortsublab/pourthru/pts_demo/index.htm. It needs low-cost equipment and instruction of people is easy. The following procedure has to be followed. The crop is irrigated thoroughly. After approximately 1 h for equilibration, a saucer is placed under each container to be tested. A sufficient number of containers have to be sampled to ensure reliability. A sufficient amount of distilled water is poured evenly on top of each pot to yield ~50 ml leachate out of the bottom of the container. The demineralized water will replace the substrate solution during drainage of the container (via the 'piston effect'). The container has to drain completely in about 5 min. The solution on the saucers is poured into beakers and its EC and pH values are measured. Some variation due to the amount of water poured onto the container can lead to variation of the data. When channeling occurs the water does not displace the substrate solution.

The results obtained using the PT method must be carefully interpreted, and calibration curves were developed for comparing pour-through and saturated media extract nutrient values (Cavins et al., 2004). Yet, pH data derived from the PT methods differ significantly from those of other methods and they require careful interpretation (Handreck, 1994; Rippy and Nelson, 2005).

7.4.2 EXCHANGEABLE, SEMI- AND NON-WATER SOLUBLE ELEMENTS

7.4.2.1 The Calcium Chloride/DTPA Method

One of the characteristics of calcium chloride as an extraction liquid is that it offers a certain salt concentration during the extraction. This results in the release of weakly adsorbed reserves, mostly concerning main elements, N, P, K and Mg. This corresponds to what happens in reality when fertilizers are supplied. The DTPA is a strong complex-forming agent, binding the trace elements Fe, Mn, Cu and Zn. The combination of calcium chloride and DTPA is therefore meant for both main and trace elements. Moreover, the activity of the roots is simulated to some extent. The method is described in CEN 13651 (2001). The extraction liquid is 0.01 M $CaCl_2$/0.002 M DTPA with an average pH of 2.6. NH_4, NO_3, P, SO_4, K, Na, Mg, Fe, Cu, Zn, Mo, Mn, Pb, Cd and B are determined in the extract. Contents are converted and presented in mass element (g or mg) per unit volume of the substrate. Determination of B may give rise to problems, depending on the detection method used, due to the strong colouring of the extract. The elements are expressed in mg L^{-1} substrate.

7.4.2.2 The 0.1 M Barium Chloride 1:1.5 Volume Method

The extraction with barium chloride is meant to determine the amount of exchangeable cations, displaced from adhesion sites by the barium ion. Since it is not naturally present in growing media, barium does not give rise to disturbances of the measurement of individual cations. The results of the determination give an estimate of the total of cations 'available' to the plant. The sample is wetted to a state equivalent to that arising from a pressure head of −32 cm and a volume of, for example, 100 ml is measured (see 1:1.5 method).

This volume of substrate is mixed with 1.5 times the volume of a 0.1 M barium chloride solution, after which the suspension is shaken intensively for 15 min. After a waiting period of 1 h, the suspension is filtered through a paper filter which gives no contamination of the extract with the elements to be determined. For some elements (particularly trace elements) filtration is done with a 0.45 μm filter. The method is being applied for coir pith, coir fibre and coir chips. In the extract pH, NH_4, K, Na, Ca, Mg, Fe, Mn, Zn and Cu are measured (Verhagen, 1999).

7.4.2.3 Active Manganese Method

Mn may occur in various forms in a growing medium: (a) water soluble, (b) adsorbed or exchangeable, (c) easily reducible, and (d) inert (Leeper, 1947). The aim of the determination of 'active' manganese is to investigate how much Mn can theoretically be released. Mn-active is the sum of the three Mn forms (a), (b) and (c) mentioned above. On the basis of these, a prediction can be made of the possible toxicity of Mn (Sonneveld, 1979).

To a weighed amount of air-dry medium, 20 times as much (by weight) hydroquinone (1.4-benzenediol) extractant containing 1 M ammonium acetate solution with a nominal pH of 7 are added. After 1 h shaking, filter and measure the Mn in the filtrate. The determination is recommended for materials which are expected to contain significant quantities of Mn (such as some barks). Mn is expressed in mg per kilogram dry material.

7.4.3 THE pH IN LOOSE MEDIA

The method is described in CEN 13037 (1999) and the volume is determined according to CEN 13040 (2006). The pH can also be measured in the above-mentioned water extraction methods.

According to CEN 13040, a weighed amount of growing medium, corresponding to a volume of 60 ml or 250 ml is put in a plastic bottle, and the necessary amount of water (300 ml or 1250 ml) is added. It is shaken for 1 h in a horizontal shaking machine. Within 1 h after shaking the pH is measured. A reading is taken if the value changes no more than 0.1 unit during 15 s. The method is suitable for all loose, non-preformed media. With materials that consist for at least 80 per cent of the volume of particles smaller than 20 mm, a weight equivalent of 60 ml is weighed out. For all other materials, a weight equivalent of 250 ml should be taken. Note also that pH, EC and nutrients can all be done on the same extract.

The same procedure is being used with 1 M KCl or 0.01 M $CaCl_2$ as extractant.

7.4.4 NITROGEN IMMOBILIZATION

Some organic growing media can immobilize a substantial amount of nitrogen during the course of the cropping and especially during the first weeks of use as a growing medium or as a component in potting soils. For these media, it is important to estimate the potential for nitrogen immobilization by the medium prior to their use. For this purpose, a method has been developed in which the degree of immobilization is determined and expressed in the so-called 'nitrogen fixation index' (NFI), modified after the nitrogen draw-down index (NDI). Methods based on Handreck (1993) are suitable for all organic materials. The NFI indicates the percentage of the nitrogen present in a substrate on day 0 and after 4 and 20 days, respectively.

To a material, 0.1 mmol NO_3 per gram dry matter is added. This content is equal to that in an average potting soil in which 10 mmol l^{-1} NO_3 occurs in the substrate solution. After 4 and 20 days, the NO_3 content in the solution is determined and the NFI is calculated. Only limited experience has been gained with this method, so that no estimate can be given as to the accuracy which can be obtained with the method.

7.4.5 CALCIUM CARBONATE CONTENT

The aim is to determine the amount of carbonates. The content of the carbonates is expressed as calcium carbonate. A certain weight of air-dry medium, for example 3 g, is placed into a closed container and an excess of 4 M HCl is added, so that all carbonates in the sample react to produce CO_2. On the basis of the increase of the amount of gas above the sample, the amount of $CaCO_3$ can be calculated. The method is meant for all growing media (NEN 5751, 1989).

The sum of the carbonates is expressed as calcium carbonate equivalent; the mode of expression is in the fraction of $CaCO_3$ on weight basis. There may be other substances (for instance organic aromatic acids) which may give off CO_2, or another gas may be formed (e.g. H_2S and/or SO_2). $FeCO_3$ does not completely dissolve in hydrochloric acid, which results in a faulty representation of the carbonate content. In most situations, this does not lead to significant practical errors in determining $CaCO_3$ content.

7.5 BIOLOGICAL ANALYSIS

The determination of the following characteristics will be discussed: stability/break-down (respiration), the presence of weeds and phytotoxicity (growing tests).

Different methods will be applicable depending on the nature of the material, such as its mineral and organic content, whether it is pre-formed material or not and its EC.

7.5.1 STABILITY (AND RATE OF BIODEGRADATION)

7.5.1.1 Definition and Units

There are different ways to look at stability (Page, 1982). There is physical, chemical and biological stability. Physical stability has to do with mechanical strength, chemical stability with the breakdown of certain carbohydrate structures and biological stability with microbiological activity. A common approach is to define stability as the microbial degradation rate of the organic matter under aerobic conditions (Haug, 1993). A lower degradation rate corresponds to a higher degree of stability. The microbial degradation of the organic matter ($C_aH_bO_cN_d$) can be represented by Eq. (8).

$$C_aH_bO_cN_d + xO_2 + yNH_4 \to aCO_2 + zH_2O\,\Delta H + X \qquad (8)$$

where

ΔH is the heat production in kJ
X represents biomass in gr L^{-1}
O_2 and CO_2 may be measured in mg g^{-1} h^{-1}.

Compost stability and maturity are terms which indicate the degree of organic matter decomposition and potential of phytotoxicity caused by insufficient composting (see Chap. 11 and Wu and Ma, 2001). When unstable materials are added to growing media, they may have a negative impact on plant growth due to reduced oxygen content and/or available nitrogen and/or the presence of phytotoxic compounds (Brodie et al., 1994; Keeling et al., 1994; He et al., 1995). Phytotoxic substances of organic origin can be broken down by microorganisms during the composting process into non-toxic organic compounds. Because of the mentioned concerns, extensive research has been conducted to study the composting process and to develop methods to evaluate the stability/maturity of compost prior to its agricultural use (Jimenez and Garcia, 1992; Mathur et al., 1993; Iannotti et al., 1994; Hue and Liu, 1995). However, at present, there is no well-accepted, official definition of compost stability/maturity within the compost industry or research community. Correspondingly, there is no universally accepted method of evaluating compost stability/maturity. However, there is an European standard in preparation under the banner of CEN after an European project called 'HORIZONTAL' failed (Cooper, 2004).

7.5.1.2 General Principle

A sample is incubated at a certain moisture content and a certain temperature with continuous replacement of carbon dioxide-free air. Carbon dioxide or oxygen use is measured. An interesting overview of the available tests is given by Cooper (2004).

7.5.1.3 Special Cases

Many tests are available, some simple to operate, others requiring quite sophisticated and expensive apparatuses. Some workers hold views that more than one test may be required.

7.5.2 POTENTIAL BIODEGRADABILITY

Knowledge of the potential biodegradability is a valuable tool, especially for material intended for landfill. Godley (2003) stated that organic compounds can be classified as either readily degradable, moderately degradable, poorly degradable or recalcitrant depending on how easily they are decomposed. The question to be answered is how to determine how readily degradable a material will be in the environment is which it will be placed. Sloot et al. (2003) has suggested that water-soluble organic matter is a simple and rapid procedure giving valuable information on the potential biodegradability of the material under test. The determination of cellulose, hemi-cellulose and lignin gives further information. Most animal nutrition laboratories undertake such a test (Van Soest, 1963; Van Soest and Wine, 1967). They may be known as acid detergent fibre – hemicellulose and acid detergent fibre – neutral detergent fibre (ADF-NDF). A method of carbohydrate fractionation as a measure of estimating the biological stability could be considered. Some laboratories are looking at the possibility of using NIR (near infra-red reflectance) to determine the degree of mineralization of wastes. Preliminary findings seem to indicate that the technique has potential but could well be waste specific.

7.5.3 HEAT EVOLUTION (DEWAR TEST)

Heat evolved during the composting process can be measured and used as an indication of microbiological activity. In the 'Dewar test', a sample is placed in a heat-retaining flask (Dewar or similar) and the heat evolved is measured over a period of up to 10 days either by a thermocouple or by a thermometer. The test is limited in the sense that it best distinguishes very immature from mature compost; it can not distinguish moderate maturity from high maturity which may be important for potting mix use of composts (Brinton, 2000). A more sensitive test is based on the very high sensitivity of modern microcalorimeters (Laor et al., 2004).

7.5.4 SOLVITA TEST™

Solvita Test is a commercially available field test kit. It employs gel-colorimetric technology in which respiration gases from composts (CO_2 and NH_3) are captured and the amounts accurately indicated by the colour change on test sticks and calibrated to a wide range of known conditions. This easy to use system can, under certain circumstances, give an interesting indication of the stability. However, Adani and Scaglia (2003) did not obtain good result by comparing a dynamic test (rate of biodegradation) and Solvita, due to the impossibility to determine a 'a priori' lag phase for Solvita.

7.5.5 RESPIRATION RATE BY CO_2 PRODUCTION

In this method the rate of biodegradation is estimated by measuring the rate of CO_2 production under specific circumstances (Cooper, 2004). The method was proposed as an European standard but the work was not completed.

7.5.6 RESPIRATION RATE BY O_2 CONSUMPTION (THE POTENTIAL STANDARD METHOD)

In this method the rate of biodegradation is estimated by measuring the rate of O_2 consumption under specific circumstances (Veeken et al., 2003). This method, formerly known as the OxiTop®, uses a standardized time of oxidation by an added surplus of microorganisms in a water-dispersed organic growing medium at standard time, temperature and pH and with prevention of nitrification. In this method also the N-mineralization and N-fixation can be measured.

A nitrification inhibitor can be added to products with a high ammonia level. The conversion of ammonia to nitrate also consumes oxygen and could be wrongly interpreted as instability. In these cases nitrification must be blocked.

7.5.6.1 Common Values for Media

Common values for respiration rates and N-mineralization rates of representative substrates are shown in Tables 7.3 and 7.4, respectively.

TABLE 7.3 Respiration Rates of Various Types of Organic Matter Under Standardized Conditions

Type of organic matter	O_2 respiration rate (mmol kg^{-1} h^{-1})
Biowaste compost	17.0 +/− 7.8
Green waste compost	10.4 +/− 3.3
Finished mushroom substrate	28.1 +/− 3.8
Peat	2.1 +/− 0.9
Bark and composted bark	13.4 +/− 9.6

Source: After Veeken et al. (2003).

TABLE 7.4 N-mineralization Determined by Standardized Oxitop Measurements

Type of organic matter	N-mineralisation (mg kg^{-1} h^{-1})
Biowaste compost	6.6 +/− 3.4
Green waste compost	5.1 +/− 2.3
Finished mushroom substrate	38.8 +/− 23.5
Peat	5.6 +/− 3.3
Bark and composted bark	−2.9 +/− 2.5

Source: After Veeken et al. (2003).

7.5.6.2 Influence of the Property on Growth

Some linear regressions were performed to find out if there is a relation between growth of a crop and stability measurements as performed in the laboratory (Silva, 2004). As materials for this research, two different composts made of three different kinds of green waste were used. Compost A: 70 per cent trimmings, 30 per cent grass, Compost B: 30 per cent trimmings, 30 per cent grass, 40 per cent leaves. They were composted and samples were taken 4 (A1 and B1), 7 (A2 and B2) and 14 days (A3 and B3) after the start of the composting process. Using these two composts (at 20 per cent and 40 per cent vol. with peat) the growth of kohlrabi (*Brassica oleracea*) was assessed. As not the mixtures but only the pure materials were measured for stability, some assumptions were made. It was assumed that the peat respiration was zero. This was also approximately found in the performed analyses with the HORIZONTAL method (method for rate of biodegradation, Cooper, 2004). For the 40 per cent addition the value as found for 100 per cent compost was used for the regression and for 20 per cent this value was divided by two. The found differences for kohlrabi could be explained to a large extent by nitrogen draw-down in the substrate. There is also a clear relation between respiration measured with the O_2 uptake method and the growth of kohlrabi (Fig. 7.11). The method as developed for HORIZONTAL in which the CO_2 release is measured gives a clear relation between the stability results and the growth of kohlrabi too. There is a good comparison between the stability methods (Table 7.5). It seems, however, that in the method according to O_2 uptake, the circumstances for respiration are better as higher values are found. The O_2 uptake is a more robust laboratory method. Criteria can be set up based on the results of O_2 uptake method. Some proposed values for biowaste and for greenwaste composts can be found in Table 7.6.

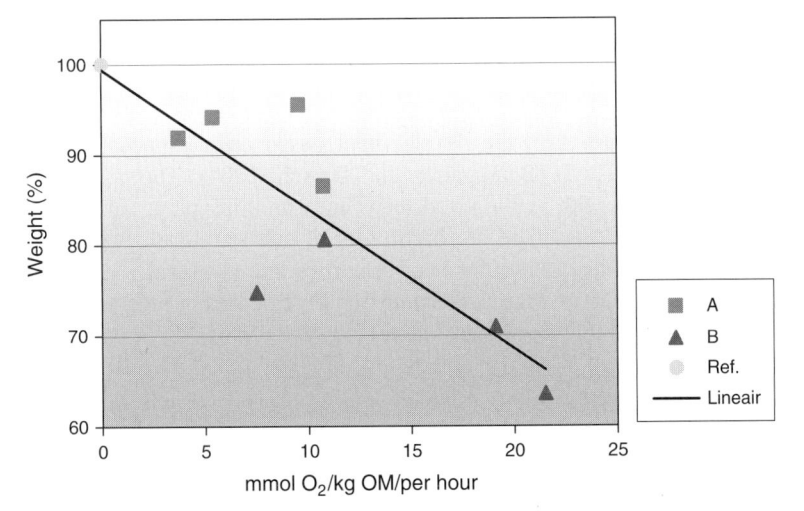

FIGURE 7.11 Relation between relative growth of kohlrabi and stability measurements using the Oxitop respiration method from compost A1, A3, B1 and B3 as in Table 7.6. $y = -1.55x + 99.4$, $R^2 = 0.74$.

TABLE 7.5 Results of the Stability Measurements

	Respiration rate by CO_2 production (dry weight)	Respiration rate by O_2 consumption (fresh weight)
	mg g^{-1} d^{-1}	mmol kg^{-1} h^{-1}
Peat	0.3	n.d.
Compost A1	7.4	19.1
Compost A2	5.1	n.d.
Compost A3	4.2	7.5
Compost B1	8.3	21.5
Compost B2	5.3	n.d.
Compost B3	5.1	10.8

n.d. – Not determined.

TABLE 7.6 Criteria Proposed for Oxygen Uptake by Biowaste and for Greenwaste Composts in mmol kg^{-1} h^{-1}, Based on Fresh Weight

Type of material	Biological oxygen demand
Very unstable (compost)	> 30
Unstable (compost)	15–30
Stable (compost)	5–15
Very stable (compost)	< 5

Source: After Veeken et al. (2003).

7.5.7 WEED TEST

7.5.7.1 Definition and Units

A European standard is in the development based on a proposal coming from the project HORIZONTAL (Baumgarten and Dersch, 2004). The contamination of horticultural substrates with viable weed seeds and sprouting plant parts (propagules) is of special importance as there are high costs of eradicating them in horticultural practice. With a weed test, the number of seedlings and propagules emerging under certain circumstances is assessed. The number of emerged weeds is given as the number per area – (number per square metre).

7.5.7.2 General Principle

To investigate whether a growing medium is infected with weeds, a substrate sample is fertilized and if necessary limed (to pH 5.5–6.0). The substrate is wetted thoroughly with clean water. In duplicate trays, cover a minimum of 0.25 m^2 with a 4-cm thick layer of substrate. Maintain a temperature of 18°–30°C and also a high humidity by covering the trays with plastic film. During the test the substrate must be kept moist, as under normal growing conditions. Care must be taken to ensure no secondary infection arises, for example from seeds blown in by the wind. The number of emerging weeds is monitored and recorded for four weeks.

7.5.7.3 Special Cases

Dilution of the tested material is necessary if its chemical properties, especially the salt content expressed as electrical conductivity, could inhibit the germination and the sprouting. In such cases, the substrate has to be diluted using suitable diluents. The same problem occurs if the water-holding capacity is low. In this case also dilution with another material must be considered. Suitable diluents can be materials with a low salt content, weed-free and a high water retention, for example peat or perlite.

Whether seeds reach germination depends on two factors: first, the dormancy must be broken, and secondly germinating conditions must be optimal. One of the consequences of this is that the results of a germination test, such as the current weed test, do not always indicate how many seeds may potentially reach germination. The physiological phenomenon of seeds dormancy may be broken partly by physical treatments, for example, by a low temperate stimulus under wet cool conditions at 4°C for three days. Next to this process of stratification, gibberellic acid treatment, KNO_3 treatment, compost washing and dilution with peat can improve germination. For instance, lettuce needs stratification as without a cool period the germination is poor. An interesting overview of the available weed tests is given by Baumgarten and Dersch (2004).

7.5.7.4 Common Values for Media

Depending on the source of the materials, different levels can be expected. Depending on the use, different levels may be acceptable. For peat substrates standards set for RHP certification are < 15 per square metre in general and < 4 per square metre for land weeds (Bos et al., 2002; Zevenhoven et al., 1997).

7.5.8 GROWTH TEST

7.5.8.1 Definition and Units

Phytotoxicity is defined as a delay of seed germination, inhibition of plant growth or any adverse effect on plants caused by specific substances (phytotoxins) or growing conditions (WRAP, 2002). In chemical research it is possible to determine a number of important parameters and also to indicate certain dangers with respect to plant growth. However, it is practically impossible to screen a material for all substances which might cause damage, and in order to assess whether there are harmful substances in a material, a growth test may be helpful. Such a test does not give definitive answers, as not all crops react similarly but it does give more confidence about the suitability of a material as a substrate for a specific crop. In some cases, it may be necessary to carry out a cropping test with the material diluted and formulated as a growing medium. This may be the result of chemical research in which unexpectedly high concentrations of certain substances were found. To test whether these substances affect plant growth, a cropping test should be undertaken. Also, when it is already known that the use of

FIGURE 7.12 Growing trial for testing phytotoxicity as used for potting soil and peat (left is reference, right is mixture with peat).

a growing medium with a certain crop gives problems and the cause is unknown, a cropping test can be informative. In certain growing media, phytosanitary problems can be expected, for example with nematodes or certain virus diseases. Specific tests for such cases are available. Occasionally it may also be necessary to test whether a process in substrate manufacturing (e.g. composting or sterilization) gives adequate killing of certain pathogens. There is a wide range of methods for examining phytotoxicity, although there are hardly any general standards for testing growing media (Fig. 7.12).

7.5.8.2 General Principle

A material can be tested as it is but it is also possible to just use an extract from the material. Testing the material *per se* offers the advantage that the plant is in direct contact with the medium. However, other factors like physical characteristics can influence growth significantly. When this is expected, an extract may be preferable. The choice of a test crop should reflect the purpose of the substrate. There is no general advice for a good indicator crop. Gossow et al., (1995) tested different crops for sensitivity to different toxic substances.

They found that if only a single species is to be used, Chinese cabbage (*Brassica chinensis*) should be the choice. Often two or three test crops from different families are used, for example Chinese cabbage, lettuce and barley.

7.5.8.3 Cropping Test for Loose Substrates

The material to be tested is mixed in various ratios with a reference substrate, for example fine peat (type 3, fraction 0–7 mm). The experiment is carried out for a certain time with a suitable transplanted crop, for example tomato seedlings which were pre-germinated in stone wool plugs. Of course, chemical characteristics should be taken into account, for example available nitrogen when assessing composts. After some time, plant weight is determined.

7.5.8.4 Testing an Extract

Make an extract of the medium to be tested with a nutrient solution, correct for EC, pH and other nutrients. Add the extract to a reference substrate, a test plate for root elongation or test in hydroculture. In the latter case adequate oxygen levels in the rooting solution must be maintained. If a nutrient immobilization is known to take place (e.g. nitrogen), necessary corrections must be carried out. After some time, germination and plant weight are determined.

7.5.8.5 Special Cases

Dilution of the tested material is necessary if chemical properties, especially the salt content, are unfavourable. An interesting overview of available growing tests is given by Baumgarten (2004b).

7.5.8.6 Common Values for Media

Common values given for media phytotoxicity are per cent germination and root elongation. Comparisons of different test methods can be found in Blok et al., (2005), Emino and Warman (2004) and Wever (2004) (Fig. 7.13).

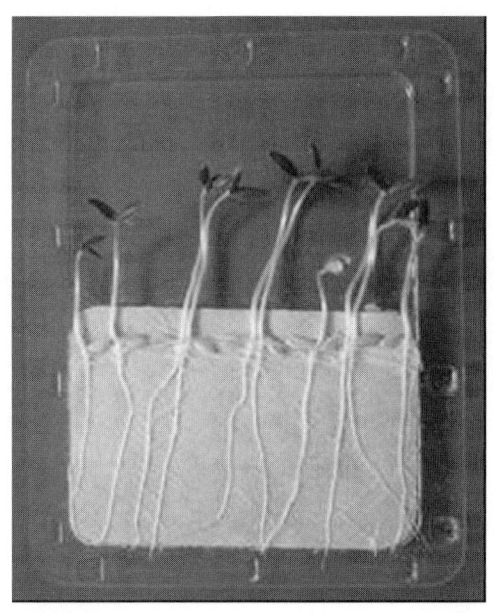

FIGURE 7.13 Transparent test plate with germinated cucumber seeds after a few days incubation in a microbiotest (after Blok et al., 2005).

REFERENCES

Abad, M., Fornes, F., Carrión, C. and Noguerra, V. (2005). Physical properties of various coconut coir dusts compared to peat. *Hort. Science*, **40**, 2138–2144.

Adani, F. and Scaglia, B. (2003). Biological stability determination in compost and waste by dynamic respiration index (DiProVe Method). Private communication.

Allaire, S.E., Caron, J., Duchesne, I., et al. (1996). Air filled porosity, gas relative diffusivity, and tortuosity: Indices of *Prunus x cistena* sp. growth in peat substrates. *J. Am. Soc. Hortic. Sci.*, **21**, 236–242.

Australian Standard (2003a). Potting mixes. AS 3743–2003.

Australian Standard (2003b). Compost, soil conditioners and mulches. AS 4454–2003.

Baas, R., Wever, G. and Koolen, A.J. (2001). Oxygen supply and consumption in soilless culture: Evaluation of an oxygen simulation model for cucumber. *Acta Hort.* (ISHS), **554**, 157–164.

Baumgarten, A. (2004a). CEN-methods (European standards) for determining plant available nutrients – A comparison. *Acta Hort.* (ISHS), **644**, 343–349.

Baumgarten, A.S. (2004b). Phytotoxicity (Plant tolerance). Report Horizontal.

Baumgarten, A. and Dersch, G. (2004). Contamination with viable weed seeds and plant propagules. Report Horizontal.

Blake, G.R. and Hartge, K.H. (1986). Particle density. In *Methods of Soil Analysis. Pt. 1. Physical and Mineralogical Methods*, 2nd edn (A. Klute, ed.). Madison, Wisconsin: Am. Soc. Agron.

Blok, C. and Cassamassimo, R.E. (2001). Non-destructive root oxygen use measurement II. Cucumber propagation in rockwool in a climate chamber. PPO Report 245. Naaldwijk, The Netherlands, Applied Plant Research.

Blok, C. and Gérard, S. (2001). Non-destructive root oxygen use measurement III. Cucumber propagation in rockwool in a climate chamber, July–August 2001. Naaldwijk, Applied Plant Research.

Blok, C., Persoone, G. and Wever, G. (in press). A practical and low cost microbiotest to assess the phytotoxic potential of growing media and soil. *Acta Hort.* (ISHS), In press.

Bos, E.J.F., Keijzer, R.A.W., Van Schie, W.L., et al. (2002). Potgrond en substraten. Naaldwijk, The Netherlands, Stichting RHP.

Brinton, W.F. (2000). Compost quality standards and Guidelines: An international view. Woods End Research Laboratory, Inc.

Brodie, H.L., Francis R.G. and Lewis E.C. (1994). What makes good compost? *Biocycle*, **35**, 66–68.

Bullens, H.P.G. (2001). Mechanische eigenschappen van tuinbouwkundige groeimedia. Wageningen, Leerstoelgroep Bodemtechnologie, Wageningen University, The Netherlands.

Caron, J. and Nkongolo, V.K.N. (1999). Aeration in growing media: Recent developments. *Acta Hort.* (ISHS), **481**, 545–551.

Caron, J. and Nkongolo, V.K.N. (2004). Assessing gas diffusivity coefficients in growing media from in situ water flow and storage measurements. *Vadose zone j.*, **3**, 300–311.

Caron, J. R., Rieviere, L.-M., Charpentier, S., et al. (2002). Using TDR to estimate hydraulic conductivity and air entry in growing media and sand. *Soil Sci. Soc. Am. J.*, **66**, 373–383.

Caron, J., Morel, P.H. and Rivière, L.-M. (2001). Aeration in growing media containing large particle size. *Acta Hort.* (ISHS), **548**, 229–234.

Cavins, T.J., Whipker, B.E. and Fonteno, W.C. (2004). Establishment of calibration curves for comparing pour-through and saturated media extract nutrient values. *Hort. Science*, **39**, 1635–1639.

CEN (1998). draft PrEN 12579: Soil improvers and growing media – Sampling.

CEN (1998). draft PrEN 13041: Soil improvers and growing media – Determination of physical properties – Dry bulk density, air volume, water volume, shrinkage and total pore space.

CEN (1999). EN 12580: Soil improvers and growing media – Determination of a quantity.

CEN (1999). EN 13037: Soil improvers and growing media – Determination of pH.

CEN (2001). draft prEN 13651: Soil improvers and growing media – Extraction of calcium chloride/DTPA soluble elements.

CEN (2001). EN 13652: Soil improvers and growing media – Extraction of water soluble nutrients and elements.

References

CEN (2006). draft PrEN 13040: Soil improvers and growing media – Sample preparation for chemical and physical tests, determination of dry matter content, moisture content and laboratory compacted bulk density.

Cooper, B.J. (2004). Stability (Biodegradability). Report Horizontal –7 WP4.

Da Silva, F.F., Wallach, R. and Chen, Y. (1995). Hydraulic properties of rockwool slabs used as substrates in horticulture. *Acta Hort.* (ISHS), **401**, 71–75.

De Kreij, C. and Wever, G. (2005). Proficiency testing of growing media, soil improvers, soils, and nutrient solutions. *Commun. Soil Sci. Plant Anal.*, **36**, 81–88.

De Kreij, C., Van Elderen, C.W., Meinken, E. and Duijvestijn, R.C.M. (2001). Extraction of growing media regarding its water holding capacity and bulk density. *Acta Hort.* (ISHS), **548**, 409–414.

De Kreij, C., Van Elderen, C.W., Meinken, E. and Fischer, P. (1995). Extraction methods for chemical quality control of mineral substrates. *Acta Hort.* (ISHS), **401**, 61–70.

Ellerbrock, R.H., Gerke, H.H., Bachmann, J. and Goebel, M.-O. (2005). Composition of organic matter fractions for explaining wettability of three forest soils. *Soil Sci. Soc. Am. J.*, **69**, 57–66.

Emino, E.R. and Warman, P.R. (2004). Biological assay for compost quality. *Compost Sci. Util.*, **12**, 342–348.

Gabriëls, R. (1995). Standardization of growing media analysis and evaluation: CEN/ISO/ISHS. *Acta Hort.* (ISHS), **401**, 555–558.

Gabriëls, R., Van Keirsbulck, W. and Verdonck, O. (1991). Reference method for physical and chemical characterization of growing media: An International Comparative Study. *Acta Hort.* (ISHS), **294**, 147–160.

Godley, A. (2003). Estimating biodegradable municipal solid waste diversion from landfill. Phase 1 Review of tests for assessing the potential biodegradability of MSW. (Draft report Project record P1-513). Swindon, Wiltshire, UK, Environment Agency R&D Dissemination Centre.

Goebel, M.-O., Bachmann, J., Woche, S.K., et al. (2004). Water potential and aggregate size effects on contact angle and surface energy. *Soil Sci. Soc. Am. J.*, **68**, 383–393.

Gossow, W., Grantzau, E. and Scharpf, H.C. (1995). Keimpflanzentest zur Substratprüfung. *Deutscher Gartenbau*, **49**, 1568–1571.

Handreck, K.A. (1993). Use of nitrogen drawdown index to predict fertilizer nitrogen requirements in soilless potting media. *Commun. Soil Sci. Plant Anal.*, **24**, 2137–2151.

Handreck, K.A. (1994). Pour-through extracts of potting media: Anomalous Results for pH. *Commun. Soil Sci. Plant Anal.*, **25**, 2081–2088.

Haug, R.T. (1993). *The Practical Handbook of Composting Engineering: Process Kinetics and Product Stability* Boca Raton: Lewis.

He, D.I., Logan, T.J. and Traina, S.J. (1995). Physical and chemical characteristics of selected U.S. municipal solid-waste composts. *J. Environ. Qual.*, **24**, 543–552.

Holtman, W., Van Duijn, B., Blaakmeer, A. and Blok, C. (2005). Optimalization of oxygen levels in root systems as effective cultivation tool. *Acta Hort.* (ISHS), **697**, 57–64.

Hue, N.V. and Liu, J. (1995). Predicting compost stability. *Compost Sci. Util.*, **3**, 8–18.

Iannotti, D.A., Grebus, M.E., Toth, B.L., et al. (1994). Oxygen respirometry to assess stability and maturity of composted municipal solid waste. *J. Environ. Qual.*, **23**, 1177–1183.

Jimenez, E.I. and Garcia, V.P. (1992). Determination of maturity indices for city refuse compost. *Agric. Ecosyst. Environ.*, **38**, 331–341.

Keeling, A.A., Paton, I.K. and Mullett, J.A.J. (1994). Germination and growth of plants in media containing unstable refuse derived compost. *Soil Biol. Biochem.*, **26**, 767–772.

Kipp, J.A. and Wever, G. (1998). Characterisation of the hydrophysical behaviour of stonewool. Applied Plant Research.

Kipp, J.A., Wever, G. and De Kreij, C. (2001). *International Substrate Manual*. Elsevier.

Klute, A. and Dirksen, C. (1986). Hydraulic conductivity and diffusivity: Laboratory methods. In *Methods of Soil Analysis. Part 1 Physical and Mineralogical Methods.* (A. Klute, ed.). Madison, Wisconsin, USA: Soil Science Society of America.

Laor, Y., Raviv, M. and Borisover, M. (2004). Evaluating microbial activity in composts using microcalorimetry. *Thermochimica Acta*, **420**, 119–125.

Leeper, G.W. (1947). The forms and reactions of manganese in the soil. *Soil Sci.*, **63**, 79–94.
Mathur, S.P., Owen, G., Dinel, H. and Schnitzer, M. (1993). Determination of compost biomaturity. I. Literature review. *Biol. Agric. Hortic.*, **10**, 65–85.
Michel, J.C., Naasz, R. and Montgermont, N. (2004). A tool for measuring the shrink/swell phenomena of peat growing media by image analysis. Proc. 12th Intl. Peat Cong. wise use of peatlands, 6–11 June, Tampere, Finland.
Michel, J.-C., Riviere, L.-M. and Bellon-Fontaine, M.-N. (2001). Measurement of the wettability of organic materials in relation to water content by the capillary rise method. *Eur. J. Soil Sci.*, **52**, 459–467.
Naasz, R., Michel, J.C. and Charpentier, S. (2005). Measuring hysteretic hydraulic properties of peat and pine bark using a transient method. *Soil Sci. Soc. Am. J.*, **69**, 13–22.
NEN (1989). NEN 5751: The determination of calcium carbonate in air-dry soil.
Page, A.L. (ed.) (1982). *Methods of Soil Analysis. Part 2. Chemical and Microbiological Properties*, Madison, Wisconsin, USA.
Raviv, M., Wallach, R., Silber, A., et al. (1999). The effect of hydraulic characteristics of volcanic materials on yield of roses grown in soilless culture. *J. Am. Soc. Hortic. Sci.*, **124**, 205–209.
Rippy, J.F.M. and Nelson, P.V. (2005). Soilless root substrate pH measurement technique for titration. *Hort. Science,* **40**, 201–204.
Silva, M. (2004). Compost stability. Relation between respiration measurements of compost and plant growth. PPO Report. Naaldwijk, PPO.
Sloot Van Der, H.A., Comans, R.N.J. and Bakker, F.P. (2003). Role of stable organic matter, degradable organic matter, combustible organic matter and dissolved organic matter in waste characterisation.
Sonneveld, C. (1979). Changes in chemical properties of soil caused by steam sterilisation. In *Soil Disinfestation* (D. Mulder, ed.). Amsterdam: Elsevier.
Sonneveld, C. and Van Elderen, C.W. (1994). Chemical analysis of peaty growing media by means of water extraction. *Commun. Soil Sci. Plant Anal.*, **25**, 3199–3208.
Sonneveld, C., Van Den Ende, J. and Van Dijk, P.A. (1974). Analysis of growing media by means of a 1:11/2 volume extract. *Commun. Soil Sci. Plant Anal.*, **5**, 183–202.
Van Soest, P.J. (1963). Use of detergents in the analysis of fibrous feeds. II. A rapid method for the determination of fiber and lignin. *J. Ass. Off. Agric. Chem.*, **46**, 829–835.
Van Soest, P. and Wine, R.H. (1967). Use of detergents in the analysis of fibrous feeds. IV. Determination of plant cell wall constituents. *J. Ass. Off. Agric. Chem.*, **50**, 50–55.
Veeken, A.H.M., De Wilde, V., Hamelers, H.V.M., et al. (2003). OxiTop® measuring system for standardised determination of the respiration rate and N-mineralisation rate of organic matter in waste material. Wageningen, WUR/NMI.
Verhagen, J.B.G.M. (1999). CEC and the saturation of the adsorption complex of coir dust. *Acta Hort.* (ISHS), **481**, 151–155.
Wever, G. and Leeuwen, van A.A. (1995). Measuring mechanical properties of growing media and the influence of cucumber cultivation on these properties. *Acta Hort.* (ISHS), **401**, 27–34.
Wever, G. (2004). Comparison of different bioassays for testing the phytotoxicity of growing media. *Acta Hort.* (ISHS), **644**, 473–477.
Wever, G. and Eymar, E. (1999). Characterisation of the hydrophysical and mechanical properties of pressed blocks for transplanting. *Acta Hort.* (ISHS), **481**, 111–120.
Wever, G. and Van Winkel, A. (2004). Interlaboratory study CEN-methods for the analysis of growing media and soil improvers. *Acta Hort.* (ISHS), **644**, 597–603.
Wever, G., Baas, R., Marquez, J. and Van Aanholt, L. (2001). Oxygen supply and gas exchange in the root environment of growing media in horticulture. *Acta Hort.* (ISHS), **554**, 149–155.
Wever, G., Nowak, J.S., De Sousa Oliveira, O.M. and Van Winkel, A. (2004). Determination of hydraulic conductivity in growing media. *Acta Hort.* (ISHS), **648**, 135–143.
Wever, G., Verhagen, J.B.G.M., Baas, R. and Straver, N. (2005). Bemestingsadviesbasis potplanten voor de Europese EN 1:% volumemethode. PPO Report 593. Naaldwijk, PPO.
Wrap, T.C.A. (2002). Public available specification 100 – Specification for composted material, Annex D: Method to assess contamination by weed propagules and phytotoxins in composted material.

Wu, L. and Ma, L.Q. (2001). Effects of sample storage on biosolids compost stability and maturity evaluation. *J. Environ. Qual.*, **30**, 222–228.

Yeager, T.H., Wright, R.D. and Donohue, S.J. (1983). Comparison of pour-through and saturated pine bark extract N, P, K, and pH levels. *J. Am. Soc. Hortic. Sci.*, **108**, 112–114.

Zevenhoven, M.A., Jacobs, L.A.J. and Keijzer, R.A.W. (1997). Planten in het veen, een selectie, Brochure. Naaldwijk, Stichting RHP.

8

NUTRITION OF SUBSTRATE-GROWN PLANTS

AVNER SILBER AND ASHER BAR-TAL

8.1 General
8.2 Nutrient Requirements of Substrate-grown Plants
8.8 Impact of N Source
8.4 Integrated Effect of Irrigation Frequency and Nutrients Level
8.5 Salinity Effect on Crop Production
8.6 Composition of Nutrient Solution
References

8.1 GENERAL

The basic principles of mineral nutrition of crops have been reviewed by many authors (Marschner, 1995; Mengel and Kirkby, 2001; Epstein and Bloom, 2005). While the theory of plant nutrition for soilless-grown plants is not different from that for soil-grown plants, some aspects, however, are different. One of the objectives of this chapter is to highlight these differences.

Universal problems, such as effects of deficiency or of toxic conditions on plant development, and the relevant diagnostic tools have been explored and widely reviewed in the literature (Marschner, 1995; Loneragan, 1997; Mengel and Kirkby, 2001; Epstein and Bloom, 2005), therefore these issues are not discussed in this chapter. The main factor that distinguishes between fertilisation management of soil-grown from that of soilless-grown plants is the limited volume of substrate in the latter case; this means lower buffer capacity for solution composition and limited supply of nutrients (capacity factor). Consequently, soilless culture methods offer unique benefits such as capabilities to control water availability, pH and nutrient concentrations in the root zone. At the same time, there are higher risks because of the smaller root system and low buffering capacity for water, nutrients and because of the increased risk of exposure to extreme ambient temperatures. These subjects are further discussed in Chap. 13. The general objective of this chapter is to exemplify nutritional issues of substrate-grown plants and to discuss differences between fertilisation/nutrition of soil-grown plants and of soilless-grown ones. The specific objectives are (i) to present the nutritional requirements of soilless-grown crops; (ii) to illustrate the impacts of the NH_4/NO_3 ratio in the irrigation water on rhizosphere-pH and on the development of soilless-grown plants; (iii) to discuss the integrated effect of irrigation frequency and nutrient availability on crop growth; (iv) to present the crop responses to changes in the electrical conductivity of the root-zone solution and to the ionic composition of the water, in soilless culture; (v) to describe the mode of fertilisation and the potential nutrient sources; and (vi) to discuss possible interactions among various fertilisers, and between the roots, the solution and the substrate, including the effects of the medium. Special attention will be devoted to pH effects on plant root functions and on availability of specific nutrients.

8.2 NUTRIENT REQUIREMENTS OF SUBSTRATE-GROWN PLANTS

8.2.1 GENERAL

Plants absorb many elements through their roots: more than 50 elements have been found in various plants. However, not all are considered to be essential elements. The essential nutrients required by green plants are exclusively inorganic, and an essential element may be defined as one that is required for the normal life cycle of a green plant and whose role cannot be assumed by another element. Twenty elements are thought to be essential to the growth of most plants, and they are usually classified as macronutrients and micronutrients. The former are those which are required in relatively large amounts: carbon, hydrogen, oxygen, nitrogen, phosphorus, calcium, sulphur, potassium and magnesium. The micronutrients are those required in small amounts, such as chlorine, iron, manganese, boron, zinc, copper, molybdenum, sodium and selenium (Marschner, 1995; Mengel and Kirkby, 2001; Epstein and Bloom, 2005). Silicon is usually not considered as essential – except in case of some algae and horsetails (*Equisetaceae*) – but it plays a significant role in plant resistance to stress conditions and to some diseases, and is, therefore, designated as 'quasi-essential' (Epstein, 1994, 1999; Epstein and Bloom, 2005). Nickel was found to be an essential element for legumes (Loneragan, 1997). The elements required in the largest quantities are the main structural elements, C (carbon), H (hydrogen), O (oxygen), N (nitrogen),

P (phosphorus), K (potassium) and S (sulphur), whereas elements required in very small amounts, such as Ni (nickel) and Mo (molybdenum), are needed for only one or very few enzymes (Loneragan, 1997). The essential macroelements were first recognised in the nineteenth century by De Saussure and Boussingault, who showed by chemical analysis that plants contain C, H, O and N. Soilless (hydroponic) culture has served as the major tool in the investigation of plant nutrition from the early studies of Sachs and Knop during the nineteenth century in Germany; they identified N, P, S, K, Ca, Mg and Fe as essential elements (Marschner, 1995). Most of the essential microelements were recognised in the twentieth century, from 1930 and onwards, thanks to the development of more highly purified chemicals and more sensitive methods for analysing trace concentrations (Loneragan, 1997). The range of concentrations of the essential elements in plant tissues and the required annual amounts for maximum yields are given in Table 8.1.

Except for carbon and oxygen, which are absorbed predominantly by the canopy from the air, the essential nutrients are taken up by the roots. Carbon is also taken up by the roots, in the form of HCO_3^-, and oxygen, both as gaseous O_2 and with hydrogen in water molecules. Hydrogen is also taken up as the ion H^+. In this chapter, following the practice of commercial horticulture, we will define as nutrients those elements that are taken up predominantly by the roots (Table 8.1). Most of these nutrients are taken up as cations or anions, except for boron, which is absorbed as boric acid or as the borate ion, depending on the pH. Thus, nutrient solutions are composed of mineral salts, acids and bases. The effects of the composition of the nutrient solution and the ratios of nutrient uptakes by plants on the pH of the solution and of the rhizosphere will be discussed later, in Sect. 8.3. Nitrogen is a unique nutrient that can be absorbed

TABLE 8.1 Ranges of the Essential Element Concentrations in Nutrient Solutions and Plant Tissues, and the Required Annual Amounts for Maximum Yields

Element	Chemical symbol	Form available to plants	Nutrient solution	Plant tissues	Annual consumption
Macroelements			mg L^{-1}	g kg^{-1}	kg ha^{-1} y^{-1}
Calcium	Ca	Ca^{+2}	40–200	2.0–9.4	10–200
Magnesium	Mg	Mg^{+2}	10–50	1.0–2.1	4–50
Nitrogen	N	NO$_3^-$, NH$_4^+$	50–200	10–56	50–300
Phosphorus	P	HPO$_4^{-2}$, H$_2$PO$_4^-$	5–50	1.2–5.0	5–50
Potassium	K	K$^+$	50–200	14–64	40–250
Sulfur	S	SO$_4^{-2}$	5–50	2.8–9.3	6–50
Micronutrients			mg L^{-1}	μg g^{-1}	g ha^{-1} y^{-1}
Boron	B	H$_3$BO$_3$, HBO$_3^-$	0.1–0.3	1.0–35	50–250
Copper	Cu	Cu$^+$, Cu^{+2}	0.001–0.01	2.3–7.0	33–230
Iron	Fe	Fe^{+3}, Fe^{+2}	0.5–3	53–550	100–4000
Manganese	Mn	Mn^{+2}	0.1–1.0	50–250	100–2000
Molybdenum	Mo	MoO$_4^{-2}$	0.01–0.1	1.0–2.0	15–30
Zinc	Zn	Zn^{+2}	0.01–0.1	10–100	50–500

as either cation or anion, NH_4^+ and NO_3^-, respectively. This characteristic of nitrogen influences plant nutrition in general and it has a strong impact on fertilisation in soilless culture.

In 1909, E.A. Mitscherlich developed an equation that related growth to the supply of plant nutrients. He observed that when plants were supplied with adequate amounts of all but one nutrient, their growth was proportional to the amount of this one limiting element, which was supplied to the soil. Plant growth increased as more of this element was added, but not in direct proportion to the amount added. Mitscherlich expressed this mathematically as

$$dy/dx = (A - y)C \qquad (1)$$

where y is the yield, x is the nutrient amount, A is the potential yield that would be obtained by supplying all growth factors in their optimum amounts and C is the proportionality constant that depends on the individual growth factor. By integration one obtains

$$y = A(1 - e^{-Cx}) \qquad (2)$$

The value of C was found by Mitscherlich to be 0.122 for N, 0.60 for P and 0.4 for K. Numerous workers since Mitscherlich have found that C is not, in fact, constant, but varies rather widely with crops grown under various climatic conditions. Therefore, the value of C has to be determined for the element of interest under the specific climatic conditions. An extension of this concept has been developed for the case in which two or more factors limit growth:

$$y = A(1 - e^{-C1 \times 1})(1 - e^{-C2 \times 2})(1 - e^{-C3 \times 3}) \cdots (1 - e^{-Cn \times n}) \qquad (3)$$

Note that in this concept the rate of application is not considered, nor are the growth rate and nutrient consumption, as functions of time.

The main issue that distinguishes between fertilisation management of substrate- and of soil-grown plants is the limited volume of substrates, which means lower buffer capacity for solution composition and limited supply of nutrients (capacity factor). These two factors can be defined as the quantity and intensity factors. The quantity factor is the demand by plant as described in the next section. The intensity factor is the solution concentration. Root and nutrient concentrations have complementary effects, since uptake rate is the integral of the [flux × root surface area (or length)] in a given soil sub-volume, over the total number of sub-volumes in the soil profile. The flux F [mole (cm root)$^{-1}$ h^{-1}] is determined by the nutrient concentration in the soil solution at the root surface, C_r (mole L^{-1}), as shown by the Michaelis–Menten equation:

$$F = F_{max} \frac{C_r}{K_m + C_r} \qquad (4)$$

F_{max} mole (cm root)$^{-1}$ h^{-1} and K_m (mole L^{-1}) are plant coefficients obtained in flowing or well-stirred nutrient solution experiments. The integrated uptake approach and the presented equation are the basis for most models, which simulate nutrient uptake and plant growth. Such models can be developed as fertigation decision tools, by comparing the computed and the required uptake rates at each time step during the simulation process. If the estimated uptake and required uptake rates do not match, a corrective action must be taken, subject to an economic optimisation procedure.

Unlike cultivation in soils, in soilless culture there is a need to supply most of the essential elements, including the micro-nutrients, continuously, because of the limited buffer capacity of the medium and its limited supply of nutrients (Savvas, 2001). Since the development of all-purpose nutrient solutions by Hoagland and Arnon (1938), many authors and organisations have published recommended tables of solution composition for different crops grown in soilless culture (Joiner et al., 1983; Steiner, 1984; Wright and Niemiera, 1987; Sonneveld and Straver, 1994; Schwartz, 1995; Nelson, 1996, 2003; Hanan, 1998; Resh, 1998; Adams, 2002; Sonneveld, 2002; Jones, 2005). However, the exact composition of the nutrient solution varies according to crop, stage of development, environmental conditions and irrigation regime. Therefore, in this chapter, we will concentrate on the main principles of nutrition in soilless culture rather than presenting recipes for growing plants in soilless culture.

8.2.2 CONSUMPTION CURVES OF CROPS

8.2.2.1 Dry Weight Accumulation

The above description of crop responses to fertilisation does not address the time factor. However, the demand for nutrients varies widely and dramatically during crop growth. There are considerable differences in uptake rate and in the time at which maximum consumption rate occurs, among crops and among varieties of the same species. In many cases, the consumption function is not monotonic, but exhibits sharp changes at critical physiological stages. Basically, the rate of nutrient requirement at each growth phase is associated with two predominant processes: (i) formation of new vegetative plant tissues; and (ii) formation of reproductive organs (flowers, fruits, seeds, etc.). The nutrient requirements for dry weight (DW) increases are primarily related to the photosynthesis rate, which is affected by various meteorological factors such as photosynthetically active radiation (PAR), air temperature and humidity, wind speed, and solar azimuth position (Thornley and Johnson, 1990; Goudriaan and van Laar, 1994). Theoretically, the net CO_2 assimilation rates by leaves, at 20°C under high light and CO_2 levels, are 1200 and 2000 $\mu g\, m^{-2}\, s^{-1}$ for C_3 and C_4 plants, respectively (Goudriaan and van Laar, 1994). In the light of these data, and assuming that the average C content in the DM of vegetative crops is 45 per cent and the C content of CO_2 is 27.3 per cent, the calculated DW production of C_4 plants throughout 10 h of daylight, at PAR values above a threshold of 200 W m^{-2} (Thornley and Johnson, 1990), is 43.6 g m^{-2} d^{-1}. However, in addition to climatic aspects, other growth factors may limit biomass production; thus, for example, the actual DW production rate in the Netherlands during the summer has been reported to be only 15–25 g m^{-2} d^{-1} (Goudriaan and van Laar, 1994). The initial growth phase is commonly characterised by an exponential rate, as each new leaf contributes to the light interception and to the biomass production, whereas during further growth phases mutual shading leads to a falling growth rate. At advanced growth phases, translocation of carbohydrates to non-photosynthetic organs such as root, flower, fruit or seed further diminishes the growth rate, and non-vegetative phases are characterised by a linear rate followed by a senescence rate. Representative rates of DW accumulation are illustrated in Figs. 8.1A–8.1D. Note that lettuce (*Lactuca sativa* L.) production, which is characterised by fast vegetative growth, is characterised by an exponential rate (Fig. 8.1A), whereas vegetable fruits such as tomato (*Lycopersicum esculentum* Mill) and bell pepper (*Capsicum annuum* L.) are characterised by initial exponential growth rates that

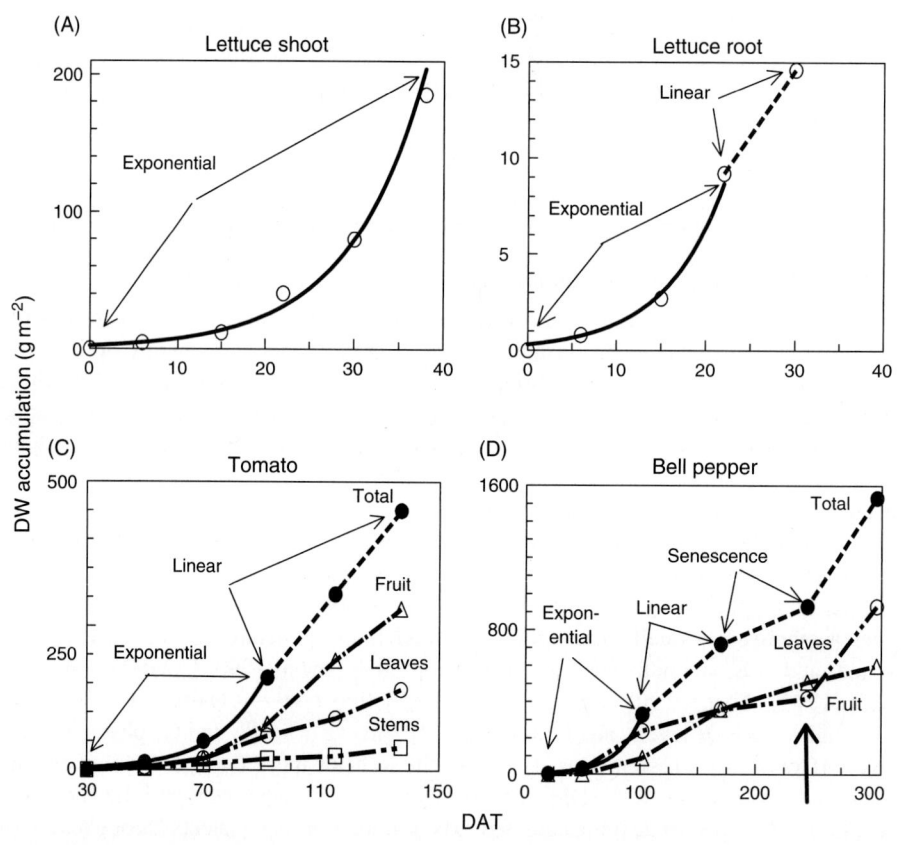

FIGURE 8.1 Dry weight (DW) accumulation as a function of time (DAT – days after transplanting) of several crop organs: (A) lettuce shoot (based on Xu et al., 2004); (B) lettuce root (based on Xu et al., 2004); (C) tomato fruits, leaves, stems and total plant (based on Tanaka et al., 1974); and (D) bell pepper fruits, leaves + stems and total plant (based on Tzipilevitz et al., 1996). Solid lines denote the initial intensive growth phases (exponential phase), and broken lines denote the subsequent growth phases (linear and senescence phases). Arrow indicates the beginning of spring season as climate became more favourable for vegetative growth.

continue until 92 and 102 DAT (days after transplanting), respectively, followed by linear growth (Figs. 8.1C and 8.1D), and even senescence in the case of bell pepper. The decrease in DW accumulation rate of bell pepper plants between 170 and 145 DAT resulted from too low temperature in the winter season (Tzipilevitz et al., 1996). In the spring season, the climate became more favourable for vegetative growth (indicated by arrow, Fig. 8.1D) and the rate of DW accumulation increased to its former value. Note that the rate of DW accumulation (slopes of the curves in Figs. 8.1A–8.1D) of lettuce in the course of the vegetative phase is significantly higher than those of tomato and bell pepper during the reproductive phase (18.5, 8 and 10.5 g m^{-2} d^{-1}, respectively).

8.2.2.2 Nutrient Accumulation

Maynard and Lorenz (1979) reviewed the advances made in vegetable mineral nutrition and fertilisation during the first 75 years of the twentieth century, and Hochmuth (2003) has described the more recent progress in the field. The main technical change in this period was the application of drip irrigation with liquid fertilisers, which offers the benefits of nutrient application to the crops in amounts and at times when it is most required by the plant, and at the location where it is most likely to be absorbed by the roots. The principles of fertigation and the solute dynamics in the root zone have been reviewed by Bar-Yosef (1999) and Mmolawa and Or (2000). Fertiliser management to combine the greatest yield with the least negative impact on the environment has become a major issue in the last two decades. A schedule for nutrient application is the key to meeting the environmental requirements and satisfying the crop demand. Consumption curves provide the basic information required for optimal fertilisation management, and fertigation is the tool with which to apply optimal management. Kläring (2001) reviewed the strategies for controlling water and nutrient supplies to greenhouse crops. The aim of the present section is to describe the specific features of nutrition of substrate-grown plants.

There are considerable differences among the shapes of consumption curves of crops and among those of varieties of the same species. In many cases the consumption curve of a nutrient exhibits sharp changes that are related to the physiological stages of plant development (Bar-Yosef, 1999). The major differences between consumption curves of different crops stem from the development stages that the crops pass through and the duration of each stage. The general nutrient consumption curve described by Hochmuth (1992) is principally analogous to that of DW accumulation illustrated in Figs. 8.1A–8.1D, that is, an exponential rate during the initial vegetative growth, followed by a linear and a senescence rate as reproductive organs develop. This consumption curve closely fits published measurements of nutrient uptake by determinant crops such as maize (*Zea mays* L.), fruit vegetables such as topped tomatoes (Tanaka et al., 1974), and melon (*Cucumis melo* L.) grown under environmental conditions that retard continuous fruit production. When non-terminating fruit tomatoes were grown continuously under well-controlled climatic conditions, the nutrients uptake rate grew steadily until production of the first fruit truss, and then became monotonic (Tanaka et al., 1974). Similar results were obtained with greenhouse tomato (Bar-Tal et al.,

1994b; Bar-Yosef, 1999) and pepper (Bar-Tal et al., 2001b). The nutrients consumption curve depends on the dry matter production curve, but there are differences between the two curves, which vary with developmental stage and between specific nutrients.

The major differences between the consumption curves of various fruit vegetables stem from differences between their patterns of production, that is, concentrated or continuous production. Daily nutrient uptake rates can be derived from the consumption curves, and those that result in optimum yield and product quality are crop specific and depend on climatic conditions, but are independent of soil characteristics and of irrigation techniques. Lack of attention to the changes in uptake rate with time may lead to periods of over-fertilisation and under-fertilisation. Over-fertilisation may enhance soil salinity and environmental contamination, whereas under-fertilisation may result in nutrient deficiency and yield reduction (Bar-Yosef, 1999). The rates of nutrient uptake by a typical vegetative crop (lettuce) and two widespread fruit crops (tomato and bell pepper) at various growth phases are illustrated in Figs. 8.2A–8.2C.

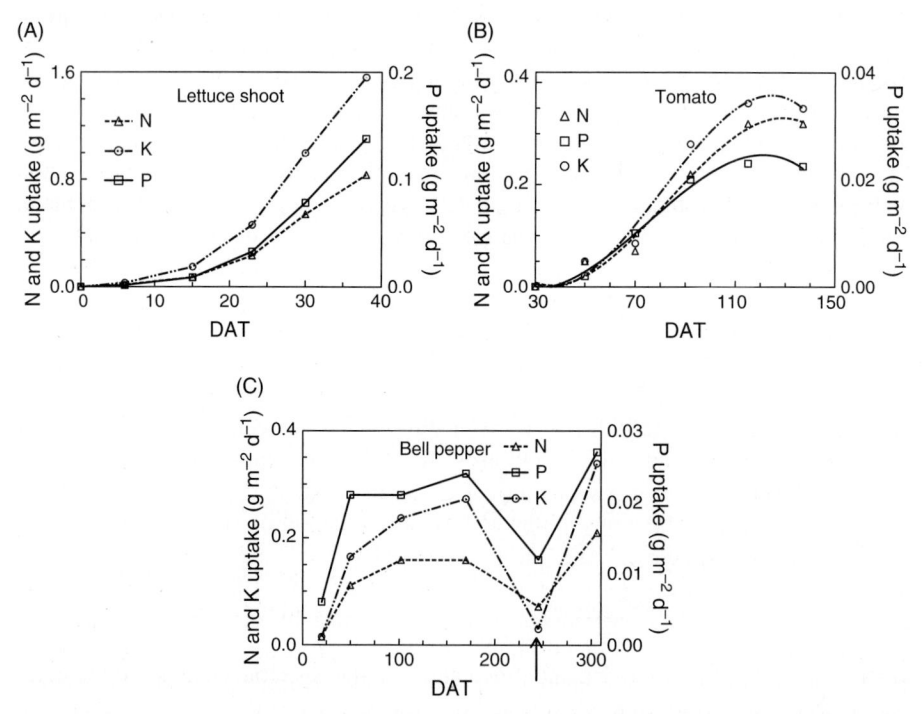

FIGURE 8.2 Rate of NPK uptake by: (A) lettuce plant (based on Xu et al., 2004); (B) terminated (eight trusses) tomato plant (based on Tanaka et al., 1974); and (C) bell pepper plant (based on Tzipilevitz et al., 1996). The arrow indicates non-optimal climatic conditions (mainly too low temperature) in the winter season that delayed fruit ripening and decreased the rate of fruit DW accumulation.

Note that the rate of nutrient uptake by the leafy vegetable (lettuce) is characterised by a purely exponential curve, and increases sharply as time proceeds (Fig. 8.2A), whereas those of tomato and bell pepper (Fig. 8.2B and 8.2C) became linear and even decreased in the course of fruit development. The sharp decrease in nutrient uptake rate of bell pepper plants (Fig. 8.2C) resulted from non-optimal climatic conditions (mainly too low temperature, Tzipilevitz et al., 1996) in the winter season, which delayed fruit ripening and decreased the rate of fruit DW accumulation (indicated by arrow, Figs. 8.1D and 8.2C). Nutrient consumption curves of non-topped and topped tomato plants were studied by Tanaka et al. (1974), who found that the rate of accumulation was slow in the early stages but accelerated as the rate of dry matter production increased (only curves for non-topped plant are presented in Fig. 8.2B). This active accumulation continued until the end of the experiment in the non-topped plants, but slowed down in the topped plants. The uptake of nutrients was continuous throughout growth, especially in non-topped plants, indicating that the physiological status of the plant did not appear to change with flowering; in other words, vegetative growth continued throughout growth without any marked effect upon fruit development. There was, however, some redistribution of N and P from vegetative organs at lower strata to the developing fruits during rapid fruit development. The amounts of N, K and Ca absorbed were greater than those absorbed by cereal crops under ordinary cultural conditions (Tanaka et al., 1974). Similar to other greenhouse crops, tomato tissues are specific in that they have a high mineral content but, nevertheless, the mineral content of the fruits is lower than that of the vegetative organs (Gary et al., 1998); a larger and more continuous supply of these elements is needed by tomato than by cereal crops. In fruits there is a continuous accumulation of carbohydrates in the vegetative organs during rapid growth, whereas, in cereals there is an apparent retranslocation of stored carbohydrates from the vegetative organs to the grains during ripening; development proceeds from the lower to the upper leaves and truss units. Each unit grows by using the currently absorbed elements and the currently produced photosynthates. The fruits grow by using the current photosynthates and the elements absorbed from the root during fruit growth.

Several examples of recommended nutrient solution compositions, adjusted for the stage of development and season in Israel, are presented in Table 8.2.

However, extrapolation of known NPK uptake data to environmental conditions different from those specified should be done carefully, and treated only as a first approximation (Bar-Yosef, 1999). For example, Xu et al. (2001) reported that the total consumption of N by pepper plants in the summer was about 2.2–2.8 times higher than that in the winter, although the nutrient solution contained the same concentration of N.

Not only can the total demand fluctuate, but the specific demand for individual nutrients can vary independently of fluctuations in total demand. The total uptake of nutrients is more or less determined by the growth and transpiration rates, but the uptake of individual nutrients depends more on the stage of growth (Voogt, 2003).

TABLE 8.2 Recommended Nutrient Solution Compositions Matched to the Growth Phase in Soilless Culture in Israel

Growth phase	N	P	K	Ca	Mg
			(mg L^{-1})		
Strawberry in greenhouse					
Transplanting	55–60	20–25	45–60	60–70	35–40
Anthesis and first fruit wave	70–85	20–25	70–90	100	45
Second fruit wave	80–85	25–30	80–90	100	45
Third fruit wave	80–85	25–30	80–90	100	45
Fourth fruit wave	55–60	20–25	55–60	80	35
Summer sweet pepper in greenhouse and net-house					
Transplanting to blooming	50–60	50–60	75–80		
Anthesis to fruit growth	80–100	80–100	100–120		
Fruit ripening and harvesting	100–120	100–120	140–160		
Fruit harvesting	130–150	130–150	180–200		
Fall-winter tomato					
Transplanting	80–90	30–40	120–140	180–220	40–50
Blooming and anthesis	120–150	30–40	180–220	230–250	40–50
Fruit ripening and harvesting	180–200	30–40	230–250	180–220	40–50
Fruit harvesting	120–150	30–40	180–220	180–220	40–50

Source: Israeli Extension Service Recommendations.

8.3 IMPACT OF N SOURCE

The nitrogen source may significantly affect the rhizosphere pH (Nye, 1981; Marschner, 1995; Marschner and Römheld, 1996; Bar-Yosef, 1999; Bloom et al., 2003), especially that of soilless-grown plants (Bar-Yosef, 1999). The common N fertiliser sources in soil systems are urea, NH_4^+ and NO_3^-. Urea is the cheapest N source (per N unit) and is the most concentrated N fertiliser (46 per cent); it is highly soluble, moves easily with the irrigation water and is, therefore, widely used in agriculture. However, urea is not taken up directly by plants and, therefore, only the derivative of urea hydrolysis – NH_4^+ ions – and the product of ammonium oxidation – NO_3^- ions – are absorbed by plant roots. The hydrolysis reactions are time-dependent processes (Lad and Jackson, 1982), therefore, urea is not commonly used in soilless culture, unless recycling systems are used (see Chap. 9), so that NH_4^+ and NO_3^- ions are the main N source (Barker and Mills, 1979). Rapid adsorption on the surface sites of the solid phases, and nitrification reactions diminish the NH_4^+ content in soils and media, and hence, rhizosphere NH_4^+ concentrations are commonly low even under irrigation with a high NH_4-N/NO_3-N ratio (R_N). However, the NH_4^+ concentrations in soilless media that have low buffer capacity and are subjected to frequent fertigation events may be high. Consequently, in soilless culture the impact of varying irrigation solution NH_4^+ concentration and R_N on rhizosphere pH and on physiological processes become crucial and controlling the R_N is of utmost importance. In addition, it is well accepted that the nitrifying bacteria exist as isolated colonies rather than being homogeneously distributed through the substrate volume (Darrah et al., 1987a; Strong et al., 1997). Consequently, the heterogeneity within low-buffered media in pH and in other chemical features driven by nitrification processes may be large.

Transformation of NH_4^+ to NO_3^- is a two-step process, in which each step is mediated by a highly specific group of bacteria. These are autotrophic bacteria that use inorganic material as energy sources (ammonia and nitrite as donors of electrons) and carbon dioxide as the carbon source for cell construction. In the first step, NH_4^+ is oxidised to NO_2^- by ammonium-oxidising bacteria (of which the most investigated are the *Nitrosomonas* group) (Eq. [5]), and in the second step, NO_2^- is oxidised to NO_3^- by nitrite-oxidising bacteria including the *Nitrobacter* group (Eq. [6]).

$$NH_4^+{}_{(aq)} + 1.5O_{2(g)} \leftrightarrow NO_2^-{}_{(aq)} + 2H^+{}_{(aq)} + H_2O \quad (5)$$

$$NO_2^-{}_{(aq)} + 0.5O_{2(g)} \leftrightarrow NO_3^-{}_{(aq)} \quad (6)$$

High levels of NH_4^+ can inhibit the nitrification processes (Focht and Verstraete, 1977; Niemiera and Wright, 1987b; Lang and Elliott, 1991; Russell et al., 2002) and it has been shown that nitrite (NO_2) oxidising bacteria are more sensitive to high NH_4^+ levels than the NH_4^+ oxidisers (Nakos and Wolcott, 1972). In addition to the specific effect of NH_4^+, high solute concentrations, in themselves, might inhibit nitrification, and the degree of inhibition was found to be related to both osmotic pressure and the specific anion that accompanied the NH_4^+ ions (Darrah et al., 1985, 1987b). The osmotic pressure threshold for inhibition was found to be 3.3 atm; however, chloride inhibited nitrification to a greater extent than its contribution to the osmotic pressure would indicate (Darrah et al., 1987b). The specific inhibitory effect of chloride must be considered carefully, because progressive chloride accumulation in the root zone as a result of high uptake of water and low frequency of irrigation is common under semi-arid conditions.

Nitrification increases with increasing temperature, within moderate limits (5–35°C), and the relationships between temperature and nitrification rates are usually described with the Arrhenius equation (Focht and Verstraete, 1977; Rodrigo et al., 1997; Russell et al., 2002) as presented in Fig. 8.3A. Nitrite-oxidising bacteria were found to be more sensitive to low temperature than the NH_4^+ oxidisers, therefore NO_2, the intermediate product of the nitrification process, may accumulate at low temperatures (Russell et al., 2002). Nitrite is well known to be toxic to plants; therefore NH_4^+ fertilisation in a cold climate should be treated with caution. However, under hot conditions the general trend of increasing nitrification rates with elevating temperature (Fig. 8.3A) changes direction, and above 30°C nitrification rates drop (Fig. 8.3B).

The pH of the medium has been reported to be an important influencing factor in nitrification processes, and the tendency for nitrification rates to decrease under acid conditions, as depicted in Fig. 8.4A and 8.4B, is generally recognised (Niemiera and Wright, 1987b; Lang and Elliott, 1991; Olness, 1999; Ste-Marie and Paré, 1999; Russell et al. 2002; Kyveryga et al., 2004). However, the shape and magnitude of the effect of pH depends on medium properties, water content and N sources (Stevens et al., 1998; Ste-Marie and Paré, 1999; Russell et al., 2002).

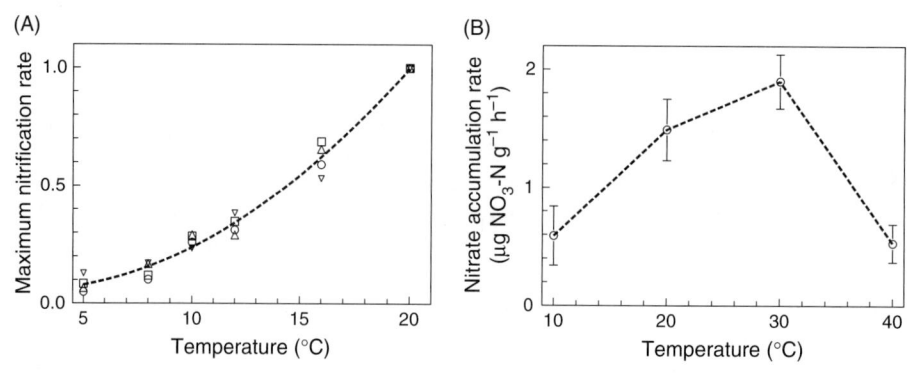

FIGURE 8.3 (A) Effect of temperature on the relative maximum nitrification rates in sandy soil amended with four N sources (based on Russell et al., 2002); (B) Influence of temperature on nitrate accumulation rate (µg NO_3^--N g^{-1} h^{-1}) in a pine bark medium (based on Niemiera and Wright, 1987a).

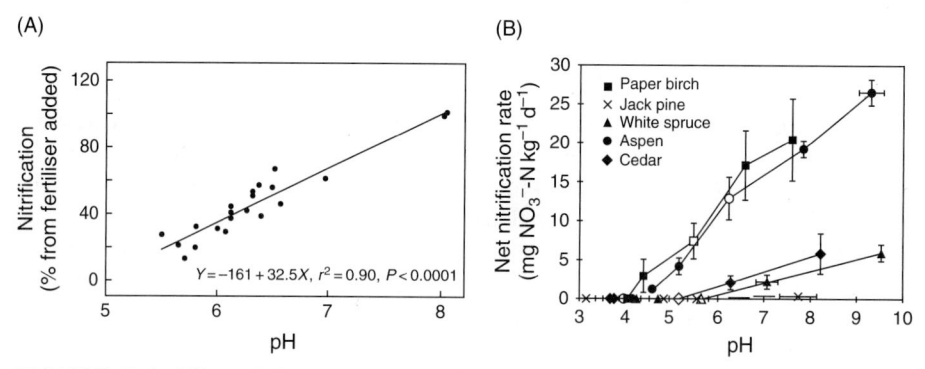

FIGURE 8.4 Effects of pH on nitrification. (A) Relationship between soil pH and percentage nitrification of fertiliser N (from Kyveryga et al., 2004, with kind permission from the Soil Science Society of America Journal); (B) pH effect on net nitrification rates in the nitrifying forest floor of different stand types (from Ste-Marie and Paré, 1999, with kind permission from Elsevier Science).

Similar to the effect of temperature, NO_2-oxidising bacteria were found to be more sensitive to acidic conditions than the NH_4^+ oxidisers, and accumulation of toxic levels of NO_2 under low pH has been reported (Russell et al., 2002). The pH has a direct impact on the oxidation process of NH_4^+ to NO_2^-, as H^+ is a product of this reaction (Eq. [5]), and consequently, accumulation of free protons in the solution inhibits further oxidation of NH_4^+. The direct effects of pH on microbial activity, or the size or composition of populations are not clear, because changes in pH frequently induce changes in additional vital features, for example membrane permeability, nutrient availability and element toxicity. Suzuki et al. (1974) showed that the change in the rate of oxidation in the pH range of 6.5–9.1 was relatively small when expressed on the basis of free ammonia rather than the ammonium ion. The relation between NH_4^+, H^+ and NH_3(aq) under equilibrium is given by (Eq. [7]):

$$NH_{4(aq)} \leftrightarrow NH_{3(g)} + H^+_{(aq)} \tag{7}$$

It is well accepted that ammonia is the real substrate for ammonia oxidisers (Suzuki et al., 1974). The ammonia molecule diffuses easily through the cytoplasm into the cells (Kleiner, 1981). When the pH is low the bacteria have to ingest ammonium actively through the cytoplasm into the cells, and to maintain a higher pH in order to obtain higher concentrations of free ammonia (De Boer and Kowalchuk, 2001). An acidic pH causes low concentrations of carbonates (Lindsay, 1979), and a lack of the carbon sources required for microbial growth could be the direct cause of the reduction of nitrification rates (Kinsbursky and Saltzman, 1990; Stevens et al., 1998).

The main effects of nitrogen source on soilless-grown plants, that is, modification of the rhizosphere pH, availability of further nutritional elements, NH_4^+ toxicity and incidence of physiological disorders such as chlorosis and blossom-end rot (BER), will be detailed and discussed below.

8.3.1 MODIFICATION OF THE RHIZOSPHERE pH AND IMPROVEMENT OF NUTRIENT AVAILABILITY

The nitrogen source may affect the rhizosphere pH via three mechanisms (Marschner, 1995; Marschner and Römheld, 1996; Bar-Yosef, 1999): (i) displacement of H^+/OH^- adsorbed on the solid phase; (ii) nitrification/denitrification reactions; and (iii) release or uptake of H^+ by roots in response to NH_4 or NO_3 uptake, respectively. Mechanisms (i) and (ii) are not associated with any plant activity, and affect the whole volume of the fertigated substrate, whereas mechanism (iii) is directly related to the uptake of nutritional elements, and may be very effective because it affects a limited volume in the immediate vicinity of the roots (Moorby et al., 1984; Gahoonia and Nielsen, 1992; Gahoonia et al., 1992; Marschner and Römheld, 1996; Taylor and Bloom, 1998; Bloom et al., 2003) as presented in Fig. 8.5. The extent of the

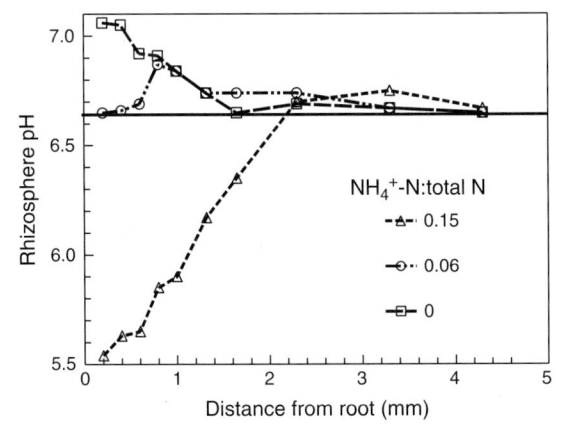

FIGURE 8.5 pH as a function of NH_4^+-N fraction (of the total N application) and the distance from the roots of rape (*Brassica napus*) plant. Solid line indicates the pH of the unplanted soil (based on Gahoonia and Nielsen, 1992).

pH alterations caused by the three mechanisms described above depends on medium properties, medium volume, plant activity and all the environmental factors that affect nitrification rate, as detailed above.

Reduction of the medium pH, driven by NH_4^+ nitrification and root excretion of protons, is commonly used for overcoming growth disorders induced by micronutrient deficiencies, such as chlorosis and 'little leaf' or 'rosette' (Tagliavini et al., 1995; Bar-Yosef et al., 2000; Savvas et al., 2003; Silber et al., 1998, 2000a,b, 2004). The effects of N level and source throughout the growth cycle on the pH of the leachate from soilless-grown rice flower (*Ozothamnus diosmifolius*, Asteraceae) are illustrated in Fig. 8.6A and 8.6B, respectively.

Increasing the NH_4^+ supply, either by increasing the total N concentration (fixed R_N) or by increasing the R_N (fixed N concentration), resulted in a sharp decrease in the pH (Silber et al., 2004). The pH during the growth period was significantly affected by the amount of irrigation solutions applied, as well as by the N concentration and source. The higher volumes of irrigation solutions that were applied during the summer (up to 120 DAT) in response to the increased transpiration rate enhanced the supply of NH_4^+, and the leachate pH fell to 4.1 at high NH_4^+ application. This trend changed during the winter (from 180 DAT) as the transpiration and irrigation rates and, consequently, the NH_4^+ doses decreased, and the leachate pH rose to close to that of the irrigation solution (Fig. 8.6). The growth of rice flower plants exposed to low-N fertilisation or low R_N was poor, and the plants exhibited growth disorders such as tip burn, and severe chlorosis and necrosis (Silber et al., 2004). The pH in the root environment was probably the most important factor affecting rice flower growth and yield. A linear regression was obtained between the total dry mass production and the average pH

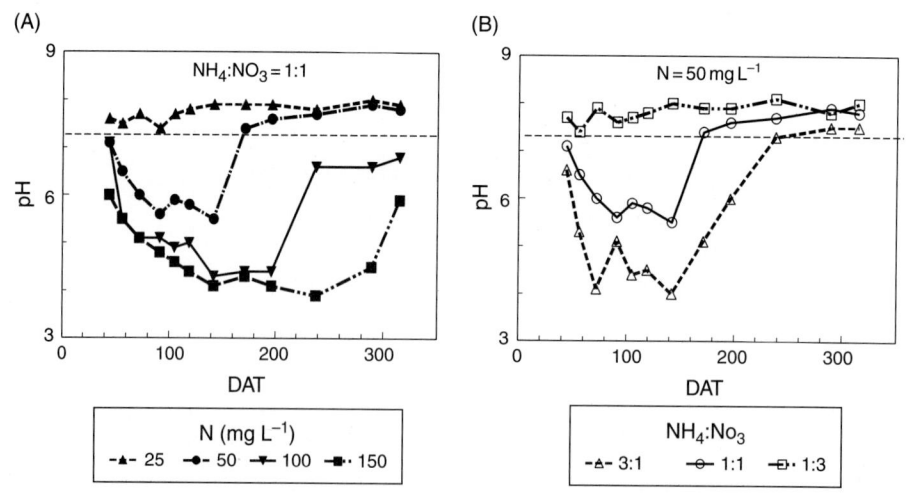

FIGURE 8.6 Effects of N level and source throughout the growth cycle on the pH of the leachate from soilless-grown rice flower. (A) Different N concentrations at constant $NH_4^+:NO_3^-$-N ratio; and (B) Different $NH_4^+:NO_3^-$-N ratio at constant N concentration (based on Silber et al., 2004).

8.3 IMPACT OF N SOURCE

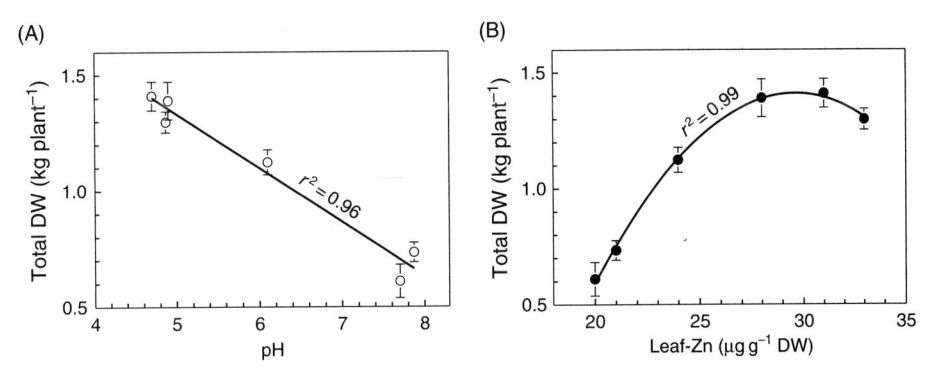

FIGURE 8.7 Relationships between total dry weight production of rice flower: (A) averaged pH in leachates during the main growth season; and (B) leaf-Zn concentrations. Vertical bars indicate the standard error (based on Silber et al., 2004).

in leachates within the pH range of 4.5–8.0 (Fig. 8.7A). It is reasonable to consider that the pH in itself did not affect rice flower growth, but did so indirectly, through its effect on nutrient availability. The only nutrient that was significantly correlated with pH and yield parameters (total fresh and dry mass production, or the number of flowering stems) was Zn. Zinc solubility is strongly pH-dependent (Barrow, 1993; Lindsay, 1979; Marschner, 1993), therefore, Zn-deficient conditions are induced by a high rhizosphere pH. A significant quadratic regression was obtained between the yield and leaf-Zn concentrations (Fig. 8.7B).

Similar effects of R_N on the drainage solution pH and on diverse growth disorders induced by high pH have been observed for other soilless-grown plants such as chrysanthemum (*Dendrathema* × *grandiflorum* Kitam), impatiens (*Impatiens walkerana* Hook, F.), 'Safari Sunset' (*Leucadendron salignum* Bergius × *L. laureolum* (Lam.) Fourc.), wax flower (*Chamelaucium uncinatum* L.) and gerbera (*Gerbera jamesonii*) (Siraj-Ali et al., 1987; Argo and Biernbaum, 1997a,b; Silber et al., 1998, 2000a,b; Bar-Yosef et al., 2000; Savvas et al., 2003).

It is important to note that decreasing the NO_3^- nutrition, by decreasing either the N level or R_N, may reduce metal mobilisation towards the roots, as a result of diminished organic acid exudation (Fig. 8.8). Formation of carboxylic anions is increased as a result of NO_3^- assimilation in plants (Fig. 8.9) and some of the carboxylates are exuded from the roots to the rhizosphere (Kirkby and Mengel, 1967; Kirkby and Knight, 1977; Touraine et al., 1990; Marschner, 1995).

Organic acids play a key role in mobilisation of metal ions towards the roots (Gardner et al., 1982a,b, 1983; Bar-Yosef, 1996), therefore, the data presented in Figs. 8.8 and 8.9 clearly indicate that R_N augmentation may play a dual role in affecting micronutrients availability. On the one hand, micronutrients availability is enhanced through rhizosphere acidification (Tagliavini et al., 1995; Silber et al., 1998, 2000a,b, 2004; Bar-Yosef et al., 2000; Savvas et al., 2003) but, on the other hand, it is diminished because of the decrease of organic acid exudation (Figs. 8.8 and 8.9).

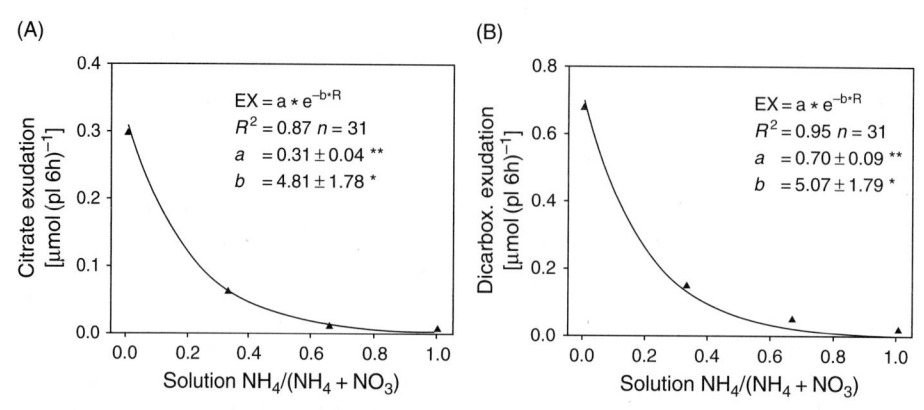

FIGURE 8.8 Relationships between R_N and carboxylic acids exuded from tomato roots. (A) Citric acid; and (B) dicarboxylic acids (from Imas et al., 1997, with kind permission from Elsevier Science).

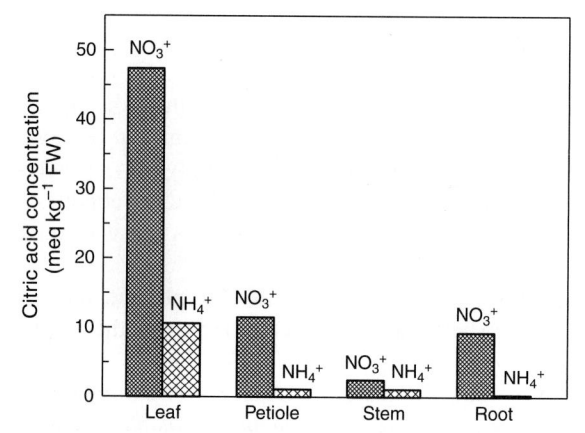

FIGURE 8.9 Influence of N source on citric acid concentration in tomato tissues (based on Kirkby and Mengel, 1967).

However, organic acids are rapidly consumed by micro-organisms in the rhizosphere, therefore their role in increasing nutrient availability in soilless culture is probably not important.

In addition to the indirect effect of R_N on nutrient availability in the rhizosphere, through pH modification as detailed above, R_N affects the apoplastic pH and, consequently, Fe^{III} reduction and mobilisation in plants (Mengel et al., 1994; Kosegarten et al., 1998, 1999a,b, 2004; Zou et al., 2001). Mengel (1994) found that inhibition of iron uptake from the root apoplast into the cytosol of root cells, rather than low iron availability in the rhizosphere, was the main cause of iron chlorosis. Kosegarten (1999b) found that the pH in the apoplast was different from that of the outer solution because of the strong H^+ buffer capacity of the apoplast, but that NO_3^- nutrition increased the apoplastic pH and depressed the ferric-chelate reductase (FC-R) activity.

Therefore, high NO_3^- or $CaCO_3$ in the rhizosphere induced chlorosis incidence even when the concentration of iron in plants was normal and not different from that in non-chlorotic plants fed with high R_N in the irrigation solution.

8.3.2 CATION-ANION BALANCE IN PLANT AND GROWTH DISORDERS INDUCED BY NH_4^+ TOXICITY

The nitrogen source may significantly affect numerous physiological processes in plants (Marschner, 1995; Forde and Clarkson, 1999; Mengel and Kirkby, 2001; Epstein and Bloom, 2005). The cation–anion balance in plant tissues is maintained by diffusible and non-diffusible organic and inorganic ions, and has been found to be notably affected by the sources of N nutrition (Marschner, 1995; Mengel and Kirkby, 2001; Epstein and Bloom, 2005). It has been widely reported that NH_4^+ nutrition depressed the uptake of cations, especially in leaves and petioles, and that NO_3^- nutrition depressed that of anions (Kirkby and Mengel, 1967; Kirkby and Knight, 1977) (Table 8.3).

Excess of cations and anions in plants is balanced by organic acids; therefore, the decrease of cation uptake followings NH_4^+ nutrition induced lower organic acid content in plants (Kirkby and Mengel, 1967; Touraine et al., 1990) and excretion from their roots (Imas et al., 1997) as depicted in Fig. 8.9.

Although N is an essential nutrient for plant growth, high NH_4^+ concentrations are toxic to most plants, especially at high root temperatures and under high salinity (Kafkafi, 1990; Forde and Clarkson, 1999; Adams, 2002; Sonneveld, 2002; Britto and Kronzucker, 2002). The detrimental effect of NH_4^+ on vegetative growth, fruit production and fruit quality (BER and flat fruit) of pepper and tomato is well documented (Ganmore-Neumann and Kafkafi, 1980a,b; Bar-Tal et al., 2001a; Claussen 2002; Aki et al., 2003; Silber et al., 2005b). The direct cause of NH_4^+ toxicity remains obscure and, in spite of numerous studies, the causative mechanism has not been identified (Britto and Kronzucker, 2002).

TABLE 8.3 Influence of N Source on Cation and Anion Concentrations in Tomato Tissues

Tissue	N source	K	Ca	Mg	SO_4^{2-}	$H_2PO_4^-$
		(meq 100 g^{-1} DW)				
Leaf	NO_3^-	58	161	30	22	13
	NH_4^+	29	62	25	35	15
Petiole	NO_3^-	176	126	38	12	15
	NH_4^+	90	61	17	29	20
Stem	NO_3^-	162	86	35	8	13
	NH_4^+	54	50	18	14	15
Root	NO_3^-	93	44	40	10	27
	NH_4^+	43	38	11	15	21

Source: Based on Kirkby and Mengel (1967).

NH_4^+ toxicity has been hypothesised to be associated with the solution concentration of $NH_3(aq)$ (Bennett and Adams, 1970a,b; Bennett, 1974). $NH_3(aq)$ can enter easily through cell membranes, and even at a very low concentration (0.17 mM in the nutrient medium), it might inhibit vital processes such as photosynthesis, respiration and enzyme activity (Bennett and Adams, 1970b; Bennett, 1974). Inhibition of root elongation as a result of NH_4^+ nutrition has been found to be pH-dependent and was directly related to the calculated solution concentrations of $NH_3(aq)$ (Bennett and Adams, 1970b), as depicted in Fig. 8.10A and 8.10B. However, Bennett's and Adams' (1970b) interpretation is not straightforward, because the severe inhibition of root growth at pH above 8 (Fig. 8.10A) could be related to deficiency of Ca (Bennett and Adams, 1970a) or of other nutrients.

It has been reported recently (Britto et al., 2001a,b) that disruption of the influx regulation and consequent accumulation of NH_4^+ in the cytosol caused useless cycling of N across the plasma membrane of root cells. This futile cycling induced an abrupt increase of respiration, with high energy cost and subsequent decline in growth (Britto et al., 2001b). Almost all the NH_4^+ ions taken up from the irrigation solution are rapidly assimilated in the roots (Raven, 1985, 1986; Marschner, 1995), therefore the demands for carbohydrates and oxygen in the roots are higher under high-NH_4-N than under high-NO_3-N treatment (Ganmore-Neumann and Kafkafi, 1983; Kafkafi, 1990). High temperature directly reduces solution-O_2 concentrations; it also indirectly reduces carbohydrate availability and solution-O_2 concentrations, through increases in plant respiration (Gur et al., 1972). Thus, growth impairment and yield reduction of plants that are susceptible to ammonium toxicity are expected to be more important under hot than under cold or temperate climates (Ganmore-Neumann and Kafkafi, 1980a, 1983; Kafkafi, 1990, 2001; Marschner, 1995; Mengel and Kirkby, 2001). It should be noted that the detrimental combination of hot temperature and high R_N occurs more frequently in soilless than in soil culture, because of the lower buffer and heat capacities of the former. On the other hand, high temperatures (up to 30–40°C, see Niemiera and Wright, 1987a; Walden and Wright, 1995) enhance the nitrification rate and therefore curtail exposure of the roots to the toxic NH_4^+-N concentrations in the rhizosphere. In soilless culture, the high air-filled porosity of most media (Raviv et al., 2002) ensures the availability of oxygen for nitrification processes that rapidly reduce the NH_4^+-N concentration in the rhizosphere.

The NH_4^+ concentration is affected by both the total N concentration and the R_N; therefore, it is more practical to express NH_4^+ activity in terms of the solution NH_4^+-N concentration. Bar-Tal et al. (2001a) found that the incidence of BER and flat fruits increased as the concentration of NH_4^+ in the nutrient solution increased, because of increases in the concentration either of total N or specifically of NH_4^+ (only the relationship between incidence of flat fruits and NH_4^+ concentration is depicted in Fig. 8.11).

However, the threshold values of NH_4^+-N concentration in the irrigation solution, as determined in field experiments, are not definitive. Nitrification is a time-dependent process and therefore the transient NH_4^+ concentration in the rhizosphere is significantly affected by the interval between successive irrigations. Thus, under irrigation

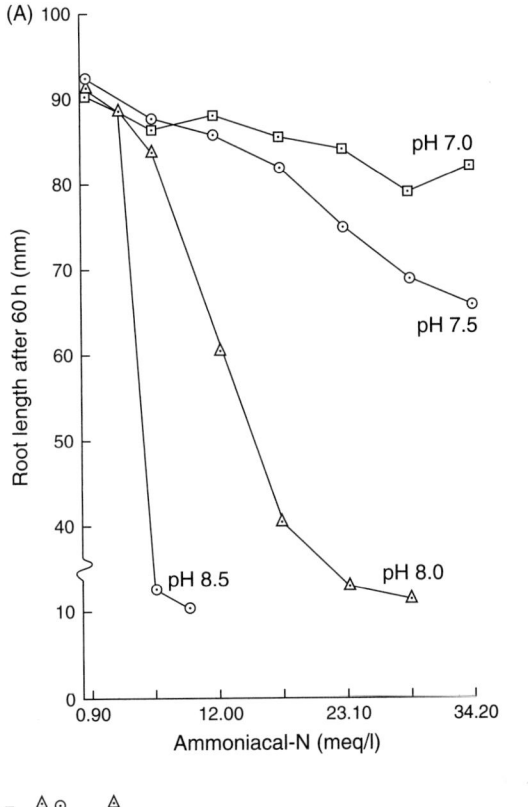

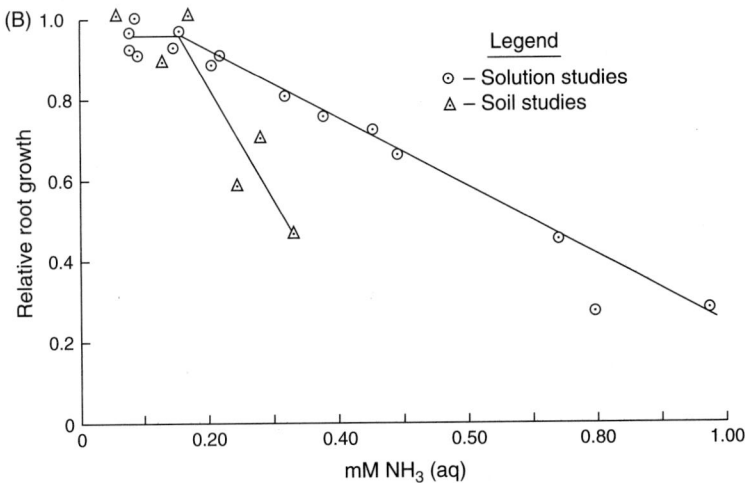

FIGURE 8.10 (A) The effect of solution concentration of NH_4^+ at different pH values, on root growth of cotton in nutrient solution during 60 h; and (B) the relationship between growth inhibition of cotton roots and the calculated concentration of NH_3(aq) in nutrient and in soil solution (from Bennett and Adams, 1970b, with kind permission from the Soil Science Society of America Journal).

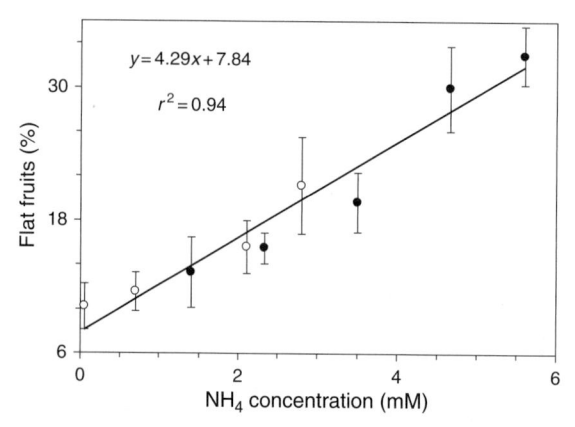

FIGURE 8.11 Relationship between incidence of flat fruit (sweet pepper) and solution concentration of NH_4^+. The empty symbols are for Expt. 1 (N varied from 0.25 to 14.0 mmol L^{-1}, constant R_N of 0.25) and the full symbols are for Expt. 2 (R_N varied from 0.25 to 2.0, constant N concentration of 7.0 mmol L^{-1}), the line is the best-fit linear regression. Vertical bars indicate the standard error (from Bar-Tal et al., 2001a, with kind permission from the American Society of Horticultural Science).

schedule of one to four daily irrigation events, roots are exposed to applied NH_4^+ only during short periods immediately after the irrigation events, whereas during most of the time between successive irrigations, NO_3^- rather than NH_4^+ is the main N source. However, under frequent fertigation, with 10–30 daily irrigation events, the transient NH_4^+ concentration in the vicinity of the roots may be high (Silber et al., 2005b). It is well documented that high NH_4^+ concentration and high R_N enhances the incidence of BER in vegetable fruits (Wilcox et al., 1973; Morley et al., 1993; Bar-Tal et al., 2001a; Silber et al., 2005b). It is generally accepted that the effect of the NH_4^+ concentration on the incidence of BER is through its effect on Ca uptake (Ho et al., 1993; Adams, 2002; Sonneveld, 2002). Since NH_4^+ has been found to depress the leaf Ca content (Table 8.3) it was suggested that in susceptible crops such as tomato and sweet pepper, NH_4-N should be restricted to 15 or even 10 per cent of the total N supplied via irrigation (Adams, 2002; Sonneveld, 2002). Claussen and Lenz (1995) reported that vegetative growth in eggplants (*Solanum melongena*) was promoted by supplying them with NH_4^+ nitrogen before bloom, and set of flowers and fruit occurred earlier than on similar plants supplied with nitrate. Furthermore, the number of flowers and the fruit yield were increased.

8.4 INTEGRATED EFFECT OF IRRIGATION FREQUENCY AND NUTRIENTS LEVEL

The mechanisms of water and nutrients transport in the rhizosphere, and the mechanisms of ion uptake by root have been widely reviewed by Barber (1995), Tinker and Nye (2000), Hopmans and Bristow (2002), Jungk (2002) and Silberbush (2002).

A quantitative approach is presented in Chap. 3 of this book. Therefore, these issues are not discussed in this chapter, and attention is focussed on the interrelationships between irrigation management, especially under non-stress conditions, nutrient availability in soilless media and several nutritional aspects of plant growth.

The beneficial effects of high-frequency irrigation were recognised some decades ago, and it is considered a useful tool for optimising the root environment. Rawlins and Raats (1975) stated that '... we should not overlook the possibilities for conserving and making most effective use of our water, land and fertilizers resources by using high-frequency irrigation', and Clothier and Green (1994) provided further support. Although these conclusions were based on studies of soil and of soil-grown plants, their basic approach is valid for soilless media as well. It is important to note that in the 1970s and until the 1990s, high-frequency irrigation was defined as having intervals of less than 7 days (Martin et al., 1990), but nowadays the time between successive irrigation events has diminished to hours, or even less. Recently, micro-drip irrigation systems have been developed that provide emitter discharges of $<0.5\,L\,h^{-1}$. These systems are mostly based on low-pressure gravity flow and have been studied mainly with soilless-grown crops. Preliminary results have indicated promising potential. The discussion of the effects of irrigation frequency on root-zone matric potential, and water and nutrients uptake by plant roots in this chapter is mainly focused on non-stress conditions, that is, the irrigation frequency does not lead to water-stress conditions, and the amount of available water in the root volume between two successive irrigation events is sufficient and does not limit uptake by the root.

8.4.1 NUTRIENT AVAILABILITY AND UPTAKE BY PLANTS

Adsorption on the solid phases and precipitation of insoluble compounds decrease the concentrations of nutrients in the root area (Chap. 6). Thus, the nutrient concentrations in the rhizosphere may be high or even excessive immediately after irrigation, and may subsequently fall to deficit levels as illustrated in Fig. 8.12A. These processes are time dependent; therefore, reducing the time interval between successive irrigations to maintain constant, optimal water content in the root zone may also reduce the variations in nutrient concentration (Fig. 8.12B), thereby increasing their availability to plants and reducing their leaching out of the root zone.

Water and nutrients acquisition by plants, and the formation of a depletion zone in the immediate vicinity of the roots, drive solute movement towards the roots. Nutrient transport to the root surface takes place by two simultaneous processes: convection in the water flow (mass flow), and diffusion along the concentration gradient (Barber, 1995; Tinker and Nye, 2000; Jungk, 2002). Medium properties, crop characteristics and growing conditions affect the relative importance of each mechanism, but the general situation is that the mobile NO_3 ion supply is taken up mainly through mass flow, whereas for less mobile elements such as P and K, diffusion is the governing mechanism (Barber, 1995; Claassen and Steingrobe, 1999; Jungk, 2002; Mmolawa

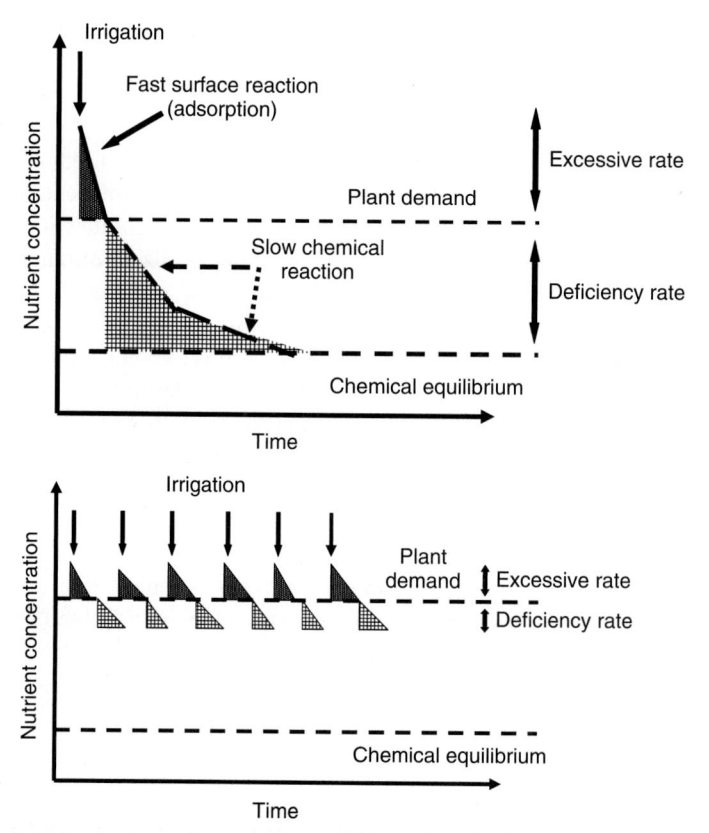

FIGURE 8.12 Schematic presentation of the time variation of nutrient concentration in the rhizosphere under conventional and frequent irrigation (A and B, respectively). Excess and deficiency rates correspond to nutrient concentration above or below plant demand, respectively; chemical equilibrium corresponds to nutrient concentration governed by equilibrium processes.

and Or, 2000; Tinker and Nye, 2000). The diffusion takes place through the aquatic phase and the diffusion coefficient (D_e) is given by Eq. (8) (Nye, 1966):

$$D_e = D_L \theta f (dC_l/dC_s) \quad (8)$$

in which D_L is the diffusion coefficient of the ion in water, θ is the volumetric water content, f is the tortuosity or the medium-impedance factor, and dC_l/dC_s is the reciprocal of the medium buffer power for the ion in the solution, C_l is the ion concentration in solution and C_s is the total diffusible amount of the ion. Nutrient diffusion takes place in water-filled pores and, therefore, θ has an important effect on D_e (Kovar and Claassen, 2005). The tortuosity factor (f) is strongly affected by θ; therefore, the effect of θ on D_e is usually described by a quadratic equation (Claassen and Steingrobe, 1999). Increasing θ from 0.1 to 0.4 resulted in increases of D_e for K (Kuchenbuch et al., 1986a) and P (Bhadoria et al., 1991) by factors of about 10. In addition, θ affects the shape of the depletion zone (water- or nutrient-depleted area surrounding plant roots), so that reducing θ increased the width of the depletion zone and consequently reduced the soil volume available for root acquisition and limited the K uptake per unit root area (Kuchenbuch et al., 1986b), as well as that of other nutrients. However, simulations demonstrated that the root was able to partially compensate for the decrease of diffusion by increasing the concentration gradient (Claassen and Steingrobe, 1999). Simulation of NO_3^- uptake showed that it was less sensitive than K uptake to changes in θ (Claassen and Steingrobe, 1999). The nitrate ion is almost not bound to the solid phases, therefore the soil buffer power (dC_l/dC_s) is significantly lower for NO_3^- than for K. In the light of the simulations by Claassen and Steingrobe (1999) and taking into account the soil buffer power for the nutritional elements, it is to be expected that P uptake will be strongly affected by θ variations. Bar-Tal et al., (1994a) used a simulation to predict that increasing the θ of the medium would compensate for low P concentration in the irrigating solution. In addition, soil-P concentration is governed by adsorption and precipitation reactions (Chap. 6); therefore, the mechanism by which increasing the nutrient concentration in the vicinity of the root compensates for the above-described decrease in diffusion is not valid for P. Indeed, altering θ significantly affected the depletion of inorganic P in the rhizosphere (Gahoonia et al., 1994). In addition to the effect of θ on D_e, frequent watering enhanced the mixing of solute between pores containing mobile and immobile water, respectively, and thereby improved P and K uptake by plants (Kargbo et al., 1991).

Claassen and Steingrobe (1999) commented that under adequate supply conditions even NO_3^- is transported mainly by diffusion, and they concluded that diffusion was a much more efficient mechanism for nutrient transport than mass flow. However, these conclusions were based mainly on experiments and model calculations in which nutrients were supplied independently of, and much less frequently than irrigation water. Furthermore, in most of those experiments, nutrients were supplied only before planting seedlings or sowing, therefore, on a time scale of days or weeks, nutrient transport from the soil solution to the root interface took place mainly by diffusion. Nevertheless, the simultaneous application of irrigation water and nutrients, together with the high values of saturated hydraulic conductivity for container media, ensure that the root surface and its vicinity are frequently supplied with fresh nutrient solution during the irrigation events and the subsequent redistributions. These frequent replenishments diminish the depletion zone formed at the root surface by uptake of nutrients during the period between successive irrigation events, decrease the concentration gradient between the medium solution and the root interface and diminish the role of diffusion in nutrient transport towards the roots (Bar-Tal et al., 1994a). Consequently, the contribution of convection to the nutrient uptake of soilless-grown plants is much more significant than that of diffusion.

Nitrate, the main N source for soilless-grown plants (Sonneveld, 2002), is hardly ever involved in adsorption or precipitation reactions; therefore, the concentration of NO_3^- in the irrigation water and its actual concentration in the vicinity of the roots are quite similar. In contrast, P availability to plant roots is time dependent, as a result of adsorption and precipitation reactions (Chap. 6). Potassium ions are hardly ever involved in precipitation reactions, but may be adsorbed on negatively charged surfaces. Therefore, the difference between the K concentrations in the irrigation solution and the rhizosphere lies between those between the respective NO_3^- and P concentrations. Consequently, it can be expected that the impact of fertigation frequency on the uptake of nutritional elements by plants will be related to both their mobility and their availability, as indeed has been reported (Mbagwu and Osuigwe, 1985; Kargbo et al., 1991; Silber et al., 2003). Although the effects of irrigation frequency on nutrient concentration in soilless-grown lettuce leaves presented in Fig. 8.13 followed the expected order of P > K > N, the magnitudes of the nutrient increases in the plant was found to be closely related to the fertilisation level.

The increases in the leaf P and K concentrations were attributed to both direct and indirect effects of irrigation frequency on the P and K concentrations at the root surface. The direct effect is the frequent elimination of the depletion zone at the root surface by the supply of fresh nutrient solution during and soon after the irrigation events. This supply was fully available to the roots soon after the irrigation events, at which times its uptake rate behaved purely in accordance with the Michaelis–Menten equation (Eq. [4]). Moreover, a higher irrigation frequency maintains higher dissolved P and K concentrations in the substrate solution, by shortening the period during which precipitation takes place. The indirect effect of irrigation frequency on nutrient availability is manifested through the higher convective and diffusive fluxes of dissolved nutrient from the substrate solution to the root surface, which increase with increasing irrigation frequency. The findings that increasing the fertiliser rate improved nutrient uptakes and plant yield, and that increased irrigation frequency resulted in systematic diminution of the nutrients uptake enhancement, may indicate that the main effect of increased fertigation frequency was related to an improvement in nutritional status, mainly with regard to P (Silber et al., 2003, 2005a; Xu et al.,

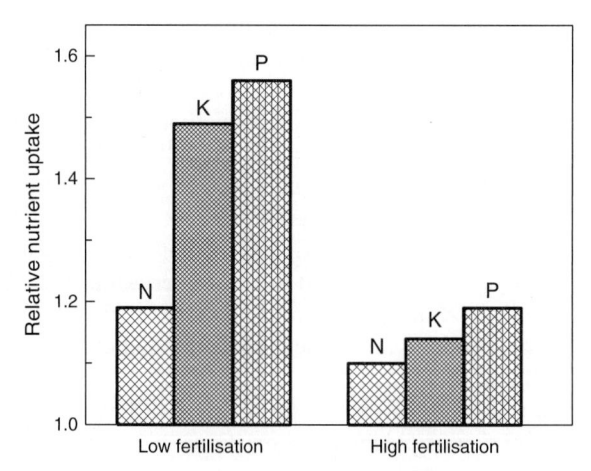

FIGURE 8.13 Relationships between fertilisation level and irrigation frequency. Relative nutrient uptake was defined as the increases of N, P and K in lettuce plants under high-frequency irrigation (18 daily irrigation events) relative to those under low-frequency irrigation (2 daily irrigation events).

2004). Thus, increasing the irrigation frequency may compensate for certain nutrient deficiencies and in the above examples the lower yields of plants fertigated at low frequency might have been a result of nutrient shortage rather than water shortage.

8.4.2 DIRECT AND INDIRECT OUTCOMES OF IRRIGATION FREQUENCY ON PLANT GROWTH

8.4.2.1 Root Growth and Root/Shoot Ratio

The adaptive changes in root growth and morphology in response to water stress, and nutrient deficiency in the rhizosphere were thoroughly described in the scientific literature (Waisel et al., 2002). Alterations of growth conditions generally lead to modifications of the root system, therefore irrigation frequency may have an effect on the root system through two main mechanisms: (i) the direct effect on the wetting patterns and water distribution in the soil volume, which modulate root distribution and growth (Phene et al., 1991; Clothier and Green, 1994; Coelho and Or, 1996, 1999); and (ii) indirect effect on nutrient availability (Lorenzo and Forde, 2001), especially that of P, which significantly modify root system efficiency (Glass, 2002; Lynch and Ho, 2005), including their effects on root hair density (Bates and Lynch, 1996; Ma et al., 2001) and root system architecture (Liao et al., 2001; Williamson et al., 2001).

Separation between the direct effect of irrigation frequency, that is, through the increases of θ in the root area, and the indirect effect of increased P availability is not straightforward. The effect of irrigation frequency on the root/shoot ratio has been reported to be smaller than that of P concentration (Xu et al., 2004; Silber et al., 2005a) and to be very sensitive to plant age (Xu et al., 2004). The reasons for the age-linked diminution of the effect of irrigation frequency on root/shoot ratio (Fig. 8.14A) could

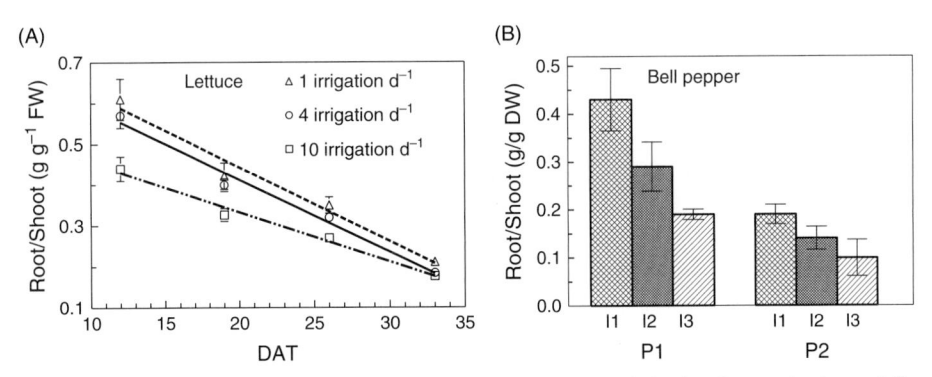

FIGURE 8.14 Root/shoot ratio of: (A) lettuce plants under three irrigation frequencies (same daily amount of water) – 1, 4 and 10 irrigation events per day – during lettuce growth (DAT – days after transplanting) (from Xu et al., 2004, with kind permission from Springer Science and Business Media); and (B) bell pepper under two water-P levels: P1 and P3 – 3 and 30 mg L^{-1}, respectively, and three irrigation frequencies (same daily amount of water): I1 (two irrigation events per day), I2 (four irrigation events per day) and I3 (1.5 min every 30 min throughout the day) (based on Silber et al., 2005a). Vertical bars indicate the standard error (not shown when smaller than the symbol).

be the following: (i) during early growth, the roots were mainly located at the top of the pots and were more sensitive to the drying and rewetting processes than later on; (ii) adsorption of the added P by the sand substrate induced stronger deficiency conditions in the early growth period than later; and (iii) the young roots in the early stages were mostly active roots, whereas in the later stages, part of the roots became inactive and probably masked the changes (Xu et al., 2004).

Similar to the findings with lettuce, observations of bell pepper plants showed high sensitivity of the root/shoot ratio to variations in irrigation frequency under low P application and a diminished response under high P application (Fig. 8.14B). Note that the root/shoot ratio under low P and high-frequency irrigation was very similar to that under high P and low-frequency irrigation, which may indicate that both treatments affect the same mechanism. Therefore, if a single mechanism is assumed and if the main impact of irrigation frequency actually arises from the increase of P availability and the consequently higher P uptake by the plant, then the root/shoot ratio should be correlated with a single plant indicator, common to both factors. Actually, irrespective of the experimental causes for leaf-P variations (P level or irrigation frequency), the values of root/shoot ratio were significantly correlated with leaf-P concentrations (Fig. 8.15).

8.4.2.2 Yield and Physiological Aspects

It has been shown that yield gained under high irrigation frequency can be primarily related to increased availability of nutrients, especially P (Mbagwu and Osuigwe, 1985; Silber et al., 2003, 2005a). The relationships between DW production of several crops and leaf-P concentrations induced by irrigation frequency are depicted in Fig. 8.16. Multiple stepwise regressions relating nutrient concentrations in the plant to yield revealed a significant correlation between the DW production and P concentration in

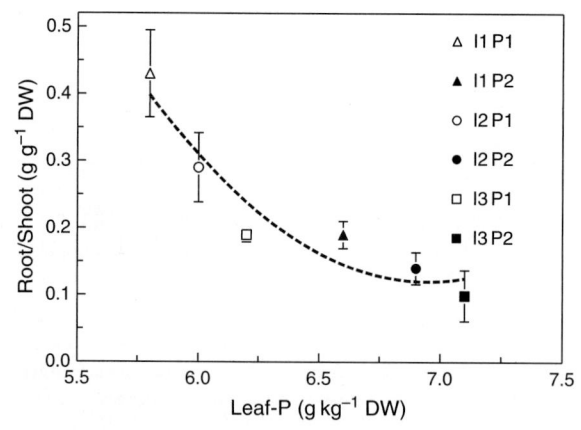

FIGURE 8.15 Relationships between root/shoot ratio and leaf-P concentration in bell pepper plants. The treatments included three irrigation frequencies (same daily amount of water): I1 (two irrigation events per day), I2 (four irrigation events per day) and I3 (1.5 min every 30 min throughout the day); and two water-P levels: P1 and P2 – 3 and 30 mg L^{-1}, respectively (based on Silber et al., 2005a). Vertical bars indicate the standard error (not shown when smaller than the symbol).

8.4 INTEGRATED EFFECT OF IRRIGATION FREQUENCY AND NUTRIENTS LEVEL

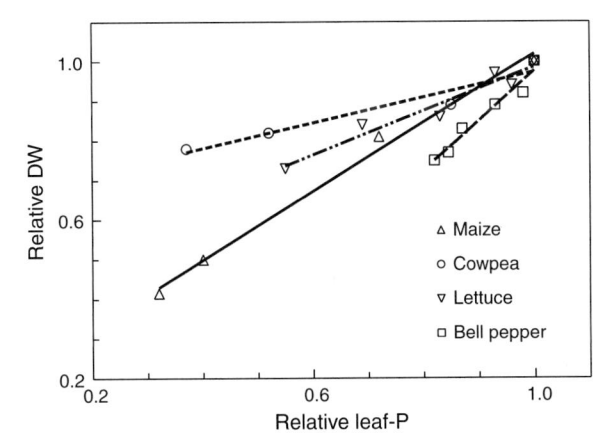

FIGURE 8.16 Relationships between DW production and leaf-P concentrations. The values are relative to the DW and leaf-P concentrations obtained in the best treatment (data for maize and cowpea were taken from Mbagwu and Osuigwe, 1985; for lettuce from Silber et al., 2003 and for bell pepper from Silber et al., 2005a).

leaves (Silber et al., 2003, 2005a; Xu et al., 2004), indicating that the main effect of fertigation frequency was related to improvements in P mobilisation and uptake.

In addition to its effects on soil-water tension and uptake, P availability, root system geometry and performance, and DW production, detailed above, irrigation frequency directly and indirectly influenced other processes in plants. Indirect effects of irrigation frequency on the concentrations of starch in the leaves and of sucrose and reducing sugars in the fruits have been reported (Silber et al., 2005a). Increased starch and reduced sucrose and hexose concentrations have previously been found in phosphorus-deficient plants (Craft-Brandner, 1992; Qiu and Israel, 1992; Paul and Stitt, 1993; Plaxton and Carswell, 1999); therefore, the differences in the concentrations of starch in the leaves and of sucrose and reducing sugars in the fruits were attributed to the variations in leaf P (Silber et al., 2005a).

A considerable and important effect of irrigation frequency on blossom-end rot (BER) incidence has been reported recently (Silber et al., 2005a; Fig. 8.17A). The cause(s) of high BER incidence under low-frequency fertigation is/are unclear, but it is generally accepted that BER incidence may be associated with water stress, for example soil water deficit, high osmotic pressure or high salinity (Ho et al., 1995; Saure, 2001; Adams, 2002). BER has also been related to Ca deficiency and, especially, to low Ca transport to the fruits, particularly to the distal fruit tissue (Ho et al., 1993, 1995; Marcelis and Ho, 1999; Ho and White, 2005). However, unlike BER incidence, the fruit Ca concentrations were almost unaffected by the fertigation frequency (Silber et al., 2005a). The discrepancy between the Ca concentrations in the fruits and BER incidence was consistent with the general remark of Saure (2001) that the role of Ca in BER (in tomato fruit) should be reassessed. Interestingly, a negative correlation was obtained between the number of fruits with BER that accumulated throughout the experiment and the fruit Mn concentrations (Fig. 8.17B).

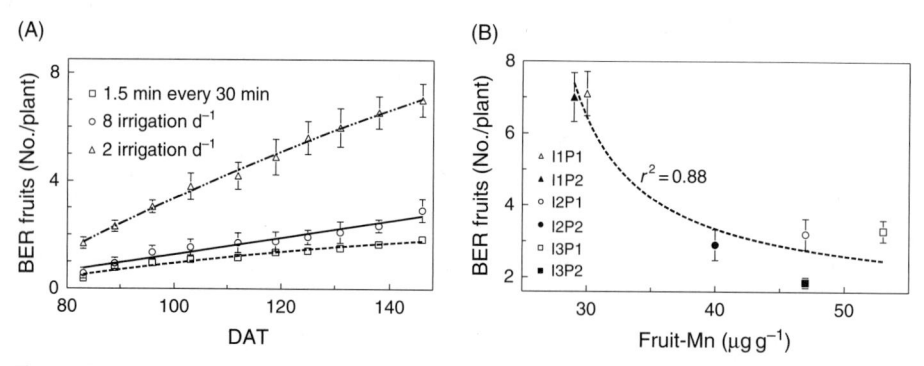

FIGURE 8.17 (A) Effect of irrigation frequency on the cumulative number BER-affected fruits per plant. (B) Relationships between cumulative number of BER-affected fruits and fruit-Mn concentration. Vertical lines indicate the standard error (not shown when smaller than the symbol) (from Silber et al., 2005a, with kind permission from Springer Science and Business Media).

It is much to be expected that precipitation of insoluble compounds might have reduced Mn solubility to a very low level and, therefore, that a high fertigation frequency would improve Mn availability. In a previous study, it was found that features of physiological disorders such as BER included the production of oxygen-free radicals, and decreases in anti-oxidative compounds and enzymatic activities in the fruit tissue (Aktas et al., 2005). Manganese has been found to play a crucial role in enzyme activities and in detoxification of oxygen-free radicals; therefore, the relationships between BER incidence and fruit Mn concentration presented in Fig. 8.17B may indicate that BER is related to Mn deficiency. However, future studies are needed to evaluate this hypothesis.

8.5 SALINITY EFFECT ON CROP PRODUCTION

8.5.1 GENERAL

High salt concentrations usually harm crops through two main mechanisms. The first is increase of the osmotic potential, both outside the roots where it depresses the external water potential, and in plant organs as a result of salt accumulation. The second mechanism comprises specific ion effects, for example ion toxicity and/or disturbed mineral nutrition (Läuchli and Epstein, 1990). Note that an increase of osmotic potential in plant organs is not necessarily deleterious and is frequently used to improve fruit quality. Growth reduction caused by salinity is a time-dependent process, and Munns (1993, 2002) proposed a two-phase model to depict the detrimental effects of salinity on plant growth. The first phase includes a rapid (minutes, hours) apparent impairment caused by the increase of external osmotic potential (the first of the above mechanisms), and is basically identical to water stress (Munns, 2002). The second phase is much slower, taking days or weeks to develop, and leads to internal injuries caused by excessive salt accumulation in transpiring leaves.

The majority of crops grown in intensive agriculture are defined as glycophyte plants, which are sensitive to salinity. The yields of such plants diminish as the salinity increases above a threshold value, which is specific for each crop (Maas and Hoffman, 1977; Maas, 1990; Shanon and Grieve, 1999; Sonneveld et al., 1999). The model of Maas and Hoffman (1977) is often used to describe the response of soil-grown crops to salinity; however, pertinent growth conditions such as temperature, humidity, irradiance, the atmospheric CO_2, concentration and irrigation frequency affect both threshold values and yield reduction gradients (Hoffman and Rawlins, 1971; Feigin et al., 1988; Salim, 1989; Zeroni and Gale, 1989; Pasternak and De Malach, 1995; Sonneveld et al., 2004). Nutritional elements added to soil-grown plants via fertilisation normally contribute only a minor part of the total ion concentration, whereas in soilless culture this contribution is relatively more significant: it has been reported that in Dutch greenhouses, N and K formed one-third of the total ion concentration (Sonneveld et al., 1990). Sonneveld et al. (2004) argued that salinity and nutrition effects would be mixed in the linear model of Maas and Hoffman (1977) and used polynomial and exponential models to describe the effect of the solution electrical conductivity (EC) on substrate-grown crops. The various osmotic effects were induced by addition of varied amounts of nutritional elements (mainly NO_3 and K); therefore, the yield reductions at low EC values reported by Sonneveld et al. (2004) probably resulted from simple nutrient deficiencies rather than real osmotic effects.

The EC in the soil solution is usually higher than that in the irrigating water (Mass and Hoffman, 1977; Sonneveld, 1988), as a result of slow and inefficient leaching processes caused by narrow and tortuous soil pores. In contrast, most of the soilless media contain wide pores that enable fast and efficient leaching (Raviv et al., 2002; Chap. 3), so that the EC in the medium solution can be controlled. Consequently, the actual EC in the root zone of soilless-grown plants may be lower than that of soil-grown plants under similar irrigation conditions. Salinity may build up in a recirculating system; this will be presented and discussed in Chap. 9. The salinity of the water is determined by the ion contents in the raw water and in the salts added as fertilisers. As the goal in soilless culture is to maintain the optimal growth conditions for the crops, it is necessary to minimise the contribution of fertilisers to the salinity of the irrigation water and the substrate solution.

Within the relevant range, there is a linear relationship between the salinity and the EC of the solution, therefore the EC is widely used as a measure of the salinity of the solution. A 1-cmol L^{-1} solution has an EC of about 1 dS m^{-1} and an osmotic pressure of ~ -0.036 MPa (at 25°C). The general recommendation for source water is that the EC should be below 1.0 dS m^{-1} (Schröder and Lieth, 2002), but higher values are common in semi-arid and arid zones.

8.5.2 SALINITY-NUTRIENTS RELATIONSHIPS

Salinity may interfere with mineral nutrition acquisition by plants in two ways (Grattan and Grieve, 1992): (i) the total ionic strength of the soil solution, regardless of its composition, can reduce nutrient uptake and translocation; and (ii) uptake competition with specific ions such as sodium and chloride can reduce nutrient uptake. These interactions may lead to Na-induced Ca and/or K deficiencies (Volkmar et al., 1998) and Cl^--induced inhibition of NO_3^- uptake (Xu et al., 2000). Navarro et al. (2002) found that the negative effects of salinity on pepper production and fruit damage by BER were more severe under exposure to chlorides than to sulphates. Application of a large amount of K might increase the plant's capacity for osmotic adjustment in saline habitats (Cerda et al., 1995), because K is the most abundant cation in the cytoplasm of glycophytes (Marschner, 1995). Chow et al. (1990) found that spinach (*Spinacia oleracea* L.) required more K for shoot growth under high-salinity conditions than under low-salinity conditions, and Kafkafi (1987) suggested that the salinity tolerance of crops could be improved by the appropriate use of nutrients. There is abundant evidence that Na and the Na/Ca ratio can affect K uptake and accumulation within plant cells and organs (Lahaye and Epstein, 1971; Kent and Läuchli, 1985; Lynch and Läuchli, 1985; Cramer et al., 1985, 1986; Ben-Hayyim et al., 1985, 1987; Chow et al., 1990), and salt tolerance seems to be correlated with selectivity for K uptake over Na (Läuchli and Stelter, 1982; Wrona and Epstein, 1985), although salinity had no effect on the leaf K content of two pepper cultivars (Chartzoulakis and Klapaki, 2000). Ben-Hayyim et al. (1987) found that growth of cultured citrus cells in solutions containing various concentrations and proportions of NaCl and $CaCl_2$ was a function of internal K concentration, irrespective of the NaCl concentration. It has been hypothesised that K application could reduce the deleterious effects of salinity on plant development (Khalil et al., 1967; Kafkafi, 1984; Ben-Hayyim et al., 1987). However, contradictory results were obtained regarding the effects of K fertilisation under saline conditions on whole plant and field crops. These results included a reduction in salinity damage to various crops when high concentrations of K were present in the media (Lagerwerff and Eagle, 1962; Helal and Mengel, 1979; Jeschke and Nassery, 1981; Chow et al., 1990; Ben Asher and Pacardo, 1997), no response to K under salinity (Jeschke and Wolf, 1988; Feigin et al., 1991), and even a negative effect of K nutrition on salt tolerance (Lunin and Gallatin, 1965; Lauter et al., 1988; Bar-Tal et al., 1991).

It has been reported that the detrimental effect of salinity was closely related to the oxygen regime in the root zone (Drew and Dikumwin, 1985; Drew and Läuchli, 1985; Drew et al., 1988). The combination of high NaCl level and anoxic conditions accelerated Na^- transport to maize shoots while depressing K^+ transport and, consequently, the shoot Na/K ratio increased significantly. Breakdown of the Na^+ exclusion mechanism under oxygen deficiency may be responsible for the salinity damage of salt-sensitive plants (Drew and Läuchli, 1985).

Antagonism between Cl^- and NO_3^- uptake by plants was demonstrated in numerous publications and was reviewed by Xu et al. (2000). This antagonism was found in various plants, including substrate-grown crops such as melon and lettuce (Feigin, 1985), strawberry (*Fragaria* × *ananassa* Duchesne) (Wang et al., 1989), tomato (Kafkafi et al., 1982; Kafkafi, 1984; Feigin et al., 1987; Heuer and Feigin, 1993). However, no positive effects were obtained in tomato by adding high levels of NO_3^- at high salinities (Heuer and Feigin, 1993). The competition of Cl^- with NO_3^- was found to be stronger in salt-sensitive than in salt-tolerant plants (Xu et al., 2000), so that Bar et al. (1997) found that when the irrigation water contained NO_3^- at about 8–16 mM, even sensitive plants such as avocado (*Persea americana* Mill.) could survive at chloride concentrations of 8–16 mM, and Cordovilla et al. (1995) found that supplying NO_3^- considerably relieved the suppressive effects of Cl salinity on faba bean (*Vicia faba* L.). In the light of the above findings, Xu et al. (2000) recommended increasing the NO_3^- supply to sensitive crops irrigated with water containing high chloride concentrations.

Much information is available on the effect of saline-water irrigation on rose (*Rosa* × *hybrida*) production (Yaron et al., 1969; Hughes and Hanan, 1978; Fernandez Falcon et al., 1986; De Kreij and van den Berg, 1990; Urban et al., 1994; Baas et al., 1995; Raviv and Blom, 2001). The range of sodium and chloride concentrations that allow optimal production of roses is well defined (Bernstein, 1964; Yaron et al., 1969; Hughes and Hanan, 1978; Fernandez Falcon et al., 1986; De Kreij and van den Berg, 1990), and damage to the plants is known to occur under exposure to Cl or Na at as little as $4\,mmol\,L^{-1}$ (Hughes and Hanan, 1978). Irrigation with saline water increased the concentration of chloride in rose tissue (Yaron et al., 1969; Bernstein et al., 1972; Hughes and Hanan, 1978; Baas and van den Berg, 1999; Cabrera and Perdomo, 2003; Bernstein et al., 2006). The reported threshold leaf chloride concentration for rose plant damage is $4.5\,g\,kg^{-1}$ (Bernstein et al., 1972; Feigin et al., 1988). Several studies found that the sodium content in rose leaves was unaffected by salinity (Sadasivaiah and Holley, 1973; Cabrera and Perdomo, 2003; Bernstein et al., 2006). However, Na exclusion is not general to all rootstocks; other reports described the stimulation of Na accumulation under salinity (Fernandez Falcon et al., 1986; Baas and van den Berg, 1999).

Ben Asher and Pacardo (1997) proposed a conceptual model in which enhanced fertilisation with KNO_3 under saline conditions might reduce the toxic effect of salinity, even if KNO_3 increased the soil osmotic potential. At two soil salinity levels ($2\,dS\,m^{-1}$ and $8\,dS\,m^{-1}$), application of KNO_3 was associated with a significant increase in the dry weight of the shoot of groundnut (*Arachis hypogaea* L.) by up to 150–300 kg KNO_3 ha^{-1}. Both the growth and the development of tomato plants irrigated with a saline solution of EC $5.5\,dS\,m^{-1}$ and containing 50 mM NaCl were improved by adding 2 mM KNO_3 to the saline solution, and to a lesser extent by adding 20 mM $Ca(NO_3)_2$, at EC of $7.5\,dS\,m^{-1}$ (Satti et al., 1994). However, an external K supply was not required for root growth of castor bean (*Ricinus communis* L.) under saline conditions (Jeschke and Wolf, 1988). Tip-burn symptoms in Chinese cabbage (*Brassica rapa*), induced by NaCl plus $CaCl_2$ salinity, were not alleviated by the addition of KNO_3 (Feigin et al., 1991).

The uptake interaction between chloride and phosphorus appears to be complex; there is no direct competition, because of the great differences in their physical and physiological properties (Xu et al., 2000). However, in several studies phosphate uptake was stimulated when chloride concentration was low and suppressed when it was high, whereas in other studies no effect was obtained (Xu et al., 2000). Only a few studies found that addition of adequate P could also help to alleviate salt stress

(Champagnol, 1979; Awad et al., 1990) and that it increased the yields of foxtail millet (*Setaria italica* L.) and clover (*Trifolium alexandrinum* Fahli) grown in a saline soil (Ravikovitch and Yoles, 1971).

Calcium plays an important role in the integrity, selectivity and permeability of plant membranes (Hanson, 1984; Grattan and Grieve, 1999; Hirschi, 2004). It has been shown that high salinity and high sodium concentration displace calcium from membrane-binding sites, and so cause disfunctioning and alteration of the membrane permeability and selectivity (Cramer et al., 1985). Several studies with various plant species demonstrated that calcium could ameliorate the adverse effects of sodium on plant membranes, roots and whole plants (Lahaye and Epstein, 1971; Kent and Läuchli, 1985; Cramer et al., 1986). Carvajal et al. (2000) reported that high-NaCl solution (50 mM) decreased the passage of water through detached roots of melon by reducing the activity of water channels (aquaporins), but pre-treatment with addition of 10 mM $CaCl_2$ solution prevented the decrease in root hydraulic conductance. A similar effect was observed when the flux of Ca^{2+} into the xylem and the Ca^{2+} concentration in the plasma membrane of the root cells were determined. The ameliorative effect of Ca^{2+} on NaCl stress could be related to the functioning of the water-channel. The negative effects of salinity (50 mM NaCl) on hydroponically grown pepper plants with respect to tissue calcium concentration and water relations were mitigated if there was a supply of Ca^{2+}, and this effect increased when the root-zone temperature was increased (Cabanero et al., 2004). Supplementary Ca partially ameliorated the negative effects of moderate salinity (35 mM NaCl) on plant growth and fruit yield of strawberry (Kaya et al., 2002), and a similar effect was obtained with tomato (Navarro et al., 2000).

Several reviewers concluded that the results reported in the literature on the interaction between salinity and nutrients were contradictory (Kafkafi, 1984; Feigin, 1985). However, part of this conflict can be removed by using the definitions of Bernstein et al. (1974), who described three different types of idealised salinity/nutrition interactions: (i) increased salt tolerance at suboptimal nutritional levels; (ii) independent effects of salinity and nutrition at optimal and suboptimal nutritional levels; and (iii) decreased salt tolerance at suboptimal nutritional levels. These definitions (Bernstein et al., 1974) were used for defining the interaction on the basis of plant performance at optimal fertilisation relative to that at suboptimal fertilisation (Maas, 1990; Grattan and Grieve, 1992) and are depicted in Fig. 8.18. This approach can be used to analyse data from a variety of publications and to find out what type of interaction(s) exist(s). Some of the conclusions on salinity/nutrition interaction published in the early literature were based on a misleading interpretation. A schematic illustration of the interactions between salinity and fertility levels is presented in Fig. 8.19. Under low salinity, nutrient-deficient control plant growth was accompanied by an increased salt tolerance function (left part of Fig. 8.19). A misleading interpretation in such case is that fertilisation relieved the salinity stress. As the salinity level increased, plant stress imposed by salinity equalled that imposed by nutrient deficiency (unaffected salt tolerance function), but as the salinity level increased further, salinity was the main yield-limiting factor, and a decreasing salt-tolerance function was obtained (right part of Fig. 8.19).

8.5 SALINITY EFFECT ON CROP PRODUCTION

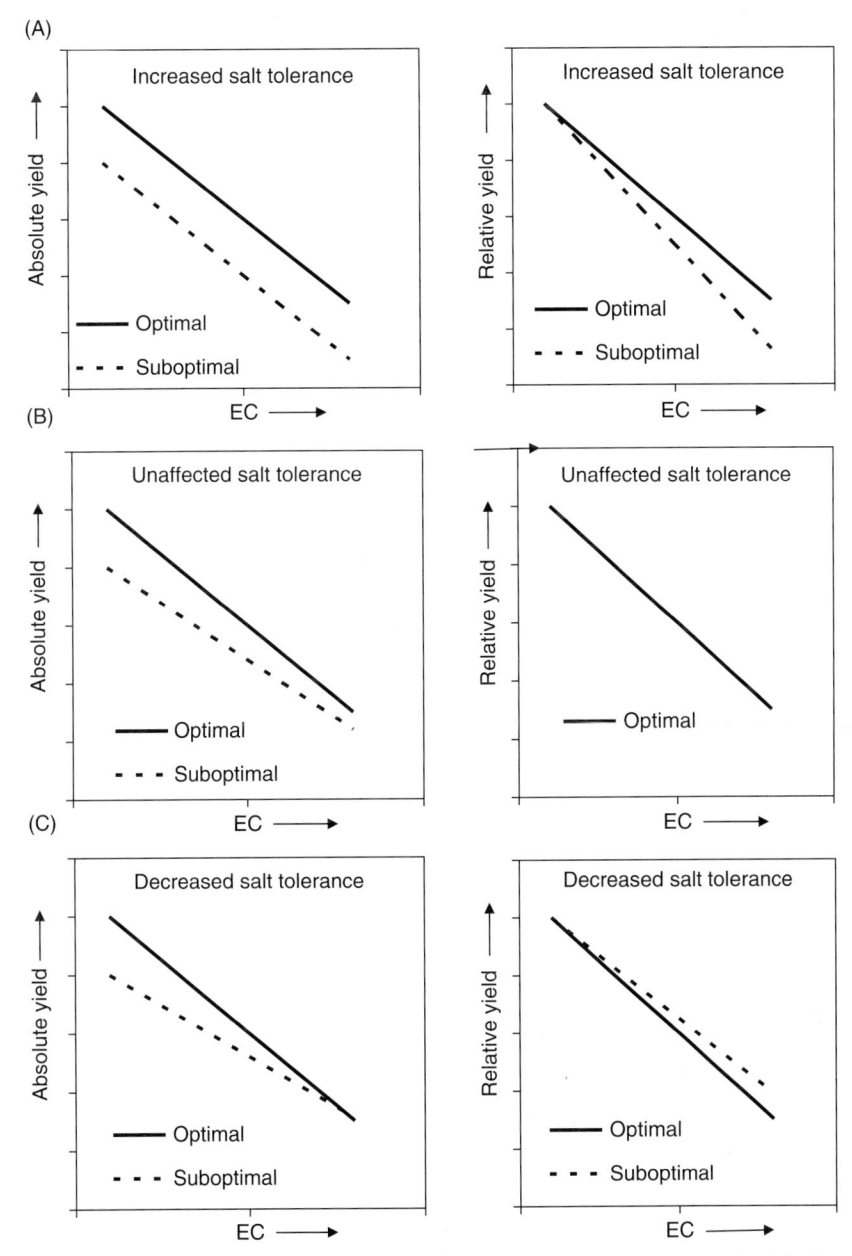

FIGURE 8.18 Types of idealised interactions between salinity and nutrients level, in their effects on absolute and relative yields. (A) increased salt tolerance under deficient nutritional levels; (B) unaffected effects of salinity and nutrition at optimal and deficient nutritional levels; and (C) decreased salt tolerance under deficient nutritional levels (based on Bernstein et al., 1974).

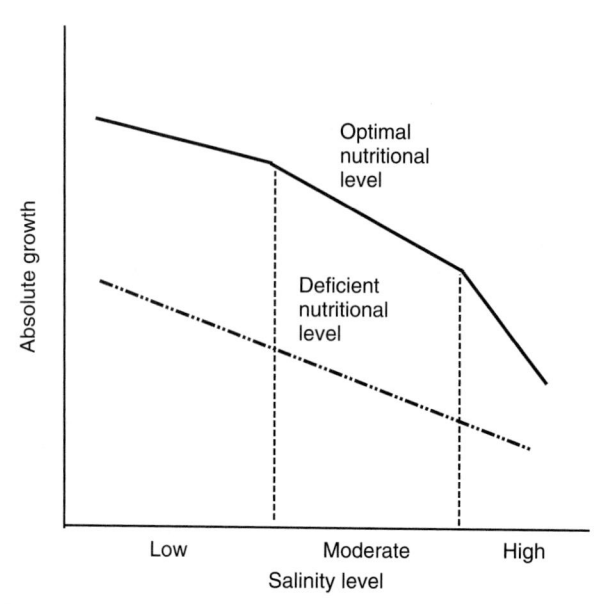

FIGURE 8.19 Effect of salinity levels on absolute yield under optimal (solid line) and deficient (broken line) nutrition levels (based on Bernstein et al., 1974 and Grattan and Grieve, 1992).

Shalhevet (1994) concluded in a review on the use of water of marginal quality for crop production that fertiliser application does not materially change the salinity response function of crops. In some cases, a better response to fertiliser was found under non-saline than under saline conditions, but only rarely could a relative advantage to fertiliser application under saline conditions be found. Grattan and Grieve (1999) concluded that, 'Despite a large number of studies that demonstrate that salinity reduces nutrient uptake and accumulation or affects nutrient partitioning within the plant, little evidence exists that adding nutrients at levels above what is considered optimal in non-saline environments, improves crop yield.'

8.5.3 YIELD QUALITY INDUCED BY SALINITY

The effects of salinity on crop quality depend mainly on the plant part that is marketed. Fruit quality is generally more tolerant to high salinity than fresh leaf yield (Xu et al., 2000). Salinity improves both the taste and the visual quality of tomato and melon fruits (Mizrahi, 1982; Mizrahi and Pasternak, 1985; Sonneveld and Welles, 1988; Adams, 1991; Li et al., 2000). The increase in quality under saline conditions was attributed to the significantly higher contents of total soluble solids, sugars and aromatic constituents (Ehret and Ho, 1986; Adams, 1991; Sonneveld and van den Burg, 1991; Li et al., 2000; Mulholland et al., 2002). Mulholland et al. (2002) found, in a split root experiment, that tomato fruit quality could be enhanced by high EC in half of the root system. In contrast, salinity reduced product quality and yield of leafy vegetables such as lettuce and cabbage (*Brassica oleracea* L.), because of tip burn (Feigin et al., 1991); under high salinity the leaves accumulated chloride, whereas both soluble sugars and vitamin C contents were significantly lower than under non-saline conditions (Xu et al., 2000).

The negative effects of salinity are often related to Ca deficiency. Salinity increases the occurrence of BER in tomato (Adams, 1991; Adams and Ho, 1992, 1993; Ho and White, 2005) and pepper (Sonneveld, 1988; Tadesse et al., 1999; Navarro et al., 2002), internal rot in eggplant (Savvas and Lenz, 1996), tipburn in lettuce and cabbage and blackheart in celery (*Apium graveolens* L.) (Feigin et al., 1991; Sonneveld, 1988). Adams and Ho (1993) found that increasing the EC of NFT solution from 3 to 17 dS m^{-1} by adding NaCl had no effect on the occurrence of BER in tomato fruits, whereas achieving the same EC increase by adding the major nutrients resulted in almost 100 per cent BER. Flower production is generally more sensitive to salinity than vegetable yield. For example, the recommended optimal EC values for anthurium (*Anthurium*) are 0.7 and 1.0 dS m^{-1} in the applied solution and the drainage water, respectively (Sonneveld and Voogt, 1993). The yield of various cut-flower crops decreases as salinity increases above a specific threshold (Sonneveld, 1988; Sonneveld et al., 1999; Adams, 2002). Baas et al. (1995) found that salinity reduced the yield and quality of carnation (*Dianthus caryophyllus* L.) and gerbera flowers. They reported that a similar yield decrease occurred in a high-EC treatment which used increased concentrations of major elements, and concluded that the yield decrease was not related to specific ion-toxic effects of sodium and chloride. Pre- and post-harvest application of calcium improved the quality and vase-life of cut flowers such as gerbera (Gerasopoulos and Chebli, 1999) and rose (Halevy and Mayak, 1981; Michalczuk et al., 1989; Starkey and Pedersen, 1997; Torre et al., 1999, 2001; Bar-Tal et al., 2001c). Maintaining low Ca/K and Ca/Na ratios in the solution can have considerable negative impacts on fruit and flower quality in soilless culture under saline conditions.

8.6 COMPOSITION OF NUTRIENT SOLUTION

The source of fertiliser to be used in fertigation has to be chosen very carefully. Solubility and compatibility with the specific substrate and irrigation water characteristics must be considered. Loss of nutrients such as P, K or NH$_4^+$, or of micronutrients, through adsorption on strongly adsorbing substrates and/or through precipitation processes detailed elsewhere, (Chap. 6) may lead to deficiency levels in the rhizosphere. Thus, nutrient deficiency is a common occurrence even if the nutrient concentration in the irrigation water is adequate. The quality of the irrigation water plays an important part in determining which kind of fertilisers should be used. An improper choice may cause difficulties by clogging the irrigation system or by corrosion, or it may lead to losses of nutrient elements. Generally, the fertilisers used should have complete solubility, quick dissolution and high nutrient content, and should be free of any toxic constituents. Methods for preparing nutrient solutions for various crops, based on target pH, EC and nutrient concentration ratios, have been methodically explored and widely reviewed in the literature (Steiner, 1961, 1984; Graves, 1983; Sonneveld and Straver, 1994; Schwartz, 1995; Resh, 1998; De Kreij et al., 1999; Savvas and Adamidis, 1999; Savvas, 2001; Adams, 2002; Sonneveld, 2002; Jones, 2005) and, therefore, are discussed only very briefly in this chapter.

8.6.1 pH MANIPULATION

The rhizosphere pH plays a key role in the determination of nutrient availability to the plant, therefore, pH management, which can lead to optimisation of the nutritional regime, attracts a lot of interest in both soil-based and soilless culture. However, manipulations of rhizosphere pH are more feasible and their accomplishment is more practicable in soilless media, characterised by low buffer capacity and subjected to frequent applications of nutrient solutions, than in soils. The optimal pH values recommended in soilless culture are usually within the range of 5.0–6.0 (Adams, 2002; Sonneveld, 2002) but, nevertheless, pH values between 6.0 and 7.0 are adequate for the majority of crops (Graves, 1983).

High pH values (>7.0) in the irrigation water are undesirable, because Ca and Mg carbonates and orthophosphates may precipitate in the irrigation lines and drippers. In addition, at pH above 7.2 the proportion of $H_2PO_4^-$ species decreases and that of HPO_4^{2-} increases (Lindsay, 1979). Since the $H_2PO_4^-$ ion is much more available to plants than HPO_4^{2-} (Hendrix, 1967), high pH may induce P-deficient conditions, even if there is adequate total P in the solution, and high substrate pH may reduce micronutrients availability to plants, because of precipitation reactions (Lindsay, 1979). However, it is possible to ensure an acceptable rhizosphere pH range by fertiliser management, with higher concentrations of micronutrients compensating for lower solubility at high pH (Smith et al., 2004a,b). To avoid precipitation at pH > 5, and to facilitate sufficient transport towards roots in soil, micronutrients are added in solutions as chelates of organic ligands. Chelating agents such as EDTA (ethylene-diamine tetraacetic acid), DTPA (diethylenetriamine pentaacetic acid) or EDDHA [ethylenediamine di (o-hydroxyphenylacetic acid)] may be used to maintain micronutrient availability for plants. The synthetic chelates lose their capacity to retain microelements at pH < 3 and, therefore, it is advisable to store chelated microelements separately from acid fertilisers, at pH > 5, and to inject them into the water after the acid stock solution. In general, it is not recommended to use any component, such as ammonia, that might raise the solution pH when injected into the irrigation water. Compounds that may reduce the irrigation water pH are nitric (HNO_3) and orthophosphoric (H_3PO_4) acids. Depending on their price and the raw water composition, these acids may be used to lower the irrigation water pH. The acid dose required to attain a target pH depends primarily on the HCO_3 concentration in the tap-water. However, at HCO_3 concentrations above 2–3 mM, it is not possible to use only H_3PO_4 to adjust the pH to 5.0–5.5 without adding excessive amounts of phosphate to the solution. For instance, if the raw water contains 3 mM HCO_3, about 2–2.5 mmol of H_3PO_4 should be added per litre of nutrient solution, in order to neutralise as much HCO_3 as is necessary to achieve the target pH. However, the H_2PO_4 concentration in the nutrient solution should normally not exceed values above 1.0 mM. The use of nitric acid is not limited by such considerations, since the target NO_3 concentration in the nutrient solutions is normally higher than 10 mM, and the HCO_3 concentrations never reach this level. It is important to note that the response of the medium pH to any chemical treatment should be expected to vary according to the components of the medium, the chemical properties of the tap-water and plant interactions (Bishko et al., 2003). Acidic pHs (below 5.0) may be detrimental to root membranes and may increase the Al and Mn concentrations in the substrate solution to toxic levels. Soilless media based on un-amended sphagnum peat and pine bark are characterised by low pH, and liming materials, that is $Ca(OH)_2$ and $Mg(OH)_2$, or carbonates, that is $CaCO_3$ and $MgCO_3$, must be added to raise the pH (Argo and Biernbaum, 1996, 1997a,b; Elliott, 1996; Bishko et al., 2002).

The approach presented by Steiner (1961, 1984), Sonneveld and Straver (1994), De Kreij et al. (1999) and Savvas and Adamidis (1999), who focussed basically on attempting to attain an optimal pH in the irrigation water (target pH), is generally accepted (Graves, 1983; Schwartz, 1995; Resh, 1998; Savvas, 2001; Adams, 2002; Sonneveld, 2002; Jones, 2005). However, there is no doubt that the target pH of the irrigation water is only a simple part of the many aspects of nutrient availability in real circumstances, that is, in the presence of a microbial population and plant activity. Even in the absence of plants, the H^+ contributed from the nitrification process should not be ignored. Two moles of H^+ are released during nitrification of one mole of NH_4^+ (Eqs. [5] and [6]). The quantity of H^+ released to the solution via nitrification per litre of substrate is 2×10^{-4} mol, under the following conditions: irrigation-NH_4^+ concentration, $1 \, mmol \, L^{-1}$ (10 per cent of the NH_4^+ that corresponds to a total N concentration of $10 \, mmol \, L^{-1}$); 10 L of substrate per plant; and irrigation dose of $1 \, L \, d^{-1}$ per plant. Thus, according to this simple calculation there could be more H^+ added via nitrification than via acidification, and this should be taken into account. Furthermore, the chemical properties of the root environment are notably different from those of the bulk medium, as a result of changes induced by the plants, as detailed elsewhere in this book (Chap. 6). Roots strongly affect their environment to a range of 1–2 mm, as depicted in Fig. 8.5 in this chapter and Fig. 6.17 in Chap. 6, by secreting H^+ and citric acid. Published data on root length (Bar-Yosef, 1999; Xu et al., 2004; Silber et al., 2005a) indicate that the volume of substrate that is strongly affected by root activity amounts to only 10–15 per cent of the total volume. Thus, the concentrations of H^+, OH^- and organic components excreted from roots into this limited volume may exceed those in the bulk substrate by an order of magnitude. The straightforward conclusion to be drawn from these remarks is that the real challenge lies in managing nutrient availability and pH in the vicinity of the roots (rhizosphere) rather than in the irrigation solution.

8.6.2 SALINITY CONTROL

According to uptake data presented in the previous section, and assuming a daily irrigation of 5 mm ($50 \, m^3 \, ha^{-1}$), N and K concentrations in the irrigation water at the time of maximum consumption rate may reach 15–20 mmol L^{-1}, which correspond to an EC of 1.5–2.0 dS m^{-1}. Under such conditions, especially when the water EC is above 1.0 dS m^{-1}, care should be taken to minimise the amount of accompanying ions added with the N or K. For example, KCl, which is a cheap source of K, should be replaced with KNO$_3$ and K$_2$HPO$_4$, and NH$_4$NO$_3$ and urea should be preferred over (NH$_4$)$_2$SO$_4$. Under high chloride concentrations, K$_2$SO$_4$ should be the preferred source of K, because of its lower salt index and absence of Cl (Zehler et al., 1981; von Braunschweig, 1986). Sodium-based fertilisers (e.g., NaNO$_3$ or NaH$_2$PO$_4$) are not advisable sources of N and P, because of the adverse effects of Na on plant functioning.

In most cases, the source of N will be NH$_4$NO$_3$, (NH$_4$)$_2$SO$_4$ and multi-nutrient fertilisers such as KNO$_3$, Ca(NO$_3$)$_2$ and (NH$_4$)$_3$PO$_4$. The multi-nutrient fertilisers should be preferred when the salt content of the water is high. Potassium application in irrigation water is relatively problem-free, thanks to the high solubility of most K salts; KCl is more soluble than K$_2$SO$_4$, and in high-Ca water there is a possibility of CaSO$_4$ precipitation. On the other hand, when the Cl concentration is high, KCl is not to be recommended for sensitive crops, but KNO$_3$ is a very efficient source of both K and N.

REFERENCES

Adams, P. (1991). Effect of increasing the salinity of the nutrient solution with major nutrients or sodium chloride on the yield, quality and composition of tomatoes grown in rockwool. *J. Hortic. Sci.*, **66**, 201–207.

Adams, P. (2002). Nutritional control in hydroponics. In *Hydroponic Production of Vegetables and Ornamentals* (D. Savvas and H.C. Passam, Eds). Athens: Embryo Publications, pp. 211–262.

Adams, P. and Ho, L.C. (1992). The susceptibility of modern tomato cultivars to blossom-end rot in relation to salinity. *J. Hortic. Sci.*, **67**, 827–839.

Adams, P. and Ho, L.C. (1993). Effects of environment on the uptake and distribution of calcium in tomato and on the incidence of blossom-end rot. *Plant Soil*, **154**, 127–132.

Aki, I.A., Savvas, D., Papadantonakis, N., et al. (2003). Influence of ammonium to total nitrogen supply ratio on growth, yield and fruit quality of tomato grown in a closed hydroponic system. *Europ. J. Hortic. Sci.*, **68**, 204–211.

Aktas, H., Karni, L., Chang, D.L., et al. (2005). The suppression of salinity-associated oxygen radicals production, in pepper (*Capsicum annuum* L.) fruit, by manganese, zinc and calcium in relation to its sensitivity to blossom-end rot. *Physiol. Plant.*, **123**, 67–74.

Argo, W.R. and Biernbaum, J.A. (1996). The effect of lime, irrigation-water source, and water-soluble fertilizer on root-zone pH, electrical conductivity, and macronutrient management of container root media with impatiens. *J. Am. Soc. Hortic. Sci.*, **121**, 442–452.

Argo, W.R. and Biernbaum, J.A. (1997a). Lime, water source, and fertilizer nitrogen form affect medium pH and nitrogen accumulation and uptake. *Hort. Science*, **32**, 71–74.

Argo, W.R. and Biernbaum, J.A. (1997b). The effect of root media on root-zone pH, calcium, and magnesium management in containers with impatiens. *J. Am. Soc. Hortic. Sci.*, **122**, 275–284.

Awad, A.S., Edwards, D.G. and Campbell, L.C. (1990). Phosphorus enhancement of salt tolerance of tomato crop. *Crop Sci.*, **30**, 123–128.

Baas, R. and van den Berg, D. (1999). Sodium accumulation and nutrient discharge in recirculation systems: A case study with roses. *Acta Hort.* (ISHS), **507**, 157–164.

Baas, R., Nijssen, H.M.C., van den Berg, T.J.M. and Warmenhoven, M.G. (1995). Yield and quality of carnation (*Dianthus caryophyllus* L.) and gerbera (*Gerbera jamesonii* L.) in a closed nutrient system as affected by sodium chloride. *Sci. Hort.*, **61**, 273–284.

References

Bar, Y., Apelbaum, A., Kafkafi, U. and Goren, R. (1997). Relationship between chloride and nitrate and its effect on growth and mineral composition of avocado and citrus plants. *J. Plant Nut.*, **20**, 715–731.

Barber, S.A. (1995). *Soil Nutrient Bioavailability: A Mechanistic Approach* (2nd edn). New York: Wiley.

Barker, A. and Mills, H.A. (1979). Ammonium and nitrate nutrition of horticultural plants. *Hortic. Rev.*, **1**, 395–423.

Barrow, N.J. (1993). Mechanisms of reaction of zinc with soil and soil components. In *Zinc in Soils and Plants* (A.D. Robson, ed.). Dordrecht: Kluwer Academic Publishers, pp. 15–31.

Bar-Tal, A., Aloni, B., Karni, L., et al. (2001a). Nitrogen nutrition of greenhouse pepper. I. Effects of nitrogen concentration and $NO_3:NH_4$ ratio on yield, fruit shape, and the incidence of blossom-end rot in relation to plant mineral composition. *Hort. Science*, **36**, 1244–1251.

Bar-Tal, A., Aloni, B., Karni, L., et al. (2001b). Nitrogen nutrition of greenhouse pepper: II. Effects of nitrogen concentration and $NO_3:NH_4$ ratio on growth, transpiration, and nutrient uptake. *Hort. Science*, **36**, 1252–1259.

Bar-Tal, A., Baas, R., Ganmore-Neumann, R., et al. (2001c). Rose flower production and quality as affected by Ca concentration in the flower. *Agronomie*, **21**, 393–402.

Bar-Tal, A., Bar-Yosef, B. and Kafkafi, U. (1994a). Modeling pepper seedling growth and nutrient uptake as a function of cultural conditions. *Agron. J.*, **85**, 718–724.

Bar-Tal, A., Feigenbaum, S. and Sparks, D.L. (1991). Potassium-salinity interactions in irrigated corn. *Irrig. Sci.*, **12**, 27–35.

Bar-Tal, A., Feigin, A., Rylski, I. and Pressman, E. (1994b). Effects of root pruning and $N-NO_3$ solution concentration on nutrient uptake and transpiration of tomato plants. *Sci. Hortic.*, **58**, 77–90.

Bar-Yosef, B. (1996). Root excretion and their environmental effects: Influence on availability of phosphorus. In *Plant Roots: Te Hidden Half*, 2nd edn (Y. Waisel, A. Eshel and U. Kafkafi, eds). New York: Marcel Dekker, pp. 581–605.

Bar-Yosef, B. (1999). Advances in fertigation. *Adv. Agron.*, **65**, 1–77.

Bar-Yosef, B., Silber, A., Markovitch, T., et al. (2000). Waxflower (*Chamelaucium uncinatum* L.) response to nitrogen trickle fertigation and iron foliar fertilization: III. Water and nutrients status in soil (in Hebrew). *Prachim, J.*, Israeli Flower Assoc. **16**, 61–75.

Bates, T.R. and Lynch, J.P. (1996). Stimulation of root hair elongation in Arabidopsis thaliana by low phosphorus availability. *Plant Cell Environ.*, **19**, 529–538.

Ben Asher, J. and Pacardo, E. (1997). K uptake by root systems grown in saline soil: A conceptual model and experimental results. In *Food Security in the WANA Region, the Essential Need for Balanced Fertilization* (A.E. Johnston, ed.). Basel: International Potash Institute, pp. 360–369.

Ben-Hayyim, G., Kafkafi, U. and Ganmore-Neumann, R. (1987). Role of internal potassium in maintaining growth of cultured citrus cells on increasing NaCl and $CaCl_2$ concentrations. *Plant Physiol.*, **85**, 434–439.

Ben-Hayyim, G., Spiegel-Roy, P. and Neumann, H. (1985). Relation between ion accumulation of salt-sensitive and isolated stable salt-tolerant cell lines of *Citrus aurantium*. *Plant Physiol.*, **78**, 144–148.

Bennett, A.C. (1974). Toxic effects of aqueous ammonia, copper, zinc, lead, boron, and manganese on root growth. In *The Plant Root and its Environment* (E.W. Carson, ed.). Virginia, Charlottesville: University, pp. 669–683.

Bennett, A.C. and Adams, F. (1970a). Calcium deficiency and ammonia toxicity as separate causal factors of $(NH_4)_2HPO_4$ injury to seedlings. *Soil Sci. Soc. Am. Proc.*, **34**, 255–259.

Bennett, A.C. and Adams, F. (1970b). Concentration of $NH_3(aq)$ required for incipient NH_3 toxicity to seedlings. *Soil Sci. Soc. Am. Proc.*, **34**, 259–263.

Bernstein, L. (1964). Salinity and roses. *Amer. Rose Ann.*, **49**, 120–124.

Bernstein, L., Francois, L.E. and Clark, R.A. (1972). Salt tolerance of ornamental shrubs and ground covers. *J. Am. Soc. Hortic. Sci.*, **97**, 550–556.

Bernstein, L., Francois, L.E. and Clark, R.A. (1974). Interactive effects of salinity and fertility on yields of grains and vegetables. *Agron. J.*, **66**, 412–421.

Bernstein, N., Bar-Tal, A., Friedman, H., et al. (2006). Application of treated wastewater for cultivation of roses (*Rosa hybrida*) in soil-less culture. *Sci. Hortic.*, **108**, 185–193.

Bhadoria, P.B.S., Kaselowski, J., Claassen, N. and Jungk, A. (1991). Phosphate diffusion coefficients in soil as affected by bulk density and water content. *Z. Pfanzenernahr. Bodenk.*, **154**, 53–57.

Bishko, A.J., Fisher, P.R. and Argo, W.R. (2002). Quantifying the pH-response of a peat-based medium to application of basic chemicals. *Hort. Science*, **37**, 511–515.

Bishko, A.J., Fisher, P.R. and Argo, W.R. (2003). The pH-response of a peat-based medium to application of acid-reaction chemicals. *Hort. Science*, **38**, 26–31.

Bloom, A.J., Meyerhoff, P.A., Taylor, A.R. and Rost, T.L. (2003). Root development and adsorption of ammonium and nitrate from the rhizosphere. *J. Plant Growth Regul.*, **21**, 416–431.

Britto, D.T. and Kronzucker, H.J. (2002). NH_4^+ toxicity in higher plants: A critical review. *J. Plant Physiol.*, **159**, 567–584.

Britto, D.T., Glass, A.D.M., Kronzucker, H.J. and Siddiqi, M.Y. (2001a). Cytosolic concentrations and transmembrane fluxes of NH_4^+/NH_3. An evaluation of recent proposal. *Plant Physiol.*, **125**, 523–526.

Britto, D.T., Siddiqi, M.Y., Glass, A.D.M. and Kronzucker, H.J. (2001b). Futile transmembrane NH_4^+ cycling: A cellular hypothesis to explain ammonium toxicity in plants. *Proc. Natl. Acad. Sci. USA*, **98**, 4255–4298.

Cabanero, F.J., Carvajal, M. and Martinez, V. (2004). Does calcium determine water uptake under saline conditions in pepper plants, or is it water flux which determines calcium uptake? *Plant Sci.*, **166**, 443–450.

Cabrera, R.I. and Perdomo, P. (2003). Reassessing the salinity tolerance of greenhouse roses under soilless production conditions. *Hort. Science*, **38**, 533–536.

Carvajal, M., A Cerda, A. and Vicente Martinez, V. (2000). Does calcium ameliorate the negative effect of NaCl on melon root water transport by regulating aquaporin activity? *New Phytol.*, **145**, 439–447.

Cerda, A.J., Pardines, M., Botella, M.A. and Martinez, V. (1995). Effect of potassium on growth, water relations and the inorganic and organic solute contents for two maize cultivars grown under saline conditions. *J. Plant Nutr.*, **18**, 839–851.

Champagnol, F. (1979). Relationship between phosphate nutrition of plants and salt toxicity. *Phosphorus Agric.*, **76**, 35–44.

Chartzoulakis, K. and Klapaki, G. (2000). Response of two greenhouse pepper hybrids to NaCl salinity during different growth stages. *Sci. Hort.*, **86**, 247–260.

Chow, W.S., Ball, C.M. and Anderson, J.M. (1990). Growth and photosynthetic responses of spinach to salinity: Implications of K^+ nutrition for salt tolerance. *Aust. J. Plant Physiol.*, **17**, 563–578.

Claassen, N. and Steingrobe, B. (1999). Mechanistic simulation models for a better understanding of nutrient uptake from soil. In *Mineral Nutrition of Crops, Fundamental Mechanisms and Implications* (Z. Rengel, ed.). New York: Haworth Press, pp. 327–369.

Claussen, W. (2002). Growth, water use efficiency, and proline content of hydroponically grown tomato plants as affected by nitrogen source and nutrient concentration. *Plant Soil*, **247**, 199–209.

Claussen, W. and Lenz, F. (1995). Effect of ammonium and nitrate on net photosynthesis, flower formation, growth and yield of eggplant. *Plant Soil*, **171**, 267–274.

Clothier, B.E. and Green, S.R. (1994). Root zone processes and the efficient use of irrigation water. *Agric. Water Manag.*, **25**, 1–12.

Coelho, F. and Or, D. (1996). A parameter model for two-dimensional water uptake by corn roots under drip irrigation. *Soil Sci. Soc. Am. J.*, **60**, 1039–1049.

Coelho, F. and Or, D. (1999). Root distribution and water uptake patterns of corn under surface and subsurface drip irrigation. *Plant Soil*, **206**, 123–136.

Cordovilla, M.P., Ocana, A., Ligero, F. and Lluch, C. (1995). Growth and macronutrient contents of faba bean plants: Effects of salinity and nitrate nutrition. *J. Plant Nutr.*, **18**, 1611–1628.

Craft-Brandner, S.J. (1992). Phosphorus nutrition influence on starch and sucrose accumulation, and activities of ADP-glucose pyrophosphorylase and sucrose-phosphate synthase during the grain filling period in soybean. *Plant Physiol.*, **98**, 1133–1138.

Cramer, G.R., Läuchli, A. and Polito, V.S. (1985). Displacement of Ca^{+2} by Na^+ from the plasmalemma of root cells. A primary response to salt stress? *Plant Physiol.*, **79**, 207–211.

Cramer, G.R., Läuchli, A. and Epstein, E. (1986). Effects of NaCl and $CaCl_2$ on ion activities in complex nutrient solutions and root growth of cotton. *Plant Physiol.*, **81**, 792–797.

Darrah, P.R., Nye, P.H. and White, R.E. (1985). Modelling growth responses of soil nitrifiers to additions of ammonium sulphate and ammonium chloride. *Plant Soil*, **86**, 425–439.

Darrah, P.R., Nye, P.H. and White, R.E. (1987a). The effect of high solute concentrations on nitrification rates in soil. *Plant Soil*, **97**, 37–45.

Darrah, P.R., White, R.E. and Nye, P.H. (1987b). A theoretical consideration of the implications of cell clustering for the prediction of nitrification in soil. *Plant Soil*, **99**, 387–400.

De Boer, W. and Kowalchuk, G.A. (2001). Nitrification in acid soils: Micro-organisms and mechanisms. *Soil Biol. Biochem.*, **33**, 853–866.

De Kreij, C. and van den Berg, T.J.M. (1990). Nutrient uptake, production and quality of *Rosa hybrida* in rockwool as affected by electrical conductivity of the nutrient solution. In *Plant Nutrition – Physiology and Applications* (M.L. van Beuichem, ed.). Dordrecht, The Netherlands: Proceedings XI Int. Plant Nutrition Coll. 1989, Kluwer, pp. 519–525.

De Kreij, C., Voogt, W. and Baas, R. (1999). Nutrient solutions and water qualities for soilless cultures. Brochure No. 196, Research Station Floriculture and Glasshouse Vegetables, The Netherlands: Naaldwijk.

Drew, M.C. and Dikumwin, E. (1985). Sodium exclusion from the shoots by roots of *Zea mays* (cv. LG 11) and its breakdown with oxygen deficiency. *J. Exp. Bot.*, **36**, 55–61.

Drew, M.C., Guenther, J. and Läuchli, A. (1988). The combined effects of salinity and root anoxia on growth and net Na^+ and K^+ accumulation in *Zea mays* grown in solution culture. *Ann. Bot.*, **61**, 41–53.

Drew, M.C. and Läuchli, A. (1985). Oxygen-dependent exclusion of sodium ions from shoots by roots of *Zea mays* (cv. Pioneer 3906) in relation to salinity damage. *Plant Physiol.*, **79**, 171–176.

Ehret, D.L. and Ho, L.C. (1986). The effect of salinity on dry matter partitioning and fruit growth in tomatoes grown in nutrient film culture. *J. Hortic. Sci.*, **61**, 361–367.

Elliott, G.C. (1996). pH management in container media. *Commun. Soil Sci. Plant Anal.*, **27**, 635–649.

Epstein, E. (1994). The anomaly of silicon in plant biology. *Proc. Natl. Acad. Sci. USA*, **91**, 11–17.

Epstein, E. (1999). Silicon. *Annu. Rev. Plant Physiol. Mol. Biol.*, **50**, 641–664.

Epstein, E. and Bloom, A.J. (2005). *Mineral Nutrition of Plants: Principles and Perspectives* (2nd edn). Sunderland, MA: Sinauer Associates.

Feigin, A. (1985). Fertilization management of crops irrigated with saline water. *Plant Soil*, **89**, 285–299.

Feigin, A., Ganmore-Neumann, R. and Gilead, S. (1988). Response of rose plants to Cl and NO_3 salinity under different CO_2 atmospheres. Proc. 7th Int. Cong. on Soilless Culture. Flevohof, the Netherlands. Published by the Secretariat of ISOSC. The Netherlands: Wageningen, pp. 135–143.

Feigin, A., Pressman, E., Imas, P. and Miltau, O. (1991). Combined effects of KNO_3 and salinity on yield and chemical composition of lettuce and chinese cabbage. *Irrig. Sci.*, **12**, 223–230.

Feigin, A., Rylski, I., Meiri, M. and Shalhevet, J. (1987). Response of melon and tomato plants to chloride-nitrate ratio in saline nutrient solutions. *J. Plant Nutr.*, **10**, 1787–1794.

Fernandez Falcon, M., Alvarez Gonzalez, C.E., Garcia, V. and Baez, J. (1986). The effect of chloride and bicarbonate levels in the irrigation water on nutrient content, production and quality of cut roses 'Mercedes'. *Sci. Hortic.*, **29**, 373–385.

Focht, D.D. and Verstraete, W. (1977). Biochemical ecology of nitrification and denitrification. *Adv. Microbiol. Ecol.*, **1**, 135–214.

Forde, B.G. and Clarkson, D.T. (1999). Nitrate and ammonium nutrition of plants: Physiological and molecular perspectives. *Adv. Bot. Res.*, **30**, 1–90.

Gahoonia, T.S., Claassen, N. and Jungk, A. (1992). Mobilization of phosphate in different soils by ryegrass supplied with ammonium or nitrate. *Plant Soil*, **140**, 241–248.

Gahoonia, T.S. and Nielsen, N.E. (1992). Control of pH at the soil-root interface. *Plant Soil*, **140**, 49–54.

Gahoonia, T.S., Raza, S. and Nielsen, N.E. (1994). Phosphorus depletion in the rhizosphere as influenced by soil moisture. *Plant Soil*, **159**, 213–218.

Ganmore-Neumann, R. and Kafkafi, U. (1980a). Root temperature and percentage NO_3^-/NH_4^+ effect on tomato plant development. I. Morphology and growth. *Agron. J.*, **72**, 756–761.

Ganmore-Neumann, R. and Kafkafi, U. (1980b). Root temperature and percentage NO_3^-/NH_4^+ effect on tomato plant development I. Morphology and growth. *Agron. J.*, **72**, 762–766.

Ganmore-Neumann, R. and Kafkafi, U. (1983). The effect of root temperature and NO_3^-/NH_4^+ ratio on strawberry plant. I. Growth, flowering, and root development. *Agron. J.*, **75**, 941–947.

Gardner, W.K., Barber, D.A. and Barbery, D.G. (1983). The acquisition of phosphorus by *Lupinus albus* L. III. The probable mechanism by which phosphorus movement in the soil/root interface is enhanced. *Plant Soil*, **70**, 107–124.

Gardner, W.K., Barbery, D.G. and Barber, D.A. (1982a). The acquisition of phosphorus by *Lupinus albus* L. I. Some characteristics of the soil/root interface. *Plant Soil*, **68**, 19–32.

Gardner, W.K., Barbery, D.G. and Barber, D.A. (1982b). The acquisition of phosphorus by *Lupinus albus* L. II. The effect of varying phosphorus supply and soil type on some characteristics of the soil/root interface. *Plant Soil*, **68**, 33–41.

Gary, C., Bertin, N., Frossard, J.S. and Le Bot, J. (1998). High mineral contents explain the low construction cost of leaves, stems and fruits of tomato plants. *J. Exp. Bot.*, **49**, 49–57.

Gerasopoulos, D. and Chebli, B. (1999). Effects of pre- and postharvest calcium applications on the vase life of cut gerberas. *J. Hortic. Sci. Biotechnol.*, **74**, 78–81.

Glass, A.D.M. (2002). Nutrient absorption by plant roots: Regulation of uptake to match plant demand. In *Plant Roots: The Hidden Half*, 2nd edn (Y. Waisel, A. Eshel, A. and U. Kafkafi, eds). New York: Marcel Dekker, pp. 571–586.

Goudriaan, J. and van Laar, H.H. (1994). *Modelling Potential Crop Growth Processes. Textbook with Exercise*. Dordrecht: Kluwer Academic Publishers.

Grattan, S.R. and Grieve, C.M. (1992). Mineral element acquisition and growth response of plants grown in saline environments. *Agric. Ecosys. Environ.*, **38**, 275–300.

Grattan, S.R. and Grieve, C.M. (1999). Salinity-mineral nutrient relations in horticultural crops. *Sci. Hortic.*, **78**, 27–157.

Graves, C.J. (1983). The nutrient film technique. *Hortic. Rev.*, **5**, 1–44.

Gur, A., Bravdo, B.A. and Mizrahi, Y. (1972). Physiological response of apple trees to supraoptimal root temperature. *Physiol. Plant.*, **27**, 130–138.

Halevy, A. and Mayak, S. (1981). Senescence and postharvest physiology of cut flowers – Part 2. *Hortic. Rev.*, **3**, 59–143.

Hanan, J.J. (1998). *Greenhouses: Advanced Technology for Protected Horticulture*. Boca Raton: CRC Press LLC.

Hanson, J.B. (1984). The function of calcium in plant nutrition. In *Advances in Plant Nutrition*, Vol. 1 (P.B. Tinker and A. Läuchli, eds). New York: Praeger, pp. 149–208.

Helal, H.M. and Mengel, K. (1979). Nitrogen metabolism of young barley plants as affected by NaCl salinity and potassium. *Plant Soil*, **51**, 547–562.

Hendrix, J.E. (1967). The effect of pH on the uptake and accumulation of phosphate and sulfate ions by bean plants. *Am. J. Bot.*, **54**, 560–564.

Heuer, B. and Feigin, A. (1993). Interactive effects of chloride and nitrate on photosynthesis and related growth parameters in tomatoes. *Photosynthetica*, **28**, 549–554.

Hirschi, K.D. (2004). Update on calcium nutrition. The calcium conundrum, both versatile nutrient and specific signal. *Plant Physiol.*, **136**, 2438–2442.

Ho, L.C., Adams, P., Li, X.Z., et al. (1995). Response of Ca-efficient and Ca-inefficient tomato cultivars to salinity in plant growth, calcium accumulation and blossom-end rot. *J. Hortic. Sci.*, **70**, 909–918.

Ho, L.C., Belda, R., Brown, M., et al. (1993). Uptake and transport of calcium and the possible causes of blossom-end rot. *J. Exp. Bot.*, **44**, 509–518.

Ho, L.C. and White, P.J. (2005). A cellular hypothesis for the induction of blossom-end rot in tomato fruit. *Ann. Bot.*, **95**, 571–581.

Hoagland, D.R. and Arnon, D.I. (1950). The water-culture method for growing plants without soil. *Circular* **347**, Agricultural Experimental Station, University of California, Berkely, CA.

Hochmuth, G.J. (1992). Fertilizer management for drip-irrigated vegetables in Florida. *HortTechnol.*, **2**, 27–31.

Hochmuth, G.J. (2003). Progress in mineral nutrition and nutrient management for vegetable crops in the last 25 years. *Hort. Science*, **38**, 999–1003.

Hoffman, G.J. and Rawlins, S.L. (1971). Growth and water potential of root crops as influenced by salinity and relative humidity. *Agron. J.*, **63**, 877–880.

References

Hopmans, J.W. and Bristow, K.L. (2002). Current capabilities and future needs for root water and nutrient uptake modelling. *Adv. Agron.*, **77**, 103–183.

Hughes, H.E. and Hanan, J.J. (1978). Effect of salinity in water supplies on greenhouse rose production. *J. Am. Soc. Hortic. Sci.*, **103**, 694–699.

Imas, P., Bar-Yosef, B., Kafkafi, U. and Ganmore-Neumann, R. (1997). Release of carboxylic anions and protons by tomato roots in response to ammonium nitrate and pH in nutrient solution. *Plant Soil*, **191**, 27–34.

Jeschke, W.N. and Nassery, H. (1981). K^+-Na^+ selectivity in roots of *Triticum, Helianthus* and *Allium*. *Physiol. Plant.*, **52**, 217–224.

Jeschke, W.D. and Wolf, O. (1988). External potassium supply is not required for root growth in saline conditions: Experiments with *Ricinus communis* L. grown in a reciprocal split-root system. *J. Exp. Bot.*, **39**, 1149–1168.

Joiner, J.N., Poole, R.T. and Conover, C.A. (1983). Nutrition and fertilization of ornamental greenhouse crops. *Hortic. Rev.*, **5**, 317–403.

Jones, B.R. (2005). *Hydroponics: A Practical Guide for the Soilless Grower* (2nd edn). Boca Raton: St.-Lucie Press.

Jungk, A.O. (2002). Dynamics of nutrient movement at the soil-root interface. In *Plant Roots: The Hidden Half*, 3rd edn (Y. Waisel, A. Eshel and U. Kafkafi, eds). New York: Marcel Dekker, Inc., pp. 587–616.

Kafkafi, U. (1984). Plant nutrition under saline conditions. In *Soil Salinity Under Irrigation* (I. Shainberg and J. Shalhevet, eds). Berlin: Springer-Verlag.

Kafkafi, U. (1987). Plant nutrition under saline condition. *Fert. Agric.*, **95**, 3–17.

Kafkafi, U. (1990). Root temperature, concentration and the ratio of NO_3^-/NH_4^+ effect on plant development. *J. Plant Nutr.*, **13**, 1291–1306.

Kafkafi, U. (2001). Root zone parameters controlling plant growth in soilless culture. *Acta Hort.* (ISHS), **554**, 27–38.

Kafkafi, U., Valoras., N. and Letay, J. (1982). Chloride interaction with NO_3 and phosphate nutrition in tomato. *J. Plant Nutr.*, **5**, 1369–1385.

Kargbo, D., Skopp, J. and Knudsen, D. (1991). Control of nutrient mixing and uptake by irrigation frequency and relative humidity. *Agron. J.*, **83**, 1023–1028.

Kaya, C., Kirnak, H., Higgs, D. and Saltali, K. (2002). Supplementary calcium enhances plant growth and fruit yield in strawberry cultivars grown at high (NaCl) salinity. *J. Hortic. Sci. Biotechnol.*, **93**, 65–74.

Kent, L.M. and Läuchli, A. (1985). Germination and seedling growth of cotton: Salinity-calcium interaction. *Plant Cell Environ.*, **8**, 155–159.

Khalil, M.A., Amer, F. and Elgabli, M.M. (1967). A salinity-fertility interaction study on corn and cotton. *Soil Sci Soc. Am. Proc.*, **31**, 683–686.

Kinbursky, R.S. and Saltzman, S. (1990). CO_2-nitrification relationships in closed soil incubation vessels. *Soil Biol. Biochem.*, **22**, 571–572.

Kirkby, E. and Mengel. K. (1967). Ionic balance in different tissues of the tomato plant in relation to nitrate, urea, or ammonium nutrition. *Plant Physiol.*, **42**, 6–14.

Kirkby, E.A. and Knight, A.H. (1977). Influence of the level of nitrate nutrition on ion uptake and assimilation, organic acid accumulation, and cation-anion balance in whole tomato plants. *Plant Physiol.*, **60**, 349–353.

Kläring, H.P. (2001). Strategies to control water and nutrient supplies to greenhouse crops. A review. *Agronomie*, **21**, 311–321.

Kleiner, D. (1981). The transport of NH_3 and NH_4^+ across biological membranes. Biochim. *Biophys. Acta*, **639**, 41–52.

Kosegarten, H., Grolig, F., Esch, A., et al. (1999b). Effects of NH_4, NO_3^-, and HCO_3^- on apoplast pH in the outer cortex of root zones of maize as measured by fluorescence ratio of fluorescein boronic acid. *Planta*, **209**, 444–452.

Kosegarten, H., Hoffmann., B. and Mengel, K. (1999a). Apoplastic pH and Fe^{3+} reduction in young sunflower leaves. *Plant Physiol.*, **121**, 1069–1079.

Kosegarten, H., Hoffmann, B., Rroco, E., et al. (2004). Apoplastic pH and Fe^{III} reduction in young sunflower (*Helianthus annuus*) roots. *Physiol. Plant.*, **122**, 95–106.

Kosegarten, H., Wilson, G.H. and Esch, A. (1998). The effect of nitrate nutrition on iron chlorosis and leaf growth in sunflower (*Helianthus annuus*). *Eur. J. Agron.*, **8**, 282–292.

Kovar, J.L. and Claassen, N. (2005). Soil-root interactions and phosphorus nutrition of plants. In *Phosphorus: Agriculture and the Environment. Agronomy Monograph No. 46* (J.T. Sims and A.N. Sharpley, eds). Madison: Soil Science Society of America, Inc., Crop Science Society of America, Inc., Soil Science Society of America, Inc., pp. 379–414.

Kuchenbuch, R., Claassen, N. and Jungk, A. (1986a). Potassium availability in relation to soil moisture. I. Effect of soil moisture on potassium diffusion, root growth and potassium uptake of onion plants. *Plant Soil*, **95**, 221–231.

Kuchenbuch, R., Claassen, N. and Jungk, A. (1986b). Potassium availability in relation to soil moisture. II. Calculation by means of a mathematical simulation model. *Plant Soil*, **95**, 233–243.

Kyveryga, P.M., Blackmer, A.M., Ellsworth, J.H. and Isla, R. (2004). Soil pH effects on nitrification of fall-applied anhydrous ammonia. *Soil Sci. Soc. Am. J.*, **68**, 545–551.

Lad, J.N. and Jackson, R.B. (1982). Biochemistry of ammonification. In *Nitrogen in Agricultural Soils* (F.J. Stevenson, ed.). Madison: Soil Science Society of America, pp. 173–228.

Lagerwerff, J.V. and Eagle, H.E. (1962). Transpiration related to ion uptake by beans from saline substrates. *Soil Sci.*, **93**, 420–430.

Lahaye, P.A. and Epstein, E. (1971). Calcium and salt toleration by bean plants. *Physiol. Plant.*, **25**, 213–218.

Lang, H.J. and Elliot, G.C. (1991). Influence of ammonium: Nitrate ratio and nitrogen concentration on nitrification activity in soilless potting media. *J. Am. Soc. Hortic. Sci.*, **116**, 642–645.

Läuchli, A. and Epstein, E. (1990). Plant response to saline and sodic conditions. In *Agricultural Salinity Assessment and Management* (K.K. Tanji, ed.). New York: American Society of Civil Engineers, pp. 113–137.

Läuchli, A. and Stelter, A. (1982). Salt tolerance of cotton genotypes in relation to K/Na selectivity. In *Biosaline Research: A Look to the Future* (A. San Pietro, ed.). New York: Plenum Press, pp. 511–514.

Lauter, D.J., Meiri, A. and Shuali, M. (1988). Isoosmotic regulation of cotton and peanut at saline concentration of K and Na. *Plant Physiol.*, **87**, 911–916.

Li, L.Y., Stanghellini, C. and Challa, H. (2000). Effect of electrical conductivity and transpiration on production of greenhouse tomato (*Lycopersicon esculentum* L.) *Sci. Hortic.*, **88**, 11–29.

Liao, H., Rubio, G., Yan, X., et al. (2001). Effect of phosphorus availability on basal root shallowness in common bean. *Plant Soil*, **232**, 69–79.

Lindsay, W.L. (1979). *Chemical Equilibrium in Soils*. New York: Wiley.

Loneragan, J.F. (1997). Plant nutrition in the 20th and perspectives for the 21st century. *Plant Soil*, **196**, 163–174.

Lorenzo, H. and Forde, B. (2001). The nutritional control of root development. *Plant Soil*, **232**, 51–68.

Lunin, J. and Gallatin, M.H. (1965). Salinity-fertility interactions in relation to the growth and composition of beans. I. Effect of N, P and K. *Agron. J.*, **57**, 339–342.

Lynch, J. and Ho, M.D. (2005). Rhizoeconomics: Carbon costs of phosphorus acquisition. *Plant Soil*, **269**, 45–56.

Lynch, J. and Läuchli, A. (1985). Potassium transport in salt-stressed barley roots. *Planta Berl.*, **161**, 295–301.

Ma, Z., Bielenberg, D.G., Brown, K.M. and Lynch, J.P. (2001). Regulation of root hair density by phosphorus availability in *Arabidopsis thaliana*. *Plant Cell Environ.*, **24**, 459–467.

Maas, E.V. (1990). Crop salt tolerance. In *Agricultural Salinity Assessment and Management* (K.K. Tanji, ed.). New York: American Society of Civil Engineers, pp. 262–304.

Maas, E.V. and Hoffman, G.J. (1977). Crop salt tolerance – current assessment. *J. Irrig. Drain. Div. ASCE*, **103 [IR 2]**, 115–134.

Marcelis, L.F.M. and Ho, L.C. (1999). Blossom-end rot in relation to growth rate and calcium content in fruits of sweet pepper (*Capsicum annuum* L.). *J. Exp. Bot.*, **50**, 357–363.

Marschner, H. (1993). Zinc uptake from soils. In *Zinc in Soils and Plants* (A.D. Robson, ed.). Dordrecht: Kluwer Academic Publishers, pp. 59–77.

Marschner, H. (1995). *Mineral Nutrition of Higher Plants* (2nd edn). New York: Academic Press.

Marschner, H. and Römheld, V. (1996). Root-induced changes in the availability of micronutrients in the rhizosphere. In *Plant Roots: The Hidden Half*, 2nd edn (Y. Waisel, A. Eshel and U. Kafkafi, eds). New York: Marcel Dekker, pp. 557–579.

Martin, D.L., Stegman, E.C. and Fereres, E. (1990). Plant-soil-water relationship. In *Management of Farm Irrigation Systems* (G.J. Hoffman, T.A. Howell and K.H. Solomon, eds). St. Joseph, MI: Amer. Soc. Agric. Eng., pp. 15–29.

Maynard, D.N. and Lorenz, O.A. (1979). Seventy-five years of progress in the nutrition of vegetable crops. *Hort. Science*, **14**, 355–358.

Mbagwu, J.S.C. and Osuigwe, J.O. (1985). Effects of varying levels and frequencies of irrigation on growth, yield nutrient uptake and water use efficiency of maize and cowpeas on a sandy loam ultisol. *Plant Soil*, **84**, 181–192.

Mengel, K. (1994). Iron availability in plant tissues – iron chlorosis on calcareous soils. *Plant Soil*, **165**, 275–283.

Mengel, K. and Kirkby, E.A. (2001). *Principles of Plant Nutrition*. Bern: IPI.

Mengel, K., Planker, R. and Hoffmann, B. (1994). Relationship between leaf apoplast pH and iron chlorosis of sunflower (*Helianthus annuus*). *J. Plant Nutr.*, **17**, 1053–1065.

Michalczuk, B., Goszczynska, D.M., Rudnicki, R.M. and Halevy, A.H. (1989). Calcium promotes longevity and bud opening in cut rose flowers. *Isr. J. Bot.*, **38**, 209–215.

Mizrahi, Y. (1982). Effect of salinity on tomato fruit ripening. *Plant Physiol.*, **69**, 966–970.

Mizrahi, Y. and Pasternak, D. (1985). Effect of salinity on quality of various agricultural crops. *Plant Soil*, **89**, 301–307.

Mmolawa, K. and Or, D. (2000). Root zone solute dynamics under drip irrigation: A review. *Plant Soil*, **222**, 163–190.

Moorby, H., Nye, P.H. and White, R.E. (1984). The influence of nitrate solution on H^+ efflux by young rape plants (*Brassica napus* cv. Emerald). *Plant Soil*, **84**, 403–415.

Morley, P.S., Hardgrave, M., Bradley, M. and Pileeam, D.J. (1993). Susceptibility of sweet pepper (*Capsicum annuum* L.) cultivars in the calcium deficiency disorder 'Blossom-end rot'. In *Optimization of plant nutrition* (M.A.C. Fragoso, M.I. Van Beusichem, and Houwers, A. eds). Dordrecht, NL: Kluwer Academic Publishers, pp. 561–567.

Mulholland, B.J., Fussell, M., Edmondson, R.N., et al. (2002). The effect of split-root salinity stress on tomato leaf expansion, fruit yield and quality. *J. Hortic. Sci. Biotechnol.*, **77**, 509–519.

Munns, R. (1993). Physiological processes limiting plant growth in saline soil: Some dogmas and hypotheses. *Plant Cell Environ.*, **16**, 15–24.

Munns, R. (2002). Comparative physiology of salt and water stress. *Plant Cell Environ.*, **25**, 239–255.

Nakos, G.G. and Wolcott, A.R. (1972). Bacteriostatic effect of ammonium to *Nitrobacter agilis* in mixed culture with *Nitrosomonas europaea*. *Plant Soil*, **36**, 521–527.

Navarro, J.M., Garrido, C., Carvajal, M. and Martinez, V. (2002). Yield and fruit quality of pepper plants under sulphate and chloride salinity. *J. Hort. Sci. Biotech.*, **77**, 52–57.

Navarro, J.M., Martinez, V. and Carvajal, M. (2000). Ammonium, bicarbonate and calcium effects on tomato plants grown under saline conditions. *Plant Sci.*, **157**, 89–96.

Nelson, P.V. (1996). Macronutrient fertilizer program. In *A Grower's Guide to Water, Media and Nutrition for Greenhouse Crops* (W.D. Reed, ed.). Batavia: Ball Publishing, p. 314.

Nelson, P.V. (2003). *Greenhouse Operation and Management* (6th edn). Upper Saddle River, NJ: Prentice Hall.

Niemiera, A. and Wright, R.D. (1987a). Influence of temperature on nitrification in a pine bark medium. *Hort. Science*, **22**, 615–616.

Niemiera, A. and Wright, R.D. (1987b). Influence of NH_4-N application rate on nitrification in a pine bark medium. *Hort. Science*, **22**, 616–618.

Nye, P.H. (1966). The measurement and mechanism of ion diffusion in soil: I. The relation between self-diffusion and bulk diffusion. *J. Soil Sci.*, **17**, 16–23.

Nye, P.H. (1981). Changes of pH across the rhizosphere induced by roots. *Plant Soil*, **61**, 7–26.

Olness, A. (1999). A description of the general effect of pH on formation of nitrate in soils. *J. Plant Nutr. Soil Sci.*, **162**, 549–556.

Pasternak, D. and De Malach, Y. (1995). Irrigation with brackish-water under desert conditions. X. Irrigation management of tomatoes (*Lycopersicon esculentum* Mill.) on desert sand dunes. *Agric. Water Manag.*, **28**, 121–132.

Paul, M.J. and Stitt, M. (1993). Effects of nitrogen and phosphorus deficiencies on levels of carbohydrates, respiratory enzymes and metabolites in seedlings of tobacco and their response to exogenous sucrose. *Plant Cell Environ.*, **16**, 1047–1057.

Phene, C., Davis, K.P., Hutmacher, R.B., et al. (1991). Effect of high-frequency surface and subsurface drip irrigation on root distribution of sweet corn. *Irrig. Sci.*, **12**, 135–140.

Plaxton, W.C. and Carswell, C. (1999). Metabolic aspects of the phosphate starvation response in plants. In *Plant Responses to Environmental Stress* (H. Lerner, ed.). New York: Marcel Dekker, pp. 350–372.

Qiu, J. and Israel, W. (1992). Diurnal starch accumulation and utilization in phosphorus-deficient soybean plants. *Plant Physiol.*, **98**, 316–323.

Raven, J.A. (1985). Regulation of pH and generation of osmolarity in vascular land plants: Costs and benefits in relation to efficiency of use of water, energy and nitrogen. *New Phytol.*, **101**, 25–77.

Raven, J.A. (1986). Biochemical disposal of excess H^+ in growing plants? *New Phytol.*, **104**, 175–206.

Ravikovitch, S. and Yoles, D. (1971). The influence of phosphorus and nitrogen on millet and clover growing in soils affected by salinity. *Plant Soil*, **35**, 555–567.

Raviv, M. and Blom, T.J. (2001). The effect of water availability and quality on photosynthesis and productivity of soilless-grown cut roses – A review. *Sci. Hortic.*, **88**, 257–276.

Raviv, M., Wallach, R., Silber, A. and Bar-Tal, A. (2002). Substrates and their analysis. In *Hydroponic Production of Vegetables and Ornamentals* (D. Savvas and H.C. Passam, eds). Athens: Embryo Publications, pp. 25–101.

Rawlins, S.L. and Raats, P.A.C. (1975). Prospects for high-frequency irrigation. *Science*, **188**, 604–610.

Resh, H.M. (1998). *Hydroponic Food Production, a Definitive Guidebook for the Advanced Home Gardener and the Commercial Hydroponic Grower* (5th edn). Santa Barbara: Woodbridge Press.

Rodrigo, A., Recous, S., Neel, C. and Mary, B. (1997). Modelling temperature and moisture effects on C-N transformations in soils: Comparison of nine models. *Ecol. Model.*, **102**, 325–339.

Russell, C.A., Fillery, I.R.P., Bootsma, N. and McInns, K.J. (2002). Effect of temperature and nitrogen source on nitrification in a sandy soil. *Commun. Soil Sci. Plant Anal.*, **33**, 1975–1989.

Sadasivaiah, S.P. and Holley, W.D. (1973). Ion balance in nutrition of greenhouse roses. Roses Inc. Bull. November, pp. 1–27.

Salim, M. (1989). Effects of salinity and relative humidity on growth and ionic relations of plants. *New Phytol.*, **13**, 13–20.

Satti, S.M.E., Lopez, M. and Al-Said, F.A. (1994). Salinity induced changes in vegetative and reproductive growth in tomato. *Commun. Soil Sci. Plant Anal.*, **25**, 2825–2840.

Saure, M.C. (2001). Blossom-end rot of tomato (*Lycopersicon esculentum* Mill.) – a calcium- or a stress-related disorder. *Sci. Hortic.*, **90**, 193–208.

Savvas, D. (2001). Nutritional management of vegetables and ornamental plants in hydroponics. In *Crop Management and Postharvest Handling of Horticultural Products*, Vol. 1 (R. Dris, R. Niskanen and S.M. Jain, eds). Enfield, N.H.: Quality Management. Science Publishers, pp. 37–87.

Savvas, D. and Adamidis, K. (1999). Automated management of nutrient solutions based on target electrical conductivity, pH, and nutrient concentrations ratios. *J. Plant Nutr.*, **22**, 1415–1432.

Savvas, D., Karagianni, V., Kotsiras, A., et al. (2003). Interactions between ammonium and pH of the nutrient solution supplied to gerbera (*Gerbera jamesonii*) grown in pumice. *Plant Soil*, **254**, 393–402.

Savvas, D. and Lenz, F. (1996). Influence of NaCl concentration in the nutrient solution on mineral composition of eggplants grown in sand culture. *Angew Bot.*, **70**, 124–127.

Schröder, F.G. and Lieth, J.H. (2002). Irrigation control in hydroponics. In *Hydroponic Production of Vegetables and Ornamentals* (D. Savvas and H. Passam, eds). Athens: Embryo Publications, pp. 263–298.

Schwartz, M. (1995). *Soilless Culture Management*. Advanced Series in Agricultural Sciences, Vol. 24. Berlin: Springer-Verlag.

Shalhevet, J. (1994). Using water of marginal quality for crop production: Major issues. *Agric. Water Manag.*, **25**, 233–269.

Shanon, M.C. and Grieve, C.M. (1999). Tolerance of vegetables crop to salinity. *Scientia Hortic.*, **78**, 5–38.

Silber, A., Ackerman, A., Mitchnick, B., et al. (2000a). The response of *Leucadendron* 'Safari Sunset' to the fertilisation regime. *J. Agric. Sci.*, **135**, 27–34.

Silber, A., Ackerman, A., Mitchnick, B., et al. (2000b). pH dominates *Leucadendron* 'Safari sunset' growth. *Hort. Science*, **35**, 647–650.

Silber, A., Ben Yones, L. and Dori, I. (2004). Rhizosphere pH as a result of nitrogen level and NH_4/NO_3 ratio and its effect on Zn availability and on growth of rice flower (*Ozothamnus diosmifolius*). *Plant Soil*, **262**, 205–213.

Silber, A., Bruner, M., Kenig, E., et al. (2005a). High fertigation frequency and phosphorus level: Effects on summer-grown bell pepper growth and blossom-end rot incidence. *Plant Soil*, **270**, 135–146.

Silber, A., Ganmore-Neumann, R. and Ben-Jaacov, J. (1998). Effects of nutrient addition on growth and rhizosphere pH of *Leucadendron* 'Safari Sunset'. *Plant Soil*, **199**, 205–211.

Silber, A., Levkovitch, I., Dinkin, I., et al. (2005b). High irrigation frequency and temporal NH_4 concentration: The effects on soilless-grown bell pepper. *J. Hortic. Sci. Biotech.*, **80**, 233–239.

Silber, A., Xu, G., Levkovitch, I., et al. (2003). High fertigation frequency: The effects on uptake of nutrients, water and plant growth. *Plant Soil*, **253**, 467–477.

Silberbush, M. (2002). Simulation of ion uptake from the soil. In *Plant Roots: The Hidden Half*, 3rd edn (Y. Waisel, A. Eshel and U. Kafkafi, eds). New York: Marcel Dekker, Inc., pp. 651–661.

Siraj-Ali, M.S., Peterson, J.C. and Tayama, H.K. (1987). Influence of nutrient solution pH on the uptake of plant nutrients and growth of *Chrysanthemum morifolium* 'Bright golden Anne' in hydroponic culture. *J. Plant Nutr.*, **10**, 2161–2168.

Smith, B.R., Fisher, P.R. and Argo, W.R. (2004a). Growth and pigment content of container-grown impatiens and petunia on relation to root substrate pH and applied micronutrient concentration. *Hort. Science*, **39**, 1421–1425.

Smith, B.R., Fisher, P.R. and Argo, W.R. (2004b). Nutrient uptake in container-grown impatiens and petunia in response to root substrate pH and applied micronutrient concentration. *Hort. Science*, **39**, 1426–1430.

Sonneveld, C. (1988). The salt tolerance of greenhouse crops. *Neth. J. Agric. Sci.*, **36**, 63–73.

Sonneveld, C. (2002). Composition of nutrient solutions. In *Hydroponic Production of Vegetables and Ornamentals* (D. Savvas and H.C. Passam, eds). Athens: Embryo Publications, pp. 179–210.

Sonneveld, C., Baas, R., Nijssen, H.M.C. and De Hoog, J. (1999). Salt tolerance of flower crops grown in soilless culture. *J. Plant Nutr.*, **22**, 1033–1048.

Sonneveld, C. and Straver, N. (1994). *Nutrients Solutions for Vegetables and Flowers Grown in Water or Substrates* (10th edn). Series: Voedingsoplossingen Glastuinbouw, No. 8, PBG, Aalsmeer: PBG Naaldwijik.

Sonneveld, C. and van den Burg, A.M.M. (1991). Sodium chloride salinity in fruit vegetable crops in soilless culture. *Neth. J. Agric. Sci.*, **39**, 115–122.

Sonneveld, C., van den Ende, J. and De Bes, S.S. (1990). Estimating the chemical composition of soil solutions by obtaining saturation extracts or specific 1:2 by volume extracts. *Plant Soil*, **122**, 169–175.

Sonneveld, C., van den Os, A.L. and Voogt, W. (2004). Modeling osmotic salinity effects on yield characteristics of substrate-grown greenhouse crops. *J. Plant Nutr.*, **27**, 1931–1951.

Sonneveld, C. and Voogt, W (1993). The concentration of nutrients for growing *Anthurium andreanum*in substrate. *Acta Hort.* (ISHS), **342**, 61–67.

Sonneveld, C. and Welles, W.H. (1988). Yield and quality of rockwool-grown tomatoes as affected by variations in EC-value and climatic conditions. *Plant Soil*, **111**, 37–42.

Starkey, R.K. and Pedersen, A.R. (1997). Increased levels of calcium in the nutrient solution improve the post-harvest life of potted roses. *J. Amer. Soc. Hortic. Sci.*, **122**, 863–868.

Steiner, A.A. (1961). A universal method for preparing nutrient solutions of a certain desired composition. *Plant Soil*, **15**, 134–154.

Steiner, A.A. (1984). The universal nutrient solution. Proc 6th Int. Congr. Soilless Culture. Luntern, Wageningen: ISOSC, pp. 633–649.

Ste-Marie, C. and Paré, D. (1999). Soil pH and N availability effects on net nitrification in the forest floors of a range of boreal forest stands. *Soil Biol. Biochem.*, **31**, 1579–1589.

Stevens, R.J., Laughlin, R.J. and Malone, J.P. (1998). Soil pH affects the processes reducing nitrate to nitrous oxide and di-nitrogen. *Soil Biol. Biochem.*, **30**, 1119–1126.

Strong, D.T., Sale, P.W.G. and Heylar, K.G. (1997). Initial soil pH affects the pH at which nitrification ceases due to self-induced acidification of microbial microsites. *Aust. J. Soil Res.*, **35**, 565–570.

Suzuki, I., Dular, U. and Kwok, S.C. (1974). Ammonia or ammonium as substrate for oxidation by *Nitrosomonas europea* cells and extracts. *J. Bacteriol.*, **120**, 556–558.

Tadesse, T., Nichols, M.A. and Fisher, K.J. (1999). Nutrient conductivity effects on sweet pepper plants grown using a nutrient film technique. 2. Blossom-end rot and fruit mineral status. *New Zealand J. Crop Hort. Sci.*, **27**, 239–247.

Tagliavini, M., Masia, A. and Quartieri, M. (1995). Bulk soil pH and rhizosphere pH of peach trees in calcareous and alkaline soils as affected by the form of nitrogen fertilizers. *Plant Soil*, **176**, 263–271.

Tanaka, A., Fujita, K. and Kikuchi, K. (1974). Nutrio-physiological studies on the tomato plant. I. Outline of growth and nutrient absorption. *Soil Sci. Plant Nutr.*, **20**, 57–68.

Taylor, R.A. and Bloom, A.J. (1998). Ammonium, nitrate, and proton fluxes along the maize root. *Plant Cell Environ.*, **21**, 1255–1263.

Thornley, J.H.M. and Johnson, I.R. (1990). *Plant Crop Modelling. A Mathematical Approach to Plant and Crop Physiology*. Oxford: Clarendon Press.

Tinker, P.B. and Nye, P.H. (2000). *Solute Movement in the Rhizosphere* (2nd edn). Oxford: Blackwell Science Publishers.

Torre, S., Borochov, A. and Halevy, A.H. (1999). Calcium regulation of senescence in rose petals. *Physiol. Plant.*, **107**, 214–219.

Torre, S., Fjeld, T. and Gislerod, R.H. (2001). Effect of air humidity and K/Ca ratio in the nutrient supply on growth and postharvest characteristics of cut roses. *Sci. Hortic.*, **90**, 291–304.

Touraine, B., Grignon, N. and Grignon, C. (1990). Interaction between nitrate assimilation in shoots and nitrate uptake by roots of soybean (*Glycine max*) plants, role of carboxylate. *Plant Soil*, **124**, 169–174.

Tzipilevitz, A., Silverman, D., Omer, S. and Reznik, D. (1996). Consumption curves of high quality greenhouse bell pepper (*Capsicum annuum* cv. Mazurka). *Gan Sade VeMeshek*, **2**, 54–62 (in Hebrew).

Urban, L., Brun, R. and Pyrrha, P. (1994). Water relations of leaves of 'Sonia' rose plants grown in soilless greenhouse conditions. *Hort. Science*, **29**, 627–630.

Volkmar, K.M., Hu, Y. and Steppuhn, H. (1998). Physiological responses of plants to salinity: A review. *Can. J. Plant. Sci.*, **78**, 19–27.

von Braunschweig, L.-Ch. (1986). Types of K fertilizers in the K replacement strategy. *Nutrient Balances and the Need for Potassium*. Basel: International Potash Institute, pp. 253–262.

Voogt, W. (2003). Nutrient management in soil and soil-less culture in the Netherlands: Towards environmental goals. Proc. 529 Int. Fert. Soc. and Dahlia Greidinger Symp., 'Nutrient, Substrate and Water Management in Protected Cropping Systems'. York, UK. pp. 225–250.

Waisel, Y., Eshel, A. and Kafkafi, U. (2002). *Plant Roots: The Hidden Half* (3rd edn). New York: Marcel Dekker.

Walden, R.F. and Wright, R.D. (1995). Interaction of high temperature and exposure time influence nitrification in a pine bark medium. *Hort. Science*, **30**, 1026–1028.

Wang, D., Guo, B.C. and Dong, X.Y. (1989). Toxicity effects of chloride on crops. *Chin. J. Soil Sci.*, **30**, 258–261.

Wilcox, G.E., Hoff, J.E. and Jones, C.M. (1973) Ammonium reduction of calcium and magnesium content of tomato and sweet corn leaf tissue and influence of blossom-end rot fruit. *J. Amer. Soc. Hort. Sci.*, **98**, 86–89.

Williamson, L.C., Ribrioux, S.P.C.P., Fitter, A.H. and Leyser, H.M.O. (2001) Phosphate availability regulates root system architecture in Arabidopsis. *Plant Physiol.*, **126**, 875–882.

Wright, R.D. and Niemiera, A.X. (1987). Nutrition of container-growth woody nursery crops. *Hortic. Rev.*, **9**, 75–101.

Wrona, A.F. and Epstein, E. (1985). Potassium transport in two tomato species, *Lycopersicon esculentum* and *Lycopersicon cheesmanii*. *Plant Physiol.*, **79**, 1068–1071.

Xu, G., Magen, H., Tarchitzky, J. and Kafkafi, U. (2000). Advances in chloride nutrition of plants. *Adv. Agron.*, **68**, 97–150.

Xu, G., Levkovitch, I., Soriano, S., et al. (2004). Integrated effect of irrigation frequency and phosphorus level on lettuce: Yield, P uptake and root growth. *Plant Soil*, **263**, 297–309.

Xu, G.H., Wolf, S. and Kafkafi, U. (2001). Effect of varying nitrogen form and concentration during growing season on sweet pepper flowering and fruit yield. *J. Plant Nutr.*, **24**, 1099–1116.

Yaron, B., Zieslin, N. and Halevy, A.H. (1969). Response of Baccara roses to saline irrigation. *J. Am. Soc. Hortic. Sci.*, **94**, 481–484.

Zehler, E., Kreipe, H. and Gething, P.A. (1981). Potassium Sulphate and Potassium Chloride. IPI-Research Topics 9. Basel: International Potash Institute.

Zeroni, M. and Gale, J. (1989). Response of 'Sonia' roses to salinity at three levels of ambient CO_2. *J. Hortic. Sci.*, **64**, 503–511.

Zou, C., Shen, J., Zang, F., et al. (2001). Impact of nitrogen form on iron uptake and distribution in maize seedlings in solution culture. *Plant Soil*, **235**, 143–149.

9

FERTIGATION MANAGEMENT AND CROPS RESPONSE TO SOLUTION RECYCLING IN SEMI-CLOSED GREENHOUSES

BNAYAHU BAR-YOSEF

9.1 System Description
9.2 Management
9.3 Specific Crops Response to Recirculation
9.4 Modelling the Crop-Recirculation System
9.5 Outlook: Model-based Decision-Support Tools for Semi-Closed Systems
Acknowledgement
References

The main reason for recycling greenhouse effluents is to maintain clean and safe environment in the greenhouse production area and local underground water. Under semiarid growth conditions, greenhouse crops grown in open irrigation systems are irrigated with ~20 000 m^{-3} water ha^{-1} y^{-1} and ~50 per cent of it is discharged to the surroundings. At a conventional N concentration of 150 g m^{-3} in the drainage the quantity of water and N disposed from 1000 ha of greenhouse area amounts to 10^7 m^3 water and 1500 ton N annually. Eliminating this discharge entirely could save $2 million in water and $82 000 in fertilizers (prices obtained from ERS, 2005). Moreover, when the discharged N reaches underground water and dissolves uniformly in a 50-m deep water body × 2000 ha (twice the greenhouse area), it can increase the NO_3-N concentration in that water volume by $1.5\,g\,m^{-3}\,y^{-1}$, and within seven years surpass the EPA permitted NO_3-N concentration in drinking water ($10\,g\,m^{-3}$). *In situ* denitrification, *ex situ* ion exchange and reverse osmosis methods to remove the dissolved nitrates are expensive. If employing membrane desalinization techniques, the nitrate removal price might be as high as desalinization, namely $\sim \$1\,m^{-3}$ (General Electric, 2005). At this price the nitrate exclusion outweighs the aforementioned potential savings in water and fertilizers combined.

The potential reductions in water and fertilizer use in closed-loop irrigation systems are significant at a national level but insufficient to persuade most growers to adopt them. While charging growers for nitrate pollution cleanup might change their attitude towards recirculation, some other factors also contribute to the slow adoption of recirculation. The first factor is the high initial investment which is required to switch from soil to detached substrate and thereof to closed-loop irrigation systems (Stanghellini et al., 2004). The second factor is the uncertainty about potential savings in water and fertilizer use and the lack of knowledge regarding optimal operation of recirculation systems. Available hydroponics books either disregarded recirculation (Romer, 1993; Schwartz, 1995; Resh, 2001; Jones, 2005), treated it inadequately (Savvas and Passam, 2002), or overlooked its potential yield and produce quality enhancement in comparison with open irrigation systems.

Water quality (salinity and ionic composition) is the main factor determining volumes of greenhouse effluent discharge. In arid zones, characterized by high evapotranspiration (ET) and salt concentration in fresh water ($\sim 1\,dS\,m^{-1}$), the main problem is the rise in osmotic potential and accumulation of Cl^- and Na^+ to toxic levels. In temperate climates (low ET and better water quality) the main concern is Na^+ accumulation in solution and resulting toxicity in certain crops (Voogt and Sonneveld, 1996; Baas and van der Berg, 1999; Sonneveld, 2000).

The second most important factor common to all growth conditions is accumulation of root excretions in recycled solutions. This pertains mainly to H^+/OH^- and carboxylates release in response to NH_4/NO_3 nutrition (Marschner, 1995), but accumulation of root derbies and secreted sugars and amino acids play also a role. Elevated pH induces carbonate- and orthophosphate-Ca precipitation that may cause emitters to clog and non-uniform water and nutrients application in comparison with open irrigation systems where variations in pH are much smaller. The unequal salt and nutrient distributions reduce water and nutrients uptake thus impairing crop yield (Sonneveld and Voogt, 1990; Sonneveld and de Kreij, 1999).

Economical and environmental constrains limit the daily quantity of water and fertilizer that can be added via open irrigation systems and the irrigation frequency. Consequently, salt and nutrient concentrations at the root surface increase and decrease, respectively, between successive irrigations thus reducing water and nutrients uptake efficiency, and rendering salt leaching from the root zone less effective. In well-planned closed-loop irrigation systems, the quantity and frequency constrains are practically non-existent, and temporary stresses stemming from reduced water and nutrient potentials at the root surface do not limit crop development and yield. It is expected therefore that transition from open- to closed-loop irrigation would result not only in safer environment and diminishing input costs but also in enhanced yield and produce quality, as some earlier studies have shown (Bar-Yosef et al., 2003, 2005).

The objectives of this chapter are to (i) describe principles related to greenhouse solution recycling; (ii) evaluate the horticultural consequences of reduced solution emission from greenhouses and the impact of salinity build-up and ample and high-frequency fertigation on yield and quality of important greenhouse crops; (iii) describe and assess efforts to simulate salt accumulation and crop response to recirculation under different environmental conditions.

9.1 SYSTEM DESCRIPTION

9.1.1 ESSENTIAL COMPONENTS

A closed-loop irrigation system is presented schematically in Fig. 9.1. The essential components of this system are (i) detached substrate (S); (ii) drainage system collecting substrate effluents by gravity flow and delivering it to drainage tank (D); (iii) submerged pump (P) pumping solution from D to either an operational water reservoir (R) or outside the greenhouse (V_{out}); (iv) booster (B) boosting solution from R into the irrigation system (IR) defined by emitters geometry and discharge rate (Q_{in}); and (v) fresh water and fertilizer replenishment device (W, NPK). Due to the potential risk of root pathogen proliferation in closed-loop irrigation systems, it is advisable to include a drainage disinfection device to treat the effluent prior to returning it to

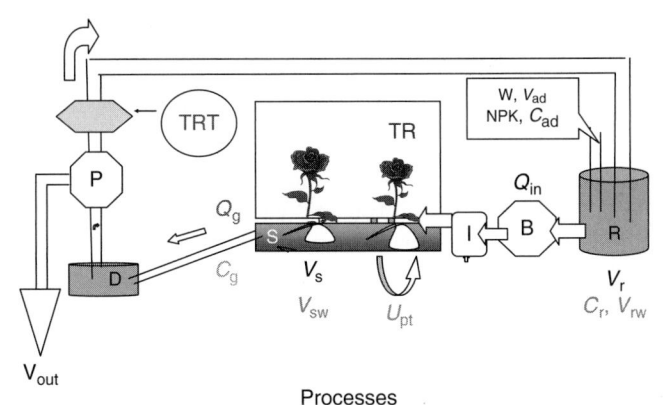

FIGURE 9.1 Schematic description of a closed-loop irrigation system. System components: D – drainage tank; P – pump (normal flow – towards water reservoir R; when receiving command to dispose – towards V_{out}); S – substrate; B – booster; I – irrigation system, discharge rate Q_{in} (L m^{-2} substrate h^{-1}); TRT – slow filtration (biofilter) effluent treatment (disinfection by chemical means is done at I); W, NPK – fresh water (quantity V_{ad}) and fertilizer (concentration C_{ad}) replenishment. Other terms are explained in the text.

the drainage tank. The device can be sand biofilter (see later), UV radiation, solution heating or chemical disinfection, for example ozonation, chlorination or hydrogen peroxide application (for review of methods, see Runia and Boonstra (2004) and Chap. 10). When a sand biofilter or UV is used, it should be inserted after the drainage disposal junction (TRT, Fig. 9.1) to avoid filtration of disposed solution. Insertion of such systems changes the volume and hydraulics of the recirculation system and probably the solution chemistry (see later). Chemical disinfectors are injected in minute quantities into the irrigation solution (I) and do not affect the recirculation hydraulics. Disinfection methods impact on root pathogens is outside the scope of this review and will be discussed in the next chapter.

9.1.2 PROCESSES AND SYSTEM VARIABLES AND PARAMETERS

9.1.2.1 Variables and Parameters

System variables: The operational water reservoir volume V_r (L m^{-2} ground) should equal to at least one daily-ET at the peak water consumption period. This requirement stems from the small substrate volume and limited water buffering capacity in greenhouses relative to the growth conditions in soil. As an example, a crop having $ET_{max} = 5$ L m^{-2} ground d^{-1} and occupying 1000 m^{-2} greenhouse ground should have V_r of 5 m^3. The added water volume reduces the rate of electrical conductivity (EC) increase and allows application of various sources of N fertilizers via the replenishment solution without affecting considerably the N composition and concentrations of the recirculated solution (see Sect. 9.1.5.3; also Chap. 8). Other system variables are V_s, the substrate's volume (L m^{-2} ground); Q_{in}, the irrigation system discharge rate (L m^{-2} ground); and Q_g, the drainage rate (L m^{-2} ground) which is determined by the substrate's hydraulic properties and volumetric water content (θ).

Dynamic variables: The target nutrient concentration in recirculated solution is crop specific but depends also on fertigation management and salinity level.

The permitted salt concentration (also called threshold EC [EC_{thr}]) in recirculated solution is crop specific and depends on the expected trade-off between yield and quality reduction as well as savings in water and nutrients input and emission. When drainage EC is greater than EC_{thr}, the effluent is discharged rather than transferred to R.

Substrate, ion and crop parameters include medium's hydraulic conductivity and water retention functions, effective ion diffusion coefficients, and crop uptake and dry matter and N partitioning parameters, all determining the system processes mentioned later.

State variables: These include concentration of elements ($i = 1, \ldots, n$) in reservoir and drainage solutions, C_{ri}, C_{gi}, respectively (mg L^{-1}); the actual water volume in R (V_{rw}) and substrate (V_{sw}) (L m^{-2} substrate); and the effluent outflow rate Q_g (L m^{-2} ground h^{-1}).

9.1.2.2 Processes

The main processes taking place in the system are evapotranspiration (ET, L m^{-2} ground h^{-1}); water and ions uptake (kg m^{-2} ground h^{-1}); ions partitioning between substrate solution and solid phase; water and ions transport in substrate; crop growth and dry matter (DM) production (kg m^{-2} ground h^{-1}), and root extension in substrate.

It is noted that solution volume in tank D (Fig. 9.1) is very small in comparison with the drainage volume and therefore will not be taken into account in the following estimations. The container can be small since pump D is operated automatically when the water head reaches a certain level and drainage volume is maintained at a minimum level.

Water replenishment, irrigation, drainage and discharge volumes are continuously measured by water meters. ET is estimated by Eq. (1), where W is the replenishment water volume and V_{out} is discharge volume.

$$\text{ET} = W - V_{out} \quad (1)$$

Note that (i) irrigation (Irr) is the amount of water applied to plants and is several-fold greater than W, and (ii) ET also equals to Irr − Drainage, but this estimate is prone to considerable measurement error when using commercial water meters (Bar-Yosef, 2003).

9.1.2.3 The Water and Salt Mass Balance

For optimal system control, salt concentration in recirculated solution (C_g) and water reservoir (C_r) and their water volumes (V_g and V_r, respectively) must be known. For a simplified system (no biofilter, known ions uptake and ET rates, and no drainage container) these variables can be estimated by solving the rate equations (2)–(5) for a single ion, for example Cl$^-$.

$$\frac{dC_r}{dt} = \frac{[Q_{ad} C_{ad} - Q_{in} C_r + (Q_g - Q_{out})C_g]}{V_r} \quad (2)$$

$$\frac{dC_g}{dt} = \frac{(Q_{in} C_r - Q_g C_g - U_{pt})}{V_g} \quad (3)$$

$$\frac{dV_r}{dt} = Q_{ad} + Q_g - Q_{in} \quad (4)$$

$$\frac{dV_g}{dt} = Q_{in} - Q_g - \text{ET} \quad (5)$$

Here, U_{pt} = Cl uptake rate by plants, g m^{-2} h^{-1}; Q_{ad} = time averaged rate of fresh water addition to system, L m^{-2} h^{-1}; C_{ad} = Cl$^-$ concentration in added water; ET, Q_{in} and Q_g were defined above; and $Q_{out} = Q_{ad} - $ ET. When Q_{ad}, Q_{in}, Q_g and ET are constants, the equations can be solved by ordinary differential equation solvers (e.g. Polymath, Cutlip et al., 2005) and used to plan the recirculation system. The mathematical description of the system processes becomes more complex when temporal and spatial solute and water distributions in the substrate are included. This is taken into account, however, in the comprehensive models described later in this review.

Solution culture (substrate-less) hydroponic systems constitute a special case of the aforementioned recirculation systems. The water volume in the growth container is maintained constant (e.g. via an external constant water head tank with feed solution) and solution is discharged from the growth container when $EC \geq EC_{thr}$ (e.g. the Ein Gedi system, Brooke, 1990). An interesting modification of deep flow (DFT) hydroponics is aero-hydroponics (Soffer and Burger, 1988), where solution is sprayed on part of the root system in the container, thus improving oxygen availability. Other hydroponic systems are nutrient film technique (NFT) and semi-deep flow techniques (see Chap. 5), where oxygen is supplied by diffusion from air and mixing occurs due to the relatively rapid solution flow rate.

9.1.3 SUBSTRATE CONSIDERATIONS

9.1.3.1 Physical Properties

The substrate is defined in terms of its volume (V_s, length × width × height, $L\,m^{-2}$ ground), water and nutrients concentrations (θ_s, $L\,L^{-1}$, and C_{si}, $g\,L^{-1}$) and hydraulic conductivity (K_h, $m\,h^{-1}$). To describe water transport in substrate, one needs to know its K_h vs. θ and θ vs. water matric potential (ϕ, m) functions. The theoretical basis of the hydraulic conductivity and water retention functions and how they should be used to predict temporal and spatial θ in substrates are reviewed in Chap. 3. Here more applicative aspects will be discussed. The first aspect is the substrate's capillary fringe, or capillary rise, that can be derived from carefully constructed water retention curves (h_c, m, Caron et al., 2002, 2005; Hillel, 2003). To avoid anoxic conditions when the bottom of the substrate is submerged in water (e.g. when rate of water supply exceeds the drainage rate), the substrate's height should equal $h_c +$ [air buffer layer], the buffer layer being usually >5 cm. Data on the capillary fringe of several substrates is summarized in Table 9.1. For a given height, the substrate width is determined by the desired substrate volume per plant (see later) and number of plants per m bed.

The second property is substrate bulk density (ρ, $kg\,L^{-1}$) from which the medium total porosity (V_a, $L\,L^{-1}$) can be estimated ($V_a = 1 - \rho/\rho_r$, where ρ_r is the particle bulk density). In soil ρ_r is about $2.65\,kg\,L^{-1}$, in organic substrates ρ_r is about $1\,kg\,L^{-1}$, and in perlite $\rho_r = 1.16\,kg\,L^{-1}$ (perlite producers product leaflet). Available data of ρ is summarized in Table 9.1.

Attempts to simulate solute transport and water drainage in substrates by employing the convection–diffusion and Richard's equations (see Hillel, 2003) were mostly unsuccessful (e.g. Wallach and Raviv, 2005). The main reason for this shortcoming is the strong dependency of hydraulic parameters on packing uniformity (Caron et al., 2002; Fermino and Kampf, 2005), variations in bulk density (Fonteno, 1993) and substrate matrix stability during crop production (Nelson et al., 2004; Wallach and Raviv, 2005). To avoid this uncertainty in simulation, Bar-Yosef et al. (2004c) used empirical drainage functions obtained in real growth containers to estimate outflow fluxes from substrate after irrigation termination. They employed a 50–60 cm long plexiglas container with adjustable side walls; drainage holes at the bottom (optionally attached to electric valves to accurately control initial water content) deliver the effluent into a container placed on electronic scale. They (i) saturate the dry substrate and record the volume of water consumed; (ii) open the drainage holes, record effluent weight (W) as a function of time, and calculate drainage fluxes (J_{ef}, $L\,m^{-2}$ substrate s^{-1});

9.1 SYSTEM DESCRIPTION

TABLE 9.1 Hydraulic Properties[a] of Representative Substrates[b]

Substrate	K_{hs} cm h^{-1}	$\theta_{95\%}$	$\theta_{1/2h}$ v v^{-1}	θ_{12h}	θ_{-1cb}	$J(\theta_s)$	J_{fd} cm h^{-1}	ρ kg L^{-1}	h_c cm
Perlite	>1000[i]	0.90	0.66	0.43	0.30	450	10	0.10	6.0[c]
Stone wool	276[d]	0.90	0.85	—	0.42	3.5	0.75	0.12	3.5[c]
Tuff	180[e]	0.70	0.34	—	—	5.0	0.9	1.1	5
Pumice	—	0.38	0.21	0.19	—	360	10	0.75	—
Coir	200[g]	0.92[c]	—	—	0.80	—	—	0.16	4.7[c]
Peat	360[h]	0.89[f]	—	—	—	—	—	0.10[f]	3.6–5[h]

[a] K_{hs} = saturated hydraulic conductivity; $\theta_{95\%s}$ = volumetric water content at 95 per cent saturation; $\theta_{1/2h} = \theta$ after 1/2 h drainage starting at saturation; $\theta_{-1cb} = \theta$ under 1 centibar suction; $J(\theta_s)$ = flux of drainage between 100 and 95 per cent saturation; J_{fd} = mean J between θ_s and $\theta_{1/2h}$; ρ = bulk density; h_c = capillary fringe.
[b] If source is absent, the parameter was derived from unpublished data by the author. All drainage data was obtained in $0.5 \times 0.2 \times 0.2$ m containers (see text).
[c] Caron et al., 2002.
[d] da Silva et al., 1995.
[e] Wallach and da Silva, 1992.
[f] Weiss et al., 1998.
[g] Raviv et al., 2001.
[h] Caron et al., 2005, 60 per cent sphagnum peat, 30 per cent composted pine bark, 10 per cent coarse sand.
[i] Marfa and Orozco, 1995, average of their coarse and fine perlite data.

(iii) transform $W(t)$ to entire substrate mean water content (θ_m) as a function of time, and then to J_{ef} as a function of θ_m. Example of functions obtained by the described tool with pumice is shown in Fig. 9.2.

It can be seen that after approximately 1/2 h the fast drainage stops and outflow during additional 10 h comprised less than ~2 per cent of the 1/2 h discharge (not presented). Drainage studies in stone wool, perlite and tuff yielded similar results, and consequently $\theta_{1/2h}$ can be used as an indicator of the substrate's moisture content when it is ready to receive a new irrigation. Data of $\theta_{1/2h}$ for the above-mentioned substrates and their saturated hydraulic conductivity (K_{hs}) as published in the literature are summarized in Table 9.1 as well. For a 1000 m^2 greenhouse, substrate dimensions of $0.4 \times 0.2 \times 600$ m, and $\theta_{1/2h} = 0.21$ or 0.66 (pumice and perlite, respectively, Table 9.1), the water quantity at the end of fast drainage (Vsw$_{bl}$) is 10.1 and 31.7 m^3, respectively. The data show that K_{hc} is considerably greater than $J(\theta_s)$. This is not surprising in view of the fact that θ in the new method could not be maintained at saturation for a sufficiently long time once drainage began, θ at a depth of 10 cm immediately dropped (see θ_{-1cb}, Table 9.1) and the decline in average θ caused a decrease in J. Another empirical parameter is the average drainage flux between θ_S and $\theta_{1/2h}$ (J_{fd} in Table 9.1) which is more stable than $J(\theta_s)$.

When planning the drip irrigation system, it is essential to choose emitter discharge rate and distance between emitters to promote complete salts leaching and avoid salt

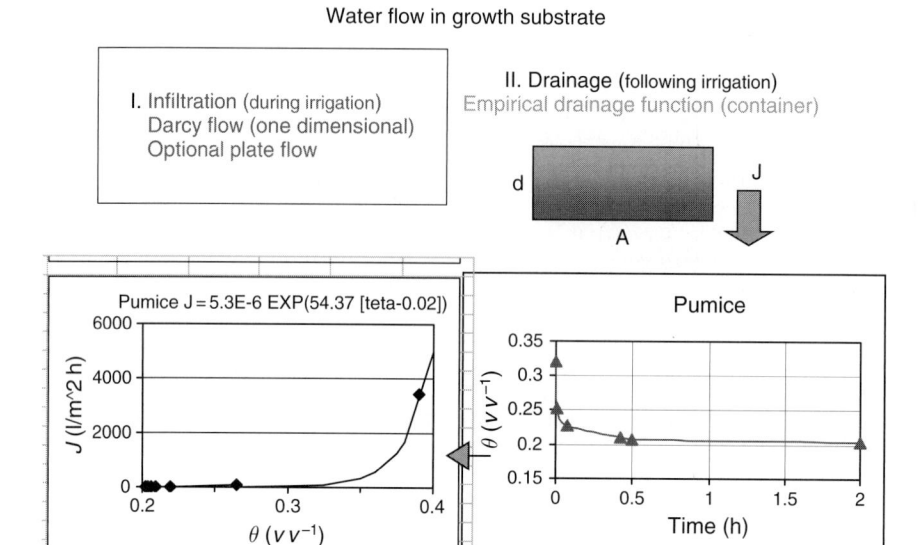

FIGURE 9.2 Obtaining the empirical pumice drainage function (for details, see Bar-Yosef et al., 1999). The substrate in the experimental container is saturated then water is allowed to drain while weighing the container. The derived water volume in substrate is divided by its volume to give average θ and drainage flux (J) as a function of time. The corresponding J and θ along the drainage process are plotted to give the $J(\theta)$ function. Water flow during infiltration (stage I) is usually described by employing the Darcy law or plate theory.

accumulation in pockets. During irrigation, a water saturated zone is formed around the emitter. In clayey soils the saturated volume is a hemisphere but in sandy soil the vertical axis is longer than the lateral axis (Bar-Yosef and Sheikolslami, 1976). In substrates, which are characterized by weaker water retention than sand, the saturated zone below emitters is expected to be cylinder-like. The relationship between emitter discharge rate (Q, L h^{-1}), irrigation time (t, h) and the saturated cylinder radius (r, cm) is given in Eq. (6), where $\theta_{1/2}$ is θ at irrigation commencement and h_s is the cylinder height (=substrate height, cm):

$$Qt = \pi r^2 h_s (\theta_s - \theta_{1/2}) \tag{6}$$

Note that θ_s cannot be accurately determined but it can be estimated from the 95 per cent θ_s ($\theta_{s95\%}$) presented in Table 9.1. In the case of stone wool ($\theta_{s95\%} = 0.90$ and $\theta_{1/2} = 0.85$, Table 9.1), $h_s = 10$ cm, $t = 0.1$ h and $Q = 2.4$ L h^{-1}, r is calculated from Eq. (6) to be 12.3 cm. This means that a distance of 20 cm between adjacent 2.4 L h^{-1} emitters should provide uniform salt leaching in this substrate and prevent salt pockets formation. Plant water uptake may reduce substrate θ below $\theta_{1/2}$ but under common irrigation frequencies this can be neglected. For example, ET of 1 L m^{-2} ground h^{-1} is a very high value in greenhouses. When irrigating every hour at $\theta_{1/2}$, the plant extracts water during the 30 minutes between fast drainage termination and new irrigation. The stone wool volume is 24 L m^{-2} ground and at $\theta_{1/2}$ it holds 20.4 L water. Correcting for the water uptake gives $\theta = (20.4 - 0.5)/24 = 0.83$ (instead of 0.85).

Experimental data of lateral EC distribution in stone wool at the end of a pepper (*Capsicum annuum*) growth period indeed indicates that no lateral salt pockets existed at the end of the season at the aforementioned drippers discharge rate and geometry (Table 9.2). However, in both presented treatments in Table 9.2 the EC in the top 2 cm was significantly higher than that in the 0–8 cm layer. This could stem from evaporation-mediated salt accumulation in the top few mm that could not be effectively leached by drip irrigation. Experimental results of Cl$^-$ distribution in perlite at the end of gypsophila (*Gypsophila paniculata*) growth season in closed-loop irrigation system also show uniform leaching at similar irrigation regime, but when irrigation rate was reduced by ~45 per cent, the Cl$^-$ concentration at a distance of 10 cm was higher than that near the emitter (Table 9.3). As in stone wool, the Cl$^-$ concentration at the top layer exceeded the concentration at the bottom.

TABLE 9.2 Distribution of EC in a Stone Wool Slab in Two Treatments Differing in Threshold EC at the End of Experiment. Test Crop was Pepper Grown at Bet Dagan, Israel

	123 days after planting			
	Distance[a] (cm)		Depth[b] (cm)	
EC$_{thr}$ (dS m^{-1})	0	10	0–2	0–10
	EC (dS m^{-1}) in slab			
2.5	2.5	2.6	3.8	2.7
4.5	4.4	4.4	6.2	4.4

Source: Data derived from Bar-Yosef et al. (1999).
[a] From emitter. Distance between emitters on lateral = 20 cm. Substrate sampled at the depth of 4–5 cm, 7–10 irrigations per day.
[b] From top. Mean of substrate samples taken at distances of 0 and 10 cm from emitter.

TABLE 9.3 Distribution of Cl in Pumice and Perlite[a] at the End of an Experiment with Gypsophila in Closed-Loop Irrigation System at Bet Dagan, Israel (EC$_{thr}$ 3.3 dS m^{-1})

	Distance from emitter (cm)					
	0	10	0	10	0	10
Depth (cm)	Pumice		Perlite 2		Perlite 2 (reduced)	
	Cl$^-$ concentration (mg kg^{-1} substrate)					
5	400	690	3240	3190	3080	3690
10	250	540	2420	2640	2660	2810
20	270	440	2120	1990	1740	1970

Source: Data derived from Bar-Yosef et al. (gypsophila 1999b).
[a] Eight to ten irrigations per day, 6 min per irrigation; in the 'perlite 2 reduced' treatment 5–6 irrigations per day, 6 min per irrigation. Distance between emitters was 20 cm.

In pumice ($\theta_{s95\%s} = 0.38$ and $\theta_{1/2} = 0.21$, Table 9.1), $h_s = 20$ cm, and $Q = 2.4$ L h^{-1}, an irrigation period of 0.1 h can saturate a cylinder of $r = 4.7$ cm, indicating possible salt pockets between emitters located 20 cm apart. Experimental data with this substrate indeed show salt accumulation at a distance of 10 cm from the emitter (Table 9.3). To obtain a saturated $r \geq 10$ cm, a higher discharge rate and/or longer irrigation period are needed. To avoid salt pockets without changing Q or irrigation time, a smaller distance between emitters should be used.

Microbes' migration in closed-loop irrigation systems depends, among other things, on the substrate's filtration capability. Microbes' migration was studied in soils and soil columns (for review, see e.g. Friedman, 1999) but meager information can be found on the mobility in substrates *per se*. Comparison between representative microbes concentration in the leachate of pumice and stone wool that received the same recycled solution via drip irrigation for 218 days indicates that pumice was considerably more effective at filtering of bacteria (total and *Pseudomonas fluorescens*), and probably also of *Pythium* (Table 9.4), out of the recirculating solution. While these results were received in pepper, similar data was obtained with gypsophila, only here pumice was shown to be better filter than perlite (Table 9.4).

Another substrate-related question is whether to grow a crop in a long, continuous trough or in individual containers at the same plant density and substrate volume per

TABLE 9.4 Comparison Between Representative Microbe Concentrations in Leachates of Pumice, Stone Wool and Perlite Under Pepper[a] and Gypsophila.[b] In Both the Experiments, the Two Test Substrates Shared the Same Recirculation System, Namely the Inflowing Solutions were Identical but the Outflow Solutions Differed due to Different Substrate Reactivity

Substrate	Crop	Total bacteria	*Pseudomonas Fluorescens*	Xanthomonas	Pythium
		cfu mL^{-1}			mL^{-1}
Pumice	Pepper	6.20 E4	3.20 E3	4.0 E3	0
Stone wool	Pepper	2.36 E6	1.24 E5	3.0 E5	0.25
		$P < 0.05$	$P < 0.05$	n.s.	n.s
		Total bacteria	*Pseudomonas Fluorescens*	CVO Erwinia	PDA[c]
		cfu mL^{-1}			mL^{-1}
Pumice	Gypsophila	1.3 E4	1.0 E4	0	26
Perlite	Gypsophila	4.0 E6	3.0 E6	0.4	97
		$P < 0.01$	$P = 0.10$	ns	ns

[a] Microbiological analyses were carried out by Dr G. Kritzman, Agricultural Research Organization, Bet-Dagan, Israel. Data were derived from Bar-Yosef et al., 1999.
[b] As footnote a, but data derived from Bar-Yosef et al. (2001).
[c] Total fungi onto Potato Dextrose Agar.

plant. In closed-loop irrigation systems, the expected root pathogens distribution is similar in both systems because container leachates are mixed and reapplied over the entire greenhouse area. The main difference between container and trough is that lateral root growth in the latter persists without interference, while in containers root contact with walls inhibits elongation (e.g. Romano et al., 2003) and promotes formation of more root tips per unit root weight. Implications of this effect are further discussed in the Sect. 9.2.5.2. Another problem with a reduced root-zone volume is the reduced oxygen supply and lower root respiration rate (Peterson et al., 1991). In conclusion, there appears to be no reason for different fertigation managements for container-and trough-grown crops as long as the substrate volume per plant and substrate height are identical in both systems.

9.1.3.2 Chemical Properties

Substrate's chemical characteristics which are pertinent to solute flow in greenhouses are (i) cation exchange and anion and cation adsorption isotherms; (ii) mineralization rate of indigenous and added organic matter; and (iii) nitrification rate. Adsorption characteristics depend on the presence of metal oxides, clay minerals and their weathering products in mineral substrates, and on carboxylic and other charged groups in organic substrates. In synthetic substrates which are manufactured at very high temperatures (e.g. perlite), the aforementioned adsorption sites collapse and the substrate becomes inert with respect to ion adsorption. Microbial activity is substrate specific due to substrate effect on nutrient availability, redox conditions and microbe mobility. The chemical properties of common substrates are discussed in Chap. 6 and are not further addressed herein except in relation to cation transport in substrates. Ions and organic compounds released from substrate surfaces may interfere with solution chemistry reactions, particularly precipitation/dissolution, but overall this effect is of minor importance.

9.1.3.3 Irrigation Systems

Irrigation solution can be applied from the bottom, top or sides of the substrate. When irrigating from below, the bottom of the substrate is submerged in irrigation water or is in liquid contact with a wet mat. If upward transport is fast enough (depending on substrate's hydraulic properties) and diffusion downward into the saturated zone is sufficiently slow, then this results in a system where the average EC of the liquid in the root zone increases slowly over days and weeks (de Kreij and Straver, 1988; Otten, 1994). The advantage of this system is that there are virtually no water and nutrients losses and the energy requirement is minimal. The disadvantage is that the salts accumulating at the top substrate layer must be actively managed resulting in some irrigation solution or substrate being discarded. Moreover, the upward capillary movement depends on the substrate's matrix and evaporation rate, and the inability to leach the unpredictable salt accumulation from the top of the root zone represents a risk of developing a layer in the root zone (at the top) that is toxic to the plant. This suggests that sub-irrigation be undertaken only if the supply water is of low salinity.

One approach to dealing with the salt accumulation risk is to equip the root zone with an overflow above the constant water table and to install a top irrigation system to wash excess salts. The excess top irrigation in case of leaching is drained out of the system via the overflow device and later recirculated or discharged, depending on EC. Such hybrid systems are used on a commercial scale in The Netherlands and California, usually in single-plant containers, where the constant water head is maintained by a pipe connecting all containers. The outflow is collected by a parallel pipe and delivered to an external water reservoir. Goodwin et al. (2003) compared overhead and capillary irrigation in containerized plants and concluded that capillary irrigation was most efficient in water and N use, but overhead irrigation gave highest yield.

The most common method in closed-loop irrigation systems is irrigation from the top of the root zone. The substrate is in containers (or bags) perforated at the bottom and placed on a rigid plastic tray at adequate slope in the direction of flow; the effluent is captured and available for recirculation. This type of irrigation is generally applied via mini-sprinklers or drip irrigation.

Irrigation systems are characterized by their effective discharge rate (J_{irs}, L m^{-2} substrate h^{-1} or J_{irg}, L m^{-2} ground h^{-1}). Under drip irrigation, J_{irs} is the product of emitter discharge rate (J_{tr}, L h^{-1}) multiplied by the number of emitters per m^{-2} substrate (n_{tr}). The substrate-to-greenhouse area ratio (f_s) is necessary to transform J_{irs} into J_{irg}:

$$J_{irs} = n_{tr} J_{tr} \qquad J_{irg} = f_s J_{irs} \qquad (7)$$

For a more detailed description of different irrigation systems and configurations, see Chap. 5.

9.1.3.4 Biofilters

Biofilters are designed to remove pathogens from recycled solutions by filtration and by exposing these pathogens for extended periods of time to unfavourable conditions reducing their viable population. The adverse growth conditions include competition with local microbes on carbon and dissolved oxygen along the flow path. The effectiveness of biofilters depends on the residence-time of effluent within the system and on the particle-size distribution in the filling material (van Os et al., 1998). Disinfection aspects of biofilters are discussed in the next chapter. From hydraulic and chemical points of view, the biofilter might affect closed-loop irrigation systems, thus a short discussion of some concerns follows.

Let us consider a typical sand biofilter as used by Kramer et al. (2002). It was packed with (bottom to top) 20 cm of coarse basalt gravel (0.6–1.0 cm), 10 cm of fine basalt gravel (0.25–0.4 cm), 30 cm of quartz sand (0.08–0.15 cm), 30 cm of quartz sand (0.06–0.08 cm) and on top a 30 cm layer of fine quartz sand (0.03–0.06 cm), altogether comprising 120 cm of substrate. When this biofilter was saturated and covered with 40 cm of water, the discharge rate from the bottom (J_{bf}) was 300 L m^{-2} h^{-1} (= 0.3 m h^{-1}). After about two months, this rate dropped to 100 L m^{-2} h^{-1}. Scrubbing

9.1 SYSTEM DESCRIPTION

the clogged top 2 cm sand layer restored J_{bf} to its initial value. Considering a time-average flow rate of $0.2 \, \text{m h}^{-1}$ results a residence time in the biofilter (1.60 m) of eight hours. The minimum biofilter surface area (S_{bf}) servicing S_g m² greenhouse ground with known ET (L m⁻² ground d⁻¹), irrigated over t hours a day and operating 24 h a day can be calculated from Eq. (8):

$$24 S_{bf} J_{bf} = (t_{ir} f_s J_{irs} - ET)S_g \qquad S_{bf} = \frac{(t_{ir} f_s J_{irs} - ET)S_g}{(24 J_{bf})} \qquad (8)$$

For $t_{ir} = 2\,\text{h}$, $J_{irs} = 50\,\text{L m}^{-2}$ substrate h⁻¹, $f_s = 0.31$, ET = 0, $S_g = 1000\,\text{m}^2$ and $J_{bf} = 200\,\text{L m}^{-2}\,\text{h}^{-1}$, the required biofilter area is 6.46 m². During day hours, when greenhouse drainage rate exceeds the biofilter flow rate, the drained solution must be stored in an auxiliary container for later (e.g. night time) filtration.

The biofilter discussed above was found to have a negligible effect on pH, NO₃ and NH₄ concentration in solutions flowing through it during spring (Table 9.5). Chloride concentration in biofilter effluent tended to be lower than that in the inflow solution due to slight dilution. This was not observed for NO_3^- and NH_4^+. A similar dilution was found for Ca at the beginning of April, but on April 25 this trend reversed (Table 9.5).

TABLE 9.5 Ion Concentration in Inflowing and Outflowing Biofilter Solutions During Two Days in April. The Reported Recirculation Treatments Differed in Their Threshold EC Value[a]

	Inflow	Outflow	Inflow	Outflow
EC_{thr} (dS m⁻¹)		4 April		25 April
	pH			
2.9	6.9	6.8	7.1	6.9
3.8	6.3	6.6	6.8	7.2
	NO₃-N (mg N L⁻¹)			
2.9	43	36	142	127
3.8	109	100	117	145
	NH₄-N (mg N L⁻¹)			
2.9	4.7	2.5	2.6	2.3
3.8	3.8	3.3	2.2	2.9
	Cl (mg L⁻¹)			
2.9	488	525	600	464
3.8	696	674	720	682
	Ca (mg L⁻¹)			
2.9	200	194	218	232
3.8	312	260	240	330

Source: Data derived from Kramer et al. (2002).
[a] Experiment conducted with roses on tuff substrate in the Arava region in Israel.

9.1.3.5 Fertilizer Addition

Fertilizers are added to the system to account for nutrients consumption by plants. The addition can be done by one of the two possible methods: (i) Introducing fertilizers into the water reservoir in the form of a ready mix solution which also replenishes the water consumption by the crop (W, NPK in Fig. 9.1). A less desired modification of (i) is adding the nutrients in the form of stock solution or solid fertilizers followed by tap water application. The disadvantage is that monitoring the added fertilizer this way is manual and inaccurate. (ii) Injecting concentrated solution of individual fertilizers in minute volumes into the irrigation water (I in Fig. 9.1) during irrigation and filling the water reservoir with water. The ready mix solution (option (i)) requires an extra reservoir, the volume of which depending on the replenishment frequency. Method (ii) requires an accurate and expensive metering pumps but it has a more rapid effect on irrigation solution chemical composition, it uncouples between water and nutrients supply, and the pumps can be programmed to add fertilizers at critical hours along the day thus offering better control of nutrients availability to plants. The daily nutrients supply should be independent of application method and matched to the target daily nutrient demand by the crop.

9.1.4 MONITORING

Circulated solutions should be monitored regularly to ensure that nutrient concentrations, EC and pH are within predefined limits. To allow comparison of solution results over time and growth conditions, solution should be sampled (via one of the emitters) at exactly the same hour of the day, or preferably, collected over a 24 h time period. The reason is that in recycled solution, EC and pH vary considerably over the day depending on water and nutrients replenishment volume and timing. Typical variations in EC and pH over the day are shown for roses (*Rosa* × *hybrida*) in Israel in Fig. 9.3.

In addition to temporal effects, there is a significant difference between ionic composition of emitter and drainage solutions, generated by the water and ion uptake from substrate and resulting in higher EC in the drainage than in the irrigation solution. Since the resulting EC difference is an important indicator of irrigation sufficiency, it is therefore recommended that both emitter and drainage solutions be monitored regularly and simultaneously.

The pH differential between irrigation and drainage solutions is caused by plant root activity. Measuring pH in both solutions is essential information needed for controlling the NH_4/NO_3 ratio in the replenishment solution (discussed in detail later).

Nutrient concentrations should be monitored in the irrigation solution to ensure that they are close to the target concentrations. Testing the drainage solution as well adds relatively little information. The frequency of analyses depends on the permitted deviation from the objective values and the daily replenishments. Quality control concerns usually require that this period is ~7 days, otherwise protracted deviation might cause irreversible damage.

Chemical analysis of plant leaves should also be performed periodically. It is particularly important in recirculation systems because nutrient uptake is potentially inhibited

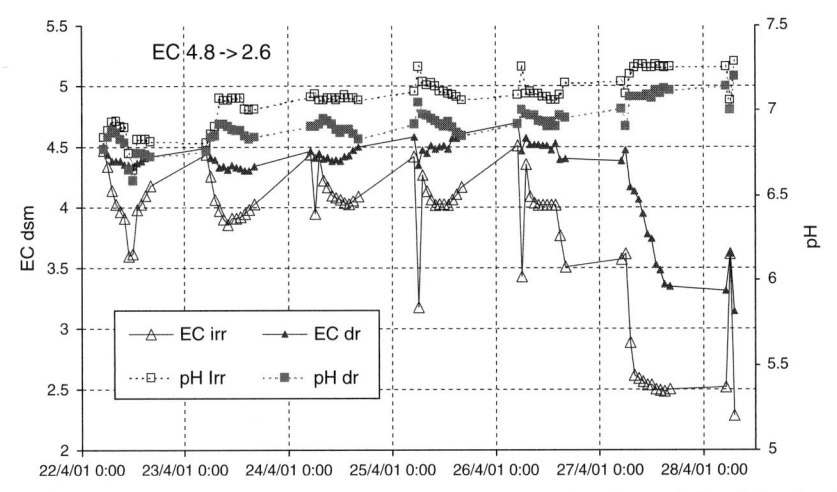

FIGURE 9.3 Hourly variations in recycled solution EC and pH at the substrate inflow (irr) and outflow (dr) points. The daily replenishment of fresh solution into the water reservoir (Fig. 9.1) occurred at 0600. One-year-old rose plants, Arava, Israel (derived from Kramer et al., 2002).

due to elevated salinity, thus it is not adequate to rely on evaluating nutrients status in plant solely from information about solution concentrations. Frequency of analysis of leaf tissue depends on plant's physiological stage, rate of nutrient consumption and average solution salinity level. Measured tissue analysis results are compared with target nutrient concentrations so that deviations can be used to identify the needed fertilizer management modifications.

Another monitoring approach is to determine crop characteristics not directly comparable with target values, but yet indicating time variations that can be interpreted as positive or adverse plant response to prevailing growth conditions. The so-called 'phytomonitoring' includes chlorophyll fluorescence imaging (Chaerle and Van Der Straeten, 2001), on the one hand, and simpler devices like stem and fruit diameter, leaf temperature and sap flow, on the other hand. For review of commercially available methods and data interpretation, see, for example, Ehret et al. (2001).

Monitoring the substrate water status using sensors is generally more important in open irrigation systems than in closed-loop systems. The reason is that in closed systems the irrigation is set to exceed the crop ET several-fold, subject to the system's constraints, and timed at an interval that assures that the root zone is never below predetermined moisture for even a few hours or minutes.

9.1.5 CONTROL

9.1.5.1 Water Replenishment

In order to avoid water spills the replenishment should take place at a certain reference condition on a daily basis. A convenient reference state is the end of substrate drainage after the last irrigation in the day, after all drained volume of leachate is collected in reservoir R. Since in most substrates, drainage stops within 1 h (Fig. 9.2) of the end of an irrigation event, the suggested timing of replenishment is between 1 h after drainage termination and the first irrigation in the morning. After replenishment the water reservoir is at full capacity. If this procedure is repeated every day, then the recharged volume is equal to ET + [disposed volume] of the previous day. The entire water volume in the system is $V_{wT} = V_r + V_{wse}$, where V_{wse} is the substrate water volume after the end of the rapid drainage (L m^{-2}). The next day first irrigation, usually taking place when ET is still negligible, should not exceed $V_{wssat} - V_{wse}$, where V_{wssat} is the substrate water volume at saturation.

Under high ET conditions, more than single water replenishment per day and continuous monitoring of EC are advised. When the EC around noon exceeds the first irrigation of the day EC by ~ 1 dS m^{-1} (ΔEC_n), then it is recommended that about 75 per cent of the morning-to-noon ET (ET$_{mn}$) and nutrients uptake be replenished immediately. The 25 per cent precaution is recommended to avoid overflowing the reservoir in case the ET is overestimated. If estimated or measured daily ET $\geq V_r$, then the midday replenishment is imperative, and if $V_r \leq$ ET$_{mn}$, two or more replenishments must take place. The decision regarding the magnitude of ΔEC_n depends on the expected crop response to fluctuations in EC along the day, and on the current EC in the system.

When the recirculation system includes a biofilter, the reference condition for replenishment should be the end of the entire drainage filtration and return of the treated solution to the water reservoir. The biofilter may add a considerable volume of water to the system. For the aforementioned 1000 m^2 greenhouse, V_r and V_{wsb} (perlite) comprise 5 and 38.4 m^3, respectively while the biofilter water volume should be 5.7 m^3 $\{= S_{bf} [h_w + h_{snd} \theta_{bfsat}]$, where S_{bf} (top surface area) $= 6.5$ m^2, h_w (constant water head above the biofilter surface) $= 0.4$ m, h_{snd} (height of biofilter substrate) $= 1.2$ m and θ_{bfsat} (porosity of biofilter substrate) $= 0.4$ v/v$\}$.

It may be concluded that early morning and midday ET replenishments are more effective in abating salinity symptoms in comparison with early morning replenishment only, but it requires a more sophisticated control system to guarantee spill-proof operation.

Water addition to the reservoir during replenishment should be automated so that it can stop when the water level reaches a certain height. The added amount of water should be recorded and stored in the greenhouse control computer. In case of several replenishments per day the fresh water (or ready mix solution) supply valve should be opened at preset times but closed automatically after the required solution volumes were implemented. Each water application must be defined in terms of volume and time at which the supply valve was opened; otherwise ET + discharge will not be smoothly estimated because certain volumes might be erroneously attributed to the previous or following day.

9.1.5.2 Solution Discharge

Any closed-loop irrigation system must support an option for solution discharge when a predetermined discharge criterion is met. EC is the simplest criterion as it can be accurately and continuously monitored and the output can be used by the greenhouse computer to open a discharge valve disposing the effluents rather than recirculating them (see Fig. 9.1).

If nutrient concentrations in the recirculated solution fluctuate dramatically, it will affect the EC independent of the build-up of undesirable solutes. For example, if N concentration drops from 140 to 14 mg L^{-1} (due to inadequate N replenishment), it reduces the EC by ~1 dS m^{-1} (roughly 10 mM(c) per 1 dS m^{-1}). Under such conditions, EC monitoring is insufficient and the accumulating ions, for example Cl$^-$ or Na$^+$, must be specifically measured. Currently no robust sensors exist to measure these ion concentrations independently and continuously in situ, so that these ions must be determined through a laboratory analysis or with a test-kit. This means that disposal cannot be performed automatically.

Regardless of the disposal criterion, three parameters must be defined for calculating discharge volumes: (i) the current time-averaged EC or ion concentration in solution (C_{av}); (ii) the permitted upper deviation from this concentration at which discharge should occur (ΔC_{up}); (iii) the desired lower deviation (ΔC_{low}) resulting from disposal and replenishment with fresh water. For example, the disposed volume per 1000 m^2 ground (V_{dis}) for C_{av}, ΔC_{up} and ΔC_{low} of 500, 50 and 50 g Cl m^{-3}, respectively, total recirculated volume V_T and fresh water Cl$^-$ concentration of 250 g m^{-3} (C_{fw}) would be (Eq. [9]):

$$V_T(\Delta C_{up} + \Delta C_{low}) = V_{dis}[(C_{av} + \Delta C_{up}) - C_{fw}]; \quad V_{dis} = V_T \frac{100}{300} \quad (9)$$

It should be noted that C_{up}, C_{low}, C_{av} and C_{fw} are approximated as the average of the continuously monitored irrigation and drainage solutions. Another assumption is that V_T is constant with time. Similar computational approaches and assumptions were used in simplified models for salt accumulation in closed-loop hydroponics (Carmassi et al., 2003; Kempkes and Stanghellini, 2003; Stanghellini et al., 2004). In practice, ion concentration can increase due to water consumption (declined V_T) and the drainage solution is discharged ($C_{drng} > C_{av}$), so the required disposal volume can be smaller than V_{dis}. To minimize unnecessary solution discharge, it is advised that: (i) disposal be limited to no more than ~75 per cent of V_{dis}, (ii) any discarding of the leachate solution occur prior to replenishment, and (iii) disposal occur in the afternoon between ~1300 and ~1500 with complete disposal (to V_{dis}) on the next day if the lower C limit was not achieved. The time between discharge events depends on ΔC, transpiration rate and C_{fw}.

Under conditions of well-controlled nutrient concentrations in solution, the variations in EC are caused mainly by Cl$^-$ (and more rarely by SO$_4^{2-}$, depending on fresh water chemical composition) and accompanying cations. Data collected in Israel from recycled solutions differing in salinity, crops, fresh water EC and climate (Fig. 9.4) indeed reveal a significant linear relationship between Cl$^-$ concentration and EC. The slope for roses, pepper and cucumber (*Cucumis sativus*) was 0.125 ± 0.006 dS m^{-1}/mM(c) while the intercept varied due to different constant nutrient concentrations in each of the three crops. Pepper gave the lowest R^2 because the nutrient concentrations at the different salinity levels were less steady than in roses

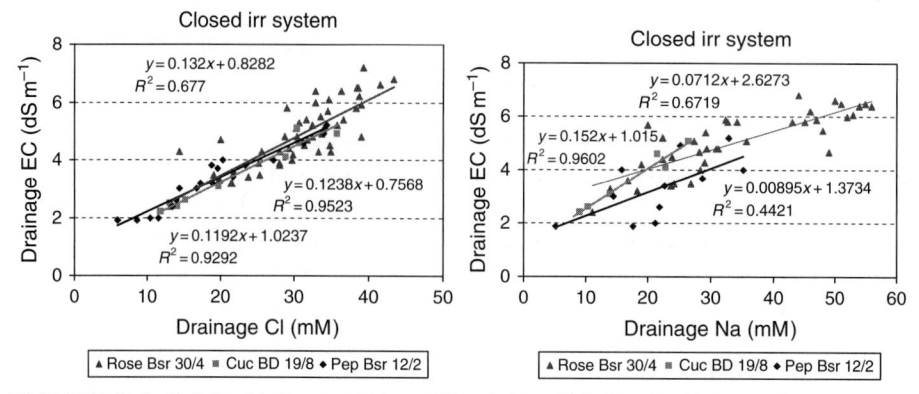

FIGURE 9.4 Relationship between drainage EC and either Cl (left) or Na (right figure) concentration in it. Data obtained in three experiments conducted during 2002 in closed-loop irrigation systems with roses and pepper (pep) at Besor and cucumber (cuc) at Bet-Dagan, Israel. Dates in legends are solution sampling dates.

and cucumber. The correlation between EC and Na^+ concentration in drainage was also high (lowest for pepper), but the slopes of the three crops differed considerably, indicating a crop-specific relationship (Fig. 9.4).

A general relationship between EC and solution ionic composition might be very useful when specific information like Fig. 9.4 is unavailable. Empirical functions of total salt concentration (C_t, mmol(c) L^{-1}) vs. EC were reviewed by Savvas (2002). An example is the relationship suggested by Sonneveld et al., (1999a) for greenhouse solutions:

$$EC = 0.095\, C_t + 0.19 \tag{10}$$

A more fundamental approach is to relate EC to the ionic strength (I, mol L^{-1}) of the solution:

$$I = K_I\, EC \tag{11}$$

$$I = 0.5\, \Sigma(|z_i|^2\, C_i) \tag{12}$$

where z_i is the ion valence and C_i is its concentration in mol L^{-1}. According to Griffin and Jurinak (1973), K_I for soil water extracts $= 0.013$ mol L^{-1}/dS m^{-1}. Reluy et al. (2004) showed that Eq. (11) can be derived theoretically, and K_I is a constant proportional to the known and tabulated specific conductivity of the ions in solution. Solution data from a rose experiment in closed-loop irrigation system in southern Israel (Kramer et al., 2002) was used to derive an equation analogous to Eq. (11) for a system where the main accumulating ions were Na^+, Ca^{2+}, Mg^{2+}, Cl^- and SO_4^{2-} and the nutrients (K^+, NH_4^+, NO_3^- and P) were maintained at constant concentrations. The formula obtained

$$I = 0.0122\, EC + 0.0117 \quad R^2 = 0.82 \tag{13}$$

has a slope which is similar to the slope of Griffin and Jurinak (1973) but with an intercept that indicates that the calculated I was overestimated because concentrations, rather than activities, were used and ion pairs were disregarded.

9.1.5.3 Nutrients Replenishment

Continuous recording of daily application of nutrients is one of the most important control tools as it allows for imposition of an application rate that is at least equal to the projected consumption rate of the crop. If cumulative supply exceeds the sum of the projected cumulative uptake of nutrients, the amount of discharged nutrients, plus the nutrients found in the recirculated solution, then the difference is likely to be found in substrate pockets located between wetting fronts of adjacent emitters. Existence of such pockets requires alteration of the substrate composition or irrigation regime to induce greater lateral water flow and more efficient solute leaching (see Chap. 3).

In a previously mentioned example of a $1000\,m^2$ greenhouse, the water volume $V_r + V_{wsb(perlite)}$ was $43.4 (=5+38.4)\,m^3$. At a N solution concentration of $100\,g\,N\,m^{-3}$ the quantity of N in the greenhouse is 4.34 kg, in comparison with a daily N uptake rate of $\sim 0.2\,kg\,d^{-1}$. When replenishing the consumed 0.2 kg by adding it in one dose into the water reservoir (R, Fig. 9.1), the N concentration there would increase to $140\,g\,N\,m^{-3}$, which will have small impact, and after the first irrigation the solution will gradually dilute to $105\,g\,N\,m^{-3}$ (uptake neglected). If instead the 0.2 kg N is injected into the irrigation water and diluted in a water dose of $1\,m^3$, the N concentration in the water pulse would be $200\,g\,N\,m^{-3}$, which is less desirable than $140\,g\,N\,m^{-3}$ due to the resulting elevated EC, but in subsequent irrigations it will also be diluted to $105\,g\,N\,m^{-3}$. Such considerable fluctuations in N concentration suggest that dividing the daily N rate into 3–5 portions seems more appropriate when injecting the nutrients directly into the irrigation water.

9.2 MANAGEMENT

9.2.1 INORGANIC ION ACCUMULATION

Ion accumulation in closed irrigation systems depends on the balance between rates of application and consumption by plants. An example of this balance for some important greenhouse crops is presented in Table 9.6. It can be seen that among the accumulating ions Na^+, Ca^{2+}, Cl^- and SO_4^{2-}, the chloride accumulation is greatest ($mg\,L^{-1}$ basis), but even this ion does not pose a problem in crops having short growth period and low transpiration rate, for example lettuce (*Lactuca sativa*). Calcium and Mg^{2+} accumulate in solution of all presented crops except lettuce and muskmelon (*Cucumis melo*). Notably the greater Ca and Mg consumption by these two crops in comparison with supply (Table 9.6) stems from the fact that the actual fresh-water Ca and Mg concentrations were higher than that specified in the table (see footnote).

In Western Europe, water application rates are lower than in semiarid and arid zones, and the fresh water salt concentration is usually lower than that cited in Table 9.6. In

TABLE 9.6 Evaluating Potential Cl, SO₄, Na, Ca and Mg Accumulation in Closed-Loop Irrigation System of Representative Greenhouse Crops [Tomato, Pepper, Lettuce, Muskmelon (M.melon), Strawberry (Strbry), Roses and Lisianthus (Lis)] Grown in Semiarid Zone. The Accumulation is Expressed as Potential Concentration (Pot. conc.)[a] of Elements in the Recirculated Solution at the End of the Growing Season. For Simplicity, All Crops were Assigned with the Same Element Concentrations in Fresh Water (in Parenthesis) Even Though There were Some Differences Between Experiments. Where Deviations From the Indicated Concentrations were Significant, They were Mentioned in Text. Uptake Data were Obtained Experimentally[b] in Treatments Owning $EC_{thr} > 4\,dS\,m^{-1}$; Under No Circumstances had the Presented Uptake Rates Caused Yield Reductions > 20 per cent of the Control ($EC_{thr} = 2.7\,dS\,m^{-1}$).

	Tomato	Pepper	Lettuce	M.melon	Strbry	Roses[c]	Lis
Fresh water (mm)	480	490	55	350	330	1293	610
Cl (conc. in fresh water 175 mg L⁻¹)							
Addition (kg ha⁻¹)	840	857	96	612	577	2263	1067
Uptake (kg ha⁻¹)	371	125	23	154	85	103	301
Pot. conc. (mg L⁻¹)	1172	1830	182	1145	1230	5400	1915
S (SO₄-S conc. in fresh water 50 mg L⁻¹)							
Addition (kg ha⁻¹)	240	245	27	175	165	646	305
Uptake[c] (kg ha⁻¹)	50	45	5	30	80	176	—
Pot. conc. (mg L⁻¹)	475	500	55	362	212	1175	—
Na (conc. in fresh water 110 mg L⁻¹)							
Addition (kg ha⁻¹)	528	539	61	385	363	1422	671
Uptake (kg ha⁻¹)	291	172	24	85	80	78	214
Pot. conc. (mg L⁻¹)	593	917	92	750	712	3360	1142
Ca (conc. in fresh water 85 mg L⁻¹)							
Addition (kg ha⁻¹)	408	416	47	298	280	1099	518
Uptake (kg ha⁻¹)	302	132	18	340	79	212	70
Pot. conc. (mg L⁻¹)	265	710	72	—	502	2217	1120
Mg (conc. in fresh water 24 mg L⁻¹)							
Addition (kg ha⁻¹)	115	118	13	84	79	310	146
Uptake (kg ha⁻¹)	13	72	9	202	34	61	75
Pot. conc. (mg L⁻¹)	255	115	10	—	112	622	177

[a] (Addition − Uptake)/total solution volume.
[b] References for reported crops, all in recycled solutions, are: tomato – Reshef et al., 2006; pepper – Bar-Yosef et al., 2005; lettuce – Bar-Yosef et al., 2002; Muskmelon – Bar-Yosef et al., 2000; strawberry – Matan et al., 2003; roses – Bar-Yosef et al., 2003; lisianthus – Dori et al., 2005.
[c] Estimate based on prevalent %S in plant as derived from Hochmuth et al. (2004) and plant DM in the above-mentioned references.

The Netherlands, for example, Cl⁻ accumulation is not regarded as a problem and the main concern is Na concentration build-up (Baas and van der Berg, 1999).

The increase in ion concentrations is equal to the accumulated quantity over time divided by the total water volume (V_{wT}). Representative examples depicting concomitant temporal increases in EC and Cl⁻ and Na⁺ concentration in recycled solution of three crops grown under threshold EC of ∼5 dS m⁻¹ is presented in Fig. 9.5.

9.2 MANAGEMENT

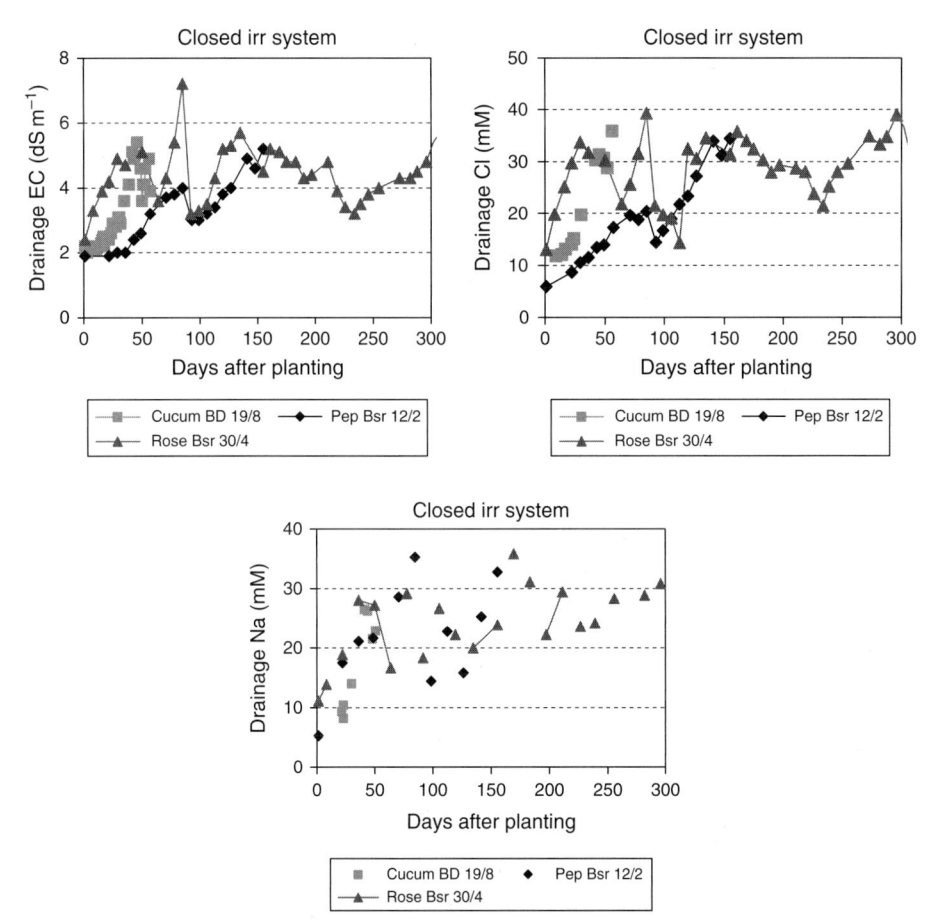

FIGURE 9.5 Increase in EC and Cl and Na concentration of recycled greenhouse solutions in the presence of cucumber (planted on 19/8 at Bet Dagan, Bar-Yosef et al., 2003), pepper (planted on 12/2 at Besor, Reshef et al., 2002a) and roses (planted on 30/4 at Besor, Bar-Yosef et al., 2003). Greenhouse water volumes were 29, 50, 30 m^3, 1000 m^{-2} greenhouse and plant stands 2500, 3500 6000 pl, 1000 m^{-2} greenhouse, respectively. NaCl concentration in fresh water was 6.6 mM in Besor and 6.0 mM in Bet Dagan.

Pepper that was planted in winter showed the slowest rate of EC and Cl and Na concentrations increase during the first three months of growth until EC$_{thr}$ was reached. Rose (planted in spring) had a faster salt accumulation rate and cucumber (planted in autumn) had the fastest rate due to its highest growth rate. The cumulative transpiration (ET$_{cum}$) which is required to obtain the threshold Cl$^-$ concentration in solution (C$_{thr}$, corresponding to EC$_{thr}$) depends on the Cl$^-$ concentration in fresh water (Cl$_{fw}$), initial Cl$^-$ concentration in recycled solution (Cl$_{in}$), Cl uptake by the crop (U$_{Cl}$) and total recycled solution volume (V$_T$):

$$ET_{cum} = \frac{[V_T(Cl_{thr} - Cl_{in}) + U_{Cl}]}{Cl_{fw}} \quad (14)$$

A similar mass balance approach was used by Carmassi et al. (2003) and Kempkes and Stanghellini (2003) (see later).

After EC_{thr} was obtained (Fig. 9.5), the solution EC of the three crops continued to fluctuate, sometimes in excess of the permitted lower and upper deviations (Eq. [9]). The fluctuations stemmed from inadequate solution discharge control stemming mainly from automation failure. The impact of such fluctuations on yield and quality is hard to evaluate; attempts to elucidate this problem are described later in this chapter.

While there was a clear similarity between EC and Cl$^-$ build-up pattern and fluctuations, Na behaved differently: its accumulation was not crop specific and it was subject to larger concentration fluctuations over time. This result agrees with the weaker correlation between EC and Na concentration in comparison with EC vs. Cl concentration (Fig. 9.4).

Temporal Na$^+$ concentration in solution (C_t) was satisfactorily simulated by Carmassi et al. (2003) and Kempkes and Stanghellini (2003), who used the ion mass balance equation (Eq. 15):

$$V_T C_t = V_T Cl_{in} + V_{in} C_{fw} - U - \Sigma V_{dis}(C_{thr} + \Delta C_{up})$$
$$C_t = C_{in} + C_{fw} V_{in}/V_T - (C_{thr} + \Delta C_{up})\Sigma V_{dis}/V_T - U/V_T \quad (15)$$

Here V_{in} is the cumulative fresh water application ($= \Sigma ET + \Sigma V_{dis}$), C_{in} is the initial ion concentration in the system solution and the other variables were defined in Eqs. (10 and 11).

EC variation over time cannot be calculated because (i) plant uptake is undefined, and (ii) occasional variations in nutrient concentrations cause variations in EC not accounted for by Eq. (15). However, as shown earlier (Fig. 9.4) the empirical relationship between C_{Cl} and EC is highly correlated, so predictions of C_{Cl} can be used as estimators of EC.

Accumulation of both Ca^{2+} and SO$_4^{2-}$ may result in gypsum (CaSO$_4$) precipitation and consequently emitters clogging. From solubility product (K_{sp}) considerations precipitation will not occur as long as (Ca^{2+}) (SO$_4^{2-}$) < K_{sp}, where parentheses denote activity and $K_{sp} = 10^{-4.5}$ (Lindsay, 1979). For (SO$_4^{2-}$) = 2 mM (concentration ~2.2 mM) which is attainable in closed-loop irrigation systems (see Table 9.6), (Ca^{2+}) should not exceed 15.8 mM, otherwise gypsum precipitation is likely to occur.

Two less common elements that may accumulate in recycled solutions are Si and Al. Si is usually found in fresh water; both may be released from substrates containing alumosilicates or their by products, for example stone wool, perlite, pumice and tuff. At pH 6.4, Al concentration in pumice is similar to its concentration in fresh water (Table 9.7). It increases to 1.3 mg L^{-1} as pH dropped to 3.8 due to increased NH$_4$ supply. The data indicate (Table 9.7) that transition from pH 4.9 to 4.2 caused a sharp increase in Al concentration from 0.32 to 0.66 mg L^{-1}. According to Adams and Moore (1983), a concentration of 1 mg L^{-1} Al^{3+} is potentially toxic to most plants, but complexation by various organic anions that are found in plant rhizosphere may reduce the activity of the free ion thus ameliorating its potential phytotoxicity (Grauer and Horst, 1992). At similar pH, perlite, pumice and stone wool did not differ considerably in their Al release to solution.

TABLE 9.7 Concentrations of Al, Si and Mn Found in Recycled Solution Under Pepper, Muskmelon and Gypsophila Grown in Perlite, Stone Wool (RW) and Pumice. Solution Sampling Time is Given as Number of Days After Planting (dap)

Ion	Pepper (68 dap) RW	Pepper (68 dap) Pumice	Muskmelon (67 dap) Perlite	Muskmelon (67 dap) Perlite	Muskmelon (67 dap) RW	Gypsophila (68 dap) Perlite	Gypsophila (68 dap) Pumice
	\multicolumn{7}{c}{Solution pH}
	5.5	6.9	6.4	4.2	4.9	6.0	5.1	4.5	3.8
	\multicolumn{7}{c}{Solution ion concentration (mg L^{-1})}
Al	0.13	0.010	0.038	0.66	0.32	0.10	0.35	0.17	1.3
Si	13.9	16.8	26.9	5.4	6.6	5.2	10.0	21	41
Mn	0.37	0.08	0.26	0.30	0.46	0.13	0.22	0.43	0.47

Sources: Data derived from Bar-Yosef et al. (pepper, 1999; muskmelon, 2000b; gypsophila, 1999b). Fresh water Al, Si and Mn concentrations were 5 µg L^{-1}, 6 ± 2 mg L^{-1} and 0, respectively.

Lindsay (1979) found that Si concentration in treatment solutions surpassed expected Si concentration in equilibrium with quartz (2.8 mg L^{-1}). Except in pumice, the Si concentrations were lower than expected in soil solution containing alumosilicates (22 mg L^{-1}); in pumice the Si concentration (27 or 41 mg L^{-1} depending on pH, Table 9.7) slightly exceeded that value. Silica concentration in perlite fluctuated with pH in the range of pH 4.2–6.9, and was more crop dependent than pH dependent. The crop-specific concentrations reflect plausible differences in Si uptake by various plants (Table 9.7). Stamatakis et al. (2003) reported that soluble Si at 2 mM (56 mg L^{-1}) in hydroponic solution ameliorated salinity-induced yield reduction in tomato (*Lycopersicon esculentum*) and alleviated incidence of blossom end rot under non-saline conditions; unfortunately the effect of lower concentrations was not investigated. On the other hand, Lieten et al. (2002) found that irrigation solution Si concentration > 0.55 mM (as K$_2$SiO$_3$) was associated with enhanced strawberry (*Fragaria* × *ananassa* Duchense) fruit albinism. The direct mechanism of this disorder is still unexplored.

Boron is prone to accumulation in recycled solutions when added in excess of plant consumption. For example, in two closed-loop irrigation system experiments in Israel, B was added continuously in the replenishment solution at concentration of 0.2 mg L^{-1} and its concentration in the recycled solution increased depending on the threshold EC (which determined discharge) and time after planting (only one date is presented in Table 9.8). In lisianthus (*Eustoma grandiflorum*) the concentration at the end of the second flush reached values of 0.4–0.8 mg L^{-1} and in cucumber the concentration varied between 0.5 and 1.1 mg L^{-1}.

Similarly, Yermiyahu et al. (2003) found no yield reduction in greenhouse pepper exposed to B concentrations of 0.5, 1.0, 2.0 and 4.0 mg L^{-1}, no toxicity symptoms were observed in the crops described in Table 9.8 either.

TABLE 9.8 Boron Accumulation in Recycled Solution as a Function of Threshold EC (EC$_{thr}$)[a] in Two Crops. Note That Solution Sampling Time was Different in the Two Experiments. Boron Concentration in the Feed Solution was 0.2 mg L^{-1}

	Lisianthus[b] Harvest 2nd flush	Cucumber[c] 45 days after planting
EC$_{thr}$ (dS m^{-1})	B (mg L^{-1})	
2.3–2.4	0.4	0.5
3.5–3.6	0.6	0.9
4.3–4.6	0.8	1.1

[a] As EC$_{thr}$ increases, discharge volume decreased.
[b] Bar-Yosef, 2003, unpublished data.
[c] Derived from Bar-Yosef et al. (2003).

Another inorganic toxicant that may accumulate in recirculated solutions is nitrite (NO$_2^-$). The problem can be acute in new synthetic substrates lacking nitrifying bacteria where NH$_4$ is a significant N source. Under such conditions, oxidation of NH$_4$ to NO$_2$ occurs at a faster rate than conversion of NO$_2$ to NO$_3$, so that nitrite accumulates in the solution. An example of this phenomenon is presented for cucumber growing in perlite in Fig. 9.6. At the end of the first week of growth, NO$_2$-N concentration was ~1 mg L^{-1} regardless of N total concentration and it sharply increased within two weeks to the peak which varied between 9 and 13 mg NO$_2$-N in the 2–8 mM N treatments (30:70 NH$_4$:NO$_3$ ratio). In the highest N treatment (16 mM) the same trend was seen, but NO$_2$ concentrations were lower probably due to lower solution pH which may have inhibited the NH$_4$ oxidation to NO$_2$. The nitrite concentration

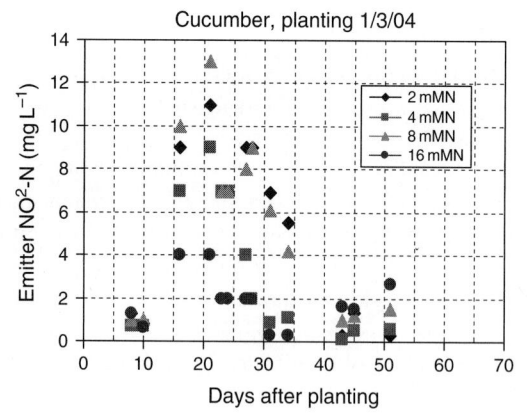

FIGURE 9.6 Nitrite (NO$_2^-$) concentration in recycled solution of cucumber plants grown in perlite 2 as a function of time and concentration of N in replenishment solution (NH$_4$:NO$_3$ ratio = 30:70). Derived from Bar-Yosef et al., (2005).

CHAPTER 9 FERTIGATION MANAGEMENT AND CROPS RESPONSE

TABLE 9.10 Concentration of Total Bacteria, *Pseudomonas Fluorescens*, and Total Fungi (on PDA, Dextrose Agar) in Circulating Solution of Given Crops Grown in Perlite and Stone Wool (RW)[a]

EC_{thr} (dS m^{-1})	Substrate	Season	Total bacteria (cfu ml^{-1})	Pseudomonas Fluorescens (cfu ml^{-1})	PDA fungus (cfu ml^{-1})
4.5	Perlite	Spring	1.9×10^6	2.3×10^5	—
2.5	Perlite	Spring	5.4×10^6	2.9×10^5	—
2.3	Perlite	Summer	0.9×10^6	1.0×10^5	160
4.5	RW	Spring	4.0×10^4	2.6×10^4	260
4.5	Perlite	Spring	7.0×10^4	1.1×10^4	66
2.0	Perlite	Spring	2.0×10^5	1.8×10^4	310
4.5	Perlite	Summer	2.4×10^5	1.2×10^3	2080

[a] Unless otherwise stated, solutions were sampled 2–3 months after planting. Analyses were done by Zman and Ben-Yeffet, Agricultural Research Organization, Israel. Total bacteria count in fresh RW is 6.2×10^4 cfu mL^{-1}. No disinfection treatments took place in presented experiments. See also data for this crop and substrate ($EC_{thr} = 2.9$ dS m^{-1}) in Table 9.4.

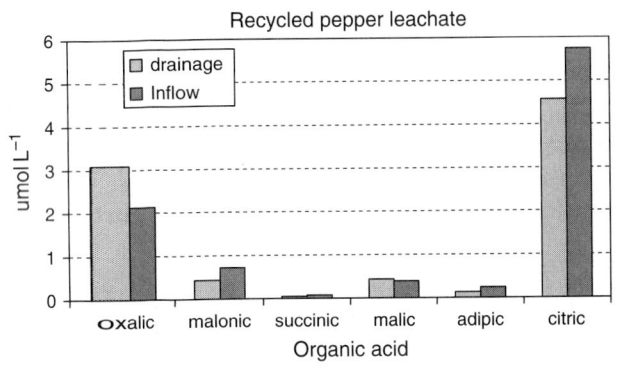

FIGURE 9.7 Concentration of organic acids found in recycled solution of pepper grown on rockwool, Dagan, Israel. The total organic carbon concentration is presented in Table 9.9. Solutions were in emitter (inflow) and in substrate drainage in late autumn (30 days after planting). Unpublished Schwarz, Erner and Bar-Yosef (1999). Formulae of acids yielding concentration > 0.5 μM are $(COOH)_2 \cdot 2H_2O$ (oxalic), $CH_2(COOH)_2$ (malonic) and $H_3C_6H_5O_7 \cdot H_2O$ (citric).

per cent of analysed organic anions and that concentrations in solution flowing into the substrate and out of it are practically equal (Fig. 9.7). The concentration of oxalate (~5 μM) is similar to the molar concentration of Zn and Mn in the same solution, therefore complexation might influence the mobility and adsorption of these microelements in the system. The OC concentration in the recycled solution of pepper was shown to be ~20 mg L^{-1} C (Table 9.9) which exceeds the ~0.5 mg L^{-1} C contributed by the organic acids presented in Fig. 9.7. This means that the OC in recycled solutions is composed mainly of more stable carbon compounds, probably stemming from microbe and root debris.

9.2 MANAGEMENT

returned to ~1 mg L^{-1} value after approximately six weeks
young and their root systems are small, the exposure to suc
render them susceptible to root pathogens (e.g. *Pythium*). N
anoxic conditions too, but in this case it stems from nitrat

9.2.2 ORGANIC CARBON ACCUMU

Organic-C (OC) accumulation is important as it serv
microorganisms and affects cations speciation due to comp
Microorganisms consume oxygen and thus compete with
well-adapted beneficial populations compete with root path
and thus suppress their potency. In systems with EC$_{thr}$ arou
small and therefore OC concentration exceeded the one in lc
which did not differ meaningfully from fresh water OC
Note that the values reported are the mean OC concentratio
season, as fluctuations with time after the first month of gro
the experimental error (±20 per cent of the mean). The stea
reflects the crop-specific balance between root excretion a
of C. Measurements in pepper indicated that OC concentra
heated to 16°C) surpassed the concentration in spring (mean
probably due to reduced microbial activity in winter (Tal
can be evaluated only for the same substrate and growing s
effects from the presented results are hard to obtain.

Data on OC constituents in recycled solutions is meag
per (same crop as in Table 9.10) show that citrate and oxa

TABLE 9.9 Steady-State^a Organic Carbon (OC) Concentration in Re
of Various Crops and Substrates at Different Growth Seasons in Israel. OC
in Fresh Water = 10–13 mg L^{-1} C

Crop	Substrate	EC$_{thr}$ (dS m^{-1})	Season	OC (mg L^{-1}
Pepper	Stone wool	4.0	Winter	38
Muskmelon	Perlite	2.5	Spring	11
		4.5	Spring	21
Gypsophila	Perlite	2.3	Spring	14
		2.9	Spring	18
Lettuce	Stone wool	2.1	Autumn	8
		2.8	Autumn	16
Cucumber	Stone wool	2.1	Spring	15
		4.8	Spring	33

^{a} In all presented systems solution samples were taken in two weeks inte
season. Between one month and harvest the OC concentrations fluctuated wi
±20 per cent of the presented mean (Bar-Yosef et al., 2003a).

It is noted that in all analysed recycled solutions (Table 9.9), dissolved O_2 concentration was higher than 90 per cent of the equilibrium concentration with air at the given solution temperature (~ 8 mg L^{-1} O_2, data not presented). This was possible under the regime of 8–16 irrigations per day exercised in the reported studies, which prevent formation of stagnant solutions in the substrate.

9.2.3 MICROFLORA ACCUMULATION

In pepper the concentrations of total bacteria and *Pseudomonas Fluorescens* in circulated solutions were two orders of magnitude smaller in pumice than in stone wool and in gypsophila two orders of magnitude smaller in pumice than in perlite (Table 9.4). In muskmelon (EC$_{thr}$ = 4.5 dS m^{-1}) the difference in microbe concentration between stone wool and perlite was relatively small and inconsistent (Table 9.10). The substrate effect plausibly stemmed from different pore-size distributions determining the filtration capacity of the medium but this effect should be further investigated before practical conclusions can be drawn. Increasing threshold EC had no clear effect on microbe population, but in muskmelon on perlite, the total bacteria concentration was meaningfully lower than in pepper and gypsophila (Table 9.10). Comparison between microbe accumulation in summer and spring (muskmelon on perlite, Table 9.10) showed small, insignificant differences. No disinfestation was used in any of the aforementioned experiments; despite this, the incidence of plant mortality was negligible in all three crops. The last observation is compatible with results by McPherson and Hardgrave (1993) and Tu et al. (1999), showing that the incidence of tomato plant mortality due to Pythium inoculation was greater in an open irrigation system than in closed one. The mechanism suggested by McPherson and Hardgrave was that some beneficial (though unidentified) compounds that are leached in open systems do accumulate in closed systems and suppress pathogens. Tu et al. (1999) suggested that a larger bacterial population that developed in their closed system suppressed the Pythium.

9.2.4 DISCHARGE STRATEGIES

Very little is known about the effects of permitted deviation from the mean threshold EC ($\Delta C_{up} + \Delta C_{low}$, Eq. [9]) on plant performance in closed-loop irrigation systems. Salinity stress (obtained by a large permitted deviation, or amplitude) has been shown to improve fruit quality (Mizrahi et al., 1988; Plaut, 1997) but possible toxicity and uptake inhibition effects might delineate this advantage, and minimal deviations around EC$_{thr}$ may be preferred. Kramer et al. (2002) investigated the EC amplitude effect in roses in the Arava valley in Israel where fresh water EC was 2.6 dS m^{-1}. Four treatments were investigated: DR4 and DR3 had EC$_{thr}$ of 3.7 dS m^{-1} and amplitudes of 1.7 and 0.3 dS m^{-1}, respectively. Treatment Dr1 had EC$_{thr}$ of 2.8 dS m^{-1} and amplitude of 0.3, and treatment Dr2 had EC that fluctuated between the maximum EC of treatment 3 and the minimum EC of treatment 1 (Fig. 9.8).

The treatments were applied for a year and results showed no significant difference in cumulative fresh flower weight and ET between low- and high-amplitude treatments

Closed-loop irrigation, Rose, Arava Israel

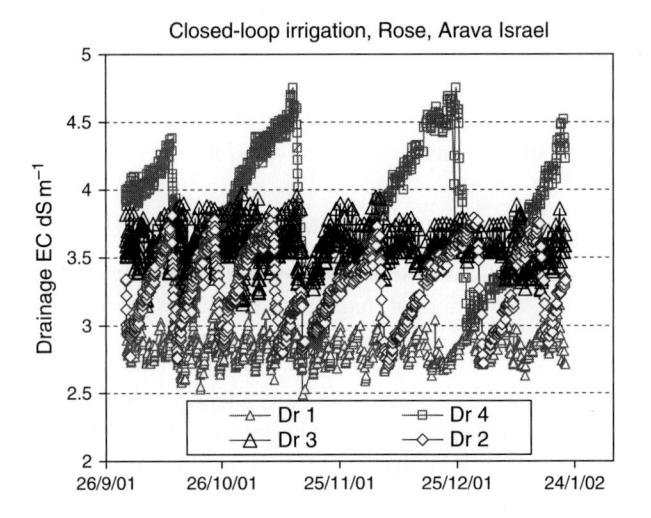

Tr	FWt	ET	Discharge Water	Discharge N
	g m^{-2}	L m^{-2}	L m^{-2}	g m^{-2}
Dr1	9070	1311	1750	131
Dr2	9284	1260	791	79
Dr3	8111	1239	678	68
Dr4	7997	1221	428	43

FIGURE 9.8 Effect of EC saw tooth amplitude ($\Delta C_{up} + \Delta C_{low}$, Eq. [10]) on total rose cut flower (CV. Long Jaguar) weight (FWt), cumulative evapotranspiration (ET) and discharge of water and N during the growth period, 31/12/99–31/8/01. The substrate was tuff, fresh water EC was 2.6 dS m^{-1}, and irrigation frequency 8–12 per day.

(Dr3 and Dr4). However, water and N discharge were 35 per cent smaller in the high than in the low amplitude (table adjacent to Fig. 9.8). The reason is that in order to dispose a certain quantity of salts, a smaller volume is needed when salt concentration is higher. When reducing the mean salinity level in solution (treatments Dr1 and Dr2) the yields and ET significantly increased but again the high-amplitude treatment resulted in significantly smaller discharge in comparison with low amplitude. The 'large saw tooth' (high amplitude) approach requires fewer discharge events (points of sharp decline in EC, Fig. 9.8), therefore its operation is expected to be cheaper than 'small saw tooth' systems.

9.2.5 SUBSTRATE AND SOLUTION VOLUME PER PLANT

9.2.5.1 Background

The substrate and solution volumes per plant affect root growth (see earlier discussion and Chap. 13), determine the buffering capacity of the closed irrigation system to water, nutrients and oxygen, and influence the salt accumulation rate in the substrate. Reduced root mass (R, g fresh weight root m^{-2} substrate) decreases the water and nutrients uptake rate (Q_W, Q_N, respectively, Eq. [16], g m^{-2} substrate h^{-1}) and reduces root cytokinin and abscisic acid (ABA) synthesis and supply to the canopy (Aiken and Smucker, 1996).

$$Q_N = F_N R; \qquad Q_W = J_W R \qquad (16)$$

The flux terms (F_N and J_W) are commonly described by the Michaelis–Menten equation (Eq. [17]) and the water flux of uptake (Eq. [18]) equations:

$$F_N = F_{maxN} C/(K_{mN} + C) \qquad (17)$$

$$J_W = K_{sr}(\phi_s - \phi_r) \qquad (18)$$

Here, F_{maxN} is the maximum F value for N, K_{mN} is the uptake efficiency constant (L g N^{-1}), K_{sr} is the combined soil-root resistance to water permeability into root (g H$_2$O bar^{-1} h^{-1}) and ϕ_s and ϕ_r (bar) are water potential in soil at the root surface and inside the root, respectively.

With regard to the hormone/root-volume interaction it is worthwhile to note that auxin, which is synthesized mainly in above-ground plant organs, is participating in root growth regulation via root proton production and movement (Mulkey et al., 1982). Through this mechanism the rapid promotion of root cell elongation by auxin is mediated by auxin-induced acid efflux into the cell wall, reducing cell wall rigidity, allowing it to expand. This proton movement is accompanied by rapid H$^+$ efflux from the elongation zone to embedding solution (Mulkey et al., 1982). It is difficult to quantify the auxin impact as it also induces biosynthesis of ethylene, an inhibitor of root elongation. The auxin flow to roots is aimed to root tips. As a result, promoting root branching (e.g. by reducing substrate volume per plant or containerizing) may reduce auxin concentration in apical root cells and thus impeding root elongation. It is possible that by increasing H$^+$ production in roots via enhanced NH$_4^+$ uptake, the auxin becomes more effective in promoting root growth, particularly under conditions enhancing root branching. Increasing the number of root tips per plant may also decrease imported gibberellin concentration in growth loci in roots and thus inhibit root growth by reducing stimulation to cell elongation.

When the target uptake rate (Q_{target}) is known, the minimal root mass which is required to sustain it under maximum flux conditions ($F = F_{max}$) is $R_{min} = Q_{target}/F_{max}$. If actual root weight is less than R_{min}, modifying the fertigation regime will not facilitate an increase in uptake rate and the only option to enhance uptake is to increase root mass, for example by increasing substrate volume per plant. Michaelis–Menten constants for some ions and crops are presented in Table 9.11.

For example, in order to satisfy Q_{NO_3} of 500 mg N m^{-2} substrate d^{-1} (= 50 mg N m^{-2} h^{-1} for 10 hours of uptake per day) by a tomato plant with $F_{max} = 25\,\mu g$ N g^{-1} fw

TABLE 9.11 Michaelis–Menten Constants[a] of Selected Ions and Crops Grown in Stirred or Unstirred Solution Culture. F_{max} in $mol\,cm^{-1}\,s^{-1}$, and K_m in μM.

Crop	Ion	F_{max}	K_m	System
Maize	NO_3	1.16	10	Well stirred
	H_2PO_4	0.50	3.0	
	K	5.02	16	
Soybean	H_2PO_4	0.10	2.0	Well stirred
Tomato[a]	NO_3	5.1	258	Well stirred
Tomato[b]	NO_3	4.6	400	Stirred
	NH_4	20.3	3000	
	H_2PO_4	2.2	350	
Tomato	NO_3–NH_4	11.0	3000	Sand
	H_2PO_4	1.8	320	Rock-wool
	K	3.0	1000	Aeropohnics
Pepper	NO_3–NH_4	14.0	550	Unstirred
	H_2PO_4	17.0	25	

[a] Units conversion: for root owning radius of 0.025 cm and bulk density of 1 g fw cm^{-3} root, F_{max} of $NO_3 = 10\,mol\,N\,cm^{-1}\,root\,s^{-1}\,10^{-13}$ is identical to $25.8\,\mu g\,N\,g^{-1}$ fw root h^{-1}; $K_m = 500\,\mu M\,NO_3$ is identical to $7\,mg\,L^{-1}\,NO_3$-N.
[b] Scaife and Bar-Yosef (1995).

root h^{-1}, $R_{min} = 50/0.025 = 2000$ g fresh weight root m^{-2} substrate. If the substrate is 0.2 m deep, this root volume occupies ~2 L (assuming root density of 1 g (mL^{-1})) out of 200 L substrate. Estimating the minimal root weight which is necessary to sustain a given transpiration rate is more difficult because the J_W parameters (Eq. [18]) are not available in the literature. The role of the Michaelis–Menten parameter K_m in planning recirculation regimes will be discussed below.

Water volume per plant determines the rate by which a nutrient concentration in solution varies with time. For the aforementioned 200 L substrate volume and predawn θ of 0.6 v v^{-1}, daily uptake of 500 mg N m^{-2} substrate reduces C_N by ~4 mg L^{-1}. This is true, of course, if no N is added via the water replenishing ET. For $f_s = 0.3$, this uptake rate is equivalent to 150 mg N m^{-2} ground. For tomato consumption rate in the high range (~350 mg N m^{-2} ground, Reshef et al., 2003), the daily reduction in C_N is 9.7 mg L^{-1} N. By applying Eqs. (16) and (17), one can estimate the expected reduction in N uptake rate due to this decline in C_N and decide what should be the frequency of N replenishment prior to exceeding the permitted deviation from Q_{target}. If f_c is 0.125 (20 cm substrate width in a bed of 160 cm, instead of 48 cm substrate above), the daily change in C_N would be 23.3 mg L^{-1} N.

The faster decline in C_N (and faster increase in salinity) associated with decreased root volume suggest that greater substrate and water volumes per plant would be better. Other factors, however, suggest smaller substrate and water reservoir volumes, among them, (i) economic reasons such as substrate acquisition and disposal expenses; (ii) reduced disinfection costs; and (iii) smaller drip irrigation systems are required.

9.2.5.2 Experimental Data of Crop Response to Substrate Volume

Pepper (CV. Cuby) grown in a closed irrigation system in Besor, Israel, was subjected to EC_{thr} of 2.7 or 4.0 dS m^{-1}. Total and export quality yields were highest in 0.50 m wide and 0.2 m high perlite beds (16.6 L pl^{-1} substrate), followed by 0.2 and 0.1 m wide beds (6.7 and 3.3 L pl^{-1}). The 0.2 and 0.1 m wide treatments did not differ significantly from each other (Fig. 9.9).

Xu and Kafkafi (2001) compared pepper growth response to container volumes of 9, 18 and 33 L pl^{-1} and in agreement with the above result reported that growth

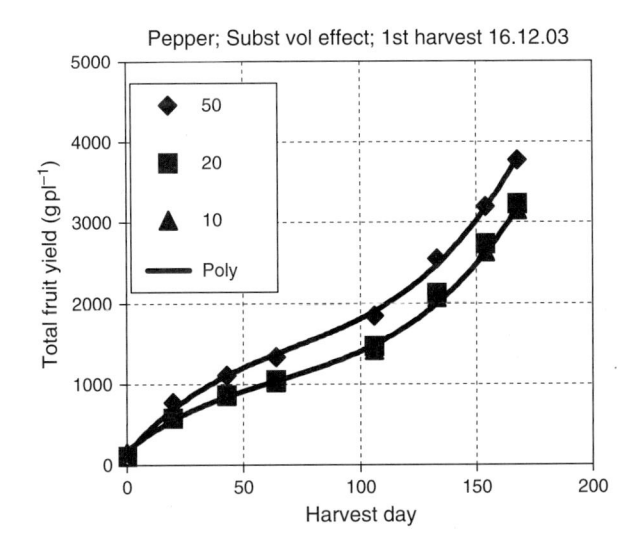

Width	Exp yld	BER	Fruit wt
cm	kg pl^{-1}	%	g frt^{-1}
50	2.68 a	39	164
20	2.08 b	41	162
10	2.10 b	41	164

FIGURE 9.9 The effect of perlite width (10, 20 or 50 cm) on total pepper (CV Cuby) fruit yield as a function of time (Figure), and high quality fruit yield (Exp yld), blossom end rot (BER) incidence and average single fruit weight (Fruit wt) (Box). The substrate volumes per plant were 3.3, 6.7 and 16.6 L pl^{-1}, respectively; substrate height was 0.20 m, $EC_{thr} = 4.0$ dS m^{-1} and irrigation frequency was 10–16 per day, depending on the season. First harvest day was Dec 16, 2003. Experiment was conducted in Besor, Israel (Bar-Yosef et al., 2004a).

and fruiting were maximal in the intermediate volume. Substrate width in the Besor experiment had no significant effect on the incidence of blossom-end rot (BER) or the average fruit weight at the beginning of June. The response was similar under the two studied EC_{thr} levels, therefore only the higher EC_{thr} is presented. The data (Fig. 9.9) indicate that the advantage of the 50 cm wide bed became apparent ~60 days after the start of harvest.

Root excavation carried out on June 10 showed that root weight (excluding >4 mm diameter roots) in 0.5, 0.2 and 0.1 m wide substrate decreased from 21 to 17 and 9 g DM pl^{-1}, respectively. This order is compatible with the treatments effect on fruit yield thus indicating that root growth restriction might have limited fruit production. The results also support the suggested hypothesis that increasing number of root tips stemming from decreasing substrate volume reduced root growth because of lower auxin and gibberellin concentrations in apical root cells. Direct measurements of root tips and hormone concentrations are needed for elucidating the obtained results.

The substrate volume effect on crop performance in closed irrigation systems was also studied in tomato. As above, yield in the 0.5 m wide perlite exceeded the yield in 0.2 or 0.1 m substrates, the latter two not differing significantly from each other (Table 9.12).

The yield increase stemmed from more and larger fruits, and lower BER incidence. Increasing the substrate width elevated P and K (but no other nutrient) concentrations in leaves in late November (0.97, 1.06 and 1.12 per cent P and 5.0, 5.2 and 5.5 per cent K, respectively) and increased transpiration rate (TR) during the spring (1.6, 2.8, 3.8 mm d^{-1}, respectively). The effect of substrate width on leaf nutrient concentrations in the spring was negligible, so it may be concluded that the yield reduction in the 0.1 and 0.2 m wide substrates resulted from water stress in the spring.

In the cited pepper experiment the maximum TR was $2 L\,pl^{-1}\,d^{-1}$. During the noon hours the corresponding TR was $0.5 L\,pl^{-1}\,h^{-1}$. In the 10 cm substrate ($3.3 L\,pl^{-1}$), water quantity after the rapid drainage was $\sim 1.5 L\,pl^{-1}$ ($\theta_{1/2h} = 0.45\,v\,v^{-1}$). Removal of additional 0.5 L water by the plant reduces θ to a value which strongly decreases the perlite's unsaturated hydraulic conductivity, thus jeopardizing water and nutrients transport to roots. In 0.2 m substrates, this problem probably did not occur.

TABLE 9.12 Effect of Substrate Width[a] on Tomato Total and Marketable Yields, Blossom End Rot (BER) Incidence and Large Fruit Yield (Besor, Israel)[b]

Substrate width (cm)	Accumulated yield (kg pl^{-1}) Total	Marketable	BER (kg pl^{-1})	Fruit size (kg pl^{-1}) >67 mm
10	9.55 b	6.00 b	1.68 a	3.12 b
20	9.46 b	6.11 b	1.58 ab	3.22 b
50	10.10 a	6.74 a	1.50 b	3.63 a

Means followed by the same letter are not significantly different according to Duncan's LSD test (0.05).

[a] $EC_{thr} = 4.5\,dS\,m^{-1}$ in all treatments; fertigation frequency 10–16 d^{-1}, depending on season.
[b] Reshef et al., (2006).

9.2.6 EFFECT OF SUBSTRATE TYPE

Substrates and their physical and chemical properties are reviewed in Chaps. 3, 6, 11 and 12. The most important substrate properties in closed-loop irrigation systems are water infiltration rate, drainage rate and, where applicable, microbes filtration capacity. The overall substrate effect on crop performance in a closed fertigation system with a discharge criterion of $EC_{thr} > 4\,dS\,m^{-1}$ was studied for pepper growing in stone wool and pumice, as well as for muskmelon growing in perlite and stone wool (Table 9.13). Under identical fertigation regime and target nutrient consumption curves, higher yields were achieved in perlite and pumice than in stone wool, but the fruit quality [BER incidence in pepper and total soluble solids (TSS) in muskmelon] did not differ significantly or was even better in stone wool (lower BER incidence). The difference in nutrients uptake (not presented) and in water and N inputs was small and within experimental error. The disadvantage of stone wool in comparison with pumice could also result from its smaller volume per plant. The comparison between stone wool and perlite was achieved with the same substrate volume per plant and indeed the difference was less pronounced than in the stone wool–pumice comparison. The slower drainage rate of stone wool in comparison with perlite or pumice (Table 9.1) may account for the poorer performance of stone wool, while the advantage of pumice may be at least partly attributed to its superior filtration capability, as evidenced by the reduced pathogen presence in the recirculated solution (Table 9.10).

TABLE 9.13 Substrate Type Effects, Under Water Recirculation Conditions, on Crop Marketable Yield, Fruit Quality (Blossom End Rot [BER] Incidence in Pepper, and Total Soluble Solids [TSS] in Muskmelon) and Water and N Inputs.[a] The Fertigation Frequency, Water Volume Per Irrigation and Nutrient Target Concentrations were Identical in all Reported Substrates[b]

Substrate	EC_{thr}	Marketable yield	Fruit quality	Water inputs	N input
			Pepper (CV. Mazurka)		
	$dS\,m^{-1}$	$kg\,m^{-2}$	BER (%)	mm	$g\,m^{-2}$
Stone wool	4.2	6.1 b	11.5 b	347	46
Pumice	4.4	7.6 a	17.7 a	380	52
			Muskmelon (CV. 5093)		
	$dS\,m^{-1}$	$kg\,m^{-2}$	TSS (%)	mm	$g\,m^{-2}$
Perlite	4.5	6.06 a	10.8	325	36
Stone wool	4.5	5.42 ab	9.2	354	38

Means followed by the same letter are not significantly different according to Duncan's LSD test (0.05).

[a] Derived from Bar-Yosef et al., 1999 (pepper) and Bar-Yosef et al., 2000 (muskmelon). Experiments carried out at Bet Dagan, central Israel.

[b] Substrate width was 100 cm in pumice, and 2×20 cm (double row) in perlite and stone wool. Irrigation frequency was $6–8\,d^{-1}$, depending on the season.

9.2.7 WATER AND NUTRIENTS REPLENISHMENT

Nutrient concentrations in fill solutions (C_f, mg L^{-1}) are designed to replenish the net loss on the previous day due to crop nutrient consumption (U) and system disposal (D). These solutions should be such as to maintain a predefined target concentration in the entire solution volume. The added water volume (V_f, L m^{-2} ground) multiplied by C_f should equal $U + D$ (mg m^{-2} ground):

$$C_f V_f = U + D; \qquad C_f = \frac{U}{V_f} + \frac{D}{V_f} \qquad (19)$$

When $D = 0$, the quotient U/V_f is the nutrient concentration in the transpiration stream (C_{ts}). In practice, U and D are not measured daily; instead, U is estimated from a crop-specific target nutrient consumption curve and D is evaluated using the estimated nutrient concentration in the drainage and the measured disposed drainage volume. Over- and under-estimation of C_f may cause accumulation or depletion of nutrients in the recirculated solution, respectively, therefore chemical analysis of the solution must be performed at least every two weeks. Under conditions of (i) $D = 0$ and (ii) real $C_{ts} = C_f$, the concentration of the nutrient in the recirculated solution is maintained at C_f.

The target nutrient concentration in solution (C_{trg}) is defined as a concentration providing an uptake rate equalling the target nutrient consumption rate by the crop at the specific growth stage. The target nutrient consumption curve is defined as a unique time function coinciding with optimal yield. Currently this function is obtained experimentally (Bar-Yosef, 1999) but in the future it may be possible to calculate it from crop model. When inserting C_{trg} in the Michaelis–Menten equation (Eq. [17]), a unique flux is obtained. Multiplying this flux by the current given (or estimated) crop root mass yields an uptake rate value that should match the target uptake rate by the plants at the appropriate growth stage. Apparently C_f (derived from quantity factors) is not necessarily equal to C_{trg} (derived from intensity factors) but the closer these values are to each other, the more balanced and stable is the recirculation system. The target concentration C_{trg} can be used as long as the nutrient quantity in the system ($C_{trg} \times V_T$) exceeds the current daily demand by the crop multiplied by a confidence factor f_c ($f_c \geq 1$). When this restriction is not met, and V_T satisfies the ET demand and cannot be increased, the nutrient concentration in solution should be elevated above C_{trg}, despite possible adverse effects on plant development. In practice, C_{trg} can seldom be calculated accurately due to scant Michaelis–Menten parameter estimates (Table 9.11) and root mass data. Consequently it is approximated as C_f and corrected during production to a steady-state concentration.

Fertigation frequency may affect the choice of C_{trg} because it is one of the factors determining the time-averaged nutrient concentration at the root surface. At higher frequency the nutrient depletion zones around roots are more often replenished by fresh solution, thus the time-averaged concentration in the root zone increases. Depletion zones are formed if the flux of uptake exceeds the rate of element transport from the bulk solution to the root. The transport is driven by diffusion and convection. The first is determined by the effective ion diffusion coefficient in the substrate, θ, and the

9.2 MANAGEMENT

nutrient concentration gradient along the pathway from the bulk solution to the root. Convection is caused by transpiration and the nutrient concentration in solution. Under similar weather and substrate conditions, a higher C_{trg} is required under low irrigation frequency (IF) than high IF because a greater concentration gradient towards the root is needed in order to provide a predefined flux of uptake.

In the Michaelis–Menten equation (Eq. [17]) the concentration term C is defined at the root surface. When adopting Michaelis–Menten parameter estimates from the literature, it is imperative that the experimental C indeed represents conditions at the root surface and not in the bulk solution (e.g. unstirred nutrient solution) or bulk substrate solution (Table 9.11).

Olsen and Kemper (1968) applied the steady-state solution of the general convection–diffusion equation in cylindrical coordinates to describe nutrient concentration at the root surface (C_a, g L^{-1}) as a function of its concentration in the bulk substrate solution (C_b, g L^{-1}), steady water uptake rate per unit root length (W, mL h^{-1} cm^{-1} root), the system's effective diffusion coefficient (Dp, cm^2 h^{-1}), root radius (a, cm) and substrate volumetric water content (θ, v v^{-1}). Simple algebraic transformation of their original solution yields,

$$C_a = -\beta + \sqrt{(\beta^2 + K_m C_b \alpha)}$$
$$\beta = 0.5\left[K_m - \left(\frac{F_{max}}{W}\right)(1-\alpha) - C_b \alpha\right]; \quad \alpha = a^{W/(2\pi\theta D_p)}; \quad D_p = D_o \exp(s\theta), \tag{20}$$

where D_o = diffusion coefficient (cm^2 h^{-1}) in water, s = substrate constant (units 1 θ^{-1}) and K_m and F_{max} are the Michaelis–Menten parameters in Eq. (17). Note that F_{max} is expressed per unit root length rather than per unit root weight.

Salts accumulate at the root surface between successive irrigations if transport rate to root exceeds the rate of uptake by the root. In this case, fresh irrigation solution displaces the salts from the root surface and reduces the osmotic potential in the root zone.

Salinity is affecting the choice of C_{trg} because it reduces mass flow towards roots and decreases the flux of nutrients uptake due to competition on ion uptake sites, or transporters (Marschner, 1995). A semi-empirical function representing inhibited uptake is the competitive Michaelis–Menten equation (Fried and Broashard, 1967):

$$F_{NO_3} = \frac{F_{maxNO_3} C_{NO_3}}{(K_{mNO_3} + C_{NO_3} + C_{Cl} K_{mNO_3}/K_{mCl})} \tag{21}$$

The function describes the flux of nitrate uptake, F_{NO_3}, in the presence of a given Cl$^-$ concentration (C_{Cl}, mg L^{-1}) by taking into account the affinity of Cl$^-$ to the uptake site (K_{mCl}, L mg^{-1}). Other variables in Eq. (21) were defined above at Eq. (17). According to Eq. (21), adding 150 mg L^{-1} Cl to a solution of 100 mg L^{-1} NO$_3^-$ reduces F_{NO_3} by tomato roots from 8 to 1.3 µg g^{-1} root NO$_3$-N (K_{mNO_3} = 3.6 10^{-3} L g^{-1} and K_{mCl} = 1.0 × 10^3 L g^{-1}). Equation (21) has been used in modelling recirculation (Bar-Yosef et al., 2004c) but has yet to be calibrated or validated for a wide range of practical applications.

Competitive uptake is also known between NH_4^+ and Mg^{2+} and K^+. For background information on this subject, see Marschner (1995) and Chap. 8; specific information for hydroponics roses can be found in Lorenzo et al. (2001).

Substrate volume is affecting chosen C_{trg} because reduced root mass calls for increase in flux of uptake if the absorption rate is to be unaltered.

9.2.7.1 Effect of Target N Concentration on Crop Performance

Bar-Yosef et al. (2000c) found that pepper fertigated 18 times per day gave similar total yield, large fruit yield and unmarketable yield under target N concentration of 70 and 140 mg L^{-1}. In this experiment, however, P and K concentrations varied concomitantly with N (Table 9.14). The expected effect of the higher NH_4 concentration in the 140 mg L^{-1} N treatment on higher incidence of unmarketable fruits (mainly blossom end rot) due to Ca uptake inhibition (Kirkby and Mengel, 1967) could not be confirmed in this study (Table 9.14).

Tomato response to N target concentration was stronger than pepper (Table 9.15) and showed a clear decline in total, marketable and large fruit yields as N concentration

TABLE 9.14 Effect of Macronutrient Concentration in Solution (C_{NPK}) on Pepper[a] Yield in a Closed-Loop Irrigation System[b] at Bet Dagan, Israel[c]

C_{NPK}[d] (mg L^{-1} N,P,K)	Total (kg m^{-2} ground)	Large (kg m^{-2} ground)	Unmarketable (%)
70,15,30	10.9	5.3 a	17.5 b
140,75,150	11.0	4.9 ab	19.1 ab

Means followed by the same letter are not significantly different according to Duncan's LSD test (0.05).

[a] CV. Cuby; planting, 30 August 1999 in perlite; first and last harvests, 20 December 1999 and 15 May 2000, respectively. The irrigation regime was identical in all treatments.
[b] The irrigation regime was identical in all treatments.
[c] Derived from a larger table by Bar-Yosef et al., 2000c.
[d] $NH_4:NO_3$ ratio in fill solution = 1:4; EC_{thr} = 3.5 dS m^{-1}; 18 irrigations d^{-1}.

TABLE 9.15 Effect of Target N Concentrations in Solution on Tomato[a] Yield and Ca and Mg Concentration in Plant Leaves. Plants Grew in Closed Irrigation System[b] Greenhouse at Besor, Israel[c]

C_N[d] (mg L^{-1} N)	Yield (kg m^{-2} ground)				Content in leaves (g 100 g^{-1})	
	Total	Marketable	Large	BER[e]	Ca	Mg
50	10.8 a	7.7 a	4.2 a	1.2 c	1.01 a	2.78 a
150	9.4 b	5.7 b	2.8 b	1.7 b	0.81 b	2.01 b
250	7.6 c	3.6 c	2.2 c	2.2 a	0.72 b	1.64 c

Means followed by the same letter are not significantly different according to Duncan's LSD test (0.05).

[a] CV. 870; planting 7 September 2004; 9–18 irrigation per day, depending on the season.

increased from 50 to 150 and 250 mg L^{-1} N. The decline in yield occurred despite a small but significant increase in N and P concentration in leaves (data not presented), and decrease in Cl$^-$ leaf concentration from 4.4 to 2.3 and 2.0 g 100 g^{-1}. The yield reduction may be related to the observed decline in Ca and Mg concentration in leaves stemming from the increased NH$_4$ in the fill solutions (fixed NH$_4$:NO$_3$ ratio) (Table 9.15). The reduction in Ca concentration was accompanied by an increase in the incidence of BER (Table 9.15). Even though the EC was identical in all treatments (EC$_{thr}$ = 4.5 dS m^{-1}) the ET in the 250 mg L^{-1}N treatment was lower than that in the 150 mg L^{-1}N (data not presented), compatible with the observed decline in yield in the two treatments.

A study of rose response to N target concentration (EC$_{thr}$ = 3.5 dS m^{-1}, 8–10 irrigation per day) showed that N concentration could be reduced to 28 mg L^{-1} without impairing total yield or increasing the number of excessively short or long stems in comparison with the concentration of 56 or 112 mg L^{-1} N (Table 9.16). It is shown that a target N concentration of 168 mg L^{-1} N caused a considerable decline in yield relative to the lower concentrations (Table 9.16). It could not be resolved in this experiment whether the yield decline stemmed from excess N uptake or NH$_4$-mediated reduced cation uptake.

As discussed above, the demonstrated crop preference for low N concentration is characteristic of irrigation involving high fertigation frequency with ample amounts of irrigation solution. It does not contradict current N concentration recommendations in open irrigation systems which are considerably higher (e.g. de Kreij et al., 1999) because irrigation frequency and dose volume are appreciably lower.

9.2.7.2 Type of Nitrogen Fertilizer (Ammonium:Nitrate Ratio)

The ammonium to nitrate ratio is an important factor in fertigation management (Bar-Yosef, 1999). For a given N dose and irrigation regime, the supply ratio affects several solution and crop factors: (i) N uptake efficiency. This stems from the higher NH$_4$ Michaelis–Menten uptake kinetics in comparison with NO$_3$ (e.g. tomato data, Table 9.11). (ii) Ca, Mg and K absorption rates decline due to competitive NH$_4$ uptake. This phenomenon is common to a variety of crops (Mengel and Kirkby, 2001), especially in inert media not adsorbing NH$_4$. (iii) NH$_4$ supply decreases solution pH while NO$_3$ tends to increase it (see, e.g., Imas et al., 1997). Several mechanisms are responsible for this phenomenon: (a) NH$_4$ nitrification which occurs in the entire substrate volume and yields two moles of H$^+$ per mole of NH$_4^+$ according to the equations:

$$NH_4^+ + 1.5O_2 = NO_2^- + H_2O + 2H^+$$
$$NO_2^- + 0.5O_2 = NO_3^- \tag{22}$$
$$\text{Overall}: NH_4^+ + 2O_2 = NO_3^- + H_2O + 2H^+ (E_0 = 348\,mV)$$

The nitrite oxidation (second step) requires a redox potential of 420–530 mV (depending on pH, Lindsay, 1979) and presence of bacteria capable of performing the two oxidation steps. In new synthetic substrates (e.g. perlite, stone wool) the bacteria are initially absent, the second stage is delayed resulting in nitrite accumulation (Fig. 9.6).

(b) Excess cation over anion uptake is neutralized by proton excretion by roots. Due to the high NH_4 uptake efficiency, H^+ release is greater in the presence of this cation in comparison with others. The incorporation of NH_4 in the synthesis of amino acids (R-NH_2) also releases H^+, and part of it is excreted in order to prevent over acidification in plant tissue. The released protons accumulate in narrow water films around roots and are thus very effective in reducing the pH at the root surface. (c) High nitrate uptake rate decreases the amount of excreted protons (charge balance considerations), thereby increasing solution pH. In addition, nitrate reduction in plant produces hydroxyls that need to be disposed of to maintain a desired pH in cells as per

$$\begin{aligned} NO_3^- + 8H^+ + 8e^- &\leftrightarrow NH_3 + 2H_2O + OH^- \\ K^+ + OH^- + CO_2 + RH &\leftrightarrow K^+ + RCOO^- + H_2O \end{aligned} \quad (23)$$

where RH is organic acid and $RCOO^-$ is carboxylate. This is done via decarboxylation, the end product of which is carboxylic acid anions (Mengel and Kirkby, 2001). Citrate and malate (Fig. 9.7) are examples of prevalent carboxylates released by greenhouse tomato (Imas et al., 1997). Since carboxylic acids are weak, carboxylates excretion causes pH increase in solution. (iv) Excess NH_4 uptake particularly at root temperature >28°C is deleterious to root development (Ganmore-Neumann and Kafkafi, 1980, 1983). The subject of nitrate/ammonium ratio is further discussed in Chap. 8.

Another source of nitrogen is urea [$CO(NH_2)_2$] which is the cheapest source of nitrogen per unit N. Due to its zero charge it is unaccompanied by ions that accumulate in closed-loop irrigation systems, it does not contribute to solution EC, and its impact on solution osmotic potential (OP) is weakest among all N fertilizers. For illustration, the osmolality (O_s kg^{-1}, where O_s = freezing point depression/1.86) of 5 g L^{-1} solution of KNO_3, $(NH_4)_2SO_4$, NH_4Cl, $NaNO_3$ and urea is 0.092, 0.093, 0.173, 0.11 and 0.083, respectively (Weast, 1977). Per g of N the smaller effect of urea on freezing point depression and osmotic potential is much more pronounced. The disadvantage of urea is that its uptake efficiency before hydrolysis is significantly lower than ammonium or nitrate (Kirkby and Mengel, 1967; Herndon and Cochlan, 2007) and therefore it is not used in greenhouses having open irrigation systems where the residence time in the substrate is too short for hydrolysis. The hydrolysis (Eq. [24]) consumes H^+ and therefore increases pH.

$$\begin{aligned} CO(NH_2)_2 + 2H_2O &\leftrightarrow CO_3^{2-} + 2NH_4^+ \\ CO_3^{2-} + H^+ &\leftrightarrow HCO_3^- \end{aligned} \quad (24)$$

According to Ikeda et al. (2001), 60 per cent of the urea applied to soilless medium is subject to hydrolysis at greenhouse normal temperature in one day. This rate is similar to NH_4 nitrification (~50 per cent in one day, Lang and Elliot, 1991; Ikeda et al., 2001), therefore the hydrolysis end product is ammonium plus nitrate. Presence of plants may change the hydrolysis rate and ratio between the ammonium and the nitrate species. There are two important constraints to using urea: (i) It should be free of biurate which can accumulate in the closed-loop irrigation system and become toxic to plants. (ii) The urea-N concentration should not exceed a certain fraction of the total N in the system due to (a) its slow uptake, and (b) urea might induce toxic ammonia

release (Eq. [25]) due to the pH increase during hydrolysis and the presence
. According to

$$NH_4^+ + OH^- \leftrightarrow NH_4OH \leftrightarrow NH_{3(g)} + H_2O$$

$$K_{hdr} = 10^{-11.04} = \frac{(PNH_{3(g)})(H^+)}{(NH_4^+)} \tag{25}$$

with $PNH_{3(g)}$ in bars, a PNH_3/NH_4 ratio of 1/1000 is expected at pH 8.04, so addition of urea at cf. pH > 7.5 is not advised.

The optimal fraction urea:total N in recirculated solutions has not yet been determined, so it is arbitrarily recommended to be less than 15 ± 5 per cent. If the total daily N uptake plus discharge is also 15 ± 5 per cent of the total N, then the entire daily depletion can be replenished by urea. Kramer et al. (2003) compared rose response to ammonium nitrate vs. urea supply in the fill solution at a target N concentration of 112 mg L^{-1} N in the recirculated solution. The cumulative number of cut flowers over 18 months of harvest and their quality were the same in both sources (Table 9.16).

TABLE 9.16 Effect of N Concentration in Solution (C_N) and N Source (NH_4NO_3 vs. urea) in Fill Solution on Rose (Long Jaguar)[a] Yield in a Closed-Loop Irrigation System[b] in Arava (Hazeva), Israel[c]

N source and target concentration (mg L^{-1})		Cut flower yield (number m^{-2} greenhouse)		
NH$_4$NO$_3$-N	Urea-N	Total	50 cm	70 cm
28	—	1080 a	260 ab	34 a
56	—	1075 a	260 ab	32 ab
112	—	1075 a	267 a	32 ab
168	—	1000 b	231 b	29 b
—	112	1020 ab	244 ab	33 a

Means followed by the same letter are not significantly different according to Duncan's LSD test (0.05).

[a] Growth period, 30 May 2002 to 31 December 2003.
[b] Tuff; total water volume in the system 30 L m^{-2} greenhouse; 9 irrigations per day, ET ~ 4–5 mm d^{-1}; minimal night temperature: 18°C.
[c] Derived from Kramer et al., 2003.

9.2.7.3 Macro-Nutrients Other Than N

Phosphorus

Three possible commercially available water-soluble P sources are H_3PO_4, KH_2PO_4 and $NH_4H_2PO_4$. These are readily incorporated in pre-mixed fertilizers. They differ in accompanying cation and acidity. Phosphoric acid (H_3PO_4) must be carefully added to stock solution containing complexed microelements as the chelates break down at pH below 3. Polyphosphates (e.g. Subbarao and Ellis, 1975) are used rarely and will not be discussed here.

In open irrigation systems, P concentration in irrigation solution is ~30 mg L^{-1} P (Hoagland and Arnon, 1950; de Kreij et al., 1999). The appropriate concentration in closed-loop irrigation systems depends on the interrelationship between P uptake and salinity (e.g. Feigin, 1985), fertigation frequency and substrate's P adsorption characteristics (empirical results are presented below).

Potassium

To avoid Cl accumulation, as much K should be added as KNO_3 and KH_2PO_4 as possible while not exceeding nitrate and P target concentrations. The rest can be added as KCl and as K_2SO_4 as dictated by Ca concentration and gypsum solubility (see Sect. 9.2.1). Potassium concentration in open-loop irrigation systems varies between 150 and 240 mg L^{-1}, depending on crop (Hoagland and Arnon, 1950; de Kreij et al., 1999). In recirculated solutions where significant part of the N is added as NH_4, the K uptake rate might be inhibited, leading to the need for higher K target concentrations (empirical results are presented below). Due to the similar N and K transport mechanism in non-adsorbing substrates, it is expected that increased fertigation frequency will allow reducing target K concentrations similarly to N.

9.2.7.4 Microelements

When Fe, Zn and Mn are found in chelated form (usually EDTA or DTPA), their target concentrations in open irrigation systems are ~1.0, 0.25 and 0.5 mg L^{-1}, respectively (de Kreij et al., 1999). Their availability to plants is pH dependent (see later). As discussed above, fertigation frequency affects recirculated solution pH, so that this in turn can be expected to affect microelements availability.

9.2.8 WATER QUALITY ASPECTS

9.2.8.1 Using Rain or Desalinized Water

Fresh water in arid and semiarid zones usually contain Ca, Mg, SO_4 and Cl in excess of crops demand (Table 9.6) so that these tend to accumulate in closed-loop irrigation systems. When using water with negligible concentration of these elements, they must be added concomitantly with N, P, K and microelements to the recycled solution. Recommended Ca, Mg, SO_4 and Cl concentrations in open and closed irrigation systems for temperate climate countries can be found for various crops in de Kreij et al. (1999). It is still unknown what the optimal concentration of these ions is in recycled solution in arid zones. Adopting recommended concentrations in Western Europe for use in arid/semiarid zones must be done carefully as transpiration-induced mass flow under higher ET conditions is greater, and lower solution concentrations might be necessary to maintain required ion activity at the root surface.

When both fresh water with high EC and desalinized (or rain) water are available, the two sources can be mixed so as to result in an EC level below a predefined threshold EC or below specific toxic ion concentration (see, e.g. Raviv et al., 1998). The optimal dilution rate should take into account water price and impact on discharge volume. The dilution strategy should also account for crop susceptibility to salinity at various growth stages. For example, tomato was shown to be more sensitive to salinity stress during the vegetative growth stage than during the reproductive growth stage (Mizrahi et al., 1988).

9.2.8.2 Reclaimed Wastewater

The main concern in using reclaimed wastewater is the presence of heavy metals, organic matter and human pathogens that may accumulate in the system and cause unpredictable damage to the crop and consumers. This subject is beyond the scope of this chapter. As far as the higher salinity level of reclaimed wastewater is concerned, the same principles discussed so far are applicable. Special attention should be paid to boron as its concentration in municipal wastewater is usually twofold higher than that recommended in nutrient solutions. In areas where residential water use involves extensive use of water softeners, it should also be anticipated that municipal waste water may contain excessive concentrations of Na.

9.2.9 FERTIGATION FREQUENCY

9.2.9.1 Principles

Maximum irrigation frequency depends on the substrate's water-holding capacity, the substrate drainage rate and the quantity of water which is applied to meet plant's demand and leaching salts from the substrate. The water quantity per m² substrate per irrigation (V_p) should be $\leq [\theta_{95\%sat} - \theta_{1/2\,h\,drng}] \times$ [substrate volume per m² substrate]. In perlite, for example, V_p should be less than $58\,L\,m^{-2}$ (see Table 9.1). The actual water dose (V_{irr}) should be smaller (~30 per cent of V_p, or ~$18\,L\,m^{-2}$) to prevent temporary water saturation and to slow down the drainage rate. Exceeding V_p results in water flow on top of the substrate rather than through it, thus impairing effective nutrients and oxygen recharge and root flushing with fresh solution.

In a greenhouse with 1.6 m wide beds and 0.4 m wide perlite container, the V_{irr} is equivalent to $4.5\,L\,m^{-2}$ ground, or 4.5 mm. The irrigation time (t_{irr}) needed to supply this dose is $t_{irr} = V_{irr}/I_{irr}$, where I_{irr} is the irrigation system discharge rate ($L\,m^{-2}$ subst h^{-1}). For a conventional $I_{irr} = 50$, each irrigation event should last for $(60 \times 18/50)$ ~20 min. The time between successive irrigations (t_{si}) should exceed $t_{irr} + t_d$, where t_d is drainage time, calculated as V_{irr}/I_d and I_d is the drainage rate in $L\,m^{-2}$ subst h^{-1}. The I_d for perlite is approximated from its J_{fd} ($cm\,h^{-1}$, Table 9.1) using the transformation $10\,cm\,h^{-1} = 100\,L\,m^{-2}$ subst h^{-1}. The t_d is therefore $(60 \times 18/100)$ ~11 min, and t_{si} $(= 11 + 20)$ ~30 min. This result indicates that two irrigations per h, each comprising ~4 mm, is a safe irrigation frequency in perlite. The permitted rate of $8\,mm\,h^{-1}$ is about 4–5-fold greater than the ET during summer middays in arid zones. The recommended V_{irr} for perlite in Israel at the aforementioned I_{irr} is 2–4 mm/irrigation event, compatible with 15–30 per cent of V_p. The recommended number of irrigations per day, n_I, is

$$n_I = \frac{[\text{daily irrigation rate}, V_d]}{V_{irr}} \quad (26)$$

For a conventional $V_d = 5\times$ [daily ET], [daily ET] $= 3$ mm and $V_{irr} = 2$ mm, the number of irrigations per day is 7–8, which is within the safe range discussed above.

9.2.9.2 Experimental Evidence

The effect of irrigation frequency in closed-loop irrigation systems was investigated for pepper at Bet Dagan, Israel and for tomato at Besor, Israel (Tables 9.17 and 9.18). In pepper, 12 irrigations per day (n_1) gave better yield and quality than $n_1 = 6$, and did not differ from 18 irrigations per day. In tomato, the irrigation frequency effect was very similar to pepper and the intermediate frequency (8–16 per day) gave best results (Table 9.18). The range in frequency in a given treatment stems from seasonal differences in ET. Baas and van der Berg (2004) studied rose (CV. Frisco) response to irrigation frequency under normal, ×2 and ×4 irrigations per day (water

TABLE 9.17 Effect of Irrigation Frequency (Same EC_{thr} and Irrigation Rate) on Pepper[a] Yield in a Closed-Loop Irrigation System[b] at Bet Dagan, Israel[c]

Fertigation frequency (Num d^{-1})	Yield Total (kg m^{-2} ground)	Marketable (kg m^{-2} ground)	Unmarketable (%)
18	11.0	4.9 a	19.1 ab
12	10.8	4.8 a	17.8 b
6	10.5	4.3 b	20.0 a

Means followed by the same letter are not significantly different according to Duncan's LSD test (0.05).
[a] CV. Cuby; planting 30 August 1999, first harvest 20 December 1999, last 15 May 2000.
[b] Threshold EC in all cases 3.5 dS m^{-1}; perlite substrate 50 cm wide; NH_4-N, NO_3-N, P, K concentrations 2, 8, 1 and, 4 mM, respectively.
[c] Derived from Bar-Yosef et al. (2000).

TABLE 9.18 Effect of Irrigation Frequency (Same EC_{thr} and Irrigation Rate) at Two N Concentrations on Tomato[a] Yield and Incidence of Blossom End Rot (BER) in a Closed-Loop Irrigation System[b] in a Greenhouse at Besor, Israel[c]

	Solution N concentration (mg N L^{-1})					
	50	150	50	150	50	150
Irrigation frequency (Num d^{-1})	Total yield (kg pl^{-1})		Marketable yield (kg pl^{-1})		BER (num pl^{-1})	
4–8	—	8.06 b	—	4.35 b	—	42.5 a
8–16	10.57	9.87 a	7.11	5.94 a	24.4	23.5 b
16–32	10.10	—	7.07	—	23.0	—

Means followed by the same letter are not significantly different according to Duncan's LSD test (0.05).
[a] CV. 870; planting 1 September 2004, end 15 July 2005.
[b] Threshold EC in all cases 4.5 dS m^{-1}; perlite, mean of 10, 20 and 50 cm wide substrate; P, K concentrations identical in all treatments (1 and 4.5 mM, respectively).
[c] Derived from Reshef et al. (2006).

rates increasing proportionally to frequency) using closed-loop irrigation system in the Netherlands. Under the prevailing growth conditions, (3 L substrate per plant and EC of 0.9 or 1.9 dS m^{-1}) irrigation frequency had no significant effect on cut-flower production.

As noted earlier, the irrigation frequency effect is interrelated with the nutrient concentration in the recycled solution. In the case of tomato, $n_1 = 16$, the yield at N concentration of 50 mg L^{-1} exceeded that at 150 mg L^{-1}, the latter being the concentration at which the frequency effect was studied (see Table 9.18).

An important constraint to irrigation dose and frequency (at fixed V_p) in recirculated systems is the biofilter or UV set-up (if used). Increasing the top surface area of the filter, or number of UV lamps, to account for greater daily irrigation, and the handling of the larger drainage volume are expensive, therefore it is essential to minimize solution recirculation. Chemical disinfection (e.g. chlorination) does not require extra equipment when drainage volume increases, therefore it is better suited for recirculation systems prone to changes in irrigation frequency over time.

9.2.10 pH CONTROL: NITRIFICATION AND PROTONS AND CARBOXYLATES EXCRETION BY ROOTS

9.2.10.1 NH$_4$/NO$_3$ Ratio Induced Effects

Solution pH can be controlled by two methods: (i) modifying the NH$_4$/NO$_3$ ratio in recycled solution. (ii) Applying the entire N as NO$_3$ and reducing pH with acid (HNO$_3$, H$_3$PO$_4$ or HCl where water Cl$^-$ is low). The advantage of method (i) over (ii) is that it reduces the salt load in the system by not requiring the added acid and it also eliminates the safety measures associated with handling strong acids. Its disadvantage is that NH$_4$ might inhibit Ca, Mg and K uptake and it may impair root development at high substrate and solution temperatures.

In order to adopt method (i), it is necessary to evaluate the following three factors: (a) the potential proton excretion by roots of a given crop, (b) the nitrification rate in the recirculation systems, and (c) the effect of nitrate uptake on carboxylates release into solution. Insight into factors (a) and (c) can be derived from the work published by Imas et al. (1997). They measured pH (Fig. 9.10) and short-term (6 h) NH$_4$ and NO$_3$ uptake and proton and citrate release by roots in solutions varying in NH$_4$:NO$_3$ ratio in the presence of young tomato plants (30–44 d after planting) (Fig. 9.11).

The data indicate that a certain threshold uptake rate (TUR) must be surpassed before exudation starts and then it proceeds at a constant rate which declines with root age. The potential accumulation of protons and citrate in real recycled solutions was estimated from data in Figs. 9.10, 9.11 and is summarized in Table 9.19. The calculations indicate that during 30 days of recirculation, tomato plants absorbing NH$_4$ from a solution of ~5 mM NH$_4^+$ may reduce an un-buffered recycled solution pH from ~7 to 3.1 (50 L solution m^{-2} ground). The effect of NH$_4$:NO$_3$ ratio on carboxylates exudation is evaluated from the data in Fig. 9.11. When all N is absorbed as nitrate, and assuming that no consumption by microbes takes place, the estimated potential citrate concentration after 30 days is 144 μM (Table 9.20). This concentration

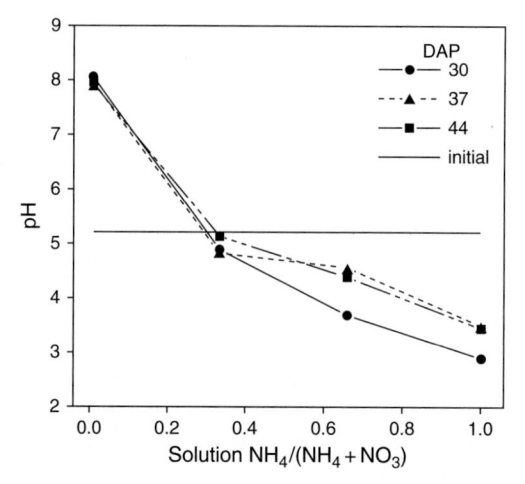

FIGURE 9.10 Solution pH after 6 h of uptake by tomato plants from solutions varying in $NH_4:NO_3$ ratio (total $N = 7.5$ mM) (derived from Imas et al., 1997).

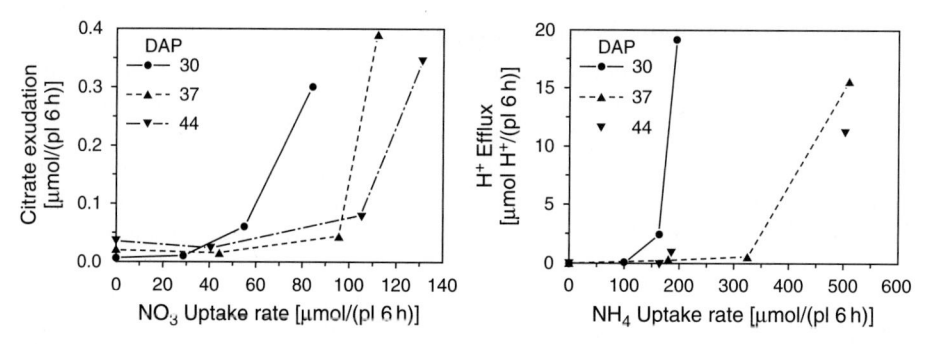

FIGURE 9.11 Citrate exudation vs. NO_3 uptake and H^+ excretion vs. NH_4 uptake in young tomato plants (derived from Imas et al., 1997).

is ~25-fold greater than the C concentration found in actual greenhouse recycled solution of 30-days old pepper plants receiving $100\,mg\,L^{-1}$ N at $NH_4:NO_3$ ratio of 15:85 (Table 9.9). The difference indicates the extent of organic acid consumption by microbes in recycled solutions (Table 9.10). The equivalent C concentration in the 144 μM citrate solution is $\sim 10\,mg\,L^{-1}$ C, which is similar to the total organic carbon (TOC) values found in recycled solutions reported above (Fig. 9.7). This indicates that the TOC consists mainly of microbial mass which consumed the exuded organic acids minus CO_2 released in respiration.

As mentioned earlier, the rate of pH evolution stemming from nitrification must also be known for pH management. In a control open-loop irrigation treatment, Reshef et al. (2002b) reported that 72 per cent of added NH_4 disappeared from substrate

TABLE 9.19 Estimated Potential H⁺ Release by Tomato Plants (40-days old) and Resulting pH in Recycled Solution of 5 mM NH₄, Zero pH Buffering Capacity and 2.5 Plants m⁻² Ground

Threshold uptake[a] (mol NH₄ pl⁻¹ d⁻¹)	H⁺ release/NH₄ uptake[b] (mol H⁺/mol NH₄)	Effective NH₄ uptake rate[c] (mol N m⁻² ground d⁻¹)	Estimated daily H⁺ release[d] (mol m⁻² ground)	Estimated daily pH[e]	30 days pH[f]
0.0012	3/10	0.004	0.0012	4.6	3.1

[a] NH₄ uptake rate below which no H⁺ is excreted.
[b] The ratio is the slope of the line in Fig. 9.11 subject to NH₄ uptake rate > threshold (average of three plant ages).
[c] Equals to [Total uptake rate] − [threshold uptake rate] × number of plants m⁻² ground. Total uptake rate = 0.007 mol N m⁻² d⁻¹.
[d] Column 2 × Column 3.
[e] Solution volume = 50 L m⁻² ground.
[f] Estimated daily H⁺ release × 30 days/50 L.

TABLE 9.20 Estimated Potential Citrate Release by Tomato Plants (40-days old) Stemming from Nitrate Uptake from Recycled Solution of 5 mM NO₃ and 2.5 Plants m⁻² Ground

Threshold uptake[a] (mol NO₃ pl⁻¹ d⁻¹)	Citrate release/ NO₃ uptake[b] (mol Citratate/mol NO₃)	Effective NO₃ uptake rate[c] (mol N m⁻² ground d⁻¹)	Estimated daily Citrate release[d] (mol m⁻² ground)	Estimated Citrate conc.[e] (μM)	30 days Citrate release (mol m⁻² ground)	30 days solution Citrate (μM)
0.0004	0.4/10	0.006	0.00024	4.6	0.0072	144

[a] NO₃ uptake rate below which no Citrate is excreted.
[b] The ratio is the slope of the line in Fig. 9.11 subject to NO₃ uptake rate > threshold (average of 3 plant ages).
[c] = [Total uptake rate] − [threshold uptake rate] × number of plants m⁻² ground. Total uptake rate = 0.007 mol N m⁻² d⁻¹.
[d] = Column 2 × Column 3.
[e] Solution volume = 50 L m⁻² ground.

solution within 24 h in the presence of plants, compared with 57 per cent in the absence of plants (mean of 6 dates in winter and spring months, Table 9.21). In recirculated solution with fill solution identical to the emitter solution in the open system, the emitter NH₄ concentration was considerably lower (~10 vs. 45 mg L⁻¹ N) and the NH₄ disappearance rate was 82 per cent in 24 h in the presence of plants and 67 per cent in their absence (Table 9.21).

In the open irrigation system the NH₄ and NO₃ concentrations were ~45 and 95 mg L⁻¹ N, respectively, and the pH indeed declined in the presence of plants from ~7 to ~5.8 in 24 h. In the absence of plants (nitrification only), the pH dropped to ~6.2. Notably the root excretion effect on pH was considerably stronger in late spring and summer than in winter and beginning of spring.

TABLE 9.21 The Daily NH$_4$ Concentration and pH in Open- and Closed-Loop Irrigation Systems in a Pepper Greenhouse (cv. Celica) in Besor, Israel. The Ionic Composition of the Fill Solution (Recirculation Treatment) was Equal to the Irrigation Solution Ionic Composition in the Open Irrigation System. Both the Emitter and the Effluent Solutions (With or Without Plants) were Collected for 24 h and Then Analysed[a]

	Open irrigation system			Closed irrigation system EC$_{thr}$ = 3.2 dS m^{-1}			
Date	Emitter	Drang + pl	Drang − pl	Fill	Emitter	Drang + pl	Drang − pl
	NH$_4$-N (mg L^{-1}) (and pH)						
12 Feb.	59 (7.6)	27 (7.3)	—	—	29 (8.3)	12 (6.3)	—
19 Mar.	19 (7.0)	10 (6.6)	19 (6.5)	44	4 (6.8)	0.2 (6.6)	1.9 (6.3)
30 Apr.	47 (7.0)	15 (5.8)	17 (6.2)	49	12 (6.2)	5.3 (5.1)	8.2 (5.6)
28 May.	46 (7.1)	17 (5.7)	18 (6.1)	30	2 (6.3)	0.3 (6.3)	0.4 (6.1)
18 Jun.	46 (7.0)	18 (5.5)	21 (6.0)	46	10 (6.7)	0.3 (6.4)	2.3 (5.6)
16 Jul.	48 (7.0)	18 (5.6)	24 (6.4)	33	9 (6.9)	0.1 (6.3)	0.8 (6.1)

[a] Derived from Reshef et al. (2003).

In the closed-loop irrigation system, the NH$_4$ and NO$_3$ concentrations were ~10 and ~130 mg L^{-1} N. Under these conditions the drainage pH in the presence of plants exceeded the pH in the absence of plants. The reason was that in the absence of plants, basidification stemming from NO$_3$ uptake did not occur (Table 9.21). In order to obtain a pH comparable to the pH in open system, the fill solution NH$_4$ concentration should exceed the one in the open system.

Recirculated solution pH is expected to be affected by fertigation frequency as well. This stems from the aforementioned observation that increasing frequency enhances ion uptake rate by plants. In solutions consisting predominantly of NO$_3^-$, higher frequency should increase the pH, while under relatively high NH$_4^+$ concentrations the change in pH should depend on the relative efficiency of protons release due to NH$_4^+$ uptake and nitrification. In the experiment where pepper response to fertigation frequency was studied (Table 9.17), the time-averaged pH in recirculated solution slightly declined (from 6.4 to 6.2) when the frequency was increased from 6 to 12 fertigations per day, while increasing the frequency to 18 fertigations per day had no further effect on pH. In tomato, (see Table 9.18) raising the frequency from 4–8 to 8–16 (at total N concentration of 150 mg L^{-1}) or from 8–16 to 16–32 (at 50 mg L^{-1} N) had no measurable effect on pH. In open irrigation systems the pH response to fertigation frequency was reported to be much stronger. For example, Bar-Yosef et al. (2000d) reported that fertigating waxflower (*Chamelaicium uncinatum* L.) in sandy soil with 7 mM N (70 per cent NH$_4^+$) at 3 irrigations per day, 1 irrigation every 3 d and 1 irrigation every 5 d (identical water and N quantities) resulted after one year in soil-root-volume pH of 5.8, 6.4 and 6.7, respectively. The interpretation was that the time-averaged NH$_4$ concentration in the soil increased as irrigation frequency increased, nitrate concentration and uptake decreased and consequently the pH declined. This

9.2 MANAGEMENT

mechanism cannot operate effectively in closed-loop irrigation systems because fresh N solution is applied only via the fill solution which comprises ~25 per cent of the total N in the system, most of which is in the form of nitrate.

Another factor that may affect recirculated solution pH is EC_{thr}. Comparing three EC_{thr} values (2.7, 4.0 and 5.5 dS m^{-1}, pepper in Besor) reveals that as long as the lowest EC threshold has not been surpassed (~30 November, Fig. 9.12A), the difference in pH between treatments was negligible. Between EC 2.7 and 4.0 dS m^{-1} (5 February) solution in treatment $EC_{thr} = 2.7$ was frequently disposed to maintain the EC within permitted boundaries. Since the fresh water pH exceeded the pH of disposed solution, the recirculated solution pH in this treatment was highest (Fig. 9.12A). Between EC 4.0 and 5.5 dS m^{-1} (25 March) solution in treatment $EC_{thr} = 4.0$ also had to be disposed

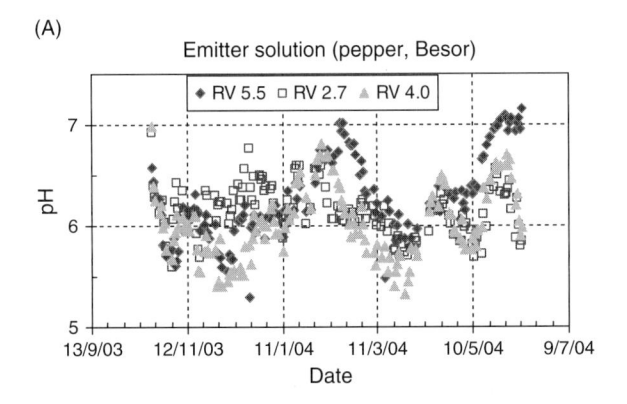

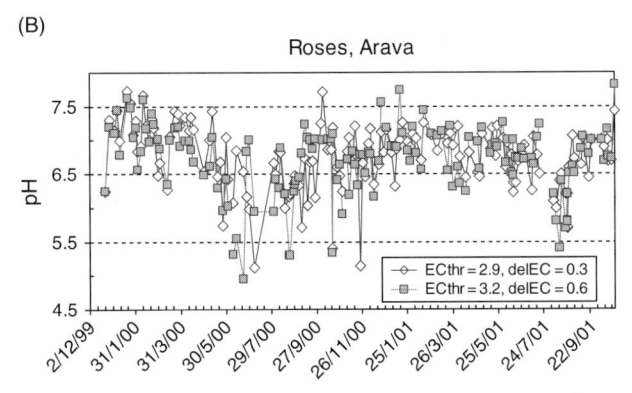

FIGURE 9.12 Effect of assigned threshold EC (EC_{thr}, 2.7, 4.0, 5.5 dS m^{-1}) (**part a.** – top) and permitted deviation from EC_{thr} (delEC, 0.3 and 0.6 dS m^{-1}) (**part b.** – bottom) on recirculated solution pH. **Part a.** – pepper in Besor (derived from Bar-Yosef et al., 2004). Note that EC as a function of time for the $EC_{thr} = 4.0$ dS m^{-1} treatment is presented in Fig. 9.4. **Part b.** – Roses in Arava (derived from Kramer et al., 2002). Note that EC as a function of time for these treatments is presented in Fig. 9.7 (Tr 1 and 2 correspond to delEC of 0.3 and 0.6, respectively).

and the recirculated solution pH became lower than in treatment $EC_{thr} = 5.5$. The reason was that at this stage the $NH_4:NO_3$ ratio in the replenishing solution (common to all treatments) was increased and more frequent solution replacements resulted in higher time-averaged NH_4 concentration in the recirculated solution and lower pH than in treatment $EC_{thr} = 5.5$. When all three treatments attained their EC_{thr} (end of March) the pH was determined by the frequency of solution displacement (lowest in $EC_{thr} = 5.5$, highest in $EC_{thr} = 2.7$) and was therefore maximal in $EC_{thr} = 5.5$.

In Fig. 9.12B the effect of ΔEC (0.3 vs. 0.6 dS m^{-1}, Eq. [9]) on recycled solution pH can be assessed. Lower ΔEC stimulates higher disposal frequency and its effect is expected to be similar to lower EC_{thr}. Careful comparison confirms this expected behaviour but in practical terms the difference between 0.3 and 0.6 dS m^{-1} was too small to generate a significant effect. The EC_{thr} impact in the pepper experiment was considerable, because the obtained pH difference of $\sim$0.5 units occurred around pH 6 which might be critical for Mn availability (see next section).

9.2.10.2 pH-Dependent Nutrient Concentrations in Solution

Nutrients whose solubility is strongly pH dependent are Ca, P, Fe, Zn and Mn. The relationship between pH, CO_2 partial pressure (PCO_2) and Ca concentration (C_{Ca}, M) in calcite ($CaCO_3$) saturated solution is (Lindsay, 1979):

$$0.5 \log(C_{Ca}^{2+}) = 4.87 - 0.5 \log(PCO_2) - pH \tag{24}$$

Under $PCO_2 = 3 \times 10^{-4}$ bar (300 ppm CO_2) and pH 7, $C_{Ca}^{2+} = 183$ mM; at pH 8.0 and 8.5 and same PCO_2, C_{Ca}^{2+} drops to 1.83 and 0.183 mM, respectively. When C_{Ca}^{2+} is controlled by gypsum its concentration was shown to be 4.8 mM Ca^{2+} regardless of pH which means that at pH ≤ 7, precipitated gypsum (if present) controls Ca solution concentration.

In addition to Ca^{2+}, two common Ca species should be mentioned: $CaHCO_3^+$ and $CaCO_3^0$. At total carbonates concentration of 10 mM, total Ca of 2 mM and pH 8, the mole fractions of Ca^{2+}, $CaHCO_3^+$ and $CaCO_3^0$ are 0.92, 0.05 and 0.03, respectively. At common carbonates concentration of 3–5 mM, the concentration of $CaHCO_3^+$ and $CaCO_3^0$ becomes negligible.

At a fixed Ca concentration, the pH determines the solution P concentration in equilibrium with Ca–P minerals. The solubility decreases in the order: $CaHPO_4 \cdot H_2O$ (DCPD) > $Ca_4H(PO_4)_3$ (OCP) > $Ca_3(PO_4)_2$ (βTCP) > $Ca_5(PO_4)_3OH$ (HA). If the activity product of Ca^{2+} and HPO_4^- exceeds the DCPD solubility product, the mineral precipitates within minutes to hours, and if the solution is maintained at $C_{Ca}^{2+} = 4.5$ mM and pH 7.0, an equilibrium P concentration of 3.5 mg L^{-1} is obtained. At pH 6.4 the concentration increases to 14.0 mg L^{-1} P (Table 9.22).

OCP precipitates slower than DCPD (within hours) and its corresponding equilibrium P concentrations are 0.9 and 9.0 mg L^{-1} P; βTCP precipitates within months and is therefore not formed within one growth season; its equilibrium concentrations at $C_{Ca}^{2+} = 4.5$ mM and pH 7.0 and 6.4 are 0.15 and 2.4 mg L^{-1} P, respectively (Table 9.22).

9.2 MANAGEMENT

TABLE 9.22 P Concentration in Equilibrium with Dicalcium Phosphate Dihydrate (DCPD), Octa Ca Phosphate (OCP) and β-Tricalcium Phosphate (bTCP) at Two pHs and One Ca Concentration[a] (C_{Ca}) Representing Conditions in Arid Zone Recycled Solutions (Ionic Strength = 10 mM)

Mineral	pH	Equilibrium concentration (mg L^{-1} P)
CaHPO$_4$ 2H$_2$O (DCPD)	6.4	14.0
	7.0	3.5
Ca$_4$H(PO$_4$)$_3$ (OCP)	6.4	9.1
	7.0	0.9
Ca$_3$(PO$_4$)$_2$ (βTCP)	6.4	2.4
	7.0	0.15

[a] $C_{Ca} = 4.35$ mM (=174 mg L^{-1}). The calculation assumes that the only Ca species was Ca^{2+}.

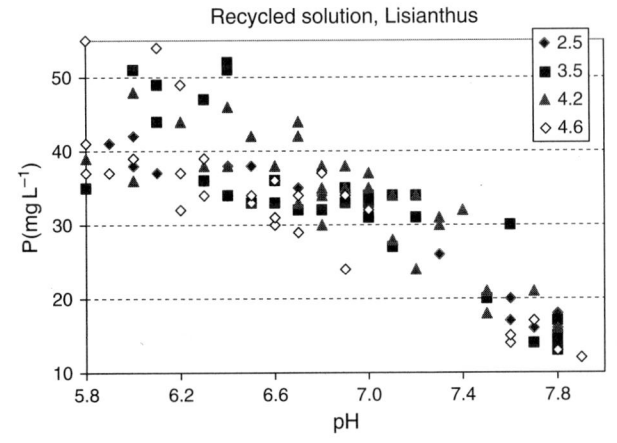

FIGURE 9.13 Inorganic P concentration in recycled solution as a function of pH. The crop was lisianthus grown in perlite in Besor. P was replenished in all threshold EC treatments (inset) to maintain 35 mg L^{-1} P in the recycled solution.

When the 4.5 mM Ca includes Ca^{2+}, CaHCO$_3^+$ and CaCO$_3^0$, the total-P equilibrium concentration can be ~20–50 per cent greater than that shown in Table 9.22, depending on specific conditions. Under high-frequency irrigation, the substrate solution is usually supersaturated with respect to the concentrations shown in Table 9.22, as the experimental results of P concentration in recycled solution vs. pH indicate (Fig. 9.13). The data clearly show a decline in P concentration as pH rises above 6.5, as theory predicts.

Iron, Zn and Mn have limited solubilities due to formation of insoluble hydroxides (Fe^{3+} + 3(OH$^-$) ↔ Fe(OH)$_{3(s)}$, Zn^{2+} + 2(OH$^-$) ↔ Zn(OH)$_{2(s)}$ and Mn^{3+} + 3(OH$^-$) ↔ MnOOH$_{(s)}$ + H$_2$O) and oxides (MnO$_2$)$_{(s)}$ at elevated pHs. These reactions decrease Fe^{3+}, Mn^{3+} and Mn^{4+} concentration to practically zero at pH > 5. The three microelements are added to fill solution as salts of the divalent cations. To precipitate as hydroxides, the Fe^{2+} and Mn^{2+} must be oxidized first. The kinetics of oxidation and precipitation in closed or open irrigation systems is as yet unknown. Zaw and Chiswell (1999) reported that the oxidation of Mn^{2+} to Mn^{4+} (the Mn^{3+} is an intermediate in this reaction) in water with pH < 8.5 is slow and takes days to weeks to complete; Scott et al. (2002) found, on the other hand, that Mn^{2-} oxidation and precipitation in a slightly acidic mountain stream was a matter of hours to days. The contrasting results can be explained by the fact that Mn^{2+} is oxidized both abiotically and biotically (Morgan, 2005). The first process is slow, as described by Zaw and Chiswell (1999). The second process is mediated by a wide variety of bacteria and is rapid (half life of about 1 to 10 h, Morgan, 2005) compatible with the rates reported by Scott et al. (2002).

The redox potentials at which Mn^{4+} and Fe^{3+} begin their reduction to Mn^{2+} and Fe^{2+} are 225 and 120 mV, respectively (redox potential decreases under anoxic conditions). These reduction reactions cannot occur before all NO_3^- in the system is depleted, as nitrate reduction starts at redox potential of 250 mV (for review of this subject matter, see, e.g. Cronk and Fennessy, 2001).

Minerals that may control Mn^{2+} concentration in solution are $MnCO_3$ (Schwab and Lindsay 1983; Luo and Millero, 2003) and $Mn_5H_2(PO_4)_4$ or $Mn_3(PO_4)_2$ (Boyle and Lindsay, 1986). Zinc concentration in solution is controlled by $Zn(OH)_2$ which is very insoluble at pH > 6, and $Zn_3(PO_4)_2$ (Lindsay, 1979). At pH 7 and $H_2PO_4^-$ of 1.0 mM, the equilibrium Mn^{2+} and Zn^{2+} concentrations are $10^{-3.4}$ and $10^{-6.1}$ M, respectively (solubility products taken from Lindsay, 1979; no carbonates or Ca^{2+} were assumed to be present in the calculation). The Mn^{2+} concentration (21 mg L^{-1}) is appreciably higher than target Mn in solution (~0.5 mg L^{-1}) while the Zn^{2+} (0.05 mg L^{-1}) is approximately fivefold smaller than needed. When Mn^{2+} is in equilibrium with 5 mM HCO_3^- (pH 7, no P or Ca present) its concentration is 2 mg L^{-1} (solubility product of $MnCO_3$ taken from Lindsay, 1979), which is still higher than the target concentration. At pH 8 the Mn^{2+} concentration in equilibrium with the above PO_4 or CO_3 minerals is 1.0 and 0.2 mg L^{-1}, respectively. The low concentration problem is solved by using chelates (Table 9.23) that can sustain desired microelement concentrations in solution

TABLE 9.23 Stability Constants (Ksc) of Several Ca, Mg, Fe, Zn and Mn Chelates and Complexes Pertinent to Recirculated Solutions

		$EDTA^{4-}$	$DTPA^{4-}$	$EDDHA^{4-}$	$Citrate^{3-}$	$Oxalate^{2-}$
Reaction[a]	Product			log Ksc		
$Ca^{2+} + L$	CaL	11.6	12.02	8.20	4.2	2.0
$Mg^{2+} + L$	MgL	9.8	10.6	9.0	—	—
$Fe^{2+} + L$	FeL	15.27	17.67	15.3	—	—
$Fe^{3+} + L$	FeL	26.5	29.2	35.4	12.5	8.9
$Zn^{2+} + L$	ZnL	17.44	19.6	17.8	5.5	4.6
$Mn^{2+} + L$	MnL	14.5	16.7	—	4.5	3.7

Sources: Lindsay (1979) and Norvel (1972).
[a] $zM + L = M_nL$; $Kc = (M_nL)/[(M)^z (L)]$.

TABLE 9.24 Microelement Concentrations as a Function of pH in Lettuce and Pepper Recirculated Solutions in Besor, Israel. Concentrations were Determined by Atomic Absorption thus they Comprise the Sum of all Species in Solution

pH	Mn	Zn	Fe	Source
	\multicolumn{3}{c}{L mg^{-1}}	
Lettuce				
5.6	0.120	0.30	1.9	
6.1	0.120	0.30	1.9	Yosef-Bar et al., 2001
6.5	0.035	0.30	2.1	
6.8	0.040	0.16	2.0	
Pepper				
5.0	0.460	0.24	2.5	
5.9	0.400	0.33	3.5	Yosef-Bar et al., 1999
6.7	0.030	0.16	1.5	
7.0	0.020	0.16	1.5	

even at high pHs. Among the aforementioned cations, Fe^{3+} forms the most stable chelates with common natural and synthetic agents. For all greenhouse fertigation purposes the stability constants (Ksc) of Fe^{3+}, Fe^{2+} and Zn^{2+} with EDTA (Table 9.23) are sufficient to warrant solubility in recirculated solution even at pH ~ 7 (Table 9.24). The Ksc of Mn^{2+} is weaker, therefore its concentration in the recycled solution decreased sharply as pH approached 6.5 (Table 9.24). Information on Ksc of Mn^{3+} or Mn^{4+} could not be found in the literature and it is also not known what the $Mn^{2+}/(Mn^{3+} + Mn^{4+})$ ratio was in the recycled solution. Using DTPA instead of EDTA might improve the Mn low-solubility problem as can be inferred from the Ksc of various cations with DTPA. The natural complexing agents citrate and oxalate are not adequate substitution to EDTA (Table 9.23).

The above discussion emphasizes the fact that maintaining the recirculated solution pH between 5 and 6, which is essential to avoid Mn deficiency (without resorting to foliage Mn application), is one of the main and most difficult tasks in closed-loop irrigation systems management. The available knowledge and efficient monitoring of solution pH warrant that this problem will not become a growth-limiting factor in closed-loop irrigation systems.

9.2.11 ROOT ZONE TEMPERATURE

The effect of root temperature on water and ions uptake has been intensively studied during the last three decades (Clarkson et al., 1986; Bowen 1991; Macduff et al., 1994; Chung et al., 1998) and will not be reviewed in detail in this chapter. The subject is relevant because solutions in closed irrigation systems are stored for one-to-several days in reservoirs which are usually exposed to outside air temperatures. It has been shown that a root temperature decline from 20 ± 2 to $14 \pm 2°C$ is associated with reduced water (Ho and Adams, 1995; Adams, 2002; Urrestarazu et al., 2003) and nutrients uptake rates (Engels, 1993; Engels et al., 1993; Adams, 2002; Urrestarazu et al., 2003) and the magnitude of the effect depends on the crop's physiological stage. Bravo and Uribe (1981) used a Q_{10} value of 2–2.5 to account for temperature effect on active nutrients uptake and stated that the Q_{10} for nutrient transport processes was considerably lower. Benoit and Ceustermans (2001) reported that stone wool cooling to 17–22°C increased Ca concentration in paprika fruit relative to the temperature of 23–33°C and attributed this effect to higher oxygen concentration in the root zone at the lower temperature range. Root zone temperature affects root growth too. For example, decreasing the maize roots temperature from 20 to 5°C reduced the root growth rate from 1.2 to 0.02 mm h^{-1} (Pritchard et al., 1990). Ganmore-Neumann and Kafkafi (1980, 1983) found that tomato and strawberry root temperature exceeding 28–30°C was detrimental to root development, particularly at high NH_4:NO_3 ratio in the feed solution. It may be concluded that root zone temperatures below ~18°C and above 28°C may seriously impair uptake and root growth. This implies that water reservoirs should be insulated to avoid temperature fluctuations exceeding this range.

Another problem related to temperature is fertilizer solubility in stock and feed solutions. Most common fertilizers are approximately 30 per cent more soluble at water temperature of 20°C than at 1°C. Potassium concentration is limited by K_2SO_4, the solubility of which at 1, 20 and 50°C is 69, 110 and 170 g L^{-1}, respectively. Nitrate solubility is minimal when added to water as KNO_3 and maximal when supplied as NH_4NO_3. The solubility of all N and P fertilizers increases with temperature between 1 and 50°C (Table 9.25). In order to be on the safe side and account for lower night temperatures, solutions should be prepared even during hot summer days according to the 20°C solubility data.

TABLE 9.25 Effect of Temperature on Solubility of Fertilizers used in Fertigation

Fertilizer	Formula	1	20	50
			g fertilizer L^{-1}	
Ammonium nitrate	NH_4NO_3	1183	1950	3440
Monoammonium phosphate	$NH_4H_2PO_4$	227	282	417
Diammonium phosphate	$(NH_4)_2HPO_4$	429	620	1030
Phosphoric acid	H_3PO_4	—	548	—
Potassium chloride	KCl	280	347	430
Potassium nitrate	KNO_3	133	316	360
Potassium sulfate	K_2SO_4	69	110	170
Monopotassium phosphate	KH_2PO_4	—	330	835[a]
Calcium nitrate	$Ca(NO_3)_2$	1020	3410	3760[a]
Magnesium nitrate	$Mg(NO_3)_2$	—	423	578[a]

Temperature range (°C) spans columns 1, 20, 50.

Sources: Hodgman (1949) and Weast (1977).
[a] Temperature of 90°C.

9.2.12 INTERRELATIONSHIP BETWEEN CLIMATE AND SOLUTION RECYCLING

Increasing the relative humidity (RH) in greenhouses lowers ET and thereby decreases salt accumulation in closed-loop irrigation systems by reducing fresh water addition (Cohen et al., 2003). It is debatable whether the reduced ET stems from decreased stomatal conductance, which is expected to reduce photosynthesis too, or from reduced water vapour gradient between leaf substomatal cavity and boundary air layer, which is expected not to decrease DM production if stomata closure is unaffected. Another unexplored question is the crop response to RH under varying EC conditions mediated by solution recirculation (Sonneveld and Welles, 1988). The aforementioned questions were studied by Bar-Yosef et al. (2003b, 2004a) with roses and pepper as test crops. In roses, three climate regimes were imposed (roof ventilation, RV, wet pad, WP, and shading, SH), each at threshold EC of 2.7 or 4.0 dS m^{-1}. The whole-season mean RH in WP, RV and SH were 64, 59 and 54 per cent, respectively. The whole-season mean air temperature in WP was 1°C lower than in RV or SH, albeit during warm hours the difference was 2–6°C, depending on the season. Global radiation in SH was 30 per cent lower than in WP or RV, which significantly decreased DM production and yield (Table 9.26). The two other climates gave similar total flowering stem yield, but RV increased the fraction of shorter (40–50 cm) stems in comparison with WP. A similar effect was obtained by increasing the threshold EC from 2.7 to 4.0 dS m^{-1} (same table). The advantage of WP over RV became apparent at EC 4.0 dS m^{-1}, where it gave a significantly higher yield and reduced discharge of water (Table 9.26) and N. Total water savings could not be ascertained since water consumption by the wet pad *per se* was not measured. However, the reduced N disposal

TABLE 9.26 Effect of Three Climate Regimes (Wet Pad, WP; Roof Ventilation, RV and Sliding Screen Shading, SH) Coupled with Two Threshold EC (EC$_{thr}$) Values for Replacing Recycles Solutions on Rose Yield (cv. Mercedes) (Total[a] and 40–50 cm Long, Expressed per m^2 Greenhouse), ET, Dry Matter (DM) Production and Water Discharge[b]

Climate[c]	EC$_{thr}$ (dS m^{-1})	Flowering stems yield (Num m^{-2}) Total	40–50 cm	Cum ET (mm)	Total DM (kg m^{-2})	Discharged water (mm)
WP	2.7	256 b	54 c	877	2.85	524
	4.0	298 a	73 b	947	3.11	162
RV	2.7	274 ab	88 ab	1091	2.66	596
	4.0	268 ab	100 a	1053	2.40	241
SH	2.7	233 c	60 c	949	2.42	556
	4.0	239 c	75 b	852	2.14	213

Means followed by the same letter are not significantly different according to Duncan's LSD test (0.05).

[a] Total yield comprised the following length groups: 40–50, 50–60, 60–70, >70 cm.
[b] Concentration of N in discharged solution was in all treatments 140–160 mg L^{-1} N.
[c] The experiment took place at Besor, Israel (Bar-Yosef et al., 2003).

($\sim$24 g m^{-2} ground in WP and $\sim$36 in RV) is significant. Two more facts should be emphasized: (i) the water use per kg DM was 306 $\pm$ 2 L in the WP and 425 $\pm$ 15 L in the RV treatment (calculated based on data in Table 2.26). This is compatible with the hypothesis that increasing RH reduces ET without affecting DM production. (ii) Yield in the WP treatment was higher at EC$_{thr}$ of 4.0 than at EC$_{thr}$ of 2.7 dS m^{-1}. This indicates that in recirculation one or more of the accumulating ions was beneficial to roses and under WP conditions its benefit outweighed its adverse effect via increased salinity. Under RV the salinity impact was stronger and no yield increase was recorded as EC$_{thr}$ rose from 2.7 to 4.0 dS m^{-1} (Table 9.26). The accumulated ions in this experiment were Ca, Mg and B, and their time-averaged concentration during the last 4 months of growth (when discharge controlled the solution EC) were 125, 100 and 0.26 mg L^{-1}, respectively, at EC$_{thr}$ of 4.0 dS m^{-1}, and 100, 65 and 0.16 at EC$_{thr}$ of 2.7. The concentrations in fresh water were 57, 28 and 0.1 mg L^{-1}. All three ions and interrelationships among them could contribute to the beneficial effect of the EC$_{thr}$ of 4.0 dS m^{-1} treatment under wet pad cooling.

The effect of climate – EC interaction on pepper is summarized in Table 9.27.

Total fruit yield was significantly higher in WP than in RV or fogging with tap water (FG) while the effect of EC was insignificant. Fogging caused leaf torch which reduced the photosynthetic leaf area and thus yield. BER incidence was significantly higher in RV than in WP or FG, and in all treatments it was lower at EC of 4.0 than 2.7. Like roses the water consumption per kg DM was lower in WP than in RV and similar to FG (218 $\pm$ 1, 244 $\pm$ 7 and 220 $\pm$ 4 L kg^{-1}, respectively); water discharge at EC of 2.7 was considerably lower in WP than in RV or FG but at EC of 4.0 dS m^{-1} the differences diminished because less water was discharged in all cooling treatments (Table 9.27).

TABLE 9.27 Effect of Three Climate Regimes (Wet Pad, WP; Roof Ventilation, RV and Fogging, FG) Coupled with Two Threshold Salinities (EC$_{thr}$) for Replacing Recycled Solutions on Total Pepper (CV. Celica) Yield (per m^2 greenhouse), Blossom End Rot (BER) Incidence (per cent of Total Fruit Number), Cumulative (Cum) ET, Dry Matter (DM) Production and Water Discharge[a]

Climate[b]	EC$_{thr}$ (dS m^{-1})	Fruit yield Total (kg m^{-2})	BER (% Num)	Cum ET (Mm)	Total DM (kg m^{-2})	Discharged water (mm)
WP	2.7	5.50 a	19.4 b	520	2.19	304
	4.0	5.26 a	15.8 c	531	2.28	125
RV	2.7	4.87 b	25.7 a	559	2.07	459
	4.0	4.83 b	24.4 a	521	1.98	146
FG	2.7	4.52 c	24.2 a	387	1.79	451
	4.0	4.44 c	14.5 c	407	1.83	105

Means followed by the same letter are not significantly different according to Duncan's LSD test (0.05).

[a] Concentration of N in discharged solution was in all treatments 140–160 mg L^{-1} N.

[b] The experiment took place at Besor, Southern Israel. (Bar-Yosef et al., 2004).

The accumulating ions in solution in the pepper experiment were Ca, Mg, SO$_4$ and B, and their concentrations at the end of the experiment were 133, 75, 207, and 0.31 mg L^{-1}, respectively, at EC$_{thr}$ = 4.0 dS m^{-1}, and 75, 45, 96 and 0.26 at EC$_{thr}$ = 2.7. The concentrations in fresh water were 48, 26, 57 and 0.1 mg L^{-1}. Improved supply and utilization of Ca is usually associated with decline in the incidence of BER (for review, see, e.g. Saure, 2001). Since BER incidence declined at EC of 4.0 in comparison with EC$_{thr}$ of 2.7, it is deduced that the accumulating beneficial ion was Ca. Indeed, for pepper the recommended Ca concentration in the irrigation solution in inert substrates in The Netherlands is ~120 mg L^{-1} Ca (de Kreij et al., 1999). The fact that there was a response to higher Ca concentration in solution (at EC of 4.0 vs. 2.7 dS m^{-1}) even though Ca concentration in treatment EC = 2.7 dS m^{-1} increased with time is not surprising as uptake is affected by competition with other cations, for example NH$_4$ and K (see, e.g. Terada et al., 1996) and enhanced Ca concentration increases its capacity to compete on common absorption sites.

Final decision whether RV or WP should be employed in closed-loop irrigation systems should be subject to economic analysis taking into account WP installation, operation and maintenance cost relative to RV. Certainly the effectiveness of WP rises as the recycled solution EC increases and in regions and seasons where high greenhouse temperatures limit production.

9.2.13 EFFECT OF N SOURCES AND CONCENTRATION ON ROOT DISEASE INCIDENCE

The nutritional status of plants has a major impact on their disease susceptibility to various pathogens (see collection of papers in Engelhard, 1989). The exact mechanism underlying this relationship is unknown but there are indications that it is associated with N concentration and NH$_4$:NO$_3$ ratio in soil or substrate solutions. Duffy and Defago (1999) reported that the severity of Fusarium crown and root rot in tomato seedlings grown in stone wool was increased by NH$_4$ and reduced by NO$_3$; intermediate NH$_4$NO$_3$ concentration (40–80 mg L^{-1} N) reduced disease, but rates > 100 mg L^{-1} N increased it. Bar-Yosef et al. (2005) investigated the solution N–root disease relationship in a closed-loop irrigation system with Pythium inoculated cucumber as test crop. The N variables were NH$_4$:NO$_3$:Urea ratio at equal total N concentration (126 ± 14 mgL^{-1} N) and N concentration at constant NH$_4$:NO$_3$ ratio of 30:70 in the fill solution. They reported that 40 days after planting, the mortality incidence increased as the NH$_4$:NO$_3$ ratio rose from 20:80 (70 per cent) to 80:20 (94 per cent mortality), and in a treatment of 50 per cent NO$_3$:50 per cent urea, the mortality dropped to 22 per cent (Table 9.28). The NH$_4$ concentration in the recirculated solution of the urea treatment was similar to the concentration in the 50 NO$_3$:50 NH$_4$ treatment and the concentration of free urea-N was less than 10 per cent of the total N. At the end of the growing season (day number 75) more than 90 per cent of the plants died in all treatments as a result of Fusarium root rot (spontaneous infection) and/or Pythium.

When maintaining a 30:70 NH$_4$:NO$_3$ ratio and increasing total N concentration from 2 to 4–8 and 16 mM, the mortality incidence after 40 days increased from 0 to

TABLE 9.28 Effect of $NH_4:NO_3$:Urea Ratio (NH:NO:U) at Constant Total N Concentration in the Fill Solution of Recirculation System (Experiment 1) and Increasing Total N Concentration at Constant $NH_4:NO_3$:Urea Ratio (Experiment 2) on per cent Cucumber Plants Mortality on Day 40 and at the End of Experiment. Also Presented is the Number of Days Required in Each Treatment to Obtain 60 per cent Mortality. Plants were Inoculated by Pythium on Planting Day. Mortality was Due to Pythium and Fusarium (Spontaneous Infection). Plants were Planted on 1 April 2004 in Closed-Loop Irrigation System at Bet Dagan, Israel[a]

Experiment	Treatment NH:NO:U		Mortality (%) End	Mortality (%) 40 days	Number of days 60% mortality
1	50:50:00	E1	94	86	32
8–10 mMN	20:80:00	E2	91	70	38
	80:20:00	E3	94	94	27
	00:50:50	E4	88	22	51
2	2	W1	0	0	—
30:70:00	4	W2	78	30	48
	8	W3	69	18	51
	16	W4	82	74	35

[a] Derived from Bar-Yosef et al. (2005).

18–30 and 74 per cent, respectively, and at 2 mM N no plant died until the end of the experiment. A similar conclusion can be drawn from the results of number of days that were required to kill 60 per cent of the initial plant population (Table 9.28). Enhanced NH_4 concentration in solution decreased pH which was linearly correlated with the per cent mortality and number of days required to kill 60 per cent of the plants (Fig. 9.14). It is unknown if the low NH_4 *per se* caused the reduced mortality

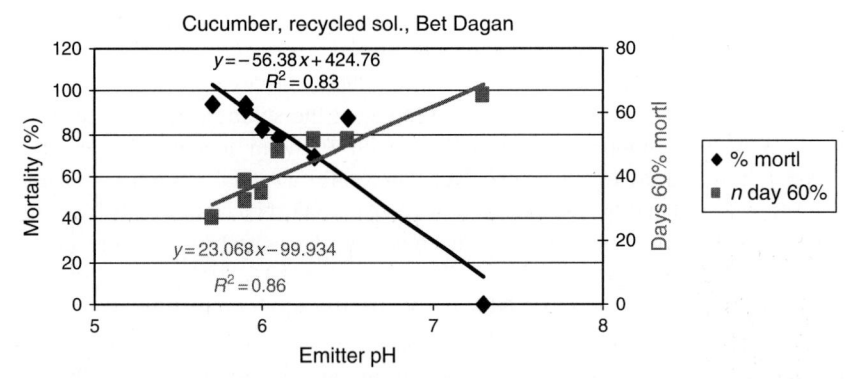

FIGURE 9.14 The effect of irrigation solution pH in closed loop-irrigation system (Emitter pH) on per cent cucumber plant mortality 30 days after planting, and on number of days (n) that were required to kill 60 per cent of the plants. Variations in pH were caused by N treatments (4 $NH_4:NO_3$ ratios at constant total N, and 4 N concentrations at constant ratio); plants were inoculated by Pythium but mortality was due to Pythium and Fusarium (spontaneous infection). Experiments were conducted in Bet-Dagan, Israel (Bar-Yosef et al., 2005).

(Duffy and Defago, 1999) or it was due to the higher pH, but it is clearly indicated that reducing the N concentration (NH$_4$:NO$_3$ ratio = 30:70) to 2 mM N in a closed irrigation system considerably abated cucumber susceptibility to Fusarium. The implications of such a low concentration on crop yield and quality must be studied in detail, but previous data indicated that at least in tomato grown in closed irrigation system C$_N$ reduction from the conventional 10 mM N to 4 mM was feasible. Another aspect of nutrition–microbial interaction is the pH impact on nitrifying bacteria. According to Lang and Elliott (1991) the NH$_4$ oxidation rate in soilless potting media slows down at pH < 6.8, and becomes insignificant at pH < 5.6. When NH$_4$ oxidation stops the nitrate uptake which increases, solution pH becomes negligible and the ammonium uptake by plants is further decreasing the pH.

9.3 SPECIFIC CROPS RESPONSE TO RECIRCULATION

This section describes and summarizes experimental evidence of specific vegetable and ornamental crop response to solution recirculation in arid and semi-arid zones where solution recycling is most problematic. Background information on the reported crops is presented in the Appendix. A conclusion common to all studied crops is that the response over time to EC$_{thr}$ was milder than the response to constant EC observed in open irrigation systems. This can be explained by the fact that all studied crops escaped salt stress during the vegetative growth stage (salt accumulation phase in closed-loop irrigation systems) which was shown (Meiri, 1984) to be more susceptible to salinity than the reproductive stage. Indeed, in all conducted experiments the exposure to EC >∼ 3.2 dS m^{-1} commenced during the reproductive growth stage, in which plants can even benefit from enhanced salinity (Mizrahi et al., 1988). Other reasons could be (i) plant adaptation to salinity during the gradually increasing EC along the vegetative growth stage (Bohnert et al., 1995); (ii) the very frequent and ample irrigation (which is possible in closed but not in open irrigation systems) that decrease the time-averaged salt concentration at the root surface; and (iii) the accumulation of a certain ion(s) or compound(s) in the closed-loop irrigation system which promote crop growth and yield in comparison with open irrigation system where it is prone to leaching out of the system.

9.3.1 VEGETABLE CROPS

9.3.1.1 Pepper

Pepper (CV. Cuby) yield response to increasing threshold salinity over the range 2.7–4.0 dS m^{-1} in perlite was presented in Table 9.27. The data showed no yield reduction and decrease in BER incidence in response to the elevated EC$_{thr}$. Pepper response to a wider range of EC$_{thr}$ was studied in stone wool at Bet Dagan, central Israel (CV. Mazurka, Table 9.29). Elevating EC$_{thr}$ from 2.3 to 4.0 and to 5.5 dS m^{-1} caused an overall 12 per cent decrease in marketable yield, or 3.7 per cent per dS m^{-1}. According to Maas and Hoffman (1977) and Feigin (1985), the expected pepper yield

TABLE 9.29 Pepper[a] (CV. Mazurka) Yield and Water and N Balance as a Function of Threshold EC for Disposing Recycled Effluent in Closed Irrigation Loop System. Note That Disposed Volume Equals to Added Amount Minus Uptake (Quantity Change in System is Negligible)

Threshold EC (dS m^{-1})	Marketable yield (kg m^{-2})	BER[b] (%)	DM[c] (kg m^{-2})	Water (mm) added	ET	N (g m^{-2}) added	uptake
2.3 ± 0.2	6.7	13.2	320	768	280	111	26
4.0 ± 0.6	6.1	11.5	295	347	285	46	23
5.5 ± 1.0	5.9	16.0	300	293	303	36	23
Signif.	ns	*	ns				

* significant difference at $p < 0.05$; ns = not significant.
[a] For details on crop and substrate, see Appendix.
[b] Blossom end rot (% of total fruits).
[c] Total dry matter production.

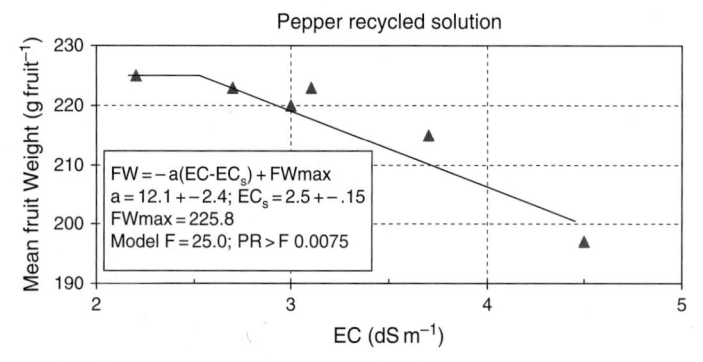

FIGURE 9.15 Effect of threshold EC on pepper (CV. Cuby) marketable fruit weight (FW) in closed-loop irrigation system in the Besor, southern Israel. EC$_s$ is the EC beyond which fruit weight started to decline. Data derived from Reshef et al., 2004. A similar normalized relationship can be found in Bar-yosef et al. (2001c).

decline in open irrigation systems is 10% ± 3% per dS m^{-1}. The yield decline in the closed-loop irrigation system was caused by reduction in marketable fruit weight, compatible with results obtained in a complementary experiment in the Besor, southern Israel (Fig. 9.15), where the relative fruit weight decline was 5.3 per cent per dS m^{-1}, greater than the aforementioned decrease in marketable yield.

Within the EC$_{thr}$ range in the Bet Dagan experiment, increasing the EC$_{thr}$ from 2.3 ± 0.2 to 4.0 ± 0.6 dS m^{-1} significantly reduced N and Mn concentration in diagnostic leaves, increased Ca and Mg concentrations, and had no significant effect on P and K concentrations. The increase in Ca and Mg concentrations stemmed from their accumulation in the recycled solution. Increasing the EC$_{thr}$ further to 5.5 ± 1.0 dS m^{-1} caused no additional change in N, Ca and Mg concentrations but reduced further the Mn concentration in the diagnostic leaves (Fig. 9.16). The decline in leaf nutrient concentrations started 90 days after planting and the difference among treatments remained fairly constant to the end of the season.

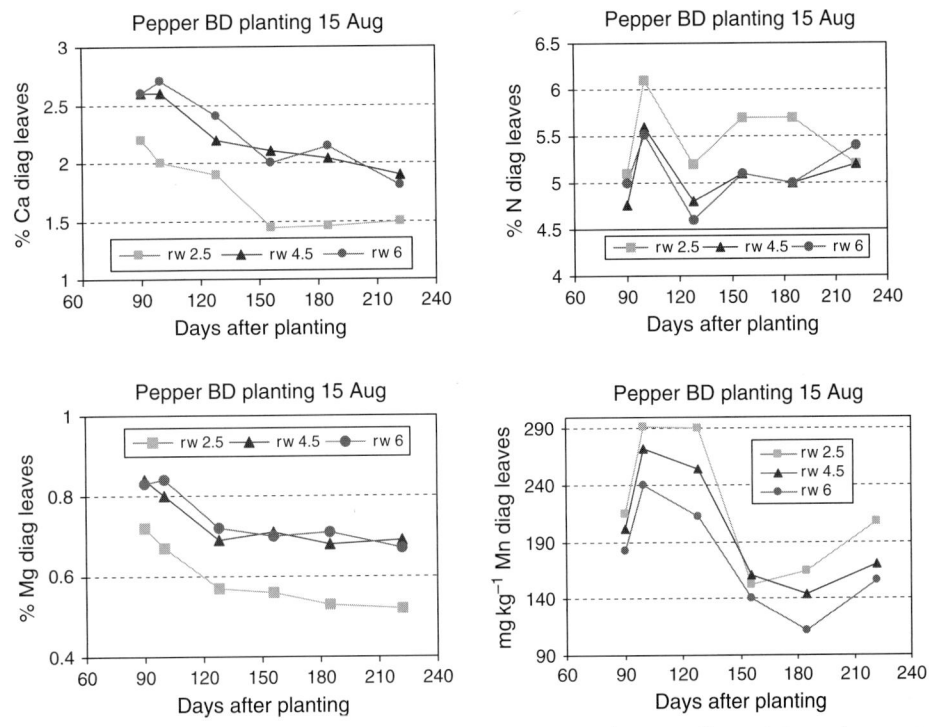

FIGURE 9.16 The concentration of reduced-N, Ca, Mg and Mn in pepper diagnostic leaves (youngest fully expanded leaf) as a function of time and threshold EC in recycled solution The pepper plants (CV. Mazurka) grew in rockwool at BD, central Israel. Other nutrients were not significantly affected by the threshold EC treatments.

The trade-off between fruit yield and quality, on the one hand, and water and N input and discharge, on the other hand, is presented in Table 9.29. Threshold EC values of ~2.3, ~4 and ~5.5 dS m^{-1} reduced marketable yield from 6.7 to 6.1 and 5.9 kg m^{-2} (insignificant difference, P = 0.05), and increased BER incidence from 13 to 16 per cent. The impact on dry matter production was negligible. However, water input declined from 768 to 347 and 293 mm, and N addition from 111 to 46 and 36 g m^2, while water and N consumption by plants were only affected slightly. Under such conditions and fertigation frequency of 12 irrigations per day during late spring and summer, the EC$_{thr}$ of 4 dS m^{-1} treatment seemed to be the best management option.

The effect of fertigation frequency (6, 12, 18 per day, perlite) on pepper yield at EC$_{thr}$ of 3.5 dS m^{-1} was shown to be small, though significant (Table 9.17) with best results in the intermediate value. The effect on total DM production and water and N consumption and discharge was negligible and inconsistent (detailed data not presented).

9.3.1.2 Muskmelon

Increasing EC$_{thr}$ from 2.2 to 3.4 and 4.6 dS m^{-1} decreased the reduced-N concentration in stems, and P and Mn concentration in leaves. The effect on %K in leaves

TABLE 9.30 Threshold EC (EC_{thr}) Effect on Nutrient Concentrations in Muskmelon Leaves and Stems (CV. 5093) at the End of the Experiment (102 Days after Planting [dap]). The Experiment was Carried out at Bet Dagan, Israel.

EC_{thr} dS m^{-1}	N	P	K	Ca	Mg	Fe	Mn	Zn
			%				mg kg^{-1}	
				Leaves				
2.2	3.0	0.69 ab	4.7	6.7 a	0.98	234	494 a	89
3.4	3.3	0.72 a	5.1	5.4 b	0.87	215	380 b	84
4.6	2.9	0.60 b	4.8	6.2 ab	0.92	268	406 b	74
Signif.[a]	ns		ns		ns	ns		ns
				Stems				
2.2	2.0 a	0.72 ab	8.2	—	—	—	—	—
3.4	1.9 a	0.76 a	8.2	—	—	—	—	—
4.6	1.8 b	0.66 b	8.1	—	—	—	—	—
Signif.[a]			ns					

Means followed by the same letter are not significantly different according to Duncan's LSD test (0.05).

For details, see Appendix (ns = not significant; otherwise significant difference at $p = 0.05$).

[a] ns = difference between treatments is not significant at $p < 0.05$; otherwise the difference was significant at this probability level.

and stems and on other nutrients in leaves was insignificant or inconsistent (day 102, end of experiment) (Table 9.30). On day 75, none of the nutrients was significantly affected by EC_{thr} probably because the exposure time to the higher EC values was too short (data not presented).

The elevated threshold EC reduced marketable fruit yield but increased fruit total soluble solids (TSS) and reduced ET without affecting DM production and N uptake (Table 9.31).

TABLE 9.31 Water and N Balance and Muskmelon[a] Yield as a Function of Threshold EC for Disposing Recycled Effluent in Closed-Loop Irrigation System. Note That Disposed Volume Equals to Addition Minus Uptake (Quantity Changes in System are Negligible)

Threshold EC (dS m^{-1})	Marketable yield (kg m^{-2})	TSS[b] (%)	DM[c] (kg m^{-2})	Water (mm) added	ET	N (g m^{-2}) added	uptake
2.1±0.2	7.6	9.1	1.14	847	381	111	29
3.4±0.5	6.7	9.6	1.08	387	327	47	29
4.6±1.0	6.1	10.8	1.04	303	292	38	26
Signif.	*	*	ns				

* Significant difference at $p < 0.05$; ns = not significant.
[a] For details on crop, see Appendix.
[b] Total soluble solids.
[c] Total dry matter production.

Due to the reduced water and N inputs and disposal, an economical analysis is required in order to decide whether the yield reduction matches the obtained inputs savings and reduced environmental pollution.

9.3.1.3 Cucumber

Threshold EC values between 2.2 and 4.5 dS m^{-1} (stone wool) had no significant effect on nutrient concentrations in plant leaves during the entire growth period (detailed data, taken from Bar-Yosef et al., 2003 not presented). This is attributed to the fact that the time exposure to the threshold salinity was too short and gradual to impede uptake. Increasing the substrate (perlite) volume per cucumber plant from 10 to 40 L significantly increased P, K and Mn concentration in plant leaves (0.68 to 0.82 per cent, 4.0 to 5.4 per cent and 236 to 369 mg kg^{-1}, respectively, all at 30 days after planting). The impact on concentration of other nutrients was insignificant (Bar-Yosef et al., 2004b).

Similarly the treatments had no significant effect on total or marketable fruit yield, total DM production and ET and N uptake (Table 9.32). However, in order to maintain EC$_{thr}$ of 2.2 or 3.2, water input had to be increased from ~200 mm in EC$_{thr}$ of 4 or 4.5 dS m^{-1} to 298 and 234 mm, respectively, with concomitant increase in N

TABLE 9.32 Water and N Balance and Cucumber[a] Yield (cv. Ringo) as a Function of Two Factors: (i) Threshold EC for Disposing Recycled Effluent; (ii) Substrate Volume per Plant (at EC$_{thr}$ of 3.2 dS m^{-1}). Disposed Volume Equals to Addition Minus Uptake (Quantity Change in System is Negligible)

Treatment EC$_{thr}$ (dS m^{-1})	Yield (kg m^{-2}) Total	A[b]	DM[c] (kg m^{-2})	Water (mm) Added	ET	N (g m^{-2}) Added	Uptake
2.2	5.0	4.3	735	298	204	39	29
3.2	5.2	4.7	741	234	214	32	28
4.0	5.1	4.4	773	196	191	31	30
4.5	4.7	4.2	723	203	203	26	26
	ns	ns	ns				

Substr[c] (L pl^{-1})	Marketable (kg m^{-2})	Fruit wt g fruit^{-1}	DM[c] (kg m^{-2})	Added mm	ET	Added kg m^{-2}	Uptake
10	11.3	82	0.57	313	306	51	20
40	11.2	81	0.62	404	384	59	23
	ns	ns	ns				

ns = not significant ($p = 0.05$).
[a] For details on crop, see Appendix.
[b] Class A (export quality).
[c] Total dry matter production.

application. This indicates that under the described growth conditions the optimal EC_{thr} for cucumber is ~4 dS m^{-1}.

9.3.1.4 Tomato

Increasing the threshold EC from 2.5 to 4.0 and 5.5 dS m^{-1} caused a decrease in reduced-N, P, Mn and K concentrations in tomato plant leaves and an increase in Cl and Na concentration (K and Na are not presented) (Table 9.33, plant sampling June 2). In mid-March a similar effect was found, but the impact on N concentration was insignificant (data not presented). The effect of EC on concentration of other nutrients in leaves (including Ca and Mg) was insignificant. In a control open system treatment (same water and nutrients rate as in Tr R2.5) the nutrient concentrations were slightly lower than in Tr R2.5 (Table 9.33). A decline in fruit yield was observed when EC_{thr} was raised from 4.0 to 5.5 dS m^{-1}. The adversely affected yield components were number of marketable fruits, fruit weight and BER incidence (Table 9.33).

The yield reduction was accompanied by a meaningful decline in ET and reduced water and N input and discharge (Table 9.34). As in previous crops, an economic evaluation of this trade-off should be performed. Note the impressive difference in water and N use between the closed-loop irrigation treatments and 'farmer's practice in an open irrigation system' in southern Israel (Tr F2.0). The low ET of the 'farmer's' treatment is additional proof that water availability in a closed-loop irrigation system exceeds the availability in an open system due to the higher irrigation frequency and amount. The whole-season water use ratio (cumulative ET/DM) in TR R2.5, R4.0 and R5.5 were 386, 389 and 342 L kg^{-1}, respectively, indicating reduced ET with minimal reduction in DM production.

TABLE 9.33 Tomato (CV. 870) Yield, Yield Components and Nutrient Concentrations in Leaves (Plant Average) as a Function of Threshold EC Values for Replacing Recirculated Solutions. Except the Presented Nutrients Treatments Effect on Other Element Concentrations was Insignificant ($p = 0.05$). Plants were sampled on June 2

Tr[a]	Yield (kg m^{-2})	Frt num (Num m^{-2})	Frt wt (g)	BER (%)	N (% in leaves)	P (% in leaves)	Cl (% in leaves)	Mn (mg kg^{-1})
R2.5	28.2 a	161 a	152 a	8 abc	3.5 a	1.1 a	1.9 c	479 a
R4.0	28.8 a	182 a	145 a	4 c	3.4 a	0.8 b	2.8 ab	342 bc
R5.5	22.7 b	137 b	131 b	12 a	2.9 b	0.7 b	2.5 b	329 bc
C2.0	31.9 a	180 a	149 a	11 a	3.1 ab	0.9 b	1.6 c	418 ab
F2.0	29.9 a	168 a	152 a	9 ab	—	—	—	—

Means followed by the same letter are not significantly different according to Duncan's LSD test (0.05).

Source: Data derived from Reshef et al. (2003).

[a] R2.5, R4.0 and R5.5 stand for EC_{thr} of 2.5, 4.0 and 5.5 dS m^{-1}, respectively; C2.0 is an open-loop irrigation system supplied with the same solution composition and volume and same irrigation frequency as Tr R2.5; F2.0 is open system 'farmer's practice' (45 per cent leaching fraction).

TABLE 9.34 Water and N Balance in Tomato (CV. 870) as a Function of Threshold EC for Disposing Recycled Effluent

	Water (mm)			N^b (g N m^{-2} ground)		
Tr[a]	Fill	Disposed	ET	Added	Disposed	Consumed
R2.5	1106	353	753	135	83	47
R4.0	889	68	821	108	15	46
R5.5	643	11	632	78	2	36
F2.0	1641	1104	537	230	166	45

Difference between N consumed and added N-disposed N stems from N remaining in system and experimental error.
[a] Treatments are explained in the footnote of Table 9.33. More details in Appendix.
[b] Added = Fill × N conc. in Fill solution mg N L^{-1}; Disposed N = Disposed water × corresponding drainage N concentration; consumed N = N in harvested fruits + N in leaves + N in stems.

9.3.1.5 Strawberry

Increasing the EC$_{thr}$ from 2.5 to 3.2 and 4.5 dS m^{-1} had no effect on nutrients concentrations in fruit and root collar at the end of the experiment (June 4, Table 9.35). One exception was Mn in fruits which declined from 139 to 101 and 62 mg kg^{-1}, and in root collar from 1040 to 690 and 464 mg kg^{-1}, respectively (detailed data not presented). The elevated EC$_{thr}$ increased the plant leaves Ca and P concentrations, decreased the concentration of Mn and had an inconsistent effect on Mg concentration (Table 9.35). The concentration of other nutrients in leaves were unaffected by EC$_{thr}$. The Mn concentration showed the greatest decline with EC$_{thr}$ but apparently it had no impact on yield (Table 9.36). In winter (February) the threshold EC effect was similar to the one in summer, only Mn leaf concentration was unaffected by EC$_{thr}$ (Table 9.35).

TABLE 9.35 Threshold EC (EC$_{thr}$) Effect on Nutrient Concentrations in Strawberry (CV. Tamar) Leaves in the Winter and Summer Growth Periods.[a] Other Nutrients (N, K, Fe, Zn) Were not Significantly Affected by EC$_{thr}$ in Both Dates and are not Presented Therefore. The Closed-Loop Irrigation System Substrate was Coir; the Experiment was Carried out at Besor, South Western Israel

		P		Ca		Mg		Mn	
Crop	EC$_{thr}$ (dS m^{-1})	Feb	June	Feb	June	Feb	June	Feb	June
		%						mg kg^{-1}	
Strawberry	2.5	0.58	0.33	0.62	1.05	0.38	0.34	707	397
	3.2	0.65	0.33	0.69	1.05	0.42	0.33	638	315
	4.0	0.63	0.39	0.69	1.25	0.44	0.41	658	125
								ns	

ns = not significant; otherwise significant at $p \leq 0.05$.
[a] Plants were grown in hanging containers 10 cm wide, 0.8 L substrate pl^{-1}. More details about the experiment in Appendix.

TABLE 9.36 Strawberry[a] Yield and Dry Matter (DM) Production and Water and N Mass Balance as a Function of Threshold EC for Disposing Recycled Effluent in Closed Irrigation-Loop System. Note That Disposed Volume Equals to Addition Minus Uptake (Quantity Change in System is Negligible)

Threshold EC (dS m^{-1})	Yield (kg m^{-2}) All	Yield (kg m^{-2}) Class A	DM[b] (kg m^{-2})	Water (mm) Added	Water (mm) ET	N (g m^{-2}) Added	N (g m^{-2}) Uptake
Open							
1.8	11.9	2.35	1.26	~1200	~375	~100	24
2.5	10.3	2.26	1.13	445	348	36	24
3.2	10.4	1.95	1.14	322	272	26	18
4.0	10.9	2.08	1.18	389	321	27	19
	ns	ns	ns				

ns = not significant at $p = 0.05$.
[a] For details on crop, see Appendix.
[b] Total dry matter production.

There was no fruit yield decline or reduced DM production due to increasing EC$_{thr}$ from 2.5 to 3.2 and 4.0 dS m^{-1} or in comparison with an open irrigation system at EC of 1.8 dS m^{-1} (Table 9.36). However, a considerable reduction in water and N application and disposal was obtained in the transition from 2.5 to 3.2 or 4.0 dS m^{-1} (the difference between 3.2 and 4.0 dS m^{-1} was insignificant). This indicates that EC$_{thr}$ of 4.0 is the preferred value for growing greenhouse strawberries under the production conditions in Israel. The mechanism explaining this response is unclear yet, but it may be related to the enhanced Ca concentration in leaves stemming from Ca accumulation in the recycled solution.

Due to the small substrate (coir) volume per strawberry plant (0.8 L pl^{-1}), the yield response to fertigation frequency is of particular importance. Matan et al. (2003) studied strawberry yield and water and N mass balances at fertigation frequencies of 5–10 and 10–20 per day (depending on growth period) under identical total daily water and N application rate. Results showed (Table 9.37) that fruit yield and quality, DM

TABLE 9.37 Strawberry[a] Yield and Water and N Mass Balance[b] as a Function of Irrigation Frequency at Threshold EC of 3.5 dS m^{-1}

Treatment Irr. freq.[c] (num d^{-1})	Yield (kg m^{-2}) Class A	Yield (kg m^{-2}) Total	DM (kg m^{-2})	Water (mm) Added	Water (mm) ET	N (g m^{-2}) Added	N (g m^{-2}) Uptake
5–10	1.95	10.4	1.35	322	272	26	18
10–20	1.98	10.1	1.31	292	268	23	20
Signif.	ns	ns	ns				

ns = not significant at $p = 0.05$.
[a] Details on experimental conditions are given in Table A1 (appendix).
[b] Disposal = added − uptake (change in substrate was negligible).
[c] Irrigation frequency in a given treatment increased with transpiration rate.

production, and water and N addition, consumption and disposal were unaffected by the increased frequency.

9.3.1.6 Lettuce

Increasing the EC_{thr} from 2.5 to 3.5 and 4.5 dS m^{-1} had no adverse effect on lettuce nutrient tissue concentrations in the spring (Bar-Yosef et al., 2002). An exception was a small (but significant) decrease in leaf K and Mg concentrations (11.8, 10.4 and 9.1 per cent K and 0.78, 0.73 and 0.67 per cent Mg, respectively). To obtain the above EC range, NaCl was added to the solution as otherwise the small ET and short growth period (~30 days to harvest) would not enable such a range in EC. The EC effect on commercial yield was negligible as old leaves are trimmed at harvest to obtain uniform stalks which are considerably smaller than the real growth.

9.3.2 ORNAMENTAL CROPS

9.3.2.1 Roses

The leaf nutrient concentrations of tuff-grown roses decreased with increasing EC_{thr} only after the exposure period exceeded ~1 year (from planting). In Besor the %N, P and K declined as EC_{thr} rose from 2.7 to 4.0 and 5.5 dS m^{-1} (plant leaves sampled in mid-March), and in Arava the same trend was observed in diagnostic leaves sampled in July (Table 9.38). The decline in %N (not including nitrate) could be partly explained by the known adverse effect of Na$^+$ salinity on nitrate reductase activity (Lorenzo et al., 2000). Despite Ca and Mg accumulation in solution, their concentration in diagnostic leaves in Arava declined somewhat (Table 9.38). Nutrients not included in Table 9.38 were not significantly affected by the EC_{thr} treatments.

Under conditions of roof ventilation in Besor, EC_{thr} values of 2.7, 4.0 and 5.5 dS m^{-1} coincided with yields of 270, 268 and 209 flowering stems per m^2 ground (Table 9.39).

TABLE 9.38 The Effect of Increasing Threshold EC (EC_{thr}) on Nutrient Concentrations in Rose Leaves at Two Experimental Sites: Besor (West Southern Israel, CV. Mercedes) and Arava[a] (Southern Israel, CV. Long Jaguar). Both Substrates were Tuff Irrigated 10–18 Times a Day. Only Results That were Significantly ($p \leq 0.05$) Affected by EC_{thr} are Presented.

	Besor (March 2003)			Arava (July 2001)				
EC_{thr}	N	P	K	EC_{thr}[a]	N	P	Ca	Mg
dS m^{-1}	% in all leaves			dS m^{-1}	% in diagnostic leaves			
2.7	2.7	0.63	2.3	2.9	2.0	0.26	0.80	0.39
4.0	2.2	0.45	2.2	3.8	1.7	0.23	0.73	0.33
5.5	2.0	0.34	2.1	4.8	1.6	0.22	0.70	0.34

Cooling by roof ventilation.
[a] Upper limit of a 'saw tooth' EC experiment described in p. 368 (Kramer et al., 2002).

TABLE 9.39 Water and N Balance and Rose[a] Yield as a Function of Threshold EC for Disposing Recycled Effluent in Closed-Loop Irrigation System in Besor. The Substrate was Tuff. Only Roof Ventilation Data is Presented. Note That Disposed Volume Equals to Addition Minus Uptake (Quantity Change in System is Negligible)

Treatment		Flowering stem yield[b]			Water (mm)		N (g m^{-2})	
Climate	EC$_{thr}$ (dS m^{-1})	All	<40 cm (num. m^{-2})	Fresh wt (kg m^{-2})	Added	ET	Added	Disposed
RV	2.7	270	18	5.86	1687	1091	276	94
RV	4.0	268	19	5.62	1293	1053	208	49
RV	5.5	209	48	4.36	1127	983	204	28
Signif.		0.05	0.05					

[a] For details on crop, see Appendix. Data cover the period 13 May 2002–13 July 2003.
[b] DM = Fresh weight × 0.23 (Kramer et al., 2002); Disposed N = disposed water × 150 mg N L^{-1}.

This is somewhat higher than the data reviewed by Raviv and Blom (2001) showing a yield decrease of 2 per cent per dS m^{-1} around EC 5 dS m^{-1}, attributed mainly to reduced specific leaf area. The elevated salinity (Table 9.39) increased the fraction of <40 cm stems, decreased average stem weight and decreased ET. The corresponding water and N disposal rates were 596, 240 and 144 mm (added – ET) and 94, 49 and 28 g N m^{-2}; again indicating an advantage of EC$_{trh}$ of 4.0 over 2.7 or 5.5 dS m^{-1} (Table 9.39). The same conclusion was drawn when comparing yield response to EC$_{thr}$ of 2.7 or 4.0 dS m^{-1} under wet pad cooling conditions in the Besor (data not presented).

The yield in Arava (Table 9.40) significantly declined (both flowers number and weight) as EC$_{thr}$ increased from 2.9 to 3.8 dS m^{-1}, but no reduction in total DM

TABLE 9.40 Rose[a] Cut Flower Yield and Water and N Mass Balance[b] as a Function of Threshold EC (EC$_{thr}$) Value[c] in Arava, Israel

Treatment EC$_{thr}$ ± dev (dS m^{-1})	Flowering stem yield			Water (mm)		N (g m^{-2})	
	Num. (>50 cm) (m^{-2})	Fresh wt (kg m^{-2})	DM produced (kg m^{-2})	Added	ET	Added	Uptake
2.9	288	9.07	2.53	3061	1311	306	175
3.8	256	8.11	2.53	1917	1239	192	124
4.8	254	8.00	2.46	1649	1221	165	122
Signif.	0.05	0.05	ns				

ns = not significant ($p = 0.05$).
[a] Details on experimental conditions are given in Appendix. Data cover the period 1 January 2001–31 December 2001.
[b] Disposal = added – uptake (change in substrate was negligible).
[c] Upper limit of tested 'saw tooth' EC. For details, see p. 368 (Kramer et al., 2002).

production was observed. This indicates that under the heat load in Arava (compare the higher ET in Arava in comparison with Besor) increasing salinity has greater impact on DM partitioning to short stems (<50 cm) in comparison with Besor (milder temperatures). The yield reduction in Arava was compensated by ~30 per cent saving in water and N application and >50 per cent saving in disposal (Table 9.40).

9.3.2.2 Lisianthus

Lisianthus response to EC_{thr} was studied in two experiments in south-western Israel (Besor). In the first experiment, EC_{thr} values of 2.3, 3.2 and 4.5 dS m^{-1} had no significant effect on N, P, K, Ca, Mg and microelement concentration in leaves sampled in mid-November (buds opening stage, first out of two annual flushes). In the second experiment (EC_{thr} 2.6, 3.2 and 4.0 dS m^{-1}), leaves were sampled at the end of the second flush (beginning of June 2003) and similarly to previous crops P and Mn concentration in leaves (out of all aforementioned nutrients) were significantly affected: P dropped from 0.48 to 0.33 and 0.30 per cent and Mn from 103 to 71 and 63 mg kg^{-1} (detailed data not presented). The decline in P and Mn concentrations had no significant effect on number of harvested flowers and total DM production, but they were associated with a decline in flowering stem weight (Table 9.41). The recirculation at $EC_{thr} = 2.6$ dS m^{-1} gave yield identical to a control open irrigation system but the water and N applications were ~50 per cent lower (Table 9.41). The disposal rate decreased as EC_{thr} increased, indicating that similarly to previous crops recirculation at EC_{thr} between 3.2 and 4.0 was optimal in lisianthus.

TABLE 9.41 Water and N Balance and Lisianthus[a] Yield and Dry Matter (DM) Production During the Second Flowering Flush (Terminating 3 June 2003) as a Function of Threshold EC for Disposing Recycled Effluent in Closed-Loop Irrigation System. Note That Disposed Volume Equals to Addition Minus Uptake (Quantity Change in System is Negligible)

Threshold EC (dS m^{-1})	Flowers yield[b] number (m^{-2}ground)	stem wt (g)	DM (kg m^{-2})	Water (mm) Added	ET[c]	N (g m^{-2}) Added	Uptake
Open							
2.0	109	106	1872	1282	682	113	43
2.6±0.3	107	101	1704	890	733	62	39
3.2±0.2	116	98	2048	971	852	50	45
4.0±1.0	111	87	1720	878	792	43	35
Signif.	ns	0.05	ns				

ns = not significant.
[a] For details on crop, see Appendix.
[b] Fl-st = flowering stems per m^2 ground; Stem wt = average flowering stem weight; mean flowering stem length (wave 2) = 86 ± 1.
[c] The ET during the first flowering wave was 78 ± 8 mm in all treatments.

9.3.2.3 Gypsophila

The response of gypsophila (CV. Yukinko) to increasing EC_{thr} (2.2, 3.4 and 4.2 dS m^{-1}) was studied at Bet-Dagan, central Israel (Bar-Yosef et al., 1999b, 2000a, 2001b). With the exception of P the EC_{thr} had no significant effect on nutrients content in plant organs in mid-July. The P content in leaves decreased with increasing EC_{thr} giving concentrations of 0.39, 0.36 and 0.34 per cent P, respectively. During autumn the EC_{thr} was reduced to 2.1, 2.6 and 3.0 dS m^{-1} but despite the decrease, plants sampled at the end of October exhibited a significant increase in Ca and Mg concentration and decrease in Mn concentration in leaves, while other nutrients were unaffected. The Ca concentrations increased from 1.3 to 1.5 and 1.8 per cent, Mg from 0.62 to 0.70 and 0.90 per cent and Mn declined from 148 to 142 and 116 mg kg^{-1}, respectively (detailed data not presented). Flower yield was unaffected by treatments either in the summer (96 ± 7 flowering stems m^{-2}bed) or in the autumn (29 ± 2 stems m^{-2}bed) harvests. A similar lack of response was obtained in flowers fresh weight (detailed data not presented). It is noted that for the given planting date (May 18) the increase in EC was quite slow and ascribed EC_{thr} values were achieved only after 55 d of growth. The slow salinity build-up may explain the fact that gypsophila yield and nutrient uptake in July were unaffected by the EC_{thr} treatments.

9.3.2.4 Solidago (*Solidago rugosa*)

The effect of EC_{thr} on solidago (CV. Tara) yield was studied in south-western Israel, at Besor, with perlite as growth substrate. Between 16 July 2001 and 31 December 2001, the increase in EC was slow and reached only 2.5 dS m^{-1} (Table 9.42). The salt accumulation rate increased during February, and in mid-April the planned threshold EC values were obtained. Treatments (open, 2.5, 3.5 and 4.2 dS m^{-1}) had no significant effect on entire year cumulative flower yield and average stem weight and length (Table 9.42). This can be explained by the relatively short exposure time to high EC and the high irrigation frequency (7 ± 2 irrigations per day) and water dose (2 mm per irrigation) that minimized salt accumulation at the root surface. As the EC_{thr} increased water and N inputs decreased, particularly in the transition from 2.5 to 3.5 dS m^{-1} (1100 vs. 786 mm, Table 9.42). The ET was unaffected by treatments and even at the highest EC_{thr} it did not differ from ET in a control open system (Table 9.42). Nitrogen uptake was lower in the recirculated solutions than in the control open system, but the EC_{thr} treatments caused only a small decrease in uptake which was within the range of experimental error (Table 9.42). While the water disposal in EC_{thr} of 3.5 and 4.2 dS m^{-1} was ~50 mm, in EC_{thr} of 2.5 dS m^{-1}, it increased to 350 mm.

The whole-year mean N concentration in the transpiration stream (N uptake/ET) was unaffected by EC_{thr} (31 ± 2 mg L^{-1} N) but this average value was smaller than the concentration in the open irrigation system (41 mg L^{-1} N) and replenishment solution (75 mg L^{-1} N). Due to this difference, another experiment was carried out (Dori et al., 2005) to determine solidago response to target N concentration in recirculated solution in Besor (CV. Tara, planting 23 October 2003, perlite). Three N concentrations (35, 70 and 140 mg L^{-1} N) were investigated under identical EC_{thr} of 3.0 dS m^{-1} and irrigation frequency of 7 ± 2 per day (depending on climate). The treatments did not exert a

TABLE 9.42 Effects of Threshold EC (EC_{thr}) on Solidago (CV. Tara) Yield, Flowering Stem Length and Weight, Water Balance and N Uptake.[a] Plants were Grown on Perlite in a Closed-Loop Irrigation System. Note That Disposed Volume Equals to Water Addition Minus ET (Water Content Change in System is Negligible)

EC_{thr} (dSm^{-1})	Maximum EC^b ($dS\,m^{-1}$) 16/7/01– 31/12/01	1/1/02– 17/7/02	Flowering stems[c] Cumulative yield (Stems m^{-2} ground)	Weight (g)	Length (cm)	Water (L m^{-2} ground) Added	ET	N^d uptake (g m^{-2} ground)
Open								
2.0	1.6	2.0	275	22.8	71	3960	710	29
2.5	2.3	3.0	273	26.2	74	1100	750	25
3.5	2.5	3.5	256	27.3	74	786	730	23
4.2	2.5	4.2	279	26.5	73	766	715	21
Signif.			ns	ns	ns			

[a] For details on crop, see Appendix. Planting 16 July 2001.
[b] Maintained for at least 2 weeks.
[c] Yield is the sum of the autumn, spring and summer harvests; Stem weight and length refer to the spring harvest only. In the summer harvest the average stem weight and length were 22.5 ± 0.8 g and 86 ± 2 cm, respectively (insignificant EC_{thr} effect).
[d] N uptake was obtained by plant tissue analyses. The approximated ($\pm 20\%$) N input to above treatments ($=$ water addition $\times$ 75 mg L^{-1} N) were 297, 82.5, 59 and 57 g m^{-2} ground N, respectively, all as NH_4NO_3. If pH dropped below 4.5, the NH_4:NO_3 ratio was decreased without changing total N concentration in the replenishment solution.

significant effect ($P=0.05$) on flower yield (74, 81, 75 flowering stems per m^2 ground, respectively, in the first flush [11 February 2004–8 March 2004]; 66, 72, 77 in the second flush [19 May 2004–26 May 2004]; and 71, 69, 72 flowers per m^2 in the third flush [29 July 2004–8 August 2004]) (no tabulated data is presented).

The results indicate that under the growth conditions described above solidago can be successfully grown in recirculated solution with a threshold EC of ~ 4 dS m^{-1} and target N concentration between 35 and 70 mg L^{-1} N without impairing yield but with considerable saving in water and N input and reduced disposal to the surroundings.

EC threshold values for crops not investigated thoroughly in closed-loop systems and not discussed above (gerbera, carnation, aster, lily) can be found in Sonneveld et al. (1999b) and used as a first approximation of permitted salt accumulation and yield reduction per unit increase in EC. More information on nutrient management in semi-closed systems where EC was determined by the nutrients themselves can be found in Voogt and Sonneveld (1996).

9.4 MODELLING THE CROP-RECIRCULATION SYSTEM

9.4.1 REVIEW OF EXISTING MODELS

Any comprehensive model describing crop growth in a climate-controlled closed-loop irrigation system greenhouse should include the following components: (i) simulator of crop DM production and partitioning between plant organs as a function of greenhouse climate and crop water and N status (for review, see Marcelis et al., 1998). Nitrogen is given higher priority than other nutrients due to its importance in plant development and specific effects of NH_4 and NO_3 on root growth, excretion of protons and carboxylates, and certain ions uptake. (ii) Simulator of water and nutrients uptake by the crop as a function of their concentration and distribution in the substrate. The specific problem of salt accumulation in closed irrigation systems requires that competitive uptake phenomena (e.g. Na-K and Cl-NO_3) and osmotic potential effects on water uptake and nutrients mass flow to roots be accounted for in the simulation. Another concern is salinity effect on plant characteristics, for example dry matter content or specific leaf area (Heuvelink et al., 2003). (iii) Simulator of water, solute and nutrients transport in substrate including ion partitioning between solid and liquid phases, mineralization of organic N and nitrification as a function of root-zone conditions. Comprehensive crop-soil-atmosphere models (e.g. DSSAT, 2004) fulfil the above requirements, but they are not calibrated to greenhouse crops, detached substrates and constrained root volumes, nor are they programmed to account for solution recirculation and salinity build-up effects. Two models specifically designed to simulate crop growth and uptake in closed-loop irrigation system greenhouses and to account for the aforementioned requirements are Greenman by Bar-Yosef et al. (2004a) and CLOSYS by Marcelis et al. (2003, 2005). The models differ in the simulation of root-controlled processes and transport phenomena but are quite similar in their approach to calculate crop net photosynthesis and DM and N partitioning. Both models still lack the capability to predict produce quality and the effect of tissue elements other than N on DM production.

Other greenhouse models simulating only part of the entire system have also been published. Heinen (1997) presented a detailed model of nutrients transport, root growth and uptake but it lacks the interaction with DM production and does not account for the different components of recirculation systems. Giuffrida et al. (2003), Stanghellini et al. (2004) and Carmassi et al. (2003) published models capable of predicting salt accumulation in closed-loop irrigation systems but these models include no feedback relations with crop response to rising EC values.

9.4.2 EXAMPLES OF CLOSED-LOOP IRRIGATION SYSTEM SIMULATIONS

9.4.2.1 Chloride Accumulation and Impact on N and Water Uptake

The threshold Cl^- concentration for replacing recycled solution determines the time it takes under given climatic conditions and irrigation regime until solution discharge must take place. As described above, threshold EC is simpler and less expensive to monitor but cannot be calculated directly, therefore it is being approximated from simulated Cl^- concentrations. Simulation results from the Greenman model (Bar-Yosef et al. 2004a) show that in order to maintain an average Cl^- concentration of 500 mg L^{-1} in the presence of tomato, the recycled solution must be partly discharged after day 27. If a concentration of 1750 mg L^{-1} is tolerated, then the discharge starts after 126 days (Fig. 9.17). A similar trend was obtained in a pepper experiment reported earlier

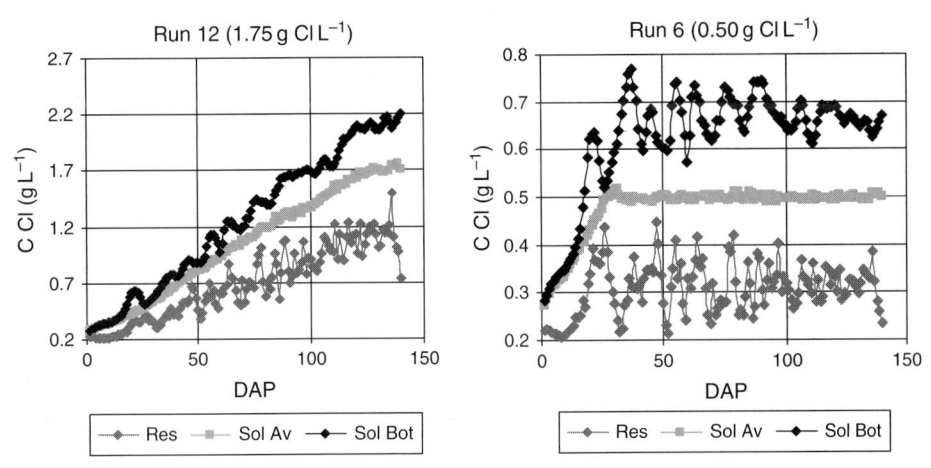

FIGURE 9.17 Simulated concentration of Cl⁻ in recirculated solution in closed-loop irrigation system as a function of time and threshold Cl concentration. Simulated crop was tomato; growth conditions and fertigation management were typical to the Besor area in southern Israel. Threshold Cl⁻ concentration = 1750 mg L^{-1} (left) or 500 mg L^{-1} (right). Res = water reservoir; Sol A = average concentration in entire water volume and Sol Bot = concentration in solution at the bottom of the substrate. Calculated by Greenman (Bar-Yosef et al., 2004c).

(Fig. 9.5) where ~140 days were required to obtain an average Cl⁻ concentration of ~1050 mg L^{-1} in recycled solution at Besor, Israel. The model also predicts that Cl concentration at the bottom of the substrate is considerably higher than in the water reservoir (= irrigation solution) and the difference is widening with increasing threshold concentration (Fig. 9.17). The simulated daily fluctuations are greater at the bottom of the substrate than in the water reservoir, which is replenished with fresh solution in order to maintain constant water volume in the system. The accumulated Cl⁻ affects plant growth via several mechanisms: (i) inhibiting nitrate absorption by competitive uptake (Eq. [21]); (ii) reducing water uptake and therefore mass flow of nutrients to roots due to osmotic potential. (iii) decreasing the specific leaf area and increasing the %DM, particularly in fruits (Heuvelink et al., 2003). The Greenman model distinguishes between effects (i) and (ii) while keeping (iii) constant. Based on calibrated parameters for tomato, the model predicts that fruit yield is affected more by N uptake inhibition than water uptake. When excluding competitive uptake, the simulated N quantity in plant was ~35 g N m^{-2} ground (day 80), while when accounted for it with a threshold Cl⁻ concentration of 500 mg L^{-1} the quantity is reduced to 10 g N m^{-2} ground (Fig. 9.18).

The N deficiency thus created induced a reduced DM production rate (expressed as 'N stress factor') that increased with time. When the Cl⁻ effect on N-uptake was disregarded, the N stress was negligible (stress factor = 1) (Fig. 9.18). The water stress factor (Fig. 9.18) was determined by the fertigation regime and weather conditions, and was negligibly affected by the reduced N uptake.

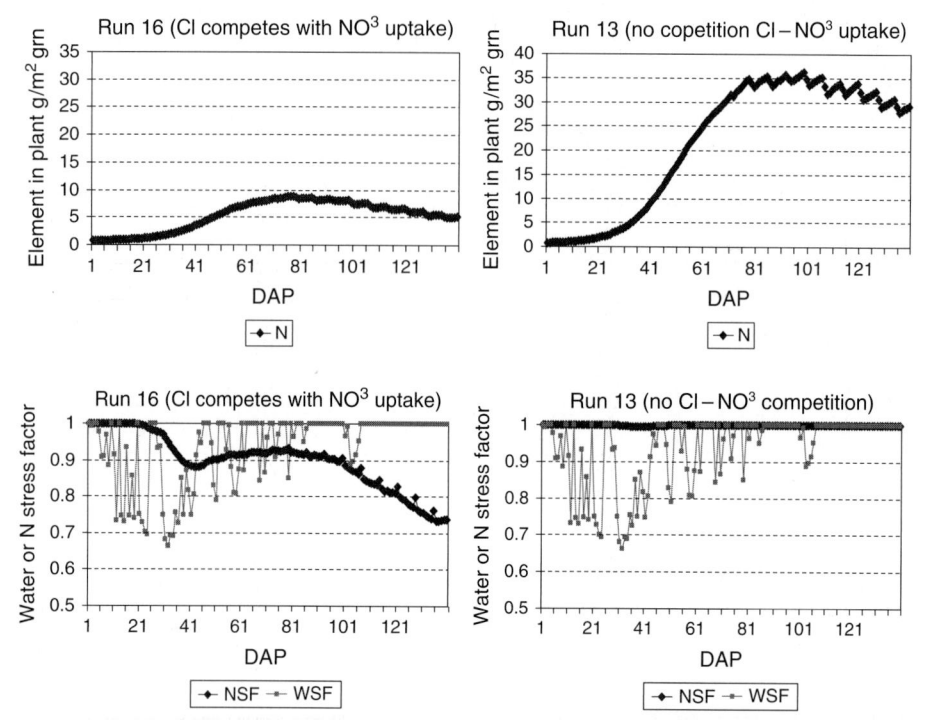

FIGURE 9.18 The effect of competitive Cl–NO$_3$ uptake on N absorption by tomato plants and N stress factor (NSF) (Cl$_{thresh}$ = 500 mg/L). The effect is visualized by comparing results when competitive uptake in accounted for and when it is disregarded. The water stress factor (WSF) was determined mainly by the assigned irrigation regime and weather conditions. Plants grew in a closed-loop irrigation system in the Besor, southern Israel. Calculations by the Greenman model (Bar-Yosef et al., 2004c).

The solution osmotic potential effect on water and N uptake and disposal was simulated for threshold Cl$^-$ concentration of 500 or 1225 mg L^{-1} (Fig. 9.19). The higher concentration created an osmotic potential of ~8 MPa. The tomato plant 'water retention curve' (water content θ vs. water potential ϕ in plant) which is used in Greenman allows very low values of ϕ (~30 MPa, Bar-Yosef et al., 1980), therefore water uptake (which is proportional to the total water potential difference inside and outside the root, Eq. [18]) could be sustained at a rate not appreciably smaller than at 500 mg L^{-1} Cl$^-$. Experimental ET and N uptake data indeed show (Table 9.34) that both values did not differ considerably when EC$_{thr}$ was 2.5 or 4.0 dS m^{-1} (EC$_{thr}$ of 4.0 corresponds to ~40 mM of monovalent electrolyte which is close to the 1225 mg L^{-1} Cl concentration). The model further predicts that in order to maintain a threshold Cl of 500 instead of 1250 mg L^{-1}, about 200 L water and 20 g N m^{-2} ground must be discarded (Fig. 9.19).

Both the model and the experimental results showed that maintaining the recirculated solution EC at 4.0 dS m^{-1} caused insignificant reduction in tomato yield (data not presented).

9.4 MODELLING THE CROP-RECIRCULATION SYSTEM

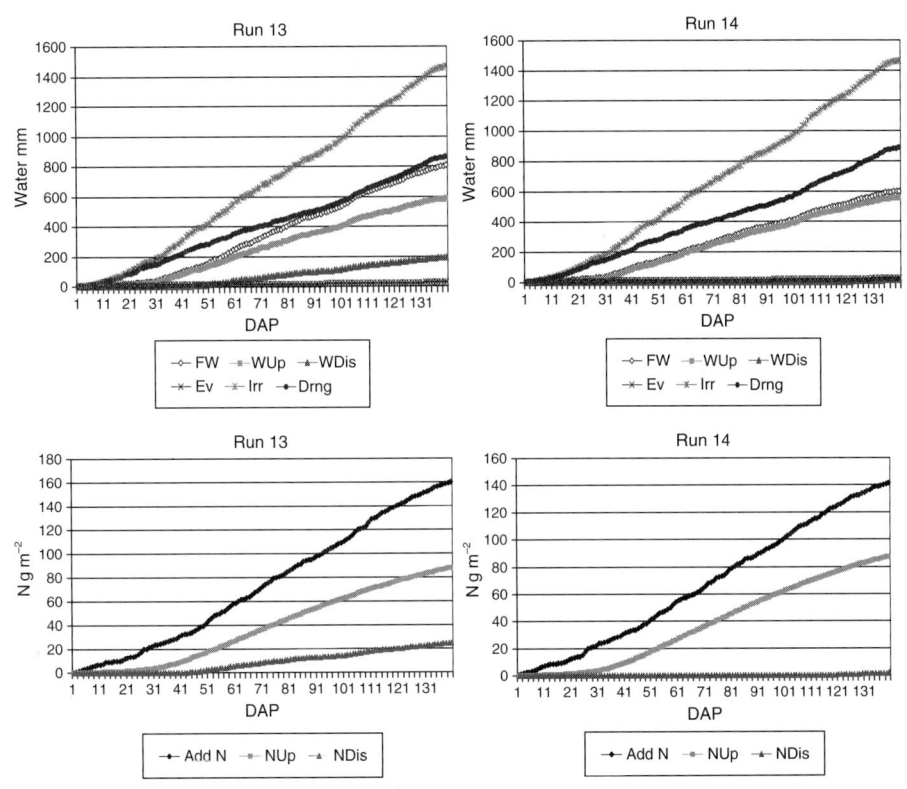

FIGURE 9.19 The simulated effect of threshold Cl concentration (*run* 13 = 500, *run* 14 = 1225 mg L^{-1} Cl) on water and N uptake (Wup and Nup) and disposal (Wdis and Ndis) by accounting for the osmotic potential mechanism and disregarding Cl–NO$_3$ competitive uptake. The applied irrigation (Irr) and simulated evaporation (Ev) and drainage (Drng) are also included in the top figures.

9.4.2.2 Effect of Target Nitrate Concentration

A simple criterion to evaluate the adequacy of a nutrient target concentration in irrigation water is to compare the ion concentration in the top substrate layer (approximately = the irrigation water concentration) with that of deeper substrate layers. This applies of course to top irrigation systems only. When loading the system with solution containing 130 mg L^{-1} NO$_3$-N and fertigating with 50 mg L^{-1} NO$_3$-N (=target concentration), the concentration at the bottom of the substrate exceeds the target concentration for the first 35 days of growth (Fig. 9.20). After that, consumption by plants increases and NO$_3$-N concentration at the bottom decreases below the target concentration. The concentration in the water reservoir is maintained at the target value because water and N are replenished to maintain the N concentration and total water volume in the system at specified values. When the target NO$_3$-N concentration is 130 mg L^{-1}, the concentrations in deep substrate layers are very similar to the target value between 50 and 100 days after planting, but later they exceed the target concentration (Fig. 9.20), indicating that target values should be adjusted with plant age.

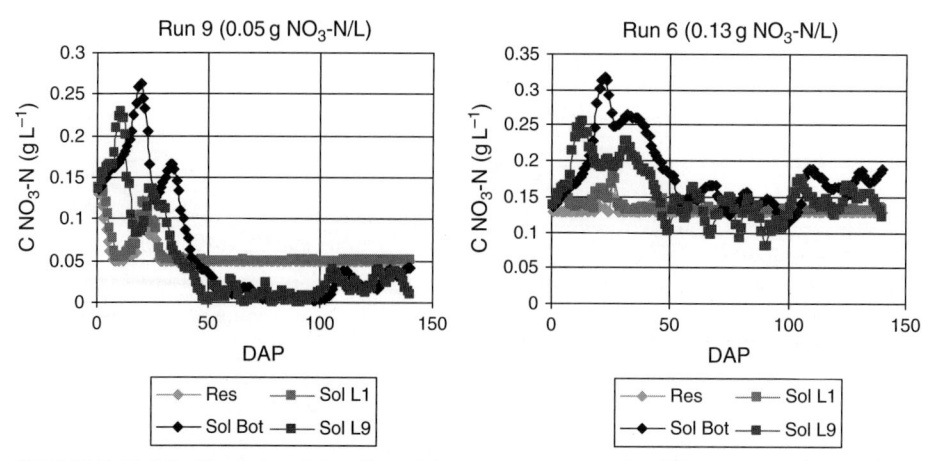

FIGURE 9.20 Simulation of the effect of nitrate target concentration (50 or 130 mg L^{-1} NO$_3$-N) on NO$_3$-N temporal concentration in recirculated solution in water reservoir, top 1 cm of substrate, mid-substrate layer (L9) and at the bottom of the substrate (Bot). Initial concentration in all parts of the system 130 mg L^{-1}.

The model predicts that increasing target concentrations (5, 25, 50 and 130 mg L^{-1}) result in elevated reduced-N concentration in plant leaves, the effect changing with plant age (Fig. 9.21). The secondary increase in N concentration on day ~80 is possible under ample nitrate supply because fruit harvesting decreases the fruit N

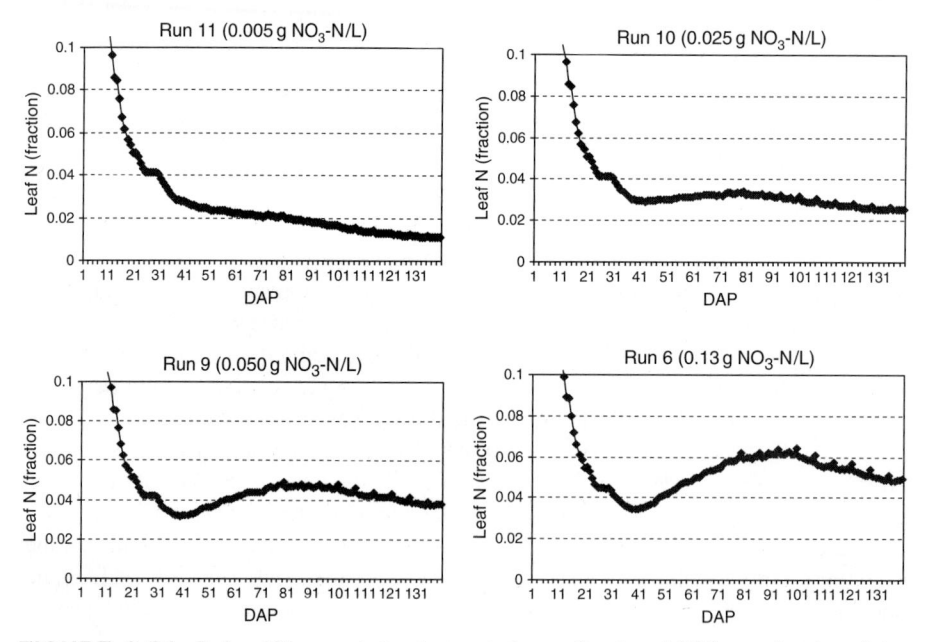

FIGURE 9.21 Reduced-N concentration in tomato leaves (fraction of DM) as a function of time and target nitrate concentration (5, 25, 50 and 130 mg L^{-1} NO$_3$) in recirculated solution. Simulation by Greenman (Bar-Yosef et al., 2004c).

9.4 MODELLING THE CROP-RECIRCULATION SYSTEM

sink power and more N is allocated to leaves. It is noted that the model accounts for nitrate reduction in the plant, so the reduced-N concentration results depend on the prescribed nitrate reduction rate constants. Frequent plant tissue analyses must be performed under realistic growth conditions to test the model predicted temporal leaf N concentrations.

9.4.2.3 Predicting Solution pH

Controlling pH of recycled solution was mentioned earlier as one of the major problems in managing closed-loop irrigation systems. Based on principles discussed earlier, the Greenman model simulates both the nitrification and the ammonium uptake effects on recirculated solution pH (Fig. 9.22). It is shown that for $NH_4:NO_3$ ratio 1:4 and 140 mg L^{-1} N the NH_4 uptake is more effective in decreasing solution pH than nitrification. The model predicts an increase in solution pH starting ~40 days after planting coinciding with the commencement of fruit filling and reduced DM allocation to roots. The short-term fluctuations in pH stem from the difference between fresh water and recirculated solution pH. Predicted variations in pH along the season resemble experimental observations from the Besor area, Israel (Fig. 9.22, bottom), so it is expected that when better calibrated the model might be helpful in choosing ammonium nitrate ratios and concentrations facilitating not only target N uptake rates but also target solution pH.

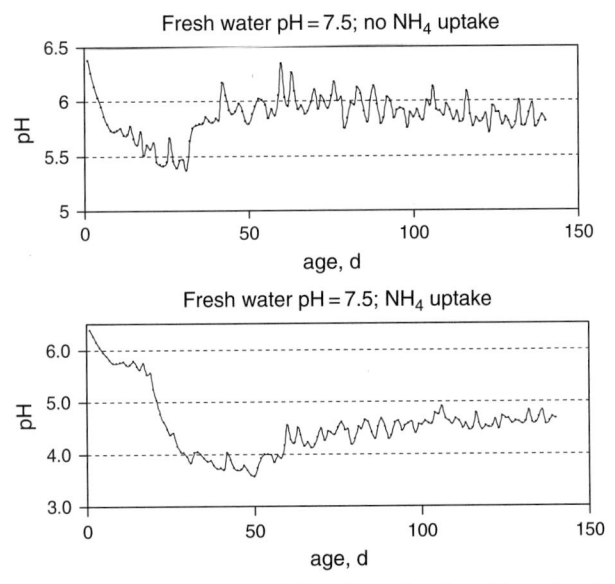

FIGURE 9.22 Simulation of recirculated solution pH as a function of days after planting (age). Initial pH is 6.5, fresh water pH is 7.5; $NH_4:NO_3 = 1:4$. Top: disregarding NH_4 uptake effect on H^+ excretion by roots (nitrification driven pH change only); middle: disregarding nitrification effect (NH_4 uptake driven pH change only). The third graph (bottom) depicts experimental weekly data obtained for tomato in closed-loop irrigation system in the Besor area, Israel (derived from Reshef et al., 2003).

9.5 OUTLOOK: MODEL-BASED DECISION-SUPPORT TOOLS FOR SEMI-CLOSED SYSTEMS

One of the major potential contributions of models is in their use for deriving management recommendations, and in the context of this chapter, optimizing the control of semi-closed greenhouses. To transform a model into a decision-support system (DSS), two essential components must be incorporated in it: (i) DM and uptake target functions according to which crop must develop in order to give an anticipated yield. (ii) An 'expert-driven interface' that analyses deviations between

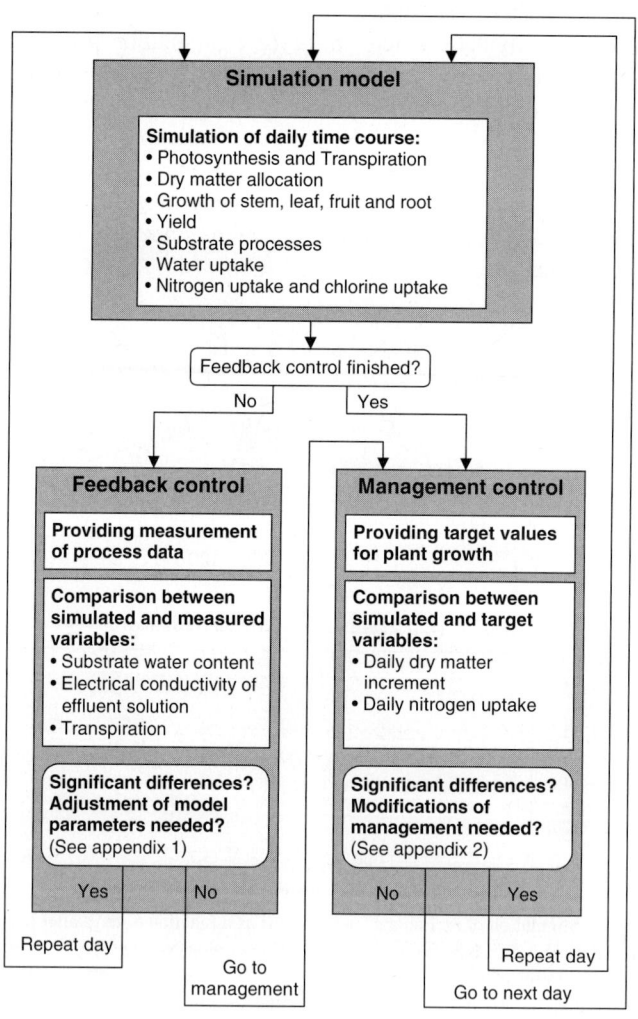

FIGURE 9.23 The Decision-Support System (DSS) used by Bar-Yosef et al., 2004.

model predictions and target values and recommends modifications in management rules to minimize the gap. Two secondary prerequisites are (i) developing a hardware–software package allowing real-time acquisition of weather and EC and pH data to the DSS. (ii) A feedback mechanism that evaluates model's predictions based on real-time comparison between calculated and measured data, for example solution pH or daily transpiration. An example of such a system is presented in Fig. 9.23 (Bar-Yosef et al., 2004). A DSS by Marcelis et al. (2003) is driven by similar considerations but it relies more on plant-related variables, for example plant height and stem expansion.

Greenhouse DSSs, like the above, are designed to achieve automatic control of cooling or heating the greenhouse, CO_2 enrichment, relative humidity, harvest and leaf pruning timing, and irrigation–fertilization management, including quantities and frequency of application and permitted threshold salinity and disposal. The DSS also allows evaluating pre-planting decisions like substrate type and volume of water reservoir. It is hard to assess how much time will be needed until currently developed DSS will become operative under real world conditions. Some engineering problems still exist in executing in real time even simple model recommendations, for example 'increase irrigation frequency by 50 per cent and report back'. Another difficulty in applying DSS is lack of validation under commercial production and feedback from growers, extension personnel and associated scientists.

ACKNOWLEDGEMENT

The cooperation, devotion and skill of the following researchers, extension specialists and technicians, who participated in the 'greenhouse drainage recycling' project supported by the Ministry of Agriculture, Israel, is highly appreciated: **S. Kramer**, Field Extension Specialist, Ministry of Agriculture, Israel; **E. Matan**, Head, The Besor Greenhouse Experiment Station and Southern Israel R&D; **D. Shmuel, Irit Dori, Hana Yehezkhel** and **S. Cohen**, The Besor Greenhouse Experiment Station researchers; **Dr G. Kritzman** and **Dr Svetlana Fishman**, phytopathology and crop modelling researchers at the Agricultural Research Organization (ARO), Bet-Dagan, Israel; **Irit Levkovich, Shoshana Soriano** and **T. Markovitch**, technicians at the ARO; and **G. Reshef** and **M. Bruner**, Vegetable and Field Extension Specialists, respectively, Ministry of Agriculture, Israel.

The partial supported by **GIF** (German-Israeli foundation for scientific research) is greatly appreciated.

APPENDIX

TABLE A.1 Summary of Studies Conducted in Closed-Loop Irrigation Systems and Cited in This Work (All Done in Israel)

Crop	Cultivar	Growth period	Location[a]	Substrate type[b] (L pl^{-1})	Variables[c]	Source
Pepper	Mazurka	15/9/97–27/4/98	Bet-Dagan	RW 4 Pumice 35	EC_{thr}	Bar-Yosef et al., 1999
	Celica	1/1/02–28/7/02	Besor	Perlite 2	EC_{thr}	Reshef et al., 2002a
Pepper	Cuby	30/8/99–15/5/00	Bet-Dagan	Perlite 14	Irr freq $C_{fertilizer}$	Bar-Yosef et al., 2000
Pepper	Celica	1/9/03–15/7/04	Besor	Perlite 3,6,15	EC_{thr} climate $V_{substrate}$	Bar-Yosef et al., 2005
Tomato	870	1/9/02–15/7/03	Besor	Perlite	EC_{thr}	Reshef et al., 2003
Tomato	870	1/9/04–15/7/05	Besor	Perlite 3,6,15	Climate N conc $V_{substrate}$	Reshef et al., 2006
Lettuce	9273	23/4/01–30/5/01	Bet-Dagan	RW 4	EC_{thr}	Bar-Yosef et al., 2002
Cucumber	Ringo 128	19/8–20/10/02	Bet-Dagan	RW 6	EC_{thr}	Bar-Yosef et al., 2003
Cucumber	AV-36	7/4/03–30/5/03	Bet-Dagan	Perlite 10, 40	$V_{substrate}$	Bar-Yosef et al., 2004b
Musk-melon	5093	8/3/99–15/6/99	Bet-Dagan	RW 12	EC_{thr}	Bar-Yosef et al., 2000
Straw-berry[d]	Tamar 328	20/9/01–15/5/02	Besor	Coco-fibre 0.8	EC_{thr} Irr freq	Matan et al., 2003
Gypsophila	Yukinko	18/5/98–4/11/98	Bet-Dagan	Perlite 17 Pumice 17	EC_{thr} Substrate	Bar-Yosef et al., 1999
Rose	Long Jaguar	1/1/01–31/12/01	Arava	Tuff 8.5	EC_{thr} $EC_{amplitude}$	Kramer et al., 2002
Rose	Mercedes	10/12/01–1/7/03	Besor	Tuff 8.5	EC_{thr}, climate	Bar-Yosef et al., 2003
Lisianthus	Eko champain	6/10/01–15/6/02	Besor	Perlite 1.6	EC_{thr}	Dori et al., 2002
Lisianthus	Eko champain	6/10/02–15/6/03	Besor	Perlite 1.6	EC_{thr}	Dori et al., 2003a
Solidago	Tara	16/7/01–16/7/02	Besor	Perlite 0.8	EC_{thr}	Dori et al., 2003b
Solidago	Tara	23/10/03–8/8/04	Besor	Perlite 0.8	Target N Concn.	Dori et al., 2005

[a] Bet-Dagan – coastal plane central Israel; Besor – South-western Israel; Arava – South-eastern Israel.
[b] RW = stone wool, produced in Israel; perlite = perlite 2 for agriculture; tuff = 0–8 M.
[c] EC_{thr} = threshold EC for solution disposal; $V_{substrate}$ = substrate volume effect; Irr freq = irrigation frequency effect.
[d] Plants were grown in hanging containers 1, 0.1 and 0.15 m long, wide and deep, respectively.

REFERENCES

Adams, P. (2002). Nutritional control in hydroponics. In *Hydroponic Production of Vegetables and Ornamentals* (D. Savvas and H. Passam. eds). Athens: Embryo Publications, pp. 211–261.

Adams, F. and Moore, B.L. (1983). Chemical factors affecting root growth in subsoil horizons of coastal plain soils. *Soil Sci. Soc. Am. J.*, **47**, 99–107.

Aiken, R.M. and Smucker, A.J.M. (1996). Root system regulation of whole plant growth. *Annu. Rev. Phytopathol.*, **34**, 325–346.

Baas, R. and van der Berg, D. (1999). Sodium accumulation and nutrient discharge in recirculation systems: a case study with roses. *Acta Hort.* (ISHS), **507**, 157–164.

Baas, R. and van der Berg, D. (2004). Limiting nutrient emission from a cut rose closed system by high flux irrigation and low nutrient concentrations? *Acta Hort.* (ISHS), **644**, 39–46.

Bar-Yosef, B. (1999). Advances in fertigation. *Adv. Agron.*, **65**, 1–79.

Bar-Yosef, B. and Sheikolslami, M.R. (1976). Distribution of water and ions in soils irrigated and fertilized from a trickler source. *Soil Sci. Soc. Am. J.*, **40**, 575–582.

Bar-Yosef, B., Fishman, S. and Klaering, H.P. (2004a). A model based decision support system for closed irrigation loop greenhouses. *Acta Hort.* (ISHS), **654**, 107–121.

Bar-Yosef, B., Levkovich, I. and Markovich, T. (2001c). Pepper response to leachate recycling in a greenhouse in Israel. *Acta Hort.* (ISHS), **548**, 357–364.

Bar-Yosef, B., Levkovich, I. and Markovich, T. (2000a). Gypsophila paniculata response to leachate recycling in a greenhouse in Israel. *Acta Hort.* (ISHS), **554**, 193–203.

Bar-Yosef, B. Levkovitch, I. and Markovitch, T. (2000b). Muskmelon response to water and fertilizer recycling in greenhouses. Gan Sadeh and Meshek Year 2000, 65–69. (Hebrew).

Bar-Yosef, B. Levkovitch, I. Markovitch, T. (2001a). Lettuce response to water and fertilizer recycling in greenhouses. Report submitted to the Chief Scientist, Ministry of Agriculture, Israel. (Hebrew).

Bar-Yosef, B., Raviv, M. and Meiri, A. (1999). Biological and chemical aspects of water and fertilizers recycling in greenhouses: pepper, muskmelon and gypsophila. Final report 1996–1998 submitted to the Chief Scientist, Ministry of Agriculture, Israel. (Hebrew).

Bar-Yosef, B., Raviv, M. and Meiri, A. (2000c). Crops response to water and fertilizers recycling in greenhouses: pepper. Annual report submitted to the Chief Scientist, Ministry of Agriculture, Israel. (Hebrew).

Bar-Yosef, B., Sagiv, B. and Eliah, E. (1980). Tomato drip fertigation in greenhouses in the Besor area. Final Report submitted to the Ministry of Agriculture, Israel (Special publication No 775).

Bar-Yosef, B., Levkovich, I., Markovich, T. and Mor, T.Y. (2001b). Gypsophila response to drainage recycling in greenhouses. *Dapey Meida*, **17**, 47–52 (Hebrew).

Bar-Yosef, B., Cohen, Y., Kritzman, G. and Matan, E. (2003a). Integrated crop response to solution recycling, greenhouse climate and root disease: cucumber. Report submitted to the Chief Scientist, Ministry of Agriculture, Israel. (Hebrew).

Bar-Yosef, B., Cohen, Y., Kritzman, G. and Matan, E. (2003b). Integrated crop response to solution recycling, greenhouse climate and root disease: roses and cucumber. Report submitted to the Chief Scientist, Ministry of Agriculture, Israel. (Hebrew).

Bar-Yosef, B., Cohen, Y., Kritzman, G. and Matan, E. (2004b). Integrated crop response to solution recycling, greenhouse climate and root disease: pepper. Second year progress Report submitted to the Chief Scientist, Ministry of Agriculture, Israel. (Hebrew).

Bar-Yosef, B., Cohen, Y., Kritzman, G. and Matan, E. (2005). Integrated crop response to solution recycling, greenhouse climate and root disease: pepper, roses and cucumber. Report submitted to the Chief Scientist, Ministry of Agriculture, Israel.

Bar-Yosef, B., Raviv, M., Meiri, A. and Ganmore, R. (2002). Crops response to solution recycling in greenhouses in Israel: pepper and lettuce. Final report 2000–2002 submitted to the Chief Scientist, Ministry of Agriculture, Israel. (Hebrew).

Bar-Yosef et al.(2004c). Integrated crop response to solution recycling, greenhouse climate and root disease: cucumber. Second year Report submitted to the Chief Scientist, Ministry of Agriculture, Israel. (Hebrew).

Bar-Yosef, B., Silber, A., Markovitch, T., et al. (2000d). Wax flower response to N fertilization via the water, irrigation interval and Fe foliar fertilization. III. Water and nutrients status in soil. *Dapey Meida* **14**, 61–65 (Hebrew).

Benoit, F. and Ceustermans, N. (2001). Impact of root cooling on blossom end rot in soilless paprika. *Acta Hort.* (ISHS), **548**, 319–326.

Bohnert, H.J., Nelson, D.E. and Jensen, R.G. (1995). Adaptations to environmental stresses. *Plant Cell*, **7**, 1099–1111.

Boyle, F.W., Jr. and Lindsay, W.L. (1986). Manganese phosphate equilibrium relationships in soils. *Soil Sci. Soc. Am. J.*, **50**, 588–593.

Bowen, G.B. (1991). Soil temperature, root growth and plant function. In *Plant Roots: The Hidden Half* (Y. Waisel, A. Eshel, and U. Kafkafi, eds). New York: Marcel Dekker Inc., pp. 309–330.

Bravo, F.P. and Uribe, E.G. (1981). Temperature dependence of the concentration kinetics of absorption of phosphate and potassium in corn roots. *Plant Physiol.*, **67**, 815–819.

Brooke, L.L. (1990). From dessert, to laboratory, to backyard. Aero-hydroponics – the hydroponic method of the future. *The Growing Edge*, **2**(1), 25.

Carmassi, G., Incrocci, L., Malorgio, M., Tognoni, F. and Pardossi, A. (2003). A simple model for salt accumulation in closed loop hydroponics. *Acta Hort.* (ISHS), **614**, 149–154.

Caron, J. and Elrick, D. (2005a). Measuring the unsaturated hydraulic conductivity of growing media with a tensiometer disc. *Soil Sci. Soc. Am J.*, **69**, 783–793.

Caron, J.E., Elrick, D., Beeson, R. and Boudreau, J. (2005b). Defining critical capillary rise properties for growing media in nurseries. *Soil Sci. Soc. Am J.*, **69**, 794–806.

Caron, J., Riviere, L.M., Charpentier, S., et al. (2002). Using TDR to estimate hydraulic conductivity and air entry in growing media and sand. *Soil Sci. Soc. Am J.*, **66**, 373–383.

Chaerle, L. and Van Der Straeten, D. (2001). Seeing is believing: imaging techniques to monitor plant health. *Biochimica et Biophysica Acta.*, **1519**, 153–166.

Chung SoonKyung, Wonhea Kim, MiHyea Park, Yun Jun Park. (1998). Effects of winter root zone warming on the productivity and quality of cut rose (Rosa hybrida L.) in rockwool culture. *J. Korean Soc. Hortic.* **38**, 766–770.

Clarkson, D.T., Hopper, M.J. and Jones, L.H.P. (1986). The effect of root temperature on the uptake of nitrogen and the relative size of the root system in *Lolium perenne*. I. Solutions containing both NH_4^+ and NO_3^-. *Plant Cell Environ.*, **9**, 535–545.

Cohen, Y., Fuchs, M., Li, Y. et al. (2003). Reducing transpiration by an evaporation pad to delay solute accumulation in closed loop fertigation systems. *Acta Hort.* (ISHS), **609**, 173–179.

Cronk, J.K. and Fennessy, M.S. (2001). *Wetland Plants Biology and Ecology*. Boca Raton, FL, USA: Lewis Press.

Cutlip, M.B., Shacham, M. and Elly, M. (2005). Efficient Solution of Numerical Problems within Polymath and Excel. 7th World Congress of Chemical Engineering, Glasgow, Scotland, 10–14 July, 2005.

da Silva, F.F., Wallach, R. and Chen, Y. (1995). Hydraulic properties of rockwool slabs used as substrates in horticulture. *Acta Hort.* (ISHS), **401**, 71–75.

de Kreij, C. and Straver, N. (1988). Flooded-bench irrigation: effect of irrigation frequency and type of potting soil on growth of Codiaeum and on nutrient accumulation in the soil. *Acta Hort.* (ISHS), **221**, 245–252.

de Kreij, C., Voogt, W. and Baas, R. (1999). Nutrient solution and water quality for soilless cultures. Brochure 196. Research Station for Floriculture and glasshouse vegetables (PBG) Naaldwijk, The Netherlands.

Dori, I., Bar-Yosef, B., Shmuel, D., et al. (2002). Crops response to solution recycling in greenhouses in the Besor area, Israel: Lisianthus (1st year). Submitted to the Chief Scientist, Ministry of Agriculture, Israel. (Hebrew).

Dori, I., Bar-Yosef, B., Shmuel, D., et al. (2003a). Crops response to solution recycling in greenhouses in the Besor area, Israel: Lisianthus (2nd year). Submitted to the Chief Scientist, Ministry of Agriculture, Israel. (Hebrew).

References

Dori, I., Bar-Yosef, B., Shmuel, D., et al. (2003b). Crops response to solution recycling in greenhouses in the Besor area, Israel: Solidago (2nd year). Submitted to the Chief Scientist, Ministry of Agriculture, Israel. (Hebrew).

Dori, I., Bar-Yosef, B., Shmuel, D., et al. (2005). Crops response to solution recycling in greenhouses in the Besor area, Israel: Solidago (3rd year). Submitted to the Chief Scientist, Ministry of Agriculture, Israel. (Hebrew).

DSSAT – The Decision Support System for Agrotechnology Transfer (V4 version). (2004). International Consortium for Agricultural Systems Application. http://www.icasa.net/dssat/

Duffy, B.K. and Defago, G. (1999). Macro and microelement fertilizers influence the severity of Fusarium crown and root rot of tomato soilless production system. *Hort. Science.*, **34**, 287–291.

Ehret, D.L., Lau, A., Bittman, S. et al. (2001). Automated monitoring of greenhouse crops. *Agronomie.*, **21**, 403–414.

Engelhard, A.W. (1989). *Soilborne Plant Pathogens: Management of Diseases with Macro- and Micro- Elements.* St. Paul, Minn: APS Press.

Engels, C. (1993). Differences between maize and wheat in growth related nutrient demand and uptake of potassium and phosphorus at suboptimal root zone temperatures. *Pl. Soil*, **150**, 129–138.

Engels, C., Munkle, L. and Marschner, H. (1993). Effect of Zone Temperature and Shoot Demand on Uptake and Xylem Transport of Macronutrients in Maize (*Zea mays* L.). *J. Exp. Bot.*, **249**, 537–547.

ERS (Economic Research Service), USDA. (2005). The economics of food, farming, natural resources and rural America. http://www.ers.usda.gov.

Feigin, A. (1985). Fertilization management of crops irrigated with saline water. *Pl. Soil*, **89**, 285–299.

Fermino, H. and Kampf, A. (2005). Considerations about the packing density of growing media prepared under increasing levels of humidity. *Acta Hort.* (ISHS), **697**, 147–151.

Fonteno, W.C. (1993). Problems and considerations in determining physical properties of horticultural substrates. *Acta Hort.* (ISHS), **342**, 197–204.

Fried, M. and Broashard, H. (1967). The soil – plant System. Academic Press, New York.

Friedman, S. (1999). Bacteria transport in saturated and unsaturated soils. *Agric. Res. Israel*, **10**, 99–146. (Hebrew).

Ganmore-Neumann, R. and Kafkafi, U. (1980). Root temperature and percentage NO_3/NH_4 effect on tomato plant development. I. Morphology and growth. *Agron. J.*, **72**, 758–761.

Ganmore-Neumann, R. and Kafkafi, U. (1983). The effect of root temperature and NO_3^-/NH_4^+ ratio on strawberry plants. I. Growth, flowering and root development. *Agron. J.*, **75**, 941–947.

General Electric. (2005). Water and Process Technologies. http://www.gewater.com

Giuffrida, F., Lipari, V. and Leonardi, C. (2003). A simplified management of closed soilless cultivation systems. *Acta Hort.* (ISHS), **614**, 155–160.

Goodwin, P.B., Murphy, M., Melville, P. and Yiasoumi, W. (2003). Efficiency of water and nutrient use in containerised plants irrigated by overhead, drip or capillary irrigation. *Aust. J. Exp. Agric.*, **43**, 189–194.

Grauer, U.E. and Horst, W.J. (1992). Modelling cation amelioration of Al phytotoxicity. *Soil Sci. Soc. Am. J.*, **56**, 166–172.

Griffin, R.A. and Jurinak, J.J. (1973). Estimation of activity coefficients from the electric conductivity of natural aquatic systems and soil extract. *Soil Sci.*, **116**, 26–30.

Heinen, M. (1997). Dynamics of water and nutrients in closed recirculating cropping system in glasshouse horticulture with special attention to lettuce grown in irrigated sand beds. PhD thesis Wageningen University, 270 pp.

Herndon, J. and Cochlan, W.P. (2007). Nitrogen utilization by the raphidophyte *Heterosigma akashiwo*: growth and uptake kinetics in laboratory cultures. *Harmful Algae* **6**(2), 260–270.

Heuvelink, E.P., Bakker, M. and Stanghellini, C. (2003). Salinity effects on fruit yield in vegetable crops: a simulation study. *Acta Hort.* (ISHS), **691**, 133–140.

Hillel, D. (2003). *Introduction to Environmental Soil Physics.* New York: Academic Press.

Ho, L.C. and Adams, P. (1995). Nutrient uptake and distribution in relation to crop quality. *Acta Hort.* (ISHS), **396**, 33–44.

Hoagland, D.R. and Arnon, D.I. (1950). The water culture method for growing plants without soil. Circulation 347. University of California Experiment Station.

Hochmuth, G., Maynard, D., Vovrina, C., et al. (2004). Plant tissue analysis and interpretation for vegetable crops in Florida. University of Florida IFAS Extension. http://edis.ifas.ufl.edu/EP081.

Hodgman, C. (1949). *Handbook of Chemistry and Physics*. Cleveland, OH: CRC Press.

Ikeda, H., Tan, X.W., Ao, Y. and Oda, M. (2001). Effects of soilless medium on the growth and fruit yield of tomatoes supplied with urea and or nitrate. *Acta Hort.* (ISHS), **548**, 157–164.

Imas, P., Bar-Yosef, B., Kafkafi, U. and Ganmore-Neumann, R. (1997). Release of carboxylic anions and protons by tomato roots in response to ammonium/nitrate ratio and pH in nutrient solution. *Pl Soil.*, **191**, 27–34.

Jones, J.B. (2005). *A Guide for the Hydroponic and Soilless Grower* (2nd edn). Portland, OR: Timber Press.

Kempkes, F. and Stanghellini, C. (2003). Modelling salt accumulation in closed system: a tool for management with irrigation water of poor quality. *Acta Hort.* (ISHS), **614**, 143–148.

Kirkby, E.A. and Mengel, K. (1967). Ion balance in different tissues of the tomato plant in relation to nitrate, urea or ammonium nutrition. *Plant Physiol.*, **42**, 6–14.

Kramer, S., Bar-Yosef, B. Tzubari, G. and Osherovitch, A. (2003). Rose response to N concentration in recycled solution and N source in fill solution of closed loop irrigation system in the Arava, Israel. Report submitted to the Research and Development, Arava (Hebrew).

Kramer, S., Bar-Yosef, B., Tzubari, G., et al. (2002). Rose response to water and fertilizer recycling in the Arava. Annual report submitted to the Research and Development Authority, Arava, Israel. (Hebrew).

Lang, H. J. and Elliott, G. C. (1991). Influence of ammonium:nitrate ratio and nitrogen concentration on nitrification activity in soilless potting media. *J. Am. Soc. Hort. Sci.*, **116**, 642–645.

Lieten, P., Horvath, J. and Asard, H. (2002). Effect of silicon on albinism of strawberry. *Acta Hort.* (ISHS), **567**, 361–364.

Lindsay, W.L. (1979). Chemical Equilibria in Soils. New York: Wiley-Interscience.

Lorenzo, H., Siverio, J.M. and Caballero, M. (2001). Salinity and nitrogen fertilization and nitrogen metabolism in rose plants. *J. Agric. Sci.* **137**, 77–84.

Lorenzo, H., Cid, M.S., Siverio, J.M. and Ruano, M.C. (2000). Effect of sodium on mineral nutrition in rose plants. *Annals Appl. Biol.* **137**, 65–72.

Luo, Y. and Millero, F.J. (2003). Solubility of rhodochrosite ($MnCO_3$) in NaCl solutions. *J. Solution Chem.*, **32**(5), 405–416.

Maas, E.V. and Hoffman, G.J. (1977). Crop salt tolerance – current assessment. *J. Irrig. Drain. Div. ASCE*, **103**, 115–134.

Macduff, J.H., Jarvis S.C. and Cockburn, J.E. (1994). Acclimation of NO_3^- fluxes at low root temperature by *Brassica napus* in relation to NO_3^- supply. *J. Exptl. Bot.*, **45**, 1045–1056.

Marcelis, L.F.M., Heuvelink, E. Goudriaan, J. (1998). Modelling biomass production and yield of horticultural crops. A review. *Sci. Hortic.*, **74**, 86–111.

Marcelis, L.F.M., De Groot, C.C., DelAmor, et al. (2003). Crop nutrient requirement and management in protected cultivation. *Proc.*, 525, New York, UK: The International Fertilizer Society, pp. 117–152.

Marcelis, L. F.M., Brajeul, E. Elings, A. et al. (2005). Modelling nutrient uptake of sweet pepper. *Acta Hort.* (ISHS), **691**, 285–292.

Marschner, H. (1995). *Mineral Nutrition of Higher Plants* (2nd edn). New York: Academic Press.

Marfa, O. and Orozco, R. (1995). Non saturated hydraulic conductivity of perlite – some effects on pepper. *Acta Hort.* (ISHS), **401**, 235–242.

Matan. E., Bar-Yosef, B., Hana, Y., et al. (2003). Crops response to solution recycling in greenhouses in Israel: strawberries (3rd year). Submitted to the Chief Scientist, Ministry of Agriculture, Israel. (Hebrew).

McPherson, M. and Hardgrave, M. (1993). HDC Progress Report. Grower (August 19). Sussex UK: Nexus Publishing, pp. 17–18.

Meiri, A. (1984). Plant response to salinity: Experimental and methodology and application to the field. Reassessment of water quality criteria for irrigation. In *Soil Salinity Under Irrigation* (I. Shainberg, and J. Shalhevet, eds). *Ecol. Stud. Anal. Synth.*, No 51, New York: Springer, pp. 284–297.

Mengel, K. and Kirkby, E. A. (2001). *Principles of Plant Nutrition*. (5th edn), New York: Springer.

Mizrahi, Y. (1982). Effect of salinity on tomato fruit ripening. *Plant Physiol.*, **69**, 966–970.

Mizrahi, Y., Taleisnik, E., Kagan Zur, V., et al. (1988). A saline irrigation regime for improving fruit quality without reducing yield. *J. Am. Soc. Hort. Sci.*, **113**, 202–205.

References

Morgan, J.J. (2005). Kinetics of reaction between O_2 and Mn(II) species in aqueous solutions. *Geochimica et Cosmochimica Acta.*, **69**(1), 35–48.

Mulkey, T.J., Kuzmanoff, K.M. and Evans, M.L. (1982). Promotion of growth and hydrogen ion efflux by auxins in roots of maize pretreated with ethylene biosynthesis inhibitors. *Plant Physiol.*, **70**, 186–188.

Nelson, P.V., Oh, Y.-M. and Cassel, D.K. (2004). Changes in physical properties of coir dust substrates during crop production. *Acta Hort.* (ISHS), **644**, 260–268.

Norvel, W.A. (1972) Equilibria of metal chelates in soil solution. In *Micronutrients in Agriculture* (J.J. Mortvedt, P.M. Giordano and W.L. Lindsay, eds.). Madison, WI: Soil Sci. Soc. Am., pp. 115–138.

Olsen, S.R. and Kemper, W.D. (1968). Movement of nutrients to plant roots. *Adv. Agron.*, **20**, 91–151.

Otten, W. (1994). Dynamics of water and nutrients for potted plants induced by flooded bench fertigation. Thesis, Agricultural University Wageningen, ISBN 90 5485 304 2, 115pp.

Peterson, T.A., Reinsel, M.O. and Krizek, D.T. (1991). Tomato (*Lycopersicon esculentum* Mill CV. Better Bush) plant response to root restriction. I. Alternation of plant morphology. *J. Exp. Bot.*, **42**, 1241–1249.

Plaut, Z. (1997). Irrigation with low quality water: effects on productivity, fruit quality and physiological processes of vegetable crops. *Acta Hort.* (ISHS), **449**, 591–597.

Pritchard, J., Barlow, P.W., Adam, J.S. and Tomas, A.D. (1990). Biophysics of the inhibition of the growth of maize roots by lowered temperature. *Plant Physiol.*, 93, 222–230.

Raviv, M., Krasnovsky, A., Medina, S. and Reuveni, R. (1998). Assessment of various control strategies for circulation of greenhouse effluents under semi-arid conditions. *J. Hortic. Sci. Biotech.* **73**, 485–491.

Raviv, M. and Blom, T. (2001). The effect of water availability and quality on photosynthesis and productivity of soilless-grown cut roses. *Scientia Hortic.*, **88**, 257–276.

Raviv, M., Lieth, J.H. and Wallach, R. (2001). The effect of root-zone physical properties of coir and UC mix on performance of cut rose (CV. Kardinal). *Acta Hort.* (ISHS), **554**, 231–238.

Raviv, M., Wallach, R., Silber, A. and Bar-Tal, A. (2002). Substrates and their analysis. In *Hydroponic production of vegetables and ornamentals* (D. Savvas, and H. Passam, eds). Athens: Embryo Publications, pp. 25–101.

Reluy, F.V., de Paz Becares, J.M., Zapata Hernandez, B.D. and Sanchez Diaz. J. (2004). Development of an equation to relate electrical conductivity to soil and water salinity in a Mediterranean agricultural environment. *Australian J. Soil Res.*, **42**, 381–388.

Resh, H.M. (2001). Hydroponic food production. *A definitive guidebook of soilless food-growing methods.* (6th edn). Beaverton, OR: Woodbridge Press Publ. Co.

Reshef, G., Bar-Yosef, B., Shmuel, D., et al. (2002a). Pepper response to solution recycling in the Besor. Report submitted to the Chief Scientist, Ministry of Agriculture, Israel. (Hebrew).

Reshef, G., Bar-Yosef, B., Shmuel, D., et al. (2002b). Pepper response to solution recycling in the Besor. South Israel Research and Development, 2001/2002 reports, p. 23. (21 p., Hebrew).

Reshef, G., Bar-Yosef, B., Shmuel, D., et al. (2003). Tomatoes response to solution recycling in the Besor. Report submitted to the Chief Scientist, Ministry of Agriculture, Israel. (Hebrew).

Reshef, G., Bar-Yosef, B., Shmuel, D., et al. (2006). Tomato response to solution recycling in the Besor. Report submitted to the Chief Scientist, Ministry of Agriculture, Israel. (Hebrew).

Romano, D., Paratove, A. and LiRosi, A. (2003). Plant density and container cell volume effects on Solanaceous seedling growth. *Acta Hort.* (ISHS), **614**, 247–253.

Romer, J. (1993). *Hydroponic Crop Production.* Kenthurst: Kangaroo Press P/L.

Runia, W.T. and Boonstra, S. (2004). UV-Oxidation Technology for Disinfection of Recirculation Water in Protected Cultivation. *Acta Hort.* (ISHS), **644**, 549–555.

Saure, M.C. (2001). Blossom-end rot of tomato (Lycopersicon esculentum Mill.) – A calcium or a stress related disorder. *Sci. Hortic.*, **90**, 193–208.

Savvas, D. and Passam, H. (eds). (2002). Hydroponic Production of Vegetables and Ornamentals. Athens: Embryo Publications.

Savvas, D. (2002). Nutrient solution recycling. In *Hydroponic Production of Vegetables and Ornamentals* (D. Savvas, and H. Passam, eds). Athens: Embryo Publications, pp. 299–343.

Scaife, A. and Bar-Yosef, B. (1995). *Fertilizing for High Yield and Quality Vegetables: Nutrient and Fertilizer Management in Field Grown Vegetables.* IPI-Bulletin No 13. Basel, Switzerland: International Potash Institute, 104 p.

Schwab, A.P. and Lindsay, W.L. (1983). Effect of redox on the solubility and availability of manganese in a calcareous soil. *Soil Sci. Soc. Am. J.*, **47**, 217–220.

Schwartz, M. (1995). *Soilless Culture Management*. Berlin: Springer Verlag.

Scott, D.T, Mcnight, D.M., Voelker, B.M. and Herncir, D.C. (2002). Redox processes controlling manganese fate and transport in a mountain stream. *Environ. Sci. Tech.*, **36**(3), 453–459.

Soffer, H. and Burger, D.W. (1988). Effects of dissolved oxygen concentrations in aero-hydroponics on the formation of adventitious roots. *J. Am. Soc. Hortic. Sci.*, **113**, 218–221.

Sonneveld, C. (2000). Effect of salinity on substrate grown vegetables and ornamentals in greenhouse horticulture. Thesis Wageningen University, Netherlands, 151 pp.

Sonneveld, C. and de Kreij, C. (1999). Response of cucumber (*Cucumis sativus L.*) to an unequal distribution of salts in the root environment. *Pl. Soil.*, **209**, 47–56.

Sonneveld, C. and Voogt, W. (1990). Response of tomatoes (*Lycopersicon esculentum*) to an unequal distribution of nutrients in the root environment. *Pl. Soil.*, **124**, 251–256.

Sonneveld, C. and Welles, G.W.H. (1988). Yield and quality of rockwool grown tomatoes as affected by variations in EC values and climatic conditions. *Pl. Soil.*, **111**, 37–42.

Sonneveld, C., Voogt, W. and Spaans, L. (1999a). A universal algorithm for calculation of nutrient solution. *Acta Hort.* (ISHS), **481**, 331–339.

Sonneveld, C., Baas, R., Nijssen, H.M.S. and de Hoog, J. (1999b). Salt tolerance of flower crops grown in soilless culture. *J. Pl. Nutr.*, **22**, 1033–1048.

Stamatakis, A., Papadantonakis, N., Savvas, D., et al. (2003). Effects of silicon and salinity on fruit yield and quality of tomato grown hydroponically. *Acta Hort.* (ISHS), **609**, 141–147.

Stanghellini, C., Kempkes, F., Pardossi, A. and Incrocci, L. (2004). Closed water loop in greenhouses: effect of water quality and value of produce. *Acta Hort.* (ISHS), **691**, 233–241.

Subbarao, Y.V. and Ellis, R., Jr. (1975). Reaction products of polyphosphates and orthophosphates with soil and influence on uptake of phosphorus by plants. *Soil Sci. Soc. Am. Proc.*, **39**, 1085–1088.

Terada, M., Goto, T., Kageyama, Y. and Konishi, K. (1996). Effect of potassium concentration in nutrient solution on growth and nutrient uptake of rose plants. *Acta Hort.* (ISHS), **440**, 366–370.

Tu, J.C., Papadopoulos, A.P., Hao, X. and Zheng, J. (1999). The relationship of Pythium root rot and rhizosphere microorganisms in a closed circulating and an open system in rockwool culture of tomato. *Acta Hort.* (ISHS), **481**, 577–583.

Urrestarazu, M., Salas, M.C., Gómez, A. and Valera, D. (2003). Cucumber Crop Response to Heated Nutrient Solution in Soilless Crop. *Acta Hort.* (ISHS), **614**, 649–651.

Van Os, E.A., van Kuik, F.J., Th. Runia, W. and van Buuren, J. (1998). Prospects of slow sand filtration to eliminate pathogens from recirculating solutions. *Acta Hort.* (ISHS), **458**, 377–382.

Voogt W. and Sonneveld, C. (1996). Nutrient management in closed growing systems for greenhouse production. In *Plant Production in Closed Ecosystems and Automation, Culture and Environment*. Dordrecht: Kluwer Academic Press, pp. 83–102.

Wallach, R. and da Silva, F.F. (1992). Unsaturated hydraulic characteristics of tuff (scoria) from Israel. *J. Am. Soc. Hort. Sci.*, **117**, 415–421.

Wallach, R. and Raviv, M. (2005). The dependence of moisture-tension relationship and water availability on irrigation frequency in containerized growth medium. *Acta Hort.* (ISHS), **697**, 293–300.

Weast, R.C. (ed.) (1977). *CRC Handbook of Chemistry and Physics*, Cleveland, OH: CRC Press.

Weiss, R., Alm, J., Laiho, R. and Laine, J. (1998). Modelling moisture retention in peat soils. *Soil Sci. Soc. Am. J.*, **62**, 305–313.

Xu, G. and Kafkafi, U. (2001). Nutrient supply and container size effects on flowering, fruiting, assimilate allocation and water relations of sweet pepper. *Acta Hort.* (ISHS), **554**, 113–120.

Yermiyahu, U., Finegold, I., Keren, R., et al. (2003). Response of pepper to boron and salinity under greenhouse conditions. *Acta Hort.* (ISHS), **609**, 149–154.

Zaw, M. and Chiswell, B. (1999). Iron and manganese dynamics in lake water. *Water Res.*, **33**(8), 1900–1910.

10

PATHOGEN DETECTION AND MANAGEMENT STRATEGIES IN SOILLESS PLANT GROWING SYSTEMS

JOEKE POSTMA, ERIK VAN OS
AND PETER J.M. BONANTS

10.1 Introduction
10.2 Detection of Pathogens
10.3 Microbial Balance
10.4 Disinfestation of the Nutrient Solution
10.5 Synthesis: Combined Strategies
Acknowledgements
References

10.1 INTRODUCTION

10.1.1 INTERACTION BETWEEN GROWING SYSTEMS AND PLANT PATHOGENS

Intensive greenhouse cropping systems face a variety of risks for the outbreak of plant diseases. In field production, crop rotation is used, in part, to deal with root-borne diseases; in soilless greenhouse production this is generally not feasible. In soil systems, methyl bromide and other soil fumigants have been used (and in some countries still are) extensively to overcome problems with root diseases. However, a phase-out of the use of methyl bromide in agriculture was assigned in the Montréal Protocol for 2005. The use of methyl bromide in greenhouse production systems was eliminated in the Netherlands in the 1980s by the adoption of soilless systems. Soilless culture systems have, indeed, proven to be a successful alternative to the use of methyl bromide for economic production of several crops (Braun and Supkoff, 1994). Other advantages of soilless systems are a higher production, energy conservation, better control of growth and the grower becomes independent of soil type and quality (van Os, 1999).

Soilless production systems afford the possibility to save water and nutrients, as well as to avoid pollution of ground and surface water by the excess of fertilizers in the drain water, by reusing the irrigation water. Systems which reuse all effluent from the root zone are 'closed systems'. One of the main barriers to recirculation of irrigation water, however, is the presence of plant disease causing-organisms in the drainage water. Failure to manage such organisms emanating from even one infected plant will subject the entire crop to significant risks. Already in 1996, an overview was given on the *Fusarium*, *Pythium* and *Phytophthora* species that could spread through the growing system by recirculation of the irrigation water (Rattink, 1996).

10.1.2 DISEASE-MANAGEMENT STRATEGIES

Chemical control of plant pathogens is a robust and effective method to control pests and diseases, but it has environmental drawbacks and human safety risks. For the safety of the consumers, a period of several days should elapse between the application of a pesticide and harvesting a food crop. In greenhouse crops, where vegetables and flowers are often harvested daily, re-entry time regulations can be a serious practical disadvantage. The ability to apply chemical crop protection agents depends on regulation by the authorities and differs by country. In general, the number of registered chemical crop protection agents is more limited for greenhouse crops than for the large field crops, since registration costs are high for the relatively small greenhouse market (Paulitz and Bélanger, 2001).

The risks of root-borne diseases can be reduced by various non-chemical strategies, such as the use of (partial) resistant varieties, sanitation and cultural practices, and biological control. In greenhouse systems, climate and water regime can be managed, offering possibilities to control diseases. A unique feature of soilless systems is the possibility of starting a production cycle completely free of pathogens, and the eradication of pathogens present in the recirculated irrigation water with various disinfestation techniques. Sanitary practices to avoid pathogen infestation should be combined with preventive measures to make crops in closed systems less susceptible to pathogen outbreaks. Furthermore, the potential of modern detection methods to identify and treat the relevant pathogens is improving continuously.

10.1.3 OVERVIEW OF THE CHAPTER

This chapter describes pesticide-free strategies to minimize the risks of root-borne diseases in soilless systems, since these strategies are unique for soilless plant growing systems. Recirculation of the nutrient solutions offers an ideal opportunity for the detection of micro-organisms. Moreover, recirculation of nutrient solutions also allows treatment of pathogens in it. Furthermore, the start with fresh and pathogen-free substrates is a challenge for the design of microbiologically optimized growth media which are more suppressive against plant diseases. Section 10.2 deals with detection of pathogens in the nutrient solution. Recent technical developments of detection techniques allow the design of specific and quantitative detection of pathogens in nutrient solutions (Bonants et al., 2005; Lievens et al., 2005; Okubara et al., 2005). The implementation of these techniques in horticulture has just started and will give many new possibilities in the near future. Section 10.3 deals with the development of a microbially balanced system where we present the concept of balancing populations of various microbes so as to keep them all under better control. In that section, we describe the current knowledge about microbes in soilless production systems, application of biocontrol agents, factors stimulating a more balanced microbial system, as well as disease suppression by various substrates.

Over the years, various methods and technologies have been developed for disinfestation of the recirculated nutrient solution which are described in Sect. 10.4, grouped by the following approaches: filtration, heat treatment, oxidation (e.g. hydrogen peroxide, ozone), electromagnetic radiation (e.g. UV), active carbon adsorption and copper ionization. The challenge is to combine the available strategies into environmentally and economically sound soilless plant production systems with low risks for pathogen outbreaks. Soilless systems have the potential of creating a balance between a pathogen-free start and a suppressive microflora. In Sect. 10.5, we provide examples of such combined strategies and present an outlook to the future.

10.2 DETECTION OF PATHOGENS

10.2.1 DISEASE POTENTIAL IN CLOSED SYSTEMS

Traditionally, the strategy in greenhouse horticulture has been to keep the growing systems as 'clean' as possible, by using pathogen-free propagation material and substrates and by using disinfestation and other sanitation techniques. Survival of soil-borne pathogens such as *Fusarium oxysporum, Verticillium dahliae,* nematodes and many others in soil, especially in the deeper soil layers from which they cannot be eradicated, can be circumvented by using soilless substrates which are regularly replaced. However, changing from soil-based production to a soilless production system with recirculating nutrient solution may be conducive to outbreaks of other types of plant pathogens. For example, zoospore-producing organisms, such as *Pythium* spp. and *Phytophthora* spp., are well adapted to life in liquids, and pose a serious threat in soilless systems. *Phytophthora* and *Pythium* species, previously called fungi, are fungus-like organism classified in the Phylum *Oomycota.* They are taxonomically more related to brown algae than to fungi, but a clear and univocally classification of the oomycetes is still unavailable. They flourish in soilless substrates which involve free water (Stanghellini and Rasmussen, 1994). Zoospores actively swim to their hosts so that infection can occur within minutes. Multiplication of these pathogens is explosive under favourable conditions.

Contamination of soilless systems with pathogens can occur during the entire period of crop growth, since they can be brought in via the water supply, by air, insects or inadvertently by grower or equipment. Many plant pathogenic species can pose a threat for plants in soilless growing systems. This is illustrated by the long list of plant pathogens which have been detected from various water resources: 17 species of *Phytophthora*, 26 species of *Pythium*, 27 genera of fungi, 8 species of bacteria, 10 viruses, and 13 species of plant parasitic nematodes (Hong and Moorman, 2005). The type of pathogens that cause economic problems in soilless greenhouse systems is largely dependent on the cultivated crop, cultural and climatic conditions, and the possibility to avoid infestations with the pathogen.

Problems with pathogens in closed systems can also be expected if infected plant material (seeds, cuttings or young plants) is introduced into a clean system where the harmful organisms have a chance to proliferate in the absence of competition with a stable microflora. With the virtual impossibility of achieving sterility, or at least complete absence of pathogens, comes a finite amount of risk of pathogen infection. Management of this risk requires the ability to detect and monitor pathogens (qualitatively and quantitatively) so as to be able to predict the occurrence of disease symptoms and to implement pest and disease management strategies.

10.2.2 BIOLOGICAL AND DETECTION THRESHOLDS

The 'biological threshold of the pathogen' is defined as the amount of inoculum of the pathogen that must be present in the growing system to result in subsequent disease development. This inoculum can be present in various forms, such as mycelium, zoospores, conidia, resting spores and so on. For *Pythium* species, concentrations of less than one zoospore of *Pythium aphanidermatum* per millilitre irrigation water resulted in cucumber (*Cucumis sativum* L.) crops showing *Pythium* root rot (Fig. 10.1)

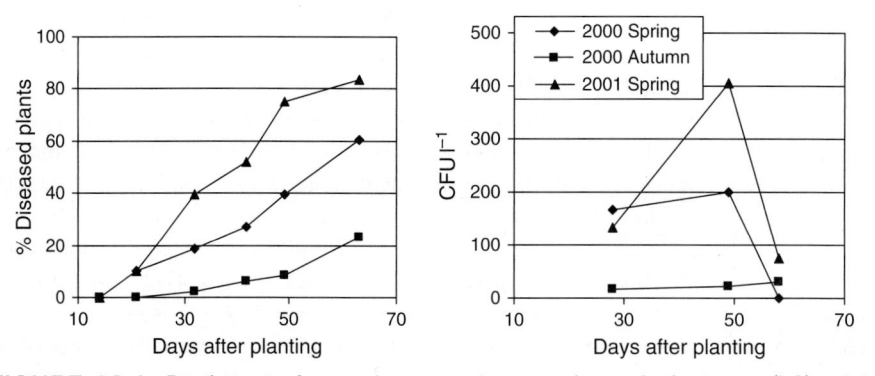

FIGURE 10.1 Development of root and crown rot in a cucumber production system (left) and the presence of *Pythium aphanidermatum* in the drain water (right). Populations were enumerated with plate counts on a specific medium (summarized data from the EU-project MIOPRODIS, FAIR CT98-4309).

(Postma et al., 2001). In this case 0.1–0.4 colony-forming units (CFU) per millilitre drain water resulted in a crop with 20–80 per cent of the plants showing crown rot symptoms (Postma et al., 2001). For other pathogens higher concentrations are needed for symptom development; for *Fusarium oxysporum*, for example, 10^3–10^4 CFU per gram potting soil is needed to obtain symptoms (Postma and Luttikholt, 1996). Moreover, concentrations of pathogens will not be evenly spread over the whole system. For example, propagules of the pathogen can be filtered out from the water by the rooting substrate or can settle by gravity to the bottom of the drain water tank (Rattink, 1996). Day-night rhythm of greenhouse temperature, crop activity and irrigation might influence pathogen growth and sporulation, causing fluctuations in spore release. Therefore, the spatial and temporal fluctuations of the pathogen should be taken into account for proper sampling.

Another complication is the fact that in horticultural practice combinations of pathogens may occur. Combined occurrence of pathogens might result in lower biological damage thresholds for each pathogen, particularly when pathogens have additive or synergistic effects. Few examples are plant pathogenic nematodes facilitating fungi such as *Verticillium*, *Fusarium*, *Pythium* and *Rhizoctonia* to penetrate the roots (Agrios, 1988), *Polymyxa* and *Olpidium* species being a vector of viruses (Agrios, 1988).

In summary, there is a lack of information on biological thresholds in greenhouse systems to predict yield loss for the specific pathogen–crop combinations. The purpose of detection and quantification of the different pathogens is to apply control measures in an early stage, but only if necessary. This could decrease pesticide use, environmental pollution and production costs. However, if a pathogen is present without causing damage, pesticide use should be avoided. Thus, biological threshold information on important host–parasite combinations in greenhouse systems, depending upon the mix of pathogens in the water, the growing conditions and the susceptibility of the crop, is needed urgently (Hong and Moorman, 2005).

The detection threshold of a method, that is the minimum quantity of a pathogen which can be detected with a certain detection method, should be lower than the biological threshold, in order to be a useful tool in risk assessment. With low biological thresholds, large volumes of samples should be analysed. For example, with aggressive pathogens such as *P. aphanidermatum*, samples should be in the range of 1 l. Such large samples can be concentrated in several ways: centrifugation, sieving or filtration, baiting with plant parts. More information on this subject can be found in the review of Hong and Moorman (2005). Samples should be taken as close as possible to the roots to obtain a representative population from the location where infections take place: that is, it is better to sample fresh leachate from the substrate, than to take a sample from the storage tank. Moreover, spores can sink to the bottom of the tank (Rattink, 1996). Shipment and storage time of the samples should be as short as possible, since living organisms are involved.

10.2.3 METHOD REQUIREMENTS FOR DETECTION AND MONITORING

Various aspects have to be considered and certain criteria have to be fulfilled for developing effective detection methods (Bonants et al., 2005):

- *Specificity*. The test should be very specific and should only detect the target pathogen and not closely related species; no false positives or false negatives should be observed.
- *Sensitivity*. The test should be sensitive in detecting low concentration of the pathogen.
- *Costs*. For routine testing, the test should be cost effective.
- *Complexity*. The availability of sufficient technical and personal expertise to perform the test should be considered.
- *Robustness*. The test should be robust, including a high repeatability and reproducibility.
- *High throughput*. Sometimes the test should be suitable for high throughput screening.
- *Rapidity*. The results obtained should be available within a short time.

10.2.4 DETECTION TECHNIQUES

Plant pathogens can be detected by several methods based on totally different strategies:

- direct observation of the pathogen in plant tissue or extracts by microscopic techniques;
- isolation of the pathogen using young plants, plant parts or specific media;
- targeting the antigenic compounds of the pathogen with serological methods;
- targeting the DNA (or RNA) of the pathogen with molecular methods.

The general principles and many examples of these methods for quantitative detection of different plant pathogens have been described in several books (e.g. Schots et al., 1994; Van Vuurde and Postma, 1996; Dehne et al., 1997). Table 10.1 gives

TABLE 10.1 Characteristics of the Basic Detection Strategies of Plant Pathogens

Method	Examples	Target	Type of organisms
Isolation	plant assays, (selective) media	infectious entities, colony-forming units	pathogens quickly showing symptoms, culturable bacteria and fungi
Microscopy	light microscope, transmission electron microscope	visual entities	nematodes and obligate pathogens, viruses
Serology	ELISA, IF, dipstick	antigens	viruses, some bacteria, few fungi
Molecular	PCR, quantitative PCR	DNA or RNA	fungi, bacteria, viruses

TABLE 10.2 Properties of Three Different Detection Methods for *Pythium* and *Phytophthora*

Criterion	Isolation: plate counts	Serology: dipstick	Molecular: PCR
Sensitivity (propagules/ml)	0.1–10	0.01–1	0.02–0.5
Specificity	part of genus	part of genus	species
Propagule type	culturable mycelium and zoospores	motile zoospores	DNA of living as well as dead targets
Quantitative	yes	partly	partly
Rapidity (days)	1–2	1	0.3

Source: Van Os and Bruins, 2004.

an overview of four basic detection strategies. As an illustration, properties of various detection methods available for *Pythium* and *Phytophthora* are compared in Table 10.2.

Isolation on selective media can be used for identification and detection of culturable pathogens. This involves inoculating sterilized culture media with extracts, concentrates of the pathogen or roots fragments, and culturing these over several days. Analysing physiological, biological and morphological characteristics requires specialized taxonomic expertise. Only culturable organisms can be detected in this way, making this methodology unsuitable for nematodes, viruses and obligate (fungal) pathogens. Moreover, these methods are usually time consuming and mostly not species specific. An alternative for the isolation on culture media is to use plant parts as baiting material, or young and sensitive plants to capture infectious plant pathogens.

Microscopy is used for direct observation of pathogens in plant material or in concentrated extracts. This method is specifically suitable for those pathogens that can not be cultured on a medium. Nematodes are usually extracted from soil or plant material, then concentrated, identified and enumerated by microscopy. For viruses, transmission electron microscopy is useful in case new types are involved for which no serological method is developed yet. Microscopic methods, in general, require specialized expertise to recognize morphological characteristics and are time consuming.

The introduction of serological (or immunological) techniques in the 1960s was a breakthrough. This detection strategy is based on the intrinsic antigenic properties of the outer surface of a pathogen. Most viruses, several bacteria and some fungi have such specific antigens, for which antibodies can be developed. At present, ELISA tests (enzyme-linked immunosorbent assay) are generally used in the agricultural sector and by commercial companies to detect the presence of bacteria and viruses in products such as potatoes, ornamentals and flower bulbs. This is a cheap, fast and reliable technique. Another serological technique applied to detect bacteria is called IF (immunofluorescence), where an antibody combined with a fluorescent dye reacts specifically with the antigens of the pathogen. A serological test which can be used on-site is the dipstick, a commercially available kit where the pathogen is first trapped on a membrane and then visualized by serological probes. Such dipsticks are available for several plant pathogens, including *Pythium* and *Phytophthora* spp. (Pettitt et al., 2002). The limitation of the use of serological methods is the presence of specific antigens on the surface of the target pathogen. Therefore, serological techniques are mainly used for plant pathogenic viruses and some bacteria. For other bacteria and most of the fungi, specific antigens are not present.

Molecular techniques, on the basis of nucleic acids (DNA or RNA), filled this gap. For all type of organisms, specific DNA or RNA fragments can be found. The introduction of amplification methods for DNA and RNA, and the increasing availability of sequence data, resulted in a large variation of new molecular detection and identification techniques since 1980s. Development of these techniques is still ongoing. In most methods, DNA or RNA amplification by the use of polymerase chain reaction (PCR) is the basis. Several examples can be mentioned (Bonants et al., 2005):

- PCR-based identification techniques of single isolates, for example ITS-RFLP, SCAR-PCR, AFLP, RAPD.
- Community profiling of mixed populations, for example DGGE or TGGE
- General detection methods using specific primers, PCR
- Quantitative detection methods using probes, for example Molecular Beacons or TaqMan probes
- Activity measurements can be based on mRNA, RT-PCR and NASBA.

For population dynamics studies, quantitative PCR (Q-PCR) is used more and more. This is a real-time PCR technique where fluorescence is used to monitor the accumulation of the PCR product after each PCR cycle. The fluorescence data are used to extrapolate the amount of target DNA present. A recent overview of this technique and its applications is given by Okubara et al. (2005).

For routine detection of a large range of pathogens, several companies and institutes are active in developing multiplex detection systems; that is the detection of different organisms in the same sample. Possibilities and drawbacks of this strategy are described in the paragraph below.

10.2.5 POSSIBILITIES AND DRAWBACKS OF MOLECULAR DETECTION METHODS FOR PRACTICAL APPLICATION

DNA macro-array technology is implemented to routinely detect plant pathogens in environmental samples by various companies (Lievens and Thomma, 2005; Lievens et al., 2005). With this test, 50 different plant pathogens including fungi and bacteria can be detected. However, this is a qualitative method and a quantitative score can hardly be given.

A disadvantage of all DNA techniques is that no distinction is made between the types of propagules (spores, cells and mycelium) of living or dead organisms. However, in an active system, dead organisms will generally disintegrate in short time. Detection of mRNA is generally accepted as a method to detect living organisms in contrast to detection of DNA (as in PCR), where both dead and living organisms are detected.

For quantitative results, verification of the resolution and accuracy of the methods, using test samples prepared with known numbers of target organisms, is necessary. In addition, the inclusion of an internal control in molecular methods for all samples is a check for the inhibition of reactions used in the molecular technique caused by chemical compounds in the sample.

Specificity of detection methods, even with the most advanced molecular techniques, is still a problem for poorly defined genera, as well as genera containing strains with different ecological behaviour. *Fusarium oxysporum* is an example of the latter: it contains pathogens causing various symptoms (wilt versus crown and root rot), as well as non-pathogenic isolates, which can even act as antagonistic organisms against Fusarium diseases.

Another relevant issue, arising due to the lower detection thresholds of the newly developed molecular techniques, is the lack of biological thresholds; that is, at what pathogen concentration is control really needed? The first experience with the new detection techniques is that increased amounts of pesticides were applied in the greenhouses where occurrence of pathogens was monitored (personal communication: Ludeking, Relab den Haan). In conjunction with the development of the new detection techniques, growers should get used to the lower detection limits of these techniques, and, simultaneously, biological thresholds of the pathogens should be studied.

10.2.6 FUTURE DEVELOPMENTS

The technological developments in the molecular detection technology are likely to bring new methods for detecting plant pathogens. Such methods will enable more precise quantitative detection of organisms or genes, as well as simultaneous detection of various plant pathogens (the so-called 'multiplex detection')

The newest development in molecular detection technology is the micro-array technology, in which various oligonucleotides can be detected in little more than one square mm (Fig. 10.2). Micro-array technology provides the next generation of DNA diagnostics, enabling the detection of several pathogens and pathogenic variants in parallel, improving specificity of the test. Several micro-array systems are available (Bonants et al., 2005) in various formats. For sensitivity reasons a pre-amplification procedure is still necessary to be able to detect small numbers of pathogen propagules.

Looking into the future, we can envision that in recirculating water systems and in air, monitoring systems will be developed which monitor water and air in real time for the presence of numerous organisms pathogenic for different crops. This will not only be in a qualitative manner, but also quantitative and multiplex. If a certain pathogen then exceeds its threshold level, an automatic alarm system will warn the grower to take appropriate action.

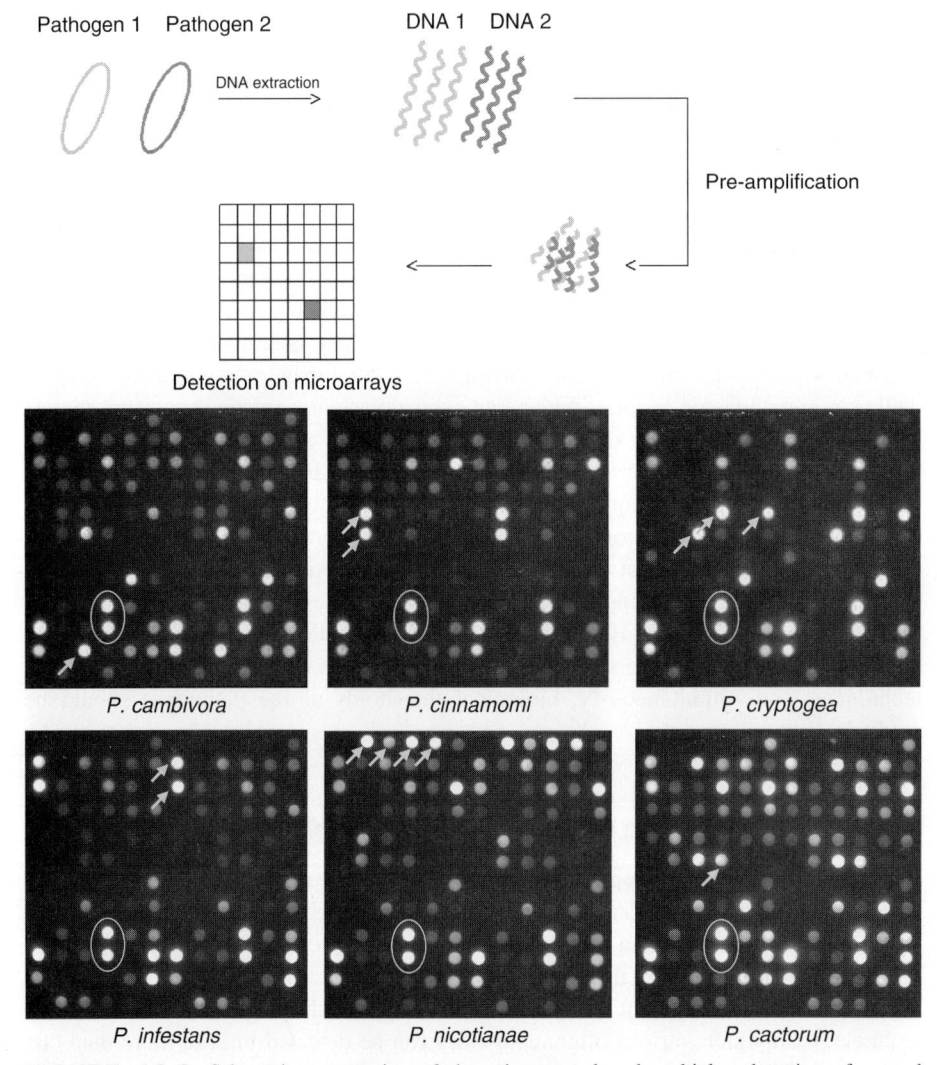

434　　　　　　　　CHAPTER 10　PATHOGEN DETECTION AND MANAGEMENT STRATEGIES

FIGURE 10.2 Schematic presentation of the micro-array-based multiplex detection of several pathogens on a single chip (upper part). Below part: Hybridization patterns for six different *Phytophthora* species on a micro-array on which species-specific probes were spotted. Examples of common *Phytophthora* DNA sequences (circle) and species-specific sequences (arrows) are indicated (see also Plate 20).

10.3　MICROBIAL BALANCE

10.3.1　MICROBIOLOGICAL VACUUM

Due to the use of new or sterilized substrates, soilless systems often start with a 'microbiological vacuum', lacking a diverse and competitive microflora. Various aspects of microbiologically balanced systems, including disease suppressiveness and the addition of biological control agents, are discussed below.

10.3.2 MICROBIAL POPULATIONS IN CLOSED SOILLESS SYSTEMS

A fundamental difference with soil is that many soilless systems that are used for closed systems do not contain a substantial organic fraction. Organic matter in soil is an important source of nutrients for micro-organisms. Soil is a microbiologically rich substrate, generally containing 10^7–10^9 culturable bacterial propagules and 10^4–10^6 fungal propagules per gram agricultural soil (Alexander, 1977).

Soilless systems are generally not sterile either, but containing lower levels of micro-organisms compared to soil. Bacterial plate counts showed that 10^5–10^7 CFU of bacteria are present per millilitre nutrient solution depending on the crop type, age and type of soilless system (Postma et al., 2000, 2003; Koohakan et al., 2004; Calvo-Bado et al., 2006). Soilless systems without organic components are a relatively poor growing medium for micro-organisms at the start of a crop due to the lack of organic matter, but once plants grow in the system, exudates from the roots, or even sloughed root material itself, provide organic substrates on which the micro-organisms can grow. Two examples illustrating that soilless systems start with a low level of micro-organisms: (1) numbers of bacteria in the solution of a tomato (*Lycopersicon esculentum* Mill.) crop were 10^3 CFU per millilitre before crop growth, and 10^6 CFU per millilitre 20 h after planting (Berkelmann, 1992); and (2) moist stone wool with young cucumber plants contained 10^6, compared with up to 6×10^7 CFU bacteria per gram moist stone wool of an older crop (Postma et al., 2000). The numbers of fungi (including *Trichoderma* spp.) and filamentous actinomycetes (like streptomycetes) are very low at the start of a crop in soilless systems: often less than 10^2 CFU per gram moist stone wool (Postma et al., 2000; Koohakan et al., 2004; Postma et al., 2005).

While the numbers of micro-organisms in soilless systems are typically low at the start of a crop, so is the diversity (Postma et al., 2000). Figure 10.3 shows the genetic profile of the bacterial population in the nutrient solution of a cucumber crop analysed with PCR–DGGE (polymerase chain reaction–denaturing gradient gel electrophoresis, a molecular method to visualize the community composition). The bands in the PCR–DGGE profile reflect the dominant bacterial species in the community. The number of bands (i.e. species) was low at the start of the experiment (just before planting, i.e. week 0) and increased in week 1 and 3 after planting. In another study, no changes in the bacterial composition were detected in a tomato crop using PCR–DGGE. In this case the microbial community was analysed for the first time 6 weeks after planting up to about 14 weeks after planting (Calvo-Bado et al., 2006). The establishment of the microflora probably occurs within the first weeks of a crop, and later treatments do not result in a distinct influence on the dominant species which are represented in the PCR–DGGE profile.

Micro-organisms that occur at the start of a crop in a soilless growing system are probably only those that can survive in water, air, seed and plant material, and on

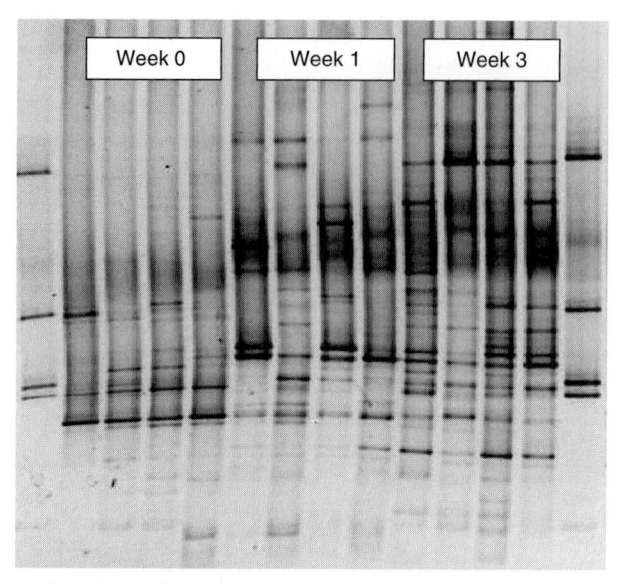

FIGURE 10.3 Composition of the bacterial population in the nutrient solution of a cucumber crop 0, 1 and 3 weeks after planting analysed with PCR–DGGE.

dry surfaces. Plant-specific micro-organisms and those adapted to the conditions in soilless systems should be introduced during crop growth or with the plant material. In agricultural soil systems, the diversity of the microflora and the presence of several plant-associated micro-organisms, such as *Rhizobium*, mycorrhizal fungi, plant growth promoting bacteria, are important characteristics of a soil (Alexander, 1977). In soilless systems, however, these plant-associated organisms will not be present in new or sterilized substrate and little research has been done on the development of the microbial populations during crop growth; that is, which species are present at various growth stages in these growing systems.

One impressive study on the occurrence of fungi is performed by Menzies et al. (2005), who identified 1250 fungal isolates present on healthy roots. They showed that there was a greater density of fungal colonies on roots grown in soil than on roots grown in other substrates, including stone wool. *Penicillium* (87.2 per cent), *Trichoderma* (4.6 per cent) and *Pythium* (3.0 per cent) were the most common genera isolated from cucumber roots. Some of the *Penicillium* isolates had positive effects on the germination of seeds as well as on early plant growth (Menzies et al., 2005).

Molecular analyses (PCR–DGGE in combination with sequencing) of the fungal community including the 'water moulds' showed that increasing numbers of zoosporic organisms were present in samples of soil, stone wool slabs and a nutrient solution (unpublished data Postma). In both soilless systems, DNA related to species of the phylum *Chytridiomycota* was detected.

10.3.3 PLANT AS DRIVING FACTOR OF THE MICROFLORA

In soil and soilless systems, a substantial part of the nutrients used by the microflora is derived from plant roots, resulting in high numbers of micro-organisms on the surface of plant roots (rhizoplane). Exudates released by the roots consist of organic compounds such as short-chain organic acids, carbohydrates, mucilage and lysates. In addition, dead cell material is accumulating with time. All the released organic material was estimated to account for 15–20 per cent of the total carbon fixed by the plant (Bolton et al., 1992). The nature of plant-derived compounds is dependent on plant species, growth conditions, rooting medium, the stage of plant development and plant root health. The microflora around the roots was shown to be altered in several characteristics (utilization capacity of carbon sources, growth rate) as a function of time and location on the root, probably because of the changes that occur in the exudation patterns of roots as plants and roots age (Folman et al., 2001).

In soil, where water transport is less extensive than in soilless substrate, the rhizosphere can extend up to 2 mm from the root surface (Bolton et al., 1992). The rhizosphere is defined as the area around the root system, where the roots influence the microflora (Alexander, 1977). In soilless substrate, a flow of nutrient solution and diffusion are transporting root exudates more easily away from the roots as compared to soil due to the free water which is present. The micro-organisms rapidly metabolize the available carbon leaking from the roots. Attachment to the surface of the root could be an important strategy of micro-organisms to compete for nutrients in soilless systems. Figures 10.4 and 10.5 show the high numbers of bacteria present on the

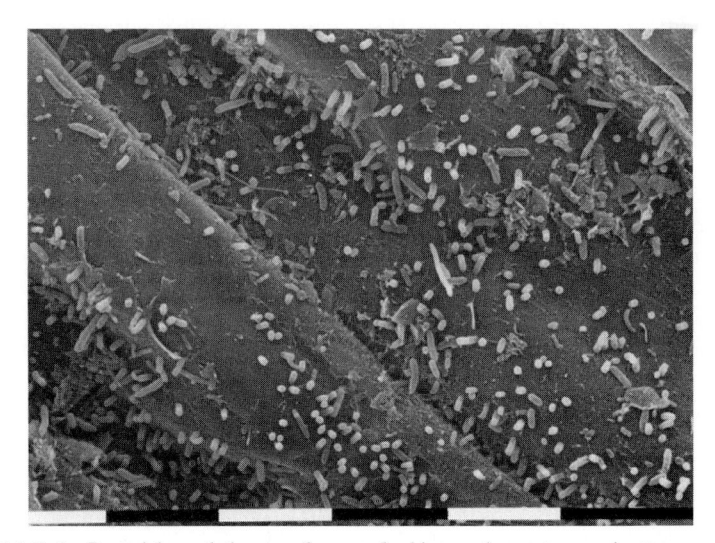

FIGURE 10.4 Bacterial population on a three-week-old cucumber root grown in stone wool visualized by scanning electron microscopy (SEM) (bar = 10 μm) (photo by Anke Clerkx, Plant Research International) (see also Plate 21).

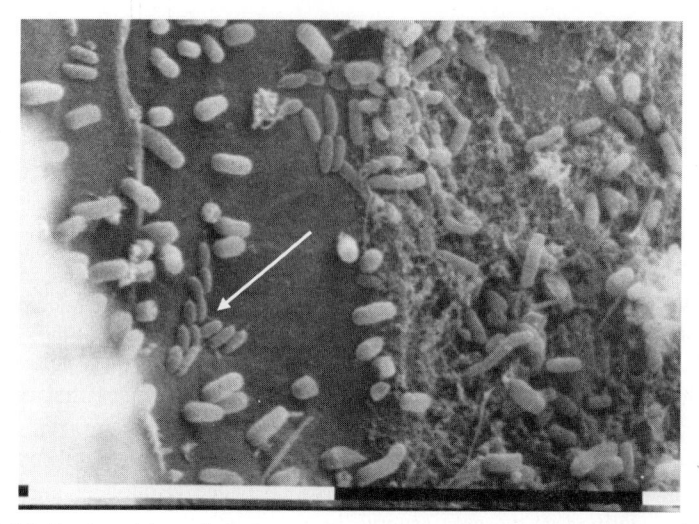

438 CHAPTER 10 PATHOGEN DETECTION AND MANAGEMENT STRATEGIES

FIGURE 10.5 Bacterial populations on a cucumber root grown in nutrient solution (bar = 10 μm). Arrow points at a micro-colony (photo by Anke Clerkx, Plant Research International) (see also Plate 22).

surface of cucumber roots grown in stone wool and nutrient solution, respectively; about 10–20 per cent of the surface is covered with bacterial cells.

Numbers and activity of the bacterial population are stimulated by the roots (the so-called 'rhizosphere effect'): higher numbers of micro-organisms are detected in the slab and the drainage water from the slabs, compared to the nutrient solution which had not yet passed the plant roots (van Os et al., 2004b). This illustrates that the rhizosphere has a significant effect in the soilless system, and the importance of the exudates and other plant-root-derived materials as influencing factor (nutrient source) for the microflora.

10.3.4 BIOLOGICAL CONTROL AGENTS

Biological control agents (also called biocontrol agents or BCA) can play an important role in suppressing root pathogens in soilless systems. Biocontrol agents are those products that control plant pathogens or pests or reduce their amount or their effect by one or more organisms other than man (i.e. viruses, bacteria, fungi, insects). An overview of biological control principles and many examples is described by Campbell (1989).

Compared to a field soil, where it is difficult to introduce biocontrol agents in sufficient concentrations at lower parts of the roots, the limited volume of the matrix around the roots in soilless systems facilitates introduction of the antagonist in the root environment. In general, soilless systems allow a good interaction among host, pathogen and antagonist. Also, the establishment of antagonists is easier in soilless systems with a new substrate having an unbalanced microflora than in a microbiologically buffered system such as soil with a tremendous competition with the micro-organisms already present (Paulitz and Bélanger, 2001). A third advantage of greenhouse systems for a successful introduction of antagonists is the regulated climate with a more uniform temperature (Paulitz and Bélanger, 2001).

Many antagonists (i.e. micro-organisms that inhibit other organisms) of root pathogens are known and have been tested in greenhouse cropping systems. However, limited numbers of antagonists are available as commercial products. The possibilities of using biocontrol agents in greenhouse systems have been reviewed by Paulitz and Bélanger (2001). Biocontrol agents of root-borne (fungal) pathogens which are developed into a product are *Trichoderma harzianum, Gliocladium virens, G. cathenulatum*, non-pathogenic *F. oxysporum, Coniothyrium minitans, Streptomyces griseoviridis* and *Bacillus subtilis*. A lot of research has also been done on *Pseudomonas* species which are good antagonists; however, a product for soilless systems is not available yet. One problem concerning several *Pseudomonas* species is the poor survival of the bacterial cells after drying and storage (so called shelf life), since it does not produce spores. An overview of recent published results on bacterial and fungal antagonists against fungal and bacterial pathogens in the rhizosphere of soil and other growing media is described by Whipps (2001).

Antagonists can be active through several mechanisms, such as (myco)parasitism, antibiosis or other inhibitory substances, competition for nutrients or space or induced resistance. A biocontrol mechanism more recently discovered is the application of micro-organisms that produce biosurfactants which cause lysis of zoospores (Postma, 1996; Stanghellini and Miller, 1997). This is an interesting mechanism for soilless systems with high water retention capacity, that, as a consequence, have problems with many different zoospore producing pathogens. Examples are *Pseudomonas aeruginosa* controlling *Phytophthora* root rot (Stanghellini and Miller, 1997), *Pseudomonas fluorescens* controlling root rot in hyacinth bulbs caused by *Pythium intermedium* (De Souza et al., 2003) and *Lysobacter enzymogenes* controlling cucumber root and crown rot caused by *P. aphanidermatum* (Folman et al., 2003).

Nevertheless, biocontrol of root diseases often shows variable results (Postma et al., 2001). This can be due to a lack of survival or activity of the biocontrol agents, or to insufficient colonization of the infection sites by the biocontrol agents. Especially, if the antagonist was originally isolated from soil or from a crop different from the one to which it is applied, it may not be fully adapted to soilless systems. For effective biocontrol, the activities of the antagonist and pathogen should be synchronized in time and in space. The influence of the location of the antagonist in relation to the location of the pathogen is illustrated by the failure of suppression of Fusarium wilt in carnation in a recirculation system, when the antagonist was added on top of the stone wool blocks, while the pathogen was introduced via the nutrient solution (Rattink and Postma, 1996). However, when introduced via the nutrient solution where the pathogen was located, the antagonist proved to be extremely effective.

A specific group of micro-organisms which has received little attention yet are endophytes, that is organisms present inside the plant tissue (Hallmann et al., 1997).

They have the advantage of being protected from environmental stresses and might be able to protect the plant in a soilless system from propagation material throughout the crop cycle. Non-pathogenic *Fusarium* species which are present in plants can, for example, protect carnation (*Dianthus caryophyllus* L.) against wilt disease caused by *F. oxysporum* (Postma and Luttikholt, 1996), or tomato against the plant-parasitic nematode *Meloidogyne incognita* (Hallmann and Sikora, 1994). An example with bacterial endophytes is the endophytic colonization of tomato tissue with a *Pseudomonas* strain that inhibits the wilting by *V. dahliae* (Sharma and Nowak, 1998).

Another group of micro-organisms are those that protect the plant through systemic-induced resistance (e.g. Chen et al., 1999). This protection is not necessarily limited to the location where the antagonist is present. Many experiments have been conducted to test if a biocontrol agent acts through induced resistance, either by using a split root system or by challenging the roots with the control agent followed by evaluating the symptoms of a leaf pathogen. Two examples of *Pseudomonas* isolates inducing resistance are described for Pythium root rot in cucumber (Zhou and Paulitz, 1994) and Fusarium wilt in tomato (Duijff et al., 1998). However, not much is known about the importance of this mechanism for an entire crop, since most research on systemic-induced resistance of root pathogens has been done with individual young plants using a split root system.

Biological control has the potential to solve problems with root diseases in soilless systems. However, there is a need for better products, and more knowledge on the optimal inoculation procedures as well as environmental conditions under which the biocontrol agent will be active. There is a need for organisms which are better adapted to soilless systems and synchronized to the pathogens occurring in these systems.

For commercial application of biocontrol agents, the need for legislation of these products is an economical barrier. To get a biocontrol agent registered, safety for the user, the consumer and the environment, as well as biocontrol efficacy need to be proven. This registration procedure is relatively expensive for products which can only be applied within a relatively small market.

10.3.5 DISEASE-SUPPRESSIVE SUBSTRATE

Whereas in soil and organic media the existence and the potential of a suppressive microflora towards root pathogens is generally accepted (see Chap. 11), suppressiveness in soilless systems has only been recently demonstrated. Suppression of *P. aphanidermatum* disease in cucumber in re-used stone wool was proven to be the result of the microflora present in this substrate, since suppressiveness of sterilized stone wool was recovered after its recolonization with the original microflora (Postma et al., 2000, 2005). In contrast to the application of singly introduced antagonists, the indigenous microflora caused reproducible suppression of Pythium crown and root rot in cucumber. Without exception, all stone wool slabs without Pythium symptoms in the previous crop were suppressive. The suppressiveness in this stone wool system correlated with bacterial diversity as well as the number of filamentous actinomycetes (mainly Streptomycetes) (Postma et al., 2000, 2005). However, growers should not re-use stone wool slabs with distinct Pythium symptoms in the previous crop, because these slabs often resulted in high disease percentages in the experiments.

Also the nutrient solution of a hydroponic system can become suppressive. This was discovered by Berkelmann (1992). The solution from a nine-week-old tomato crop inhibited mycelial growth of *F. oxysporum* f.sp. *lycopersici* very strongly in an *in vitro* test. This was due to the living microflora, since heat or filter sterilization destroyed the inhibition completely. A fresh nutrient solution, which had not been in contact with the crop, was only weakly suppressive. It would be very interesting for the use of such suppressiveness to know at what time after planting this suppressiveness in the system develops.

In general, substrates are selected on the basis of their horticultural productivity and not on the basis of disease suppressive properties. Therefore, not many substrates have been compared properly as to their influence on disease development. A complicating factor is that substrates can vary in physical and chemical properties, besides the differences in their biological characteristics. Using substrate with low water content, that is perlite instead of stone wool, is also a strategy to avoid the development of Pythium in a crop (van der Gaag and Wever, 2005). This suppressiveness is probably due to less available free water for the transport of Pythium zoospores or to different root morphology and it is not expected to lead to a more suppressive microflora. Volcanic scoria (tuff) was also found to be more suppressive to Pythium than stone wool (Kritzman et al., 2000). However, it is unclear if this is due to the presence of a suppressive microflora or the physical and chemical properties being less favourable for Pythium.

The composition of microflora in root samples of two soilless substrates has been shown to differ in tomato production (Khalil and Alsanius, 2001). Lower numbers of total aerobic bacteria and pseudomonads were present in peat than in stone wool slabs, while the numbers of filamentous actinomycetes and fungi were higher. Similarly Koohakan et al. (2004) found lower numbers of pseudomonads, higher numbers of fungi, but similar numbers of bacteria in the nutrient solution, as well as on tomato roots, grown in coconut fibre versus stone wool. However, it is not known if these differences in composition of the microflora are reflected in differences in suppressiveness.

In two independent studies it was suggested that a disease-suppressive microflora was built up in a closed system due to recirculation of the nutrient solution: (1) against *Phytophthora cryptogea* in tomato (McPherson, 1998), and (2) against *Pythium* sp. in tomato (Tu et al., 1999). The suppressiveness was lost if the nutrient solution was not recycled anymore, that is the system was changed into run-to-waste.

The role of composts in soil-borne disease suppressiveness is described in detail in Chap. 11 and therefore will not be treated here.

10.3.6 CONCLUSIONS

Several examples of soilless systems being suppressive towards certain root diseases prove the existence of beneficial micro-organisms in soilless systems. The question arises, how can suppressiveness be used as a tool and perhaps be increased? In soil, the occurrence of disease suppression is fully accepted and has led to the isolation of antagonistic species. In soilless systems it is not yet known, which species or properties within the microflora are causing disease suppression. Hopefully, antagonists that are better adapted to soilless systems will be found by analysing the microbial composition of suppressive soilless systems.

For practical applications, it is important to understand how growers or soilless substrate suppliers can enhance the abundance or activity of a suppressive microflora. The increase of the suppressiveness of a soilless system is most likely by introducing well-adapted antagonistic organisms at the start of the crop or by influencing the microflora through the crop itself or its growing conditions. If soilless systems can be created with a disease suppressiveness comparable to natural soils, they will combine the advantages of soil systems and soilless substrates, that is a microbiologically well-buffered culture system, as well as the possibility to have a pathogen-free start by renewing the substrate. The exploitation of the indigenous microflora to suppress diseases in soilless systems might be a new trend, which is breaking with the 'sterility' concept commonly used in soilless systems.

10.4 DISINFESTATION OF THE NUTRIENT SOLUTION

10.4.1 RECIRCULATION OF DRAINAGE WATER

In closed systems, where the drainage water is collected and re-used, soil-borne pathogens released by plants can rapidly be dispersed among all the plants in the production system. To minimize such risks of disease spreading, the solution should be treated before re-use. The use of pesticides for such a treatment is limited: (1) effective pesticides are not available for all such pathogens; (2) due to the direct contact of pesticides with the roots, phytotoxic levels may be achieved at concentrations that typically do not cause problems in soil; and (3) environmental legislation restricts release of water with pesticides into the environment.

Moreover, the capture of the drainage from the production system provides opportunities for treatments to eliminate soil-borne pathogens without the use of pesticides. Such methods consist of several approaches including: filtration, subjecting the effluent to high temperatures, oxidizing reactions or radiation. The efficacy of disinfestation can be measured as percent reduction of viable disease propagules. Generally a 99.9 per cent reduction of the original population is targeted (also called a log 3 reduction) since less drastic reductions result in too many pathogen propagules that may survive and infect healthy roots.

10.4.2 VOLUME TO BE DISINFECTED

The amount of drainage water that needs to be treated depends on the design and style of the installed closed system. In a traditional stone wool or perlite

10.4 DISINFESTATION OF THE NUTRIENT SOLUTION 443

system used for growing greenhouse tomato, cucumber or sweet pepper (*Capsicum annum* L.), about 30 per cent of the supplied nutrient solution is expected to return (see Chap. 5). The absolute amount of liquid to be processed per day varies with water use of the crop and irrigation frequency. A full grown tomato crop transpires in full production about 3–5 mm per day in northwest Europe to 6–9 mm in sub-tropical areas, consequently, 30 per cent or 1–3 mm has to be disinfected during a 24-h period if the system is operating under best-management standards. For a greenhouse of 1000 m^2, it means a disinfestation capacity of about 1–3 m^3 per day. If the supply water might be infected too, which might be the case if water is stored outside in open basins or obtained from surface water, all the supply water has to be disinfected as well (4–12 mm per day). If an NFT-system (Nutrient Film Technique) is used, the disinfestation capacity has to be increased dramatically to about 100 m^3 per day (supply and drain water has to be disinfected and flow rate is much higher compared to drip irrigation; see Chap. 5). Generally such a capacity is considered to be uneconomical to disinfect (Ruijs, 1994). Therefore, the risk of soil-borne diseases in NFT systems is rather high and, generally, only short-term crops such as lettuce (*Lactuca sativa* L.) and chrysanthemums (*Dendrathema* × *grandiflorum* Kitam.) are grown in NFT, but without disinfestation of the nutrient solution. A grower has to take into account all variables for calculating the required capacity. Because of the variable return rate of drain water, a sufficiently large catchment tank for drain water is needed in which the water is stored before it is pumped to the disinfestation unit (Fig. 10.6). After disinfestation, another tank is required to store the clean water before adjusting EC and pH and blending with new water to supply to the plants.

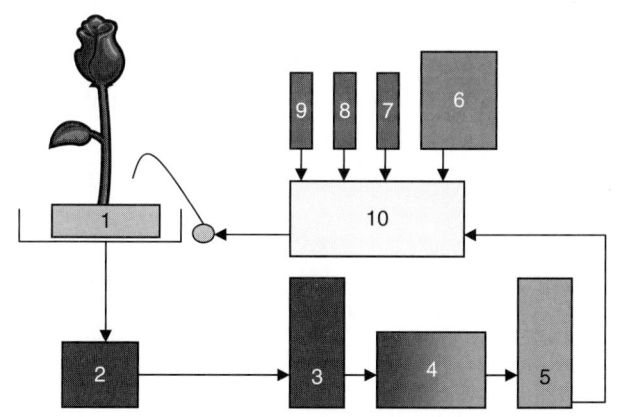

FIGURE 10.6 Closed system scheme focusing on disinfestation. Surplus water of the plants (1) flow by gravity to a central recatchment tank (2), from where it is irregularly pumped to a day storage tank (3). Twenty-four h continuous disinfestation (4) takes the nutrient solution from (3) to a tank (5), where it is stored to be used for watering. In the mixing container (10) disinfected drain water (5) is mixed with fresh water (6) and A and B nutrients (7, 8) and if needed acid or lye are added (9). After controlling EC and pH, the water is supplied to the plants (see also Plate 23).

10.4.3 FILTRATION

Filtration can be used to remove any undissolved material out of the nutrient solution (see Table 10.3). Various types of filters are available relative to the range of particle sizes. Rapid sand filters are often used to remove large particles from the drain water before adding, measuring and control of EC, pH and application of new fertilizers. After passing the fertilizer unit often a fine synthetic filter (50,000–80,000 nm) is built in the water flow to remove undissolved fertilizer salts or precipitates to avoid clogging of the irrigation emitters. These synthetic filters are also used as pre-treatment for disinfestation methods such as heat treatment, ozone treatment or UV radiation. Rapid sand and synthetic filters are placed in the water flow and can be cleaned manually or automatically.

With declining pore size, the flow is inhibited, so that removal of very small particles requires a combination of a large array of filters or higher pressures and frequent cleaning of the filter(s) to prevent clogging. Removal of pathogens requires relatively small pore size (see Table 10.3).

Various membrane filtration technologies are available where water under high pressure is pressed through a membrane. The water is separated into the desired clean water (filtrate) and the remaining water with concentrated salts (the so-called 'brine'). The brine has to be eliminated and depending on the country and situation there might be strict rules as the amount of brine might represent as much as 40 per cent of the water being treated. Membrane filters can be classified depending on the pore size of the membrane or size of particle that they allow through. Filtration method is used extensively in food industry and water purification. The present generation of membrane filters is more reliable and cheaper than has been the case in the past so that using this technology in horticulture has become feasible. As the investment is still rather high, it is only used as a supplemental method for the removal of pathogens. All over the world, use of these reverse osmosis systems are coming into use to desalinate seawater or other 'grey' water so as to be used as supply water for the plants. Simultaneously, pathogens are removed from the supply water. A big plant was built near Almería (Spain) to supply local growers with cheap water to enable

TABLE 10.3 Overview of Filtration Processes

Principle	Pore size (nm)	Removal of:
Rapid sand filter	>100,000	Algae, roots, leaves
Mechanical synthetic filters	10,000–100,000	Small particles, sand grains
Slow sand filtration	1000–10,000	Roots, pathogens
Membrane filtration:		
Microfiltration	100–1000	Colloidal and suspended particles, bacteria
Ultrafiltration	10–100	Proteins, fungi, viruses
Nanofiltration	1–10	Sugars, two valued ions, larger mono valued ions
Reverse osmosis	<1	Small ions

10.4 DISINFESTATION OF THE NUTRIENT SOLUTION

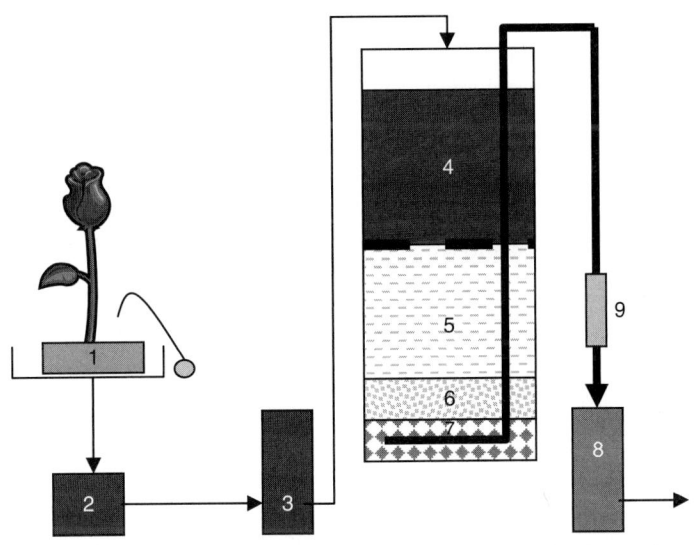

FIGURE 10.7 Scheme of slow sand filtration. Nutrient solution drains from substrate (1) to the recatchment tank (2). From there it is pumped to the day storage tank (3) and into the top of a large container or metal silo (4), from which it drips into a sand layer of 1 m thickness (5). The layer between (4) and (5) is called the Schmutz-decke or filter skin. (6) and (7) are a 10 cm fine and a 15 cm coarse gravel layer, respectively. The filtrate is pumped out of the gravel layer to container (8). In a metal silo it is done via the top, in a synthetic filter it is possible to drain via the bottom of the filter. For initial filling of the filter water is pumped from (8) into the gravel layers (7) and (6) and to above the sand layer. Flow meter (9) controls the filtration rate. From container (8) the filtrate will be mixed with fresh water to a new nutrient solution for the plants (see also plate 24).

them to use (closed) soilless culture systems. In the Netherlands, water from reverse osmosis is used in addition to rainwater supply.

Slow sand filtration is considered to be a reliable, low-cost solution to eliminate soil-borne pathogens (Wohanka, 1995; Runia et al., 1997; Van Os et al., 1997b; Ehret et al., 2001) in greenhouse horticulture (Fig. 10.7). It has been used for a long time to purify drinking water. The method disappeared but in Europe recently it has become more popular as polishing of treated drinking water to replace a chlorination process. *Phytophthora* and *Pythium* can be eliminated completely by this method, but *Fusarium spp.*, viruses and nematodes are only partly (90–99.9 per cent) removed by this method. The principle is based upon a supernatant water layer, which trickles slowly through a sand layer. Experiments proved that a flow rate of 100 l/m²/h increases the performance compared to higher flow rates and so does the selection of finer sand (grain size 0.15–0.35 mm; $D_{10} < 0.4$ mm) compared to coarser sand (Van Os et al., 1997a, 1997b). Satisfactory performances can also be obtained when either the grain size increases to 1 or 2 mm or the filtration rate increases to 300 l/m²/h (Wohanka et al., 1999). The mechanism of elimination is not only filtering (mechanical) as the size of the pores is generally larger than the pathogens eliminated. The formation of a biological active layer upon top of the sand in the filter appeared to be of great importance (Wohanka et al., 1999). Commercial installations were already in use before much was known on the limiting conditions and the working mechanism. Now, the existing commercial installations and the newly designed ones are improved, based on the results mentioned above.

10.4.4 HEAT TREATMENT

Heating the drainage water to lethal temperatures is one of the most reliable methods for disinfestation. Each type of organism has its own lethal temperature. Non-spore forming bacteria have lethal temperatures between 40 and 60°C, fungi between 40 and 70°C, with some exceptions to 85°C, nematodes between 45 and 55°C and viruses between 80 and 95°C (Runia et al., 1988) at an exposure time of 10 s. Generally the temperature setpoint (95°C) is high enough to kill most of the organisms that are likely to cause disease during the period of time that the liquid is at these killing temperatures (minimal 10 s). While this may seem very energy intensive, it should be noted that the energy is recovered and reused with a heat exchanger. The practical system operates as follows: The excess nutrient solution returns from the plants and is collected in a tank (Fig. 10.8). From this tank the solution is pumped into a pasteurization unit where it first enters a heat exchanger, where it is preheated to a temperature of about 80°C by heat recovery from the hot water exiting the unit. In the next part of the unit the solution is heated to the disinfestation temperature between 90 and 97°C, using an external heat source. From experiments, it appeared that an exposure time

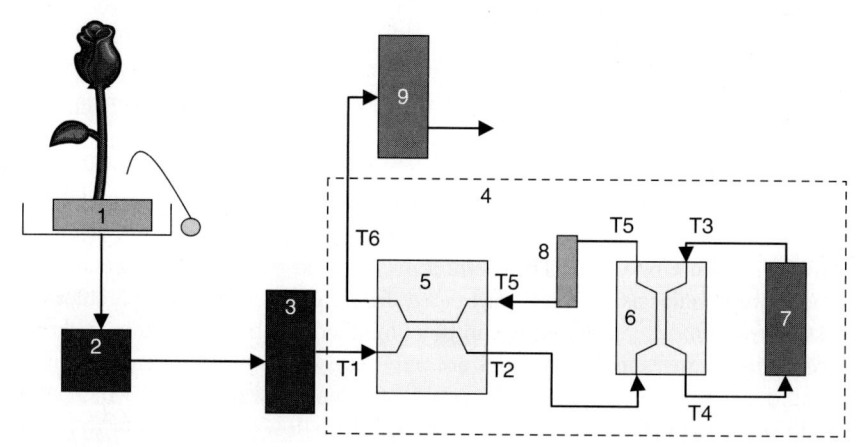

FIGURE 10.8 Scheme of heat treatment. Nutrient solution drains from substrate (1) to the recatchment tank (2). From there it is pumped to the day storage tank (3) and into heat treatment unit (4). The solution is pumped with temperature T1 (about 17–25°C) into heat exchanger (5) and preheated to T2 (70–80°C). In heat exchanger (6), it is further heated to T5 (85–97°C) by an external heat source (boiler, 7) which has an incoming temperature of T3 (95–105°C) and an outcoming temperature T4 (85–95°C). The water is kept on temperature T5 in unit (8) for an exposure time of about 30s and cooled down to temperature T6 (22–30°C) and stored in container (9) to be mixed with fresh water and nutrients for watering the plants (see also Plate 25).

of 10 s is sufficient, but for commercial purposes an exposure time of 30 s has been recommended (Runia et al., 1988). The disinfected solution flows back to the first heat exchanger to be cooled down and subsequently it is stored in a clean water tank. There has been much discussion about the disinfestation temperature, especially in relation to the elimination of viruses. In the first trials, a 99.9 per cent killing of various pathogens (log 3 reduction of the initial concentration) was achieved at 95°C for 10 s. Runia (1998) showed that similar elimination efficacies could be obtained by decreasing the disinfestation temperature and increasing the exposure time: 90°C and 2 min, 85°C and 3 min for a complete disinfestation (elimination of all pathogens including viruses) and 60°C for 2 min for a selective disinfestation (only elimination of fungi, bacteria and nematodes) (Runia and Amsing, 2000). The use of these lower temperatures in combination with longer exposure times means that no special heaters are needed to deliver the needed high temperatures in the second heat exchanger (Fig. 10.8, T3 and 7).

The temperature of the nutrient solution leaving the pasteurization unit (Fig. 10.8, T6) depends on the size of the heat exchanger (Fig. 10.8, no. 5). Generally it is designed that T6 is maximum 5°C higher than T1 (the incoming drain water). If the disinfested drain water is mixed with fresh water in a proportion of one part of drain water and two parts of fresh water, which is commonly the case in European situations growing tomato and sweet pepper with drip irrigation, there will be no problem. On the other hand, if all disinfested drain water is sent to the plants without much refreshing, which will be the case using an NFT system, the temperature of the nutrient solution will rise to unacceptable levels.

The amount of fuel needed to heat the water amounts to about 1 m^3 natural gas per 1 m^3 water for disinfestation temperatures of 95°C; this declines to 0.6 m^3 gas per 1 m^3 water while heating up to 60°C. In the Netherlands, where natural gas is available at all nurseries, this energy source is generally used for heat treatment. Other energy sources are used in places where this is not the case. If a heating system is available for the greenhouse, then its fuel type is also suitable for use in disinfestation although the type of heater determines the maximum temperature that can be achieved (Fig. 10.8, T3).

Another aspect of realizing high temperatures is the precipitation of scale (CaCO$_3$) on the plates of the heat exchanger at temperatures around 100°C. Practical experience showed that decreasing the pH of the solution to about 4.5 prevents this problem. Generally the plates of the heat exchangers are made of stainless steel; a titanium-coated heat exchanger is recommended if the chloride level in the solution is higher than 4 mmol l^{-1}.

10.4.5 OXIDATION

Several disinfestation methods use chemical oxidation as the mechanism for killing micro-organisms. An oxidizing agent reacts with the organic matter present in the solution or in the pipes through which the solution is passing. As such, it reacts with pathogens as well as non-threatening organisms (such as algae) as well as other organic matter. Prefiltration in such systems is a must.

10.4.5.1 Hydrogen Peroxide

Hydrogen peroxide (H_2O_2) is a strong, unstable oxidizing agent that reacts to form H_2O and an O-radical. The latter reacts with any type of organic material, including pathogens. If insufficient organic matter is present it forms O_2 which releases to the air. Commercially the so-called 'activators' are added to the solution to stabilize the original solution and to increase the efficacy. Activators are mostly formic acid or acetic acid, which decrease pH in the nutrient solution. Various dosages are recommended (Runia and Paternotte, 1993): 0.005 per cent against Pythium, 0.01 per cent against other fungi as Fusarium and 0.05 per cent against viruses. As weak acids are supplied to the solution, the pH decreases by one, two or three units, respectively. Before reuse of the nutrient solution, the pH has to be adjusted again. The 0.05 per cent concentration is also harmful for plant roots. Currently H_2O_2 is especially helpful for cleaning the watering system, while the use for disinfestation has been taken over by other methods. The method is inexpensive, but not efficient.

10.4.5.2 Ozone

Ozone (O_3) is a powerful oxidizing agent which reacts with all organic matter with which it comes in contact. It also reacts with iron chelates. It can kill all organisms in the water, depending on exposure time and concentration. Ozone itself is produced from dry air and electricity using an ozone-generator (converting $3O_2 \rightarrow 2O_3$). The ozone-enriched air is injected into the water that is being sanitized and stored for a period of 1 h. From several trials, Runia (1994b, 1996) concluded that an ozone supply of 10 g per hour per m^3 drainage water with an exposure time of 1 h is sufficient to eliminate all pathogens, including viruses. Supply water can be treated with 5 g per hour per m^3, because the organic load is much lower. Human exposure to the ozone that vents from the system or the storage tanks should be avoided since even a short exposure time of a concentration of 0.1 mg/l of ozone may cause irritation of mucous membranes. Therefore, in Europe, ozone treatment is not very popular anymore (expensive, strict rules) although it works technically well. In the United States, it is gaining popularity. A disadvantage is the inability to process large quantities of water at the same time. Another drawback of the use of ozone is that it reacts with iron chelate, precipitating it from the solution, depositing the iron in the irrigation system, breaking down the chelate and rendering the iron unavailable for uptake by the plant. Consequently, higher dosages of iron are needed and measures need to be taken to deal with iron deposits in the system.

10.4.5.3 Sodium Hypochlorite

Sodium hypochlorite (NaOCl) is a compound having different commercial names with different concentrations but with the same chemical structure. It is widely used for water treatment, especially in swimming pools. This material is relatively inexpensive due to this widespread use. When added to water, NaOCl decomposes to HOCl and $NaOH^-$ and depending on the pH to OCl^-, the latter decomposes to Cl^- and $O^.$ for strong oxidation. HOCl reacts strongly with cell walls at a pH of 5.5–7.5. It reacts directly with other substances and if there is enough hypochlorite it also reacts with pathogens. Le Quillec et al. (2003) showed that the tenability of hypochlorite depends on the climatic conditions and the related decomposing reactions. High temperatures and contact with air causes rapid decomposition, at which $NaClO_3$ is formed with phytotoxic properties. Runia (1984, 1988) showed that hypochlorite is not effective for eliminating viruses. Chlorination with a concentration of 1–5 mg Cl/l and an exposure time of 2 h achieved a reduction of log 1–2 of *F. oxysporum*, but some spores survived at all concentrations (Runia, 1988). Ehret et al. (2001) showed that 15–25 ppm was adequate to control bacterial wilt in sweet pepper while minimizing phytotoxicity. Safety measures have to be taken for safe storage and handling.

Hypochlorite might work against a number of pathogens, not all, but at the same time Na^+ and Cl^- concentration is increased in a closed growing system which will also lead to levels which decrease productivity of the crop and at which the nutrient solution has to be leached. Despite the above-mentioned drawbacks, the product is used in the Mediterranean area as a cheap and useful method, recommended by sellers.

10.4.5.4 Iodine

The principle of iodine disinfestation was based on the flow of the nutrient solution along cartridges which contain resins which release iodine. Getting an exact dose of iodine during a longer period is rather difficult. At the specific equipment iodine content meters were applied. Runia (1994c) proved that the efficacy against viruses and fungi was insufficient.

10.4.6 ELECTROMAGNETIC RADIATION

A number of methods to eliminate pathogens are based on electromagnetic (EM) radiation. EM radiation is characterized by its frequency and its wavelength. The lower the wavelength, the higher the energy content of a radiation, resulting in a temperature effect on the tissues and at longer exposure times, degeneration of the protein chains. Most radiation can not penetrate very deep into organisms, tissues, substrates or even turbid liquids. EM energy with a frequency of 3 GHz (wavelength < 30 cm) can penetrate only 10 cm, while waves with a frequency lower than 1 GHz (wavelength < 10 cm) can penetrate much deeper. In experiments for disinfection of the nutrient solution and substrates, microwaves (2450 MHz or longer; limited penetration) and radio waves (13 or 27 MHz, deep penetration; Rasing and Jansen, 2005; Vegter, 2005) are used. Both methods might be technically effective, but energy input and investment are very high and, consequently, not yet suitable for disinfestation of a nutrient solution.

Gamma radiation can be created with a nuclear source and has a very high efficacy in the killing of micro-organisms. As the precautions around a nuclear source are so high, it is not feasible to place a source at each greenhouse. On the other hand, you can bring a crop to the nuclear source to eliminate bacteria and other pathogens. For example, strawberries and mushrooms have been treated in such a way. Ultraviolet radiation is another method which makes use of radiation of specific wavelength and which can be applied in soilless systems.

Ultra-violet radiation (UV radiation) is electromagnetic radiation with a wavelength between 200 and 400 nm. Wavelengths between 200 and 280 nm (UV-C), with an optimum at 254 nm, have a strong killing effect on micro-organisms, because it minimizes the multiplication of DNA chains. The energy dose of the radiation is expressed in mJ/cm^2 (= mW s/cm^2), at which mW is the quantity of UV-C energy, s is the duration of the energy and cm^2 is the area to which the energy attenuated. From experiments it is known that different levels of radiation are needed for different organisms so as to achieve the same level of efficacy. Runia (1988, 1994a, 1996) recommends a dose which varies from 100 mJ/cm^2 for eliminating bacteria and fungi to 250 mJ/cm^2 for eliminating viruses. Ehret et al. (2001) showed some successful results with lower dosages. Relatively high doses are needed to compensate for variations in water turbidity and variations in penetration of the energy into the solution due to low turbulence around the UV lamp or variations in output from the UV lamp. For commercial purpose, the high dosages are recommended.

Several types of lamps are on the market to create the required dose. Most distinctive are low- and high-pressure lamps. Both lamp types do a proper job if the demanded dose can be achieved. Low-pressure lamps have a low capacity (110–300 W), produce relatively less heat and are less expensive, but more lamps are needed to disinfect a certain amount of water. High-pressure lamps are mainly high-powered lamps (3–8 kW) for large quantities of water to be disinfected. The temperature of the disinfected water will be increased compared to the incoming solution. The temperature difference is less than 1°C for low-pressure lamps and less than 2°C for high-pressure lamps. Depending on the disinfestation capacity of the installation, there are one or more lamps. The nutrient solution flows in a thin layer (<1.5 cm) in a turbulent way around the UV-lamp in a specified time to give it a good exposure.

Good installations measure the transmittance of the water with a light cell. If the transmittance is too low, too little UV-C can penetrate through the solution resulting in less elimination of pathogens. Low transmittance often results from too much organic load in the solution. An extra rapid sand filter can solve this problem. If the initial capacity is sufficient, it can be mixed with cleaner source water (rainwater, tap water). Another option is to increase the dose of UV-C, which is automatically possible in more sophisticated installations with high-pressure lamps. Those installations measure transmittance automatically and adapt the dose to the transmittance. Because of the reaction of UV-C with organic particles, frequent cleaning is needed to remove the deposits that tend to develop on the lamp. Some installations conduct an automatic frequent cleaning of the lamp. Another aspect of UV radiation is the breakdown of the used iron chelate (about 10–20 per cent). It results in a higher deposit on the lamp and an additional fertilizer demand to the nutrient solution.

10.4.7 ACTIVE CARBON ADSORPTION

Active carbon is specially produced to achieve a large internal surface area (500–1500 m^2/g) for adsorption of especially organic, non-polar substances. Also halogenated substances, odours and tastes can be adsorbed. For this, the water to be cleaned is lead through a column (Lenntech, 2005) which contains active carbon. Water flows constantly through the column realizing an accumulation of substances in the filter. Regeneration of the filter has to take place when it looses 5–10 per cent of its efficiency. The method is used for drinking water treatment but not very much for the removal of pathogens. The method is too expensive, while performance is insufficient. However, active carbon filtration pops up every few years for disinfestations of the nutrient solution. In fact it is an expensive filter, while cheaper solutions are available. It also removes a substantial amount of dissolved fertilizer, which makes fertilization much more expensive.

10.4.8 COPPER IONISATION

Electrolysis of water by silver and copper electrodes releases positive-charged free Cu^+ ions into the water, which react with membranes of micro-organisms. Runia (1988) did not see a log 3 reduction for tomato mosaic virus and for *F. oxysporum* after a treatment of 2 h, one or four days. Later equipment on the commercial market claim disinfestation of the nutrient solution with an adjustable input of Cu ions. At the same time, the Cu input in the nutrient solution is much higher than the plant needs, which will lead to toxic levels in closed systems. However, pot plant growers claim a better growth and less loss of plants when using the apparatus (Kamminga, 2004). Besides, the release of heavy metals (silver, copper) into the environment is restricted by law in many countries.

10.4.9 CONCLUSIONS

Various methods are available for disinfesting nutrient solution. Some methods are not particularly effective in their present implementations: iodine, sodium hypochlorite, active carbon and copper ionization did not sufficiently eliminate bacteria, fungi and/or viruses. Sodium hypochlorite has the disadvantage in closed systems of accumulation of chloride and sodium. However, it is popular in Mediterranean countries against several bacteria and fungi and to decrease disease pressure, because of its low cost. Partly effective are hydrogen peroxide and slow sand filtration. Hydrogen peroxide is a cheap method, but concentrations to eliminate viruses can damage plant roots. It can be used to clean the drip irrigation lines. Slow sand filtration does not eliminate all pathogens, but it eliminates the generally occurring *Pythium* and *Phytophthora* species for 99.9 per cent and decrease disease pressure of other pathogens dramatically. Besides, the method is inexpensive and can also being built by the grower. Effective methods are ozone treatment, membrane filtration, heat treatment and UV radiation. Ozone and membrane filtration eliminate pathogens effectively, but these methods often demand high investments. Heat treatment and various versions of UV radiation resulted in 99.9 per cent elimination of pathogens including viruses. They are reliable and their performance can be checked. Both methods are economically feasible when quantities of nutrient solution to be disinfected are less than about 50 m^3 per hectare per day, which is similar to about 30 per cent drain water coming from a soilless tomato system which is watered by drip irrigation. However, if there is no natural gas available, heat treatment might be too expensive.

Further it is important to realize that disinfestation of the nutrient solution is like insurance: you have to continually pay the costs for disinfestation, but without disinfestation there is a chance that a pathogen can devastate the entire crop at some unexpected time. It is the risk that the grower needs to manage by deciding if disinfestation is prudent or not.

10.5 SYNTHESIS: COMBINED STRATEGIES

10.5.1 COMBINING STRATEGIES

For proper crop management, all environmental factors should be taken into account. Crop growth depends on a suitable substrate, climatic conditions, cultural and sanitation measurements. Also for effective disease control, several practices should be combined; that is growing strong plants, reduce numbers of pathogens, improving the disease suppressiveness of the system. Chemical pesticides can be applied if they are available and allowed to be used. In the previous sections, various methods were described for reducing the pathogen incidence in the irrigation water, improving suppressiveness of a particular system, and the technical possibilities for monitoring pathogens. The following examples provide combinations of these strategies.

10.5.2 COMBINING BIOLOGICAL CONTROL AGENTS AND DISINFESTATION

Biological control agents should be added as early as possible so as to achieve a stable microbial community with a maximum of beneficial organisms before pathogens are present. If biological control agents have to be added again at later stages, the addition into a disinfested solution has the advantage that the number of microorganisms in the disinfested solution is lower, and nutrients derived from dead cells are present.

An example of combining biological control agents with a disinfestation treatment is described by Garibaldi et al. (2003): biocontrol strains of *Trichoderma* and *Fusarium* showed good efficacy in reducing *Phytophthora cryptogea* root rot on gerbera (*Gerbera jamesonii* Bolus ex Hooker f.) in combination with slow sand filtration.

10.5.3 NON-PATHOGENIC MICROFLORA AFTER DISINFESTATION

The microflora was clearly changed in the nutrient solution immediately after the disinfestation treatments itself. Disinfestation by UV radiation eliminated 90–99.9 per cent of the micro-organisms and only a few bacterial species survived (van Os et al., 2004a). After slow sand filtration, over 90 per cent of the total aerobic bacterial population was still present. However, a shift in the population occurred, since the numbers of fluorescent pseudomonads, filamentous actinomycetes and fungi decreased more drastically than the total number of bacteria (van Os et al., 2004b). Also the potential of the microflora to utilize various carbon sources and the genetic profile of the bacterial population changed (van Os and Postma, 2000; van Os et al., 2004b). Therefore it was suggested that slow sand filtration would result in higher disease suppressiveness in the growing system compared to an almost complete sterilization such as UV (van Os et al., 2004a). However, this was not the case: a decrease of the disease suppressiveness due to the disinfestation treatments was never detected. Probably, the plant-driven microflora is not disturbed. Measurements showed that the microflora in the nutrient solution around the roots was changed only little or not at all. This is explained by the fact that only part of the solution, and, as a consequence, also part of the microflora is removed from the substrate around the roots.

10.5.4 ADDITION OF BENEFICIAL MICROBES TO SAND FILTERS

Slow sand filters always needs some time before they become highly effective. The addition of proper micro-organisms, which can be obtained from already effective filters, can enhance this process. The biological activation of a filter unit was very significantly enhanced by the addition of three *Pseudomonas putida* and two *Bacillus cereus* strains (Déniel et al., 2004). The bacteria-amended filter became effective for eliminating *Fusarium oxysporum* after one month, whereas the non-inoculated filter needed six months to become effective.

10.5.5 DETECTION OF PATHOGENIC AND BENEFICIAL MICRO-ORGANISMS

The development of techniques for simultaneous detection of many different micro-organisms in the nutrient solution will allow the detection of all kinds of potential pathogens and biocontrol agents. The big advantage will be that the balance between the presence of the added biocontrol agents and the development of the pathogen can be detected. Often for effective biocontrol, the biocontrol agent should be present in 10–100 times higher concentrations than the pathogen. At too low levels of the biocontrol agent, or too high levels of the pathogen, the grower can decide either to re-apply the biocontrol agent or to control the pathogen with other control treatments.

Also the numbers of beneficial micro-organisms involved in disease suppression, such as streptomycetes, can possibly be enumerated in combination with the development of a pathogenic population.

10.5.6 FUTURE

Future development will lead to monitoring systems for both pathogenic and beneficial organisms. Based upon specific DNA sequences, multiplex detection methods will be developed, for qualitative as well as quantitative detection and monitoring. If a certain pathogen is found to exceed its threshold level, an automatic alarm system would warn the grower to take appropriate action (i.e. application of chemical, physical or biological control, altered climate or water management). This gives the grower the possibility to act directly on the levels of different micro-organisms. Detection methods will be combined with high-tech developments in the greenhouse.

A more ecological strategy using suppressive properties of competing micro-organisms could also be implemented. If soilless systems can be created with disease suppressiveness comparable to natural soils, they will combine the advantages of soil systems and soilless substrates, that is a microbiologically well-buffered culture system, as well as the possibility to have a pathogen-free start by renewing the substrate. The exploitation of the indigenous microflora to suppress disease in soilless systems might be a new trend, which is breaking with the 'sterility' concept commonly used in soilless systems. For the practical application of this strategy, it is important to understand how growers or soilless substrate suppliers can enhance the abundance or activity of a suppressive microflora aiming at a microbiologically well-buffered culture system.

ACKNOWLEDGEMENTS

The information described in this chapter was obtained in research projects supported by the Dutch Ministry of Agriculture, Nature and Food Quality and the Commission of the European Communities, Agriculture and Fisheries (MIOPRODIS, FAIR CT98-4309).

REFERENCES

Agrios, G.N. (1988). *Plant Pathology* (3rd edn). San Diego, USA: Academic Press, Inc., 803pp.

Alexander, M. (1977). *Introduction to Soil Microbiology* (2nd edn). New York, USA: John Wiley & Sons, 467pp.

Berkelmann, B. (1992). Characterisierung der Bakterienflora und des antagonistischen Potentials in der zirkulierenden Nährlösung einer Tomatenkultur (*Lysopersicon esculatum* MILL) in Steinwolle. *Geisenheimer Berichte* Band 10.

Bolton, H. Jr., Fredrickson, J.K. and Elliot, L.F. (1992). Microbial ecology of the rhizosphere. In *Soil Microbial Ecology: Application in Agriculture and Environmental Management* (F.B. Metting, ed.). New York, USA: Marcel Dekker Inc., pp. 27–63.

Bonants, P.J.M., Schoen, C.D., Szemes, M., et al. (2005). From single to multiplex detection of plant pathogens: PUMA, a new concept of multiplex detection using microarrays. *Phytopathol. Pol.*, **35**, 29–47.

Braun, A. and Supkoff, D. (1994). "Options to Methyl Bromide for the Control of Soil-borne Diseases and Pests in California with Reference to the Netherlands", Pest Management Analysis and Planning Program, California Environmental Protection Agency, Department of Pesticide Regulation, Sacremento, CA, July 1994.

Calvo-Bado, L.A., Petch, G., Parsons, N.R., et al. (2006). Microbial community responses associated with the development of Oomycete plant pathogens on tomato roots in soil-less growing systems. *J. Appl. Microbiol.*, **100**, 1194–1207.

Campbell, R. (1989). *Biological Control of Microbial Plant Pathogens*. Cambridge, UK: Cambridge University Press, 218pp.

Chen, C., Bélanger, R.R., Benhamou, N. and Paulitz, T.C. (1999). Role of salicylic acid in systemic resistance induced by *Pseudomonas* spp. against *Pythium aphanidermatum* in cucumber roots. *Eur. J. Plant Pathol.*, **105**, 477–486.

De Souza, J.T., de Boer, M., de Waard, P., et al. (2003). Biochemical, genetic, and zoosporicidal properties of cyclic lipopeptide surfactants produced by Pseudomonas fluorescens. *Appl. Environ. Microbiol.*, **69**, 7161–7172.

Dehne, H.-W., Adam, G., Diekmann, M., et al. (1997). *Diagnosis and Identification of Plant Pathogens: Developments in Plant Pathology*, Vol. 11. Dordrecht, The Netherlands: Kluwer Academic Publishers, 556pp.

Déniel, F., Rey, P., Chérif, M., et al. (2004). Indigenous bacteria with antagonistic and plant growth-promoting activities improve slow-filtration efficiency in soilless cultivation. *Can. J. Microbiol.*, **50**, 499–508.

Duijff, B.J., Pouhair, D., Olivain, C., et al. (1998). Implication of systemic induced resistance in the suppression of fusarium wilt of tomato by *Pseudomonas fluorescens* WCS417r and by nonpathogenic *Fusarium oxysporum* Fo47. *Eur. J. Plant Pathol.*, **104**, 903–910.

Ehret, D.L., Alsanius, B., Wohanka, W., et al. (2001). Disinfestation of recirculating nutrient solutions in greenhouse horticulture. *Agronomie*, **21**, 323–339.

Folman, L.B., Clerkx, A.C.M., Postma, J. and Van Veen, J.A. (2001). Ecophysiological characterization of rhizosphere bacterial communities at different root locations and plant developmental stages of cucumber grown on stone wool. *Microb. Ecol.*, **42**, 586–597.

Folman, L.B., Postma, J. and Van Veen, J.A. (2003). Characterization of *Lysobacter enzymogenes* (Christensen and Cook, 1978) strain 3.1T8, a powerful antagonist of fungal diseases of cucumber. *Microbiol. Res.*, **158**, 107–115.

Garibaldi, A., Minuto, A., Graso, V. and Gullino, M.L. (2003). Application of selected antagonistic strains against Phytophthora cryptogea on gerbera in closed soilless systems with disinfection by slow sand filtration. *Crop Prot.*, **22**, 1053–1061.

Hallmann, J., Quadt-Hallmann, A., Mahaffee, W.F. and Kloepper J.W. (1997). Bacterial endophytes in agricultural crops (review). *Can. J. Microbiol.*, **43**, 895–914.

Hallmann, J. and Sikora, R.A. (1994). Occurrence of plant-parasitic nematodes and nonpathogenic species of fusarium in tomato plants in kenya and their role as mutualistic synergists for biological-control of root-knot nematodes. *Int. J. of Pest Manag.*, **40**, 321–325.

Hong, C.X. and Moorman, G.W. (2005). Plant pathogens in irrigation water: Challenges and opportunities. *Crit. Rev. Plant Sci.*, **24**, 189–208.

Kamminga, H. (2004). Gemagnetiseerd water vermindert uitval potplanten. *Vakblad voor de Bloemiserij*, **4**, 38–39 (in Dutch); (www.aqua-perl.dk).

Khalil, S. and Alsanius, B.W. (2001). Dynamics of the indigenous microflora inhabiting the root zone and the nutrient solution of tomato in a commercial closed greenhouse system. *Gartenbauwissenschaft*, **66**, 188–198.

Koohakan, P., Ikeda, H., Jeanaksorn, T., et al. (2004). Evaluation of the indigenous microorganisms in soilless culture: Occurrence and quantitative characteristics in the different growing media. *Sci. Hortic.*, **101**, 179–188.

Kritzman, G., Katan, J., Silverman, D., et al. (2000). Disinfestation of recycled irrigation water in closed soilless growing systems. In World Congress on Soilless Culture on *Agriculture in the Coming Millennium*, Israel, p. 50 (Abstract).

Le Quillec, S., Fabre, R. and Lesourd, D. (2003). Phytotoxicité sur tomate et chlorate de sodium. *Infos-Ctifl*, **197**, 40–43.

Lenntech (2005). www.lenntech.com.

Lievens, B., Brouwer, M., Vanachter, A.C.R.C., et al. (2005). Quantitative assessment of phytopathogenic fungi in various substrates using a DNA macroarray. *Environ. Microbiol.*, **7**, 1698–1710.

Lievens, B. and Thomma, B.P.H.J. (2005). Recent developments in pathogen detection arrays: Implications for fungal plant pathogens and use in practice. *Phytopathology*, **95**, 1374–1380.

McPherson, G.M. (1998). Root diseases in hydroponics – their control by disinfection and evidence for suppression in closed systems. In Proceedings 7th International Congress of *Plant Pathology*, Edinburgh (Abstract 3.8.1S).

Menzies, J.G., Ehret, D.L., Koch, C., et al. (2005). Fungi associated with roots of cucumber grown in different greenhouse root substrates. *Can. J. bot.*, **83**, 80–92.

Okubara, P.A., Schroeder, K.L. and Paultiz, T.C. (2005). Real-time polymerase chain reaction: Applications to studies on soilborne pathogens. *Can. J. Plant Pathol.*, **27**, 300–313.

Paulitz, T.C. and Bélanger, R.R. (2001). Biological control in greenhouse systems. *Annu. Rev. Phytopathol.*, **39**, 103–133.

Pettitt, T.R., Wakeham, A.J., Wainwright, M.F. and White, J.G. (2002). Comparison of serological, culture, and bait methods for detection of *Pythium* and *Phytophthora* zoospores in water. *Plant Pathol.*, **51**, 720–727.

Postma, J. (1996). Mechanisms of disease suppression in soilless cultures. *IOBC/WPRS Bulletin*, **19**, 85–94.

Postma, J., Alsanius, B.W., Whipps, J.M. and Wohanka, W. (2003). La microflora nei sistemi di coltivazione fuori suolo. *Informatore Fitopathligico*, **3**, 35–39.

Postma, J., Bonants, P.J.M. and van Os, E.A. (2001). Population dynamics of *Pythium aphanidermatum* in cucumber grown in closed systems. *Med. Fac. Landbouww. Gent*, **66/2a**, 47–59.

Postma, J., Geraats, B.P.J., Pastoor, R. and van Elsas, J.D. (2005). Characterization of the microbial community involved in the suppression of *Pythium aphanidermatum* in cucumber grown on stone wool. *Phytopathology*, **95**, 808–818.

Postma, J. and Luttikholt, A.J.G. (1996). Colonization of carnation stems by a nonpathogenic isolate of *Fusarium oxysporum* and its effect on *Fusarium oxysporum* f.sp. *dianthi*. *Canadian Journal of Botani*, **74**, 1841–1851.

Postma, J., Willemsen-de Klein, M.J.E.I.M. and van Elsas, J.D. (2000). Effect of the indigenous microflora on the development of root and crown rot caused by *Pythium aphanidermatum* in cucumber grown on stone wool. *Phytopathology*, **90**, 125–133.

Postma, J., Willemsen-de Klein, M.J.E.I.M., Rattink, H. and van Os, E.A. (2001). Disease suppressive soilless culture systems: Characterisation of its microflora. *Acta Hort.* (ISHS), **554**, 223–229.

Rasing, F.B. and Jansen, W.J.L. (2005). Diëlektrisch ontsmetten van substraten, een goed alternatief voor stomen? *Kema rapport* 50351992-KPS/MEC 04-7173, 74p. (in Dutch).

Rattink, H. (1996). Root pathogens in modern cultural systems: Assessment of risks and suggestions for integrated control. *IOBC wprs Bulletin*, **19**(6), 1–10.

Rattink, H. and Postma, J. (1996). Biological control of fusarium wilt in carnations on a recirculation system by a nonpathogenic *Fusarium oxysporum* isolate. *Med. Fac. Landbouww. Univ. Gent*, **61/2b**, 491–498.

Ruijs, M.N.A. (1994). Economic evaluation of closed production systems in glasshouse horticulture. *Acta Hort.* (ISHS), **340**, 87–94.

Runia, W. (1996). Disinfection of recirculation water from closed production systems. In Proceedings of the Seminar on *Closed Production Systems*, (E.A. van Os, ed.). IMAG-DLO report 96-01, pp. 20–24.

Runia, W. (1998). Recirculatiewater verhitten met lagere temperatuur. *Groenten & Fruit (Glasgroenten)*, **10**(juli), pp. 6–7.

Runia, W.Th. (1984). Naar verantwoord hergebruik drain water in substraatsystemen. *De Tuinderij*, **18** (oktober), 36–39 (in Dutch).

Runia, W.Th. (1988). *Elimination of Plant Pathogens from Soilless Cultures*. ISOSC Proceedings, 429–441.

Runia, W.Th. (1994a). Elimination of root-infecting pathogens in recirculation water from closed cultivation systems by ultra-violet radiation, *Acta Hort.* (ISHS), **361**, 361–369.

Runia, W.Th. (1994b). Disinfection of recirculation water from closed cultivation systemswith ozone, *Acta Hort.* (ISHS), **361**, 388–396.

Runia, W.Th. (1994c). Disinfection of recirculation water from closed cultivation systems with iodine. *Reports Faculty Agriculture University Gent*, **59/3a**, 1065–1070.

Runia, W.Th. and Amsing, J.J. (2000). Selectieve verhitting recirculatiewater bespaart energie. *Groenten and Fruit (Glasgroenten)*, **23**(June), pp. 34–35.

Runia, W.Th., Michielsen, J.M.G.P., van Kuik, A.J. and van Os, E.A. (1997). Elimination of root infecting pathogens in recirculation water by slow sand filtration. Proceedings 9th International Congress on soilless cultures, Jersey, pp. 395–408.

Runia, W.Th., van Os, E.A. and Bollen, G.J. (1988). Disinfection of drain water from soilless cultures by heat treatment. *Neth. J. of Agric. Sci.*, **36**, 231–238.

Runia, W.Th. and Paternotte, S.J. (1993). Activators zetten waterstofperoxide aan het werk. *Vakblad voor de Bloemisterij*, **7**, pp. 44–45 (in Dutch).

Schots, A., Dewey, F.M. and Oliver, R. (eds) (1994). *Modern Assays for Plant Pathogenic Fungi: Identification, Detection and Quantification*. Oxford, UK: CAB International, 267 pp.

Sharma, V.K. and Nowak, J. (1998). Enhancement of verticillium wilt resistance in tomato transplants by in vitro co-culture of seedlings with a plant growth promoting rhizobacterium (*Pseudomonas* sp. strain PsJN). *Can. J. Microbiol.*, **44**, 528–536.

Stanghellini, M.E. and Miller, R.M. (1997). Biosurfactants; their identity and potential efficacy in the biological control of zoosporic plant pathogens. *Plant Dis.*, **81**, 19–27.

Stanghellini, M.E. and Rasmussen, S.L. (1994). Hydroponics: A solution for zoosporic pathogens. *Plant Dis.*, **78**, 1129–1138.

Tu, J.C., Papadopoulos, A.P., Hao, X. and Zheng, J. (1999). The relationship of a pythium root rot and rhizosphere microorganisms in a closed circulating and an open system in stone wool culture of tomato. *Acta Hort.* (ISHS), **481**, 577–583.

Van der Gaag, D.J. and Wever, G. (2005). Conduciveness of different soilless growing media to pythium root and crown rot of cucumber under near-commercial conditions. *Eur. J. Plant Pathol.*, **112**, 31–41.

Van Os, E.A. (1999). Closed soilless growing systems: A sustainable solution for Dutch greenhouse horticulture. *Water Sci. Technol.*, **39**, 105–112.

Van Os, E.A., Amsing, J.J., van Kuik, A.J. and Willers, H. (1997b). Slow sand filtration: A method for the elimination of pathogens from a recirculating nutrient solution. Proceedings 18th Annual Conference Hydroponic Society of America, Windsor, Ontario, Canada, pp. 169–180.

Van Os, E.A. and Bruins, M.A. (2004). MIOPRODIS: Prevention of root diseases in closed soilless growing systems by microbial optimisation, a replacement for methyl bromide. Final report 01-03-1999 to 28-02-2003. Agrotechnology and Food Innovations, Report No. 089.

Van Os, E.A., Bruins, M.A., Van Buuren, J., et al. (1997a). Physical and chemical measurements in slow sand filters to disinfect recirculating nutrient solutions. Proceedings 9th International Congress on *Soilless Culture*, Jersey, pp. 313–328.

Van Os, E.A. and Postma J. (2000). Prevention of root diseases in closed soilless growing systems by microbial optimisation and slow sand filtration. *Acta Hort.* (ISHS), **532**, 97–102.

Van Os, E.A., Postma, J., Bruins, M. and Willemsen-de Klein, M.J.E.I.M. (2004a). Investigations on crop developments and microbial suppressiveness of *Pythium aphanidermatum* after different disinfection treatments of the circulating nutrient solution. *Acta Hort.* (ISHS), **644**, 563–570.

Van Os, E.A., Postma, J., Pettitt, T.R. and Wohanka, W. (2004b). Microbial optimisation in soilless cultivation: A replacement for methyl bromide. *Acta Hort.* (ISHS), **635**, 47–58.

Van Vuurde, J.W.L. and Postma, J. (1996). Methods for quantitative and in situ detection of low levels of viable soilborne bacteria and fungi. In *Principles and Practices of Managing Soilborne Plant Pathogens* (R. Hall, ed.). St. Paul, USA: APS Press, 330pp.

Vegter, B. (2005). Met "magnetron" goedkoper substraat ontsmetten. *Vakblad voor de Bloemisterij*, **8**, 46–47.

Whipps, J.M. (2001). Microbial interactions and biocontrol in the rhizocsphere. *J.Exp. Bot.*, **52**, 487–511.

Wohanka, W. (1995). Disinfection of recirculating nutrient solutions by slow sand filtration. *Acta Hort.* (ISHS), **382**, 246–255.

Wohanka, W., Lüdtke, H., Ahlers, H. and Lübke, M. (1999). Optimization of slow filtration as a means for disinfecting nutrient solutions. *Acta Hort.* (ISHS), **481**, 539–544.

Zhou, T. and Paulitz, T.C. (1994). Induced resistance in the biocontrol of Pythium aphanidermatum by Pseudomonas spp. on cucumber. *Journal of Phytopathology-Phytopathologische Zeitschrift*, **142**, 51–63.

11

ORGANIC SOILLESS MEDIA COMPONENTS

Michael Maher, Munoo Prasad and Michael Raviv

11.1 Introduction
11.2 Peat
11.3 Coir
11.4 Wood Fibre
11.5 Bark
11.6 Sawdust
11.7 Composted Plant Waste
11.8 Other Materials
11.9 Stability of Growing Media
11.10 Disease Suppression by Organic Growing Media
References

11.1 INTRODUCTION

Horticulturists have long used organic additives such as composted leaf material and animal manure to improve the physical and chemical properties of potting substrates. More recently organic components have replaced soil completely in growing media. This chapter deals with the most important organic materials used in soilless production: peat, coir, bark, wood products and compost. It describes their physical and chemical properties and their effect on plant performance. We also describe the composting process since many organic materials require composting before they can be used as a growing medium component. The biological stability of growing media and disease suppression in growing media are also reviewed.

Soilless culture of plants in containers involves a restricted root system and a reduced root-zone volume when compared with soil-based production. While details related to this are covered in Chaps. 2 and 13, it is important here to note that under these conditions the physical and chemical properties of the growing medium must be such as to provide adequate storage of water and nutrients for the plant, while maintaining good aeration. Peat has long been used as a component of potting mixes and has become the most widely used growing medium for containers as a complete growing medium by itself.

The use of peat in horticulture has recently been questioned from an environmental standpoint, since peat is a non-renewable resource and since it plays a major role in atmospheric CO_2 sequestration. It also has a major role in ascertaining the quality of groundwater in many parts of the world. In addition, peat bogs serve as a special habitat for wild plants and animals so that there is a significant interest in conserving peat bogs (Scott and Bragg, 1994). In Southern European countries (which typically have little indigenous peat), authorization to mine peat-lands is restricted so as to protect these ecosystems (Rivière and Caron, 2001). Even in peat-rich countries, there are trends towards greater restrictions on peat extraction with some forcing the ecological restoration of peat-lands.

Also, the public pressure to recycle industrial by-products is increasing. Alternative organic substrates in organic–inorganic media mixes include waste organic by-products (e.g. wood industry wastes, urban wastes, cork, wood fibres, livestock manure composts, coconut wastes etc.). While some of these have been in use for a long time, others have been tried more recently, sometimes with promising results (e.g. coir and coconut wastes). Alternative organic substrates that are well characterized and corrected by suitable blending with inorganic components make it possible to produce high-quality horticultural plants (Calkins et al., 1997) and at the same time contribute to the reduction of overexploitation of natural peat-lands. As such the trend towards using peat-free media is likely to continue (Carlile, 1999, 2003) and the new organic growing media components have to be studied on a scientific and technical basis while evaluating the environmental and social impact linked to their large-scale use in horticultural media mixes.

In addition, the dilution or substitution of peat with renewable materials is expected to extend the life of present peat resources. Chief among replacement materials has been bark and wood fibre from forestry and the wood industry. There is also increasing interest in coconut fibre (coir dust), green waste and other plant and animal residues.

11.2 PEAT

Peat is formed by the slow decomposition of mosses, such as sphagnum moss, and sedges in wet acidic ecosystems where such biomass accumulates because the conditions of low pH and low oxygen content are not conducive to microbial activity. The harvesting of peat from bogs is relevant to the peat products that end up in use in horticulture. In the past, once the layer of mosses and heathers had been removed from the top of the bog, sod cutting machines were used to remove brick-like sods from the sides of a trench. The sods were then laid out to dry before being transported away. More commonly, nowadays a strip of bog is 'milled' with a rotary cultivator to a depth of 10–20 mm. This is allowed to dry and then it is either swept to the side of the strip to form a linear pile or else it is vacuumed into a mobile hopper. The milling process is then repeated as often as possible during the harvesting season.

As a mined product it was the primary organic component of the first standardized growing media for plants in containers (Lawrence and Newell, 1939). Their mixtures consisted of loam (7 parts), peat (3 parts) and sand (2 parts). Baker and co-workers (Baker, 1957) developed the University of California or UC growing media based on peat/sand mixes. Further work in Germany (Penningsfeld and Kurzmann, 1966), Finland (Puustjarvi, 1973) and Ireland (Woods and Kenny, 1968) showed that peat could be used as a growing medium in its own right both for container plants and for vegetable and cut-flower production.

Peat became the major component of growing media throughout Europe and North America because of its ready availability and its excellent properties. It has high porosity, provides good aeration, and has high water-holding capacity. Its low pH and nutrient content make it easy to attain the desired levels through the addition of lime and fertilizers. It is practically free of plant pathogens, although in many cases it is conducive for their rapid development (Hoitink et al., 1977; Borrero et al., 2004). It is light and therefore inexpensive to transport. It is relatively stable physically, when used as a growing medium.

An estimate of the areas of deep peatland (>30 cm) is shown in Table 11.1 (Lappalainen, 1996). Peatlands occur over a wide range of climates but a major proportion occurs in the higher middle latitudes (50–60°N) in Canada, Russia and Finland. These peat deposits developed in the post-glacial period after the recession of the ice sheets 14,000–11,000 years ago.

Peat bogs can be classified as either ombrogenous (raised and blanket bogs), where growth of the peat layer is controlled by rainfall, or topographic (basin bogs), where deposition of new peat is controlled by topography and the groundwater table (Hammond, 1975). The initial formation of peat bogs occurred in post-glacial lakes or locally wet depressions. Thus peat bogs are formed initially thorough aquatic plants and subsequently through sedges and reeds (*Carex* and *Phragmites* spp.). The accumulation of plant debris and the lack of decomposition of senesced plant materials cause the bog to slowly expand which allows the growing plants to encroach into the lake with the upward displacement of the water. This gives rise to higher water tables and allows the development of peat-forming plants beyond the original lake margins. As peat formation continues, the groundwater effect diminishes and there is an increasing dependence on rainfall to supply nutrients. If this is sufficient, then there is a transition to a raised bog and species such as *Spahgnum* and *Hypnum* become dominant.

TABLE 11.1 Area of Peatland and Peat Production for Horticulture in Different Countries

	Area of peatland (km^2) over 30 cm deep[a]	Production for horticulture (000' m^3) in 1997[b]
Canada	1,113,270	7,250
Russia	568,000[c]	2,540
USA[d]	105,000	2,201
Finland	89,210	1,626
Sweden	45,940	1,203
Belarus	23,967	272
Norway	23,700	140
UK	17,549	2,500
Germany	14,205	9,000
Ireland	13,570	1,616
Poland	12,050	680
Estonia	10,091	3,497
Ukraine	6,932	85
Latvia	6,691	650
Lithuania	4,826	1,250

[a] Lappalainen (1996).
[b] Hood and Sopo (2000).
[c] Peatland area over 70 cm deep.
[d] Excluding Alaska.

The properties of the peat depend on the nature of the plant remains and their degree of decomposition. A method for measuring the degree of decomposition was devised by von Post (1922). A small quantity of peat is squeezed in the palm of one hand and the colour, consistency and proportion of water and peat exuded is expressed as a point on a scale of 1–10. Where clear colourless water is exuded, the peat is virtually undecomposed and has a value of H1 (Table 11.2). At the other end of the scale (H10) all the peat exudes between the fingers.

TABLE 11.2 The von Post Scale for Assessing Degree of Decomposition of Peat (von Post, 1922)

Degree of decomposition (H)	Quality of water exuded	Proportion of peat exuded
1	Clear, colourless	None
2	Almost clear, yellow brown	None
3	Slightly turbid, brown	None
4	Turbid brown	None
5	Very turbid, contains a little peat in suspension	Very little
6	Muddy, much peat in suspension	One-third
7	Very muddy	One-half
8	Thick mud, little free water	Two-thirds
9	No free water	Almost all
10	No free water	All

The International Peat Society has proposed a simplified classification system for peat based on its botanical composition, degree of decomposition and nutrient status (Kivinen, 1980) as follows.

Botanical composition	1. moss peat (predominantly sphagnum and other mosses). 2. sedge peat (sedges, grasses, herbs) 3. wood peat (remains of trees and woody shrubs)
Degree of decomposition	1. weakly decomposed (H1–H3) 2. medium decomposed (H4–H6) 3. strongly decomposed (H7–H10)
Trophic status	1. oligotrophic (low in nutrients) 2. mesotrophic 3. eutrophic (high in nutrients)

Peat for use in horticulture is often classified as light peat (H1–H3 on the von Post scale), dark peat (H4–H6) and black peat (H7–H10) (Bunt, 1988). The dark peats have higher lignin content and are less prone to biodegradation during cropping. Younger peats have lower lignin content and a higher microbial activity.

11.2.1 CHEMICAL PROPERTIES

The cation exchange capacity (CEC) is a measure of the ability of the growing medium to adsorb exchangeable cations which are available to the plant and will resist the leaching of nutrients during watering. It is usually expressed in terms of centimoles per kg of dry material ($cmol\,kg^{-1}$). For further discussion on CEC, see Chap. 11.6. The CEC is pH dependant. The CEC of peat in its natural state may be as low as $50\,cmol\,kg^{-1}$. After liming to a pH over 5.5, values over $100\,cmol\,kg^{-1}$ are obtained (Puustjarvi and Robertson, 1975). The CEC of some peat types is shown in Table 11.3. CEC values for peat are high but they vary with species and degree of decomposition. Sedge peat has lower CEC than those based on Sphagnum. The highly decomposed black peat with a high proportion of humic acids has the highest value.

The high CEC values mean that peat has a good capacity to store cationic nutrients for plants. However, by contrast, the anion exchange capacity is very low. This means,

TABLE 11.3 Cation Exchange Capacities of Different Peat Types (Puustjarvi and Robertson, 1975)

Species or peat type	Cation exchange capacity (CEC)	
	$cmol\,kg^{-1}$	$meq\,L^{-1}$
Undecomposed sphagnum moss peat	130	80
Sphagnum sedge peat	110	60
Sedge peat	80	40
Highly decomposed black peat	160	240

for instance, that, unlike in most soils, phosphate can easily be leached from peat growing media as will N present in nitrate form.

In general, the pH values for raw peat (untreated, as harvested from a bog) range from 3.5 to 4.1. This means it has to be limed to be suitable for plant growth. This is an advantage as any desired pH can be provided by adding the right amount of lime. It also means that peat is a suitable growing medium for calcifuge plants such as *Azalea*.

The quantity of available nutrients in peat is normally very low and can usually be ignored. This is in contrast to many other organic materials. This too can be regarded as an advantage because it affords better control in production since it is possible to achieve any desired balance of nutrients, and high initial fertility or EC levels are never a problem. Substrates that already contain a substantial fraction of the required nutrient create difficult problems in fertigation management, as it is difficult to sustain an optimal fertigation program.

11.2.2 PHYSICAL PROPERTIES

The physical properties of a peat vary with species composition of the originating bogs as this determines the particle size distribution and degree of decomposition (Puustjarvi and Robertson, 1975). Properties of some peats are shown in Table 11.4. In general, the bulk density is low and the total porosity is high. These are important advantages of most peats because it means that the growing medium can be well aerated and yet contain an adequate reservoir of water for the plant. The low bulk density helps to reduce transport costs of the material to the nursery, an important commercial consideration.

Sorting harvested peat into various size fractions by sieving affords the possibility of providing standardized peat growing media with a wide range of physical properties

TABLE 11.4 Some Physical Properties of Peat

Organic matter (per cent)	Bulk density (g L^{-1})	Pore space (per cent vol.)	Ave. Air-filled porosity[a]	Water content[a]	EAW[b] (per cent vol)	Reference
89–100	47–290	81–97	22	50–96	ND*	Kipp et al. (2000); Limits
96	113	93	14	79	ND	Mean
96	153	90.1	23.8	66.4	19.2	Maher et al. (2001) for Irish H5 peat
ND	177	87.8	14.6	ND	29.6	De Boodt and Verdonck (1972) for 'white' (H3) peat
ND	203	86.0	13.5	ND	16.0	Black (H7) peat

[a] Air-filled porosity and water content as per cent of volume at a tension of -1 kPa.

[b] EAW – Easily Available Water content – the difference between the water content at a tension of -1 kPa and that at -5 kPa.

* ND – not determined.

TABLE 11.5 Physical Properties of Fractionated Peat

Particle size (mm)	Pore space (per cent volume)	Air-filled porosity (per cent volume)	EAW (per cent volume)
	Prasad and Maher (1993)		
0–3	94.4	13.7	37.3
6–12	92.0	36.3	17.8
10–25	91.5	40.9	14.5
	Verdonck and Demeyer (2004)		
0–1	ND*	6.9	35.5
1–2	ND	37.5	27.4
3–5	ND	50.4	13.1

* ND – not determined.

on a predictable basis. Prasad and Maher (1993) studied the physical properties of fractionated peat (Table 11.5). They found air-filled porosity values in the 6–12 mm and the 10–25 mm fractions were well above those recommended for pot plants grown in flood benches (Wever, 1991) and for nursery stock crops (Aendekerk, 1989). The water-holding capacity was reduced although the "easily available water" (EAW) value of 17.8 per cent is close to the range recommended by De Boodt and Verdonck (1972). The air-filled porosity of substrate consisting of just the fine fraction was low but the water-holding capacity was high. The data of Verdonck and Demeyer in Table 11.5 illustrate the important effect of the particles below 1 mm in reducing the air-filled porosity.

Verhagen (1997) mixed milled peats from a number of sources. He found a strong relationship between the amount of particles smaller than 1mm in the raw peat components and the air-filled porosity of the mixes. Moreover, the relationship was the same for all the types of peat included in the research. The proportion of fine particles in the peats was a better predictor of the air-filled porosity of the peat mixtures than the air-filled porosity values of the individual peats. Prasad and Ni Chualáin (2004) similarly found a good relationship between particle size <1 mm and the air-filled porosity of various peat mixes. This work could be used as a tool to compose mixtures of any given air-filled porosity. Prasad and Ni Chualáin (2004) found similar relationships with non-peat materials but the slope of the regression line was different from that for peat and also varied between materials. Thus prediction of air-filled porosity of mixtures of peat and non-peat materials was less reliable.

Scharpf (1997) used mixtures of sieved peat fractions to create substrates with air-filled porosity varying from 6 to 35 per cent. The growth of New Guinea impatiens (*Impatiens* × *hawkeri*), Elatior begonia (*Begonia* × *hiemalis*) and gloxinia (*Sinningia speciosa*) was not affected. Of the pot plants examined, only Saintpaulia (*Saintpaulia ionantha* J.C. Wendt) showed growth depressions caused by oxygen deficiency. The root growth of bean seedlings was affected at air-filled porosities below 12 per cent. Prasad and Maher (1993) found that a coarse peat fraction with an air-filled porosity of 33 per cent performed as well as stone wool for tomato (*Lycopersicon esculentum*) production in a re-circulating soilless production system. In this case, where there was intermittent irrigation with a total nutrient solution, the ability of the growing medium to store nutrients and water was less important and adequate aeration was the critical factor. Fine peat fractions have been used in seedling production systems in cell trays where substrate fluidity and water-holding capacity are more important.

11.2.3 NUTRITION IN PEAT

The low pH and very low basic level of fertility of peat requires the addition of lime and nutrients to support good plant growth. The amount of nutrients needed depends on the species being grown and the stage of development. Seeding and propagation require only a low level of nutrients, greater amounts are required for rapidly growing plants. Table 11.6 shows additions of lime and N, P and K recommended by various scientists. These rates should be taken as indicative for general purpose mixes for growing plants in pots, different rates would be ideal for different purposes. Thus for seed sowing and for growing salt-sensitive plants, lower rates would be recommended. Traditionally recommendations for additions of N have been for around 200 mg/L, of which at least 50 per cent is in the form of nitrate (Schrock and Goldsberry, 1982), for general purpose growing media. In general, these fertilizer additions are in water-soluble form such as ammonium nitrate, superphosphate and potassium sulphate. Through the life of a pot plant, depending on the length of the cultivation period, supplementary feeding will be necessary to maintain plant growth. The base levels in Table 11.6 should maintain satisfactory growth of many seedlings for 4–5 weeks. If supplementary feeding is given earlier, then the base dressing can be reduced accordingly.

Lime is most commonly applied as dolomitic lime which contains both Ca and Mg. Some formulae use ground limestone but then Mg must be added separately, often in the form of magnesium sulphate (kieserite or epsom salts). The rate of lime varies widely as this is influenced by the peat type and the species to be grown. Mixes designed for calcifuge plants such as *Azalea* would receive a dressing of $1–1.5\,\text{kg}\,\text{m}^{-3}$ (Maher et al., 2000)

TABLE 11.6 Recommendations by Researchers in a Number of Countries for Base Levels of Fertilizer and Lime Addition to Peat Based Mixes

Country	Dolomitic lime ($\text{kg}\,\text{m}^{-3}$)	N ($\text{g}\,\text{m}^{-3}$)	P ($\text{g}\,\text{m}^{-3}$)	K ($\text{g}\,\text{m}^{-3}$)	Reference
Finland	8.0	110	105	182	Puustjarvi (1973)
Germany	2.0–5.0	180	80	200	Penningsfeld (1962)
Ireland	5.6	192	112	294	Gallagher (1975)
Netherlands	7.0	120	62	199	Klapwijk and Mostert (1992)
UK	2.25	230	120	290	Bunt (1988)
USA	5.0	192	143	317	White (1974)

Penningsfeld and Kurzmann (1966) classified various ornamental plants according to nutrient requirement and susceptibility to salt levels and varied his nutrient additions as follows:

Salt-sensitive plants (fertilized at 30–60 per cent of that shown in Table 11.6)

Pot plants: *Adiantum, Anthurium scherzerianum, Asparagus plumosus, Camellia, Erica gracilis, Gardenia,* orchid, *Primula obconica, Rhododendron simsii.*
Bedding plants: *Aquilegia, Begonia semperflorens, Callistephus, Dianthus hedwigii, Godetia, Verbena.*
Propagation: As seedling compost and for rooting cuttings.

Moderately sensitive plants (fertilized as shown in Table 11.6)

Pot plants: *Aechmea fasciata, Anenome, Anthurium andreanum, Aphelandra squalosa, Cyclamen, Euphorbia fulgens, Freesia, Gerbera, Gloxinia, Hydrangea, Monstera,* rose (*Rosa spp.*), *Sanservia,* sweetpea (*Lathyrus odoratus*), *Vriesia splendens.*
Bedding plants: *Campanula, Medium, Dianthus, Matricaria, Penstemon, Petunia, Salpiglossis,* Sweet William (*Dianthus barbatus*), *Tagetes,* wallflower (*Cheiranthus allionii*), *Zinnia.*

Salt tolerant plants (fertilized at 200 per cent as shown in Table 11.6)

Pot plants: *Asparagus sprengeri, Chrysanthemum, Pelargonium,* poinsettia (*Euphorbia pulcherrima*), *Saintpaulia.*

Micronutrients must also be added to peat-based mixes. These can be added as individual inorganic salts, as components of compound fertilizers or in slow-release form, for example as fritted trace elements (FTE). For example, in Gallagher (1975) the trace elements are applied as follows; B–1.0 g m^{-3} as Borax, Cu–3.5 g m^{-3} as copper sulphate, Mn–3.5 g m^{-3} as manganese sulphate, Zn–3.2 g m^{-3}, Mo–1.2 g m^{-3} Fe–7.0 g m^{-3} as ferrous sulphate and 2.5 g m^{-3} as iron chelate (EDTA). Alternatively these could be substituted by fritted trace elements (FTE 253A) applied at 400 g m^{-3}. The advantages of the fritted form are greater convenience, fewer problems in ensuring good uniformity during mixing and potentially greater persistence. The disadvantage is that of greater expense.

The rate of macronutrient addition prior to planting must also be adjusted to account for any supply of nutrients during production in the form of liquid feed. In many cases, an initial low rate of nutrients applied as base dressing will maintain adequate growth for a short period. As this begins to be depleted, liquid feeds must be used to supply nutrients for plant growth. This gives the grower more control over the rate of growth and is particularly important in the raising of vegetable seedlings for transplanting. Growing media marketed to the non-professional market, where the use of liquid feeds is not as developed, tend to have higher levels of nutrients in the media. This promotes active plant growth over a longer period where liquid feed is not given.

Controlled release fertilizers (CRFs) consist of prills of inorganic fertilizers coated with resin or polymers. Water enters the prill through small pores in the coating and the nutrients dissolve and diffuse outwards into the substrate solution. The rate of release of nutrient from the capsules depends on the thickness of the coating, the number of pores, the formulation of the fertilizer compounds inside the prill and the temperature. They are formulated to release nutrients over periods from 3–4 months to over 2 years with the approximate release characteristics described on the product label. CRFs allow the grower to apply sufficient nutrients for the life of the crop as an initial base dressing without the danger of high salt levels occurring as long as the temperatures are relatively moderate and the prills are not damaged. They thus simplify crop fertilization as the crop needs to only be irrigated with water. Their adoption has been greatest in the hardy nursery stock industry where their use has facilitated the development of containerized production of shrubs at the expense of field production. Producers of crops under protection with greater capacity to use liquid feeding have not adopted them to the same extent deterred by their increased cost and the greater control over plant development that liquid feeding affords growers.

11.3 COIR

11.3.1 PRODUCTION OF COIR

Coir is the name given to the fibrous material that constitutes the thick mesocarp (middle layer) of the coconut fruit (*Cocos nucifera*). The husk of the coconut contains approximately 75 per cent fibre and 25 per cent fine material, the so-called 'coir pith'. Husks are often soaked in water to soften them and facilitate grinding. Often the water used is brackish and this can increase the Na and Cl levels. When the coconut husks are being processed, the coco dust is separated from the fibre. The long fibres of coir extracted from the coconut husk are used in the manufacture of industrial products, for example mats or ropes. Traditionally the dust and small fibres were left behind and accumulated as a waste product. From the late 1980s, this material has been used as a growing medium or as a component of a growing medium. Most of the coir used is exported from Sri Lanka, Vietnam, India, Philippines, Mexico and Ivory Coast. Some manufacturers store the coir pith for 6 months or alternatively compost it for that period to get a stable physical product. It is important that at this stage there should be no contamination from weeds and potential human pathogens.

Since it is a natural product that is processed in a variety of ways, the chemical properties can vary considerably. In addition, the cultivation conditions of the coco palms can also have an effect on chemical properties.

When sodium, chloride and potassium levels are high in the coir, these elements have to be leached from the substrate before it can be used as a growing medium. When coir first became available as a horticultural substrate, this leaching was done with water; later it was found that some of the K and Na were in exchangeable form so that today it is leached with water containing a cation, usually calcium nitrate. This allows these excess elements to be reduced considerably. This material is referred to as 'treated coir' (de Kreij and van Leeuwen, 2001). The resulting growing medium is then dried and compressed into bricks or blocks to minimize shipping costs. This has facilitated the export of this material over long distances.

Prior to use, the blocks of compressed treated coir are broken apart, moistened and fertilized. The volume to volume expansion ratio on reconstitution of compressed coir is about six with 1 kg expanding to about 14 L when moistened. There are usually at least 3 or 4 different products produced from coir based on particle size. These products include chips of various sizes made by slicing husks into cubes, fibre of various lengths by itself and/or mixed with coir dust in various proportions, or coir dust material by itself.

11.3.2 CHEMICAL PROPERTIES

There is great variation in pH of the coir growing media with values ranging from 4.8 to 6.9 (Table 11.7). Some of the values are not directly comparable as the ratio of the substrate to extractant can vary from 1:1.5 to 1:5, but the change of ratio of this magnitude does not affect the pH significantly (Prasad unpublished)

There is also a great deal of variation in electrical conductivity (Table 11.7). Here again there are problems in comparing values as different extraction ratios were used by different researchers. The values have been converted from saturated media extract to 1:1.5 water extract for EC using the equation $y = 0.38x + 0.012$ (Bik and Boertje, 1975). In most cases, the relationship is linear between the ratios. Mc Lachlan et al., (2004) have also found a strong correlation between saturated extract values and various ratios of water extract, 1:1,1:2 1:5 and so on. Thus it is possible to compare EC values obtained with different extraction methods.

The EC of coir can range from low to very high, depending on how the coir was processed, whether it was washed in saline or fresh water, and whether it has been in outdoor storage for a long time and thus leached of Na, K and Cl. This will result in a lower EC. Elevated levels of EC are caused mainly by naturally occurring high levels of Na, Cl and K in water-soluble form. There appears to be great deal of variation in water-soluble and exchangeable K and Na in various sources of coir (Table 11.8).

There is not a great deal of information comparing water-soluble with exchangeable cations in coir, but the values of exchangeable cations are significantly higher than the

TABLE 11.7 Levels of pH and EC in Coir

pH	EC (mS cm^{-1})	Reference
4.9–6.4	0.17–2.32	Noguera et al. (2003b)
5.6–6.9	0.13–1.26	Evans et al. (1996)
4.8–6.8	0.32–0.97	Meerow (1994)
5.5–5.7	0.80–1.90	Handreck (1993)
5.0–5.7	0.12–1.51	Prasad (1997b)
4.9–6.6	0.32–0.41	Smith (1995)
6.0–6.7	0.2–0.4	Kipp et al. (2000) (coir dust)
5.9–6.1	0.2–0.9	Kipp et al. (2000) (coir chips)

Values of Noguera et al. (2003b), Evans et al. (1996), Meerow (1994) are based on saturated media extract, Smith (1995) on 1:5 water extract, and the others on 1:1.5 water extract. EC values have been converted to 1:1.5 water extract (see text).

TABLE 11.8 Chemical Properties of Coir (Water-Soluble Macronutrients) (mg L^{-1})

NH$_4$-N	NO$_3$-N	P	K	Ca	Mg	Cl	SO$_4$	Na	Reference
<1	<1	3–34	59–926	10–51	8–28	6–753	1–31	18–180	Noguera et al. (2003b)
–	–	0–25	10–368	7–10	–	5–724	–	10–46	Evans et al. (1996)
<1	<1.0	2.0–4.0	84–149	8–9	8–9	–	11–23	34–54	Meerow (1994)
0.9	0	–	–	–	–	–	–	–	Handreck (1993)
–	–	–	4–468	–	–	5–246	–	49–105	Prasad (1997b)
<10	–	1	60–102	<5.5	<2	91–169	–	31–81	Smith (1995)
–	–	–	63–112	–	–	110–276	–	17–47	Konduru et al. (1999)
–	–	5–31	59–922	9–51	8–25	11–76	0–9	18–123	Abad et al. (2002)
0	0–24	0–3	32–62	0–4	0–2	25.74	0–10	18–32	Kipp et al. (2000) (coir dust)
0	0–2	0–9	23–90	0–4	0–144	0–56	–	21–116	Kipp et al. (2000) (coir chips)
5	11	2	101	–	–	35	29	–	Van Doren (2001)

In this Chapter, substrate nutrient concentrations are expressed as weight per litre of substrate (not extract) unless otherwise stated.
Values of N, P, Cl, SO4, of Noguera Evans, Meerow and Abad have been converted from saturated media extract to 1:1.5 water extract using the following equation: $y = 0.378x - 4.8$. $R^2 = 0.908$, n = 96: for P $y = 0.377x - 0.1$, $R^2 = 0.913$ n = 147 and for K, Na, Ca Mg $y = 0.446 + 7.3$ $R^2 = 0.935$, n = 96. (unpublished Prasad)

TABLE 11.9 Chemical Properties of Coir (CaCl$_2$/DTPA Extractable Macronutrients) (mg L^{-1})

P	K	Ca	Mg	Na	Reference
–	183–222	100–172	36–58	85–92	Kipp et al. (2000) (coir dust)
–	47–98	56–60	31–79	30–78	Kipp et al. (2000) (coir chips)
8–17	304–720	6–15	8–28	110–114	Handreck (1993)
	69–128				Prasad (1997b)

water-soluble values (Tables 11.8 and 11.9) (Prasad, 1997b; Kipp et al., 2000). Prasad (1997b) also found that although there was a close correlation between water-soluble K and exchangeable K, the relationship was logarithmic at low-water soluble K. High exchangeable K values were recorded.

These data (Tables 11.8 and 11.9) indicate that coir generally has low P values; however, there can be samples with moderate levels of P. Nitrogen levels are very low and only occasionally traces of nitrate have been recorded with values less than 5 mg/L. Ca, Mg and Chloride values can also vary a great deal. It is not known whether the high Ca values recorded (Kipp et al., 2000) are due to its treatment with a cation, for example CaNO$_3$. Coir also contains significant amounts of micronutrients and these can also vary a great deal. As was the case with the macronutrients, extractable micronutrients are higher than water-soluble micronutrients (Tables 11.10 and 11.11).

There is some variation in the CEC of coir (Table 11.12). Evans et al. (1996) quote values from 39 and 60 cmol kg^{-1} and Noguera et al. (2003b) from 32 to 95 cmol kg^{-1}.

TABLE 11.10 Chemical Properties of Coir (Water Soluble Micronutrients) (µg L^{-1})

Fe	Mn	B	Zn	Cu	Reference
400–900	100	100–1000	70–110	30–1100	Meerow (1994)
40–340	30–150	210–320	20–140	20–140	Noguera et al. (2003b)
29–67	6	61–280	0–39	0–12	Kipp et al. (2000) coir dust
17–605	0–44	0–165	20–241	237	Kipp et al. (2000) coir chips
229	336	71	7	6	Van Doren (2001)

TABLE 11.11 Chemical Properties of Coir (CaCl$_2$/DTPA Extractable Micronutrients) (µg L^{-1})

Fe	Mn	B	Zn	Cu	Reference
	1100–1500	120–180	700–1300	170–2200	Handreck (1993)
79–157	814–1540	66–77	429–527	0–6	Kipp et al. (2000) coir dust
45–140	484–561	66–154	364–552	240–448	Kipp et al. (2000) coir chips
4100–7700	900–5000	200–400	500–1100	100–300	Prasad and Maher*

* Prasad and Maher unpublished data.

TABLE 11.12 Other Chemical Properties of Coir

Lignin (per cent)	Cellulose (per cent)	Hemi Cellulose (per cent)	C:N Ratio	FT-IR ratio 1600/1050	CEC (cmol kg^{-1})	Reference
35–54	23–43	3–12	75–186		32–95	Noguera et al. (2003b)
65–70	25–30					Meerow (1994)
					39–60	Evans et al. (1996)
39.5						Prasad & Maher (2004)
42				0.70		(Coir fine)*
37				0.64		(Coir fibre)*
					35–95	Abad et al. (2002)

* Prasad and Maher unpublished data.

These values are considered high and are similar to peat. Lignin content is between 35 and 54 per cent except for Meerow (1994), who found much higher percentages. Lignin content in fibre is slightly lower than that in the fines. The Fourier Transform Infrared ratio (FT-IR) 1600 cm^{-1}/1060 cm^{-1}, which is strongly correlated to lignin, shows the same trend. (Prasad and Maher, 2004)

11.3.3 PHYSICAL PROPERTIES

It is obvious from the data that there is a wide variation in values for air-filled porosity and easily available water in coir (Table 11.13). This is probably because

TABLE 11.13 Physical Properties of Coir

Total porosity	Air-filled porosity[a]	EAW[b]	BD[c] (g L^{-1})	Reference
	11.8			Smith (1995)
94–96	13.5–29.4	23–36	74–81	Prasad (1997b)
94–96	34.0–42.6	21–23	63–82	Martinez et al. (1997)
95	12	37.0		Raviv et al. (2001)
88.0	20.0	–	–	Konduru et al. (1999)
86–90	10–12		40–80	Evans et al. (1996)
93–95	18–34	25–33	73–116	Kipp et al. (2000) (coir dust)
95	55	3	72–82	Kipp et al. (2000) (coir chips)
94	20		93	Van Doren (2001)
95.1±0.1	7.7±1.9	41.1±2.5	88±6	Raviv*, new coir
95.2±0.1	7.8±1.0	39.9±2.3	82±5	Raviv*, used coir

[a] Air-filled porosity as per cent of volume at a tension of −1 kPa.

[b] EAW – Easily Available Water content- the difference between the water content at a tension of −1 kPa and that at −5 kPa.

[c] Bulk Density on dry matter basis.

* Unpublished. Used material was tested after 3 years of rose cultivation. Average of 6 samples ± SD.

particle size, and especially the proportion less than 1 mm diameter, is strongly correlated with aeration and water availability (Noguera et al., 2003a, Prasad and Ni Chualáin, 2004). The relationship between airspace and particle size is highly significant ($R^2 = 0.94$, $n = 8$) but different from peat, with coir having a more air-filled porosity than peat for a given percentage of fine material (Prasad and Ni Chualáin, 2004). Fornes et al. (2003) ascribed differences in physical properties between coir and peat to microstructure and porosity characteristics, with coir particles having a much higher relative surface porosity (41 per cent) than peat (12 per cent). In some cases, coir decomposition in the medium was rather rapid (Raviv, unpublished). However, well-stabilized coir maintains a reasonable physical stability (Raviv, unpublished, Table 11.13). The leaching of some of the fine fractions resulted with a slight decrease in bulk density and a slight decrease in EAW, both had no effect on plant performance. Nelson et al. (2004) showed that coir was more stable physically than sphagnum peat moss.

11.3.4 PLANT GROWTH IN COIR

The production of a number of crops has been tested in coir, and compared with conventional growing media. For instance, Raviv et al. (2001) found that the number of rose flowers was 19 per cent higher in coir than in the standard UC mix. Tomatoes grown in coir compared very well with those grown in stone wool, particularly under temperature stress, namely 30°C and 35°C compared with 25°C (Islam et al., 2002). It has been shown that a number of potted plants such as Calendula (*Calendula officinalis* LINN.), coleus (*Solenostemon scutellarioides*), fuchsia (*fuchsia* × *hybrida* L.), impatiens, schefflera (*Schefflera* spp. J.R. Forst. and G. Forst.), kalanchoe (*Kalanchoe* spp. Adans.), gerbera, begonia, Dendrathema and Primula (*Primula* spp.) performed as well in coir as in peat or better (Smith, 1995; de Kreij and van Leeuwen 2001). Nursery stock plants, for example, *Pentas lanceolata*, Ixora (*Ixora coccinea* L.) (Meerow, 1994), as well as Australian native plants (e.g. Grevillea) performed as well in coir as in peat (Offord et al., 1998). Quicker root development of tomatoes in coir and consequently better yield was found by Garcia and Deverde (1994). Rooting of cuttings has been found to be better in coir than in peat (Maher and Prasad, unpublished). Strawberry (*Fragaria* × *ananassa* Duchense) grown in coir performed better than in perlite (Lopez-Medina et al., 2004). Roses are widely grown in coir in Europe and North America.

11.4 WOOD FIBRE

11.4.1 PRODUCTION OF WOOD FIBRE

Wood fibre is a renewable resource produced from wood using specialized techniques. The source of wood can be fresh or waste wood (e.g. pallets) that is free from harmful substances such as heavy metals. Usually wood from spruce (*Picea* spp.) or pine trees are used. For example, in France, maritime pine (*Pinus pinaster*) is used as the main raw material. In Germany, the important timber species is spruce.

Wood fibre is generally made from lumber and as a substrate generally contains little or no bark. Bark is a separate substrate category (discussed below) largely because timber operations generally debark logs during lumber production. A final wood fibre product generally contains bark only when the wood raw material has been debarked incompletely or not at all. The wood fibre production technique usually consists of pressing the material through an aperture which results in high pressure and high temperature. Two screw presses constitute the heart of the wood fibre friction and this heats and shreds the material. In some cases, it is treated with steam (100–120°C). Nitrogen is often added and impregnated into the wood. At this stage, natural wood colorant, for example, coal dust, can be added to dye the wood. Due to the friction involved in the processing, the product is heated to 80–90°C and is therefore free from plant pathogens. Wood fibre can be produced as coarse or fine grades. Source material and fibre structure is not very important for wood fibre substrate. The most well-known wood fibre products available in the market today are Toresa, Hortifibre, Cultifibre, Piatel and Torbella. Each of these materials generally has similar production methods. Often wood fibre is used as a component of a peat growing media rather than on its own.

11.4.2 CHEMICAL PROPERTIES

The pH levels can vary from 3.8 to 6.6. Since the wood fibres are not well buffered, the addition of acid- or alkali-forming fertilizer can lead to big swings in pH. EC values are very low except where it is pre-treated with fertilizer at the manufacturing stage (Table 11.14).

Nitrogen concentration in wood fibre is typically very low (Table 11.15), so that a supplementary nitrogen source is added either at the manufacturing stage or by the user. Nutrient levels, particularly N, can thus vary considerably depending on the manufacturing process. When N is added during manufacturing, the water-soluble N level can increase during storage. K, Na, Mg and Ca levels can also vary considerably.

TABLE 11.14 pH and EC Levels of Wood Fibre

Material	pH	EC ms/cm^{-1}
Toresa (Gumi, 2001)	5.0–6.5	0.7–1.3
Torbella (Fischer et al., 1993)	6.5	–
Hortifibre (Lemaire et al., 1988)	4.5–5.1	–
Hortifibre*	3.9–5.1	
Hortifibre*	5.21	0.15
Toresa special*	5.6	0.55
Holsfasser*	5.6	0.23

*Prasad and Maher unpublished data.

11.4 WOOD FIBRE

TABLE 11.15 Chemical Properties of Wood Fibre (Water Soluble Nutrients) (mg L^{-1} Kipp et al., 2000)

NO$_3$-N	K	Na	Ca	Mg	Cl	SO$_4$	P	CEC
3	35	2	17	7	4	2	3	10[a]

[a] Lemaire et al. (1988).

Some wood fibres were shown to have somewhat elevated levels of P. Extractable nutrient levels are higher, indicating the presence of exchange sites in wood fibre (Table 11.16).

Occasionally Mn and Zn levels can vary over a wide range, but typically the concentrations and availabilities of other micronutrients are very low. Generally, extractable micronutrient levels are higher than water-soluble micronutrient levels (Table 11.17).

Lignin levels in wood fibre, except for Toresa, are lower than peat and the C/N ratio is high. (Table 11.18).

TABLE 11.16 Chemical Properties of Wood Fibre (Extractable Macro-Nutrients) mg L^{-1}

Source	N	K	Na	Ca	Mg	P	SO$_4$
Kipp et al. (2000)	–	0–200	0–8	379	0–122	–	–
Hortifibre (Lemaire et al., 1988)	34–35	5–7	3–5	3–5		5–8	
Torbella (Fischer et al, 1993)	6	131				24	
Torbella-Toresa (Grantzau, 1991)	<1–320	20–80	–	–	–	10–150	–
Pietal*	2.1	78	20		26	5	8
Hortifibre*	16	28	6		8	2	4
Toresa Special*	86	42	19		56	0.2	142
Toresa Holzfasser*	0.5	59	66		30	0.2	11

* Prasad and Maher unpublished data (CaCl$_2$/DTPA extractable).

TABLE 11.17 Chemical properties of Wood Fibre (Extractable Micro-Nutrients) (mg L^{-1})

	Fe	Mn	Zn	B	Cu
Kipp et al. (2000) (WS)	0–0.94	0–1.96	0–96	0–0.77	0–0.026
Kipp et al. (2000) (CAT)	0–3.12	0–10.5	0–5.15	0–0.44	0.044
Pietel (CAT)*	3.15	16.7	1.4	0.18	0.15
Hortifibre (CAT)*	1.05	3.6	0.38	0.05	0.06
Toresa Special (CAT)*	11.7	11.8	1.73	0.12	0.11
Toresa (Holzfasser) (CAT)*	4.65	10.1	2.99	0.20	0.20

* Prasad and Maher unpublished data.
WS = water soluble; CAT = CaCl$_2$/DTPA extractable.

TABLE 11.18 Chemical Composition of Wood Fibre

	Lignin (per cent dry matter)	C:N Ratio	FT-IR ratio 1600 cm^{-1}/1050 cm^{-1}
Hortifibre*	26.0	1020	0.35
Piatel*	29.4	89	0.54
Toresa*	35.1	80	0.64

* Prasad and Maher unpublished data.

11.4.3 PHYSICAL PROPERTIES

Wood fibre has high level of total porosity and in most cases a very high level of air-filled porosity and a rather low level of easily available water (Table 11.19). It also has a higher oxygen diffusion rate compared to peat (Clemmenson, 2004). In addition, as a result of mechanical compression, the physical properties of wood fibre can change considerably. Gruda and Schnitzler (2004) reported volume losses of 17–28 per cent, depending on the level of compression. Unlike composted bark, wood fibre has no issues with regard to phytotoxicity.

TABLE 11.19 Physical Properties of Wood Fibre

Material	Total porosity (per cent volume)	Air-filled porosity (per cent volume)	EAW (per cent volume)	BD (g L^{-1})
Wood fibre (Kipp et al., 2000)	95	47–78	6–21	87
Toresa (Gumi, 2001)	92–97	50–65		100–180
Cultifibre (Clemmenson, 2004)	94.2	65.6		89.8
Hortifibre[a]	93.7	58.6		96.7
Piatel[a]	92.4	42		117.7
Toresa[a]	90.1	64.1		153
Toresa Special[b]	93.2	51.2	14.6	83
Toresa Nova[b]	91.4	24.9	27.3	163

[a] Prasad and Maher, unpublished data.
[b] Gruda and Schnitzler, 2004.

11.4.4 NITROGEN IMMOBILIZATION

N immobilization of wood fibre growing media from various suppliers was studied by Prasad (1997a). Nitrogen immobilization induced by microbial activity was very high and approximately 150 mg L^{-1} of N disappeared during 5 weeks of incubation. Wood fibre treated with N showed release of N, depending on the nature of the treatment. For instance, where urea was added, the release was much faster than those treated with an organic fertilizer (hoof and horn meal) after 5 weeks (Prasad, 1997a).

Gruda et al. (2000) found nitrogen immobilization of similar order of magnitude in untreated wood fibre while in N-treated wood fibre the immobilization was similar to peat. Users of wood fibre substrates need to be aware of the manufacturing process involved specifically whether N has been added. Where no N is added, up to 400 mg L^{-1} of water-soluble N may need to be added to counteract N immobilization.

11.4.5 CROP PRODUCTION IN WOOD FIBRE

A number of crops have been grown successfully in 100 per cent wood fibre or in mixtures of wood fibre and other growing media such as peat. These have included young vegetable seedlings for transplant such as lettuce (*Lactuca sativa* L.), cabbage (*Brassica oleracea*), parsley (*Petroselinum crispum*) and sweet basil (*Ocimum basilicum* L.). Other crops that have been grown successfully to maturity in production systems using predominately wood fibre include tomatoes, pepper (*Capsicum annuum*), cucumber (*Cucumis sativus*) and nursery stock plants (Bohne, 2004). Generally, N-impregnated wood fibre performed better than a non-impregnated one (Gruda and Schnitzler, 1999). It was generally found that watering had to be done more frequently than in peat. For instance, watering at −6 kPa was ideal but not at −9 kPa which reduced growth in wood fibre and wood fibre mixtures. Flowering was also earlier in Pelargonium grown in wood fibre (Gerber et al., 1999). Growers need to be aware that extra N may need to be added to the standard feeds early in the production period.

11.4.6 THE COMPOSTING PROCESS

The composting process is the biological decomposition of organic materials under controlled conditions into a stable humus-like product (Golueke, 1972). The process is aerobic and part of it is carried out under thermophilic conditions. Thorough composting of organic materials achieves a number of important objectives. It reduces phytotoxicity, pathogens and weed seeds, and stabilizes the material with respect to N and oxygen demand. It results in an attractive, easily handled material that can be used for soil conditioning and as a component of a soilless growing medium. The subject of compost production for horticultural purposes was recently reviewed by Raviv (2005).

The C to N ratio is an important variable in composting, and although various organic waste materials have been successfully composted with C:N ratios ranging between 17 and 78, the desired range for the source material is between 25 and 35 (Day and Shaw, 2001). If the value is much higher than this, then the rate of decomposition will be very slow and the resulting compost will not be properly stabilized. At lower C:N ratios the loss of N in the form of ammonia (NH$_3$) to the air will be high. The desired C:N ratio can be achieved by analysing the feedstock and correcting its C or N contents. Green waste may typically have a C:N ratio of 45–50 and would need to be combined with a material with a high N content such as food waste or poultry manure to results in a satisfactory compost. The N content is measured as total Kjeldahl N, while the C content can be estimated from the ash content (Golueke, 1972) by

$$\text{per cent C} = (100 - \text{per cent ash})/1.8$$

It should be noted that while the C:N ratio is routinely used as an indicator of compostability, there are significant differences in the biodegradability of various organic materials that are unrelated to the C:N ratio (Day and Shaw, 2001). Other elements are generally assumed not to be limiting the composting process except that P may need to be added when composting materials with high cellulose content such as municipal solid waste (Brown et al., 1998).

The C:N ratio falls during composting as CO_2 is liberated and a final value in the range of 15–20 is generally aimed for (Kayhanian and Tchobanoglous, 1993).

Material to be composted is first shredded to reduce particle size and to promote microbial invasion into the substrate to accelerate the composting process. Shredding also improves handling of the material and facilitates pre-screening and mixing (Rynk and Richard, 2001). Immediately after the material has been ground and any additions of N and water are made, the temperature begins to rise as microbiological activity increases. After this initial period of a few days, the temperature will typically be over 60°C. This is the thermophilic stage. It lasts until the easily decomposable organic matter has been broken down. It is important for uniform composting that all of the material be exposed to similar temperatures. This is accomplished in many systems by regular turning of the composting pile. Where turning is not done as in static piles, temperature uniformity throughout the pile is an important consideration in the design of the composting system. Once the easily decomposable matter is decomposed, the temperature starts to decrease and as it falls below 40°C the compost will be re-invaded by mesophilic micro-organisms. In the last stage the active composting phase gives way to the curing period where the compost matures and becomes ready for use as a component of a growing medium.

Compost stability and maturity is an important characteristic in relation to its use as a growing medium component. The terms 'stability' and 'maturity' are used interchangeably by some, others relate maturity to readiness primarily to support plant growth in horticulture or agriculture. If active decomposition continues after incorporation into a growing medium, then this will affect plant growth due to reduced levels of available oxygen and nitrogen or the presence of phytotoxic compounds. Consequently tests have been developed to evaluate the stability maturity/phytotoxicity of composts. Common stability and maturity tests are respirometric tests based on O_2 consumption and CO_2 evolution (Lasaridi and Stentiford, 1998; Scaglia et al., 2000) or self-heating tests (Kehre and Thelen-Jungling, 2006). Rate of heat production using microcalorimeter was suggested as a very sensitive measure of compost maturity (Laor et al., 2004). Chemical tests such as the nitrate/ammonium ratio, C:N ratio and content of volatile organic acids are used in various countries to assess compost maturity (Hogg et al., 2002). Inbar et al. (1989) used solid-state ^{13}C nuclear magnetic resonance and infrared spectroscopy to determine the degree of composting. Phytotoxic tests generally involve a bioassay to screen for negative impacts on seed germination and plant growth.

11.5 BARK

Ground bark has being used as a component in growing media for decades. On rare occasions (e.g. orchid production), bark is the sole component of the medium. The horticultural uses for bark have been thoroughly reviewed (Aaron, 1976; Raviv et al., 1986). Much of the bark comes from coniferous trees, although bark from hardwood trees is sometimes used particularly in North America. Prior to the 1970s, bark was generally considered to be a waste product and was burned for energy generation in sawmills. Since then it has become widely used in horticulture, mostly as decorative mulch and as a growing medium or as a component of growing media. The use has become so widespread that there are occasionally shortages of bark.

"Bark" is a term which includes the inner bark (phloem) and the outer bark (rhytidone). Bark is removed by lumber companies by using debarkers or by water friction. To be suitable as a growing medium component, bark is hammer-milled and screened to achieve a desirable particle size (usually 10–20 mm) and to achieve a consistent product. Much of the coniferous bark consists of pine bark from *Pinus radiata, Pinus taeda, Pinus pinaster, Pinus nigra* and from spruce bark.

11.5.1 CHEMICAL PROPERTIES

The composition of bark varies according to species, growth conditions, age of the trees, and even time of the year it was harvested (Solbraa, 1979). Generally, lignin content is higher in pine bark than spruce bark. Birch has the lowest levels of lignin (Table 11.20).

Pine bark can be used as a growing medium without composting (Prasad, 1980). However, it needs to be aged to reduce phytotoxicity, and some adjustments have to be made, for example, by adding extra N to overcome nitrogen immobilization, either by adding extra N as a base dressing in crop production or through a controlled release fertilizer or in the liquid feed (Prasad, 1980).

Details of bark composting have been described by Solbraa (1979). When bark is composted, supplemental N is provided to overcome N immobilization and reduce toxicity. Other elements are typically not added during composting (Maher and Prasad, unpublished) although sometimes phosphorus is also recommended (Solbraa, 1979).

TABLE 11.20 Chemical Composition of Bark

Material	Lignin	Cellulose/Hemicellulose
Pine (Morel et al., 2000)	43	42
Spruce (Solbraa, 1979)	20–35	45–55
Pine (Solbraa, 1979)	30–50	20–50
Pine*	35–36	
Birch (Solbraa, 1979)	20	35–40

* Prasad and Maher unpublished data.

Other vital parameters during composting are moisture, gaseous oxygen and temperature. A rate of urea addition of 3 per cent of the initial bark dry weight was considered to be optimum. As a result of composting there is a decrease in easily decomposable organic compounds (Solbraa, 1979). Bark may contain substances, such as terpenes, which can be inhibitory to growth of plants; composting the bark prior to use as a substrate eliminates these compounds. Some of these compounds are also water soluble so that leaching with water can reduce the toxic compounds.

Nitrogen levels are lowest in *Pinus radiata*, while bark of spruce and birch has higher levels of N; also, spruce has higher levels of K than pine (Solbraa, 1979, Table 11.21). pH is much lower in pine bark (*Pinus radiata*) than in spruce bark. Most nutrients are low to moderate except for potassium (Table 11.22). Bark has a relatively high level of iron, and some bark can have very high levels of Mn. This can cause iron deficiency and/or manganese toxicity. Maher and Thomson (1991) found that the Mn content of the bark was closely related to the exchangeable rather than total Mn level in the soil in which the tree was grown. Water-soluble and $CaCl_2$/DTPA extractable micronutrients in Irish bark are shown in Table 11.23. This shows that there are considerable amounts of extractable Mn and Zn in Irish bark. In growing

TABLE 11.21 Total Nutrient Composition of Bark (Per cent Dry Matter)

Species	N	P	K	Ca	Mg
Spruce	0.37	0.06	0.28	1.02	0.09
Pine	0.26	0.02	0.1	0.33	0.02
Birch	0.48	0.03	0.19	0.19	0.06

TABLE 11.22 Physico-Chemical Characteristics and Chemical Composition of Bark (Water Soluble) mg L^{-1}

Bark source	pH	EC (μS/cm)	N	P	K	C:N	CEC (cmol kg^{-1})
Pine (Morel et al., 2000)	6.6	90	42	11	238	170	198
Pine (Bunt, 1976)	4.0–4.3 (US)						52–57
	5.0–5.2 (Europe)						

TABLE 11.23 Extractable Micronutrients in Irish Spruce (*Picea sitchensis*) Bark, Water Soluble and $CaCl_2$/DTPA (Maher, Unpublished Data) (mg L^{-1})

Extractant	Zn	Cu	Mn
Water	0.6	0.2	0.2
$CaCl_2$/DTPA	13	0.5	100

trials (Maher and Prasad, unpublished) there was no response to the addition of Mn and Zn. Consequently it was decided not to add Mn or Zn to bark-based growing media in Ireland.

11.5.2 NITROGEN IMMOBILIZATION

Properly composted bark shows little release or N immobilization (Prasad, 1997a) while uncomposted spruce bark showed very high N immobilization. Uncomposted pine bark showed lower N immobilization. This is probably related to higher lignin content. *Pinus radiata* bark showed reduction in water-soluble P (Prasad, 1980). The mechanism of this is not clear.

11.5.3 PHYSICAL PROPERTIES

Bark mixes generally have a high level of air-filled porosity and low water-holding capacity. For a given percentage of particles with size less than 1 mm, there is a higher level of air-filled porosity than would be the case with peat which may be due to the internal porosity of pine bark (Pokorny et al., 1984). Mixing bark with peat can result in improved water-holding capacity (Beardsell et al., 1979). The proportion of each results in differences in the physical properties (Table 11.24). Bark addition to a substrate increases the bulk density markedly and reduces the pore space slightly. As the proportion of bark is increased from 50 to 100 per cent, the air-filled porosity increases from 26.7 to 31.5 per cent and the EAW shrinks from 13.8 to 7.2 per cent.

TABLE 11.24 Effect of Addition of Bark to Peat on the Physical Properties of the Mixes (Maher et al., 2001)

Rate of bark (per cent)	Bulk density ($g\,L^{-1}$)	Pore space (per cent)	Air-filled porosity (per cent)	EAW (per cent)
0	153	90.1	23.8	19.2
12.5	172	89.5	24.7	16.8
25	185	88.6	24.4	16.4
50	210	87.6	26.7	13.8
100	263	85.5	31.5	7.2

11.5.4 PLANT GROWTH

Bark should be composted prior to use in a growing medium and there are numerous publications showing its successful application. However, there are cases where non-composted bark had been used successfully (Lemaire et al., 1980; Prasad, 1980). Bark is also suitable as propagation medium both under high humidity conditions and under intermittent mist (Pokorny and Austin, 1982; Lorenzo et al., 1982) and as a growing medium for various epiphytic plants. As such, bark is very widely used as a growing medium for species of orchids and bromeliads.

11.6 SAWDUST

Sawdust has been used as a growing medium component or by itself (Raviv et al., 1986). When used by itself it is mainly as a replacement for stone wool in bags. It is mainly used in Australia, Canada, New Zealand and the Pacific coast of the United States (Maas and Adamson, 1975; Worrall, 1978; Prasad, 1979). Usually uncomposted sawdust is not suitable due to phytotoxicity (Maas and Adamson, 1975; Worrall, 1978). Sawdust generally has a high percentage of air-filled porosity and a low content of available water (Prasad, 1979; Beardsell et al., 1979) but this is also affected by particle size (Prasad and Ni Chualain, 2004). The low available water content can result in water stress during active plant growth (Allaire et al., 2005; Dorais et al., 2005). Due to the low moisture retention, irrigation must be applied frequently and in small quantities (Favaro et al., 2002). The bulk density of dry sawdust is low at 124–154 g L^{-1} (Prasad, 1979). Chemically it has a neutral or moderately low pH (6.3–7.7) and very low EC (Prasad, 1979). One problem has been the high N immobilization rate (Prasad, 1997a) and the low lignin content. Greenhouse vegetables such as tomatoes and cucumbers have been grown using sawdust with as good results as with stone wool (Maree, 1994; Parks et al., 2004). Worrall (1981) found that the growth rates of foliage plants in media containing 50–80 per cent of composted hardwood sawdust were equivalent to, or significantly better than, growth in media containing an equal percentage of sphagnum peat and receiving the same level of liquid or slow-release fertilizer. However, some negative results have also been reported by Worrall (1978) and Prasad (1980a). Prasad and Maher (2006) reported that with a number of nursery stock species, compost produced from a mixture of sawdust and spent brewers grain could successfully substitute for peat in a growing medium up to a rate of 100 per cent. Where negative results have been reported, they are often associated with no or incomplete composting (Prasad, 1980a) or insufficient neutralization of phytotoxic effects (Worrall, 1978).

11.7 COMPOSTED PLANT WASTE

The physical and chemical properties of composted materials depend on the source materials used. In an international study on disease suppressiveness in compost (Tables 11.25–11.27) (Noble, 2005), N contents of spent mushroom compost (SMC) samples from various countries were similar. The P and K content of SMC were higher than the other materials except for French SMC which had a low P content. Irish biowaste had twice the N content of green waste. Onion waste had the lowest lignin content but the highest water-soluble carbohydrate content probably as a result of a shorter composting period. Compost made from grape marc was more lignified than those made from other wastes. SMC and onion waste were the wettest materials, except for the Greek SMC which had high dry matter content.

Composted green wastes (CGW) from various countries were similar. They were also drier and had lower total N, P and K levels than SMC. Composts prepared

TABLE 11.25 Chemical Properties of Some Composted Materials Found by Researchers from Various Countries Participating in an International Project on the Disease Suppressive Properties of Composts (Noble, 2005)

Composted materials	Country	Dry matter (per cent)	N	P	K	WSC[a]	Lignin (per cent of DM)
			\multicolumn{3}{c}{(g kg^{-1} DM)}		
Spent mushroom compost (SMC)	UK (fresh)	41.0	26.7	9.6	26.0	1.8	22
	UK (mature)	37.3	20.8	7.2	12.1	1.3	22
	Ireland	32.7	24.5	7.7	29.7	2.2	19
	France	48.7	17.8	1.8	18.1	1.2	15
	Greece	83.4	20.3	4.0	12.2	1.7	21
Green waste	UK	72.8	–	–	–		24
	Ireland	61.3	11.2	1.9	9.8	0.9	20
	France	56.9	17.1	4.1	10.8	1.0	21
	Netherlands	60.7	–	–	–		20
Biowaste	Ireland	74.6	24.6	5.9	11.0	1.9	28
	UK	68.3	–	–	–	–	–
Onion waste	UK	25.5	18.4	3.3	14.0	30.4	12
Olive leaves	Greece	48.7	21.6	1.4	5.6	1.0	17
Grape marc	Greece	73.4	–	–	–	0.7	39
SCM[b]+orange peels+straw	Israel	55.0	23.0	2.1	10.0	–	–
SCM[b]+pepper plants	Israel	55.0	20.0	3.6	33.1	–	–

[a] WSC – Water-soluble carbohydrate.
[b] SCM – Separated Cattle Manure.

TABLE 11.26 Water-Soluble Nutrient Levels in Composted Materials (mg L^{-1}) Found by Researchers from Various Countries Participating in an International Project on the Disease Suppressive Properties of Composts (Noble, 2005)

Composted material	Country	pH	EC[a]	P	K	NH$_4$-N	NO$_3$-N
SMC	UK (F)	7.4	549	75	2300	0	440
SMC	UK (M)	6.9	209	35	1105	0	250
SMC	Ireland	7.1	348	61	2961	42	7
SMC	France	7.8	361	55	2260	350	0
SMC	Greece	8.0	359	82	2145	275	70
Green waste	Ireland	8.0	151	14	1009	0	119
Green waste	France	7.8	122	61	780	0	90
Biowaste	Ireland	8.4	232	36	1234	28	67
Onion waste	UK	7.1	94	70	2560	105	10
Grape marc	Greece	7.1	213	105	280	15	0
OPC+OLW[b]	Greece	6.7	319	140	561	21	0
SCM[c]+orange peel+straw	Israel	5.7	685	6	780	3	125
SCM+pepper plants	Israel	7.1	713	2	1296	2	149

[a] mS/m in a 1:5 water extract.
[b] OPC – Olive Press Cake; OLW – Olive Leaf Waste;
[c] SCM – Separated Cattle Manure.

TABLE 11.27 Physical Properties of Some Composted Materials Found by Researchers from Various Countries Participating in an International Project on the Disease Suppressive Properties of Composts (Noble, 2005)

Composted material	Country	Dry bulk density, (g L^{-1})	Pore space (per cent)	Air content (per cent)	Water content (per cent)	Particle size, per cent w/w >10 mm	4–10 mm	<4 mm
SMC	UK (F)	171	91	33	58	39	42	19
	UK (M)	263	87	20	68	33	31	36
	Ireland	149	92	42	50	21	53	26
	France	275	87	31	56	5	75	20
	Greece	197	90	53	37	14	40	46
Green waste	UK	554	73	9	65	3	18	79
	Ireland	419	81	31	50	13	31	56
	France	410	81	24	57	5	75	20
	Neths	216	90	26	53	9	46	45
Biowaste	UK	439	78	30	48	3	81	16
	Ireland	244	87	47	40	–	–	–
Onion waste	UK	328	85	17	68	77	13	0
Grape marc	Greece	444	82	45	37	1	15	84
OPC[a]+OLW	Greece	232	85	48	37	18	27	55
SCM[b]+orange peels+straw	Israel	461	76	35	40	24	34	42
SCM+pepper plants	Israel	329	82	48	35	16	28	56

[a] OPC – Olive Press Cake; OLW – Olive Leaf Waste;
[b] SCM – Separated Cattle Manure.

from vegetable wastes were moister and had higher K levels than other composts (Table 11.25). The Israeli composts based on separated cattle manure (SCM) and crop waste had the highest electrical conductivities.

The pH in all composts except for one based on SCM, orange peel and straw was high. The EC levels indicate that even if physical properties were ideal, these composts could not be used as growing media by themselves without dilution with a low base material such as peat or after thorough leaching. Alternatively, they could be used for salinity-resistant crops such as tomato (Raviv et al. 2004). The K level in the SMC, biowaste and onion waste composts are very high. The composts based on SCM had the highest EC levels (Table 11.26).

All the materials with the exception of some SMCs had much higher bulk densities than typical peats (Table 11.27) and this is a common problem with many of the organic peat substitutes. The organic matter contents of the composts prepared from vegetable wastes and olive industry wastes were higher than those of other composts. Water-holding capacities (water content at 10 cm tension) and pore space of the composted materials were greater than those of a sandy loam soil, but generally lower than those of peat. The pore space of SMC was generally greater than that of other composted wastes. Air-filled porosity values are high except for one green waste sample which had the highest content of fine particles (<4 mm), accounting for the low air content.

The use of compost as a component of growing media has been reviewed for nursery stock by Fitzpatrick (2001) and for vegetable transplant production by Sterret (2001). Pronk (1995) found that composted vegetable, fruit and garden waste could be substituted for up to 15 per cent of the peat in the potting medium, if fertilization is adjusted. Larger percentages of compost resulted to high pH levels.

The effect of adding CGW to a peat growing medium was to increase bulk density and reduce pore space (Prasad and Maher, 2001). Rates of up to 20 per cent of volume had minor effects on the physical properties but using 50 per cent CGW reduced the easily available water content significantly. Including CGW consistently increased the pH of the growing medium. In one experiment, a rate of 50 per cent CGW did not adversely affect early growth of tomato seedlings but reduced subsequent growth due to decreased availability of N. In another experiment where three CGWs were included, a 50 per cent rate resulted in a reduction in early growth. This early growth reduction was associated with high K levels. Subsequent growth was limited by the available N content. Reduced N availability was associated with high rates of CGW in the growing medium. It was concluded that a rate of 20 per cent CGW could be included without adversely affecting plant performance. In other studies, higher rates (up to 60 per cent V:V) of improved compost did not have any negative effect on plant growth. The process that led to this improvement is a wet sieving of the raw material prior to its composting. The increased organic matter and decreased salt content of the compost allow for significantly higher substitution rates of peat by compost. This was shown for cattle manure compost (Raviv et al., 1998) and Biowaste compost (Veeken et al., 2005). N immobilization was noted as a problem with CGW by Jauch and Fischer (1993), who also reported a reduction in the fresh weight of pot plants when it was used at a rate of 50 per cent. Hartz et al. (1996) also noted N immobilization by composted green waste.

Maher and Prasad (2001) investigated the effects on the composting of green waste of two N sources: calcium ammonium nitrate (CAN) and poultry manure. The inclusion of peat along with CAN was also studied. Green waste without any addition was the control treatment. The four treatments involved composting over a 20-week period and some material from each treatment was set aside and cold-stored at 2°C at 4-week intervals. During composting there was an increase in pH, N content, CEC and FT-IR peak ratio, 1485 cm^{-1}/1056 cm^{-1}. The C/N ratio and organic matter content declined over time as did available N levels in the treatments where N was added. At the end of the composting period, samples from the four treatments from the five sampling dates were analysed using a water extract, and the composts were supplemented with N to bring the available N level to 200 mg/L. They were mixed with peat, to which a standard dressing of lime and fertilizer had been added, to provide 0, 12.5, 25 and 50 per cent by volume of compost in the growing medium. The effects of these treatments on the growth of tomato seedlings were studied. At the 50 per cent rate, where the green waste had been composted without additional N and where the composting period was less than 16 weeks, plant weight was reduced. At lower rates of compost incorporation, these effects were much less. Plant growth was positively related to total N content and CEC of the green waste and negatively to the C/N ratio.

The main problems in substituting compost for peat are phytotoxicity due to incomplete composting, insufficient availability of N if the source materials are not properly formulated, and elevated pH and salt levels. Water-holding capacity is also affected as the proportion of compost in a growing medium rises.

11.8 OTHER MATERIALS

The solid fraction from animal manure can serve as an alternative to peat in container media. This may be particularly relevant in parts of Europe as the Nitrates Directive (Council of the European Communities, 1991) curtails the application of animal manure to farmland. It has been shown that composts prepared from leached separated cattle manure after a process of methanogenic fermentation can serve as a peat substitute and that its air and water content are similar to peat. Seedling growth in these media was as good as in peat mix (Chen et al., 1983). The bulk density is similar to peat. The water release curve of separated cattle manure compost shows porosity similar to that of peat but the volumetric water content within the range 1–10 kPa tension is considerably lower than peat. This is due to high air-filled porosity and higher content of residual water held at tensions greater than 10 kPa. This has implications for scheduling of irrigation (Raviv and Medina, 1997). Data on storage of composted material made from slurry fibre plus peat, slurry fibre and sawdust and on its own showed that it is quite stable as evidenced by lack of changes in N content after storage of over one year. Tomato plants grown in peat combined with up to 30 per cent of these materials were found to be satisfactory (Prasad, unpublished data).

Rice hulls, both fresh and decomposed, are often used as an amendment to peat or can be used as a replacement for perlite (Papafotiou et al., 2001). Information on the physical and chemical properties is given by Kang et al. (2004), and it has been shown that the physical and chemical properties depend on the processing. The addition of rice hull to peat decreased porosity and available water (Papafotiou et al., 2001). Lee et al. (2000) found that growth of pepper seedlings decreased with increasing rate of fresh rice hulls added to a peat-based growing medium. Where the rice hulls were composted, the best growth was obtained with 40 per cent rice hulls.

Chipboard waste can also be milled, composted and used as a substitute for peat in growing media (Molitor et al., 2004). This material can contain up to 10 per cent urea formaldehyde. It has been found to have poor storage properties due to release of N and subsequent low pH and high levels of NH_4-N and NO_3-N (Dickinson and Carlile, 1995).

11.9 STABILITY OF GROWING MEDIA

In soilless production, plants are grown over lengths of time that range from a few weeks to several years. In general, it is desired that the substrate stay static during this time, so as to not modify the growing environment in a manner that will negatively affect productivity. It is, however, important to note that the root zone is a microcosm where organisms other than the plant also prevail. There is considerable biological activity with both negative and positive impacts on the production system. The predominant negative impact of this biological activity is the physical breakdown of the substrate (i.e. lack of physical stability). However, simple destruction of the microbes that cause this is not the answer, as there are also numerous beneficial organisms in the root zone. From this perspective, it is important to have a root environment that includes disease suppressive organisms with population dynamics that are relatively stable.

11.9.1 PHYSICAL AND BIOLOGICAL STABILITY

Physical breakdown of the growing medium can have a detrimental effect on crops which have a long production period such as woody nursery stock. Crops that have a high aeration requirement cannot be grown well under wet conditions (e.g. ebb and flood systems). In this situation, the physical stability of the growing medium becomes important in maintaining favourable growing conditions for the whole period (Aendekerk, 1997; 2001; Maher and Prasad, 2004). The various peat types vary considerably in their capacity to resist physical degradation. Peats with higher degree of decomposition (H5–H6) are more structurally stable than younger peats (H2–H3) (Aendekerk, 1997). It is also known that some materials which have been suggested as substitutes for peat (e.g. spent mushroom compost) can undergo a large and rapid loss of structure (Lemaire et al., 1998). However, raw coir, for example, was found to be a very stable substrate and was found to be more stable than peat in one study (Lemaire et al., 1998 and Table 11.13).

Prasad and O'Shea (1999) studied peats of varying degrees of decomposition from bogs in Ireland, Finland, Germany, Baltic States and Sweden. Particle sizes of the peat had a major effect on the breakdown, with fine peat breaking down more than coarse peat. The degree of breakdown was inversely related to decomposition (von Post scale) and particle size. The degree of volume loss varied from 47 per cent for H2 (Finnish) peat to 5 per cent for coarse H5 (Irish) peat. Airspace decreased from 21 per cent to as low as 3.5 per cent over the 15-month incubation period for fine-textured H2 Peat. Reduction in air-filled porosity was related to volume reduction. In another trial, peat from bogs in Baltic States, Sweden, Ireland and Germany of two sizes (0–3 mm and 5–10 mm) were compared, and the loss in volume after nearly 2 years incubation was in the order of 13–20 per cent for H4 and H5 peats and 29–40 per cent for H2 and H3 peats. Wood fibres showed an even greater volume loss of the order of 50 per cent. Steaming of H2 or H5 peats did not appear to have a significant effect on break down, as evidenced by changes in particle size. FT-IR analysis of peat taken before and after incubation showed a trend towards enrichment of lignin and decrease in cellulose.

Prasad (1997b) found that volume reduction of coir as a result of biodegradation was 24 per cent over 24 months, indicating it is a relatively stable material. Shrinkage of 19–21 per cent was reported by Nelson et al. (2004) just after 8–10 weeks under cropping. Differences in biodegradation of coir batches could be related to differences in lignin content which could again be related to the age of the coir and fibre content.

Volume reduction of between 36 and 54 per cent over a 2-year period was found with various types of wood fibre (Fischer et al., 1993; Prasad and O'Shea, 1999). In another study with coir and three wood fibres on their own or mixed with H5 peat, Prasad and Maher (2004) found that the wood fibres generally break down rapidly over time compared with peat. Generally the shrinkage was much less in Toresa than in other wood fibre products (Prasad and Maher, 2004) probably due to differences in the lignin content of the materials (Table 11.19). Coir was found to be relatively stable in this study, as well. The addition of H5 peat to all the non-peat materials led to a reduction in the rate of breakdown while the addition of lime accelerated breakdown. Breakdown of the materials was strongly and negatively related to the initial lignin content. Samples were analysed by FT-IR spectroscopic analysis at the start of the experiment. A strong negative relationship was found between the FT-IR peak ratio $1600\,cm^{-1}/1056\,cm^{-1}$ (lignin to cellulose) and substrate shrinkage. Fischer et al. (1993) also found volume loss of similar magnitude with wood fibres. In a review comparing peat-based and peat-free media mixes, Carlile (2003) attributed the stability of peat-based media to the high lignin (resistant to microbial attack) content of peat.

Fitzpatrick and Verkade (1991) have reported that some compost products used in growing media start with satisfactory porosity but undergo compaction during the growing period. Maher and Prasad (2002) studied the effect of adding CGW to two types of peat on physical properties and on physical degradation of the growing medium. The peats were relatively undecomposed (H2) and moderately decomposed peat (H5). A young CGW, composted for 12 weeks with no addition of N, and a mature CGW, composted for 20 weeks with the addition of $500\,g\,m^{-3}$ of N prior to composting, were compared. Addition of CGW at 30 per cent by volume increased bulk density and air content and reduced pore space and easily available water content. Physical degradation was measured by monitoring height loss of the peat/CGW mixes over a 2-year period. H2 peat degraded at a much greater rate than the H5 peat. Addition of young CGW increased degradation of both peats whereas the mature CGW only increased degradation in the H5 peat. Also in this experiment, FT-IR spectroscopic analysis of the mixes was carried out and again the peak ratio $1600\,cm^{-1}/1056\,cm^{-1}$, which represents the aromatic/aliphatic ratio (lignin to carbohydrate ratio), was negatively correlated to physical degradation of the mixes.

Harrelson et al. (2004) studied the physical properties of fresh pine bark and aged pine bark over one year in an unprotected location in the presence and absence of coarse builders sand (bark:sand = 8:1; 11 per cent sand by volume). Container capacity and available water in the aged pine bark–sand substrate were significantly greater than in fresh pine bark–sand substrate. The authors concluded that fresh pine bark–sand substrate would require frequent irrigation with small quantities of water due to limited available water content. Over the course of the year, available water in the fresh pine bark–sand substrate increased from 15.8 to 22.3 per cent while air-filled porosity decreased from 36.3 to 24.9 per cent. Similarly, the aged pine bark–sand substrate showed an available water increase from 22.7 to 30.0 per cent with an air-filled porosity decrease from 25.9 to 17.0 per cent. The decrease in air-filled porosity for both mixes over time can present challenges to the growers for their long-term use in production.

Biological stability is an important consideration as part of N nutrition management. Some coir has been found to immobilize nitrogen (Prasad, 1997a). In some cases, immobilization over 5 weeks was around 50–60 mg N L^{-1}, while in others the N immobilization was negligible. In growing media derived from composted materials, if the composting process has not been thorough, the resulting compost may have high cellulose or hemicellulose contents which are readily degraded by micro-organisms. Thus many peat-free media mixes have an inherently higher microbial population and activity than peat-based media mixes (Dickinson, 1995). This often results in a very quick depletion of inorganic nutrients (microbial immobilization), particularly nitrogen (Dickinson and Carlile, 1995), with a simultaneous change in pH. Such changes can be ameliorated by the addition of a buffering agent such as expanded clay or other clay minerals in the media mixes (Bragg, 1998).

11.9.2 PATHOGEN SURVIVAL IN COMPOST

It is possible that plant, human or animal pathogens could contaminate waste material destined for composting. In a review of human and animal pathogen survival in CGW, Jones and Martin (2003) concluded that most pathogens are inactivated by the composting process and a composting procedure with a residence time of 3 days at a temperature higher than 55°C results in sanitized compost. There is, however, a danger that *Escherichia coli* and *Salmonella enterica* may grow in the final compost if the process has been inefficient and the organic matter remains poorly stabilized (Hess et al., 2004; Cekmecelioglu et al., 2005). Viruses are considered more heat tolerant. However, a recent study revealed that composting is capable of completely eradicating human-pathogenic enteroviruses (Monpoeho et al., 2004). Similarly, helminth ova of *Ascaris suum*, normally occurring in sewage sludge, were completely inactivated by composting (Paulsrud et al., 2004).

Noble and Roberts (2003, 2004) reviewed temperature–time effects and other sanitizing factors of composting on 60 plant pathogen and nematode species. For all of the bacterial plant pathogens and nematodes, the majority of fungal plant pathogens, and a number of plant viruses, a compost temperature of 55°C for 21 days was sufficient for ensuring eradication. Shorter periods may be satisfactory but these were not always examined. The fungal plant pathogens *Plasmodiophora brassicae*, the causal agent of clubroot of Brassicas, and *Fusarium oxysporum* f. sp. lycopersici, the causal agent of tomato wilt, were more temperature tolerant. A compost temperature of at least 65°C for up to 21 days was required for eradication. However, in a more recent study, Fayolle et al. (2006) showed a complete eradication of *Plasmodiophora brassicae* in both bench- and large-scale composting procedures at temperatures as low as 50°C provided that the moisture content was sufficient (60 per cent W:W). Several plant viruses, particularly Tobacco Mosaic Virus (TMV), were temperature tolerant. However, there is evidence that TMV is degraded over time in compost, even at temperatures below 50°C.

Compost temperatures in excess of 60°C can be achieved in a range of composting systems using a wide range of source materials. However, there is insufficient information on the survival risks of pathogens in cooler zones of the compost, particularly in in-vessel systems where the compost is not turned.

11.10 DISEASE SUPPRESSION BY ORGANIC GROWING MEDIA

11.10.1 THE PHENOMENON AND ITS DESCRIPTION

Although most growing media are initially pathogen-free, they are frequently infested with disease-causing pathogens soon after planting. Low biological activity and lack of microbial population diversity, typical of most unused media, was suggested as one of the main reasons for this potentially destructive phenomenon (Chapter 11.10 as well as Postma et al., 2000; Menzies et al. 2005). Although several peat moss types suppressed soil- borne diseases such as *Alternaria brassicicola*, *Leptosphaeria maculans*, *Rhizoctonia solani* and *Fusarium oxysporum* f.sp. *lycopersici* (Tahvonen, 1982), peat moss is still used in many cases as a control in compost suppressiveness studies (Gaag et al., 2004; Noble and Coventry, 2005; Raviv et al., 2005) due to its conduciveness to rapid infestation by a variety of pathogens (Hoitink et al., 1977; Borrero et al., 2004; Yogev et al., 2006).

Compost has been shown to provide some level of disease suppression in several production systems. As discussed above, most organic substrates undergo some sort of composting as part of the processing of the raw materials into growing media component. While the focus of this has largely been on the beneficial changes of physical and chemical properties, there is also evidence of disease suppressiveness as a result of the composting process. In addition to a substantial body of research results on suppression of fungal diseases, examples for inhibition of disease-causing nematodes (Oka and Yermiyahu, 2002; Raviv et al., 2005) and bacteria (Schonfeld et al., 2003; Yogev et al., in preparation) also exist.

11.10.2 SUGGESTED MECHANISMS FOR SUPPRESSIVENESS OF COMPOST AGAINST ROOT DISEASES

The most dramatic and immediate effect of amending an unused growing medium with compost is the large and rapid increase in microbial and fungal activities and microbial population diversity (Diab et al., 2003). In many cases, compost sterilization considerably reduces or eliminates its suppressiveness, indicating that the mechanism of disease suppression is predominantly biological (Hadar and Mandelbaum, 1986; Trillas-Gay et al., 1986; Reuveni et al., 2002; Noble and Roberts, 2003). However, in some cases a fast reintroduction of diverse microbial populations soon after sterilization resulted in resumed suppressiveness (Serra-Wittling et al., 1996; Yogev et al., 2006).

The suppressiveness phenomenon consists of a complex and intricate set of mechanisms. Understanding these is important since it is desirable in soilless production to make use of this phenomenon and to maximize its effectiveness. Hoitink and Boehm (1999) suggested several mechanisms which are likely to be factors in disease suppression:

- Competition for nutrients, space and occupation of infection sites by other micro-organisms: Beneficial effects of competition for nutrients and space were shown for a variety of pathosystems (Chen et al., 1988 and Serra-Wittling et al., 1996). Another example is the antagonistic effect towards *Fusarium oxysporum* f. sp. *radicis-lycopersici*, *Pyrenochaeta lycopersici*, *Pythium ultimum* and *Rhizoctonia solani*, all known pathogens of tomato, observed after compost was added to soil. These effects were associated with marked increases in the percentage of siderophore producers within the root-zone of tomato (de Brito-Alvarez et al., 1995). Siderophore producers such as fluorescent pseudomonads are known to compete with fusarium through suppression of their chlamydospores' germination (Elad and Baker, 1985). Massive production of siderophores leads to reduced levels of iron which is essential for successful germination of the pathogen and penetration of the host. In soil-grown plants, the proliferation of pseudomonads is dependent on continued supply of carbon through root exudation; when compost is applied, ample supply of carbon is provided, along with diverse communities of such beneficial microbes. Another type of competition among fungi is for colonization sites on the root surface (Olivain and Alabouvette, 1999; Benhamou and Garand, 2001). For example (Pharand, et al. (2002) showed that compost-mediated resistance of tomato to crown and root rot caused by *Fusarium oxysporum* f. sp. *radicis-lycopersici* was associated with non-pathogenic strains of *Fusarium oxysporum* and that this was induced through the formation of physical barriers on the roots to fungal penetration, at sites of attempted penetration by the pathogen (Pharand et al. 2002).
- Hyperparasitism followed by lysis was shown to be the main mechanism of soil suppression of *Rhizoctonia solani*, caused by *Trichoderma harzianum* (Chet and Baker, 1980). *T. harzianum* can be found frequently in suppressive composts (Kuter et al., 1983). However, Nelson et al. (1983) showed that amending immature compost with high dose of *T. harzianum* did not confer suppressiveness to this medium, thus concluded that the production of container media that were consistently suppressive to *R. solani* requires both the existence of antagonists and an environment that favours their activity. Direct hyperparasitism of *Sclerotium rolfsii* propagules was shown by Gorodecki and

Hadar (1990) in composts based on grape marc and separated cattle manure but not in peat- based media. Another example of lysis was shown for hyphae of *Pythium aphanidermatum* (Mandelbaum and Hadar, 1990). They demonstrated that hyphae of *P. aphanidermatum* grown on nylon fabric and buried in container media were rapidly lysed in compost medium based on separated cattle manure as compared with peat-based media. Light and SEM microscopic observations of hyphae retrieved from container media showed no evidence of direct parasitism by other soil fungi; however, bacteria were associated with the lysing hyphae. In general, it was concluded that pathogen propagules can be destroyed after incubation in suppressive organic substrates. This process is relatively fast, as shown by Yogev et al. (2006) for microconidia of *Fusarium oxysporum* f. sp. *radicis-lycopersici* and for *Fusarium oxysporum* f. sp. *basilici* in different compost types, while the same conidia survived in peat.

- Antibiosis: The role of production of antibiotics within composts has not proven yet but there are indications that this may be one of the mechanisms. *Gliocladium virens* produced detectable levels of gliotoxin in composted mineral soil, populated with natural micro-biota and inoculated with *G. virens*. This medium effectively controlled damping-off of zinnia (*Zinnia elegans*) seedlings caused by *Pythium ultimum* and *Rhizoctonia solani*. Aqueous extracts containing gliotoxin were as effective for control of damping-off of zinnias as the intact *G. virens*-amended medium (Lumsden et al., 1992).
- Futile pathogen germination: One potential mechanism for compost- related disease suppression is the compost's role in mimicking root exudates. Normally, a pathogen propagule will not germinate in the absence of a host as signalled by root or seed exudates (Lockwood, 1990) since these chemical signals are required for the host identification by the pathogen (Chen et al., 1988). Induction of germination of pathogens in the absence of a host may occur, however, in composted material. In compost-containing medium, germination of pathogens may be triggered before the pathogen comes in contact with living plants, so that the existing inoculum is spent. This is probably the reason why plant-derived composts are usually so effective in suppressing soil-borne diseases (Cheuk et al., 2005) and especially in reducing the inoculum level, even without the existence of plants (Yogev et al., 2006).
- Adsorption of chemical signals that enhance the germination and root colonization by pathogens on the high specific surface area of composts is another putative mechanism for disease suppression. Adsorption reduces the amount of free signal, so the germination of a smaller number of propagules can be realized. Later, part of these adsorbed molecules can be decomposed by compost micro-organisms, leading to further decrease of the signal's concentration. However, conclusive evidences to support this mechanism are still unavailable.
- Induced systemic resistance (ISR; or systemic acquired resistance, SAR) is 'a state of enhanced defensive capacity developed by a plant when appropriately stimulated' (Bakker et al., 2003). The resistance responses are

usually systemic, but localized types also exist, and may be effective against a broad range of pathogens. ISR is probably one of the most important mechanisms through which compost induces disease resistance to plants. Many rhizosphere bacteria and fungal isolates have been reported to turn on ISR in plants (van Loon et al., 1998). In addition, microbial metabolites such as salicylic acid and lipopolysaccharides have been implicated in microbially mediated ISR (Bakker et al., 2003). Some of the changes triggered by composts and its microbial communities were of anatomical nature (as described above, Pharand et al., 2002). Others are related to enzymatic activity. For example, composted pine bark was suppressive to *Pythium* root rot of cucumber (Zhang et al., 1996), whereas dark peat was not suppressive to the disease. Cucumber and *Arabidopsis* plants grown in composted pine bark expressed higher levels of β-1,3-glucanase (Zhang et al., 1998) and peroxidase (Zhang et al., 1996) than those grown in peat. Split root experiments suggested that the resistance mechanism in cucumber was systemic (Zhang et al., 1996). Compost- induced expression of pathogenesis-related genes was recently shown by Kavroulakis et al. (2006). They concluded that suppressive composts are able to elicit consistent expression of certain PR genes in the roots of tomato plants, even in the absence of any pathogen. The expression of the PR genes may be triggered by the microflora of the compost or could be associated with abiotic factors of the compost.

- In this context, it is important to note that the ISR phenomenon is not limited to the root system and inhibition of foliar diseases was also detected. Although Krause et al. (2003) and Hoitink et al. (2006) reported that systemic suppression of foliar diseases induced by compost amendments is a rare phenomenon and is mainly related to the effect of the level of compost inoculation with *Trichoderma hamatum* strain 382, other examples for foliar disease suppression were recently demonstrated (Bobev et al., 2004; Horst et al., 2005; Kavroulakis et al., 2005) suggesting that this phenomenon is far less rare than previously thought. Recently, Yogev et al. (Personal communication) working with cucumber plants grown in either peat or compost based on orange peels and separated cattle manure deliberately inoculated the leaves with spores of *Botrytis cinerea*, the causal agent of grey mould – a very severe greenhouse foliar disease of many crops. After germination, the fungus grows initially as hyphae, spreading within the leaves. The growth rate of the hyphae in compost-grown plants was 93 per cent slower as compared to their peat-grown counterparts. Similar results were obtained with melon (*Cucumis melo*).

The organisms responsible for disease suppression colonize the compost during the curing stage when the temperature drops below 40°C (Scheuerell and Mahaffee, 2005). *Bacillus* spp., *Enterobacter* spp., *Flavobacterium balustinum*, *Pseudomonas* spp., other bacterial genera and *Streptomyces* spp. as well as *Penicillium* spp., several *Trichoderma* spp., *Gliocladium virens* and other fungi have been identified as biocontrol agents in compost-amended substrates (Hoitink et al., 2001).

11.10.3 HORTICULTURAL CONSIDERATIONS OF USE OF COMPOST AS SOILLESS SUBSTRATE

In commercial soilless production, it is important to find economical and reliable methods to reduce the risk of yield loss due to soil-borne diseases. Inclusion of compost in growing media as a method to suppress a wide variety of soil-borne plant pathogens was first suggested by Hoitink et al. (1975) and is now a well-established commercial practice, corroborated by a large body of scientific evidence. This subject has been reviewed by Hoitink and Fahy (1986), Hoitink and Boehm (1999), Hoitink, et al. (2001), Noble and Roberts (2003), Noble and Coventry (2005) and Termorshuizen, et al. (2006). Some representative examples for this use are shown in Table 11.28.

Hadar and Mandelbaum (1986) found that the degree of decomposition of compost has a strong effect on the rate of disease suppressiveness. They showed that immature compost could not suppress damping-off (*Pythium aphanidermatum*) of cucumber seedlings, while mature compost could. Another example can be found in *Rhizoctonia solani*, a pathogen which is highly competitive as a saprophyte and *Trichoderma*, an effective biocontrol agent of *R. solani*. In fresh, undecomposed organic matter, biological control of *R. solani* does not occur because both organisms grow as saprophytes and *R. solani* remains capable of causing disease. The synthesis of lytic enzymes involved in parasitism of pathogens by *Trichoderma* is repressed in fresh organic matter due to high glucose concentrations. In mature compost, where concentrations of nutrients such as glucose are low, sclerotia of *R. solani* are killed by the parasite and biological control prevails (Hoitink et al., 2001).

On the other hand, if the compost is excessively stabilized, then it will not support microbiological activity so that disease suppression is lost (Widmer et al., 1998). Similarly, Tahvonen (1993) found lightly decomposed peats (H2–H3) to be suppressive of soil-borne diseases. Darker, more decomposed peats are low in microbial activity and are conducive to *Pythium* and *Phytophthora* root rots (Boehm and Hoitink, 1992).

An important practical consideration of the use of compost for disease prevention is the longevity of the suppressiveness capacity. It was shown that effectiveness of several compost types against the dollar spot disease of creeping bentgrass (*Agrostis palustris* Pencross) caused by *Sclerotinia homoeocarpa* was maintained for 1 year, under shaded, uncontrolled storage conditions (Boulter et al., 2002). This subject is now under investigation by others and preliminary results also suggest a full capacity for 1 year, with subsequent gradual decline, with only 50 per cent of the original capacity after 2 years (Raviv et al., in preparation).

In conjunction with the extensive European compost research described above (Noble, 2005), the effect of 18 composts was tested in 7 pathosystems [*Verticillium dahliae* – eggplant (*Solanum melongena*), *Rhizoctonia solani* – cauliflower (*Brassica oleracea var. botrytis*), *R. solani* – pine, *Phytophthora nicotianae*- tomato, *P. cinnamomi* – lupin (*Lupinus spp.*), *Cylindrocladium spatiphylli* – spatiphyllum, *F. oxysporum* – flax (*Linum usitatissimum*] (Termorshuizen et al., 2006). Applying 20 per cent of compost into potting soil or sand, 54 per cent of the tested combinations were significantly more disease suppressive, 3 per cent showed significant enhancement of the disease, and 43 per cent of the tested combinations did not result

TABLE 11.28 Typical Examples of Commercially Feasible Uses of Composts for Soil-borne Disease Mitigation

Pathogen	Host	Compost feedstock	Reference
F. oxysporum f. sp. *radicis-cucumerinum*	Cucumber (*Cucumis sativus*)	Greenhouse waste	Yogev et al., 2006
F. oxysporum f. sp. *radicis-lycopersici*	Tomato (*Lycopersicon esculentum* Mill.)	Greenhouse waste	Cheuk et al., 2005; Raviv et al., 2005
F. oxysporum f. sp. *basilici*	Sweet basil (*Ocimum basilicum*)	Separated cattle manure	Reuveni et al., 2002
F. oxysporum f. sp. *chrysanthemi*	*Dendrathema grandiflorum*	Hard wood bark	Chef et al., 1983
F. oxysporum f. sp. *melonis*	Melon (*Cucumis melo*)	Greenhouse waste; prunings	Raviv et al., 2005; Ros et al., 2005
F. oxysporum f. sp. *dianthi*	Carnation (*Dianthus caryophyllus*)	Hard wood bark, olive pomace	Pera and Filippi, 1987; Pera and Calvet, 1989
Meloidogyne incognita	Tomato (*Lycopersicon esculentum* Mill.)	MSW	Ryckeboer and Coosemans, 1996
Meloidogyne javanica	Tomato (*Lycopersicon esculentum* Mill.)	Cattle manure; Greenhouse waste	Oka and Yermiyahu, 2002; Raviv et al., 2005
Pythium ultimum	Cucumber (*Cucumis sativus*)	Hard wood bark	Chen et al., 1988
P. ultimum	Poinsettia *Euohorbia pulcherrima*	Hard wood bark	Daft et al., 1979
Pythium aphanidermatum	Cucumber (*Cucumis sativus*)	Licorice compost, sugarcane residues	Hadar and Mandelbaum, 1986; Theodore and Toribio, 1995
Phytophthora nicotianae	Various citruses at the nursery stage	MSW	Widmer et al., 1998
Phytophthora cinnamomi	*Rhododendron catawbiense*	Hard wood bark	Hoitink et al., 1977
P. cinnamomi	*Aucuba japonica*	Soft and hard wood bark	Spencer and Benson, 1981
Phytophthora citricola	*Aucuba japonica*	Soft and hard wood bark	Spencer and Benson, 1981
Rhizoctonia solani	Poinsettia (*Euohorbia pulcherrima*)	Hard wood bark	Daft et al., 1979
R. solani	Radish (*Raphanus sativus*)	Separated cattle manure, grape marc	Gorodecki and Hadar, 1990
R. solani	Cucumber (*Cucumis sativus*)	MSW	Tuitert et al., 1998
Sclerotinia minor	Lettuce (*Lactuca sativa*)	Sewage sludge	Lumsden et al., 1986
Sclerotium rolfsii	Radish (*Raphanus sativus*)	Cattle manure, grape marc	Gorodecki and Hadar, 1990

in significant differences compared to the control without compost. The mean disease suppressiveness per compost ranged from 14 to 61 per cent (Termorshuizen et al., 2006). Thus, generally speaking, compost has a positive effect on suppressiveness of potting soil. However, none of the compost types could improve the suppressiveness of the potting soil for all tested pathosystems. Another disadvantage is the variability and unpredictability of compost: apparently similar compost types may give different results, and compost quality is often not stable in time and among batches of the same producer. This can be changed if strict production parameters are adopted by producers, as suggested by Raviv (2005).

Consequently, it may be feasible to partially replace peat in potting mixes with well-prepared, mature compost, whenever disease suppression is desired. Normally, values of 20–30 per cent of the total resulting mix are quoted (Termorshuizen et al., 2006; Prasad, unpublished data). Higher rates can also be used, subjected to testing with the target crop.

REFERENCES

Aaron, J.R. (1976). Conifer bark: its properties and uses. Forestry Commission Record No. 110., U.K.

Abad, M., Noguera, P., Puchades, R., Maquieira, A. and Noguera, V. (2002). Physio-chemical properties of some coconut coir dusts for use as a peat substitute for containerised ornamental plants. *Bioresour. Technol.*, **82**, 241–245.

Aendekerk, T.G.L. (1989). Annual Report, Research Station for Nursery Crops, *Boskoop*, 42.

Aendekerk, T.G.L. (1997). Decomposition of peat substrate in relation to physical properties and growth of *Chamaecyparis*. *Acta Hort.* (ISHS), **450**, 191–198.

Aendekerk, T.G.L. (2001). Decomposition of peat substrates in relation to physical properties and growth of *Skimmia*. *Acta Hort.* (ISHS), **548**, 261–268.

Allaire, S.E., Caron, J., Ménard, C. and Dorais, M. (2005). Potential replacements for rockwool as growing substrate for greenhouse tomato. *Can. J. Soil Sci.*, **85**, 67–74.

Baker, K.F. (ed.). (1957). The UC system for producing healthy container grown plants. Manual 23, University of California, Agricultural Experiment Station Extension Service.

Bakker, P.A.H.M., Ran, L.X., Pieterse, C.M.J. and van Loon, L.C. (2003). Understanding the involvement of rhizobacteria-mediated induction of systemic resistance in biocontrol of plant diseases. *Can. J. Plant Pathol.*, **25**, 5–9.

Beardsell, D.V., Nichols, D.C. and Jones, D.L. (1979). Physical properties of nursery potting mixes. *Sci. Hort.*, **11**, 1–8.

Benhamou, N. and Garand, C. (2001). Cytological analysis of defense-related mechanisms induced in pea root tissues in response to colonization by nonpathogenic *Fusarium oxysporum*. *Phytopathology*, **91**, 730–740.

Bik, A.R. and Boertje, G.A. (1975). Fertilizer standards for potting composts based on 1:1.5 volume extraction method of soil testing. *Acta Hort.* (ISHS), **50**, 153–156.

Bobev, S., Willekens, K., Goeminne, G., et al. (2004). Resistance to air-borne diseases and pests induced by compost in substrate cultivation of strawberry. *Commun. Agric. Appl. Biol. Sci.*, **69**, 591–593.

Boehm, M.J. and Hoitink, H.A.J. (1992). Sustenance of microbial activity in potting mixes and severity of Pythium root rot of poinsettia. *Phytopathology*, **82**, 259–264.

Bohne, H. (2004). Growth of nursery crops in peat reduced and peat free substrates. *Acta Hort.* (ISHS), **644**, 103–106.

Borrero, C., Trillas, M.I., Ordovas, J., et al. (2004). Predictive factors for the suppression of Fusarium wilt of tomato in plant growth media. *Phytopathology*, **94**, 1094–1101.

Boulter, J.I., Boland, G.J. and Trevors, J.T. (2002). Evaluation of composts for suppression of dollar spot (*Sclerotinia homoeocarpa*) of turfgrass. *Plant Dis.* **86**, 405–410.

Bragg, N. (1998). The commercial development of a sustainable peat alternative substrate from locally derived industrial by-products. *Acta Hort.* (ISHS), **469**, 61–70.

de Brito-Alvarez, M.A.., Gagne, S. and Antoun, H. (1995). Effect of compost on rhizosphere microflora of the tomato and on the incidence of plant growth-promoting rhizobacteria. *Appl. Environ. Microbiol.*, **61**, 194–199.

Brown, K.H., Bouwkamp, J.C. and Gouin, F.R. (1998). The influence of C:P ratio on the biological degradation of municipal solid waste. *Compost Sci. Util.*, **6**(1), 53–58.

Bunt, A.C. (1976). *Modern Potting Composts*. London: George & Unwin Ltd.

Bunt, A.C. (1988). *Media and Mixes for Container Grown Plants*. London: Unwin Hyman Ltd.

Calkins, J.B., Jarvis, B.R. and Swanson, B.T. (1997). Compost and rubber tire chips as peat substitutes in nursery container media: Growth effects. *J. Environ. Hortic.*, **15**(2), 88–94.

Carlile, W.R. (1999). The effects of the environment lobby on the selection and use of growing media. *Acta Hort.* (ISHS), **481**, 587–596.

Carlile, W.R. (2003). Growing media and the environment lobby in the UK 1997–2001 ISHS *Acta Hort.* (ISHS), **644**, 107–113.

Cekmecelioglu, D., Demirci, A., Graves, R.E. and Davitt, N.H. (2005). Optimization of windrow food waste composting to inactivate pathogenic microorganisms. *Transactions ASAE*, **48**, 2023–2032.

Chef, D.G., Hoitink, H.A.J. and Madden, L.V. (1983). Effects of organic components in container media on suppression of Fusarium wilt of chrysanthemum and flax. *Phytopathology*, **73**, 279–281.

Chen, Y., Inbar, Y. and Raviv, M. (1983). Slurry produced by methanogenic fermentation of cow manure as a peat substitute in horticulture. In *Proc. 2nd Intl. Symp. Peat in Agr. and Hort.* pp. 297–317.

Chen, W., Hoitink, H.A.J., Schmitthenner, A.F. and Tuovinen, O.H. (1988). The role of microbial activity in suppression of damping off caused by *Pythium ultimum*. *Phytopathology*, **78**, 1447–1450.

Chet, I. and Baker, R. (1980). Induction of suppressiveness to *Rhizoctonia solani* in soil. *Phytopathol.* **70**, 994–998.

Cheuk, W., Lo, K.V., Copeman, R., et al. (2005). Disease suppression on greenhouse tomatoes using plant waste compost. *J. Environ. Sci. Health Part B, Pestic., Food Contam. Agric. Wastes*, **40**, 449–461.

Clemmenson, A.W. (2004). Physical characteristics of Miscanthus compost compared to peat and other growth substrates. *Compost Sci. Util.*, **12**, 219–224.

Council of the European Communities, (1991). Council Directive of 12 December 1991 concerning the protection of waters against pollution caused by nitrates from agricultural sources (91/676/EEC), O.J. No L 375, 31.12.1991, p. 1 (http://ec.europa.eu/environment/water/water- nitrates/index_en.html).

Daft, G.C., Poole, H.A. and Hoitink H.A.J. (1979). Composted hardwood bark: A Substitute for Steam Sterilization and Fungicide Drenches for Control of Poinsettia Crown and Root Rot. *Hort. Science*, **14**, 185–187.

Day, M. and Shaw. K. (2001). Biological, chemical and physical processes of composting. In *Compost Utilization in Horticultural Cropping Systems* (P.J. Stofella and B.A. Kahn, eds.). Boca Raton, Florida, USA: CRC Press LLC, p. 17–50.

De Boodt, M. and Verdonck, O. (1972). The physical properties of the substrates used in horticulture. *Acta Hort.* (ISHS), **26**, 37–44.

De Kreij, C. and van Leeuwen, G.J.L. (2001). Growth of pot plants in treated coir dust as compared to peat. *Comm. Soil Sci. Plant Anal.*, **32**, 2255–2265.

Diab, H.G., Hu, S. and Benson, D.M. (2003). Suppression of Rhizoctonia solani on impatiens by enhanced microbial activity in composted swine waste-amended potting mixes. *Phytopathology*, **93**, 1115–1123.

Dickinson, K. (1995). Plant growth, nutrient status and microbial activity in peat free growing media during storage. PhD. Thesis, Nottingham Trent University, UK, p. 175.

Dickinson, K. and Carlile, W.R. (1995). The storage properties of wood based peat free growing media. *Acta Hort.* (ISHS), **401**, 89–96.

Dorais, M., Caron, J., Bégin, G., et al. (2005). Equipment performance for determining water needs of tomato plants grown in sawdust based substrates and rockwool. *Acta Hort.* (ISHS), **691**, 293–304.

Elad, Y. and Baker, R. (1985). The role of competition for iron and carbon in suppression of chlamydospore germination of *Fusarium* spp. by *Pseudomonas* spp. *Phytopathology*, **75**, 1053–1059.

Evans, M.R., Konduru, S. and Stamps, R.H. (1996). Source variation in physical and chemical properties of coconut coir dust. *Hort. Science*, **6**, 965–967.

Favaro, J.C., Buyatti, M.A. and Acosta, M.R. (2002). Evaluation of sawdust-based substrates for the production of seedlings. *Investigación Agraria Producción y Protección Vegetales*, **17**(3), 367–373.

Fayolle, L., Noble, R., Coventry, E., et al. (2006). Eradication of *Plasmodiophora brassicae* during composting of wastes. *Plant Pathol.*, **55**(4), 553–558.

Fischer, P., Meinken, E. and Kalthoff, F. (1993). Holzfasserstoffe im Test. *Gb GW*, **26**, 1220–1222.

Fitzpatrick, G.E. (2001). Compost utilization in ornamental and nursery stock crop production systems. In *Compost Utilization in Horticultural Cropping Systems* (P.J. Stofella and B.A. Kahn, eds). Boca Raton, FLA., USA: CRC Press LLC, pp. 135–150.

Fitzpatrick, G.E. and Verkade, S.D. (1991). Substrate influence on compost efficacy as a nursery growing medium. *Proc. Florida State Hortic. Soc.*, **104**, 308–310.

Fornes, F., Belda, R.M., Abad, M., et al. (2003). The micro structure of coconut coir bust for use as alternative to peat soilless growing media. *Aust. J. Exptl. Agric.*, **43**, 1171–1179.

Gaag, D.J. van der., Rijn, E. van and Termorshuizen, A. (2004). Disease suppression in potting mixes amended with Dutch yard waste composts. *Bulletin OILB/SROP*, **27**, 291–295.

Gallagher, P.A. (1975). Peat in protected cropping. In *Peat in Horticulture* (D.W. Robinson and J.G.D. Lamb, eds). London: Academic Press, pp. 133–145.

Garcia, M. and Deverde. C. (1994). Le residu des fibres de coco a nouveau substrat pour la culture hors sol. PHM. *Rev. Hortic.*, **348**, 7–12.

Gerber, T., Steinbacher, S. and Hauser, B. (1999). Holzfasersubstrat zur Kultur von Pelargonium-Zonale-Hybriden–biophysicalishe und pflanzenbauliche Untersuchung. *Angewandte Botanik*, **73**, 217–221.

Golueke, C.G. (1972). *Composting, a Study of the Process and its Principles*. Emmaus, PA., USA: Rodale Press.

Gorodecki, B. and Hadar, Y. (1990). Suppression of *Rhizoctonia solani* and *Sclerotium rolfsii* disease in container media containing composted separated cattle manure and composted grape marc. *Crop Prot.*, **9**, 271–274.

Grantzau, E. (1991). Holzfasersubstrate im Zierpflanzenbau – Eigenschaften und Qualitatsforderungen. *Gartenbau Magazin*, **38**(9), 44–47.

Gruda, N. and Schnitzler, W.H. (1999). The influence of organic substrates on growth and physiological parameters of vegetables seedling. *Acta Hort.* (ISHS), **450**, 487–494.

Gruda, N. and Schnitzler, W.H. (2004). Suitability of wood fibre substrate for production of vegetable transplants.1 Physical properties of wood fibre substrates, *Sci. Hortic.*, **100**, 309–322.

Gruda, N., von Tucher, S. and Schnitzler, W.H. (2000). N-Immobilisierung in Holsfasersubstraten bei der Anzucht von Tomatenjungpflanzen (*Lycopersicon lycopersicium* (L) Karst.ex Farw.). *Angewandte Botanik*, **74**, 32–37.

Gumi, N. (2001). Toresa and other woodfibre products: advantages and drawbacks when used in growing media. In *Proc. Int. Peat Symposium, Peat in Horticulture, Peat and it's alternatives in growing media* (G. Schmilewski, ed.). Dutch National committee, International Peat Society, pp. 39–44.

Hadar, Y. and Mandelbaum, R. (1986). Suppression of *Pythium aphanidermatum* damping-off in container media containing composted liquorice roots. *Crop Prot.*, **5**, 88–92.

Hammond, R.F. (1975). The origin, formation and distribution of peatland resources. In *Peat in Horticulture* (D.W. Robinson and J.G.D. Lamb, eds). London: Academic Press, pp. 1–22.

Handreck, K.A. (1993). Properties of coir dust, and its use in the formulation of soilless potting media. *Comm. Soil Sci. Plant Anal.*, **24**, 349–363.

Harrelson, T., Warren, S.L. and Bilderbark, T.E. (2004). How do you manage aged versus fresh pine bark?. Proc. Southern Nursery Association Annual conference, 49th Annual Report, 63–65.

Hartz, T.K., Costa, F.J. and Schrader, W.L. (1996). Suitability of composted green waste for horticultural uses. *Hort Sci.*, **31**(6), 961–964.

Hess, T.F., Grdzelishvili, I., Sheng, H.Q. and Hovde, C.J. (2004). Heat inactivation of E. coli during manure composting. *Compost Sci. Util.*, **12**, 314–322.

Hogg, D., Barth, J., Favoino, E., et al. (2002). Comparison of compost standards within the EU, North America and Australasia. The Waste and Resources Action Programme, The Old Academy, 21 Horsefair, Banbury, Oxon OX16 0AH, (www.wrap.org.uk).

Hoitink, H.A.J. and Boehm, M.J. (1999). Biocontrol within the context of soil microbial communities: A Substrate-dependent Phenomenon. *Ann. Rev. Phytopathol.*, **37**, 427–446.

Hoitink, H.A.J. and Fahy, P.C. (1986). Basis for the control of soilborne plant-pathogens with composts. *Ann. Rev. Phytopathol.*, **24**, 93–114.

Hoitink, H.A.J., Krause, M.S. and Han, D.Y. (2001). Spectrum and mechanisms of plant disease control with composts. In *Compost Utilization in Horticultural Cropping Systems* (P.J. Stofella and B.A. Kahn, eds). Boca Raton, FLA., USA: CRC Press LLC, pp. 263–273.

Hoitink, H.A.J., Madden, L.V. and Dorrance, A.E. (2006). Systemic resistance induced by Trichoderma spp.: Interactions between the host, the pathogen, the biocontrol agent, and soil organic matter quality. *Phytopathology*, **96**, 186–189.

Hoitink, H.A.J., Schmitthenner, A.F. and Herr, L.J. (1975). Composted bark for control of root rot in ornamentals. *Ohio Report on Research and Development*, **60**(2), 25–26.

Hoitink, H.A.J., VanDoren, D.M. Jr. and Schmitthenner, A.F. (1977). Suppression of *Phytophthora cinnamomi* in a composted hardwood bark potting medium. *Phytopathology*, **67**, 561–565.

Hood, G. and Sopo, R. (2000). World Peat Production in 1997. Report to IPPA Meeting of Commission II. Quebec, Quebec. August, 2000.

Horst, L.E., Locke, J., Krause, C.R., et al. (2005). Suppression of Botrytis blight of begonia by Trichoderma hamatum 382 in peat and compost-amended potting mixes. *Plant Dis.*, **89**, 1195–1200.

Inbar, Y., Chen, Y. and Hadar, Y. (1989). Solid state carbon 13 nuclear magnetic resonance and infra red spectroscopy of composted organic matter. *Soil Sci. Soc. Am. J.*, **53**, 1695–1701.

Islam, M.S., Kahn, T., Ito, T., et al. (2002). Characterisation of the physio-chemical properties of environmentally friendly organic substrates in relation to rock wool. *J. Hort. Sci. Biotech.*, **77**, 1462–1465.

Jauch, M. and Fischer, P. (1993). Kompost in Substrat. *Deutscher Gartenbau*, **46**, 2888–2891.

Jones, P. and Martin, M. (2003). A review of the literature on the occurrence and survival of pathogens of animals and humans in green compost. The Waste and Resources Action Programme, The Old Academy, 21 Horsefair, Oxon OX16 0AH, UK.

Kang, J.Y., Lee, H.H. and Kim, K.H. (2004). Physical and chemical properties of organic horticultural substrates used in Korea. *Acta Hort.* (ISHS), **644**, 231–235.

Kavroulakis, N., Ehaliotis, C., Ntougias, S., et al. (2005). Local and systemic resistance against fungal pathogens of tomato plants elicited by a compost derived from agricultural residues. *Physiol. Mol. Plant Pathol.*, **66**, 163–174.

Kavroulakis, N., Papadopoulou, K.K., Ntougias, S., et al. (2006). Cytological and other aspects of pathogenesis-related gene expression in tomato plants grown on a suppressive compost. *Ann. bot.*, **98**, 555–564.

Kayhanian, M. and Tchobanoglous, G. (1993). Characteristics of humus produced from the anaerobic composting of the biodegradable fraction of municipal solid waste. *Environ. Technol.*, **14**, 815–829.

Kehre, B. and Pohle, A. (Eds) (1994). Determination of degree of stability by self heating tests (in German) In *Methodenbuch zur Analyse von Kompost (Methods for analysis of compost)*. Germany: Bundesgütegemeinschaft Kompost e.v. Köln.

Kehres, B. and Thelen-Jungling, (2006). Methodenbuch zur Analyse organischer Dungemittel, Bodenverbesserungmittel und substrate: IV. Al. (5th edn). Bundesgütegemeinschaft Kompost, Koeln-Gremberghoven, Germany.

Kipp, J.A., Wever, W. and de Kreij, C. (eds) (2000). International substrate manual, Elsevier International Business Information, PO Box 4, 7000 BA, Doetinchem, The Netherlands.

Kivinen, E. (1980). Proposal for a general classification of virgin peatlands. Proc. 6th International Peat Congress, Duluth, pp. 47–51.

Klapwijk, D. and Mostert, J. (eds) (1992). Richtlijnen voor de produktie van potgronden en substraten. Informatiereeks no. 73. Proefstation voor Tuinbouw onder Glas, Postbus 8, 2670 Naaldwijk, The Netherlands.

Konduru, S., Evans, M.R. and Stamps, R.H. (1999). Coconut husk and processing effect on chemical and physical properties of coconut coir dust. *Hort. Science*, **34**, 88–90.

Krause, M.S., De Ceuster, T.J.J., Tiquia, S.M., et al. (2003). Isolation and characterization of rhizobacteria from composts that suppress the severity of bacterial leaf spot of radish. *Phytopathology*, **93**, 1292–1300.

Kuter, G.A., Nelson, E.B., Hoitink, H.A.J. and Madden, L.V. (1983). Fungal populations in container media amended with composted hardwood bark suppressive and conducive to Rhizoctonia damping-off. *Phytopathol.* **73**, 1450–1456.

Laor, Y., Raviv, M. and Borisover, M. (2004). Evaluating microbial activity in composts using microcalorimetry. *Thermochimica Acta*, **420**, 119–125.

Lappalainen, E. (ed.) (1996). Global Peat Resources. International Peat Society, Kuokkalantie 4, 40420 Jyskä, Finland.

Lasaridi, K.E. and Stentiford, E.I. (1998). A simple respirometric technique for assessing composts stability. *Water Res.*, **32**, 3717–3723.

Lawrence, W.J.C. and Newell, J. (1939). *Seed and Potting Composts*. London: Allen and Unwin.

Lee, J.W., Lee, B.Y., Lee, Y.B. and Kim, K.S. (2000). Growth and inorganic element content of hot pepper seedlings in fresh and decomposed expanded rice hull based substrates. *J. Korean Soc. Hortic. Sci.*, **41**, 147–151.

Lemaire, F., Dartiques, A. and Riviere, L.M. (1980). Properties of substrates with ground pine bark. *Acta Hort.* (ISHS), **99**, 67–80.

Lemaire, F., Dartiques, A. and Riviere, L.M. (1988). Physical and chemical characteristics of a lingo-cellulose material. *Acta Hort.* (ISHS), **238**, 9–22.

Lemaire, F., Riviere, L.M., Stievenard, S., et al. (1998). Consequences of organic matter biodegradability on the physical and chemical parameters of substrates. *Acta Hort.* (ISHS), **469**, 129–138.

Lockwood, J.L. (1990). Relation of energy stress to behaviour of soilborne plant pathogens and to disease development. In *Biological Control of Soilborne Plant Pathogens* (D. Hornby, ed.). Wallingford, UK: CAB International, pp. 197–214.

Lopez-Medina, J., Perablo, A. and Flores, F. (2004). Closed soilless system growing: A Sustainable Solution to Strawberry Crop in Huelva (Spain). *Acta Hort.* (ISHS), **649**, 213–215.

Lorenzo, P., Trillas, M.I., Sant, M.D. and Aguilà, J.F. (1982). Effects of physical media properties on *Codiaeum variegatum* rooting response. *Acta Hort.* (ISHS), **126**, 293–302.

Lumsden, R.D., Millner, P.D. and Lewis, J.A. (1986). Suppression of lettuce drop caused by *Sclerotinia minor* with composted sewage sludge. *Plant Dis.*, **70**, 197–201.

Lumsden, R.D., Locke, J.C., Adkins, S.T., et al. (1992). Isolation and localization of the antibiotic gliotoxin produced by Gliocladium virens from alginate prill in soil and soilless media. *Phytopathol.*, **82**(2), 230–235.

Maas, E.F. and Adamson, R.M. (1975). Peat, bark and sawdust mixtures for nursery substrates. *Acta Hort.* (ISHS), **50**, 147–152.

Maher, M.J. and Prasad, M. (2001). The effect of N source on the composting of green waste and its properties as a component of a growing medium. In *Proceedings of the International Conference, Orbit 2001, on* Biological Processing of Waste, Seville, May 9–12, 299–305.

Maher, M.J. and Prasad, P. (2002). The Effect of addition of composted green waste to peat on the physical stability of growing media. In *Proceedings of the International Symposium, 2002*, Composting and Compost Utilization, Columbus, Ohio, May 6–8.

Maher, M.J. and Prasad, M. (2004). The effect of peat type and lime on growing medium pH and structure and on growth of *Hebe pinguifolia* 'Sutherlandii'. *Acta Hort.* (ISHS), **644**, 131–137.

Maher, M.J. and Thomson, D. (1991). Growth and manganese content of tomato (*Lycopersicon esculentum*) seedlings grown in Sitka spruce (*Picea sitchensis*) (Bong.) (Carr.) bark substrate. *Sci. Hortic.*, **48**, 223–231.

Maher, M.J., Prasad, M., Campion, J. and Mahon, M.J. (2000). Optimising nutrition of containerised nursery stock. End of Project Report, Project 4461, Teagasc, Kinsealy Research Centre, Malahide Road, Dublin 18. 22pp.

Maher, M.J., Prasad, M., Mahon, M.J., et al. (2001). Utilisation of compost in horticultural growing media. End of project report, Project 4751, Teagasc, Kinsealy Research Centre, Malahide Road, Dublin 17, Ireland.

Mandelbaum, R. and Hadar, Y. (1990). Effects of available carbon source on microbial activity and suppression of Pythium aphanidermatum in compost and peat container media. *Phytopathol.*, **80**, 794–804.

Maree, P.C.J. (1994). Using bio-degradable material as a growing media in hydroponics in the Republic of South Africa. *Acta Hort.* (ISHS), **361**, 141–158.

Martinez, F.X., Sepo, N. and Valero, J. (1997). Physical and physiochemical properties of peat-coir mixes and the effect of clay-material addition. *Acta Hort.* (ISHS), **450**, 39–46.

Mc Lachlan, K., Chong, C., Voorney, R.P., et al. (2004).Variability of soluble salts using different extraction methods on composts and other substrates. *Compost Sci. Util.*, **12**, 180–184.

Meerow, A.W. (1994). Growth of two sub-tropical ornamental plants using coir (coconut mesocarp pith) as a peat substitute. *Hort. Sci.*, **29**, 1484–1486.

Menzies, J.G., Ehret, D.L., Koch, C., et al. (2005). Fungi associated with roots of cucumber grown in different greenhouse root substrates. *Can. J. bot.*, **83**, 80–92.

Molitor, H.D., Faber, A., Marutzky, R. and Springer, S. (2004). Peat substitutes on the basis of recycled wood chipboard. *Acta Hort.* (ISHS), **644**, 123–130.

Monpoeho, S., Maul, A., Bonnin, C., et al. (2004). Clearance of human-pathogenic viruses from sludge: Study of Four Stabilization Processes by Real-time Reverse Transcription-PCR and Cell Culture. *Appl. Environ. Microbiol.*, **70**, 5434–5440.

Morel, P., Poncet, L. and Riviere, L.M. (2000). Les supports de culture horticoles 2000. Institut National de la Recherche Agronomique, Paris, France, p. 87.

Nelson, E.B., Kuter, G.A. and Hoitink, H.A.J. (1983). Effects of fungal antagonists and compost age on suppression of Rhizoctonia damping-off in container media amended with composted hardwood bark. *Phytopathology*, **73**, 1457–1462.

Nelson, P.V., Oh, Y.-M. and Cassel, D.K. (2004). Changes in physical properties of coir dust substrates during crop production. *Acta Hort.* (ISHS), **644**, 261–268.

Noble, R. (Ed.) (2005). Recycling horticultural wastes to produce pathogen suppressant composts for sustainable vegetable production. Final Report of EU FP5 project QLRT-CT-01458 RECOVEG. Warwick HRI, University of Warwick, UK.

Noble, R. and Coventry, E. (2005). Suppression of soil-borne plant diseases with composts: A Review. *Biocontrol Sci. Technol.*, **15**, 3–20.

Noble, R. and Roberts, S.J. (2003). A review of the literature on eradication of plant pathogens and nematodes during composting, disease suppression and detection of plant pathogens in compost. The Waste and Resources Action Programme, The Old Academy, 21 Horsefair, Oxon OX16 0AH, UK.

Noble, R. and Roberts, S.J. (2004). Eradication of plant pathogens and nematodes during composting: A review. *Plant Pathol.*, **53**, 548–568.

Noguera, M., Abad, R., Pucchades and Maquiera, A. (2003a), Coconut coir waste, a new and viable ecologically friendly peat substitute. *Acta Hort.* (ISHS), **517**, 279–286.

Noguera, P., Abad, M., Puchades, R., et al. (2003b). Influence of particle size on physical and chemical properties of coconut coir dust in container medium. *Comm. Soil Sci. Plant Anal.*, **34**, 593–605.

Offord, C.A, Muir, S. and Tyler, J.C. (1998). Growth of selected Australian plants in soilless media using coir as a peat substitute. *Aust. J. Agric.*, **38**, 879–887.

Oka, Y. and Yermiyahu, U. (2002). Suppressive effects of composts against the root-knot nematode *Meloidogyne javanica* on tomato. *Nematology*, **4**, 891–898.

Olivain, C. and Alabouvette, C. (1999). Process of tomato root colonization by a pathogenic strain of *Fusarium oxysporum* f. sp. *lycopersici* in comparison to a pathogenic strain. *New Phytologist*, **141**, 497–510.

Parks, S., Newman, S. and Golding, J. (2004). Substrate effects on greenhouse cucumber growth and fruit quality in Australia. *Acta Hort.* (ISHS), **648**, 129–133.

Papafotiou, M., Chronopoulos, J., Kargas, G., et al. (2001). Cotton gin trash compost and rice hull as growing medium components for ornamentals. *J. Hortic. Sci. Biotechnol.*, **76**, 431–435.

Paulsrud, B., Gjerde, B. and Lundar, A. (2004). Full scale validation of helminth ova (*Ascaris suum*) inactivation by different sludge treatment processes. *Water Sci. Technol.*, **49**, 139–146.

Penningsfeld, F. (1962). *Die Ernährung im Blumen- und Zierplanzenbau*. Berlin und Hamburg: Paul Parey.

Penningsfeld, F. and Kurzmann, F. (1966). *Hydrokultur und Torfkultur*. Stuttgart: Eugen Ulmer Verlag.

Pera, A. and Calvet, C. (1989). Suppression of Fusarium wilt of carnation in a composted pine bark and a composted olive pomace. *Plant Dis.*, **73**, 699–700.

Pera, A. and Filippi, C. (1987). Controlling of Fusarium wilt in carnation with bark compost. *Biol. Wastes*, **22**, 219–228.

Pharand, B., Carisse, O. and Benhamou, N. (2002). Cytological aspects of compost-mediated induced resistance against Fusarium crown and root rot in tomato. *Phytopathology*, **92**, 424–438.

Pokorny, F.A. and Austin, M.E. (1982). Propagation of blueberry by softwood terminal cuttings in pine bark and peat media. *Hort. Science*, **17**, 640–642.

Pokorny, F.A. and Wetzstein, H.Y. (1984). Internal porosity and root penetration of pine bark particles *Hort. Science*, **19**, 447–449.

Postma J., Willemsen-de Klein, M.J.E.I.M. and van Elsas, J.D. (2000). Effect of the indigenous microflora on the development of root and crown rot caused by *Pythium aphanidermatum* in cucumber grown on rockwool. *Phytopathology*, **90**, 125–133.

Prasad, M. (1997a). Nitrogen fixation of various materials from a number of European countries by three nitrogen fixation tests. *Acta Hort.* (ISHS), **450**, 353–362.

Prasad, M. (1997b). Physical, chemical and biological properties of coir dust. *Acta Hort.* (ISHS), **450**, 21–30.

Prasad, M. (1979). Physical properties of media for container-grown crops. 1. New Zealand peats and wood wastes. *Sci. Hortic.*, **10**, 317–323.

Prasad, M. (1980). Retention of nutrients by peats and woodwastes, *Sci. Hortic.*, **12**, 203–209.

Prasad, M. (1980a). Evaluation of woodwastes as a substrate for ornamental crops watered by capillary and drip irrigation. *Acta Hort.* (ISHS), **99**, 93–104.

Prasad M. and Maher, M.J. (1993). Physical and chemical properties of fractionated peat. *Acta Hort.* (ISHS), **342**, 257–264.

Prasad, M. and Maher, M.J. (2001). The use of composted green waste (CGW) as a growing medium component. *Acta Hort.* (ISHS), **549**, 107–114.

Prasad, M. and Maher, M.J. (2004). Stability of peat alternatives and use of moderately decomposed peat as a structure builder in growing media. *Acta Hort.* (ISHS), **648**, 145–151.

Prasad, M. and Maher, M.J. (2006). Evaluation of composted botanic materials as components of reduced peat growing media for nursery stock. In *Proceedings of the International Conference ORBIT*. Biological waste management – From local to global. Vol. 2, Germany: Weimar, pp. 401–407.

Prasad, M. and Ni Chualáin, D. (2004). Relationship between particle size and airspace of growing media. *Acta Hort.* (ISHS), **648**, 161–176.

Prasad, M. and O'Shea, J. 1999. Relative breakdown of peat and non peat growing media. *Acta Hort.* (ISHS), **481**, 121–128.

Pronk, A.A. (1995). Composted vegetable, fruit and garden waste as a substitute for peat in container-grown nursery stock. *Acta Hort.* (ISHS), **401**, 473–480.

Puustjarvi, V. (1973). Peat and its use in horticulture. Turveteollisuusliitto ry., Publication 3.

Puustjarvi, V. and Robertson, R.A. (1975). Physical and chemical properties. In *Peat in Horticulture* (D.W. Robinson and J.G.D. Lamb, eds). London: Academic Press Inc., pp. 23–38.

Raviv, M. (2005). Production of high-quality composts for horticultural purposes – a mini-review. *HortTechnology*, **15**, 52–57.

Raviv, M. and Medina, S. (1997). Physical characteristics of separated cattle manure compost. *Compost Sci. Util.*, **5**(3), 44–47.

Raviv, M., Chen, Y. and Inbar, Y. (1986). Peat and peat substitutes as growth media for container grown plants. In *The role of organic matter in modern agriculture* (Y. Chen and Y. Avnimelech, Y. eds). Dordrecht, The Netherlands: Martinus Nijhoff Publishers, pp. 255–288.

Raviv, M., Zaidman, B.Z. and Kapulnik, Y. (1998). The use of compost as a peat substitute for organic vegetable transplants production. *Compost Sci. Util.*, **6**, 46–52.

Raviv, M., Lieth, J.H., Burger D.W. and Wallach, R. (2001). Optimization of transpiration and potential growth rates of Kardinal rose with respect to root zone physical properties. *J Am. Soc. Hort. Sci.*, **126**, 638–643.

Raviv, M., Oka, Y., Katan, J., et al. (2004). High-nitrogen compost as a medium for organic container-grown crops. *Bioresour. Technol.*, **96**, 419–427.

Raviv, M., Oka, Y., Katan, J., et al. (2005). High-nitrogen compost as a medium for organic container-grown crops. *Bioresour. Technol.*, **96**, 419–427.

Reuveni, R., Raviv, M., Krasnovsky, A., et al. (2002). Compost induces protection against Fusarium oxysporum in sweet basil. *Crop Prot.*, **21**, 583–587.

Rivière, L.M. and Caron, J. (2001). Research on substrates: state of the art and need for the coming 10 years. *Acta Hort.* (ISHS), **548**, 29–41.

Ros, M., Hernandez, M.T., Garcia, C., et al. (2005). Biopesticide effect of green compost against fusarium wilt on melon plants. *J. Appl. Microbiol.*, **98**, 845–854.

Ryckeboer, J. and Coosemans, J. (1996). The influence of biowaste composts on *Pythium* and rootknot nematodes (*Meloidogyne incognita*) on tomato plants (*Lycopersicon esculentum* Mill.) in vivo. Mededelingen Faculteit Landbouwkundige en Toegepaste Biologische Wetenschappen Universiteit Gent (Belgium). **61**(2b), 567–576.

Rynk, R. and Richard, T.L. (2001). Commercial compost production systems. In *Compost Utilization in Horticultural Cropping Systems* (P.J Stofella, and B.A. Kahn, eds). Boca Raton, FLA., USA: CRC Press LLC, pp. 51–93.

Scaglia, B. Tambone, F., Genevini P.L. and Adani, F. (2000). Respirometric index determination, dynamic and static approaches. *Compost Sci. and Util.*, **8**, 90–98.

Scharpf, H.C. (1997). Physical characteristics of peat and the growth of pot plants. In *Proceeding of the International Peat Symposium, Peat in Horticulture – Its Use and Sustainability*, Amsterdam, pp. 43–52.

Scheuerell, S.J. and Mahaffee, W.F. (2005). Microbial recolonization of compost after peak heating needed for the rapid development of damping-off suppression. *Compost Sci. Util.*, **13**, 65–71.

Schonfeld, J., Gelsomino, A., Overbeek, L.S. et al. (2003). Effects of compost addition and simulated solarisation on the fate of *Ralstonia solanacearum* biovar 2 and indigenous bacteria in soil. *FEMS Microbiol. Ecol.*, **43**, 63–74.

Schrock, P.A.T. and Goldsberry, K.L. (1982). Growth responses of seed geranium and petunia to N source and growing media. *J. Am. Soc. Hort. Sci.*, **101**, 34–37.

Scott, M. and Bragg, N. (1994). Digging into composts. *Combined Proceedings of the International Plant Propagators' Society*, **43**, 150–156.

Serra-Wittling, C., Houot, S. and Alabouvette, C. (1996). Increased soil suppressiveness to Fusarium wilt of flax after addition of municipal solid waste compost. *Soil Biol. Biochem.*, **28**, 1207–1214.

Smith, C. (1995). Coir: A Viable Alternative to Peat for Potting. *The Horticulturist*, **4**, 12–28.

Solbraa, K. (1979). Composting of bark. I. Different bark qualities and their uses in plant production. Med. fra Norsk Institutt for. *Skogforskning*, **34**, 285–323.

Spencer, S. and Benson, D.M. (1981). Root rot of *Aucuba japonica* caused by *Phytophthora cinnamomi* and *P. citricola* and suppressed with bark media. *Plant Dis.*, **65**, 918–921.

Sterret, S.B. (2001). Composts as horticultural substrates for vegetable transplant production. In *Compost utilization in horticultural cropping systems* (P.J. Stofella, and B.A. Kahn, eds). Boca Raton, FLA., USA: CRC Press LLC, pp. 227–240.

Tahvonen, R. (1982). The suppressiveness of Finnish light coloured Sphagnum peat. *J. Sci. Agric. Soc. Finland*, **54**(5), 345–356.

Tahvonen, R. (1993). The disease suppressiveness of light-coloured sphagnum peat and biocontrol of plant diseases with *Streptomyces* sp. *Acta Hort.* (ISHS), **342**, 37–42.

Termorshuizen, A.J., van Rijn, E., van der Gaag, D.J., et al. (2006). Suppressiveness of 18 composts against 7 pathosystems: Variability in Pathogen Response. *Soil Biol. Biochem.*, **38**, 2461–2477.

Theodore, M. and Toribio, J.A. (1995). Suppression of *Pythium aphanidermatum* in composts prepared from sugarcane factory residues. *Plant Soil*, **177**, 219–223.

Trillas-Gay, M.I., Hoitink, H.A.J. and Madden, L.V. (1986). Nature of suppression of Fusarium wilt of radish in a container medium amended with composted hardwood bark. *Plant Dis.*, **70**, 1023–1027.

Tuitert, G., Szczech, M. and Bollen, G.J. (1998). Suppression of *Rhizoctonia solani* in potting mixtures amended with compost made from organic household waste. *Phytopathology*, **88**, 764–773.

Van Doren, J.J.M. (2001). Coir pith potential In *Proc. Int. Peat Symposium, Peat in Horticulture, peat and it's alternatives in growing media* (G. Schmilewski, (ed.). Dutch National Committee, International Peat Society, pp. 23–29.

van Loon, L.C., Bakker, P.A.H.M. and Pieterse, C.M.J. (1998). Systemic resistance induced by rhizosphere bacteria. *Ann. Rev. Phytopathol.*, **36**, 453–483.

Veeken, A.H.M., Blok, W.J., Curci, F., et al. (2005). Improving quality of composted biowaste to enhance disease suppressiveness of compost-amended, peat-based potting mixes. *Soil Biol. Biochem.*, **37**, 2131–2140.

Verdonck, O. and Demeyer, P. (2004). The influence of particle sizes on the physical properties of growing media. *Acta Hort.* (ISHS), **644**, 99–101.

Verhagen, J.G.M. (1997). Particle size distribution to qualify milled peat; a prediction of air content of ultimate mixes. In *Proceeding of the International Peat Symposium, Peat in Horticulture – Its Use and Sustainability*. Amsterdam, pp. 53–56.

von Post, L. (1922). Sveriges Geologiska Undersöknings torvinventering och nogra av dess hittils vunna resultat (SGU peat inventory and some preliminary results). *Svenska Mosskulturforeningens Tidskrift, Jonköping, Sweden*, **36**, 1–37.

Wever, G. (1991). Guide values for physical properties of peat substrates. *Acta Hort.* (ISHS), **294**, 41–47.

White, J.W. (1974). Dillon Research Fund, Progress Report on Research at Penn State. *Penn. Flower Growers Bulletin*, **273**, 3–4.

Widmer T.L., Graham, J.H. and Mitchell, D.J. (1998). Composted municipal waste reduces infection of citrus seedlings by Phytophthora nicotianae. *Plant Dis.*, **82**, 683–688.

Woods, M.J. and Kenny, T. (1968). Nutritional and cultural aspects of peat as a growing medium for tomatoes. *Proc. 6th Coll. Int. Potash Inst.*, 342–351.

Worrall, R.J. (1978). The use of composted wood waste as a peat substitute. *Acta Hort.* (ISHS), **82**, 79–86.

Worrall, R.J. (1981). Comparison of composted hardwood and peat- based media for the production of seedlings, foliage and flowering plants. *Sci. Hortic.*, **15**, 311–319.

Yogev, A., Raviv, M., Hadar, Y., et al. (2006). Plant waste-based composts suppressive to diseases caused by pathogenic *Fusarium oxysporum. Euro. J. Plant Pathol.*, **116**, 267–278.

Zhang, W., Han, D.Y., Dick, W.A., et al. (1998). Compost and compost water extract-induced systemic acquired resistance in cucumber and *Arabidopsis. Phytopathology*, **88**, 450–455.

Zhang, W., Hoitink, H.A.J. and Dick, W.A. (1996). Compost-induced systemic acquired resistance in cucumber to *Pythium* root rot and anthracnose. *Phytopathology*, **86**, 1066–1070.

12

INORGANIC AND SYNTHETIC ORGANIC COMPONENTS OF SOILLESS CULTURE AND POTTING MIXES

ATHANASIOS P. PAPADOPOULOS, ASHER BAR-TAL,
AVNER SILBER, UTTAM K. SAHA
AND MICHAEL RAVIV

12.1 Introduction
12.2 Most Commonly Used Inorganic Substrates in Soilless Culture
12.3 Most Commonly Used Synthetic Organic Media in Soilless Culture
12.4 Substrates Mixtures — Theory and Practice
12.5 Concluding Remarks
Acknowledgements
References

12.1 INTRODUCTION

The components of soilless growing media and potting mixes used in horticulture are primarily selected based on their physical and chemical characteristics and, in particular, their superior ability to provide simultaneously sufficient levels of oxygen and water to the roots. The substrates should be porous and hold the water by weak matric forces, so that plants can easily extract water and experience a lower risk of oxygen deficiency. Various soilless substrates in commercial use, such as stone wool, polyurethane, perlite, vermiculite and others are virtually pest-free and can easily be disinfested between growth cycles in case of disease contamination. They also enable relatively easy control over the pH and the availability of nutrients in the root zone. An inherent drawback of soilless cultivation is the fact that root volume is restricted. The limited root volume increases root-to-root competition for water, oxygen and nutrients since there are more roots per unit volume of medium. This subject is discussed in Chap. 13.

Description of substrates used in horticulture, as well as mode of soilless culture management, has been widely reviewed by Bunt, 1988; Schwartz, 1995; Hannan, 1998; Resh, 1998; Kipp et al., 2000; Raviv et al., 2002; Nelson, 2003; and Jones, 2005. The purpose of this chapter is to describe the main inorganic and synthetic organic substrates used in horticulture and to identify their main advantages and drawbacks.

12.2 MOST COMMONLY USED INORGANIC SUBSTRATES IN SOILLESS CULTURE

There is a lot of variability in the origin, physical and chemical characteristics of the substrates used by the horticultural industry. Moreover, new sources of natural and artificial by-products are being introduced as growing media from time to time. Therefore, it is impossible to describe all types of substrates and only the most representative and commonly used will be described. In this chapter, substrates are divided into inorganic and synthetic organic materials. The organic materials comprise synthetic substrates (like phenolic resin and polyurethane) and natural organic matter (peat, coconut coir and composted organic wastes); the natural organic substrates were discussed in Chap. 11. The inorganic substrates can be classified as natural unmodified materials (sand, tuff, pumice) and processed materials (stone wool, perlite, vermiculite, and expanded clay and zeolite). Important properties of the substrates include their chemical activity and surface charge. Therefore, substrates are characterized as active (e.g. peat, tuff) or semi-inert (e.g. stone wool and sand) materials. The difference between these two groups is quantitative, rather than a qualitative one. For a more thorough discussion of this issue, see Chap. 6.

The description of each substrate includes information on its production and origin, plus general information on its applications as a growth medium, or for other purpose. The physical characteristics – bulk density (BD), water retention and hydraulic conductivity – are given, as these properties are essential for proper irrigation management. The chemical characteristics – composition, stability as affected by pH, cation exchange capacity (CEC), pH and salinity – are given as these basic data are required for the proper management of fertilization and irrigation. Information on substrate sterilization is given, as disease control is a major factor in the successive use of substrates. Information on waste treatment is also presented since the potential for environmental contamination is becoming a central issue in intensive soilless cultivation.

12.2.1 NATURAL UNMODIFIED MATERIALS

12.2.1.1 Sand

Production, Origin and General Information – Sand is the coarse fraction of soil minerals. It is defined by the International Society of Soil Science as particles above 0.02 mm in diameter, and it is further separated into: (i) coarse sand, 0.2–2.0 mm; and (ii) fine sand, 0.02–0.2 mm. Pure sand is widely used in deserts and coastal plains because it is a cheap, local, natural source. It is often used as a growing bed situated on the ground above a polyethylene film that separates it from the soil. As a natural deposit, the particle size and distribution are often variable. The required depth of the sand layer depends on the range of particle diameters. The finer the sand, the deeper the required layer of sand to avoid water logging and poor aeration. Sand is also used as a component of various substrates, usually forming the heaviest constituent, serving as ballast for top-heavy plants and in outdoor nurseries against wind.

Physical Characteristics – The BD of sand is high relative to other growing media components, 1.48 and 1.80 g cm^{-3} for fine and coarse sand, respectively. The total porosity is relatively low, 0.45 and 0.30 per cent for fine and coarse sand, respectively, and the water content at saturation is somewhat low, 0.39 and 0.27 per cent, respectively. Sand has a narrow pore-size distribution; therefore the small pore fraction retains almost a constant water volume over increasing suction (from 0 to 10 cm water for coarse sand, or from 0 to 20 cm water for fine sand). A further increase in water suction results in a steep decline in water content. The hysteresis phenomenon is negligible. Its retention curve (da Silva, 1991; Wever et al., 1997) indicates that aeration problems are expected when using fine sand in common pots or beds. Bunt (1991) showed that the mean oxygen diffusion rate in the profile of a fine sand bed was 10–100 times lower than that of peat, perlite, redwood bark and different mixtures. The saturated hydraulic conductivity of coarse and medium sand was relatively high, 5.1 and 7.1 cm min^{-1}, respectively (da Silva, 1991). However, the unsaturated hydraulic conductivity of coarse and medium sand was reduced sharply as the water suction increased above 10 and 20 cm, respectively (da Silva, 1991).

Chemical Characteristics – Quartz (SiO_2) is the most common component of the sand fraction in soils. It is the second most common mineral in the earth's crust and is highly resistant to weathering (Drees et al., 1989). Quartz particle density is high, 2.6–2.65 g cm^{-3}, with a relatively low specific surface, 2 m^{2}g^{-1} (Drees et al., 1989). Quartz is a stable mineral with a low solubility of 3–7 mg Si L^{-1}, independent of pH in the pH range of 2.5–9.0 (Drees et al., 1989). It is one of the purest minerals known with a very low substitution of Si by Al, Fe and other trace elements. Thus, the charge deficiency that plays a major role in the physical–chemical activity of other soil minerals is very low in Quartz (Drees et al., 1989). Therefore, sand is a chemically inactive substrate that can serve as diluent to more reactive components in potting media. It must be emphasized that beach sand and road de-icing sand contain high amount of salts and therefore should not be used for horticultural purposes.

Sterilization, Re-use, and Waste Disposal – Sand can be steam-sterilized. With thin layers, the pores may rapidly fill with water, which disturbs the steaming process (Kipp et al., 2000). Sand is very durable because it is neither chemically nor biologically altered during the course of its use as a growing medium. Sand waste can be used in infrastructure and construction, thus it does not raise environmental pollution problems. On the other hand, the protection of natural sand dunes limits the use of sand as a growing medium.

12.2.1.2 Tuff

Production, Origin and General Information – Tuff is a common name for pyroclastic (Greek pyro 'fire', and klastos 'fragment') volcanic material, characterized by high porosity and surface area. Volcanic rocks are classified according to their silica content as follows (silica percentage in the solid phase): rhyolitic (more than 65), andesitic (50–65) and basaltic (less than 50). Rhyolitic lavas are formed during eruption at relatively low temperatures (800–1000°C), and therefore they contain predominantly light elements such as Si and Al, whereas their Fe, Mn, Ca and Mg contents are low. The viscosity of rhyolitic rocks is high and they have a light colour. Basaltic lavas are formed at high temperatures (above 1000°C), and therefore have a higher Fe, Mn, Ca and Mg content, which induces low viscosity and dark colour. Andesitic lavas are formed at intermediate temperatures and have intermediate colour and a chemical composition between that of rhyolitic and basaltic lavas.

Rapid cooling of magma during eruption prevents the formation of primary minerals and, therefore, pyroclastic materials contain mainly vesicular, volcanic glass. The physical and chemical properties of tuff are determined mainly by its mineralogical composition and weathering stages (Silber et al., 1994; 1999). Three tuffs erupted from the same volcano, and having almost the same chemical composition, but a differing primary mineral composition and weathering stage, they had different surface charge characteristics, P adsorption capacity and dissolution kinetics (Silber et al., 1994; 1999). In addition, grinding and sieving processes may affect tuff's physical and chemical properties.

Physical Characteristics – Tuff has a BD of 0.8–1.5 g cm^{-3}, and a total porosity of 60–80 per cent, depending on its origin and the sieving/grinding process. Retention curve and hydraulic characteristics of tuff are discussed in detail in Chap. 3.

Chemical Characteristics – Tuffs have permanent and variable charge surfaces, resulting mainly from amorphous materials. The chemical composition and the surface characteristics of three tuffs from Northern Israel differing in their weathering stage are shown in Table 12.1 (After Silber et al., 1994).

Tuff possesses a buffering capacity and may adsorb or release nutrients, especially P, during the growth period (Silber et al., 1999; Silber and Raviv, 1996). The chemical stability of tuffs depends on their mineralogical composition. Volcanic glass dissolution is very rapid while that of secondary minerals, such as kaolinite and halloysite, is slower. Hence, non-weathered materials containing a high concentration of volcanic glass, like black tuff, are unstable and dissolve easily in solution below pH 6, while red and yellow tuffs are more stable (Silber et al., 1999). Introducing plants to black or even red tuff, after equilibration with an acidic electrolyte (pH below 5), may be risky due to Al and Mn toxicity. More specific chemical characteristics of tuff are discussed in detail in Chap. 6.

TABLE 12.1 Chemical Composition, Expressed as Oxides (%) and Surface Characteristics of Three Tuff Types from Northern Israel

Oxide	SiO$_2$	Al$_2$O$_3$	Fe$_2$O$_3$	CaO	MgO	Na$_2$O	K$_2$O
Black	45.8	16.3	12.1	8.7	7.3	3.9	1.6
Red	43.3	13.3	12.2	10.9	4.3	4.2	1.7
Yellow	39.6	20.4	15.5	4.9	5.3	0.2	0.2

Tuff type	Black	Red	Yellow
Specific surface area (m^2 g^{-1})[a]	7	28	174
pH in H$_2$O[b]	9.3	8.1	7.9
pH in 1 M-KCl[b]	7.8	6.6	6.6
CEC in pH 7 (cmol kg^{-1})	10.7	28.5	60.1

[a] N$_2$ adsorption; [b] 90 min shaking 0.5 g 25 mL^{-1}.

Sterilization, Re-use, and Waste Disposal – Tuff is a stable material, which can last for many years. Growing plants may even improve the chemical properties of tuff due to the accumulation of organic matter and low-molecular-weight fulvic acid (Silber and Raviv, 1996). Steaming, solarization or chemical treatments can be used for disinfestation. Disinfestation treatments of tuff after rose (*Rosa* × *hybrida*) cut flower production resulted in yields which were superior to those of unused tuff (Raviv et al., 1998). The disinfestation process *per se* did not affect the solid phase or stability of tuff, but significantly affected the solubilization of organic matter. The soluble organic compounds released during disinfestation were adsorbed later by the tuff surfaces, thereby changing its surface charge (Silber and Raviv, 2001). Thus, beyond their beneficial effects on pathogen populations, the disinfestation treatments enhanced nutrient availability.

12.2.1.3 Pumice

Production, Origin and General Information – Pumice, like tuff, is a product of volcanic activity and usually forms from silicic lavas developed in rhyolitic composition, rich in gases and volatiles (Challinor, 1996). Rapid releases of pressure during volcanic eruptions lead to gas expansion and the formation of low-density materials composed of highly vesicular volcanic glass. Pumice is common in areas rich in volcanic activity, such as the Portuguese Azores, Greek islands, Iceland, Japan, New Zealand, Russia, Sicily, Turkey and the United States. The raw material is mined from quarries, ground and sieved to specification. Its physical and chemical properties are affected by its aggregate size. Pumice has been used since Roman times as lightweight aggregate in construction, stonewashing in the clothing industry, polishing and cleaning metal, wood and glass, and as filler in the paper and plastic industries. The physical properties of pumice are very similar to those of tuff, while the chemical properties, especially surface activity and charge, are very different. While tuffs have permanent and variable charge surfaces, and possess a buffering capacity for nutrients, pumice is an inert aluminosilicate material.

Physical Characteristics – Pumice is a lightweight aggregate, having a low BD of 0.4–0.8 g cm^{-3}, and a total porosity of 70–85 per cent (Boertje, 1994; Challinor, 1996; Raviv et al., 1999), depending on its origin and the sieving/grinding process. Pumice possesses large pores and consequently its volumetric water content decreases sharply as water tension increases (Boertje, 1994; Raviv et al., 1999). The water-holding capacity of pumice is relatively low compared with stone wool, perlite or organic substrates and may limit water and nutrient uptake by plants, especially in hot climates (Raviv et al., 1999). The volumetric total porosity, air-filled porosity and easily available water (defined at a water suction of: 0, 0–1.0 and 1.0–5.0 kPa, respectively) of three pumices are detailed in Table 12.2.

The unsaturated hydraulic conductivity of Sicilian and Greek pumice decreased by almost 4 and 6 orders of magnitude, respectively, as water tension increased from 0 to 10 kPa (Raviv et al., 1999). Therefore very frequent irrigation is required for plants grown in pumice.

Chemical Characteristics – Pumice is an inert aluminosilicate material composed primarily of silica and Al-oxide, but may also contain metal oxides, calcite or salts. The chemical composition of pumice is shown in Table 12.3.

Pumice has no buffering capacity and possesses a very low surface charge, derived mainly from impurities of carbonate and metal content (Silber, unpublished data). The material is stable even at a pH of 2.5 (Silber, unpublished data). However, caution is recommended when using new pumice material because high concentrations of Na are leached out at the beginning of use.

TABLE 12.2 Physical Properties of Pumice from Several Sources

Pumice source	Bulk density g cm^{-3}	Total porosity %	Air-filled porosity %	Easily available water %
Iceland[a]	0.4	85	40	5
Sicily[b]	0.7	70	27	4
Greece[b]	0.6	75	41	2

[a]-from Boertje, 1994; [b]- from Raviv et al., 1999.

TABLE 12.3 Chemical Composition of Pumice, Expressed as Oxides (%)

Oxide	SiO$_2$	Al$_2$O$_3$	CaO	MgO	Fe$_2$O$_3$	Na$_2$O	K$_2$O
%	70–75	12–14	1–3	0.1–0.6	0.8–2.0	3–6	4–5

Sterilization, Re-use, and Waste Disposal – Pumice is biologically inert and contains no pathogens or weeds (Challinor, 1996). It is stable and can be re-used practically indefinitely. Being a natural product, it can be disposed of without causing environmental pollution.

12.2.2 PROCESSED MATERIALS

12.2.2.1 Perlite

Production, Origin and General Information – Perlite is a glassy volcanic rock with a rhyolitic composition and 2–5 per cent of combined water (Alkan and Doğan, 1998; Doğan and Alkan, 2004). The main known world's perlite reserves (about 70 per cent) are located along the Aegean coast in Turkey. The commercial product is produced by heating the ground, sieved material to 760–1100°C. The combined water in the perlite is converted to gas at high temperature in the oven and subsequently the volume expands 4–20 times its original volume, resulting in a lightweight high porosity material. Perlite is frequently used in potting soil mixtures and as a stand-alone growing medium (Grillas et al., 2001; Gül et al., 2005). It is produced in various grades, the most common being 0–2 and 1.5–3.0 mm in diameter. The various grades differ in their physical characteristics.

Physical Characteristics – Expanded perlite is very light with a particle and BD of 0.9 and 0.1 g cm^{-3}, respectively. It is very porous, has a strong capillary action and can hold 3–4 times its weight in water. Burés et al. (1997a) reported that water retained at 10 kPa is much higher for the coarse fraction (0.5–1.0 mm diameter) than for the fine fraction (0.25–0.50 mm diameter) of expanded perlite. This difference in water-holding capacity between the coarse and fine fractions indicates that most of the water is held by the coarse particles in internal pores. However, it is not explained by the volume of internal porosity alone (Burés et al., 1997a). The slope of the reduction in water content as the water tension increases is moderate relative to sand and stone wool (Burés et al., 1997a,b). The available and non-available water in commercial perlite of 0–4 mm diameter was 13.6 and 36.5 per cent of its volume, respectively (Burés et al., 1997b). The water retention curve of perlite shows moderate hysteresis (Burés et al., 1997b; Wever et al., 1997). Wever et al. (1997) reported that the saturation of perlite was very rapid, independent of its initial moisture. The saturated hydraulic conductivity depends on particle diameter (Burés et al., 1997a). For commercial perlite of 0–4 mm diameter, having 50 per cent of the particles smaller than 0.5 mm, saturated hydraulic conductivity was 0.3 cm min^{-1} (Burés et al., 1997b). A reduction of 2 orders of magnitude in the hydraulic conductivity was obtained as the water suction increased from 0 to 30 cm water (Burés et al., 1997a). This change is moderate in comparison to sand.

Chemical Characteristics – Perlite is neutral with a pH of 7.0–7.5, but it has no buffering capacity and contains no mineral nutrient. When the pH is low there is a risk of toxic Al release into the solution. The chemical composition of the material, as analysed by Olympios (1992), is given in Table 12.4.

Sterilization, Re-use, and Waste Disposal – Perlite is a sterile product as it is produced at a very high temperature. Chemically, perlite is a stable material, which

TABLE 12.4 Chemical Composition of Perlite, Expressed as Oxides (%)

Oxide	SiO_2	Al_2O_3	CaO	MgO	Fe_2O_3	Na_2O	K_2O
per cent	73.1	15.3	0.8	0.05	1.05	3.65	4.5

can last for several years; its stability is not greatly affected by acids or microorganisms. Being an inert material, recycling perlite poses no environmental problems. Re-using perlite without processing to grow successive crops is potentially risky because of media compaction, salt build-up, and pest contamination (Hanna and Smith, 2002). Replacing used perlite with new media to raise successive crops is costly, and recovering the expense by selling the product at a higher price may not work well in a competitive market. Steam sterilization of used perlite before planting a new crop has been recommended to safeguard against pathogen contamination (Wilson, 1988). However, this treatment requires expensive steam generators and may not be adequate to restore perlite's loose structure and to reduce media salt (Hanna, 2005). In a study conducted to determine if raising tomatoes (*Lycopersicon esculentum* Mill.) in cleaned and disinfected used perlite would be more economical than new perlite and without any negative effect on yield, Hanna (2005) recycled perlite twice a year for commercial tomato production in bag (18.9 L) culture. After the separation of roots from the previous crops, used perlite was then treated with hot water (13.25 L water/18.9 L perlite) at temperatures reaching 93.3°C to leach out excess salt and disinfect the medium. This treatment raised media temperatures above limits necessary to kill several fungi and nematodes and significantly reduced media salt (EC, NO_3–N, and K were reduced by 43, 50, and 47 per cent respectively) with no noticeable change in physical condition (particle size distribution) over 8 years (which was contrary to common belief that being soft, perlite is sensitive to mechanical compression, which may grind particles to powder). Cleaning and disinfecting used perlite for recycling saved 56 per cent of the cost to replace the media and gave greater marketable yield and heavier tomato fruit than new perlite. The observed yield benefit of perlite recycling was ascribed to the collective effect of salt reduction, media disinfection and the presence of an optimum level of nutrients; it usually takes time to build nutrients to an optimum level in new perlite. Thus, it was concluded that used perlite can be cleaned and disinfected as needed and recycled for many years because it is not organic in nature and physically and chemically stable (Hanna, 2005). Marfa et al. (1993) also found that perlite retains its physical properties for successive crops.

12.2.2.2 Vermiculite

Production, Origin and General Information – The raw material for vermiculite is a natural clay mineral that has a layered structure with water in between the layers. The substrate, named expanded Vermiculite, is produced in a similar way to perlite by heating the grinded and sieved material to 1000°C. The water is converted to vapour at high temperature in the oven and pushes the layers away from each other. This process is termed 'exfoliation'. As a result, expanded Vermiculite consists of granules with an accordion shape, light weight and high porosity (Kipp et al., 2000). Vermiculite is used as a sowing medium and as a component of potting soil mixtures. Fine grades are used mainly as a mulch in transplant production while coarse grades are frequently used in rooting media (Wright, 1989).

Physical Characteristics – Expanded Vermiculite is produced in various grades, the most common being 0–2, 2–4 and 4–8 mm in diameter. All grades are very light with a particle density of $0.9\,g\,cm^{-3}$ and a BD range of $0.07-0.1\,g\,cm^{-3}$, for the coarse and fine grades, respectively. It is very porous, has strong capillary action and can hold 3–4 times its weight in water.

Chemical Characteristics – Vermiculite is neutral clay, with a pH of 7.0–7.5 and low EC. This clay type contains two tetrahedral sheets for every one octahedral sheet. Like the raw material, it has a permanent negative charge. Consequently, it has a CEC value of $15-21\,cmol\,kg^{-1}$ and a buffering capacity for pH and cations. It also adsorbs ions like phosphate due to its high surface area and some positive-charged sites on the edges of the clay. The chemical composition of the material, expressed as oxides (per cent), is given in Table 12.5.

When the pH is low, there is a risk of toxic Al release into the solution. Mg is an important component of the mineral structure and is the dominant adsorbed cation.

Sterilization, Re-use, and Waste Disposal – Vermiculite is a sterile product as it is produced at very high temperatures. However, it cannot be steam-sterilized as it disintegrates during heating. Vermiculite expanded structure collapses easily. Therefore, it is not suitable for long period of use. It is sensitive to mechanical compression, which may grind particles to powder. Disposal of vermiculite is not hazardous to the environment. In the past, asbestos was found in some vermiculite mines. All those mines were closed and now vermiculite is considered safe.

TABLE 12.5 Chemical Composition of Vermiculite, Expressed as Oxides (%)

Oxide	SiO_2	Al_2O_3	MgO	Fe_2O_3
%	20–25	5–10	35–40	32–35

12.2.2.3 Expanded Clay Granules

Production, Origin and General Information – Expanded clay is a granular product with a crystalline structure with surface and internal porosity. It is produced by forming the clay into pellets and then fired in rotary kilns at 1200°C; at this temperature, gas is released and expands the clay (Oekotau Easy Green GmbH, 2006; ESCSI, 2006). The raw material must have a low content of soluble salts so that substances, like lime, are not added during the process. Otherwise, salt may be leached during cultivation. The grade size used in horticulture is 4–8 or 4–16 mm diameter (Oekotau Easy Green GmbH, 2006). Expanded clay granules have been used in horticulture since 1936. They are used in hydro-culture and in containers for various crops.

Physical Characteristics – Expanded clay granules are light with a BD of 0.28–0.63 g cm^{-3} and water content of 11–24 per cent at a pressure head of -10 cm (de Kreij et al., 1995). Wever and van Leeuwen (1995) reported high BD, 1.2 g cm^{-3}, and 53 per cent porosity. They found that the water content was reduced from 48 to 44 per cent as the water suction increased from 3 to 50 cm, thus the amount of easily available water is very small, just 4 per cent. By contrast, the material contains a large amount of air. It has been shown that there were no changes in physical characteristics after 5 years of intensive cropping. The saturation process of the substrate from a dry condition is relatively slow, but if it started from a pressure head of 100 cm it is rapid (Wever et al., 1997). The water retention curve shows insignificant hysteresis. There is no available information on the hydraulic conductivity of expanded clay granules, but they are relatively resistant to compression (Wever and Leeuwen, 1995).

Chemical Characteristics – Expanded clay granules are neutral, with a pH of about 7.0. The EC is low if low salt clay was used and no salts were added during the baking process; otherwise, washing is required. It has been classified as an inert material with no cation exchange or buffering capacity. However, Meinken (1997) and Meinken and Fischer (1994) demonstrated that nutrients may accumulate by diffusion into the granules and they can be released back to the solution.

Sterilization, Re-use, and Waste Disposal – Expanded clay granules is a sterile product as it is produced at a very high temperature. After use, it can be washed and sterilized without deleterious effect. Expanded clay granules are very stable and can last for many years. Waste material can be used in the construction industry.

12.2.2.4 Zeolite

Production, Origin and General Information – Zeolites are crystalline hydrated aluminosilicates of alkali and alkaline cations that possess infinite, three-dimensional crystal structures (Ming and Mumpton, 1989; Mumpton, 1999). Zeolites are usually formed by metamorphism of volcanic rocks, but may also be formed from non-volcanic materials in marine deposits or aqueous environments (Ming and Mumpton, 1989). Due to their ion exchange, adsorption, hydration–dehydration and catalysis properties, zeolites are widely used in agriculture and in numerous industries for the removal of pollutants from waste and drinking water (Ming and Mumpton, 1989; Mumpton, 1999). Amelioration of the chemical and physical properties of natural zeolite is achieved by producing synthetic minerals. Sherman (1999) stated that 508 different types of synthetic zeolites with improved physicochemical properties are produced all over the world. Zeolites (mainly clinoptilolite) are used in soil remediation to adsorb nuclear waste or heavy metals (Kapetanios and Loizidou, 1992; Rosen, 1996; Chlopecka and Adriano, 1997; Paasikallio, 1998; Krutilina et al., 1999, 2000). They are used in agriculture as soil amendments for: (i) a source of P, K and NH$_4$ nutrients in infertile soils and substrates (Hershey et al., 1980; Chen and Gabelman, 1990; Notario del Pino et al., 1994; Allen et al., 1995; Williams and Nelson, 1997; Dwairi, 1998); (ii) reducing N losses and nitrate contamination (Weber et al., 1983; Ferguson and Pepper, 1987; Huang and Petrovic, 1994; Ando et al., 1996; Kithome et al., 1999); and, (iii) improving water availability (Huang and Petrovic, 1995; Yasuda et al., 1995).

A sophisticated example for the use of zeolite in growing media is by pre-charging it with aluminum in order to produce blue hydrangea (Opena and Williams, 2003). Additional agricultural uses of zeolites were described by Pond and Mumpton (1984). Currently zeolites (mainly clinoptilolite) are widely used as components of growing media, especially in Mediterranean countries such as Greece and Turkey (Gül et al, 2005; Samartzidis et al., 2005).

Zeolites possess extremely high CEC values (220–460 cmol kg^{-1}) as well as a relatively high BD (1.9–2.3 g cm^{-3}) (Ming and Mumpton, 1989) and therefore the use of zeolite as a single component growing substrate is not recommended. However, in mixed substrates, which include organic (peat and compost) or inorganic materials (sand and perlite), zeolites are used widely for flower and vegetable production all over the world (Bulgaria, China, Cuba, Italy, Japan, Jersey, Korea, Russia, USA and Yugoslavia). The experimental use of zeolite as a single growing substrate has been reported for several crops, such as carnation (*Dianthus caryophyllus* L.) (Challinor et al., 1995), sweet pepper (*Capsicum annum* L.) (Harland et al., 1999), tomato (Rivero-Gonzales and Rodriguez-Fuentes, 1988), gerbera (*Gerbera jamesonii* Bolus ex Hooker f.) (Papadopoulos et al., 1995) and lettuce (*Lactuca sativa* L.) (Gül et al., 2005).

Clinoptilolite is the principal zeolite mineral used in agriculture and therefore only its physical and chemical properties will be detailed below.

Physical Characteristics – Clinoptilolite has a particle density of 2–2.1 g cm^{-3} (Ming and Mumpton, 1989) and saturation water content of 34 per cent (Mumpton, 1999). However, grinding and sieving processes affect the chemical and physical properties of the product. The physical properties of Turkish zeolites (clinoptilolite) differed by aggregate size were almost identical to that of known materials used for substrates (basaltic and rhyolitic tuff) at the same size (Unver et al., 1989). The volumetric total porosity, air-filled porosity and easily available water of these zeolites are detailed in Table 12.6 (from Unver et al., 1989).

Chemical characteristics – Clinoptilolite is very stable, but dissolves at pH 2, or lower in a period of time (Ming and Mumpton, 1989). Clinoptilolite has a unit-cell formula of $[(Na_3K_3)(Al_6Si_{30}O_{72})]24H_2O$ and CEC of 220 cmol kg^{-1} (Ming and Mumpton, 1989). For additional discussion on the potential nutritional effect of zeolites, see Chap. 6.

Sterilization, Re-use, and Waste Disposal – Zeolites are not sensitive to mechanical compression and so their physical properties do not change with handling; this offers possibilities for zeolite re-use. Harland et al. (1999) compared zeolite (clinoptilolite) as

TABLE 12.6 Physical Properties of Zeolite Fractions

Fraction size (mm)	Bulk density (g cm^{-3})	Total porosity (per cent)	Air-filled porosity (per cent)	Easily available water (per cent)
2–6	0.69	50	28	4
6–10	0.65	40	20	10
10–14	0.64	35	10	1

a new and recycled substrate under closed (recirculating) and open (non-recirculating) fertigation systems for growing sweet pepper over a period of 4 years (three recyclings of the substrate). In the recycling system, at the end of each cropping period, the plants were removed and roots were sieved from the substrate and then it was steam-sterilized at 90°C for 3 hours, before being re-bagged and used as the substrate for the succeeding crop. Over the period of the trial, there was no loss of yield or quality attributable to the re-use of zeolite substrate (Harland et al., 1999).

12.2.3 MINERAL WOOL

12.2.3.1 Stone Wool and Glasswool

Production, Origin and General Information – Mineral wool is a light, artificial material, originally produced for thermal and acoustic insulation in the construction industry. Stone wool products for a variety of purposes began to be produced by the Danish company I/S H.J. Henriksen and V. Kähler in 1909 (see http://www.rockwool.com). The application of modified forms of stone wool as a substrate for horticulture began in Denmark in 1969 (Smith, 1987). Through various corporate transitions the Grodan Group emerged and by the 1980s its flagship product 'rockwool' became widely synonymous with stone wool. The stone wool that is used in horticulture is mainly used as slabs or blocks of bonded fibres, but is also available in granulated form as a component of potting mixtures. Stone wool is manufactured by heating a mixture of three natural raw materials: 60 per cent diabase (a form of basalt rock, dolerite), 20 per cent limestone and 20 per cent coke. These materials are melted together at a high temperature, with the coke acting as a fuel in the form of a blast furnace through which air is forced to raise the temperature to 1600°C. The molten mixture is then spun at a high speed into thin fibres of about 0.5 mm diameter, which are cooled by an air stream. The fibres are heated with certain additives (a phenolic resin and wetting agents) to bind them together and lower the natural hydrophobicity of the material. It is then pressed into blocks or slabs of various sizes (Smith, 1987). Glasswool is made by melting quartz sand in an electric oven at 1200°C. The process of production of glasswool fibres and slabs is similar to that of stone wool. In general, both stone wool and glasswool (mineral wool) are sterile, easily managed and consistent in performance. Stone wool has become one of the major greenhouse substrates in The Netherlands, Belgium, Germany, Denmark and other Western European countries, and other regions (Smith, 1987) with current uses in virtually all areas where greenhouse vegetables are produced (Canada, US, Netherlands and other countries). It is an effective growth medium for horticultural crops, in which the grower can easily manipulate the ratio between water and air and between each of the nutrients in the root zone (Papadopoulos et al., 1999). On the other hand, it is 'unforgiving' to management errors because it lacks a buffering capacity for nutrients, pH and water, and due to the low volume of the common slabs.

Physical Characteristics – Stone wool is a lightweight substrate with a low BD of about $0.05-0.1\,g\,cm^{-3}$ and a pore volume of 92–98 per cent, depending on the producer (Smith, 1987, 1998; Kipp et al., 2000). It shows high water retention at low water tension and the water content declines sharply as the water tension increases so that virtually no water is retained at suctions higher than 5 kPa (da Silva, 1991; da Silva et al., 1995). As a result, the water buffer capacity is low and a steep gradient in water content occurs from the top to the bottom of the slab following irrigation and free drainage. The air volume is low at the bottom, just 4 per cent at a height of 1 cm above the base, while the upper layers are dry. The recommended height of a slab for optimal water to air ratio is 7.5–10 cm (Smith, 1987). However, van der Gaag and Wever (2005) recommended using deeper slabs when growing cucumbers in order to reduce stone wool conduciveness to *Pythium aphanidermatum*, the causal agent of Pythium root and crown rot, as a result of stone wool's high water content in shallow slabs. A strong hysteresis phenomenon was observed in the first cycle of wetting and drying, but it was reduced in the following cycles (da Silva et al, 1995). The hydraulic conductivity at saturation is very high (4.6 cm min^{-1}) but is sharply reduced with increased suction (da Silva et al, 1995). Such reductions in hydraulic conductivity may lead to poor uptake of water and nutrients and to rapid development of water stress if irrigation is delayed. Stone wool is a soft and elastic substance that is compressed under pressure and retains its original height following relaxation, but after use it becomes softer and less elastic (Wever and van Leeuwen, 1995).

Glasswool is very light and can contain a lot of water and air. In contrast to stone wool, the fibre diameter of glass wool can vary, which affects the water-holding capacity. By having finer fibres in the upper part of the slab than the lower part, it is possible to obtain better water content distribution over the height of the slab.

Chemical Characteristics – The chemical composition of stone wool, expressed as oxides (per cent), is given (Verwer, 1976) in Table 12.7.

The main chemical characteristic of stone wool is that it is relatively inert, when expressed on a volume basis, except for some minor effects on pH. The initial pH of commercial stone wool products was rather high in the past (7.0–8.0) and values of up to 9.5 have been recorded (Smith, 1987). However, nowadays values of 6–6.5 are the industry standard. Below pH 5.0, stone wool starts to dissolve, so low pH values should be avoided. Stone wool itself has no effect on crop nutrition and all nutrients must be supplied with the water. However, it was demonstrated that a considerable amount of Fe could be taken up from stone wool by rose rootstocks (Rupp and Dudley, 1989) and some vegetables (Sonneveld and Voogt, 1985). Moreover, Rupp and Dudley (1989) reported that considerable concentration of Fe, Mn, Cu and Zn could be extracted with DTPA from various stone wool products.

Sterilization, Re-use, and Waste Disposal – Being a sterile product, new stone wool provides root environment which is initially less vulnerable to pathogenic attack,

TABLE 12.7 Chemical Composition of Stone Wool, Expressed as Oxides (Per cent)

Oxide	SiO$_2$	Al$_2$O$_3$	CaO	MgO	Fe$_2$O$_3$	Na$_2$O	K$_2$O	MnO	TiO$_2$
per cent	47	14	16	10	8	1	1	1	1

especially in closed fertigation culture system (Tu et al., 1999). Stone wool can be steamed before re-use.

After one or two crop cycles, stone wool is usually discarded, producing a high volume of waste (approximately, 125 m^3 per hectare of plant production). This environmental issue of the waste is one of the major problems in the horticultural use of mineral wool, as it can not easily be returned back to nature. Disposal of used horticultural stone wool in landfill sites has been carried out in all countries where crops have been grown on stone wool; such disposal method is either increasingly less available or unsustainable (Bussell and McKennie, 2004).

In the last decade, various methods of stone wool recycling have been developed. Research on the re-use of stone wool has been and continues to be conducted by stone wool manufacturers, particularly Grodan A/S (Denmark) and Grodan BNF (The Netherlands). The Grodan studies summarized by Bussell and McKennie (2004) included several suggestions of re-using horticultural stone wool such as recycling by a stone wool factory, soil amendments (to improve poor physical properties of clay soils), mixing with other substrates (e.g. peat and compost), brick production, landfill cover, mining cover/recultivation and mushroom casing. Re-use of stone wool and glasswool as a raw material for the re-manufacture of both horticultural and insulation mineralwool and in brick production are now used widely in the United Kingdom and The Netherlands; where nearly all of the used stone wool from crops is collected and processed at large facilities (Neefjes, 2001). In areas where this type of recycling is not feasible, stone wool is commonly disposed of in a landfill.

One of the potential uses of used stone wool slabs is their incorporation (as chopped small particles) into peat- or compost-based media for horticultural production; this provides an opportunity for used stone wool recycling and for reducing the consumption of a non-renewable resource such as peat.

In New Zealand, a pilot study explored the possibility of utilizing the used stone wool slabs (after removal of plastic sleeves) as mulch around avocado trees; the mulch successfully suppressed weed growth for 2 years without affecting the growth of avocado trees (Bussell and McKennie, 2004). The bottlenecks to the direct re-use of stone wool slabs for growing succeeding crop(s) include their susceptibility of being broken when moved, the possibility of incomplete steam sterilization between crops and the occurrence of variable amounts of root matter in the used slabs making water content and EC control much more difficult than in new slabs (Bussell and McKennie, 2004).

12.3 MOST COMMONLY USED SYNTHETIC ORGANIC MEDIA IN SOILLESS CULTURE

12.3.1 POLYURETHANE

Production, Origin and General Information – Mineral oil is the raw material for the production of various foams, slabs and granules used for the furniture and construction industries. Mineral oil products, like di-isocyanate, may be mixed with glycol to produce polyurethane. The product is a polymer that contains excess di-isocyanate groups, which react with water to release CO_2 and induce foaming of the polymer. The resulting foams are used in the furniture industry and the cutting residues provide the raw material for substrates. They are ground to granules, which are pressed together with additives to form a slab. Steam at 140°C is blown in during the process. Foam is used as a growing medium for vegetables and flowers in Belgium and The Netherlands. The granules are often used as a component in mixtures. Polyurethane is considered as a substitute for perlite in potting mixes (Cole and Dunn, 2002).

Physical Characteristics – Polyurethane is a very light material, with a low particle density and a very low BD, 1.19 and 0.078 g cm^{-3} respectively. Consequently, the pore volume porosity is very high, 0.95. As most of the pores are relatively large, it holds a lot of air and very little water under low suction (Kipp et al., 2000). The available water in the range of 10–100 cm suction is about 2 per cent of its volume. There are no available data on its hydraulic conductivity characteristics.

Chemical Characteristics – Polyurethane is a chemically inert substrate with a low EC and a pH of 6. It does not contain or release any important nutrient, except Fe, Zn and B. In the past, the foam contained harmful organic substances, but this problem has now been eliminated.

Sterilization, Re-use and Waste Disposal – Polyurethane is resistant to acids, is not decomposed during production by the action of microbial enzymes, and hence cannot be composted after use. It is a sterile product that can be steamed between crops for several growing cycles. Waste can be re-used in the production of polyurethane slabs for construction, or it can be burnt. Polyurethane substrates can be re-used for as long as 10–15 years, and they are easy to handle (Benoit and Ceustermans, 1995). The superficial roots can be brushed away mechanically, and the slabs/mats can be pressed dry, which facilitates the passage of steam, so that sterilization is made more efficient. Benoit and Ceustermans (1995) showed that the roots left in polyurethane mats in 4 years of cultivation increased water retention at 1 KPa from 8 to 16 per cent, whereas air volume decreased from 85 to 74 per cent, suggesting that some change in air/water relations occurred. In a 10-year-long study, Benoit and Ceustermans (1994) observed that in the recycled polyurethane substrate, water retention at 1.0, 1.7 and 2.0 KPa increased and air volume decreased with the years of its re-use; after the 10-year period of re-use with eight steam treatments, one each between two successive years, the air/water retention properties of polyurethane substrate were comparable to those of new stone wool. However, owing to its low density, the substrate was susceptible to compression without being able to recover its original volume; this might be a disadvantage when the substrate is re-used and steamed repeatedly. In tomato production, earliness and total production were not affected by the re-use of polyurethane substrate over 10 years (Benoit and Ceustermans, 1994). In cucumber (*Cucumis sativum* L.) production, repeated use of the same polyurethane slab for two consecutive years gave comparable marketable yield to that recorded with new perlite or organic substrate each year (best yielding media), but polyurethane slab demanded more water and resulted in higher leaching percentage (Böhme, 1995). Hardgrave (1995) reported that when cucumbers were grown on steam-sterilized polyurethane foam slabs used previously, the productivity and fruit quality were equal or better to those of new slabs of polyurethane foam or stone wool; such beneficial effects on crop performance were ascribed to the improved water-holding capacity due to the presence of root materials within the recycled slab. Tests also showed that a polyurethane foam slab was stronger, more flexible, and hence suffered much less physical damage than a stone wool slab during the steaming process between two successive crops (Hardgrave, 1995). Because of labour saving and its long-lasting nature with repeated use, polyurethane is considered an advantageous substrate economically, agronomically and environmentally. Benoit and Ceustermans (1995) suggested that higher initial purchase price of polyurethane could be depreciated over a minimum of 10 years, and the price of steam sterilization was only half the price of recycling stone wool slabs.

12.3.2 POLYSTYRENE

Production, Origin and General Information – Flakes or beads of expanded polystyrene foam are by-products of polystyrene processing. These are sometimes added to substrate mixes to improve aeration and drainage. Polystyrene is one of the best-known synthetic polymers (others include polyethylene, polypropylene, and polyester), a very commonly found, mass-produced plastic. The individual building block of polystyrene is styrene, a liquid hydrocarbon, derived in the late nineteenth century from storax balsam, which comes from a tree in Asia Minor called the oriental sweet gum (*Liquidambar orientalis* Mill.). Polystyrene was invented in 1950 primarily to provide thermal insulation. At present, polystyrene includes any grade, class or type of thermoplastic polymer, copolymer, interpolymer, alloy or blend composed primarily of polymerized styrene. Polystyrene is made in a process known as suspension polymerization. Styrene (C_8H_8) is first derived from petroleum or natural gas through a two-step reaction between ethylene (C_2H_4) and benzene (C_6H_6): (1) ethylene reacts with benzene (C_6H_6) in the presence of a catalyst such as aluminum chloride to form ethylbenzene (C_8H_8); (2) which is then dehydrogenated (hydrogen is removed) at 600–650°C to form styrene. Styrene is subsequently polymerized either by heat or by an initiator such as benzoyl peroxide. Degree of polymerization is regulated by stopping the polymerization process at some desired point using polymerization inhibitors (also called terminating agents) such as oxygen, sulphur or quinol. To form the low-density, loosely attached cells, polystyrene is then suspended in water to form droplets. A suspension agent, such as specially precipitated barium-sulphate or copolymers of acrylic- and methacrylic-acid and their esters (organic product formed by the reaction between an acid and an alcohol), is then added to the water. The suspension agents are viscous and serve to hold up the droplets, preventing them from sticking together. The beads of polystyrene produced by suspension polymerization are tiny and hard. To make them expand, special blowing agents are used, including propane, pentane, methylene chloride and the chlorofluorocarbons. The resulting polystyrene beads are then cleaned, and anomalous beads filtered out.

During the 1980s, polystyrene was incorporated in growing media by many growers. It was generally found that the polystyrene did not do much that could not be better provided by other materials. Since the air in the beads is sealed from the gas phase in the root zone and since it has no water-holding or ion-exchange capacity, all it really does is reduce the effective size of the root zone for root exploration. The biggest problem for virtually all growers was the environmental contamination around the nursery due to polystyrene beads which had blown or washed out of containers and were clogging up drains and drainage pipes. Virtually all growers have since abandoned the use of polystyrene.

Physical Characteristics – Polystyrene beads have many of the physical characteristics of perlite. They are pearl like spheres with a particle density of $1.03\,g\,cm^{-3}$. Polystyrene beads are approximately 98–99 per cent gas by volume; the gas trapped in the cells is air. They are extremely light-weighted with a BD of around $0.024\,g\,cm^{-3}$. When added to growing mixtures, polystyrene increases aeration and drainage, and decreases the BD; it has almost no water-holding capacity. In a study comparing the addition of $0.38\,m^3$ of polystyrene beads or perlite to a growing mix ($0.38\,m^3$ sphagnum peat moss $+2.27\,Kg.$ calcitic lime $+1.36\,Kg$ of dolomitic lime), Matkin (2005) measured maximum volumetric water retention as $0.27\,cm^3\,cm^{-3}$ for the polystyrene mix as compared to $0.42\,cm^3\,cm^{-3}$ for the perlite mix.

Chemical Characteristics – Chemically, polystyrene beads are composed of 94–96% polystyrene, 4–6 per cent blowing agent, <0.4 styrene monomer and <0.5 per cent water. They are chemically inert; they supply no nutrients and are devoid of CEC or anion exchange capacity (AEC). The pH is neutral.

Sterilization, Re-use, and Waste Disposal – Polystyrene beads are highly resistant to microbial decomposition because of their large particle size and hydrophobic nature (Anonymous, 1994). They may be broken down by high temperature and by certain chemical disinfecting agents. They are also unstable to gasoline and similar petroleum hydrocarbon liquids. They should not be steam heated, which would cause the beads to migrate to the top of the growing media with a resultant nuisance, if dispersed by water or wind. They can safely be incinerated and will yield only carbon dioxide and water, if the procedure is handled correctly. Handling polystyrene should be done with special care from environmental standpoint, because it contains chlorofluorocarbons (CFCs); CFCs are inert, and harmless to humans and the environment upon their release. However, long after the first use of CFCs, scientists realized that they contribute to the depletion of the ozone layer of the atmosphere. Therefore, and since recycling of polystyrene mixed with other components is not possible, its use is not recommended anymore.

12.3.3 POLYESTER FLEECE

Production, Origin and General Information – Polyester fleece is a soft, fuzzy fabric generally used for clothing or blankets, and in any other applications where a warm, wool-like material is needed. It is a two-sided pile material, meaning that both the front and the back surface of the fabric consists of a layer of cut fibres, similar to corduroy or velvet. Polyester fleece is made from pure polyester fibres, which are mixed and thermally fixed. The raw material for polyester fleece, polyester, is produced by heating terephthalic acid (a petroleum derivative) with ethylene glycol (another petroleum derivative, commonly known as antifreeze) at 150–210°C. This first reaction produces dihydroxydiethyl terephthalate, which is then pumped into an autoclave, where it is heated under pressure to about 280°C. At this temperature the chemical transforms into a polymer called polyethylene terephthalate. As the polymer cools, it becomes thick syrup (a viscous liquid). This liquid is then extruded through a showerhead-like nozzle (tiny holes in a metal disk called a spinneret). On contact with air, the streams of liquid polymer become dry and hard. In England, this polymer was called terylene; Du Pont secured exclusive US rights to the polymer in 1946, calling it polyester. If the polymer, polyethylene terephthalate, is not extruded into fibres, it can be formed into the plastic commonly used for soda bottles. Interest in recycling plastics in the 1980s led to the development of polyester fibre made from used soda bottles. Many polyester fleece garments on the market today are made from a combination of recycled and virgin polyester. The slabs of polyester fleece covered with plastic foils are used as horticultural medium. Polyester fleece is an economically attractive substrate; the amount of substrate required can be as low as 1.7–3.6 t. ha^{-1} for polyester fleece as compared to 8.8–12.5 t. ha^{-1} for stone wool.

Physical Characteristics – The water–air–solid ratio in polyester fleece medium at water retention of 0.5 KPa has been reported as 59:39:2 (Schroeder and Forester, 2000), which could be considered as optimum for growing horticultural crops. The pore volume in polyester fleece slab is 98 per cent, very close to that in stone wool slab. It is an extremely lightweight medium. Depending on the quality of the fleece and structure of the fibres (compactness), and the height of the slabs, the BD may vary considerably. Schroeder and Forester (2000) used polyester slabs with three different heights (2, 4 and 8 cm) at the same length (100 cm) and breadth (20 cm), which had BD of 0.045, 0.045 and 0.026 g cm^{-3}, respectively. The water retention in the polyester fleece slabs in the range of 1–3 kPa is much lower than that in stone wool slabs. As reported by Schroeder and Forester (2000), for an increase in suction from 0 to 1.0 kPa, water-holding capacity of fleece dropped down sharply (from 98 per cent) to <10 per cent, when water retention in stone wool was still around 70 per cent.

Chemical Characteristics – The polyester fleece fibres are pH neutral and pH-stable from pH 2 up to 11, thus providing opportunities for easy control over nutrient regimes in the rhizosphere. The CEC is close to 0 cmol kg^{-1}; it is an absolutely chemically inert medium.

Sterilization, Re-use, and Waste Disposal – After use, fleece can be steam heated for sterilization and it is re-usable for several subsequent crop(s); it is resistant to temperatures as high as 160°C. Due to paucity of research results, the performance of used fleece slabs remains unknown. Fleece can be disposed to landfill sites and used for manufacturing new slabs. It might have a potential for recycling as a soil amendment (to improve physical properties of clay soils) or it could be used in substrate mixes to improve the balance of air/water retention. A key advantage of fleece products is that dried waste fleece can also be burned in furnaces.

12.4 SUBSTRATES MIXTURES — THEORY AND PRACTICE

Individual components of mixed substrates are often chosen considering their properties so that they complement each other and the resultant medium possesses most of the desirable attributes for good plant growth and production. Soil and organic components used in substrate mixes, like peat or compost, often lack coarse particles necessary for adequate aeration and hold moisture relatively tightly around the particles, predominantly by adsorptive forces (Bilderback and Jones, 2001). In contrast, milled bark, sand and most aggregates hold moisture between particles, therefore air and water retention characteristics of the media mixes largely result from the interactions among the individual components. During aging, decomposition and softening of the particles of organic components result in a shift of particle-size distribution, which in turn results in a shift in media hydrophysical and chemical properties. Therefore, mixtures of peat or compost with coarse particles components materials are common to obtain the ideal physical properties of the substrate for the specific plant. Zeolite possesses unique adsorption capacity with a high cation exchange and dehydration-rehydration ability (Mumpton, 1999); perlite, on the other hand, is rather inert (low buffering and CEC) with the majority of its water being held superficially and released at relatively low tension, providing excellent drainage of the medium and aeration of rhizosphere (Marfa et al., 1993). Diatomite is a chalk-like, soft, friable, earthy, very finely grained, finely porous, light-coloured siliceous sedimentary rock with low density. Diatomite is essentially inert. A mixture of chemically active substrate, such as zeolite (which has poor water retention capacity) with chemically inert perlite or fine diatomite (with good air/water retention properties), is expected to yield a good substrate, because of the complementary properties of the individual components. Thus, perlite or diatomite, when mixed with zeolite, would improve the hydrophysical properties of zeolite (Grillas et al., 2001). On the other hand, the high ion-exchange capacity of zeolite would impart buffering property to the media mixes, thus restricting the fluctuations of pH and nutrient concentration in the root environment. Zeolite, perlite and diatomite, all have a relatively stable structure and could, therefore, be re-used for several successive crops.

Spomer (1974) stated that 'The optimum amount of amendment creates soil physical conditions which closely match plant requirements. The optimum is usually selected by the 'trial-and error' method of growing plants in a series of mixtures. Elimination of this inefficient and inaccurate procedure would be a significant advance in the development of soil mixtures.' We adopt this approach in the following review on physical and chemical properties of media prepared of mixtures of different substrates.

12.4.1 SUBSTRATE MIXTURES — PHYSICAL PROPERTIES

Bulk density – Bulk Density (BD) is an important physical property which has an influence on the other physical properties of a medium as described in Chap. 3. When two components are mixed and each component contributes weight in proportion to its volume fraction in the blend, the BD of the mixture can be calculated by Eq. (1) (Pokorny et al., 1986):

$$D_m = ((V_t C_1)D_1 + (V_t C_2)D_2)/V_t \qquad (1)$$

where D_m = predicted BD (g cm^{-3}) of volume mixture; V_t = total sample volume (cm^3); C_1 = volume fraction of component 1; C_2 = volume fraction of component 2; D_1 = BD (g cm^{-3}) of component 1; D_2 = BD (g cm^{-3}) of component 2. Linear increase in BD associated with increasing the fraction of sand in the medium was used to predict the fraction by volume of sand in unknown mixtures of sand and bark of the same particle size (Pokorny and Henny, 1984a).

Shrinkage or reduction in bulk volume occurs when two or more loose components, differing in particle-size distribution, are mixed and the final volume of the mixture is less than the sum of the volumes of the individual components (Spomer, 1974). For example, when the media components have greatly different particle sizes, such as pine bark and fine sand, the final volume of the mix is not additive, if they are blended together (0.0283 m^3 plus 0.0283 m^3 result in 0.0425–0.0495 m^3 rather than 0.0566 m^3); the result is an increase in BD and lower total porosity and air-filled porosity, resulting in higher container capacity and available water (Bilderback et al., 2005; Brown and Pokorny, 1975). Pokorny and Henny (1984b) reported that substrates with the same components and ratio were not identical in their physical properties, even though they were assumed to be the same. Differences in physical properties in their study were attributed to the shrinkage and particle-size differences of the components blended. Variation in identical components occurs if dry components are blended compared to blending moist components. Dry components when mixed tend to fit together tightly and increase the BD of the substrate compared to when moist components are blended; consequently, air-filled porosity is reduced. Spomer (1974) described this phenomenon schematically when small particles from the fine component occupy large pores of the coarse component (Fig. 12.1).

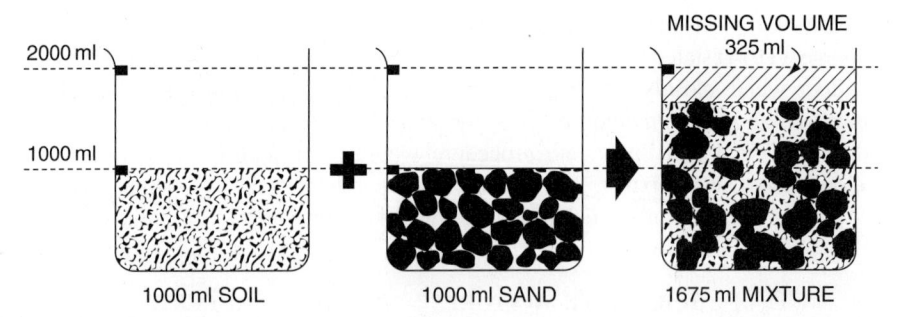

FIGURE 12.1 Sample data illustrating the decrease in total bulk volume of soil mixtures components upon mixing. The total component bulk volume is 2000 ml prior to mixing and 1675 ml after mixing (sample results rounded to nearest 5 ml) (Spomer, 1979. Hort. Science 14:75–77, with permission from the American Society for Horticultural Science).

12.4 SUBSTRATES MIXTURES — THEORY AND PRACTICE

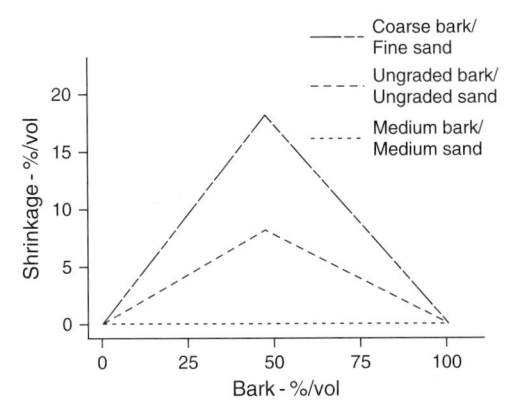

FIGURE 12.2 Shrinkage curves for potting media prepared from three grades each of pine bark and sand. Coarse bark-fine sand: 0–50 per cent bark: Y = 0.53 + 0.34 B, df = 28, r = 0.98; 50–100 per cent bark: Y = 0.73 + 0.33 S, df = 28, r = 0.96. Medium bark-medium sand: 0–50 per cent bark: Y = 0.36 + 0.00 B, df = 28, r = 0; 50–100 per cent bark: Y = 0.36 + 0.00 S, df = 28, r = 0. Ungraded bark-ungraded sand: 0–50 per cent bark: Y = 0.80 + 0.14 B, df = 28, r = 0.93; 50–100 per cent bark: Y = 0.42 + 0.15 S, df = 28, r = 0.94. B = per cent bark, S = per cent sand. (Nash and Pokorny, 1990. Hort. Science 25:930–931, with permission from the American Society for Horticultural Science).

Nash and Pokorny (1990) presented the shrinkage in volume resulting from mixtures of two components (Fig. 12.2). The maximum shrinkage was obtained when coarse particles of bark fine were mixed with fine particles of sand whereas no shrinkage was observed when medium particles of bark were mixed with medium particles of sand.

Pokorny et al., (1986); Burés and Pokorny (1991); Martinez et al., (1991); Burés and Pokorny (1991) and Burés et al., (1993a,b,c,d), in what are apparently the first attempts to predict physical properties of mixes, based on the properties of their ingredients, showed that experimentally determined physical properties (particle-size distribution, BD, particle density, total porosity, and water-release patterns) of binary volume mixtures of different media ingredients agreed reasonably well with those predicted by theoretical calculations on the corresponding characteristics of the individual components. They demonstrated the usefulness of their proposed theoretical approach in precisely selecting the appropriate mixing ratios for a desired set of properties of the media mixes. Burés et al. (1993b,c) proposed a model based on the Monte-Carlo method to predict the BD and porosity of mixtures of spherical particles of various diameters. The simulated data for a mixture of pine bark and sand was in good agreement with observations done with scanning electron microscopy (Burés et al., 1993a). Another approach was developed by Pokorny et al., (1986) who modified Eq. (1) to include the shrinkage effect as follows:

$$D_m = ((V_t C_1) D_1 + (V_t C_2) D_2) / (V_t - S) \qquad (2)$$

where terms are defined as above and S = shrinkage fraction. Pokorny et al., (1986) applied Eq. (2) successfully in combination with measured shrinkage values for predicting the BD of mixtures of sand and pine bark. Pokorny et al. (1986) modified Eq. (2),

$$D_m = \sum_{i=1,n} (V_t C_i) D_i / (V_t - S) \tag{3}$$

for predicting the BD of mixtures of three or more components; here i = indicator of number of components from 1 to n. Equations (2) and (3) enable prediction of the BD of mixtures but they include the unknown variable S. Shrinkage curves have been graphically characterized for several binary mixes (Pokorny et al., 1986; Nash and Pokorny, 1990) and a mathematical expression for a binary system has been proposed by Burés and Pokorny (1991):

$$S = 2q_s S_{max} \tag{4}$$

where q_s = component with volume <50 per cent, and S_{max} = maximum shrinkage. It was found in several binary systems that S_{max} is obtained at a 1:1 ratio (v/v) (Pokorny et al., 1986; Nash and Pokorny, 1990; Burés and Pokorny, 1991), as demonstrated in Fig. 12.2. Equation (4) was validated successfully for binary mixtures of peat moss/perlite, Pine bark/sand and Peat moss/sand (Burés and Pokorny, 1991). Thus, BD of any mixture of two components can be calculated from the BD of each component and measuring the S_{max}. Pokorny et al. (1986) obtained an excellent agreement between actual and predicted bulk densities of binary mixtures of pine bark and sand. Burés et al. (1993d) extended Eq. 1.4.4 for ternary system:

$$\begin{aligned} S &= 2(x_2 S_{max12} + x_3 S_{max13}) \quad \text{for} \quad x_1 > 0.5 \\ S &= 2(x_1 S_{max12} + x_3 S_{max23}) \quad \text{for} \quad x_2 > 0.5 \\ S &= 2(x_1 S_{max13} + x_2 S_{max23}) \quad \text{for} \quad x_3 > 0.5 \\ S &= (1 - 2x_1) S_{max23} + (1 - 2x_2) S_{max13} + (1 - 2x_3) S_{max12} \quad \text{for} \quad x_1, x_2 \text{ and } x_3 < 0.5 \end{aligned} \tag{5}$$

They validated it successfully for medium composed of a large range of fractions of pine bark, sand and calcined clay.

Jones and Or (1998) combined a fractional packing model of Koltermann and Gorelick (1995) with an empirical relation of Furnas (1931) for predicting the porosity of a binary mixture of fine and coarse glass beads of different diameter ratio as a function of the coarse component fraction. Measured porosities for coarse component mixtures of 0.33, 0.5 and 0.67 were slightly overestimated by the model ($d/D = 0.5$) (Fig. 12.3). Jones and Or (1998) found that the geometric mean produced closer estimates of the porosity than the harmonic for a New Jersey soil and for a mixture of coarse and fine sand particles (Fig. 12.3).

Water Retention – The retention curve of any substrate is a very important characteristic for water availability and aeration for growing plants and it is governed by the pore-size distribution (PSD) of the substrate as presented in Chap. 3. The relation

12.4 SUBSTRATES MIXTURES — THEORY AND PRACTICE

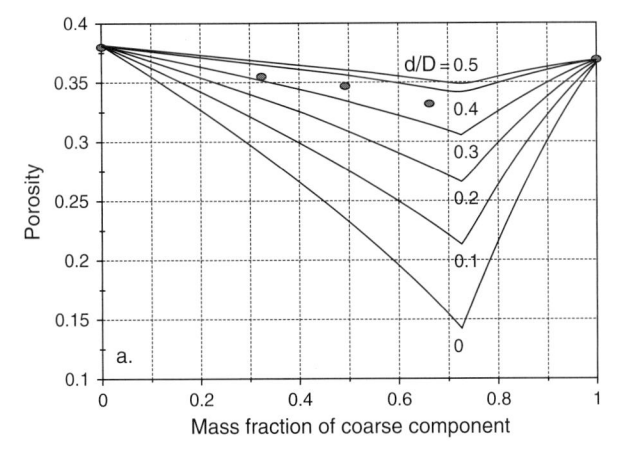

FIGURE 12.3 Estimated porosity of mixtures of glass bead sizes of 100 and 200 μm of different mass fractional contents of the coarse (200 μm) component as s function of the mass fraction of the coarse component, D, of binary mixtures whose diameter ratio, d/D, vary from 0 to 0.5. Data points are measurements of porosity for the glass bead mixtures; h is matric head (Jones and Or, 1998. *Soil Sci. Soc. Am. J.* 62:563–573, with permission from the American Society of Soil Science).

between the BD and the PSD (particle-size distribution) is also presented in Chap. 3. In mixtures of two or more materials, the retention curve is governed by the distribution of pores of each component and the new pores formed by the mixing of the particles of the fine components in the large pores of the coarse component (Bunt, 1974; Burés et al. 1993a,b,c,d; Spomer, 1974). A schematic representation of decrease in the total water retention and air-filled porosity of soil and sand mixture has been drawn by Spomer (1974) (Fig. 12.4).

Sahin et al. (2002) found that the water retention property at low tension (pF < 2.52 kPa) and pore-size distribution of 1:1 (v/v) binary mixes of perlite, pumice and creek sand were quite predictable based on the corresponding properties of the components involved. At higher tension, the water retention capacity of a mixture of perlite + pumice is higher than that of mixtures of perlite + sand and pumice + sand. The proportion of macro-pores (>100 mm) representing air-filled porosity as well as mesopores (30–100 mm) plus micropores (3–30 mm) representing available water-holding capacity was comparable for perlite–sand and perlite–pumice, but higher than pumice–sand; the perlite–pumice mix had a higher ultramicroporosity (<3 mm) than the other two mixes. In general, moisture retention characteristics of media mixes is a compromise of its individual component corresponding characteristics. However, even when the same components are blended in identical ratios, physical properties may vary due to difference in particle size that cause shrinkage in the volume of the mixture (Bilderback et al., 2005).

Nash and Pokorny (1992) derived the following model that includes the shrinkage effect:

$$\text{WRx}_x = \sum_{i=1,n} (\text{WR}_{1x} C_i)/(V_t - S) \tag{6}$$

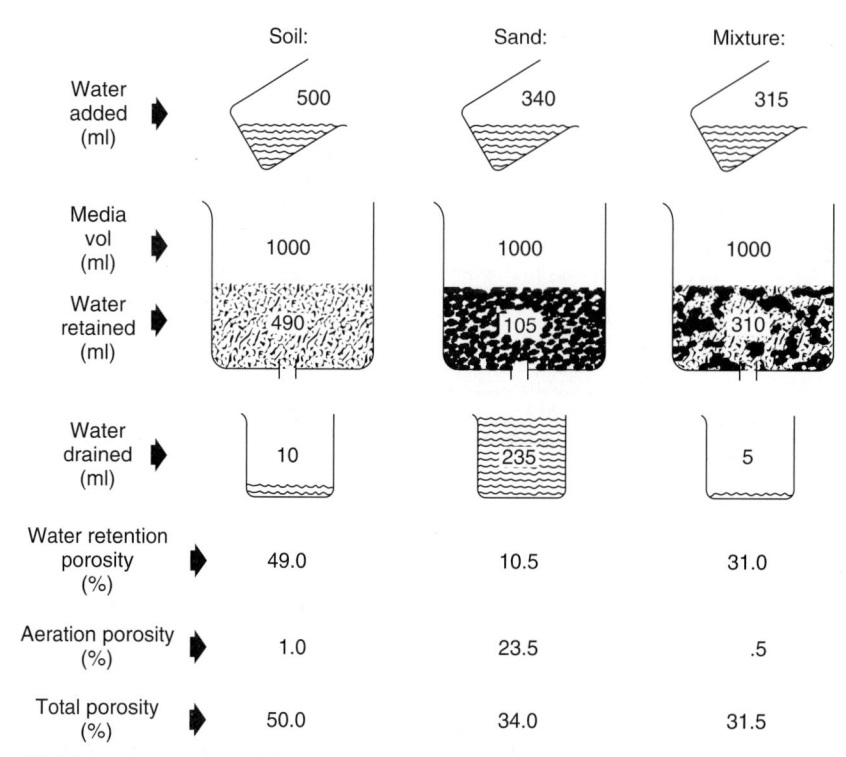

	Soil:	Sand:	Mixture:
Water added (ml)	500	340	315
Media vol (ml)	1000	1000	1000
Water retained (ml)	490	105	310
Water drained (ml)	10	235	5
Water retention porosity (%)	49.0	10.5	31.0
Aeration porosity (%)	1.0	23.5	.5
Total porosity (%)	50.0	34.0	31.5

FIGURE 12.4 Sample data illustrating the decrease in total porosity of soil mixture components upon mixing. The mixture total porosity of 31 per cent is less than either the soil (50 per cent) or sand (34 per cent) alone (Spomer, 1979. *Hort. Science* 14:75–77, with permission from the American Society for Horticulture).

where WR_x, WR_{1x}, WR_{2x}, WR_{nx} = estimated water retained in the mixture and in the components at water tension X (g), C_i = volumetric water fraction of the ith component. V_t = total volume of the mix; S = shrinkage fraction. They found close agreement between measured and calculated water retention of wide range of pine bark and sand mixtures (0–100 per cent pine bark) at several water tensions levels (0, 10, 50 and 100 cm of water). However, Eq. (6) has not been validated by other researchers for different mixtures and its generality is unproven.

The physico-empirical model of Arya and Paris (1981) for prediction of water retention curves using the PSD was adopted for substrate mixtures by Jones and Or (1998). The measured retention curve fits the optimized retention curve of mixtures of sand particles very closely, but the prediction for mixtures of glass beads deviated from the measured in the near saturation and near dry ranges for unknown reasons (Fig. 12.5).

No attempts have been made to evaluate this for other types of substrate mixtures. The retention curves of binary mixtures of fine (100 μm) and coarser (200 μm) glass beads fall between the one component curves of the fine and coarse beads for wide range of water tension from 0.4 to 0.8 m (Fig. 12.6).

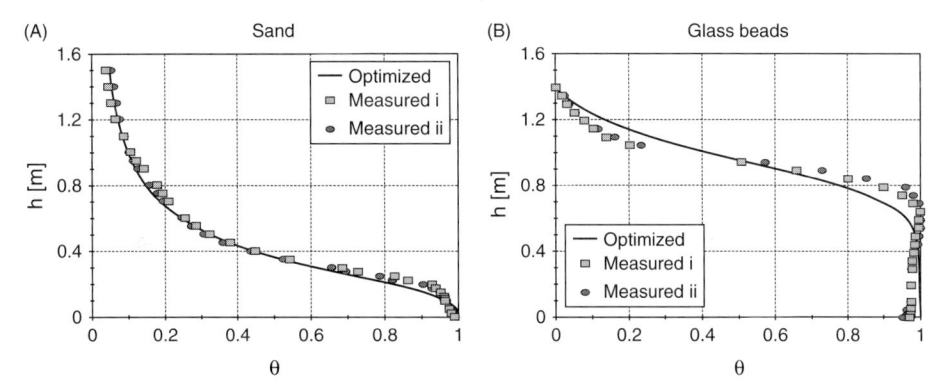

FIGURE 12.5 Comparison of optimized and measured substrate-water characteristics for (a) sand and (b) glass beads, where h is matric head and θ is the relative water content. (Jones and Or, 1998. *Soil Sci. Soc. Am. J.* 62:563–573. With permission from the American Society of Soil Science).

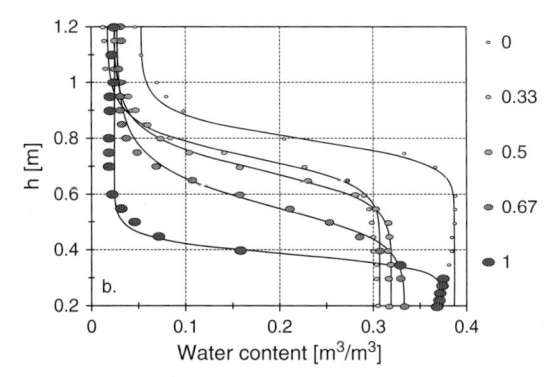

FIGURE 12.6 Substrate water characteristic curves for mixtures of glass bead sizes of 100 and 200 μm of different mass fractional contents of the coarse (200 μm) component, as s function of the mass fraction of the coarse component, D, of binary mixtures whose diameter ratio, d/D, varies from 0 to 0.5 (Jones and Or, 1998. *Soil Sci. Soc. Am. J.* 62:563–573. With permission from the American Society of Soil Science).

This pattern of the retention curves of binary systems is due to the shrinkage mechanism. Weiss et al. (1998) found that the values of the empirical parameters α and n in the van Genuchten model (Eq. (7)) for various peat soils are dependent on the BD and the percentage of botanical components *Sphagnum* and *Carex*. They suggested an alternative semi-empirical model for the retention curve which includes only one parameter, k, which was dependent on the BD, the percentage of botanical component *Sphagnum* and lignin content. Thus, the retention curves of mixtures of peat from different sources can be predicted based on their components.

Hydraulic Conductivity – The saturated (Ks) and unsaturated hydraulic conductivity (Kh) of substrates determine water and salts movement through the growing media.

The decrease of Kh in porous substrates as the water content (θ) decreases is much sharper than in soil (Raviv and Blom, 2001). A minute decrease in θ (e.g. of several percents) may decrease Kh by an order of magnitude and thus greatly affect water availability to the roots (Wallach et al., 1992a, 1992b; Raviv et al. 1999, 2001). It was shown that in substrates, the main limiting factor to water uptake is the Kh and not the matric potential (da Silva et al., 1993; Raviv et al., 2001). This concept of water availability and its dependence on the momentary hydraulic conductivity of the growing substrate is presented and discussed in Chapter 3 (Section 3.4.5).

The value of Ks of a mixture can be estimated by the Kozeny–Carman (Bear, 1972) expression:

$$Ks = (p_w g_o/\mu)(d_m^2 \eta)/(180(1-\eta)^2) \tag{7}$$

where p_w = liquid density (Mg m^{-3}), g_o = acceleration due to gravity (m s^{-2}), μ = dynamic viscosity (Pa s), η = total porosity (m^3 m^{-3}), d_m = the mean grain diameter (m) which is calculated from either the geometric or harmonic mean of d and D as described by Koltermann and Gorelick (1995). This method of estimating Ks in soils produces estimates generally within one order of magnitude (Koltermann and Gorelick, 1995). Employing this method with the geometric mean so as to produce better estimates, Jones and Or (1998) calculated Ks, of 8.6×10^{-5} m s^{-1} for a mixture of sand particles of wide range of diameters values, where the measured Ks was 4.6×10^{-4} m s^{-1}.

When the value of Ks is determined, Kh curves for mixes can be calculated by Eq. [3.3.9] based on the retention curve as shown for composted agricultural wastes and their mixtures with tuff (Wallach et al., 1992b), sphagnum peat moss mixtures with tuff (da Silva et al., 1993), sphagnum peat moss mixtures with coarse perlite (Heiskanen, 1995a, 1995b, 1999) and sphagnum peat moss mixtures with perlite and sand (Heiskanen, 1999).

We started this section with the statement of Spomer dated three decades ago about the need to eliminate the inefficient and inaccurate procedure of trial and error in developing mixtures with required physical properties. Since then there has been considerable progress in the understanding of the physical properties of substrates and our ability to estimate the BD, porosity and retention curve of mixtures based on the properties of their components and measurements of the maximal shrinkage of 1:1 (volume basis) mixtures (Eqs. (2–6)). Actual measurements of these properties are required because the models assume ideal spheres, whereas the actual shape of substrate particles varies from spheres to rods and fibres. These properties are the main physical aspects of mixtures described in the excellent review by Bunt (1988). However, with accepting the concept of water availability and its dependence on the momentary hydraulic conductivity of the growing substrate, there is a need for accurate prediction of Ks and Kr. In this area there was some progress in rough estimation of Ks (Klotermann and Gorelick, 1995; Jones and Or, 1998) as presented above but an empirical approach is still required for determining Ks. Nevertheless, the knowledge gained on the relationship between Ks, the water retention curve and Kh is useful in saving laborious measurement of Kh of mixtures under a relevant range of water content (tension).

12.4.2 SUBSTRATE MIXTURES — CHEMICAL PROPERTIES

Some of the substrate materials described in the preceding sections, such as stone wool, perlite, pumice, polyurethane foam and others are chemically inert, whilst others, such as organic materials, tuff and zeolite, exhibit a considerable ion-exchange capacity and buffering properties to the fluctuation of pH and nutrient availability. With stone wool, polyurethane slabs and pure sand, when used as substrates in soilless culture, the root zone hydrophysical properties are well defined. When the growing medium is a mixture of components (chosen to provide water, air, anchorage, CEC, buffering capacity, and nutrients), the resultant chemical environments are the result of complex interactions among the components.

When a substrate is prepared from two components that do not react with each other, the chemical properties of the substrate are expected to be in proportion to the values of the individual component fractions. Thus, the CEC of milled pine bark with sand declined with increasing percentage of sand (Brown and Pokorny, 1975). CEC was most accurately determined on a volume basis because of the large difference in BD between the media components, milled pine bark and sand (Brown and Pokorny, 1975). CEC of two component (pine bark and sand) container media was predicted from laboratory analysis of each component and applying a model similar to Eq. (2) for the BD (Nash and Pokorny, 1990):

$$CEC_m = (C_1\ CEC_1 + C_2\ CEC_2)/(C_1\ V_t + C_2\ V_t - S\ V_t) \qquad (8)$$

Increasing the percentage of sand in the potting medium mixtures of milled pine bark and sand from 0 to 100 per cent resulted in a linear increase in the pH from 4.1 to 5.4 (Brown and Pokorny, 1975). However, there are no published models on the effect of chemical reactions between components on the combined chemical properties. The research is mainly empirical on methods to modify chemical properties of substrates by adding another substitute to fit the requirements of specific plants. For example, fly ash can raise pH and alter chemical properties of the mixes (Bilderback et al., 2005). The organic acids produced and microbial enzymes involved in the decomposition of organic components may also interact both physically (through mere physical adsorption) and chemically (through sorption, complex formation, and dissolution) with the inorganic components like expanded clay granules, vermiculite, pumice and tuff. Clays are also known to abiotically modify several reactions of organic matter decomposition through catalytic action (Huang, 2000); thus, clay can affect the chemical properties of media mixes. The interactions of organic compounds with inorganic mineral, especially clays, are described in the soil literature and presented briefly in Chap. 6 (Chemical characteristics). Organic compounds, especially the active and soluble fractions (humic and fulvic acids), may cause decomposition of the solid phase (Tan, 1986) and/or modification of the surface charge (Tate and Theng, 1980; Huang and Violante, 1986). The effects of organic exudation and residues of roots on chemical properties of mineral substrates are described in Chap. 6. The main effect is the increase of negatively charged surface of tuff or other mineral substrate due to organic acids. Chap. 6 presents the contrasting effects of low weight, soluble organic acids exudation by roots on the pH of intact substrate and a substrate with a high surface charge; similar effects have been reported for the reaction of soluble organic acids with various inert minerals and those with charged surfaces. In mixtures of organic and inorganic substrate, both mechanisms of decomposition of the solid phase and modification of the surface charges of the mineral component may occur simultaneously. Moreover, the possible effect of the inorganic component may have impact on the structure of the organic component and/or its decomposition; therefore it is very difficult to predict the chemical properties of mixtures. The principles of creating substrate mixtures to regulate nutrient availability in the rhizosphere and to develop substrates that might contribute nutrients during the growth period of horticultural crops are described in Chap. 6.

12.4.3 SUBSTRATE MIXTURES — PRACTICE

Despite the importance of the hydraulic conductivity there are few published works that present the effect of mixing substrates on this parameter. Spomer (1974) suggested that drainage in organic substrates could be improved either by increasing media depth or by amending the medium with coarse inorganic components. Combining sand with organic components is a common practice in preparing media mixes. This addition of sand results in slower infiltration rate of irrigation water as it moves through the media profile, particularly in fresh or less decomposed organic components (Bilderback and Jones, 2001). Percolation rate declined as the percentage of sand was increased in a milled pine bark and sand mixture (Brown and Pokorny, 1975). The slower infiltration rate in sand-amended media mixes promoted more thorough wetting of the substrate, compared with straight organic particles, through which water can channel rapidly to the bottom (Bilderback and Jones, 2001). Da Silva (1991) found that the retention curve of 1:1 (v:v) mixture of tuff and peat was slightly closer to the curve of the peat than that of the tuff while the hydraulic conductivity of the mixture was very close to that of peat in the studied range of water height, 0–25 cm (Fig. 12.7).

Peat moss-based substrates are popular because of their high nutrient buffering capacity, good aeration and suitability to various crops (Caron et al., 2005; Papadopoulos and Tan, 1991; Steidman, 1988; Sheldrake, 1989). When lightweight organic materials such as peat are used in media mixes, a fine sand or clay are often added to increase BD so as to improve stability. Mixes with very low BD are less suitable as growing media in outdoors nurseries due to stability concerns. Sand and clay also increase the wettability of the mix (Guttormsen, 1974). However, undesirable effects of adding sand or clay are a reduction in total porosity and air-filled porosity (Bunt, 1974); it is well known that water and nutrient management becomes difficult when air-filled porosity is low. Bunt (1983) determined the principal physical properties of a mixture of peat (100, 75, 50, 25 and 0 per cent, v/v) and minerals of fine or coarse grades (0, 25, 50, 75 and 100 per cent v/v). The minerals used included perlite, vermiculite, calcined clay and sand. Generally, total porosity was inversely related to BD, and there was no interaction between the particle size of the minerals and total porosity. Except for perlite–peat, mixtures of fine grade minerals and peat had lower air-filled porosity values than those of individual components, the mineral or peat in

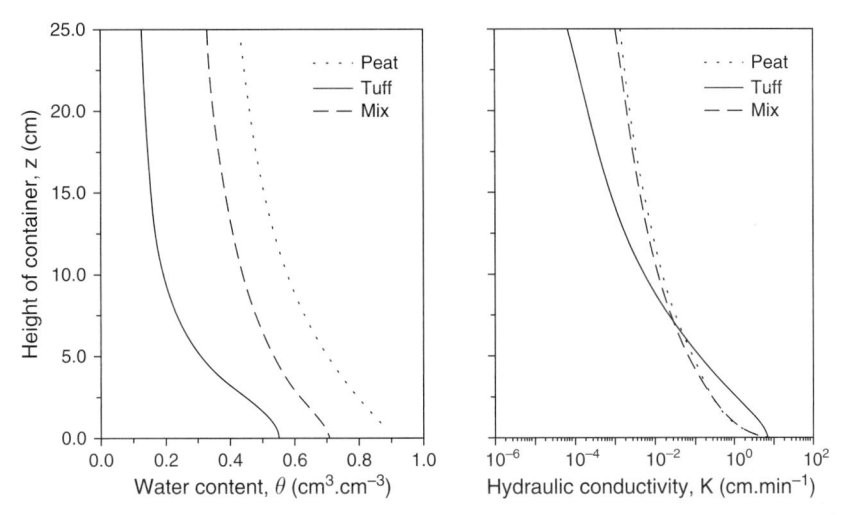

12.4 SUBSTRATES MIXTURES — THEORY AND PRACTICE   533

FIGURE 12.7 Water content and hydraulic conductivity distributions along a container under drainage equilibrium conditions for peat, tuff and a mix of the two (After da Silva, Ph.D. Thesis 1991).

agreement with the above explanation on the shrinkage effect. In peat–perlite (fine grained) mixture, the air-filled porosity remained constant for all the mixing ratios. On the contrary, addition of coarse grade minerals progressively increased the air-filled porosity of the mix as the mineral content was increased; this was observed irrespective of the mineral type. Easily available water was not significantly affected by the addition of fine-grained minerals. However, the coarse grade minerals reduced the easily available water, especially when it was present at more than 25% by volume in the mix. Addition of fine and coarse grades of vermiculite or perlite to the mix resulted in much higher easily available water than other minerals (Bunt, 1983).

Bunt (1991) observed a remarkable increase in BD and a decrease in total porosity and air-filled porosity as a consequence of blending fine sand (25–50%, v/v) or coarse sand (25 per cent) with peat; however, fine perlite (25 per cent) in the peat-based media caused only slight increase in BD without affecting total porosity or air-filled porosity. Brűckner (1997), on the other hand, did not observe any conspicuous change in the hydro-physical properties of the media mixes prepared by adding various proportions of clay materials (10–35 per cent), perlite (15–30 per cent), or sand (20 and 50 per cent) to peat in the presence or absence of other organic substrates (wood chips, pine bark, and/or rice husks). Media mixes prepared by blending peat moss with the inorganic components perlite, pumice, or creek sand at 1:1 volumetric ratio had BD, water retention characteristics and pore-size distribution intermediate between those of the two individual components; similar results were also obtained when peat or sawdust was mixed with perlite, pumice or creek sand at the same ratio (Sahin et al., 2002).

Verhagen (2004) observed that when coarse- and fine-graded clay products (5–25 per cent, v/v) were blended with peat to prepare media mixes, there was a large influence of the clay on the water uptake and retention characteristics of the mix.

When compared with pure peat, fine-graded clay products showed a stronger effect than the coarse-graded ones in increasing BD, and reducing organic matter content, total porosity and easily available water. These properties were, however, less affected by the peat:clay ratio. The fine-grade clays caused a large increase in the unsaturated hydraulic conductivity at relatively low addition. At high addition of clay, pores might have been blocked by the clay restricting capillary rise of water (Verhagen, 2004). It was suggested that the effectiveness of clay products in media mixes could not always be ascribed to specific chemical characteristics of the clay; physical effects and interaction with cultivation systems (irrigation frequency and fertilization) also play an important role (Verhagen, 2004).

Riga et al. (2003) evaluated the physical properties of four mixtures of peat and blended used stone wool (from 0 to 60 per cent, v/v) as a culture medium for geranium (*Pelargonium peltatum* L.) production. The production was studied with two values of substrate water potential as the initial set point for fertigation control (-15 and -10 kPa). Results showed that used blended stone wool, when combined with peat, had certain properties which resulted in an increased air space and decreased volumetric water content. When water potential decreased from 0 to -10 kPa, the blended stone wool lost about 85 per cent of its water content, whereas only 62 per cent of the water content of saturated peat was lost at the same water potential. However, volumetric easily available water was affected slightly by the incorporation of used stone wool to the peat substrate. Used stone wool showed higher content of soluble minerals than peat. Nevertheless, only those plants grown in a mixture containing the highest volumetric stone wool content (i.e. 60 per cent) resulted in inadequate quality for market. In contrast, when supplying fertigation at -10 kPa, all parameters were similar to those of the control, except for those plants grown in the mixture containing the highest volumetric stone wool content (i.e. 60 per cent), where a lower biomass production and similar production of inflorescences per plant and market quality were observed. It was concluded that blended recycled stone wool can be used successfully as a component of peat-based media for geranium production as long as the fertigation schedule is modified according to the physical properties of the media. Some previous studies also showed the viability of some media mixes containing previously unused stone wool blended with wood chips, peat moss, coconut fibre and perlite (Kim et al., 2000) or municipal compost (Bilderback and Fonteno, 1993); all of these studies reported that adding stone wool to culture media led to an improvement of their physical characteristics, resulting in a good hydro-physical environment for crop growth and production.

Stone wool incorporation into compost, peat or other substrates results in improved aeration and increased water-holding capacity of the mixture (Fonteno and Nelson, 1990; Choi et al., 1999; 2000). Korean research (Jeong, 2000; Kim et al., 2000; Lim and Jeong, 2000; Shin and Jeong, 2000; Jeong and Hwang, 2001; Kim and Jeong, 2003) emphasized (1) to grind used stone wool and other components of the mixture to a suitable grade, (2) to have certain substrates, particularly bark and wood chips, weathered or soaked to remove chemicals harmful to plant growth and (3) to steam sterilize the ground, used stone wool before its incorporation in the mixture.

Granulated or powdered clay materials are recommended in the media mixes to impart buffering properties against fluctuations of pH, EC and nutrient status. Martinez et al. (1997) added a granulated (2–8 mm) and a powdered commercial clay material to peat–coir mixes and studied the changes in physical and physicochemical properties due to interaction of the components. Clay material addition to the peat–coir mixes resulted in significant increases in BD, easily available water, available water and pH, and significant decreases in total porosity, air-porosity, water retention at 1 kPa, organic matter and EC; but, no significant effect on water buffering capacity and water retention at 2 KPa. The effects of two different types of clay on air/water parameters and EC were similar, but pH was significantly higher in powdered than granulated clay materials (Martinez et al., 1997). When the proportion of coir increased (from 3:1 to 1:3; peat:coir, v/v) in the mixes, significant decreases were noted on BD and organic matter (OM) content, and increases on total porosity, air-filled porosity and pH, in relation to those properties of the basic materials, peat and coir. Increased proportion of coir in the mix caused significant increases in easily available water and available water. Overall, the air/water parameters of the mix exhibited an important and non-linear bulking effect of coir when mixing with peat that seemed to be saturated when coir was one-half of the mix volume. This effect was attributed to an interaction between the rigidity of coir fibres and the flexibility of peat fibres; coir fibres could furnish a relatively rigid framework which would promote a more open rearrangement of peat fibres. Mixes made with organic materials such as peat, pine bark and coir can be difficult to wet initially, and also to rewet after they have been dried-out in the container. Clay and most generally mineral material can be used to enhance the wettability (i.e., the ability of dry media to rapidly absorb water when moistened) of peat-based substrates (Bunt, 1988). Airhart et al. (1978) reported that an air-dry peat-vermiculite substrate required 5 days to reach 70–78 per cent of moisture saturation while milled pine bark required 48 days to achieve 58–78 per cent saturation.

Issa et al. (2001) studied seasonal and diurnal photosynthetic responses and flower yield of two gerbera cultivars to perlite, zeolite and a 1:1 (volume based) mixture. They concluded that gerbera plants grown in the 1:1 mixture exhibited the best physiological responses. Samartzidis et al. (2005) studied productivity and physiological responses of rose to zeolite (z) and perlite (p) substrates at various ratios (25z:75p, 50z:50p, 75z:25p and 100z:0p). Increasing the fraction of perlite in the mixture reduced water uptake and net photosynthesis (A_{net}) values of rose at various stages. This observation was explained by the increased amount of easily available water as the fraction of zeolite increased. Both zeolite and perlite used in this study were coarse. During mixing and production phase, an alteration (fragmentation) of perlite usually occurs, whereas zeolite is not affected because of its hardness (Orozco and Marfa, 1995). It was believed that perlite particles with sizes smaller than 3 mm in the medium decreased the percentage of pores with large diameters (60 μm) and increased the percentage of those with a diameter between 30 and 60 mm (Samartzidis et al., 2005). Consequently, an increase in the total available water content of the growing medium with high zeolite fraction occurred, improving water status, improving water and nutrient uptake, and avoiding a pronounced midday water stress. The above described works (Issa et al., 2001; Samartzidis et al., 2005) demonstrate that physical and chemical characteristics of the substrate mixture may play an important role in plant growth and crop production. However, conclusions from such comparisons have to be drawn carefully, because the optimal irrigation and fertilization for each substrate is different as has been shown by Riga et al. (2003). The methods of irrigation as well as the height of the container are of crucial factors in constructing the 'ideal' mixtures (Jones and Or, 1998; Caron et al., 2005). High fraction of sphagnum peat to coarse pine bark and sand provided the best capillary rise and best growth of nursery trees grown in pots with subirrigation system (Caron et al., 2005).

12.5 CONCLUDING REMARKS

Selection of an appropriate growing medium for a given plant species along with appropriate management of nutrients and water usually have large impacts on both quantity and quality of the production in soilless culture. As reported by Olympios (1992), inert media do not react with nutrient ions and they do not posses adsorption complex; this permits a precise adjustment of the nutrient levels to the actual plant requirements (Sonneveld and Spaans, 1990). Frequently, the role of a growing medium is considered to be the mechanical support of the root system; it preserves optimal physical conditions, particularly the air–water relations. That means that the hydrophysical properties of substrates are a more important consideration than their chemical properties for optimal plant growth. As the hydrophysical properties of substrates can not be tuned during the growing period, they should be optimal from the beginning. In contrast, the chemical properties can be adapted to optimal conditions during the growth period. An important advantage of using soilless growing media over soil cultivation is that the former enable an easy control over nutrient status in the root zone and their hydrophysical characteristics pertaining to their ability to provide simultaneously sufficient levels of oxygen and water to the roots. Furthermore, there are important advantages in terms of plant health, if sterile materials are used or if they are steam-sterilized or fumigated between crops. A growing medium can even be used several times after its previous disinfection (van Os et al., 1991).

Despite the worldwide trend of growing plants in various soilless media rather than in soil (see Chap. 1), the selection of proper plant medium for horticulture remains an empirical endeavour often focusing on locally available materials and economics rather than on the physical principles that govern retention and flow of water and gas (Jones and Or, 1998). Recurrent water- and oxygen-induced stresses affecting plant growth have prompted the need for the refinement of the selection criteria for optimal growing media (see additional discussion in Chap. 13). Furthermore, it is interesting to note that even after some 40 years of research, understanding the full effect of the physical properties of the growing medium on plant performance is far from complete (Raviv et al., 2004). The most important question is how to define optimal physical characteristics of media in terms of both water and oxygen availability to plant roots. There are no distinct universally accepted standards for physical properties of horticultural growing media. However, Yeager et al. (1997) suggested the following ranges for easiest irrigation and nutrient management with most substrate mixes utilized in commercial production of horticultural crops: total porosity (50–85 per cent), air-filled porosity (10–30 per cent), container capacity (45–65 per cent), available water (25–35 per cent), unavailable water (25–35 per cent) and BD (0.19–0.7 g cm^{-3}). Other researchers questioned the usefulness of this approach because they vary with crop types, growing conditions and targets of the grower. The approach of Jones and Or (1998) seems to hold a lot of promise; they treated this issue as a non-linear least squares multi-objective optimization problem. They used an objective function formulated to maximize liquid and gaseous content and fluxes to plant roots, while matching media hydraulic properties with the control system design. This approach has been tested only for mixtures of sand particles and of glass beads. A challenge for the future is to test and modify this approach for more mixtures of materials that are commonly used as substrate in horticulture.

ACKNOWLEDGEMENTS

The generous support of Ms. Lorraine Smith in extensive literature searches during the preparation of this chapter is gratefully acknowledged.

REFERENCES

Airhart, D.L., Natarella, N.J. and Pokorny, E.A. (1978). Influence of initial moisture content on the wettability of pine bark medium. *Hort. Science*, **13**, 432–434.
Alkan, M. and Doğan, M. (1998). Surface titrations of perlite suspensions. *J. Colloid Interface Sci.*, **207**, 90–96.
Allen, E.R., Ming, D.W., Hossner, L.R., et al. (1995). Growth and nutrient uptake of wheat in clinoptilolite-phosphate rock substrates. *Agron. J.*, **87**, 1052–1059.
Ando, H., Mihara, C., Kakuda, K.I. and Wada, G. (1996). The fate of ammonium nitrogen applied to flooded rice as affected by zeolite addition. *Soil Sci. Plant Nutr.*, **42**, 531–538.
Anonymous, (1994). *Fate of Polystyrene in the Environment: A Review of Scientific Literature*. Edmonton, Alberta, Canada: Department of Soil Science, University of Alberta.
Arya, L.M. and Paris, J.F. (1981). A physicoempirical model to predict the soil moisture characteristic from particle-size distribution and bulk density data. *Soil Sci. Soc. Am. J.*, **45**, 1023–1030.
Bear, J. (1972). *Dynamics of Fluids in Porous Media*. New York, USA: Elsevier.
Benoit, F. and Ceustermans, N. (1994). A decade of research on polyurethane foam (PUR) substrates. *Plasticulture*, **104**, 47–53.
Benoit, F. and Ceustermans, N. (1995). A decade of research on ecologically sound substrates. *Acta Hort.* (ISHS), **408**, 17–29.
Bilderback, T.E., and Fonteno, W.C. (1993). Improving nutrient and moisture retention in pine bark substrates with rockwool and compost combinations. *Acta Hort.* (ISHS), **342**, 265–272.
Bilderback, T.E. and Jones, R.K. (2001). Horticultural practices for reducing disease development. In *Diseases of Woody Ornamental Trees in Nurseries* (R.K. Jones and D.M. Benson eds). St. Paul, Minn., USA: APS Press, pp. 387–400.

Bilderback, T.E., Warren, S.L., Owen J.S. Jr. and Albano, J.P. (2005). Healthy substrates need physicals too. *HortTechnology*, **5**, 747–751.

Boertje, G.A. (1994). Chemical and physical characteristics of pumice as a growth medium. *Acta Hort.* (ISHS), **401**, 85–87.

Böhme, M. (1995). Evaluation of organic synthetic and mineral substrates for hydroponically grown cucumber. *Acta Hort.* (ISHS), **401**, 209–217.

Brown, E.F. and Pokorny, F.A. (1975). Physical and chemical properties of media composed of milled pine bark and sand. *J. Am. Soc. Hort. Sci.*, **100**, 119–121.

Brückner, U. (1997). Physical properties of different potting media and substrate mixtures-especially air- and water capacity. *Acta Hort.* (ISHS), **450**, 263–270.

Bunt, A.C. (1974). Some physical and chemical characteristics of loamless pot-plant substrates and their relation to plant growth. *Acta Hort.* (ISHS), **37**, 1954–1965.

Bunt, A.C. (1983). Physical properties of mixtures of peats and minerals of different particle size and bulk density for potting substrates. *Acta Hort.* (ISHS), **150**, 143–153.

Bunt, A.C. (1988). Physical aspects. In *Media and Mixes for Container-grown Plants*. London, UK: UNWIN HYMANN, p. 48.

Bunt, A.C. (1988). Media and Mixes for Container-grown Plants. London, UK: UNWIN HYMANN.

Bunt, A.C. (1991). The relationship of oxygen diffusion rate to the air-filled porosity of potting substrates. *Acta Hort.* (ISHS), **294**, 215–224.

Burés, S., Farmer, A.M., Landau, D.P., et al. (1993a). Container media characterization by scanning electron-microscopy and comparison with a Monte-Carlo computer-simulated medium. *Comm. Soil Sci. Pl. Anal.*, **24**, 2649–2659.

Burés, S., Gago, M.C. and Martinez, F.X. (1997a). Water characterization in granular materials. *Acta Hort.* (ISHS), **450**, 389–396.

Burés, S., Landau, D.P., Ferrenberg, A.M. and Pokorny, F.A. (1993b). Monte-Carlo computer-simulation in horticulture — A model for container media characterization. *Hort. Science*, **28**, 1074–1078.

Burés, S., Marfa, O., Perez, T., Tebar, J.A. and Lloret, A. (1997b). Measure of substrates unsaturated hydraulic conductivity. *Acta Hort.* (ISHS), **450**, 297–304.

Burés, S. and Pokorny, F.A. (1991). Equation for estimating shrinkage in binary mixtures of container media. *Hort. Science*, **26**, 1087.

Burés, S., Pokorny, F.A., Landau, D.P. and Ferrenberg, A.M. (1993c). Computer-simulation of volume shrinkage after mixing container media components. *J. Am. Soc. Hort. Sci.*, **118**, 757–761.

Burés, S., Pokorny, F.A. and Ware, G.O. (1993d). Estimating shrinkage of container media mixtures with linear and/or regression models. *Comm. Soil Sci. Pl. Anal.*, **24**, 315–323.

Bussell, W.T. and McKennie, S. (2004). Rockwool in horticulture, and its importance and sustainable use in New Zealand. *New Zealand J. Hort. Sci.*, **32**, 29–37.

Caron, J., Elrick, D.E., Beeson R. and Boudreau, J. (2005). Defining critical capillary rise properties fro growing media in nurseries. *Soil Sci. Soc. Am. J.* **69**, 794–806.

Challinor, P.F. (1996). The use of pumice in horticulture. In *Proceedings of 9th International Congress Soilless Culture*, St Helier, Jersey. ISOSC, Wageningen, The Netherlands. pp. 101–104.

Challinor, P.F., La Pivert, J. and Fuller, M.P. (1995). The production of standard carnation on nutrient loaded-zeolite. *Acta Hort.* (ISHS), **401**, 293–299.

Chen, J.L. and Gabelman, W.H. (1990). A sand–zeolite culture system for simulating plant acquisition of potassium from soils. *Plant Soil*, **26**, 169–176.

Chlopecka, A. and Adriano, D.C. (1997). Influence of zeolite, apatite and Fe-oxide on Cd and Pb uptake by crops. *Sci. Total Environ.*, **207**, 195–206.

Choi, J.M., Chung, H.J., Seo, B.K. and Song, C.Y. (1999). Improved physical properties of rice hull, saw dust and wood chip by milling and blending with recycled rockwool. *J. Korean Soc. Hort. Sci.*, **44**, 755–760.

Cole, J.C. and Dunn, D.E. (2002). Expanded polystyrene as a substitute for perlite in rooting substrate. *J. Environ. Hort.* **20**, 7–10.

da Silva, F.F. (1991). Static and dynamic characterization of container media for irrigation management. M.Sc. Thesis, Faculty of Agriculture, The Hebrew University of Jerusalem (in English).

da Silva, F.F., Wallach, R. and Chen, Y. (1993). A dynamic approach to irrigation scheduling in container media. In: Proceedings of 6th International Conference on Irrigation. Agritech, Ministry of Agriculture, Tel Aviv, Israel. pp. 183–198.

da Silva, F.F., Wallach, R. and Chen, Y. (1995). Hydraulic properties of rockwool slabs used as substrates in horticulture. *Acta Hort.* (ISHS), **401**, 71–75.

de Kreij, C. van Elderen, C.W., Meinken, E. and Fischer, P. (1995). Extraction methods for chemical quality control of mineral substrates. *Acta Hort.* (ISHS), **401**, 61–70.

Doğan, M. and Alkan, M. (2004). Some physiochemical properties of perlite as an adsorbent. *Fresenius Environ. Bull.*, **13**, 252–257.

Drees, L.R., Wilding, L.P., Smeck, N.E. and Senkayi A.L. (1989). Silica in soils: Quartz and disordered silica polymorphs. In *Minerals in Soil Environment* (J.B. Dixon, and S.B. Weed eds). Madison, WI, USA: Soil Sci. Soc. Am., pp. 913–974.

Dwairi, I.M. (1998). Evaluation of Jordanian zeolite tuff as a controlled slow-release fertilizer for NH_4^+. *Environ. Geol.*, **34**, 1–4.

ESCSI. (2006). Expanded shale, clay and slate. www.escsi.org

Ferguson, G.A. and Pepper, I.L. (1987). Ammonium retention in sand amended with clinoptilolite. *Soil Sci. Soc. Am. J.*, **51**, 231–234.

Fonteno, W.C. and Nelson, P.V. (1990). Physical properties of and plant responses to rockwool-amended media. *J. Amer. Soc. Hort. Sci.*, **115**, 375–381.

Furnas, C.C. (1931). Grading aggregates: I — Mathematical relations for beds of broken solids of maximum density. *Ind. Eng. Chem.*, **23**, 1052–1058.

Grillas, S., Lucas, M., Bardopoulou, E., et al. (2001). Perlite based soilless culture systems: current commercial applications and prospects. *Acta Hort.* (ISHS), **548**, 105–113.

Gül, A., Erogul, D. and Ongun, A.R. (2005). Comparison of the use of zeolite and perlite as substrate for crisp-head lettuce. *Sci. Hort.*, **106**, 464–471.

Guttormsen, G. (1974). Effects of root medium and watering on transpiration, growth and development of glasshouse crops. I. Effects of compression at varying water levels on physical state of root media and on transpiration and growth of tomatoes. *Plant Soil*, **40**, 1, 65–81.

Hanna, H.Y. and Smith. D.T. (2002). Recycling perlite for more profit in greenhouse tomatoes. *Louisiana Agr.*, **45**, 9.

Hanna, H.Y. (2005). Properly recycled perlite saves money, does not reduce greenhouse tomato yield, and can be re-used for many years. *HortTechnology*, **15**, 342–345.

Hannan, J.J. (1998). *Greenhouses: Advanced Technology for Protected Horticulture*. Boca Raton, FL: CRC Press LLC 684 pp.

Hardgrave, M. (1995). An evaluation of polyurethane foam as a reusable substrate for hydroponic cucumber production. *Acta Hort.* (ISHS), **401**, 201–208.

Harland, I., Lane, S. and Price, D. (1999). Further experiences with recycled zeolite as a substrate for the sweet pepper crop. *Acta Hort.* (ISHS), **481**, 187–194.

Heiskanen, J. (1995a). Water status of sphagnum peat and peat-perlite mixtures in containers subjected to irrigation regimes. *Hort. science*, **30**, 281–284.

Heiskanen, J. (1995b). Physical properties of two-component growth media based on sphagnum peat and their implications for plant-available water and aeration. *Plant and Soil*, **172**, 45–54.

Heiskanen, J. (1999). Hydrological properties of container media based sphagnum peat and their potential implications for availability for water to seedlings after outplanting. *Scand. J. For. Res.*, **14**, 78–85.

Hershey, D.R., Paul, J.L. and Carlson, R.M. (1980). Evaluation of potassium–enriched clinoptilolite as a potassium source for potting media. *Hort. Science*, **15**, 87–89.

Huang, P.M. (2000). Abiotic catalysis. In *Handbook of Soil Science* (M.E. Sumner ed.). Boca Raton, FL: CRC Press, pp. 303–332.

Huang, P.M. and Violante, A. (1986). Influence of organic acids on crystallization of precipitation products of aluminium. In *Interactions of Soil Minerals with Natural Organics and Microbes* (P.M. Huang, M. Schnitzer, eds). SSSA, Spec. Pub. No. 17. Madison: Soil Science Society of America, pp. 159–222.

Huang, Z.T. and Petrovic, A.M. (1994). Clinoptilolite zeolite influence on nitrate leaching and nitrogen use efficiency in simulated sand based golf greens. *J. Environ. Qual.*, **23**, 1190–1194.

Huang, Z.T. and Petrovic, A.M. (1995). Physical properties of sand as affected by clinoptilolite zeolite particle size and quantity. *J. Turfgrass Management*, **1**, 1–15.

Issa, M., Ousounidou, G., Maloupa, H. and Constantinidou, H.A. (2001). Seasonal and diurnal photosynthetic responses of two gerbera cultivars to different substrates and heating systems. *Sci. Hort.*, **88**, 215–234.

Jeong, B.R. (2000). Current status and perspectives of horticultural medium re-use. *Korean J. Hort. Sci. Technol.*, **18**, 876–883.

Jeong, B.R. and Hwang, S.J. (2001). Use of recycled hydroponic rockwool slabs for hydroponic production of cut roses. *Acta Hort.* (ISHS), **554**, 89–94.

Jones, B.R. Jr. (2005). *Hydroponics: A Practical Guide for the Soilless Grower* (2nd edn). Boca Raton, FL: St.-Lucie Press. 423 pp.

Jones, S.N. and Or, B. (1998). Design of porous media for optimal gas and liquid fluxes to plant roots. *Soil Sci. Soc. Am. J.*, **62**, 563–573.

Kapetanios, E.G. and Loizidu, M. (1992). Heavy metal removal by zeolite in tomato cultivation using compost. *Acta Hort.* (ISHS), **302**, 63–71.

Kim, O.I., Cho, J.Y. and Jeong, B.R. (2000). Medium composition including particles of used rockwool and wood affects growth of plug seedling of petunia 'Romeo'. *Korean J. Hort. Science Technol.*, **18**, 33–38.

Kim, G.H. and Jeong, B.R. (2003). Hydroponic culture of a pot plant *Ficus benjamina* 'King' using mixtures of used rockwool slab particles and chestnut woodchips. *J. Korean Soc. Hort. Sci.*, **44**, 251–254.

Kipp, J.A., Wever, G. and de Kreij, C. (2000). *International Substrate Manual*. The Netherlands: Elsevier.

Kithome, M., Paul, J.W. and Bomke, A.A. (1999). Reducing nitrogen losses during simulated composting of poultry manure using adsorbents or chemical amendments. *J. Environ. Qual.*, **28**, 194–201.

Koltermann, C.E. and Gorelick, S.M. (1995). Fractional packing model by hydraulic conductivity derived from sediment mixtures. *Water Resour. Res.*, **31**, 3283–3297.

Krutilina, V.S., Goncharova, N.A., Panov, N.P. and Letchamo, W. (1999). Effect of zeolite and phosphogypsum on yield, plant uptake, and content of strontium in soil. *Comm. Soil Sci. Plant. Anal.*, **30**, 483–495.

Krutilina, V.S., Polyanskaya, S.M., Goncharova, N.A. and Panov, N.P. (2000). Growth, photosynthesis, and uptake of heavy metals by barley and corn plants influenced by different methods of zeolite and phosphogypsum application. *Comm. Soil Sci. Plant. Anal.*, **31**, 1287–1298.

Lim, M.Y. and Jeong, B.R. (2000). Effect of medium composition including chestnut woodchips and granular rockwool on growth of plug seedlings. *Korean J. Hort. Sci. Technol.*, **18**, 508–512.

Marfa, O., Martinez, A., Orozco, R., et al. (1993). The use of fine grade perlites in lettuce bag cultures. II. Physical properties, rheologic effects and productivity. *Acta Hort.* (ISHS), **342**, 339–348.

Martinez, F.X., Burés, S., Blanca, F., et al. (1991). Experimental and theoretical air/water ratios of different substrate mixtures at container capacity. *Acta Hort.* (ISHS), **294**, 241–248.

Martinez, F.X., Sepo, N. and Valero, J. (1997). Physical and physicochemical properties of peat-coir mixes and effects of clay-material addition. *Acta Hort.* (ISHS), **450**, 39–46.

Matkin, O.A. (2005). Comparative growth studies perlite vs. polystyrene media. http://www.schundler.com/compare.htm (Accessed 13 February 2006).

Meinken, E. (1997). Accumulation of nutrients in expanded clay used for indoor planting. *Acta Hort.* (ISHS), **450**, 321–327.

Meinken, E. and Fischer, P. (1994). Hydrokulturen in Wohnräumen. Weiter Optimalbereich bei der Ernährung. *Deutscher Gartenbau*, **48**, 1348–1350.

Ming, D.W. and Mumpton, F.A. (1989). Zeolites in soils. In *Minerals in Soil Environments* (J.B. Dixon, and S.B. Weed eds) (2nd edn). SSSA Book Series No.1, Madison, WI, USA: Soil Sci. Soc. Am., pp. 873–909.

Mumpton, F.A. (1999). La roca magica: uses of natural zeolites in agriculture and industry. *Proc. Nat. Acad. Sci. USA*, **96**, 3463–3471.

Nash, M.A. and Pokorny, F.A. (1990). Shrinkage of selected two-component container media. *Hort. Sci.*, **25**, 8, 930–931.

Nash, M.A. and Pokorny, F.A. (1992). Prediction of water-retention of milled pine bark-sand potting media from laboratory analyses of individual components. *Commun. Soil Sci. Plant Anal.*, **23**, 929–937.

Neefjes, H. (2001). Recycling rockwool reaches the top. *Vegetables and Fruits*, week 25, 2001. 2p.

Nelson, P.V. (2003). *Greenhouse Operation and Management* (6th edn). Upper Saddle River, NJ: Prentice hall, 692 pp.

Notario del Pino, J.S., Artega-Padron, I.J., Gonzales-Martin, M.M. and Garcia-Hernandez, J.E. (1994). Response of alfalfa to a phillipsite-based slow-release fertilizers. *Comm. Soil Sci. Plant. Anal.*, **25**, 2231–2245.

Oekotau Easy Green GmbH. (2006). Hydroton. www.oekotau.de/en/pdf/en_hydro.pdf.

Olympios, C.M. (1992). Soilless media under protected cultivation: rockwool, peat, perlite and other substrates. *Acta Hort.* (ISHS), **323**, 215–234.

Opena, G.B. and Williams, K.A. (2003). Use of precharged zeolite to provide aluminum during blue hydrangea production. *J. Pl. Nutr.*, **26**, 1825–1840.

Orozco, R. and Marfa, O. (1995). Granulometric alteration, air entry potential and hydraulic conductivity in perlites used in soilless cultures. *Acta Hort.* (ISHS), **408**, 147–161.

Paasikallio, A. (1998). Effect of biotite, zeolite, heavy clay, bentonite and apatite on the uptake of radio-caesium by grass from peat soil. *Plant Soil*, **206**, 213–222.

Papadopoulos, A.P., Hao, X., Tu, J.C. and Zheng, J. (1999). Tomato production in open or closed rockwool culture systems with NFT or rockwool nutrient feedings. *Acta Hort.* (ISHS), **481**, 89–91.

Papadopoulos, A., Maloupa, E. and Papadopoulos, F. (1995). Seasonal crop coefficient of gerbera soilless culture. *Acta Hort.* (ISHS), **408**, 81–90.

Papadopoulos, A.P. and Tan, C.S. (1991). Irrigation of greenhouse tomatoes grown in 'Harrow' peat-bags. *Can. J. Plant Sci.* **71**, 947–949.

Pokorny, F.A. and Henny, B.K. (1984a). Construction of a milled pine bark and sand potting medium from component particles: I. Bulk density: A tool for predicting component volumes. *J. Am. Soc. Hort. Sci.*, **109**, 770–773.

Pokorny, F.A. and Henny, B.K. (1984b). Construction of a milled pine bark and sand potting medium from component particles: II. Medium synthesis. *J. Amer. Soc. Hort. Sci.*, **109**, 774–776.

Pokorny, F.A., Gibson, P.G. and Dunavent, M.G. (1986). Prediction of bulk density of pine bark and/or sand potting media from laboratory analyses of individual components. *J. Am. Soc. Hort. Sci.*, **111**, 8–11.

Pond, W.G. and Mumpton, F.A. (1984). Zeo-Agriculture. *Use of Natural Zeolites in Agriculture and Aquaculture*. Boulder, CO: Westview Press.

Raviv, M. and Blom, T.J. (2001). The effect of water availability and quality on photosynthesis and productivity of soilless-grown cut roses. *Sci. Hort.*, **88**, 257–276.

Raviv, M., Lieth, J.H., Burger, D.W. and Wallach, R. (2001). Optimization of transpiration and potential growth rates of 'Kardinal' rose with respect to root-zone physical properties. *J. Am. Soc. Hort. Sci.* **126**, 638–643.

Raviv, M., Silber, A. and Medina, S. (1998). The effect of medium disinfestation on cut rose productivity and on some chemical properties of tuff. *Sci. Hort.*, **74**, 285–293.

Raviv, M., Wallach, R., Silber, A. (1999). The effect of hydraulic characteristics of volcanic material on yield of roses grown in soilless culture. *J. Amer. Soc. Hort. Sci.* **124**, 205–209.

Raviv, M., Wallach, R., Silber, A. and Bar-Tal, A. (2002). Substrates and their analysis. In *Hydroponic Production of Vegetables and Ornamentals* (D. Savvas, and H. Passam eds). Athens, Greece: Embryo Publications, pp. 25–101.

Raviv, M., Wallach, R. and Blom, T.J. (2004). The effect of physical properties of soilless media on plant performance – A review. *Acta Hort.* (ISHS), **644**, 251–259.

Resh, H.M. (1998). *Hydroponic Food Production, a Definitive Guidebook for the Advanced Home Gardner and the Commercial Hydroponic Grower* (5th edn). Santa Barbara, CA: Woodbridge Press Publishing Company, 528 pp.

Riga, P., Alava, S., Uson, A., (2003). Evaluation of recycled rockwool as a component of peat-based mixtures for geranium (*Pelargonium peltatum* L.) production. *J. Hort. Sci. Biotechnol.*, **78**, 213–218.

Rivero-Gonzales, L.A. and Rodriguez-Fuentes, G. (1988). Cuban experience with the use of natural zeolite substrates in soilless culture. In Proceedings of 7th Internat. Congress Soilless Culture, ISOSC, Wageningen, The Netherlands. pp. 405–416.

Rosen, K. (1996). Field studies on the behaviour of radiocaesium in agricultural environments after the Chernobyl accident. Rapport Institutionen for Radioekology Sveriges Lantbe No.78.

Rupp, L.A. and Dudley, L.M. (1989). Iron availability in rockwool may affect rose nutrition. *Hort. Science*, **24**, 258–260.

Sahin, U., Anapali, O. and Ercisli, S. (2002). Physico-chemical and physical properties of some substrates used in horticulture. *Gartenbauwissenschaft*, **67**, 55–60.

Samartzidis, C., Awada, T., Maloupa, E., Radoglou, K., Constantinidou, H.-I.A. (2005). Rose productivity and physiological responses to different substrates for soilless culture. *Sci. Hort.*, **106**, 203–212.

Schroeder, F.-G., and Forester, R. (2000). Polyester fleece as a substrate for soilless culture. *Acta Hort.* (ISHS), **554**, 67–73.

Schwartz, M. (1995). *Soilless Culture Management*. Advanced Series in Agricultural Sciences Vol. 24. Berlin: Springer-Verlag.

Sheldrake, R. (1989). Tomato profits are in the bags. *Amer. Vegetable Grower*, **37**, 24–28.

Sherman, J.D. (1999). Synthetic zeolites and other microporous oxide molecular sieves. *Proc. Nat. Acad. Sci. USA*, **96**, 3471–3478.

Shin, W.G. and Jeong, B.R. (2000). Growth of plug seedlings of petunia 'Madness Rose' and pansy 'Majestic GT' in various mixtures of recycled horticultural media. *Korean J. Hort. Sci. Technol.* **18**, 523–528.

Silber, A., Bar-Yosef, B., Singer, A. and Chen, Y. (1994). Mineralogical and chemical composition of three tuffs from northern Israel. *Geoderma*, **63**, 123–144.

Silber, A., Bar-Yosef, B. and Chen, Y. (1999). pH-dependent kinetics of tuff dissolution. *Geoderma*, **93**, 125–140.

Silber, A. and Raviv, M. (1996). Effects on chemical surface properties of tuff by growing rose plants. *Plant Soil*, **186**, 353–360.

Silber, A. and Raviv, M. (2001). Disinfestation effects on the chemical properties of tuff. *Acta Hort.* (ISHS), **554**, 41–50.

Smith, D.L. (1987). *Rockwool in Horticulture*. London, UK: Grower Books.

Smith, D. (1998). *Growing in Rockwool*. London, UK: Grower Books.

Sonneveld, C. and Spaans, L. (1990). Computer program calculates best nutrient solution plan. *Groenten-en-Fruit*, **45**, 38–39. (In Dutch).

Sonneveld, C. and Voogt, W. (1985). Studies on the application of iron to some glasshouse vegetables grown in soilless culture. *Plant Soil*, **85**, 55–64.

Spomer, L.A. (1974). Optimizing container soil amendment: the threshold proportion and prediction of porosity. *Hort. Science*, **9**, 532–533.

Steidman, B. (1988). So far peat bags do the jobs. Greenhouse Canada. p. 10.

Tan, H.K. (1986). Degradation of soil minerals by organic acids. In *Interactions of Soil Minerals with Natural Organics and Microbes* (P.M. Huang, M. Schnitzer, eds). SSSA, Spec. Pub. No. 17, Madison: Soil Science Society of America, pp. 1–27.

Tate, K.R. and Theng, B.K.G. (1980). Organic matter and its interactions with inorganic soil constituents. In *Soils with Variable Charge* (B.K.G. Theng, ed.). New Zealand: Lower Hutt, pp. 225–249.

Tu, J.C., Papadopoulos, A.P., Hao, X. and Zheng, J. (1999). The relationship of Pythium root rot and rhizosphere microorganisms in a closed circulating and an open system in rockwool culture of tomato. *Acta Hort.* (ISHS), **481**, 577–583.

Unver, I., Ataman, Y., Canga, M.R. and Munsuz, N. (1989). Buffering capacities of some mineral and organic substrates. *Acta Hort.* (ISHS), **238**, 83–97.

Van der Gaag, D.G. and Wever, G. (2005). Conduciveness of different soilless growing media to Pythium root and crown rot of cucumber under near-commercial conditions. *Eur. J. Pl. Pathol.*, **112**, 31–41.

van Os, E.A., Ruijs, M.N.A. and van Weel, P.A. (1991). Closed business systems for less pollution from greenhouses. *Acta Hort.* (ISHS), **294**, 49–57.

Verhagen, J.B.G.M. (2004). Effectiveness of clay in peat based growing media. *Acta Hort.* (ISHS), **644**, 151–155.

Verwer, F.L. (1976). Growing horticultural crops in rockwool and nutrient film. Proceedings of 4th International Congress Soilless Culture. ISOSC, Wageningen, The Netherlands, pp. 107–119.

References

Wallach, R. da Silva F.F. and Chen. Y. (1992a). Hydraulic characteristics of Tuff (Scoria) used as a container medium. *J. Am. Soc. Hort. Sci.*, **117**, 415–421.

Wallach, R., da Silva F.F. and Chen. Y. (1992b). Unsaturated hydraulic characteristics of composted agricultural wastes, Tuff and their mixtures. *Soil Sci.* **153**, 434–441.

Weber, M.A., Barbarick, K.A. and Westfall, D.G. (1983). Ammonium adsorption by a zeolite in a static and dynamic system. *J. Environ. Qual.*, **12**, 549–552.

Weiss, R., Alm, J., Laiho, R. and Laine, J. (1998). Modeling moisture retention in peat soils, *Soil Sci. Soc. Am. J.*, **62**, 305–313.

Wever, G. and van Leeuwen, A.A. (1995). Measuring mechanical properties of growing media and the influence of cucumber cultivation on these properties. *Acta Hort.* (ISHS), **401**, 27–34.

Wever, G., van Leeuwen, A.A. and van der Meer, M.C. (1997). Saturation rate and hysteresis of substrates. *Acta Hort.* (ISHS), **450**, 287–295.

Williams, K.A. and Nelson, P.V. (1997). Using precharged zeolite as a source of potassium and phosphate in a soilless container medium during potted chrysanthemum production. *J. Am. Soc. Hort. Sci.*, **122**, 703–708.

Wilson, G.C.S. (1988). The effect of various treatments on the yield of tomatoes in reused perlite. *Acta Hort.* (ISHS), **221**, 379–382.

Wright, R. (1989). Evaluation of propagation mediums trough rooting response of *Hedera helix* 'Ivalace'. *Ivy J.*, **15**, 28–32.

Yasuda, H., Takuma, K., Mizuta, N. and Nishide, H. (1995). Water retention variety of dune sand due to zeolite addition. *Bulletin Fac. Agric.* Japan: Tottori University, 48, 27–34 (in Japanese with English abstract).

Yeager, T.H, Gilliam, C., Bilderback, T.E., et al. (1997). *Best Management Practices, Guide for Producing Container-grown Plants*. Atlanta, GA, USA: Southern Nursery Association.

13

Growing Plants in Soilless Culture: Operational Conclusions

Michael Raviv, J. Heinrich Lieth, Asher Bar-Tal and Avner Silber

13.1 Evolution of Soilless Production Systems
13.2 Development and Change of Soilless Production Systems
13.3 Management of Soilless Production Systems
References

13.1 EVOLUTION OF SOILLESS PRODUCTION SYSTEMS

The previous chapters present various specific facets of soilless plant production, representing the state of the art and including some perspectives on the direction in which the field is moving. In some cases, gaps in knowledge were pointed out and required future research was suggested. Generally the chapters focus on particular facets of a system whose full complexity was beyond the scope of each specific chapter. In some cases lack of appreciation of the complexity imposed by the interaction of all the factors may lead to problems and failures for practitioners. The aim of this final chapter is to integrate various concepts from throughout the book and from other sources, to provide practitioners with practical operational tools, allowing them to optimize crop production. The focus is to develop a better understanding of the intricate processes taking place within the system along the root zone–plant–atmosphere continuum, as affected by the interactions among the growing substrate, the liquid and gaseous phases held in its matrix, and its nutritional status. Aerial conditions, although exerting major effects on plant performance, are beyond the scope of this chapter.

One concept that was discussed in Chap. 8 was the historical perspective on nutritional limitations to plant growth. The work by Mitcherlish suggested that plant productivity is limited by the nutritional factor that is limiting. This concept has been verified and is appropriate if all factors except one are non-limiting. As such this concept is also true in a broader sense when considering all factors that impact soilless crop production. However, when multiple factors are limiting, we must anticipate interacting effects which are more complex than simply suggesting causality of suboptimal production to the most-limiting factor. This is particularly important in practice because it is extremely rare that all production factors are simultaneously optimized.

A further consideration regarding this issue is the fact that the root zone is a dynamic microenvironment where many of the factors are in continual change. Many factors can change second-to-second or minute-by-minute, while some change over the course of days or weeks. Management of such dynamic systems requires constant attention, and the advent of computer technology coupled with sensor technology provides a situation where we can anticipate significant changes in the field as control technology improves.

13.1.1 MAJOR LIMITATION OF SOILLESS- VS. SOIL-GROWING PLANTS

In the first chapter, we described the historical reasons for the advent of soilless cultivation and its further expansion to the level it has reached to date, and we briefly suggested reason why its share in global food and plant production will continue to grow in the foreseeable future. It is particularly important to note the speed with which soilless production has changed over the past 50 years. Anyone considering soilless crop production should be aware that the field as a whole is dynamic, with horticultural practices changing continuously. One consequence of this is that in various chapters, methods and technologies are presented as being in widespread use or commonly accepted. Within this rapidly changing field, such methods may be quickly replaced by alternate methods as scientific understanding and economic feasibility change. Perhaps the most dramatic change is currently occurring in relation to management of recirculation systems for various soilless production systems.

Various reasons persist for the continued growth of soilless production as part of agriculture. Some of these reasons can be inferred from the chapters dealing with the physical and chemical characteristics of substrates, emphasizing their superiority over soil cultivation, ease of control of water, oxygen and nutrient availabilities and the resulting improved crop performance. Another advantage is the relative freedom from soil-borne pathogens and improved possibility of disinfestation of the medium among growing cycles. However, growing crops without soil also imposes some limitations which test the grower's skill to avoid losing the potential benefits offered by the advantages. The extent of these limitations and how to practically deal with them is discussed in this chapter.

13.1.2 THE EFFECTS OF RESTRICTED ROOT VOLUME ON CROP PERFORMANCE AND MANAGEMENT

One of the biggest contrasts between soil-less and soil-based production is the spatial confinement of the roots into a specific, well-defined root zone. The smaller the root zone, the more intensive the production system needs to be to manage this volume.

Sonneveld (1981) reported that the volume of medium available to a tomato (*Lycopersicon esculentum*, L.) plant grown in a soil bed in a greenhouse is approximately 200 L, while the corresponding volumes for production in substrates are typically an order of magnitude smaller. He calculated that the volume of water (and hence nutrients) in the root zone at any given moment to a tomato plant would be an order of magnitude greater in soil than in soilless production. Thus in soilless production involving such substrates as peat and stone wool, irrigation needs to be carried out much more frequently than in soil-based production if the nutrient and water storage aspects were the same. In fact, with Nutrient Film Technique (NFT) systems, where the water volume may well be an order of magnitude smaller than substrate-based soilless production, application of water needs to be continuous so as to avoid nutrient deficiencies and water stress.

Numerous experiments and simulation studies demonstrated that a continuous growth process of the root system is essential to ensure efficient water and nutrients uptake from soil (Williams and Yanai, 1996; Tinker and Nye, 2000; Sadanal et al., 2005). Root growth is especially important in the case of low mobility nutrients such as P, Mn, Zn and more so when their concentrations in the root-zone solution are low since in this case ion diffusion towards the depletion zone is minimal. Plants growing in soil typically exhibit fine root growth so as to gain access to water and nutrients from less-explored regions of the root zone. In fact, active, vital growth by such plants depends on continued formation of new roots (Williams and Yanai, 1996; Tinker and Nye, 2000; Sadanal et al., 2005). In frequently flushed soilless root zones, the near-absence of clear depletion zones somewhat diminishes the need by the plant for such active 'foraging'. However, since the process of root growth is, by nature, a persistent one (Bengough et al., 2006), root systems of container-grown plants are usually very dense. Optimization of soilless production systems means that irrigation and fertilization must be carried out with precise timing and location, and that such systems suffer from minimal tolerance for error.

Under normal horticultural conditions of sufficient supply of water and nutrients, substrate volume has little or no effect on root/shoot ratio (NeSmith et al., 1992). Therefore, to satisfy the canopy needs for water and nutrients, root density must increase with decreasing substrate volume (Keever et al., 1985; Boland et al., 2000). The increased root density involves greater oxygen and nutrient consumption per unit volume of root zone, resulting in intense root-to-root competition for oxygen and nutrients, leading to more rapid declines in the concentration of dissolve oxygen (DO) and available nutrients. At the same time, the reduced DO levels can negatively affect root function, increase their susceptibility to diseases and eventually cause their death. Lack of sufficient DO may also inhibit ammonium oxidation, leading to pH decrease, accumulation of toxic levels of ammonium in the liquid phase and even to toxic levels of ammonia in the gaseous phase of the root zone. This phenomenon may be aggravated by the consumption of oxygen by micro-organisms that decompose organic matter. Even in inorganic media, the decomposition of dead roots and root exudates contributes to oxygen consumption. The continued mineralization of organic matter may also be accompanied by compaction of the medium, resulting in decreased oxygen diffusion rate (ODR), leading to slow flow rate of gaseous oxygen into the medium gaseous phase. Frequent solution replenishment with DO-saturated water can minimize the severity of these problems. Another method to prevent low and decreasing ODR is through the choice of medium's components.

13.1.3 THE EFFECTS OF RESTRICTED ROOT VOLUME ON PLANT NUTRITION

Two effects of restricted root volume on plant nutrition are that: (1) it can physically restrict root growth, so that volume, length and surface area of the roots are reduced and (2) limited reservoir size within the root zone and small buffer capacity of inert media may lead to restricted nutrient supply. It is not possible to raise the nutrient supply without limit by elevating their concentration, as this involves a potentially detrimental increase in osmotic potential, which interferes with water uptake (Sonneveld, 1981). A partial solution to the limited availability of water and nutrients is through their frequent supply through fertigation. But even constant supply (as is practical in closed systems) cannot overcome all the constraints posed by small root volume, as discussed in Chap. 9, and by Bar-Tal et al. (1990, 1995), Bar-Tal and Pressman (1996) and Rieger and Marra (1994). It is therefore important to select the appropriate container size for each specific crop, based on its growth rate and final desired size and growing conditions.

Bar-Tal et al. (1995) investigated the effect of root size on the uptake of nutrients and water by tomato plants, to investigate the effects of root restriction in relation to nutrients supply on transpiration and nutrients uptake. Restriction of the root system resulted in a decrease in root volume compared with unrestricted roots. Severe and mild root restriction nearly stopped the increase in root volume 70 and 130 days after transplanting, respectively. The growth curves of the unrestricted and restricted root systems in both nutrient solutions had a smooth 'S' shape, which is described well by the Gompertz growth model (Eq. [1]) (Fig. 13.1).

$$V_t = V_f^* e^{-b^* e - k^* t} \tag{1}$$

where V_t is the root volume at time t (cm^3), V_f is the maximal root volume (cm^3) and b and k are dimensionless parameters. The estimated parameters k and V_f were significantly affected by the solution nitrate concentration (Bar-Tal et al., 1995). The rate coefficient value (k) of the intact root, 0.0228 d^{-1}, was in good agreement with published values of 0.035–0.039 d^{-1} (Bar-Tal et al., 1994). Root restriction significantly reduced V_f, k (the rate coefficient, d^{-1}) and b (dimensionless). The resulting V_f of the restricted root was in good agreement with the bag volume.

Root restriction induced an increase in F_N, the nitrogen uptake rate per root unit volume. This indicated an adaptation of the smaller root system to the shoot demand.

13.1 EVOLUTION OF SOILLESS PRODUCTION SYSTEMS

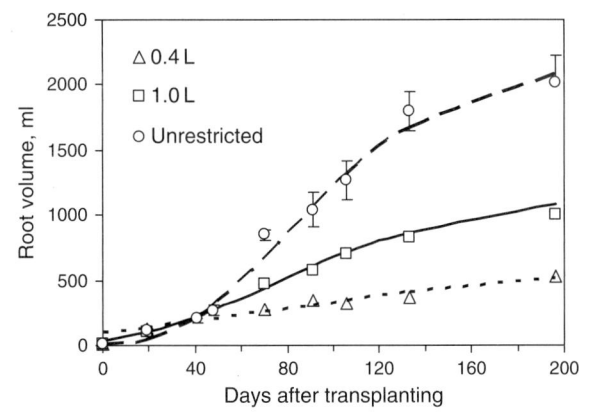

FIGURE 13.1 Root growth as affected by root restriction. Each point represents a mean of two N—NO$_3$ solution concentrations (1.0 and 9.0 mmol L^{-1}), except for the mild restriction that was studied only in 9.0 mmol L^{-1} solution. The vertical bars represent ±LSD$_{0.05}$. The curves were calculated with the Gompertz growth model. The values of the parameters were fitted with the NLIN procedure of SAS (based on Bar-Tal et al., 1995).

Thus, the smaller root system was more efficient in uptake per unit root weight, in agreement with published data (Jungk, 1974; Jungk and Barber, 1975; Edwards and Barber, 1976; de Willigen and van Noordwijk, 1987; Bar-Yosef et al., 1988). Eshel et al. (2001) found that the efficiency of absorption of K and P by the root increased following removal of part of the roots; however, this was not true for other nutrients. Consequently, the mass or the surface area of the root may affect nutrient uptake of plants grown in substrates flushed frequently with nutrient solution (Fig. 13.2). Root

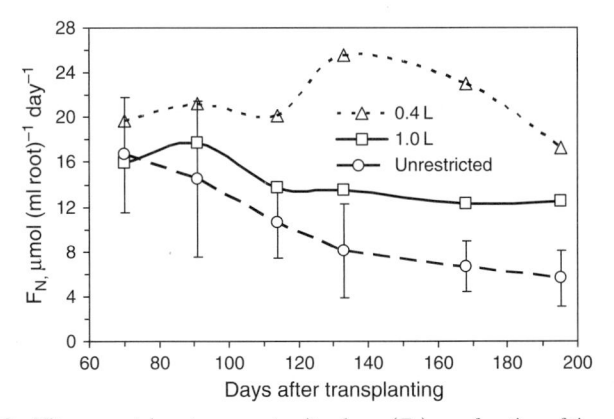

FIGURE 13.2 Nitrogen uptake rate per root unit volume (F_N) as a function of time and root volume. Each point represents a mean of two N—NO$_3$ solution concentrations (1.0 and 9.0 mmol L^{-1}), except for the mild volume that was studied only in 9.0 mmol L^{-1} solution. The vertical bars represent ±LSD$_{0.05}$ (based on Bar-Tal et al., 1995).

restriction increased F_N; values of 8.0, 13.5 and 25.6 μmol ml^{-1} d^{-1} were obtained at 133 days after transplanting (DAT) for the control and for mild and severe root restriction, respectively.

Root restriction significantly reduced the potassium uptake rate per plant but it had no effect on F_K, the potassium uptake rate per root unit volume. The values of the parameter F_{max} (Michaelis–Menten uptake model, Eq. (4) in Chap. 8) (24.9 and 8.8 μmol g^{-1} d^{-1}, for the young and old plants, respectively) fell within the range of published data for four different species (Wild et al., 1974; Bar-Yosef et al., 1992).

In the study of Bar-Tal and Pressman (1996) in the aero-hydroponics system, root restriction decreased Ca uptake rate per plant, but there was a trend of increasing F_{Ca} when root restriction was combined with low K concentration, which is consistent with the ability of the restricted-root plants to maintain Ca concentrations in plant organs similar to those in non-restricted plants. Choi et al. (1997) reported that root restriction by small containers (190 ml) strongly suppressed transport of ^{45}Ca ions to new leaves and apices relative to the plants in the control containers (20 L). Water transport, expressed on leaf area basis, was marginally reduced by root restriction, an indication that calcium transport was more severely limited than water transport. Karni et al. (2000) also obtained a decline in Ca uptake by pepper root due to removal of three quarters of the root system, while the effect on water uptake and status in the plant was marginal.

In conclusion from the above, root restriction reduces both nutrient uptake per plant and nutrient uptake per unit root when the buffer capacity and the rate of replenishment of the growing medium are the limiting factors. When the growing system enables maintaining a constant concentration of each nutrient at the root surface (as is the case in NFT, DFT, aeroponics and with very frequent irrigation pulses), the ability of the restricted root system to meet the plant requirements is not the primary limiting factor. The minimal root size to meet plant nutritional requirement for nitrogen and other mobile nutrients is much smaller than that required for water supply, aeration and other physiological demands. However, the supply of phosphate and calcium may be the limiting factors as a result of restricted root system.

13.1.4 ROOT CONFINEMENT BY RIGID BARRIERS AND OTHER CONTRIBUTING FACTORS

As previously mentioned, regeneration of new roots is essential for normal plant development (Stevenson, 1967). Root growth is directed by environmental cues, including touch (thigmotropism) and gravity. Gravity sensing occurs mainly in the columella cells of the root cap. Downward root growth is a natural response to gravi- and hydrotropism, typical to all active roots. A recent study by Massa and Gilroy (2003) suggests that normal root tip growth requires the integration of both gravity and touch stimuli. However, strong mechanical impedance such as the one that results from compact soil layer affects root growth direction (see Chap. 2) and causes clear anatomical and ultrastructural changes in the root apex (Wilson and Robards, 1979). These changes reflect several mechanisms that help root elongation in hard soils and even penetration through compacted layers. The root cap is generally smaller and thus cannot confer the same degree of protection to the root meristem as caps grown without mechanical impedance. Other changes include increased sloughing of border cells; exudation from the root cap that decreases friction and marked increase in root diameter. All these changes help to relieve stress in front of the root apex and to decrease buckling. Whole root systems may also grow preferentially in loose versus dense soil (Bengough et al., 2006). Concomitantly, mechanical impedance considerably reduces the root elongation rate and final root cell size and affects root distribution (Croser et al., 1999). Some of those energy-consuming changes may occur when roots of container-grown plant reach a rigid wall. As these roots can only change course, staying within the confined root zone, root density increases within the root zone. Eventually, root confinement within a limited volume results with reduced root growth (Hameed et al., 1987), similar to what can usually be found in compacted soils (Hoffmann and Jungk, 1995). This reduced root growth can explain the results that were described earlier, connecting root-zone volume with plant performance even under ample supply of water and nutrients.

In container-grown plants, gravi- and hydrotropism frequently results with the accumulation of root mat on the bottom (Martinez et al., 1993). This part of the root system frequently accounts for a major part of the total root biomass and may be exposed to oxygen deficiency both due to the respiration of an extensive mass of dense roots, and as a result of the existence of a perched water layer on the bottom of the container.

The main practical steps that could be applied at the grower level to ease DO deficiency are to optimize container depth, according to the crop size and growth duration and to frequently refresh the DO level using irrigation. Another useful approach is the use of liquid culture or aeroponics, where mechanical impedance is practically avoided. However, even in these conditions, increased root density leads, eventually, to growth inhibition (Peterson et al., 1991a; Bar-Tal and Pressman, 1996).

Another important limitation that sometimes occurs in soilless growing systems is the accumulation of regions of high EC within the medium, where irrigation water reaches only through diffusion and in the top layer experiencing a substantial amount of evaporation. This phenomenon limits the useful volume available to the plants. Salt pockets can be formed when drippers are too far from each other and hydraulic conductivity is low. Plants irrigated exclusively with subirrigation will generally experience high EC near the medium's surface as a result of evaporation-derived salt accumulation (Kent and Reed, 1996). It is important to plan both the irrigation system(s) and the water discharge rate, so that a frequent downward piston-like water movement will prevent this type of salt accumulation (Schwarz et al., 1995). Thus with subirrigation it may be necessary to use a second irrigation system or to plan on hand-watering occasionally so as to leach such excess salts from the upper layer, and to then provide adequate irrigation solution to push these salts completely out of the root zone. The subject of selecting desirable root-zone volume is discussed in Sect. 13.3.

13.1.5 ROOT EXPOSURE TO AMBIENT CONDITIONS

As mentioned in Chap. 2, roots growing in containers are more exposed to extreme ambient temperatures than soil-grown roots, while temperature fluctuations in deep soil layers are minimal. In addition, the relative effect of ambient low vapour pressure deficit (VPD) on water evaporation from the root zone is high, due to the typically high surface to depth ratio, as compared to this of soil-grown plants. The evaporation issue has limited horticultural significance, as normally container-grown plants are well-irrigated, the medium surface is shaded by the plant canopy and the coarseness of the medium results in low capillary movement. It may, however, be important in outdoor nurseries and in newly planted media, where the plant cover is minimal and the effect of advective heat from the surroundings is significant (Seginer, 1994).

Moderate root zone warming (20–25°C) may have a beneficial effect in many cases and is routinely practiced in many propagation nurseries during the rooting stage of stem and leaf cuttings. Elevated temperatures, on the other hand, lead to a significant increase in root respiration rates. Since this is coupled with a sharp decrease in DO levels in the root-zone solution, the result may be detrimental. The combination of high root zone temperature with ammonium as the source of N has a negative effect on the root system and whole plant growth and development (Kafkafi, 1990). Anaerobic processes may occur within the root tissue, significantly lowering its water and nutrient uptake and its growth rate. High temperature may, in some cases, negatively affect the activity of nitrifying bacteria which may, in turn, lead to toxic ammonium levels. Even under Northern Europe conditions, within the stone wool slabs in greenhouse vegetable production, temperatures can reach levels of 30–35°C, greatly affecting the incidence of, and susceptibility to, pathogens (van der Gaag and Wever, 2005). Martin et al. (1991) showed a marked effect of the size of the container on the temperature in the middle of the root zone. Summer-time temperatures were 5–6°C lower in 57-L than in 10-L containers.

Practical methods that are used to mitigate high temperatures include adequate irrigation control, possibly with occasional pulse irrigation, the use of mulch, a careful choice of the container colour and dimensions.

13.1.6 ROOT ZONE UNIFORMITY

One thing that should not be overlooked when contrasting in-ground field production with soilless production is the matter of uniformity. With in-ground field production, there is frequently variability in soil characteristics within the field. Many times farmers ignore this and simply treat the entire field as having the same soil, but the plants will generally grow and yield differently in the various parts of the field.

With soilless production this is generally not the case. With a properly assembled and mixed substrate, using standard best-management practices, each plant experiences essentially the same growing medium as all other plants in the crop, as long as drainage and supply of irrigation water is uniform. As such, in soilless production a uniform management strategy for a particular crop is ideal, while in field production, better productivity would be achieved by customizing irrigation and fertilization for each plant depending on the variations in soil within the field. With innovations in precision agriculture, it can be expected that such optimization will be feasible, yet currently this type of optimization is not yet feasible in field production. With soilless production, on the other hand, such customized optimization is possible although it is questionable how economically feasible it is. As yet, in both soil-based and soilless production, the general strategy is to control irrigation and fertilization for all plants in a crop so as to avoid stress.

13.2 DEVELOPMENT AND CHANGE OF SOILLESS PRODUCTION SYSTEMS

13.2.1 HOW NEW SUBSTRATES AND GROWING SYSTEMS EMERGE (AND DISAPPEAR)

One aspect that has had a major influence on the subject of this book is the notion by non-experts that one can grow plants in virtually any granular or porous material. This notion is fuelled by the fact that plants will grow in a variety of substrates, as long as one waters and fertilizes them reasonably well. This notion has led to a nearly routine exploration of various waste materials as growing media. This has resulted in testing of many materials, some of which have been adapted as growing media. Shredded rubber tires, for instance, have been tested. Unfortunately, this material has not yielded a product suitable for widespread use as substrate in the soilless production. On the other hand, when testing coconut husk debris in a few tropical areas of the world, this waste material was found to be horticulturally useful so that coconut coir is now a well-established component of some growing media and even as a growing medium by itself. In the same way, in California, the use of redwood sawdust, a waste product from the large lumber industry in the state, was discovered to create a product that behaved quite different from other sawdust materials because its slow decomposition results in a root zone that is very slow to collapse. This makes it an excellent substrate for growing outdoor nursery products that require many months to produce. Thus happenstance and trial-and-error have played a major role in the development of soilless growing media components. And while this approach has also led to somewhat random testing of various combinations of materials in various forms, it is today possible to engineer systems (consisting of various media components, containers, control systems and management practices) that optimize the particular application. Horticulturalists already understand some of these principles. For instance, it is well known that if one transplants a plant from a small pot in which it is growing well to a larger, deeper pot, then specific adjustments have to be made, either in the composition of the growing medium or by installing drainage elements, to continue to get the plant to grow. What the horticulturalist is doing is optimizing for one or two variables based on experience, even if not based on scientific knowledge of the underlying principles.

Several attempts have been made over the years to model and predict the characteristics of growing media consisting of well-defined components. Factors such as particle-size distribution, bulk and particle density were taken into account. Water release curves and rate of shrinkage were successfully modelled (Brown and Pokorny, 1975; Pokorny and Henny, 1984a; Pokorny et al., 1986; Bures et al., 1993a,b). Such models enable the prediction of several important properties of new combinations of ingredients.

13.2.2 ENVIRONMENTAL RESTRICTIONS AND THE USE OF CLOSED SYSTEMS

As discussed in Chap. 9, intensive greenhouse production involves several environmental risks. Environmental protection agencies in a growing number of countries require growers to capture and reuse the effluent from soilless production systems so as to minimize these risk. Simultaneously, effluent recycling leads to significant water and fertilizer savings. Moreover, the complicated issue of both temporal and spatially uniform supply of water and nutrients is not relevant if the system is closed so that it enables frequent or even constant flushing.

In view of the dwindling global reservoirs of water and energy (oil) and as a result of environmental regulation, it appears mandatory that most above-ground systems (that allow collection of discharged drainage water) will need to be modified to capture any effluent. As described in Chap. 9, many of the problems involved with such operation have been addressed and many such systems are already in operation. There are a variety of issues that still require solutions, but the primary hurdle for many commercial growers is economic. Even this may be true only on the basis of the individual grower, and not on a national or global basis. In order to promote fast adoption of this practice it is imperative that the true cost of water treatment be known and integrated into their final cost to the costumer. In addition, the future cost for cleaning soil and water resources should be considered. Thus adoption of a closed recirculation system involves careful study of the economic factors and a complete analysis over the long run as to handling of water when it reaches a level of salinity due to non-nutritive ions that makes it impossible to use for in commercial soilless production. As such 100 per cent recirculation is as yet still not possible in areas where the supply water contains ions that the plant cannot use. It should be noted, however, that semi-closed systems normally reduce the amount of effluent by $\sim$90 per cent over a system where no recirculation is practised. It is important that agencies regulating this issue be aware that growers generally must have the flexibility to discard water where some deleterious ion (e.g. Sodium) has accumulated to a level that renders the water toxic to plants. While some argue that water can be cleaned up through water treatment, the result of such treatment is always, in addition to clear water a brine solution that must be discarded. At present there is no easy solution to this problem, and regulators that attempt to enforce 100 per cent recirculation are imposing a condition that will ultimately put the grower out of business.

13.2.3 SOILLESS 'ORGANIC' PRODUCTION SYSTEMS

While 'organic agriculture' has very specific meaning to growers and regulators, the general public is rarely knowledgeable about the details that define this area of agriculture and what is involved in certification. The strict rules of most of the Organic Agriculture certification bodies state that in order to be considered as an organically certified crop, it should be grown in soil; yet the attractiveness of this type of agriculture to the general public is based on the desirability of having foods that were not subjected to chemical fertilizers or pesticides. As such it is entirely feasible to accomplish this with soilless production. This notion recently became clear to scientists (Dresboll, 2004) and may, eventually, affect the rules. This is especially true for areas of the world where fertile soil is limited, while food production lags behind the needs.

There is a growing trend of growing food crops without the use of chemical fertilizers and pesticides in soilless culture. This trend, especially developed in Australia and New Zealand, is driven by the public demand for pesticide-free produce and for decreasing the use of oil-derived products (such as chemical fertilizers). Thus, while the requirement of growing crops in soil makes certification of soilless production systems as 'organic' unlikely, the forces that drive the demand for organic products may well force changes to the definition or force the emergence of alternatives that are equally attractive to the general public. Such changes have already been noted with the advent of 'biorational' pesticides which have changed the way integrated pest management (IPM) is currently practised in conventional field agriculture.

Discussion of the crop protection aspect of these soilless production systems is beyond the scope of the present book but it should be noted that with the advent of biological control methodologies, the certification issue may ultimately be solvable. The subject of plant nutrition as part of an 'organic' soilless production system will be briefly described below.

In general, such soilless 'Organic Systems' can be based on either substrate or liquid culture. In principle, the aim of fertigation in both types is to provide completely soluble nutrients, but to have them derived from organic sources, rather than synthetic soluble salts. A variety of commercial products exist in the market, mainly based on water extraction of fermented fish, meat, compost (compost tea) or manure residues to make such liquid fertilizers available. The two main drawbacks of these materials are their extremely high cost per a unit of specific nutrient (especially nitrogen) and the fact that the ratio among the various nutrients is usually different from what plants require. Several attempts have been made to formulate an accurate combination of off-the-shelf organic fertilizers, with various degrees of success (Atkins and Nichols, 2002, 2004; Liedl et al., 2004a,b; Jarecki et al., 2005). Significant work in this area is continuing as evidenced by the frequent appearance of new products by various companies.

Development of substrates that can supply nutrients for extended periods of time is driven by the fact that some compost types can provide much of the needed nutrients for extended periods of time (Dresboll, 2004), while simultaneously suppressing many root pathogens (Raviv, 2005). More nutrients can be supplied once the compost's reservoir

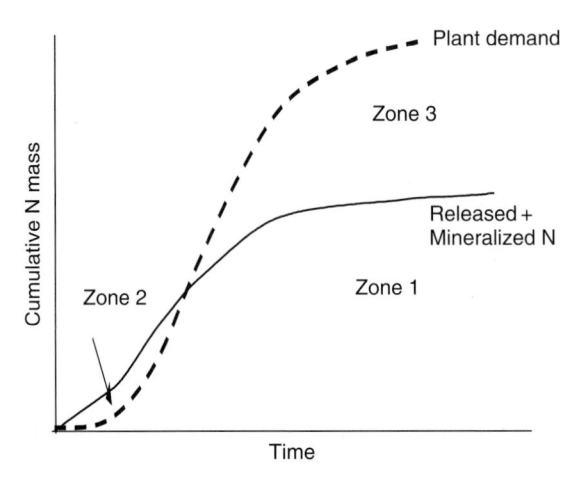

FIGURE 13.3 Graphical presentation of plant demand for nitrogen *vs.* compost mineralization rate (Thanks to Dr Yael Laor).

has been exhausted in the form of guano, feather meal or other concentrated organic fertilizers (Raviv et al., 2005). This is schematically depicted in Fig. 13.3. The difficult part of this nutritional program is, on the one hand, how to minimize nutrient leaching from the medium when the plants are young and their consumption is low (zone 2 in Fig. 13.3) while, on the other hand, accelerating the compost mineralization once the plants attained full size (zone 3), so as to maximize compost-derived nutrient uptake by the plant (zone 1). Nutrient leaching can be minimized by effluent recirculation. The decomposition process itself renders the substrate somewhat unstable (see Chap. 11) so that long-term (more than a year) crop production in such systems may not be feasible. Most concentrated organic fertilizers are very expensive, in comparison with chemical fertilizers. For example, guano-derived nitrogen cost ~$5 per kilogram, more than five times its cost when derived from chemical fertilizers. Compost, on the other hand, is a relatively cheap source of nutrients. However, most of the nitrogen exists within the compost in a recalcitrant form that resists mineralization. The challenge for future researchers is to find ways to accelerate nitrogen mineralization from compost, if possible, in a controlled manner, at times dictated by the grower. The work of Dresboll (2004) elegantly set the stage for this needed effort.

One approach to developing an organic soilless nutrient source is to use a separate reactor in which compost is aerobically brewed to form liquid extract called compost tea. Such systems are already being tested in commercial production, and considerable research is needed to develop methods for sustained uniform delivery of particular nutrients to the plants.

To date, soilless 'organic' production has not reached a significant share of the market, but there is considerable development in this area to attempt to solve the hurdles to making such systems economically feasible. In addition to the economical and professional obstacles mentioned above, the current regulations are the main stumbling block, effectively requiring the development of an entirely new system.

13.2.4 TAILORING PLANTS FOR SOILLESS CULTURE: A CHALLENGE FOR PLANT BREEDERS

In addition to its importance to the understanding of basic root growth processes, the study of the response of roots to various types of physical stress (such as oxygen deficiency and mechanical impedance) has clear horticultural relevance. Although some root traits such as enhanced ability to capture phosphate (Gahoonia et al., 2001; Bates and Lynch, 2002) and the nature of root biomass, length and architecture (McPhee, 2005; Fita et al., 2006) are starting to receive increased interest, in general the intentional selection of particular root traits has been largely neglected. The above examples were studied with field crops such as barley (*Hordeum vulgare* L.), melon (*Cucumis melo* L.) and pea (*Pisum sativum* L.). When more such responses are elucidated, especially under conditions of soilless culture, that is ample water supply and nutrition, better manipulation of root growth responses to the specific characteristics of these growing conditions will be attainable via appropriate plant breeding or genetic engineering technologies. Some examples for root characteristics sought for soilless culture conditions are increased branching and fine roots formation, so as to continuously provide the plant with actively absorbing root surfaces. In view of the increased importance of closed systems, better sodium and chloride exclusion efficiency is a trait of paramount importance. Root systems that can tolerate higher external concentrations of these ions allow the grower to set a higher threshold before effluent discharge becomes mandatory. As explained in Chap. 9, any such small increase leads to a very significant decrease in the required leaching fraction. Similarly, root systems that can tolerate higher osmotic potential in the rhizosphere can lead to the above-mentioned result, while greatly improving water and nutrient use efficiencies.

Another facet of soilless plant production that could be addressed by breeders is the plant's response to root zone temperatures. These are typically higher than the temperatures to which plant roots have adapted through evolution. Accompanying higher temperatures is a lower level of dissolved oxygen, so that selecting for plants with low susceptibility to hypoxia could be particularly advantageous in soilless production.

13.2.5 CHOOSING THE APPROPRIATE MEDIUM, ROOT VOLUME AND GROWING SYSTEM

Choosing the appropriate medium for a specific combination of crop and environmental conditions is the subject of numerous scientific papers and several books (Bunt, 1988; Handreck and Black, 2002; Savvas and Passam, 2002). Kipp et al. (2000) provided detailed tabulated information about substrate's characteristics. Most of the papers dealing with comparisons among substrates or growing systems do not involve explicit economic consideration. Moreover, the results of most of these studies are often affected by arbitrary choices, made by the researchers. For example, if various substrates are irrigated with identical irrigation regimes ('control', usually is based on the 'traditional' medium and is based on time-controlled irrigation), in spite of their differing physical characteristics, an unfair advantage is given to some of them, while others may be irrigated in a frequency that does not match their properties. Generally researchers do not make the extra effort to compare substrates using irrigation frequencies based on physical parameters such as water tension, unsaturated hydraulic conductivity, water content or real-time replenishment of transpired water based on weighing lysimeter (some of the few who did were Raviv et al., 2001; Morel and Michel, 2004; Wallach and Raviv, 2005). The use of measurable plant parameters that reflect plant water status is even less common. Irrigation control using methods such as leaf water potential (with pressure chamber); stomatal resistance for water (with diffusion porometer); canopy temperature (with thermocouples or infrared thermometers); flow of water in the stem (with the heat pulse method); and changes in stem diameter (with dendrometer) can be found in research studies but only rarely in horticultural studies. Caron et al. (1998) recommended the use of plant-based measures such as xylem water potential to assess water availability and to control irrigation. However, these methods have not gained popularity in commercial settings. As a general rule it can be stated that reports aimed at comparing media while using uniform irrigation treatments have only limited usefulness. In addition to avoiding biased conclusions, comparative research conducted while fitting the irrigation regime to the medium's physical characteristics also provides relevant information about optimal irrigation management.

Another factor that may mask media effect is their potential contribution (positive or negative) to nutrient availability. As described in Chap. 6, various media have different rates of absorption and release of nutrients. If this is not taken into account while comparing growing media, false conclusions may be reached. Unlike the case of irrigation regime, overcoming this heterogeneity among substrates is rather difficult since real-time monitoring of all nutrients is not practical even under experimental conditions.

Unfortunately we cannot suggest clear answers about the optimal choice of media for specific crop/conditions combinations except to caution the readers about the nature of this kind of pitfall. Researchers should not justify such experimental design based on the fact that such flawed method is frequently found in the scientific literature. Historically, when a large number of growers adopt a specific medium (e.g. stone wool in Northern Europe) it is accompanied by a wide assortment of peripheral elements (structural, knowledge and customized instrumentation) that constitute a substantial economic investment. This results in an inhibition to replace one medium by a newer one, even if proved as somewhat superior. Adoption of a new substrate is more common when a substrate must be replaced occasionally, when a new industry emerges, or when the use of a certain substrate is not free of problems. An example of the former is the replacement by many growers of peat moss by coir due to the fact that peat moss shrinks over the years requiring to be either top-dressed or replaced (Scagel, 2003, and see Chap. 11).

Economics and availability play a major role in substrate component selection. Tuff, for example, is readily available in Israel and in the Canary Islands (under the name Picon) and its cost in these countries and their neighbours is competitive with other substrates; therefore it found a wide use in these countries (Plaut et al., 1973; Perez-Melian et al., 1977; Silber et al., 1994). Similarly, bark compost is widely used in temperate countries, where forestry is a major industry and a large source of wood waste (Pokorny, 1982). In some cases, the need to find solutions for used media lead to the development of yet another new media by their recycling. An example is the incorporation of recycled stone wool as a component in peat-based media (Riga et al., 2003).

As mentioned above, root volume rarely affects shoot/root ratio (NeSmith et al., 1992). It can be assumed that shoot growth is regulated by root growth (Hoffmann and Jungk, 1995; Eshel et al., 2001) while, in an interactive manner, root growth is regulated by the shoot through carbohydrate supply. Modern fertigation techniques enabled growers to save costly greenhouse space and to reduce the expense of substrates by growing plants in small root volumes. High growth rate and high shoot/root ratio are common under these conditions. There is a limit, however, to the shoot/root ratio, as at a certain point the physiological ability of the root to take up water and nutrients to meet the demands of the shoot becomes a limiting factor. In well-controlled experiments, root restriction has been shown to limit plant growth in various species, as manifested in reductions of such growth parameters as leaf area, leaf number, plant height and biomass production (Cooper, 1972; Richards and Rowe, 1977; Carmi and Heuer, 1981; Carmi et al., 1983; Carmi and van Staden, 1983; Ruff et al., 1987; Robbins and Pharr, 1988; Peterson and Krizek, 1992; Choi et al., 1997; van Iersel, 1997; Karni et al., 2000; Haver and Schuch, 2001; Xu et al., 2001; Dominguez-Lerena et al., 2006). In many cases, the initial volume in which transplants are grown affects their performance after transplanting (Bar-Tal et al., 1990; di Benedetto and Klasman, 2004; Dominguez-Lerena et al., 2006). The volume in which the root can expand may affect plant growth either through plant nutrition and transpiration (Brouwer and de Wit, 1968; Hameed et al., 1987; Bar-Tal et al., 1990; Choi et al., 1997) or via root and shoot physiology (Aung, 1974; Jackson, 1993; Haver and Schuch, 2001). In fact, in some cases, the effects of root restriction are not implemented through nutrient deficiency (Carmi and Heuer, 1981; Robbins and Pharr, 1988; Izaguirre-Mayoral and de Mallorca, 1999) or water stress (Krizek et al., 1985; Ruff et al., 1987; Izaguirre-Mayoral and de Mallorca, 1999). In such cases it has been suggested that root restriction stress induces a reduction in the supply of growth substances from roots to shoots and an imbalance in root and shoot hormones (Carmi and Heuer, 1981; Peterson et al., 1991a,b; van Iersel, 1997; Hurley and Rowarth, 1999; Haver and Schuch, 2001). Peterson et al. (1991a,b) showed that root respiration was reduced by root confinement, and concluded that the decline in root respiration capacity represents a decline in root metabolism. Haver and Schuch (2001) found that severe root restriction caused apical dominance that was associated with low ethylene production by the root. Moreover, they showed that exposure of roots to ethylene overcame the apical dominance induced by root restriction. The transfer of growth substances from roots to shoots, hormonal balance and root metabolism are out of the scope of this chapter and of the main subject of this book and will not be discussed further. However, when the crop in question is a finished plant of well-balanced and compact dimensions, a relatively large root volume is advisable.

It has been shown that the depth of the growing container has a more significant effect than the volume on successful plant development (Dominguez-Lerena et al., 2006). Generally the deeper the medium depth, the larger is the ratio of air to water spaces in it. In growing media of all depths, there will be a saturated zone after drainage at the bottom due to the formation of perched water table. In shallow media, this saturated zone constitutes a larger percentage of the total root zone volume, thus affecting air/water relations, possibly leading to water logging problems. However, this is not always true. For example, when the development of a fibrous root system of tree transplants is required for subsequent optimal development after transplanting, shallow and wide containers are considered preferable (Milbocker, 1991).

The minimal desirable root-zone volume is usually determined by the grower before starting the crop. Even in cases where one or more transplanting stages are planned, it is essential to plan them in advance, to ensure the crop's success. Factors that affect final required root-zone volume include the type of crop, its growing duration and season, nature of the chosen substrate, economical consideration such as the cost of various containers, substrate and greenhouse area ('rent'). Logistical considerations, such as transportation costs, may also affect the decision in cases where the potted plant itself is the final product. Some typical examples are listed in Table 13.1.

TABLE 13.1 Typical Examples of Required Size of Root Volumes for Various Crops

Crop type	Growing duration	Special considerations	Standard volume/plant	Reference
Bedding plants and vegetable transplants	<60 days	Densely grown	<125 cm^3	Kemble et al., 1994; di Benedetto and Klasman, 2004
Lettuce and other leafy vegetables	<60 days from transplanting		<0.5 l	Gysi and Allmen, 1997
Small-size pot plants	3–6 months	Transportation required	0.5–2.0 l	
Medium-size pot pants	6–9 months	Transportation required	1–4 l	
Large-size pot plants	1–2 years	Transportation required	10–20 l	
Short-season vegetables and herbs	2–6 months		2–5 l	Baas et al., 2001
Long season cut flowers (gerbera, carnations)	9–12 months		3–6 l	
Long season vegetables (tomato, pepper)	9–12 months		~10 l	Tuzel et al., 2001
Long season vegetables (tomato, pepper)	9–12 months	Hydroponics	1–2 l	Peterson et al., 1991a
Perennial cut flowers (rose)	3–5 years		5–10 l	Raviv et al., 1999; Raviv et al., 2001
Tree saplings	one year	Transportation required	~5 l	Salifu et al., 2006

One common practical consideration for growers that dictates the root-zone volume is the interval between irrigation events. While it would be ideal to tailor the irrigation schedule to the needs ot the plants, many growers set the irrigation schedule for convenience or to take advantage of available capacity for irrigation. In such circumstances, the root-zone volume should be matched to this interval to assure that the plant does not run out of water.

13.3 MANAGEMENT OF SOILLESS PRODUCTION SYSTEMS

13.3.1 INTERRELATIONSHIPS AMONG VARIOUS OPERATIONAL PARAMETERS

One thing that should be clear from this book is that the large number of variables that interact to form the production system make it nearly impossible for humans to consider (much less to optimize) all these factors simultaneously. As such there is still a lot of room in the marketplace for growing media formulation that are specifically suited for particular applications. Much of what is available today is really a raw material that still requires substantial engineering by the end-user to create an ideal growing situation for plants.

It is important to note that growing media of today are not simply haphazard combinations of a few convenient media components; rather they are elements of engineered horticultural systems tested by marketing forces and adapted by horticulturalists to particular purposes.

All media provide some plant anchorage, while nutrients are often provided by added fertilization. Water and air are provided in the pore spaces of the media. The principal factors affecting air and water status in the growing media are: (1) the media components and ratios; (2) media depth; (3) media handling; and (4) watering practices.

Each of the generally used media components is unique in their properties pertaining to their air/water relations. For example, peat has a comparatively higher content of unavailable water at a given matric tension compared to vermiculite. This variability in the availability of water in different types of media components means no two components are exactly alike in terms of providing water to plants. Similar variability does exist among various media components with regards to wettability (i.e. the ability of dry media to rapidly absorb water when moistened) or drainage characteristics. Airhart et al. (1978) reported that an air dry peat–vermiculite substrate required 5 days to reach 70–78 per cent of moisture saturation while milled pine bark required 48 days to achieve 58–78 per cent saturation. Thus, the kind and ratio of the components in the media mixes have important bearings on the hydrophysical environment in the root zone, which in turn largely determines the performance of the plants grown thereon. In general, moisture retention characteristics of media mixes is a compromise of its individual component corresponding characteristics. However, even when the same components are blended in identical ratios, physical properties may vary due to difference in particle size (Bilderback et al., 2005). Pokorny and Henny (1984b) reported that substrates with the same components and ratio were not identical in their physical properties, even though they were assumed to be the same. Differences in physical properties in their study were attributed to the shrinkage and particle-size differences of the components blended. Variation in identical components occurs if dry components are blended compared to blending moist components. Dry components when mixed tend to fit together tightly and increase bulk density of the substrate compared to when moist components are blended; consequently, air-filled porosity is reduced.

In handling growing media, proper care is necessary to avoid compaction; air space can be drastically reduced by compaction. The moisture content of the media prior to filling containers or bags may also be important. For example, adding water to soilless media mixes filling plug trays causes the media to swell and helps create more aeration. Moistening the media before filling larger containers does not have much benefit.

Apart from the air/water relations of the media mixes imparted by the individual and interactive effects of their components, watering practices have a tremendous influence on their air/water relations. This is due to the fact that both water and air compete for the same pore spaces in the media, where they are nearly mutually exclusive.

13.3.2 DYNAMIC NATURE OF THE SOILLESS ROOT ZONE

Soilless production has been in transition ever since it was conceived (see Chap. 1). Initially this type of production was carried out by mimicking traditional methods based on production in soil. The wide array of soils on the planet has led farmers to adapt methods so as to deal with the many physical and chemical properties that abound. The production methods in highly porous soils were assumed to be an excellent initial step for managing production in soilless growing media. However, as more and more artificial components were tested and drawn into use, the physical and chemical properties of growing media became more and more different from soils, so that the methods required to be commercially successful changed dramatically. Yet even now, most practitioners view the key elements of soilless production to be driven by water in the limited root zone and avoidance of stress related to water status and plant nutrition.

While this stress-avoidance approach has been successful in attaining excellent commercial results with soilless production, the growth of the industry has resulted in strong commercial competition which continues to push growers to achieve greater efficiencies. This has resulted in the various specialized greenhouse production systems described in Chap. 5 as well as a huge outdoor nursery industry. As a segment of agriculture, it continues to increase both in size and in percentage. If one adds to this the fact that virtually every home owner owns and pampers a number of potted plants, it becomes clear that there is considerable interest in managing the root zone for objectives ranging from survival or stress avoidance to maximal growth.

While the irrigation and fertilization factors described above are certainly ones that cannot be ignored, there are a multitude of other factors that are of equal importance in soilless production. The ones that must be taken into account are those which have the capacity to change rapidly, and especially where the rate of change towards a suboptimal condition can occur within a few days under conditions that may readily occur.

As such it is worthwhile to explore the various root-zone factors discussed throughout this book in the context of how readily each can change from a condition that is generally optimal to one that is suboptimal (or vice versa). Ideally a quantitative approach would be best here, but the numerous circumstances in the real world make this impossible. For this discussion, it is important to separate out effects that occur over the course of a day versus ones that occur faster or slower. There is no need to consider a time frame longer than one week in soilless production since there are no root zone factors that do not have the capacity to change on at least a weekly basis. And therein lies an operational conclusion: while some factors in soils take multiple weeks to change significantly, every factor in soilless production has the capacity to change much more rapidly, even with substrates that are highly buffered.

The major consequence of this capacity for rapid change is that control must involve accurate and rapid sensing of all these conditions in conjunction with implementation of mitigating measures to keep the system as homeostatic as possible. In cases where accurate and effective control is not feasible at the timescales needed, we still attempt to implement some sort of buffering to slow the rapidity of change to a manageable level. In fact, the sole reason why we seek well-aerated media is to provide a passively buffered oxygen situation by allowing oxygen to diffuse into the root zone and ethylene and CO_2 to leave. If we were to force bulk movement of these gases (as we force bulk movement of all other materials through irrigation) then the aeration buffering provided by well-aerated materials would be unnecessary. In fact, in water culture methods, we generally force oxygen into the water by either bubbling air or spraying, raising the dissolved oxygen level to saturation.

A major difference between systems that are operated as open versus closed or semi-closed systems is in the moisture regime. While moisture content can change with equal rapidity in both, closed systems are generally characterized by more frequent irrigations, preventing moisture levels from ever declining to levels that would induce even hidden moisture stress in the plants. Thus in a mixture that combines an inert medium such as sand or perlite with one or two organic substrates, the moisture conditions will typically decline to levels that are dryer in open systems than in closed systems. As was pointed out in Chap. 3, one consequence of this is that the unsaturated hydraulic conductivity in open systems is likely to be much lower at times than it is in closed systems. Thus irrigation in the two systems leads to more rapid change in bulk movement of water in closed systems so that during normal operation, the practitioner can count on lateral bulk movement of water to be faster in closed systems than in open systems. Since this bulk movement carries with it dissolved gasses and nutrients, the net flow of these into plants is typically faster in closed systems, even without consideration of the bulk movement due to the more frequent irrigation events.

One stark contrast between soilless and soil-based production is the rapidity in soilless substrates of change during irrigation events in relation to the rates of change between irrigation events. Clearly it is possible to induce very significant rapid change in most of the root-zone variables with irrigation. In fact, one logical conclusion is that dissolved oxygen concentration can be raised significantly through an irrigation event. Thus noting that oxygen declines at an hourly pace between irrigation events in both open and closed systems, a logical conclusion is that irrigation on an hourly basis (practiced in many greenhouse production settings) prevents hypoxic conditions. The same is, of course, true for various nutrients, but the rate of decline to deficiency conditions typically takes longer for nutrients than for oxygen. In many soilless production techniques involving stone wool, irrigation intervals of less than 2 h are needed even when there is still ample fertilizer content in the root-zone solution. It can be argued that the current practices have stabilized around a set of water and nutrient supply recommendations that match this time interval and that it is dictated not by water and nutrients, but rather by oxygen.

In large part there is still a lot that is not known about oxygen in the root zone. Some studies have been carried out but scientific research was hampered by lack of sensors that could be used to make *in situ* measurements at specific points in the root zone of plants growing in soilless growing media. Most such methods in the past have involved taking a sample (which may well result in changing the oxygen concentration of the sample) or measuring a flowing liquid as it passes through a membrane. Most early methods used membrane technology that was consumptive, so that very small changes could not be detected because they were smaller than the oxygen utilization of the instrumentation. Recently new spectrophotometric devices have emerged (Wang et al., 1999) which allow non-consumptive measurement of dissolved oxygen at the tip of a probe that can be inserted in liquid or solid substrates. This has led to the finding that the oxygen concentration in the root zone of plants in soilless substrates is quite variable within the root zone and that it changes very rapidly. We have measured root zones of roses in 3–5-year-old coir where the oxygen concentration declined in the middle of the root zone from 8 ppm (near saturation) to less than 0.5 ppm over a period of 2 h (Flannery, 2007). The degradation level of this particular substrate was not thought to be an issue, but carrying out the same test with brand new coir resulted in diffusion of oxygen into the root zone that was fast enough to compensate for oxygen consumption. It may also be possible that this rapid decline in old coir might have been due to the microbial activity. Regardless of the cause, the operative conclusion was that irrigation was the only feasible way to restore oxygen to normoxic conditions in this particular root zone.

Earlier in this chapter, it was shown that the confined root zone has in and of itself a significant effect on the plants and their roots. In general, there is a notion that we can talk about the root zone as being of a particular composition that is uniform throughout the root-zone volume. And while this may well be the case to some extent, one should not jump to the conclusion that the root zone is entirely uniform. In fact, there are significant gradients from top to bottom and from inside to outside, as well as from surfaces on which the sun is shining to those in the shade. From the standpoint of water, there may be a perched water table at the base with declining wetness levels with height. In substrates with steep moisture retention curves, this can mean that the root zone is too wet at the bottom and too dry at the top for optimal root function. And for all substrates, this dictates the minimum and maximum depth to which the substrate can be used. Temperature gradients are quite common with the surface of the substrate being involved in active evaporative cooling even as the sun shining on the black plastic side of the container heats the nearby substrate to temperature that are too hot for humans to handle. Clearly, there is extensive heterogeneity with temperature and this will lead to substantial variation in dissolved oxygen concentration because oxygen solubility ranges from hypoxic to normoxic levels over the temperature range that roots of container-grown plants experience. These temperatures are also linked to the rate of microbial activity and population growth, further contributing to the volatility of the situation with regard to any element for which these microbes compete with the growing plant. It was shown in Chaps. 8 and 9 that utilization of nutrients involves exchange processes that affect pH and EC. Yet despite these dramatic changes in temperature, oxygen, pH, EC and moisture content, the *status quo* in the industry today does not involve any attempt to sense these variables. Clearly technical developments are needed in this arena so as to develop sensor-based methods for optimizing the root zone.

In fact, closed production systems are really only sustainable by measuring these variables dynamically. Greenhouse soilless production of vegetables is already innovating in this area (see Chap. 5) and there is a significant growth in the use of sensors and instrumentation, coupled with decision-support software, to attempt to stabilize optimal conditions for the plants. For plants growing in liquid culture, this type of control is already in use but even here, commercial settings still rely mostly on buffering by involving a lot of liquid in the system. As such, DFT, with its large amount of water, has less rapid changes (i.e. is more stable) than aeroponics, which is in turn more stable than NFT. As was pointed out in Chap. 5, NFT has not emerged as a widespread commercial system, due to this instability. As we develop instrumentation and control technologies that can replace buffering with active manipulation, we may well find that NFT can be feasible in commercial settings.

Another facet that is frequently ignored by the practitioner is the presence and activity of microbes. In fact, the prevailing notion in commercial production is that it is best to eradicate these as they are viewed to be in competition with or even antagonists to the plants. Thus there is nearly indiscriminate use of fungicides with the purpose of eradicating pathogens, with little consideration as to whether the effect to beneficial microorganisms might not result in greater problems. Chapters 10 and 11 in this book touched on the concept of disease suppressiveness in substrates that is mediated or imparted by micro-organisms. The fact that the complete absence of microbes is decidedly sub-optimal has been well known by practitioners; many will chose to pasteurize rather than sterilize growing media for this reason; or an attempt will be made to re-inoculate sterilized growing media so as to prevent the 'biological vacuum'.

We now see further evidence that we must treat the microbes as partners in production, rather than tolerable antagonists. In Chap. 9, it is shown that as systems evolve to more closed systems, we must account for the activity of microbes in the conversion of nitrogen forms and that such conversion plays a major role in the dynamics of the pH. Considerably more research is needed to assist growers with methods that manage this pH. Research in this area needs to evolve so that the microbial population is not treated as a difficult unknown, but rather as a set of dynamic populations. We do not yet know what levels of these populations are optimal, exactly which species are ideal partners for soilless plant production, and how to control the root-zone variables so that these subsystems are managed in an ideal fashion. The two chapters in this book that deal with this issue present the state of the art in this area.

There are, of course, various other horticultural actions that subject the plant to sudden changes which translate into rapid changes in the root zone. Soilless culture generally involves conditions where the environment can be controlled dynamically around the plant (e.g. greenhouse production) or where the plant can be moved to another location without disrupting its growth (e.g. container nursery production). Many times such changes are implemented so as to improve lighting conditions for the plants. These actions do have an effect on the plants that translates to the root zone as well. Improving lighting conditions can enhance photosynthetic activity which yields more sugars in the plant for the growth of tissues such as roots. The resulting increase in exudates, microbe activity, nutrient uptake by roots and microbes will result in a more dynamic situation where all variables change more rapidly. Thus a grower needs to be aware that such a horticultural action will have an impact and may require more attention to pH and fertigation management. Also, such a change may result in more rapid increase in leaf biomass and increase the need for air movement around the leaves. This will result with a faster transpiration stream (Chap. 4) which will, in turn, affect water removal from the root zone.

One operational conclusion that can be drawn from this with regard to open systems is that there is a need to maximize transpiration as part of the overall production strategy. This is something that is frequently not understood because our attention to the ever increasing value of water has led us to seek ways to reduce water use by agriculture. However, in soilless production, maximizing transpiration will assure that a steady supply of nutrients reaches the surface of the roots of plants growing in substrate (as opposed to liquid culture) and thus will increase water use efficiency (Raviv and Blom, 2001). Without this, the plant would be dependent on diffusion processes or the bulk movement during irrigation events. With diffusion of nutrient molecules being very slow, and irrigation in open systems being relatively infrequent (a few times per week, driven by water use of the plant), the boundary layer surrounding the roots would be mostly in a nutrient-depleted state.

13.3.3 SENSING AND CONTROLLING ROOT-ZONE MAJOR PARAMETERS: PRESENT AND FUTURE

With the advent of the use of computers for control of many processes in agriculture came the possibility of developing production systems that combine sensor technologies and intelligent control to create very specific production systems that marry particular plants (or even specific cultivars) to particular soilless production elements. The chapters in this book dealing with irrigation and fertigation explicitly highlight such systems as being common and dynamically developing to include new sensors, mathematical models and horticultural approaches. For example, integration of technologies to estimate the amounts of each nutrient ion in the root zone is resulting in greater dynamic control over plant nutrition, ultimately allowing greater customization of specific production systems to specific crops.

Much of the technological developments in this area have been driven by opportunity. Moisture sensors are available and significant research has been done to incorporate them into irrigation control systems. Robust-specific ion electrodes for measuring ions in the root zone have been lacking, so that research in this area has been limited and consequently there are currently no widely used control systems that obtain input about specific ions from the root zone. Once robust versions of such sensors become available, they will be integrated into production systems to dynamically optimize plant nutrition. Currently, some prototype systems exist, but these measure the leachate of soilless production systems rather than the status of the root zone directly.

Similarly, sensing dissolved oxygen concentration in the root zone has been (and continues to be) very difficult so that using this variable as a control variable for manipulating the root zone is as yet not possible. As such there are currently no irrigation systems (or aeration systems) that explicitly control oxygen in the root zone. Therefore our current recommendations involve the use of substrates with high aeration. Should it become feasible to monitor and control oxygen concentration directly, then this recommendation may well change, possibly resulting in new and revolutionary soilless production systems.

While opportunistic exploitation of sensor and controller technology is important in advancing horticultural technology, it should be noted that soilless agricultural production has advanced to such a point, with ever-increasing percentage as part of agriculture, that it is now reasonable for research agencies to explicitly set research priorities to focus on developing sensor technologies that can be incorporated specifically in soilless production systems. In fact, throughout agriculture there are numerous opportunities for advances in 'Precision Agriculture'. For soilless crop production, development of sensor technologies in conjunction with simultaneous control (and optimization) of moisture, nutrients and oxygen are likely to pay large dividends, pushing production per unit production area to efficiencies never seen before. It could be argued that if human population continues to grow as it has been, and especially in areas that are prime agricultural production areas, then soilless precision agriculture will be needed to avert food shortages.

Thus, one of the recommendations that can be extracted from this information is that horticultural engineering has significant opportunities in the area of integration of sensor technology with dynamic control over the root-zone variables that are the primary constraints to productivity. Over the past 50 years, improvements in this control have resulted in a variety of production systems, each more productive than the previous generation. It is important to note that significant advances in this area are yet possible.

REFERENCES

Airhart, D.L., Natarella, N.J. and Pokorny, E.A. (1978). Influence of initial moisture content on the wettability of pine bark medium. *Hort. Science*, **13**, 432–434.

Atkins, K. and Nichols, M. (2002). Organic hydroponics: The Massey experience. *Grower*, **57**(5), 27–32.

Atkins, K. and Nichols, M. (2004). Organic hydroponics. *Acta Hort.* (ISHS), **648**, 121–127.
Aung, H.L. (1974). Root-shoot relationships. In *The Plant Root and its Environment.* (E.W. Carson, ed.). Charlottesville: University Virginia, pp. 29–61.
Baas, R., Wever, G., Koolen, A.J., et al. (2001). Oxygen supply and consumption in soilless culture: Evaluation of an oxygen simulation model for cucumber. *Acta Hort.* (ISHS), **554**, 157–164.
Bar-Tal, A., Bar-Yosef, B. and Kafkafi, U. (1990). Pepper transplant response to root volume and nutrition in the nursery. *Agron. J.*, **82**, 989–995.
Bar-Tal, A., Bar-Yosef, B. and Kafkafi, U. (1993). Modeling pepper seedling growth and nutrient uptake as a function of cultural conditions. *Agron. J.*, **85**, 718–724.
Bar-Tal, A., Feigin, A., Rylski, I. and Pressman, E. (1994). Root pruning and N-NO$_3$ solution concentration effects on tomato plant growth and fruit yield. *Sci. Hortic.*, **58**, 91–103.
Bar-Tal, A., Feigin, A., Sheinfeld, S., et al. (1995). Root restriction and N-N-NO$_3$ solution concentration effects on nutrient uptake, transpiration and dry matter production of tomato. *Sci. Hortic.*, **63**, 195–208.
Bar-Tal, A. and Pressman, E. (1996). Root restriction and K and Ca solution concentration affect dry matter production, cation uptake, and blossom end rot in greenhouse tomato. *J. Am. Soc. Hortic. Sci.*, **121**, 649–655.
Bar-Yosef, B., Matan, E., Levkovitz, I., et al. (1992). Response of two cultivars of greenhouse tomato (F 144 and 175) to fertilization in the Besor area. Final Report 301–170–92, submitted to the Chief Scientist of the Ministry of Agriculture, Israel (in Hebrew).
Bar-Yosef, B., Schwartz, S., Markovich, T., et al. (1988). Effect of root volume and nitrate solution concentration on growth, fruit yield, and temporal N and water uptake rates by apple trees. *Plant Soil*, **107**, 49–56.
Bates, T.R. and Lynch, J.P. (2002). Root hairs confer a competitive advantage under low phosphorus availability. *Plant Soil*, **236**, 243–250.
Bengough, A.G., Bransby, M.F., Hans, J., et al. (2006). Root responses to soil physical conditions: Growth dynamics from field to cell. *J. Exp. Bot.*, **57**, 437–447.
Bilderback, T.E., Warren, S.L., Owen, J.S. Jr. and Albano, J.P. (2005). Healthy substrates need physicals too. *HortTechnology*, **5**, 747–751.
Boland, A.M., Jerie, P.H., Mitchell, P.D., et al. (2000). Long-term effects of restricted root volume and regulated deficit irrigation on peach: I. Growth and mineral nutrition. *J. Am. Soc. Hortic. Sci.*, **125**(1), 135–142.
Brouwer, R. and de Wit, C.T. (1968). A simulation model of plant growth with special attention to root growth and its consequences. In *Root Growth* (W.J. Whittington, ed.). Proceeding 15th Easter School in Agr. Sci., University of Nottingham, London, UK: Butterworths, pp. 224–242.
Brown, E.F. and Pokorny, F.A. (1975). Physical and chemical properties of media composed of milled pine bark and sand. *J. Am. Soc. Hortic. Sci.*, **100**, 119–121.
Bunt, A.C. (1988). *Media and Mixes for Container-grown Plants.* London: Unwin Hayman, 309pp.
Bures, S., Landau, D.P., Ferrenberg, A.M. and Pokorny, F.A. (1993a). Monte Carlo computer simulation in horticulture: A model for container media characterization. *Hort. Science*, **28**, 1074–1078.
Bures, S., Pokorny, F.A., Landau, D.P. and Ferrenberg, A.M. (1993b). Computer simulation of volume shrinkage after mixing container media components. *J. Am. Soc. Hortic. Sci.*, **118**, 757–761.
Carmi, A., Hesketh, J.D, Enos, W.T. and Peters, D.B. (1983). Interrelationships between shoot growth and photosynthesis as affected by root growth restriction. *Photosynthetica*, **17**, 240–245.
Carmi, A. and Heuer, B. (1981). The role of roots in control of bean shoot growth. *Ann. Bot.*, **48**, 519–527.
Carmi, A. and van Staden, J. (1983). The role of roots in regulating the growth rate and cytokinin content in leaves. *Plant Physiol.*, **73**, 76–78.
Caron, J., Xu, H.L., Barnier, P.Y., Duchesne, I. and Tardif, P. (1998). Water availability in three artificial substrates during Prunus X cistena growth: Variable threshold values. *J. Am. Soc. Hortic. Sci.*, **123**, 931–936.
Choi, J.H., Chung, G.C., Suh, S.R., et al. (1997). Suppression of calcium transport to shoots by root restriction in tomato plants. *Plant Cell Physiol.*, **38**, 495–498.
Cooper, A.J. (1972). The influence of container volume, solution concentration, pH and aeration on dry matter partition by tomato plants in water culture. *J. Hortic. Sci.*, **47**, 341–347.

Croser, C., Bengough, A.G. and Pritchard, J. (1999). The effect of mechanical impedance on root growth in pea (*Pisum sativum*). I. Rates of cell flux, mitosis, and strain during recovery. *Physiol. Plantarum.*, **107**, 277–286.

de Willigen, P. and van Noordwijk, M. (1987). *Roots, Plant Production and Nutrient Use Efficiency*. Ph.D. Thesis Agricultural University Wageningen, the Netherlands, p. 282.

di Benedetto, A.H. and Klasman, R. (2004). The effect of plug cell volume on the post-transplant growth for *Impatiens walleriana* pot plant. *Europ. J. Hortic. Sci.*, **69**, 82–86.

Dominguez-Lerena, S., Herrero-Sierra, N., Carrasco-Manzano, I., et al. (2006). Container characteristics influence *Pinus pinea* seedling development in the nursery and field. *For. Ecol. Manage.*, **221**, 63–71.

Dresboll, D.B. (2004). *Optimisation of Growing Media for Organic Greenhouse Production*. Ph.D. dissertation submitted to The Royal Veterinary and Agricultural University, Copenhagen, Denmark, 115pp.

Edwards, J.H. and Barber, S.A. (1976). Nitrogen flux into corn roots as influenced by shoot requirement. *Agron. J.*, **68**, 973–975.

Eshel, A., Srinivas Rao, Ch., Benzioni, A. and Waisel, Y. (2001). Allometric relationships in young seedlings of Faba bean (*Vicia faba* L.) following removal of certain root types. *Plant Soil*, **233**, 161–166.

Fita, A., Pico, B. and Nuez, F. (2006). Implications of the genetics of root structure in melon breeding. *J. Am. Soc. Hortic. Sci.*, **131**, 372–379.

Flannery, R.J. (2007). Oxygen dynamics in the root zone of hydroponically-grown rose and chrysanthemum Ph.D. Dissertation, University of California Davis.

Gahoonia, T.S., Nielsen, N.E., Joshi, P.A. and Jahoor, A. (2001). A root hairless barley mutant for elucidating genetic of root hairs and phosphorus uptake. *Plant Soil*, **235**, 211–219.

Gysi, C. and von Allmen, F. (1997). Lamb's lettuce in different growing media in closed recirculating soilless culture. *Acta Hort.* (ISHS), **450**, 155–159.

Hameed, M.A., Reid, J.B. and Rowe, R.N. (1987). Root confinement and its effects on the water relations, growth and assimilate partitioning of tomato (*Lycopersicon esculentum* Mill). *Ann. Bot.*, **59**, 685–692.

Handreck, K.A. and Black, N.D. (2002). *Growing Media for Ornamental Plants and Turf* (3rd edn). Sydney: University of New South Wales Press, 542pp.

Haver, D. and Schuch, U. (2001). Influence of root restriction and ethylene exposure on apical dominance of petunia (*Petunia x hybrida* Hort. Vilm.-Andr.). *Plant Growth Regul.*, **35**, 187–196.

Hoffmann, C. and Jungk, A. (1995). Growth and phosphorus supply of sugar beet as affected by soil compaction and water tension. *Plant Soil*, **176**, 15–25.

Hurley, M.B. and Rowarth, J.S. (1999). Resistance to root growth and changes in the concentrations of ABA within the root and xylem sap during root-restriction stress. *J. Exp. Bot.*, **50**, 799–804.

Izaguirre-Mayoral, M. and de Mallorca, M.S. (1999). Responses of rhizobium-inoculated and nitrogen-supplied *Phaseolus vulgaris* and *Vigna unguiculata* plants to root volume restriction. *Aust. J. Plant Physiol.*, **26**, 613–623.

Jackson, M.B. (1993). Are plant hormones involved in root to shoot communication? *Adv. Bot. Res.*, **19**, 104–187.

Jarecki, M.K., Chong, C. and Voroney, R.P. (2005). Evaluation of compost leachates for plant growth in hydroponic culture. *J. Plant Nutr.*, **28**, 651–667.

Jungk, A. (1974). Phosphate uptake characteristics of intact root systems in nutrient solution as affected by plant species, age and P supply. In *Plant Analysis and Fertilizer Problems*. Proceeding of the 7th International Colloquium, **1**, 185–196.

Jungk, A. and Barber, S.A. (1975). Plant age and the phosphorus uptake characteristics of trimmed and untrimmed corn root systems. *Plant Soil*, **42**, 227–239.

Kafkafi, U. (1990). Root temperature, concentration and the ratio of NO_3^-/NH_4^+ effect on plant development. *J. Plant Nutr.*, **13**, 1291–1306.

Karni, L., Aloni, B., Bar-Tal, A., et al. (2000). The effect of root restriction on the incidence of blossom-end rot in bell pepper (*Capsicum annuum* L.). *J. Hortic. Sci. and Biotech.*, **75**, 364–369.

Keever, G.J., Cobb, G.S. and Reed, R.B. (1985). Effects of container dimension and volume on growth of three woody ornamentals. *Hort. Science*, **20**(2), 276–278.

Kemble, J.M., Davis, J.M., Gardner, R.G. and Sanders, D.C. (1994). Spacing, root cell volume, and age affect production and economics of compact-growth-habit tomatoes. *Hort. Science*, **29**, 1460–1464.

Kent, M.W. and Reed, D.W. (1996). Nitrogen nutrition of New Guinea impatiens 'Barbados' and Spathiphyllum 'Petite' in a subirrigation system. *J. Am. Soc. Hortic. Sci.*, **121**, 816–819.

Kipp, J.A., Wever, G. and de Kreij, C. (2000). *International Substrate Manual*. The Netherlands: Elsevier, 94pp.

Krizek, D.L., Carmi, A., Mirecki, R.M., et al. (1985). Comparative effects of soil moisture stress and restricted root zone volume on morphogenetic and physiological responses of soybean [*Glycine max* (L.) Merr.]. *J. Exp. Bot.*, **36**, 25–38.

Liedl, B.E., Cummins, M., Young, A., et al. (2004a). Liquid effluent from poultry waste bioremediation as a potential nutrient source for hydroponic tomato production. *Acta Hort.* (ISHS), **659**, 647–652.

Liedl, B.E., Cummins, M., Young, A., et al. (2004b). Hydroponic lettuce production using liquid effluent from poultry waste bioremediation as a nutrient source. *Acta Hort.* (ISHS), **659**, 721–728.

Martin, C.A., Ingram, D.L. and Nell, T.A. (1991). Growth and photosynthesis of Magnolia grandiflora 'St Mary' in response to constant and increased container volume. *J. Am. Soc. Hortic. Sci.*, **116**, 439–445.

Martinez, F.X., Canameras, N., Fabregas, F.X., et al. (1993). Water exploitation in substrate profile during culture. *Acta Hort.* (ISHS), **342**, 279–286.

Massa, G.D. and Gilroy, S. (2003). Touch modulates gravity sensing to regulate the growth of primary roots of *Arabidopsis thaliana*. *Plant J.*, **33**, 435–445.

McPhee, K. (2005). Variation for seedling root architecture in the core collection of pea germplasm. *Crop Sci.*, **45**, 1758–1763.

Milbocker, D.C. (1991). Low-profile containers for nursery-grown trees. *Hort. Science*, **26**, 261–263.

Morel, P. and Michel, J.C. (2004). Control of the moisture content of growing media by time domain reflectometry (TDR). *Agronomie*, **24**, 275–279.

NeSmith, D.S., Bridges, D.C. and Barbour, J.C. (1992). Bell pepper responses to root restriction. *J. Plant Nutr.*, **15**(12), 2763–2776.

Perez-Melian, G., Luque-Escalona, A. and Steiner, A.A. (1977). Leaf analysis as a diagnosis of nutritional deficiency or excess in the soilless culture of lettuce. *Plant Soil*, **48**, 259–267.

Peterson, T.A. and Krizek, D.T. (1992). A flow-through hydroponic system for the study of root restriction. *J. Plant Nutr.*, **15**, 893–911.

Peterson, T.A., Reinsel, M.D. and Krizek, D.T. (1991a). Tomato (*Lycopersicon esculentum* Mill., cv. 'Better Bush') plant response to root restriction. 1. Alteration of plant morphology. *J. Exp. Bot.*, **42**, 1233–1240.

Peterson, T.A., Reinsel, M.D. and Krizek, D.T. (1991b). Tomato (*Lycopersicon esculentum* Mill., cv. 'Better Bush') plant response to root restriction. 2. Root respiration and ethylene generation. *J. Exp. Bot.*, **42**, 1241–1249.

Plaut, Z., Zieslin, N. and Arnon, I. (1973). The influence of moisture regime on greenhouse rose production in various growth media. *Sci. Hortic.*, **1**, 239–250.

Pokorny, F.A. (1982). Horticultural uses of bark: A bibliography. Research Report. College of Agric., Experiment Station Georgia University, No. 402.

Pokorny, F.A., Gibson, P.G. and Dunavent, M.G. (1986). Prediction of bulk density of pine bark and/or sand potting media from laboratory analyses of individual components. *J. Am. Soc. Hortic. Sci.*, **111**, 8–11.

Pokorny, F.A. and Henny, B.K. (1984a). Construction of a milled pine bark and sand potting medium from component particles. I. Bulk density: A tool for predicting component volumes. *J. Am. Soc. Hortic. Sci.*, **109**, 770–773.

Pokorny, F.A. and Henny, B.K. (1984b). Construction of a milled pine bark and sand potting medium from component particles: II. Medium synthesis. *J. Am. Soc. Hortic. Sci.*, **109**, 774–776.

Ran, Y., Bar-Yosef, B. and Erez, A. (1992). Root volume influence on dry matter production and partitioning as related to nitrogen and water uptake rates by peach trees. *J. Plant Nutr.*, **15**, 713–726.

Raviv, M. (2005). Production of high-quality composts for horticultural purposes – a mini-review. *HortTechnology*, **15**, 52–57.

Raviv, M. and Blom, T. (2001). The effect of water availability and quality on photosynthesis and productivity of soilless-grown cut roses. *Sci. Hortic.*, **88**, 257–276.

Raviv, M., Lieth, J.H., Burger, D.W. and Wallach, R. (2001). Optimization of transpiration and potential growth rates of 'Kardinal' rose with respect to root-zone physical properties. *J. Am. Soc. Hortic. Sci.*, **126**, 638–645.

Raviv, M., Oka, Y., Katan, J., et al. (2005). High-nitrogen compost as a medium for organic container-grown crops. *Bioresour. Technol.*, **96**, 419–427.

Raviv, M., Wallach, R., Silber, A., et al. (1999). The effect of hydraulic characteristics of volcanic materials on yield of roses grown in soilless culture. *J. Am. Soc. Hortic. Sci.*, **124**, 205–209.

Richards, D. and Rowe, R.N. (1977). Effects of root restriction, root pruning and 6-benzylaminopurine on the growth of peach seedlings. *Ann. Bot.*, **41**, 729–740.

Rieger, M. and Marra, F. (1994). Responses of young peach trees to root confinement. *J. Am. Soc. Hortic. Sci.*, **119**, 223–228.

Riga, P., Alava, S., Uson, A., et al. (2003). Evaluation of recycled rockwool as a component of peat-based mixtures for geranium (*Pelargonium peltatum* L.) production. *J. Hortic. Sci. Biotechnol.*, **78**, 213–218.

Robbins, N.S. and Pharr, D.M. (1988). Effect of restricted root growth on carbohydrate metabolism and whole plant growth of *Cucumis sativus* L. *Plant Physiol.*, **87**, 409–413.

Ruff, M.S., Krizek, D.L., Mirecki, R.M. and Inouye, D.W. (1987). Restricted root zone volume: Influence on growth and development of tomato. *J. Am. Soc. Hortic. Sci.*, **112**, 763–769.

Sadanal, U.S., Sharma1, P., Ortiz, N.C., et al. (2005). Manganese uptake and Mn efficiency of wheat cultivars are related to Mn-uptake kinetics and root growth. *J. Plant Nutr. Soil Sci.*, **168**, 581–589.

Salifu, K.F., Nicodemus, M.A., Jacobs, D.F. and Davis, A.S. (2006). Evaluating chemical indices of growing media for nursery production of Quercus rubra seedlings. *Hortic. Sci.*, **41**, 1342–1346.

Savvas, D. and Passam, H. (eds) (2002). *Hydroponic Production of Vegetables and Ornamentals*. Athens: Embryo Publications, p. 463.

Scagel, C.F. (2003). Growth and nutrient use of ericaceous plants grown in media amended with sphagnum moss peat or coir dust. *Hort. Science*, **38**, 46–54.

Schwarz, D., Heinen, M. and van Noordwijk, M. (1995). Rooting characteristics of lettuce grown in irrigated sand beds. *Plant Soil*, **176**, 205–217.

Seginer, I. (1994). Trnaspirational cooling of a greenhouse crop with partial ground cover. *Agric. For. Meteor.*, **71**, 265–281.

Silber, A., Bar-Yosef, B., Singer, A. and Chen, Y. (1994). Mineralogical and chemical composition of three tuffs from northern Israel. *Geoderma*, **63**, 123–144.

Sonneveld, C. (1981). Items for application of macro-elements in soilless culture. *Acta Hort.* (ISHS), **126**, 187–195.

Stevenson, D.S. (1967). Effective soil volume and its importance to root and top growth of plants. *Can. J. Soil Sci.*, **47**(3), 163–74.

Tinker, P.B. and Nye, P.H. (2000). *Solute movement in the Rhizosphere*, 2nd edition. Oxford: Blackwell Science Publishers, 464pp.

Tuzel, I.H., Tuzel, Y., Gul, A., et al. (2001). Effect of different irrigation schedules, substrates and substrate volumes on fruit quality and yield of greenhouse tomato. *Acta Hort.* (ISHS), **548**, 285–291.

van der Gaag, D.J. and Wever, G. (2005). Conduciveness of different soilless growing media to Pythium root and crown rot of cucumber under near-commercial conditions. *Eur. J. Plant Pathol.*, **112**(1), 31–41.

van Iersel, M. (1997). Root restriction effects on growth and development of Salvia (*salvia splendens*). *Hort. Science*, **32**, 1186–1190.

Wallach, R. and Raviv, M. (2005). The dependence of moisture-tension relationships and water availability on irrigation frequency in containerized growing medium. *Acta Hort.* (ISHS), **697**, 293–300.

Wang, W., Reimers, C.E., Wainright, S.C., et al. (1999). Applying fiber-optic sensors for monitoring dissolved oxygen. *Sea Technol.*, **40**(3), 69–74.

Wild, A., Skarlou, V., Clement, C.R. and Snaydon, R.W. (1974). Comparison of potassium uptake by four plant species grown in sand and in flowing solution culture. *J. Applied Ecol.*, **11**, 801–812.

Williams, M. and Yanai, R.D. (1996). Multi-dimensional sensitivity analysis and ecological implications of a nutrient uptake model. *Plant Soil*, **180**, 311–324.

Wilson, A.J. and Robards, A.W. (1979). Some observations of the effects of mechanical impedance upon the ultrastructure of the root caps of barley. *Protoplasma*, **101**, 61–72.

Xu, G.H., Wolf, S. and Kafkafi, U. (2001). Interactive effect of nutrient concentration and container volume on flowering, fruiting, and nutrient uptake of sweet pepper. *J. Plant Nutr.*, **24**, 479–501.

Index of Organism Names

COMMON NAMES

Actinomycetes, 435, 440, 441, 452
Alstroemeria, 184, 185, 193, 197
Amaryllis, 193, 199
Anthurium, 198, 325
Asparagus, 183, 193, 196
Aster, 193, 198, 409
Aubergine, 193
Avocado, 321, 518

Barley, 284, 557
Begonia, 199, 465, 473
Bell pepper, 296–9, 315–17
Bouvardia, 197
Brassica, 489
Bromeliad, 481
Butterhead lettuce, 194

Cabbage, 17, 26, 194, 284, 321, 325, 477
Calendula, 473
Carnation, 184, 197, 325, 409, 439, 440
Carrot, 17, 26
Castor bean, 321
Cauliflower, 494
Celery, 325
Chickpea, 20, 21, 32
Chinese cabbage, 26, 284, 321

Chrysanthemum, 165, 166, 184–6, 193, 198–9, 201, 202, 305, 443
Clover, 322
Coleus, 473
Cotton, 16, 19, 20, 32
Courgette, 193
Creeping bentgrass, 494
Cucumber, 26, 27, 28, 29, 180, 186, 187, 193, 202, 361, 363, 364, 395, 397, 401–402, 428, 435, 436, 439, 440, 443, 477, 482, 493, 494, 517, 519
Cyclamen, 199
Cymbidium, 198

Dollar spot disease, 494
Dracaena, 200
Durum wheat, 19

Eggplant, 310, 325, 494
Elatior begonia, 465
Endive, 183, 196
Erwinia, 350

Faba bean, 321
Ficus, 200

Figleaf gourd, 29
Flax, 494
Fluorescent pseudomonads, 452
Foxtail millet, 322
Freesia, 184, 185, 199, 202
French beans, 193
Fuchsia, 199, 473

Geranium, 534
Gerbera, 187, 190, 198, 305, 325, 409, 473, 515, 535
Gloxinia, 465
Grevillea, 473
Grey mould, 493
Groundnut, 321
Gypsophila, 349, 350, 367, 408

Horsetail, 292
Hot pepper, 193
Hyacinth, 193, 199, 439

Iceberg lettuce, 193
Impatiens, 305, 465, 473
Iris, 199
Ixora, 473

Kalanchoe, 199, 473
Kohlrabi, 281

Legume, 24, 33, 292
Lettuce, 165, 166, 180, 182, 185–8, 194, 195, 201, 283, 284, 296–9, 314, 316, 320, 325, 359, 405, 443, 477, 515
Lily, 185, 409
Lisianthus, 198, 363, 364, 389, 407
Lupin, 24, 494

Maize, 17, 185, 297, 320, 392
Mangroves, 22
Maritime pine, 474
Melon, 27, 29, 30, 297, 320, 322, 324, 493, 557
Monstera, 193, 200
Muskmelon, 359, 365, 366, 367, 373, 399–401
Mycorrhiza, 16, 33, 103, 436

Nematode, 284, 427, 428, 429, 430, 431, 440, 445, 447, 489, 512

New Guinea impatiens, 465
Nitrobacter, 32

Orchid, 198, 199, 200, 467, 479, 481
Oriental sweet gum, 520

Palm, 200, 462, 468
Parsley, 477
Pentas, 473
Phalenopsis, 200
Pigeon pea, 18
Pine, 64, 91, 214, 222, 327, 474, 479, 480, 481, 488, 489, 493, 494, 524, 526, 528, 531, 532, 533, 535, 536, 561
Poinsettia, 467, 495
Primula, 473
Pseudomonads, 441, 452, 491

Radish, 195, 495
Rice, 22, 26, 185, 245, 274, 486, 533
Rice flower, 304, 305
Rose, 57, 75, 85, 93, 95, 102, 180, 187, 190, 191, 197, 198, 202, 226, 231, 321, 325, 357, 358, 367, 377, 393–4, 405–407

Saintpaulia, 193, 199, 200, 465
Schefflera, 473
Sedge, 460, 461, 463
Sorghum, 26
Spatiphyllum, 494
Spinach, 26, 195, 320
Spruce, 473, 474, 479, 480, 481
Strawberry, 30, 180, 182, 187, 300, 320, 322, 363, 392, 403–405, 473
Streptomycetes, 435, 440, 453
Sweet basil, 477, 495
Sweet clover, 322
Sweet pea, 467
Sweet pepper, 163, 194, 202, 443, 447, 449, 515, 516
Sweet William, 467

Tobacco Mosaic Virus (TMV), 490
Tomato, 26, 28, 31, 165, 180, 186, 187, 193, 194, 201–202, 237–8, 297–9, 306–307, 320, 370, 376, 377, 382, 383, 384, 402–403, 440, 441, 443, 491, 547

INDEX OF ORGANISM NAMES

Tulip, 193, 199

Wallflower, 467
Watermelon, 18, 19
Wax flower, 305
Wheat, 18–19, 23–5, 238
White cabbage, 17

Witloof, 183, 193, 196, 199

Xanthomonas, 350

Zinnia, 492

SCIENTIFIC NAMES

Adiantum spp., 467
Aechmea fasciata, 467
Agrostis palustris Pencross, 494
Alstroemeria aurantiaca L., 184
Alternaria brassicicola, 490
Anenome spp., 467
Anthurium andreanum, 467
Anthurium scherzerianum, 467
Aphelandra squalosa, 467
Apium graveolens L., 325
Aquilegia spp., 467
Arabidopsis spp., 16, 24, 493
Arachis hypogaea L., 321
Ascaris suum, 489
Asparagus plumosus, 467
Asparagus sprengeri, 467
Aucuba japonica, 495
Azalea spp., 464

Bacillus cereus, 453
Bacillus spp., 493
Bacillus subtilis, 439
Begonia semperflorens, 467
Begonia x hiemalis, 465
Botrytis cinerea, 493
Brassica chinensis, 284
Brassica oleracea var. *botrytis*, 494
Brassica oleracea L. convar. *Capitata* (L.) Alef. var *alba* DC, 17
Brassica rapa, 26, 321

Cajanus cajan L, 18
Calendula officinalis LINN, 473
Callistephus spp., 467
Camellia spp., 467

Campanula spp., 467
Capsicum annum L., 31, 244, 443, 515
Carex spp., 461, 529
Chamelaucium uncinatum L., 305
Cheiranthus allionii, 467
Chicorium endivia L., 183
Chytridiomycota, 436
Cicer arietinum L., 20
Citrulus lanatus (Thunb.) Matsum. & Nakai, 18
Cocos nucifera, 468
Coniothyrium minitans, 439
Cucumis melo L., 27, 297, 557
Cucumis sativum L., 26, 428, 519
Cucurbita ficifolia Bouché, 29
Cyclamen spp., 467
Cylindrocladium spatiphylli, 494

Daucus carota L., 17
Dendrathema × *grandiflorum* Kitam, 305, 443
Dianthus barbatus, 467
Dianthus caryophyllus L., 325, 440, 515
Dianthus hedwigii, *Godetia*, 467

Enterobacter spp., 493
Equisetaceae, 292
Erica gracilis, 467
Escherichia coli, 489
Euphorbia fulgens, 467
Euphorbia pulcherrima, 467, 495
Eustoma grandiflorum, 363

Flavobacterium balustinum, 493
Fragaria × *ananassa* Duchense, 30, 363, 473
Freesia spp., 184

Fuchsia X hybrida L., 473
Fusarium oxysporum, 427, 429, 433, 489, 490, 491, 492
Fusarium oxysporum (non-pathogenic), 491
Fusarium oxysporum f. sp. *basilica,* 492, 495
Fusarium oxysporum f. sp. *chrysanthemi,* 495
Fusarium oxysporum f. sp. *lycopersici,* 489, 490
Fusarium oxysporum f. sp. *melonis,* 495
Fusarium oxysporum f. sp. *radicis-cucumerinum,* 495
Fusarium oxysporum f. sp. *radicis-lycopersici,* 491, 492, 495
Fusarium oxysporum f.sp. *dianthi,* 495

Gardenia spp., 467
Gerbera jamesonii Bolus ex Hooker f., 452, 515
Gliocladium cathenulatum, 439
Gliocladium virens, 439, 492, 493
Gloxinia spp., 465, 467

Hydrangea spp., 467, 515
Hypnum spp., 461
Gossypium hirsutum, 16
Gypsophila paniculata, 349

Impatiens x *hawkeri,* 465
Impatiens walkerana Hook, F., 305
Ixora coccinea L., 473

Kalanchoe spp. Adans, 473

Lactuca sativa L., 296, 443, 477, 515
Lathyrus odoratus, 467
Leptosphaeria maculans, 490
Leucadendron salignum Bergius × *L. laureolum* (Lam.) Fourc., 305
Linum usitatissimum, 494
Liquidambar orientalis Mill., 520
Lupinus albus L., 24
Lupinus spp., 494
Lycopersicon esculentum Mill., 228, 435, 495, 512
Lysobacter enzymogenes, 439

Matricaria, 467
Medium, 467
Melilotus officinalis L., 24
Meloidogyne incognita, 440, 495

Meloidogyne javanica, 495
Monstera, 193, 200, 467

Nitrobacter, 301
Nitrosomonas, 301

Ocimum basilicum L., 477
Olpidium, 429
Oomycota, 427
Oryza sativa, 26
Ozothamnus diosmifolius, 304

Pelargonium, 467, 477, 534
Pelargonium peltatum L., 534
Penicillium spp., 493
Penstemon, 467
Pentas lanceolata, 473
Persea americana Mill., 321
Petunia, 467
Phaseolus vulgaris L., 18
Phragmites spp., 461
Phytophthora cinnamomi, 495
Phytophthora citricola, 495
Phytophthora cryptogea, 441, 452
Phytophthora nicotianae, 494, 494
Phytophthora spp., 427, 432
Picea sitchensis, 480
Picea spp., 473
Pinus nigra, 479
Pinus pinaster, 474, 479
Pinus radiate, 479, 480, 481
Pinus taeda, 479
Pisum sativum L., 34, 557
Plasmodiophora brassicae, 489, 490
Polymyxa, 429
Primula obconica, 467
Primula spp., 473
Pseudomonas aeruginosa, 439
Pseudomonas fluorescens, 350, 366, 367, 439
Pseudomonas putida, 453
Pseudomonas spp., 493
Pyrenochaeta lycopersici, 491
Pythium aphanidermatum, 428, 492, 494, 495
Pythium intermedium, 439
Pythium spp., 427
Pythium ultimum, 491, 492, 495

Raphanus sativus, 495
Rhizobium, 436

Index of Organism Names

Rhizoctonia, 429, 490, 491, 492, 494, 494, 495
Rhizoctonia solani, 490, 491, 492, 494, 495
Rhododendron catawbiense, 495
Rhododendron simsii, 467
Ricinus communis L., 321
Rosa X *hybrida*, 222, 509

Saintpaulia, 465, 467
Saintpaulia ionantha J.C. Wendt, 465
Salmonella enterica, 489
Salpiglossis, 467
Sanservia, 467
Schefflera spp. J.R.Forst. & G.Forst., 473
Sclerotinia homoeocarpa, 494
Sclerotinia minor, 495
Sclerotium rolfsii, 491, 495
Setaria italica L., 322
Sinningia speciosa, 465
Solanum melongena, 310, 507
Solenostemon scutellarioides, 473
Sorghum bicolor L., 26
Sphagnum, 529
Spinacia oleracea L., 26, 320

Streptomyces griseoviridis, 439
Streptomyces spp., 493

Tagetes, 467
Trichoderma hamatum, 493
Trichoderma harzianum, 439, 491
Trichoderma spp., 435, 493
Trifolium alexandrinum Fahli, 322
Triticum aestivum L., 18
Triticum turgidum L., 19

Verbena, 467
Verticillium dahliae, 427, 494
Verticordia plumosa L., 24
Vicia faba L., 321
Vriesia splendens, 467

Zea mays L., 17, 79, 297
Zinnia, 467, 492
Zinnia elegans, 492

Subject Index

Abscisic acid (ABA), 33, 369
Acetic acid, 33, 224, 326, 448
Active carbon adsorption, 450–1
Adhesion, 50, 276
Adhesive force, 52, 53
Adsorption, 99–101
Adsorption isotherm, 100, 104
Advection, 105
Aeration, 14–15, 45, 104–106, 254–255, 460, 473, 507, 520–1, 567
Aerohydroponics, 165–6
Aeroponics, 165–6, 201
AFLP, 432
Aggregation, 48, 53, 98, 101, 107, 429, 509–10
Air-filled porosity, 42, 45, 107, 265, 308, 465–6, 472–3, 481–2, 487, 489, 524, 527, 532–3, 562
Algae, 133, 138, 185, 292, 427, 447
Alkali, 124–5, 169, 474, 514
Allophane, 211, 218, 225
Aluminium, 14, 19, 199
Aluminosilicate, 225, 510, 514
Alumosilicate, 362–3
Ammonia, 26, 32, 301–303, 378–9, 477
Ammonium acetate, 271, 276
Ammonium nitrate, NH_4NO_3, 328, 377–9, 395, 415
Amyloplast, 16
Anchorage, 15, 531, 561
Andesitic lava, 508
Anion, 217–18, 293–4, 307–10
Anion adsorption isotherm, 214, 220

Anion exchange capacity (AEC), 211, 213, 237, 521
Anoxia, 267
Antagonist, 438–40, 442, 491, 565
Antibiosis, 439, 492
Antibody, 431
Antigen, 430–2
Apical dominance, 559
Apoplast, 226, 306
Aquaporins, 20, 322
Arrhenius equation, 301
Atmospheric demand (water), 7, 90–4
Auger, 249
Autotrophic bacteria, 301
Auxin, 34, 369, 372

Bacteria, 300–303, 364, 367, 377–378, 390, 397, 428, 431–2, 435–8, 440–1, 447, 449–53, 492–3
Barium chloride, 276
Bark, 64, 91, 93, 214, 254–255, 273–4, 479–81
Basaltic lava, 508
Bedding plants, 7, 45, 186, 467
Benzene, 520
β-1,3-Glucanase, 493
β-Tricalcium phosphate, $Ca_3(PO_4)_2$ (bTCP), 209, 218, 389
Biodegradability, 279, 478
Biodegradation, 278–9

Biological control:
  agent, 438–40, 452
Biomass production, 128, 295–6, 534, 559
Biurate, 378
Blackheart, 325
Blossom-end rot (BER), 303, 307, 308, 317–18, 325, 372, 373, 376, 395, 399, 402
Bonsai, 14
Boom system, 134
Borax, 467
Boric acid, 293
Boron, 292–3, 363, 381
Boundary condition:
  Dirichlet, 73
  Neumann, 73
Boundary layer, 84, 119, 235, 566
  resistance, 84
Bulk density (BD), 15, 17, 19, 42, 44, 47, 50, 134, 237, 250–3, 274, 346, 464, 481, 482, 488, 506–508, 521–7, 529–35

C/N ratio, 214–15, 238, 485–6
Calcined clay, 526, 532
Calcite ($CaCO_3$), 125, 248, 277, 327, 388–9, 447
Calcium, 160, 322, 325, 359, 550
Calcium carbonate, 277
Calcium chloride, 275–6
Calcium nitrate, $Ca(NO_3)_2$, 328, 468
Capillarity, 50–8
Capillary:
  fringe, 346–7
  mat, 130–1, 137–9, 190
  rise, 53, 69, 259, 346, 534, 536
Carbohydrate, 17, 26, 30, 31, 33, 119, 278, 299, 308, 488, 559
Carbon dioxide (CO_2):
  assimilation, 295
  concentration (atmospheric), 105, 319
  partial pressure (PCO_2), 388
  sequestration, 460
Carbon, 33, 228, 292–3, 301, 352, 365–7, 437, 450–1
Carbonate, 277, 342, 388, 390, 510
Carboxylate, 305, 342, 378, 383, 410
Cation:
  adsorption isotherm, 351
  cation exchange capacity (CEC), 25, 210–16, 237, 463, 485–6, 521–3, 531
Cavitation, 79
Cell:
  membrane, 20, 29, 103, 308
  wall, 27, 30–1, 33–4, 226, 369, 448

Cellulose, 185, 279, 478, 487–9
Chelate, 326, 379, 380, 390–1, 448
Chemical signal, 492
Chlamydospore, 491
Chloride, 275–6, 301, 320–1, 468, 520, 557
Chlorination, 344, 383, 445, 449
Chlorine, 160, 182, 292
Chlorophyll fluorescence, 355
Chlorosis, 303, 304, 306–307
Citrate, 228, 233–6, 365–6, 383–5, 391
Citric acid, $H_3C_6H_5O_7H_2O$, 227–9, 233–6, 366
Clay pellets, 250
Clay, 20–2, 25, 44, 49, 255, 489, 512–14, 531–5
Clinoptilolite, 514–15
Closed pores, 44–5, 47
Closed systems, 180–1, 201, 202, 426, 427–8, 442, 451, 554, 563
Coal cinder, 185
Coconut coir,5 93, 506, 553
Coconut fibre, 441, 460, 534
Cohesion, 50, 77, 269
Cohesive force, 52
Coir, 271–3, 468–73, 487–9
Coke, 516
Colloids, 42, 99
Colonization, 439–40, 491–3
Colour automated grading, 200
Columella, 16, 550
Competition, 8, 14–15, 102, 194, 202, 234–5, 320–1, 352, 375, 439, 491
Complexation, 99, 226, 362, 365–6
Compost:
  biowaste compost, 482–4, 485
  cattle manure compost, 484–6, 492
  maturity, 478
  spent mushroom compost (SMC), 482–4
  tea, 555, 556
Composted grape marc, 482, 492
Composted green waste (CGW), 482, 485, 488, 489
Composted pine bark, 64, 347, 481, 493
Composting, 214–16, 238, 278, 279, 281, 460, 477–8
Cone penetrometer, 268
Conidia, 428, 492
Connectivity, 45–6
Consumption curve, 295–300, 374
Contact angle, 50–3, 64, 259
Container capacity, 7, 73–6, 91, 121–2, 135, 140, 144–5, 271, 274, 524, 537
Controlled release fertilizer (CRF), 467–8
Convection, 95–9
Convection-dispersion equation (CDE), 99–104
Copper, 226, 292, 451

Copper ionisation, 427, 451
Copper sulphate, 467
Cork, 460
Culturable pathogens, 431
Cuticle, 119
Cytokinin, 34, 369
Cytoplasm, 26–27, 30, 32, 303, 320

Darcy's equation, 261
Darcy's law, 65–7, 71–2, 81
Decision support system (DSS), 203, 416–17
Deep Flow Technique (DFT), 165–6, 185, 195, 201–202, 346, 565
Dendrometer, 150, 558
Denitrification, 20, 217, 230, 303, 342
Depletion zone, 311–14, 374, 547
Desalinization, 342, 380–1
Desorption, 46, 99, 217, 230
Detoxification, 26, 30, 31, 318
DGGE, 435
Di-isocyanate, 518–19
Diabase, 516
Diammonium phosphate, $(NH_4)_2HPO_4$, 392
Diatomite, 523
Dicalcium phosphate dehydrate, $CaHPO_4$ $2H_2O$ (DCPD), 388–9
Dielectric capacitance, 150–1
Dielectric constant, 48–9, 170
Dielectric permittivity, 150
Dielectric sensor, 151, 170–1
Diffuse double layer, 220
Diffusion:
  oxygen, 264–7
Diluter/dispenser unit, 168, 174, 176
Disease:
  control, 4, 452, 506
  suppression, 442, 490–6
Disinfestation, 442–53
Dispersion, 95–9, 197
Dispersivity, 96, 98–100, 104
Dissolution, 210–11, 225–6, 228–30, 235, 237, 508
Dissolve oxygen (DO), 547–8, 551, 552
Diversity, 270, 435, 440, 490
DNA, 432, 433, 436, 453
Dolerite, 516
Dolomitic lime, 466, 521
Dormancy, 283
Drainage, 55–6, 63–4, 73–6, 167–8, 179–82, 343–8, 354–8, 442
Drip irrigation, 123, 136–7, 163–4, 193, 196–9, 201, 311, 349–50, 451–2
DTPA, 275–6, 391, 517

Easily available water (EAW), 90–1, 465, 472, 488, 510, 514, 533–5
Ebb-and-flood, 130, 139
Ebb and Flow, 139, 167, 186, 190, 193, 198–201
EC sensor, 169, 203
Economical feasibility, 129, 159, 161, 186, 194, 451–452, 553, 556–7
EDDHA, 224, 326
EDTA, 224, 326, 380, 390, 467
Effective pore space, 46
Effluent, 97–8, 341–4, 346, 352–3, 357, 426, 554
Electrical conductivity (EC), 48, 120, 124, 150, 169, 283, 292, 319, 344, 469
Electrical resistance sensor, 150, 151
Electromagnetic radiation, 449–50
Electrostatic potential, 220
ELISA, 431
Elongation zone, 15, 17, 30, 369
Emitter, 130–1, 133, 136, 149, 154, 347–50, 354, 385–6
Endodermis, 20, 29
Endophyte, 439
Enzyme, 103, 293, 308, 318, 431, 494, 519, 531
Ethylene, 34, 224, 326, 369, 520, 522, 559, 563
Evaporation, 21, 74, 76, 77, 85, 119, 123, 138, 151 2, 552
Evapotranspiration (ET), 74, 82, 151, 222, 342
Exfoliation, 513
Expanded clay, 271, 274, 489, 506, 513–14, 531
Extensibility, 33–34
Extraction, 58, 73, 80–2, 84, 88–9, 226, 270–2, 469
Exudation, 77, 228, 305, 437–8, 491, 531, 551, 566

Feather meal, 556
Ferrous sulphate, 467
Fertigation, 95, 157–8, 167–78
Fertilization, rate of, 228
Fick's first law, 96
Filtration, 161, 273, 276, 344, 350, 352–3, 444–6
Fix bench, 189–201
Flooded concrete floor, 186
Flooded floor, 130–1, 140
Fluidity, 67, 466
Fog system, 133–4
Foliage, 132–4, 143, 145, 162
Foliage plant, 200–201, 482
Foliar application, 24–25
Formic acid, 448
Fourier Transform Infrared ratio (FT-IR), 472, 485, 487, 488

Fractal, 16
Fractionation, 250, 279
Free radicals, 318
Frequency domain (FD), 150–1, 170–1
Freundlich isotherm, 100
Fritted trace elements, 467
Fulvic acid, 226, 509, 531
Fungi, 427–8, 429–32, 435–6, 446–52, 491–3, 565

Gamma radiation, 449
Gas exchange, 45–6, 50
Germination, 18–19, 179, 283, 285, 436, 478, 491–3
Gibberellic acid (GA), 34, 283
Gibberellin, 34, 369, 372
Glasswool, 516–18
Gliotoxin, 492
Glucose, 494
Glycol, 518, 522
Glycophyte, 22, 319, 320
Goethite, 227, 234
Granulated substrate, 178, 180, 185, 197
Gravel, 6, 200, 201, 352
Gravimetric potential, 119–20, 122
Gravitropism, 14–15
Greenhouse:
  climate, 410
Growing medium, 5, 15–16, 45, 47–51, 53–6, 76–8, 80–2, 96–8, 170–1, 245–6, 271–2, 283–4, 460–1, 468, 478–9, 481, 482, 485–8, 519, 531
Guano, 556
Guttation, 77
Gypsum (CaSO$_4$), 30, 149–50, 275, 328, 362, 388
Gypsum block sensor, 149

H$^+$-ATPase, 29
Halloysite, 218, 225, 508
Hardness, 269–70
Heat capacity, 48
Heat-pulse method, 84
Heavy metals, 14, 217, 271, 381, 451, 473–4, 514
Hemicellulose, 279, 489
Hexokinase, 24
Hexose, 24, 317
Hoof and horn meal, 476
Hopeite, 223
Hormone (plant), 33–4
Humic acid, 463
Humus, 51, 477
Husk, 185, 274, 468–9, 533, 553
Hydathode, 77

Hydraulic conductivity:
  saturated, 261–2, 346–7, 372, 507, 514, 517, 529–30, 532–4, 559, 563
Hydraulic diffusivity, 72
Hydrochloric acid, 277
Hydrodynamic dispersion, 98–9
Hydrogen bonding, 50
Hydrogen peroxide, 29, 344, 427, 448, 451
Hydrophobicity, 259–60
Hydroponics, 6, 185
Hydroquinone, 276
Hydrostatic pressure, 52
Hydrotropism, 15, 18, 550–1
Hydroxide, 211, 389–90
Hydroxyapatite, 218
Hygroscopic water, 47
Hyperparasitism, 491
Hyphae, 492, 493
Hysteresis, 58–65, 122, 255, 257, 259, 507, 511, 517

Immunofluorescence, 431
Imogolite, 211
Indole acetic acid (IAA), 33–34
Induced systemic resistance (ISR), 492–3
Infection site, 439, 491
Infiltration, 45, 51, 135, 146, 148, 163, 373, 532
Injector, 128, 130, 158, 175–8
Inoculum, 428, 492
Integrated pest management (IPM), 64, 555
Interstitial filling, 255
Intrinsic permeability, 67
Iodine, 449, 451
Ion-Selective Electrode (ISE), 169–70, 203
Ionic strength, 214, 220–1, 226–7, 231–4
Iron, 234, 292, 306, 389, 448, 450, 480
Irradiance, 319
Irrigation:
  control, 143, 146, 147–9, 151–5, 158, 171, 558
  management, 58, 85, 88, 127–8, 145–6, 152–3, 311, 506, 558
Irrigation frequency (IF), 45, 56, 59, 74, 93, 104, 143–4, 223, 291, 310–318, 375, 386, 408, 417, 431, 443, 534
ITS-RFLP, 432

Kaolinite, 508

Langmuir isotherm, 100, 218
Leachate, 144, 304–305, 350–1, 357, 567

Subject Index

Leaching fraction, 144–5, 148, 238, 557
Leaf analysis, 355
Leaf area index (LAI), 203
Leaf area, 29–30, 93, 119, 153, 394, 406, 410–11, 550, 559–560
Leaf water potential, 78–9, 558
Lethal temperature, 446
Lignin, 214–15, 279, 463, 472, 475, 479, 481–2, 487–8, 529
Lime, 461, 464, 466, 485, 488, 521
Limestone, 466, 516
Lipopolysaccharide, 493
Lipoxygenase, 28
Load cell, 76, 146, 171, 269
London-van der Waals force, 50
Lysis, 439, 491, 492

Macro-array, 432–3
Macronutrient, 14, 292, 467, 471
Magnesium nitrate, $Mg(NO_3)_2$, 392
Magnesium sulphate, 466
Magnesium, 24, 292
Magnetic susceptibility, 48
Malate, 378
Malonic acid, $CH_2(COOH)_2$, 366
Manganese sulphate, 467
Manganese, 276, 480
Manure:
  chicken manure, 185
  poultry manure, 477, 485
Mass flow, 24–5, 54, 102–103, 118, 124, 311, 313, 375, 380, 410–11
Matric potential, 50, 54–9, 62, 67, 72, 88–9, 93–4, 103, 107, 119–21, 530
Mechanical dispersion, 97–9, 105
Mechanical impedance, 17, 33–34, 550–1, 557
Mesocarp, 468
Metering pump, 175, 176, 354
Methyl bromide, 425
Michaelis-Menten equation, 294–5, 369
Microcalorimeter, 279, 478
Microfiltration, 444
Microflora, 7, 235–6, 367, 434–42, 452–3
Micronutrient, 32, 217–18, 224, 292–3, 304–305, 325, 326, 471, 475
Microorganism, 7, 15, 51, 278, 280, 365, 435, 453–4, 512, 565
Microscopy, 79–81, 431
Microwave, 48, 449
Mineral weathering, 16
Mineralization, 25, 237–8, 273–5, 280, 410, 548, 556

Mist system, 133–4
Mixing tank, 172, 175–8
Moisture content, 48–9, 56, 59–60, 65, 73–6, 80, 84, 88–91, 93–4, 97, 121–2, 145, 147, 149–50, 271–2, 274
Molybdenum, 25, 292
Monoammonium phosphate, $NH_4H_2PO_4$, 379
Monopotassium phosphate, KH_2PO_4, 379, 380, 392
Montmorillonite, 21
Moss, 71, 91–2, 460–1, 490, 521, 530, 532–4, 558
Movable bench, 189–91, 198, 200
MRNA, 432–3
Mucilage, 17–18, 437
Mulch, 74, 479, 513, 518, 552
Mycelium, 428, 432
Mycoparasitism, 439
Mycorrhiza, 16, 33, 103, 436

Nanofiltration, 444
Near infra-red reflectance (NIR), 273
Necrosis, 304
Nernst equation, 169
Net photosynthesis, 410, 535
NH_4/NO_3 ratio, 383–8, 395–7, 415
Nickel, 292–3
Nitrate:
  reduction, 365, 378, 390, 415
Nitric acid, 327
Nitrification, 20, 217, 230, 280, 300–304, 308, 327, 377–9, 383–91, 415
Nitrite, 25, 32, 301, 364–5, 377
Nitrogen:
  immobilisation, 305
Nitrogen draw-down Index (NDI), 277
Nitrogen fixation index (NFI), 277
Nuclear magnetic resonance, 478
Nutrient availability, 303–307, 311–15
Nutrient Film Technique (NFT), 142, 164–5, 346, 443, 547
Nutrition, 26–34, 216–17, 291–328

Octa Ca phosphate, $Ca_4H(PO_4)_3$ (OCP), 209, 219, 389
Ohm's law, 77
Open systems, 180, 202, 367, 385–6, 402, 408, 563, 566
Orangery, 1,2 3, 4
Organic:
  acids, 24, 228, 230, 233, 235, 305–307, 365–6, 378, 384, 437, 478, 531–2
  agriculture, 239, 555

Organic (*Continued*)
  fertilizer, 467, 476, 555–6
  hydroponics, 8
Organic carbon (OC), 365–6, 384
Orthophosphate, 225, 227, 326, 342
Orthophosphoric acid, 326, 379
Osmotic potential, 33–34, 54–5, 58, 120, 318, 321, 342, 375, 378, 410–13, 548, 557
Osmotic pressure, 10, 55, 271–2, 301, 317
Outdoor nurseries, 157, 507, 552
Overhead irrigation, 132–4, 141, 246, 352
Oxalate, 365, 391
Oxalic acid, $(COOH)_2 2H_2O$, 230, 366
Oxidation, 447–9
Oxygen:
  deficiency, 7, 14–15, 25, 42, 123–4, 148, 157, 179, 320, 551, 557
Oxygen diffusion rate (ODR), 7, 265, 476, 507, 548
Oxytropism, 14
Ozonation, 344
Ozone, 448, 451, 521

Parasitic weed, 16
Particle density, 253, 507, 513, 515, 519, 521, 525, 554
Particle size distribution (PSD), 42–4, 255, 352, 464, 512, 523–5, 554
Pasteurization, 446–7
Pathogen:
  eradication, 426, 489–90
Pathogenesis-related genes, 493
PCR, 432–3, 435–6
Peat:
  black peat, 258, 260, 463
  sedge peat, 463
  sphagnum moss peat, 460
Peat bog, 460–1
Penetrability, 267–9
Perched water table, 45, 560, 564–565
Permanent wilting point, 90
Permeability, 29, 33–34, 67, 77, 302, 322, 369
Perlite, 43–4, 74–5, 93, 367, 372, 408, 511–12
Peroxidase, 31, 493
Peroxide, 29, 344, 427, 448, 451, 520
Pesticide, 8, 118, 426, 429, 433, 442, 452, 555
pH, 218–20, 222, 276–7, 388–91
Phenolic resin, 506, 516
Phosphate, 22–23, 24–25, 235, 237, 321, 326, 464, 550, 557
Phosphoric acid (H_3PO_4), 326, 379, 383
Phosphorus, 19, 213, 218–21, 379–80
Photosynthate, 299

Photosynthetically active radiation (PAR), 295
Phytomonitoring, 355
Phytotoxicity, 277, 283–5, 449, 476–9, 482, 486
Pine bark, 64, 91, 214–22, 327, 479, 481, 488–9, 493, 525–6, 532–3, 535–6
Piston flow, 97
Plant disease, 124, 142, 162, 179, 186, 425–6
Plant Plane Hydroponic, 185
Poiseuille's law, 65
Polarization (of electrodes), 169
Pollution, 8, 160, 170, 180, 182, 201, 236–237, 239, 342, 401, 426, 429, 508, 511, 514
Polyester fleece, 521–2
Polyethylene, 136, 165, 181–2, 185, 507, 520, 522
Polypropylene, 165, 181–2, 185–7, 520
Polysaccharide, 31, 34, 493
Polystyrene, 165–6, 185, 188, 195, 520–1
Polyurethane, 7, 165, 178, 185, 249, 253, 256, 506, 518–20, 531
Polyurethane foam, 165, 185, 256, 520, 531
Polyvinylchloride (PVC), 165, 181–2, 187
Pore size distribution, 46, 59, 63, 68, 107, 258, 367, 507, 527, 533
Porometer:
  diffusion porometer, 558
Porosity, 44–5, 61–3, 105–107, 253–4
Pot plant, 6–7, 187, 189, 200–201, 257, 451, 465–7, 485
Potassium chloride, KCl, 277, 328, 380
Potassium nitrate, KNO_3, 283, 321, 328, 378, 380, 392
Potassium sulphate, K_2SO_4, 328, 380, 392, 466
Potassium, 22, 165, 217, 292–3, 328, 380, 392, 466, 468, 480, 550
Potentiometric titration, 211–14
Precipitation, 103, 159, 168, 172, 174–5, 211, 217–18, 222–4, 311, 313–14, 325–6, 362, 390
Precision agriculture, 153, 553, 567
Preferential flow, 51
Pressure:
  chamber, 558
  head, 50, 53, 58, 78, 80, 130, 257, 271, 273, 276, 514
  potential, 50, 53, 55
Propagule, 180, 282, 428, 429, 432–3, 435, 442, 491–2
Proton extrusion, 23, 33–34
Pulse irrigation, 141–2, 143, 552
Pumice, 15, 44, 71, 93, 237, 245, 274, 347–50, 362–3, 367, 373, 509–11, 531, 533
Pycnometer, 253

Q-PCR, 432
Quality control, 246, 248
Quartz, 352, 363, 507, 516

Rain water, 125, 133, 380
RAPD, 432
Re-use, 440–2, 507, 509, 511, 513–19, 521–3
Recirculation:
　crops, 397, 409–15
Recycling of climate, 393–5
Redox potential, 18, 377–378, 390
Reducing sugars, 317
Redwood bark, 507
Rehydration, 255, 257–9, 523
Relative humidity (RH), 119, 133, 143, 153, 166, 393
Reproductive stage, 174, 397
Reservoir of water, 343–5, 356
Respiration, 29–30, 45, 105, 228, 277, 280–2, 308, 551, 559
Response curve, 23–4, 27, 124–5
Retention curve, 58–65, 88, 91, 93, 121, 222, 346, 507–508, 511, 514, 526–30, 532, 564
Reverse osmosis, 161, 342, 444–5
Rewetting, 18, 122, 246, 255, 257–8, 316
Reynolds number, 66–7
Rhizoplane, 437
Rhizosphere, 10, 18, 32, 225–6, 300, 303–307, 326–7, 437–8
Rhizotron, 15, 16
Rhyolitic lava, 508
Rice hulls, 245, 486
Rice husks, 185, 274, 533
Richards equation, 71–3, 80, 104
RNA, 430, 432
Rock phosphate, 24
Rockwool, 516
Rolling bench, 189, 191, 199
Root:
　cap, 16–18, 550–1
　density, 19, 76–7, 79, 82–4, 88–9, 210, 228, 230, 236, 547, 551
　disease, 395–7, 439–40, 490–3
　excretion, 226, 235, 304, 342, 365, 385
　exudates, 103, 226, 437, 492, 548
　radius, 20, 104, 375
　respiration, 29–30, 45, 105, 228, 351, 552, 559–560
　shrinkage, 20–1
　tip, 14–15, 23, 28, 34, 351, 369, 372, 550

　volume, 8, 22, 76, 80, 311, 369–70, 386, 410, 506, 547–50, 557–61
Root-borne disease:
　adventitious root, 24
　lateral root, 17–19, 23
　primary root, 17, 19, 23
Root zone temperature, 26, 153, 322, 391–2
Rootstock, 185, 199, 321, 517
RT-PCR, 432
Run-off, 8–10, 132, 138, 141, 148, 160, 161, 203, 237

Salicylic acid, 493
Salinity, 30–1, 120, 127, 149–50, 158, 160, 291, 318–25
Sand, 4–5, 15–17, 62, 253, 255–6, 352–353, 444–5, 507–508, 531
Sand filter, 444, 450, 453
Sap flow, 21–22, 29, 79, 84, 355
Saprophyte, 494
Saturated hydraulic conductivity, 261–2, 346–7, 372, 507–508, 514, 517, 529–30, 532–4, 559, 563
Saturation, 44, 47–8, 63, 68, 76, 257, 271–2, 535
Sawdust, 482
SCAR-PCR, 432
Sclerotia, 494
Selenium, 292
Serology, 430–2
Sewage sludge, 489
Shrinkage, 20–1, 64, 258, 260–1, 488, 524–30, 554, 562
Siderophore, 491
Sieving, 254, 429, 464, 485, 508, 510, 515
Silica, 3, 25, 363, 508, 510
Silicon, 224–5, 292
Slab, 180–1, 187–8, 192–3, 197–8, 441, 516–20
Sodium, 22, 158, 193, 271, 292, 320–2, 325, 448–9
Sodium hypochlorite, 448–9, 451
Soil-borne pathogen, 193, 442, 445, 546
Soil dispersion, 197
Soil heating, 196
Soilless culture, 1–10, 505–37, 545–67
Solarization, 179, 182, 509
Solid substrate, 6, 178, 564
Solubility, 217–23, 230–1, 325–6, 390–2
Sorption site, 101, 234, 351, 395
Specific leaf area, 406, 410–11
Specific surface, 30, 47, 67, 492, 507
Sponge substrate, 185
Spore, 429
Sporulation, 429

Sprinkler, 20, 127, 133–4, 137, 162–3, 166, 185, 201
Sprouting (bud), 197, 282, 283
Stability(rate of biodegradation), 278–9
Starch, 16, 317
Static pile, 478
Statocyte, 16
Statolith, 16
Steam sterilization, 179, 512, 518, 520
Steaming, 487, 507, 509
Sterilization, 8, 441, 490–1, 507, 509, 511–15, 517–22
Stickiness, 269–70
Stomata, 21, 77–8, 84, 119, 127, 393, 558
Stomatal conductance, 393
Stomatal patchiness, 84
Stomatal resistance, 78, 558
Stone wool, 268–72, 347–50, 373, 435–42, 516–18
Stratification, 283
Strengite, 218, 231
Stress, 31–2, 82, 120, 127, 143, 145–6, 322, 411–12, 557, 559, 562–3
Styrene, 520, 521
Substrate, 217–20, 236–9, 274, 292–9, 346–54, 369–72, 494–6, 506, 523–36
Subsurface systems, 131–132, 137
Sucrose, 317
Suction, 54, 56, 58–60, 62–3, 119, 255–6, 517, 519, 522
Suction table, 59, 255–6
Sulphate, 25, 172, 320, 466, 467, 520
Sulphur, 292–3, 520
Superphosphate, 466
Suppressiveness, 440–2, 453, 454, 490–3, 494, 496, 565
Surface charge, 7, 30, 210–14, 225–6, 231, 506, 508–509, 510, 531–2
Surface systems, 131–132, 134–7
Surface tension, 50, 52, 260
Surface water, 79, 158, 160–1
Symplast, 104
Systemic acquired resistance (SAR), 492

Tensiometer, 55–9, 68, 74–6, 88, 147–51
Tension, 50–2, 56, 58–60, 62, 67, 80–2, 91, 93
Texture, 42–3, 48–9, 53, 58, 61, 66, 106, 151, 274, 487
TGGE, 432
Thermogravimetry, 48
Thigmotropism, 14, 550
Time-controlled irrigation, 557

Time domain reflectometry (TDR), 47–50, 59, 84, 127, 150–1, 170–171
Tipping bucket, 74, 144
Titanium, 447
Tomato crown and root rot, 395, 440
Tortuosity, 46, 96, 236, 263, 312
Tortuosity factor, 96–7, 236, 312
Total organic carbon (TOC), 384
Total pore space, 44–5, 252–5
Total porosity (TP), 42, 44, 46, 63, 106, 253, 346, 476, 508, 510, 524, 530, 532–5
Total soluble solids (TSS), 324, 373, 400
Toxicity, 14, 24, 26, 31, 32, 276, 302–303, 307–10, 367, 479–80
Translocation, 31–32, 33, 296, 299, 320
Transmittance, 54, 77, 140, 450
Transpiration, 74, 77–9, 82, 84–9, 91–5, 118, 126–127, 128, 343, 566
Transpiration stream, 118, 128, 374, 408, 566
Trough, 130–1, 139, 164–5, 171, 179–82, 187–8, 191, 193–8, 201, 350–1
Tuff (Scoria, Pouzolane, Picon), 7, 43, 253, 441, 558–559
Turbidity, 124, 450

UC mix, 5, 71, 93, 121, 148, 473
Ultrafiltration, 444
Underground water, 341–2
Uniformity, 129–31, 136, 144–5, 163, 185, 552–3
Unsaturated hydraulic conductivity, 63, 67–8, 93, 123, 137, 139, 142, 167, 248, 255, 257–8, 262–4, 372, 507, 534, 558, 563
Urea formaldehyde, 245, 486
Urea, 32, 300, 378, 395, 486
UV radiation, 179, 344, 444, 450, 451, 452–453

Van Genuchten model, 529–30
Vapour pressure, 54, 87, 153, 552
Vapour pressure deficit (VPD), 87–8, 93–4, 144, 153, 552
Variable charged surface, 213, 218, 226, 233
Variscite, 218, 231
Vegetative stage, 23, 31, 272, 295–9, 307, 310, 397
Venturi, 175, 176–177
Vermiculite, 506, 512–13, 531–3, 561–562
Virus, 428–32, 445–9, 451, 489–90
Viscosity, 67, 105, 508, 530
Volcanic glass, 225, 226, 508–509

Walking Plant System, 191–2
Waste disposal, 8, 507–509, 511, 513–15, 519, 521
Water availability, 76–104, 530
Water buffering capacity (WBC), 91, 122, 344, 535
Water content, 46–50, 58, 60, 62–3, 79–82, 91–2, 96–9, 105, 122, 137, 138, 145, 150–1, 153–4, 273, 311–12, 346–7, 375, 441, 482, 484–6, 488–9, 507, 517, 534–5
Water filled pore fraction (WFP), 263
Water holding capacity, 42, 84, 137, 141, 142, 146, 283, 381, 461, 465–6, 481, 511, 517, 520–2
Water repellency, 51, 259–60
Water retention, 44, 46, 52, 58–65, 246, 252, 254–7, 344, 346, 439, 506, 511, 514, 519, 521–3, 526–8, 535
Water tension, 82, 510, 511, 517, 528, 558
Water uptake, 18–21, 76–88, 92, 103–104, 119, 348, 375, 533, 535, 548, 550

Water use efficiency (WUE), 10, 128, 142, 201, 566
Weathering, 16, 212, 351, 507, 508
Weed, 16, 179, 201, 277–8, 282–3, 518
Weighing lysimeter, 73, 558
Wettability, 51, 64, 532, 535, 561–562
Wetting agent, 260, 516
Wind speed, 119, 152–3, 295
Wood chips, 533–4
Wood fibre, 273, 473–8, 488

Xylem, 20–2, 76–8, 322, 558
Xylem water potential, 558

Zeolite, 61, 224, 237
Zinc, 24–25, 218, 223–224, 292, 305, 390
Zoospore, 427, 428, 439, 441